From Classical to Quantum Mechanics

This book provides a pedagogical introduction to the formalism, foundations and applications of quantum mechanics. Part I covers the basic material that is necessary to an understanding of the transition from classical to wave mechanics. Topics include classical dynamics, with emphasis on canonical transformations and the Hamilton–Jacobi equation; the Cauchy problem for the wave equation, the Helmholtz equation and eikonal approximation; and introductions to spin, perturbation theory and scattering theory. The Weyl quantization is presented in Part II, along with the postulates of quantum mechanics. The Weyl programme provides a geometric framework for a rigorous formulation of canonical quantization, as well as powerful tools for the analysis of problems of current interest in quantum physics. In the chapters devoted to harmonic oscillators and angular momentum operators, the emphasis is on algebraic and group-theoretical methods. Quantum entanglement, hidden-variable theories and the Bell inequalities are also discussed. Part III is devoted to topics such as statistical mechanics and black-body radiation, Lagrangian and phase-space formulations of quantum mechanics, and the Dirac equation.

This book is intended for use as a textbook for beginning graduate and advanced undergraduate courses. It is self-contained and includes problems to advance the reader's understanding.

GIAMPIERO ESPOSITO received his PhD from the University of Cambridge in 1991 and has been INFN Research Fellow at Naples University since November 1993. His research is devoted to gravitational physics and quantum theory. His main contributions are to the boundary conditions in quantum field theory and quantum gravity via functional integrals.

GIUSEPPE MARMO has been Professor of Theoretical Physics at Naples University since 1986, where he is teaching the first undergraduate course in quantum mechanics. His research interests are in the geometry of classical and quantum dynamical systems, deformation quantization, algebraic structures in physics, and constrained and integrable systems.

GEORGE SUDARSHAN has been Professor of Physics at the Department of Physics of the University of Texas at Austin since 1969. His research has revolutionized the understanding of classical and quantum dynamics. He has been nominated for the Nobel Prize six times and has received many awards, including the Bose Medal in 1977.

FROM CLASSICAL TO QUANTUM MECHANICS

An Introduction to the Formalism, Foundations and Applications

GIAMPIERO ESPOSITO, GIUSEPPE MARMO

INFN, Sezione di Napoli and
Dipartimento di Scienze Fisiche,
Università Federico II di Napoli

GEORGE SUDARSHAN

Department of Physics,
University of Texas, Austin

CAMBRIDGE UNIVERSITY PRESS
Cambridge, New York, Melbourne, Madrid, Cape Town, Singapore,
São Paulo, Delhi, Dubai, Tokyo

Cambridge University Press
The Edinburgh Building, Cambridge CB2 8RU, UK

Published in the United States of America by Cambridge University Press, New York

www.cambridge.org
Information on this title: www.cambridge.org/9780521143622

First published 2004
This digitally printed version 2010

A catalogue record for this publication is available from the British Library

Library of Congress Cataloguing in Publication data

Esposito, Giampiero.
From classical to quantum mechanics / Giampiero Esposito, Giuseppe Marmo,
George Sudarshan.
p. cm.
Includes bibliographical references and index.
ISBN 0 521 83324 8
1. Quantum theory. I. Marmo, Giuseppe. II. Sudarshan, George, 1931– III. Title.
QC174. 12.E95 2004 530.12–dc21 2003053212

ISBN 978-0-521-83324-0 Hardback
ISBN 978-0-521-14362-2 Paperback

For Michela, Patrizia, Bhamathi, and Margherita, Giuseppina, Nidia

Contents

Preface

The present manuscript represents an attempt to write a modern monograph on quantum mechanics that can be useful both to expert readers, i.e. graduate students, lecturers, research workers, and to educated readers who need to be introduced to quantum theory and its foundations. For this purpose, part I covers the basic material which is necessary to understand the transition from classical to wave mechanics: the key experiments in the development of wave mechanics; classical dynamics with emphasis on canonical transformations and the Hamilton–Jacobi equation; the Cauchy problem for the wave equation, the Helmholtz equation and the eikonal approximation; physical arguments leading to the Schrödinger equation and the basic properties of the wave function; quantum dynamics in one-dimensional problems and the Schrödinger equation in a central potential; introduction to spin and perturbation theory; and scattering theory. We have tried to describe in detail how one arrives at some ideas or some mathematical results, and what has been gained by introducing a certain concept.

Indeed, the choice of a first chapter devoted to the experimental foundations of quantum theory, despite being physics-oriented, selects a set of readers who already know the basic properties of classical mechanics and classical electrodynamics. Thus, undergraduate students should study chapter 1 more than once. Moreover, the choice of topics in chapter 1 serves as a motivation, in our opinion, for studying the material described in chapters 2 and 3, so that the transition to wave mechanics is as smooth and 'natural' as possible. A broad range of topics are presented in chapter 7, devoted to perturbation theory. Within this framework, after some elementary examples, we have described the nature of perturbative series, with a brief outline of the various cases of physical interest: regular perturbation theory, asymptotic perturbation theory and summability methods, spectral concentration and singular perturbations. Chapter

8 starts along the advanced lines of the end of chapter 7, and describes a lot of important material concerning scattering from potentials.

Advanced readers can begin from chapter 9, but we still recommend that they first study part I, which contains material useful in later investigations. The Weyl quantization is presented in chapter 9, jointly with the postulates of the currently accepted form of quantum mechanics. The Weyl programme provides not only a geometric framework for a rigorous formulation of canonical quantization, but also powerful tools for the analysis of problems of current interest in quantum mechanics. We have therefore tried to present such a topic, which is still omitted in many textbooks, in a self-contained form. In the chapters devoted to harmonic oscillators and angular momentum operators the emphasis is on algebraic and group-theoretical methods. The same can be said about chapter 12, devoted to algebraic methods for the analysis of Schrödinger operators. The formalism of the density matrix is developed in detail in chapter 13, which also studies some very important topics such as quantum entanglement, hidden-variable theories and Bell inequalities; how to transfer the polarization state of a photon to another photon thanks to the projection postulate, the production of statistical mixtures and phase in quantum mechanics.

Part III is devoted to a number of selected topics that reflect the authors' taste and are aimed at advanced research workers: statistical mechanics and black-body radiation; Lagrangian and phase-space formulations of quantum mechanics; the no-interaction theorem and the need for a quantum theory of fields.

The chapters are completed by a number of useful problems, although the main purpose of the book remains the presentation of a conceptual framework for a better understanding of quantum mechanics. Other important topics have not been included and might, by themselves, be the object of a separate monograph, e.g. supersymmetric quantum mechanics, quaternionic quantum mechanics and deformation quantization. But we are aware that the present version already covers much more material than the one that can be presented in a two-semester course. The material in chapters 9–16 can be used by students reading for a master or Ph.D. degree.

Our monograph contains much material which, although not new by itself, is presented in a way that makes the presentation rather original with respect to currently available textbooks, e.g. part I is devoted to and built around wave mechanics only; Hamiltonian methods and the Hamilton–Jacobi equation in chapter 2; introduction of the symbol of differential operators and eikonal approximation for the scalar wave equation in chapter 3; a systematic use of the symbol in the presentation of the Schrödinger equation in chapter 4; the Pauli equation with time-dependent magnetic fields in chapter 6; the richness of examples in chapters 7 and 8; Weyl quantization in chapter 9; algebraic methods for eigenvalue problems in chapter 12; the Wigner theorem and geometrical phases in chapter 13; and a geometrical proof of the no-interaction theorem in chapter 16.

So far we have defended, concisely, our reasons for writing yet another book on quantum mechanics. The last word is now with the readers.

Acknowledgments

Our dear friend Eugene Saletan has kindly agreed to act as merciless reviewer of a first draft. His dedicated efforts to assess our work have led to several improvements, for which we are much indebted to him. Comments by Giuseppe Bimonte, Volodya Man'ko, Giuseppe Morandi, Saverio Pascazio, Flora Pempinelli and Patrizia Vitale have also been helpful.

Rosario Peluso has produced a substantial effort to realize the figures we needed. The result shows all his skills with computer graphics and his deep love for fundamental physics. Charo Ivan Del Genio, Gabriele Gionti, Pietro Santorelli and Annamaria Canciello have drawn the last set of figures, with patience and dedication. Several students, in particular Alessandro Zampini and Dario Corsi, have discussed with us so many parts of the manuscript that its present version would have been unlikely without their constant feedback, while Andrea Rubano wrote notes which proved very useful in revising the last version.

Our Italian sources have not been cited locally, to avoid making unhelpful suggestions for readers who cannot understand textbooks written in Italian. Here, however, we can say that we relied in part on the work in Caldirola *et al.* (1982), Dell'Antonio (1996), Onofri and Destri (1996), Sartori (1998), Picasso (2000) and Stroffolini (2001).

We are also grateful to the many other students of the University of Naples who, in attending our lectures and asking many questions, made us feel it was appropriate to collect our lecture notes and rewrite them in the form of the present monograph.

Our Editor Tamsin van Essen at Cambridge University Press has provided invaluable scientific advice, while Suresh Kumar has been assisting us with TeX well beyond the call of duty.

Part I

From classical to wave mechanics

1
Experimental foundations of quantum theory

This chapter begins with a brief outline of some of the key motivations for considering a quantum theory: the early attempts to determine the spectral distribution of energy density of black bodies; stability of atoms and molecules; specific heats of solids; interference and diffraction of light beams; polarization of photons. The experimental foundations of wave mechanics are then presented in detail, but in a logical order quite different from its historical development: photo-emission of electrons by metallic surfaces, X- and γ-ray scattering from gases, liquids and solids, interference experiments, atomic spectra and the Bohr hypotheses, the experiment of Franck and Hertz, the Bragg experiment, diffraction of electrons by a crystal of nickel (Davisson and Germer), and measurements of position and velocity of an electron.

1.1 The need for a quantum theory

In the second half of the nineteenth century it seemed that the laws of classical mechanics, developed by the genius of Newton, Lagrange, Hamilton, Jacobi and Poincaré, the Maxwell theory of electromagnetic phenomena and the laws of classical statistical mechanics could account for all known physical phenomena. Still, it became gradually clear, after several decades of experimental and theoretical work, that one has to formulate a new kind of mechanics, which reduces to classical mechanics in a suitable limit, and makes it possible to obtain a consistent description of phenomena that cannot be understood within the classical framework. It is now appropriate to present a brief outline of this new class of phenomena, the systematic investigation of which is the object of the following sections and of chapters 4 and 14.

(i) In his attempt to derive the law for the spectral distribution of energy density of a body which is able to absorb all the radiant energy falling upon it, Planck was led to assume that the walls of such a body consist of harmonic oscillators, which exchange energy with the electromagnetic field inside the body only via integer multiples of a fundamental quantity ε_0. At this stage, to be consistent with another law that had been derived in a thermodynamical way and was hence of universal validity, the quantity ε_0 turned out to be proportional to the frequency of the radiation field, $\varepsilon_0 = h\nu$, and a new constant of nature, h, with dimension [energy] [time] and since then called the Planck constant, was introduced for the first time. These problems are part of the general theory of heat radiation (Planck 1991), and we have chosen to present them in some detail in chapter 14, which is devoted to the transition from classical to quantum statistical mechanics.

(ii) The crisis of classical physics, however, became even more evident when attempts were made to account for the stability of atoms and molecules. For example, if an atomic system, initially in an equilibrium state, is perturbed for a short time, it begins oscillating, and such oscillations are eventually transmitted to the electromagnetic field in its neighbourhood, so that the frequencies of the composite system can be observed by means of a spectrograph. In classical physics, independent of the precise form of the forces ruling the equilibrium stage, one would expect to be able to include the various frequencies in a scheme where some fundamental frequencies occur jointly with their harmonics. In contrast, the Ritz combination principle (see section 1.6) is found to hold, according to which all frequencies can be expressed as differences between some spectroscopic terms, the number of which is much smaller than the number of observed frequencies (Duck and Sudarshan 2000).

(iii) If one tries to overcome the above difficulties by postulating that the observed frequencies correspond to internal degrees of freedom of atomic systems, whereas the unknown laws of atomic forces forbid the occurrence of higher order harmonics (Dirac 1958), it becomes impossible to account for the experimental values of specific heats of solids at low temperatures (cf. section 14.8).

(iv) Interference and diffraction patterns of light can only be accounted for using a wave-like theory. This property is 'dual' to a particle-like picture, which is instead essential to understanding the emission of electrons by metallic surfaces that are hit by electromagnetic radiation (section 1.3) and the scattering of light by free electrons (section 1.4).

(v) It had already been a non-trivial achievement of Einstein to show that the energy of the electromagnetic field consists of elementary quantities $W = h\nu$, and it was as if these quanta of energy were localized in space

(Einstein 1905). In a subsequent paper, Einstein analysed a gas composed of several molecules that was able to emit or absorb radiation, and proved that, in such processes, linear momentum should be exchanged among the molecules, to avoid affecting the Maxwell distribution of velocities (Einstein 1917). This ensures, in turn, that statistical equilibrium is reached. Remarkably, the exchange of linear momentum cannot be obtained, unless one postulates that, if spontaneous emission occurs, this happens along a well-defined direction with corresponding vector $\vec{u}$, so that the linear momentum reads as

$$\vec{p} = \frac{W}{c}\vec{u} = \frac{h\nu}{c}\vec{u} = \frac{h}{\lambda}\vec{u}. \tag{1.1.1}$$

In contrast, if a molecule were able to emit radiation along all possible directions, as predicted by classical electromagnetic theory, the Maxwell distribution of velocities would be violated. There was, therefore, strong evidence that *spontaneous emission is directional.* Under certain circumstances, electromagnetic radiation behaves as if it were made of elementary quantities of energy $W = h\nu$, with speed c and linear momentum $\vec{p}$ as in Eq. (1.1.1). One then deals with the concept of energy quanta of the electromagnetic field, later called *photons* (Lewis 1926).

(vi) It is instructive, following Dirac (1958), to anticipate the description of polarized photons in the quantum theory we are going to develop. It is well known from experiments that the polarization of light is deeply intertwined with its corpuscular properties, and one comes to the conclusion that photons are, themselves, polarized. For example, a light beam with linear polarization should be viewed as consisting of photons each of which is linearly polarized in the same direction. Similarly, a light beam with circular polarization consists of photons that are all circularly polarized. One is thus led to say that each photon is in a given *polarization state.* The problem arises of how to apply this new concept to the spectral resolution of light into its polarized components, and to the recombination of such components. For this purpose, let us consider a light beam that passes through a tourmaline crystal, assuming that only linearly polarized light, perpendicular to the optical axis of the crystal, is found to emerge. According to classical electrodynamics, if the beam is polarized perpendicularly to the optical axis $\mathcal{O}$, it will pass through the crystal while remaining unaffected; if its polarization is parallel to $\mathcal{O}$, the light beam is instead unable to pass through the crystal; lastly, if the polarization direction of the beam forms an angle α with $\mathcal{O}$, only a fraction $\sin^2 \alpha$ passes through the crystal.

Let us assume, for simplicity, that the incoming beam consists of one photon only, and that one can detect what comes out on the other side of the crystal. We will learn that, according to quantum mechanics, in a number of experiments the whole photon is detected on the other side of the crystal, with energy equal to that of the incoming photon, whereas, in other circumstances, no photon is eventually detected. When a photon is detected, its polarization turns out to be perpendicular to the optical axis, but under no circumstances whatsoever shall we find, on the other side of the crystal, only a fraction of the incoming photon. However, *on repeating the experiment a sufficiently large number of times*, a photon will eventually be detected for a number of times equal to a fraction $\sin^2 \alpha$ of the total number of experiments. In other words, the photon is found to have a *probability* $\sin^2 \alpha$ of passing through the tourmaline, and a probability $\cos^2 \alpha$ of being, instead, absorbed by the tourmaline. A deep property, which will be the object of several sections from now on, is then found to emerge: when a series of experiments are performed, one can only predict a set of possible results with the corresponding probabilities.

As we will see in the rest of the chapter, the interpretation provided by quantum mechanics requires that a photon with oblique polarization can be viewed as being in part in a polarization state parallel to $\mathcal{O}$, and in part in a polarization state perpendicular to $\mathcal{O}$. In other words, a state of oblique polarization results from a 'superposition' of polarizations that are perpendicular and parallel to $\mathcal{O}$. It is hence possible to decompose any polarization state into two mutually orthogonal polarization states, i.e. to express it as a superposition of such states.

Moreover, when we perform an observation, we can tell whether the photon is polarized in a direction parallel or perpendicular to $\mathcal{O}$, because the measurement process makes the photon be in one of these two polarization states. Such a theoretical description requires a sudden change from a linear superposition of polarization states (prior to measurement) to a state where the polarization of the photon is *either* parallel *or* perpendicular to $\mathcal{O}$ (after the measurement).

Our brief outline has described many new problems that the general reader is not expected to know already. Now that his intellectual curiosity has been stimulated, we can begin a thorough investigation of all such topics. The journey is not an easy one, but the effort to understand what leads to a quantum theory will hopefully engender a better understanding of the physical world.

1.2 Our path towards quantum theory

Unlike the historical development outlined in the previous section, our path towards quantum theory, with emphasis on wave mechanics, will rely on the following properties.

(i) The photoelectric effect, Compton effect and interference phenomena provide clear experimental evidence for the existence of photons. 'Corpuscular' and 'wave' behaviour require that we use both 'attributes', therefore we need a relation between wave concepts and corpuscular concepts. This is provided for photons by the Einstein identification (see appendix 1.A)

$$\left(\vec{k} \cdot d\vec{x} - \omega \, dt\right) = \frac{1}{\hbar}\left(\vec{p} \cdot d\vec{x} - p_0 \, dx_0\right). \tag{1.2.1a}$$

More precisely, light has a corpuscular nature that becomes evident thanks to the photoelectric and Compton effects, but also a wave-like nature as is shown by interference experiments. Although photons are massless, one can associate to them a linear momentum $\vec{p} = \hbar\vec{k}$, and their energy equals $\hbar\omega = h\nu$.

(ii) The form of the emission and absorption spectra, and the Bohr hypotheses (section 1.6). Experimental evidence of the existence of energy levels (section 1.7).

(iii) The wave-like behaviour of massive particles postulated by de Broglie (1923) and found in the experiment of Davisson and Germer (1927, diffraction of electrons by a crystal of nickel). For such particles one can perform the de Broglie identification

$$\left(\vec{p} \cdot d\vec{x} - p_0 \, dx_0\right) = \hbar\left(\vec{k} \cdot d\vec{x} - \omega \, dt\right). \tag{1.2.1b}$$

It is then possible to estimate when the corpuscular or wave-like aspects of particles are relevant in some physical processes.

1.3 Photoelectric effect

In the analysis of black-body radiation one met, for the first time, the hypothesis of quanta: whenever matter emits or absorbs radiation, it does so in a sequence of elementary acts, in each of which an amount of energy ε is emitted or absorbed proportional to the frequency ν of the radiation: $\varepsilon = h\nu$, where h is the universal constant known as Planck's constant. We are now going to see how the ideas developed along similar lines make it possible to obtain a satisfactory understanding of the photoelectric effect.

The photoelectric effect was discovered by Hertz and Hallwachs in 1887. The effect consists of the emission of electrons from the surface of a solid when electromagnetic radiation is incident upon it (Hughes and DuBridge 1932, DuBridge 1933, Holton 2000). The three empirical laws of such an effect are as follows (see figures 1.1 and 1.2; the Millikan experiment

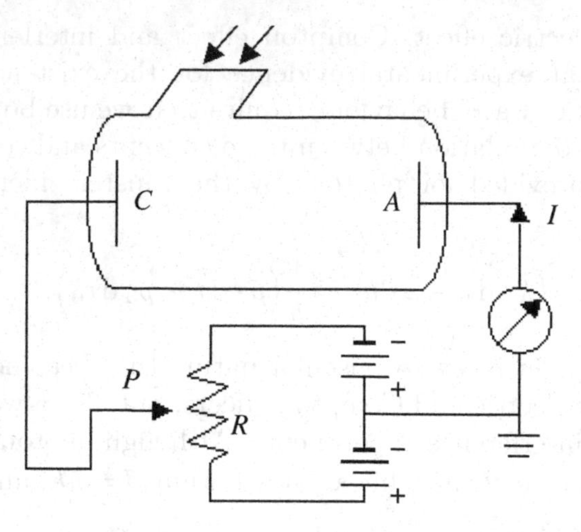

Fig. 1.1. The circuit used in the Millikan experiment. The energy with which the electron leaves the surface is measured by the product of its charge with the potential difference against which it is just able to drive itself before being brought to rest. Millikan was careful enough to use only light for which the illuminated electrode was photoelectrically sensitive, but for which the surrounding walls were not photosensitive.

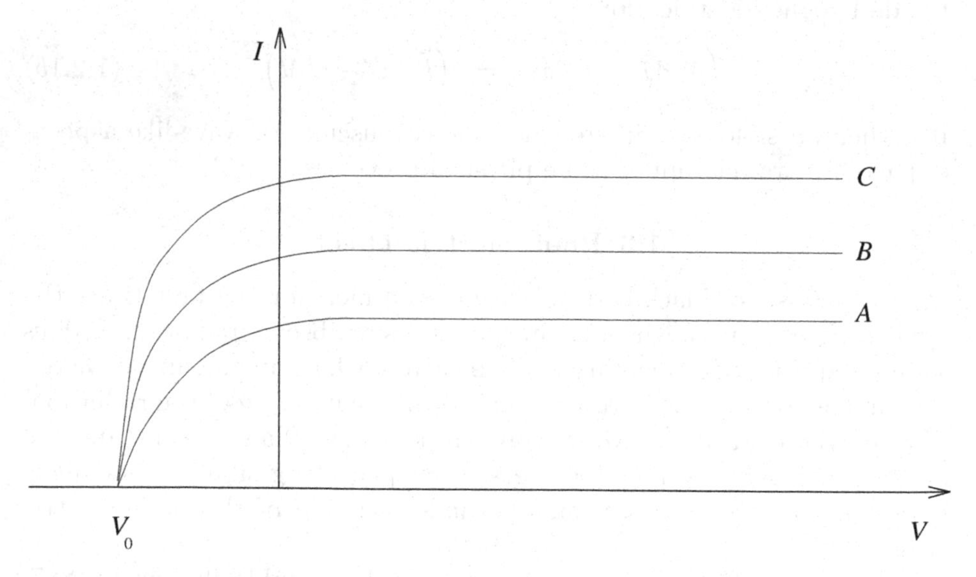

Fig. 1.2. Variation of the photoelectric current with voltage, for given values of the intensity.

quoted therein should not be confused with the measuremt of the electron charge, also due to Millikan).

(i) The electrons are emitted only if the frequency of the incoming radiation is greater than a certain value ν_0, which is a peculiar property of the metal used in the experiment, and is called *the photoelectric threshold.*

(ii) The velocities of the electrons emitted by the surface range from 0 to a maximum value of $v_{\rm max}$. The kinetic energy corresponding to $v_{\rm max}$ depends linearly on the frequency ν: $T_{\rm max} = k(\nu - \nu_0), k > 0$. $T_{\rm max}$ does not depend on the intensity of the incoming radiation.

(iii) For a given value of the frequency ν of the incoming radiation, the *number* of electrons emitted per cm^2 per second is proportional to the intensity.

These properties cannot be understood if one assumes that classical electromagnetic theory rules the phenomenon. In particular, if one assumes that the energy is uniformly distributed over the metallic surface, it is unclear how the emission of electrons can occur when the intensity of the radiation is extremely low (which would require a long time before the electron would receive enough energy to escape from the metal). The experiments of Lawrence and Beans showed that the time lag between the incidence of radiation on a surface and the appearance of (photo)electrons is less than 10^{-9} s.

However, the peculiar emission of electrons is naturally accounted for, if Planck's hypothesis is accepted. More precisely, one has to assume that the energy of radiation is quantized not only when emission or absorption occur, but can also travel in space in the form of elementary quanta of radiation with energy $h\nu$. Correspondingly, the photoelectric effect should be thought of as a collision process between the incoming quanta of radiation and the electrons belonging to the atoms of the metallic surface. According to this quantum scheme, the atom upon which the photon falls receives, all at once, the energy $h\nu$. As a result of this process, an electron can be emitted only if the energy $h\nu$ is greater than the work function W_0:

$$h\nu > W_0. \tag{1.3.1}$$

The first experimental law, (i), is therefore understood, provided one identifies the photoelectric threshold with $\frac{W_0}{h}$:

$$\nu_0 = \frac{W_0}{h}. \tag{1.3.2}$$

If the inequality (1.3.1) is satisfied, the electron can leave the metallic plate with an energy which, at the very best, is $W = h\nu - W_0$, which implies

$$W_{\max} = h(\nu - \nu_0). \tag{1.3.3}$$

This agrees completely with the second law, (ii). Lastly, upon varying the intensity of the incoming radiation, the number of quanta falling upon the surface in a given time interval changes, but from the above formulae it is clear that the energy of the quanta, and hence of the electrons emitted, is not affected by the intensity.

In the experimental apparatus (see figure 1.1), ultraviolet or X-rays fall upon a clean metal cathode, and an electrode collects the electrons that are emitted with kinetic energy $T = h\nu - W_0$. If V_0 is the potential for which the current vanishes, one has (see figure 1.3)

$$V_0 = \frac{h\nu}{e} - \frac{W_0}{e}. \tag{1.3.4}$$

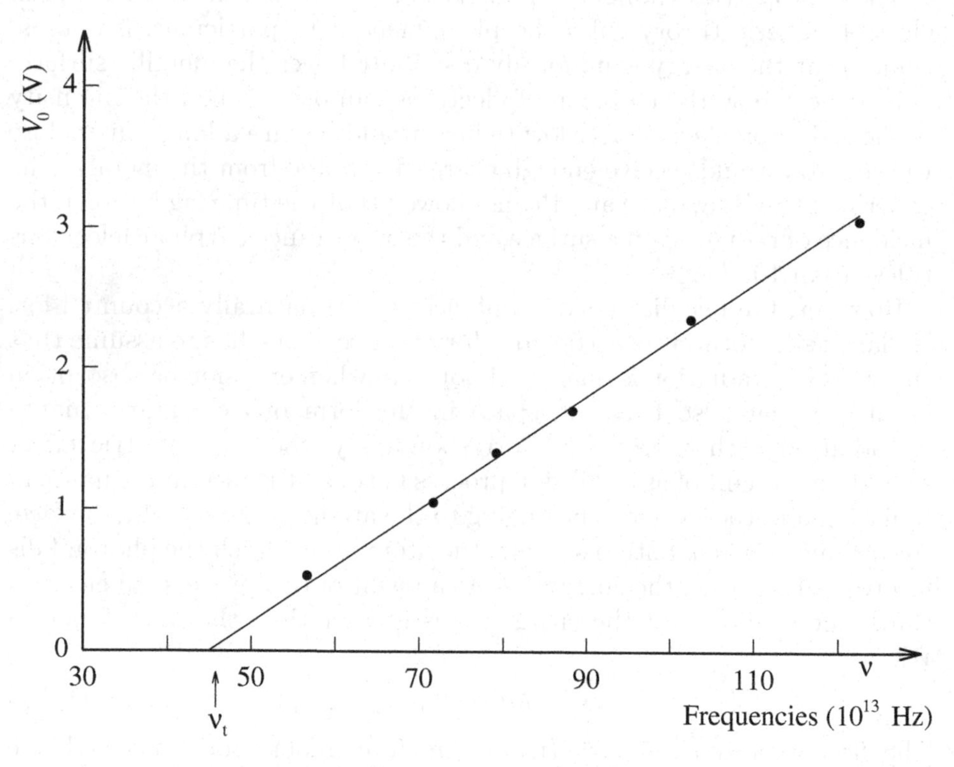

Fig. 1.3. Results of the Millikan experiment for the retarding potential V_0 expressed as a function of frequency (Millikan 1916, © the American Physical Society). A linear relation is found between V_0 and ν, and the slope of the corresponding line is numerically equal to $\frac{h}{e}$. The intercept of such a line on the ν axis is the lowest frequency at which the metal in question can be photoelectrically active.

The plot of $V_0(\nu)$ is a straight line that intersects the ν-axis when $\nu = \nu_0$. The slope of the experimental curve makes it possible to measure Planck's constant (for this purpose, Millikan used monochromatic light). The value of the ratio $\frac{h}{e}$ is 4.14×10^{-15} V s, with $h = 6.6 \times 10^{-27}$ erg s.

Einstein made a highly non-trivial step, by postulating the existence of elementary quanta of radiation which travel in space. This was far more than what Planck had originally demanded in his attempt to understand black-body radiation. Note also that, strictly, Einstein was not aiming to 'explain' the photoelectric effect. When he wrote his fundamental papers (Einstein 1905, 1917), the task of theoretical physicists was not quite that of having to understand a well-established phenomenology, since the Millikan measurements were made 10 years after the first Einstein paper. Rather, Einstein developed some far-reaching ideas which, in particular, can be applied to account for all known aspects of the photoelectric effect. Indeed, in Einstein (1905), the author writes as follows.

. . . The wave theory of light, which operates with continuous spatial functions, has worked well in the representation of purely optical phenomena and will probably never be replaced by another theory. It should be kept in mind, however, that the optical observations refer to time averages rather than instantaneous values. In spite of the complete experimental confirmation of the theory as applied to diffraction, reflection, refraction, dispersion, etc., it is still conceivable that the theory of light which operates with continuous spatial functions may lead to contradictions with experience when it is applied to the phenomena of emission and transformation of light.

It seems to me that the observations associated with blackbody radiation, fluorescence, the production of cathode rays by ultraviolet light, and other related phenomena connected with the emission or transformation of light are more readily understood if one assumes that the energy of light is discontinuously distributed in space. In accordance with the assumption to be considered here, the energy of a light ray spreading out from a point source is not continuously distributed over an increasing space but consists of a finite number of energy quanta which are localized at points in space, which move without dividing, and which can only be produced and absorbed as complete units.

1.4 Compton effect

Classically, a monochromatic plane wave of electromagnetic nature carries momentum according to the relation $p = \frac{E}{c}$. Since E is quantized, one

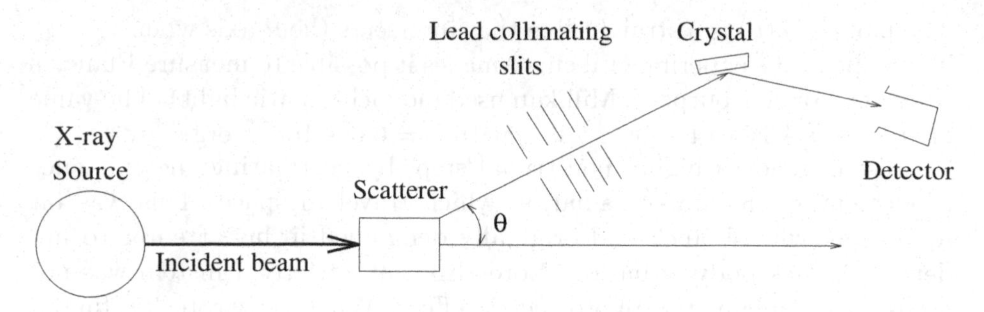

Fig. 1.4 Experimental setup for the Compton experiment.

is naturally led to ask whether the momentum is carried in the form of quanta with absolute value $\frac{h\nu}{c}$. The Compton effect (Compton 1923a,b) provides clear experimental evidence in favour of this conclusion, and supports the existence of photons. For this purpose, the scattering of monochromatic X- and γ-rays from gases, liquids and solids is studied in the laboratory (see figure 1.4). Under normal circumstances, the X-rays pass through a material of low atomic weight (e.g. coal). A spectrograph made out of crystal collects and analyses the rays scattered in a given direction. One then finds, jointly with the radiation scattered by means of the process we are going to describe, yet another radiation which is scattered without any change of its wavelength. There exist two nearby lines: one of them has the same wavelength λ as the incoming radiation, whereas the other line has a wavelength $\lambda' > \lambda$. The line for which the wavelength remains unaffected can be accounted for by thinking that the incoming photon also meets the 'deeper underlying' electrons of the scattering material. For such processes, the mass of the whole atom is involved, which reduces the value of the shift $\lambda' - \lambda$ significantly, so that it becomes virtually unobservable. We are now going to consider the scattering process involving *the external electron only.*

Let us assume that the incoming radiation consists of photons having frequency ν. Let $m_{\rm e}$ be the rest mass of the electron, $\vec{v}$ its velocity after collision with the photon and let ν' be the frequency of the scattered photon. The conservation laws that make it possible to obtain a theoretical description of the phenomenon are the conservation of energy and momentum, and the description has to be considered within a relativistic setting. We denote by $\vec{l}$ the unit vector along the direction of the incoming photon, and by $\vec{u}$ the unit vector along the direction of emission of the scattered photon (see figure 1.5).

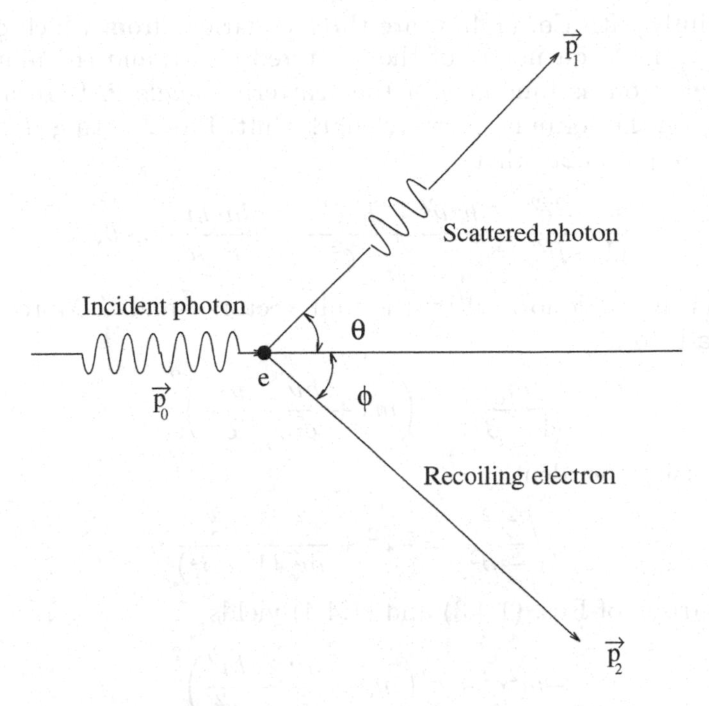

Fig. 1.5. A photon with linear momentum $\vec{p}_0$ collides with an electron at rest and is scattered with momentum $\vec{p}_1$, while the electron recoils with momentum $\vec{p}_2$.

The energy conservation reads, in our problem, as

$$m_e c^2 + h\nu = \frac{m_e c^2}{\sqrt{1 - \frac{v^2}{c^2}}} + h\nu'. \tag{1.4.1}$$

Moreover, taking into account that the momentum of the electron vanishes before the scattering takes place, the conservation of momentum leads to

$$\frac{h\nu}{c}\vec{l} = \frac{m_e \vec{v}}{\sqrt{1 - \frac{v^2}{c^2}}} + \frac{h\nu'}{c}\vec{u}. \tag{1.4.2a}$$

If Eq. (1.4.2a) is projected onto the x- and y-axes it yields the equations (see figure 1.5)

$$\frac{h\nu}{c} = \frac{h\nu'}{c}\cos\theta + p\cos\phi, \tag{1.4.2b}$$

$$\frac{h\nu'}{c}\sin\theta = p\sin\phi, \tag{1.4.2c}$$

which, jointly with Eq. (1.4.1), are three equations from which one may evaluate ϕ, the frequency ν' of the scattered X-ray, and the momentum p of the electron as functions of the scattering angle θ. Here attention is focused on the formula for wavelength shift. First, setting $\beta \equiv \frac{v}{c}$, one finds from Eq. (1.4.2a) that

$$\frac{m_{\rm e}^2 \beta^2 c^2}{(1-\beta^2)} = \frac{h^2\nu^2}{c^2} + \frac{h^2\nu'^2}{c^2} - 2\frac{h\nu}{c}\frac{h\nu'}{c}\cos\theta, \tag{1.4.3}$$

where θ is the angle formed by the unit vectors $\vec{l}$ and $\vec{u}$. Moreover, Eq. (1.4.1) leads to

$$\frac{m_{\rm e}^2}{(1-\beta^2)} = \left(m_{\rm e} + \frac{h\nu}{c^2} - \frac{h\nu'}{c^2}\right)^2. \tag{1.4.4}$$

Thus, on using the identity

$$\frac{\beta^2 c^2}{(1-\beta^2)} = -c^2 + \frac{c^2}{m_{\rm e}^2}\frac{m_{\rm e}^2}{(1-\beta^2)}, \tag{1.4.5}$$

the comparison of Eqs. (1.4.3) and (1.4.4) yields

$$-m_{\rm e}^2 c^2 + c^2\left(m_{\rm e} + \frac{h\nu}{c^2} - \frac{h\nu'}{c^2}\right)^2$$
$$= \frac{h^2\nu^2}{c^2} + \frac{h^2\nu'^2}{c^2} - 2\frac{h\nu}{c}\frac{h\nu'}{c}\cos\theta. \tag{1.4.6}$$

A number of cancellations are now found to occur, which significantly simplifies the final result, i.e.

$$\nu - \nu' = \frac{h\nu\nu'}{m_{\rm e}c^2}(1-\cos\theta). \tag{1.4.7}$$

However, the main object of interest is the formula for $\lambda' - \lambda$, which is obtained from Eq. (1.4.7) and the well-known relation between frequency and wavelength: $\nu/c = 1/\lambda, \nu'/c = 1/\lambda'$. Hence one finds

$$\lambda' - \lambda = \frac{h}{m_{\rm e}c}(1-\cos\theta), \tag{1.4.8}$$

where

$$\frac{h}{m_{\rm e}c} = 0.0024\,{\rm nm}. \tag{1.4.9}$$

Interestingly, the wavelength shift is maximal when $\cos\theta = -1$, and it vanishes when $\cos\theta = 1$. In the actual experiments, the scattered photons are detected if in turn, they meet an atom that is able to absorb them (provided that such an atom can emit, by means of the photoelectric effect, an electron, the passage of which is visible on a photographic plate).

We can thus conclude that photons behave exactly as if they were particles with energy $h\nu$ and momentum $\frac{h\nu}{c}$. According to relativity theory, developed by Einstein and Poincaré, the equation $p = \frac{E}{c}$ is a peculiar property of massless particles. Thus, we can say that photons behave like massless particles.

The frequency shift is a peculiar property of a quantum theory which relies on the existence of photons, because in the classical electromagnetic theory no frequency shift would occur. To appreciate this, let us consider the classical description of the phenomenon. On denoting the position vector in $\mathbf{R}^3$ by $\vec{r}$, with Cartesian coordinates (x, y, z), and by $\vec{k}$ the wave vector with corresponding components (k_x, k_y, k_z), the electric field of the incoming plane wave of frequency $\nu = \frac{\omega}{2\pi}$ may be written in the form

$$\vec{E} = \vec{E}_0 \cos\left(\vec{k} \cdot \vec{r} - \omega t\right), \tag{1.4.10}$$

where the vector $\vec{E}_0$ has components $\left(E_{0_x}, E_{0_y}, E_{0_z}\right)$ independent of (x, y, z, t). Strictly, one has then to build a wave packet from these elementary solutions of the Maxwell equations, but Eq. (1.4.10) is all we need to obtain the classical result. The electric field which varies in space and time according to Eq. (1.4.10) generates a magnetic field that also varies in space and time in a similar way. This is clearly seen from one of the vacuum Maxwell equations, i.e. (we do not present the check of the Maxwell equations for the divergences of $\vec{E}$ and $\vec{B}$, but the reader can easily perform it)

$$\text{curl}\, \vec{E} + \frac{1}{c}\frac{\partial \vec{B}}{\partial t} = 0, \tag{1.4.11}$$

which can be integrated to find

$$\vec{B} = -c \int \text{curl}\, \vec{E}\ \text{d}t. \tag{1.4.12}$$

Now the standard definition of the curl operator, jointly with Eq. (1.4.10), implies that

$$\left(\text{curl}\, \vec{E}\right)_x \equiv \frac{\partial E_z}{\partial y} - \frac{\partial E_y}{\partial z} = \left(k_z E_{0_y} - k_y E_{0_z}\right) \sin\left(\vec{k} \cdot \vec{r} - \omega t\right), \tag{1.4.13}$$

$$\left(\text{curl}\, \vec{E}\right)_y \equiv \frac{\partial E_x}{\partial z} - \frac{\partial E_z}{\partial x} = \left(k_x E_{0_z} - k_z E_{0_x}\right) \sin\left(\vec{k} \cdot \vec{r} - \omega t\right), \tag{1.4.14}$$

$$\left(\text{curl}\, \vec{E}\right)_z \equiv \frac{\partial E_y}{\partial x} - \frac{\partial E_x}{\partial y} = \left(k_y E_{0_x} - k_x E_{0_y}\right) \sin\left(\vec{k} \cdot \vec{r} - \omega t\right). \tag{1.4.15}$$

The coefficients of the sin function on the right-hand sides of Eqs. (1.4.13)–(1.4.15) are easily seen to be minus the components along the

X, Y, Z axes, respectively, of the vector product $\vec{k} \wedge \vec{E}_0$, and hence one finds

$$\mathrm{curl}\, \vec{E} = -\vec{k} \wedge \vec{E}_0 \, \sin\left(\vec{k} \cdot \vec{r} - \omega t\right). \tag{1.4.16}$$

By virtue of Eqs. (1.4.12) and (1.4.16) one finds

$$\vec{B} = \vec{B}_0 \cos\left(\vec{k} \cdot \vec{r} - \omega t\right), \tag{1.4.17}$$

where we have defined

$$\vec{B}_0 \equiv \frac{c}{\omega} \vec{k} \wedge \vec{E}_0. \tag{1.4.18}$$

The force acting on the electron of charge e is, therefore,

$$\begin{aligned} \vec{F} = m \frac{\mathrm{d}}{\mathrm{d}t} \vec{v} &= e \left(\vec{E} + \frac{1}{c} \vec{v} \wedge \vec{B} \right) \\ &= e \left(\vec{E}_0 + \frac{1}{c} \vec{v} \wedge \vec{B}_0 \right) \cos\left(\vec{k} \cdot \vec{r} - \omega t\right), \end{aligned} \tag{1.4.19}$$

where, by virtue of Eq. (1.4.18), one finds

$$\begin{aligned} \left(\vec{v} \wedge \vec{B}_0\right)_x &\equiv v_y B_{0_z} - v_z B_{0_y} \\ &= \frac{c}{\omega} \left[v_y \left(k_x E_{0_y} - k_y E_{0_x} \right) - v_z \left(k_z E_{0_x} - k_x E_{0_z} \right) \right] \\ &= \frac{c}{\omega} \left[k_x \left(\vec{v} \cdot \vec{E}_0 \right) - E_{0_x} \left(\vec{v} \cdot \vec{k} \right) \right]. \end{aligned} \tag{1.4.20}$$

Analogous equations hold for the other components of $\vec{v} \wedge \vec{B}_0$ so that, eventually,

$$\vec{v} \wedge \vec{B}_0 = \frac{c}{\omega} \left[\vec{k} \left(\vec{v} \cdot \vec{E}_0 \right) - \vec{E}_0 \left(\vec{v} \cdot \vec{k} \right) \right], \tag{1.4.21}$$

which implies (see Eqs. (1.4.10) and (1.4.19))

$$m \frac{\mathrm{d}}{\mathrm{d}t} \vec{v} = e \left\{ \vec{E} + \frac{1}{\omega} \left[\vec{k} \left(\vec{v} \cdot \vec{E} \right) - \vec{E} \left(\vec{v} \cdot \vec{k} \right) \right] \right\}, \tag{1.4.22}$$

with

$$\frac{\mathrm{d}}{\mathrm{d}t} \vec{r} = \vec{v}. \tag{1.4.23}$$

The magnetic forces are negligible compared with the electric forces, so that the acceleration of the electron reduces to $\frac{\mathrm{d}\vec{v}}{\mathrm{d}t} = \frac{e}{m} \vec{E}$. By virtue of its oscillatory motion, the electron begins to radiate a field which, at a distance R, has components with magnitude (Jackson 1975)

$$|\vec{E}'| = |\vec{B}'| = \frac{e}{c^2 R} \ddot{r} \sin\phi, \tag{1.4.24}$$

where c is the velocity of light and ϕ is the angle between the scattered beam and the line along which the electron oscillates. Substituting for the acceleration, one finds

$$|\vec{E}'| = |\vec{B}'| = \frac{e^2 E \sin\phi}{mc^2 R}. \tag{1.4.25}$$

1.4.1 Thomson scattering

To sum up, in a classical model, the atomic electrons vibrate with the same frequency as the incident radiation. These oscillating electrons, in turn, radiate electromagnetic waves *of the same frequency*, leading to the so-called Thomson scattering. This is a non-relativistic scattering process, which describes X-ray scattering from electrons and γ-ray scattering from protons. For a particle of charge q and mass m, the total Thompson scattering cross-section (recall that the cross-section describes basically the probability of the scattering process) reads as (Jackson 1975)

$$\sigma_{\rm T} = \frac{8\pi}{3}\left(\frac{q^2}{mc^2}\right)^2. \tag{1.4.26}$$

For electrons, $\sigma_T = 0.665 \times 10^{-24}$ cm^2. The associated characteristic length is

$$\frac{q^2}{mc^2} = 2.82 \times 10^{-13} \text{ cm} \tag{1.4.27}$$

and is called the classical electron radius.

1.5 Interference experiments

The wave-like nature of light is proved by the interference phenomena it gives rise to. It is hence legitimate to ask the question: how can we accept the existence of interference phenomena, if we think of light as consisting of photons? There are, indeed, various devices that can produce interference fringes. For example, a source S of monochromatic light illuminates an opaque screen where two thin slits, close to each other, have been produced. In passing through the slits, light is diffracted. On a plate L located a distance from the slits, interference fringes are observed in the area where overlapping occurs of the diffraction patterns produced from the slits A and B, i.e. where light is simultaneously received from A and B (see figure 1.6).

Another device is the Fresnel biprism (Born and Wolf 1959): the monochromatic light emitted from S is incident on two coupled prisms P_1

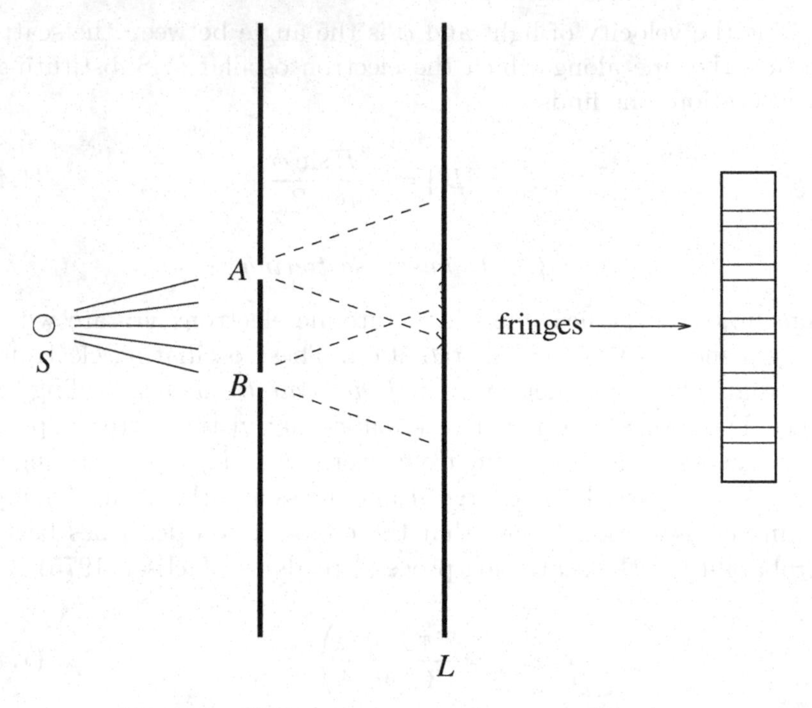

Fig. 1.6 Diffraction pattern from a double slit.

and P_2; light rays are deviated from P_1 and P_2 as if they were emitted from two (virtual) coherent sources S' and S''. As in the previous device, interference fringes are observed where light emitted both from P_1 and P_2 is collected (see figure 1.7).

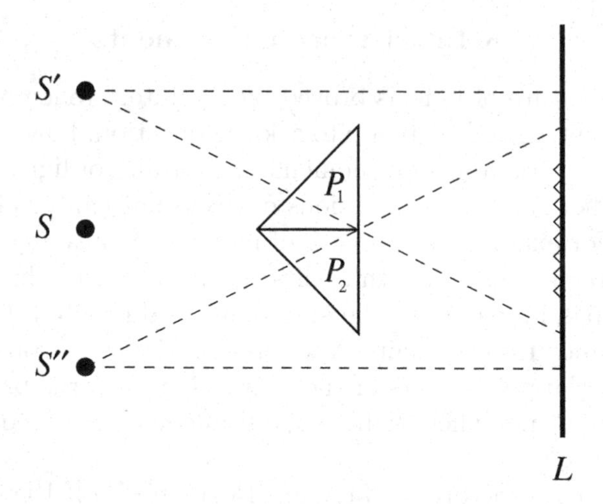

Fig. 1.7 Diffraction from a biprism.

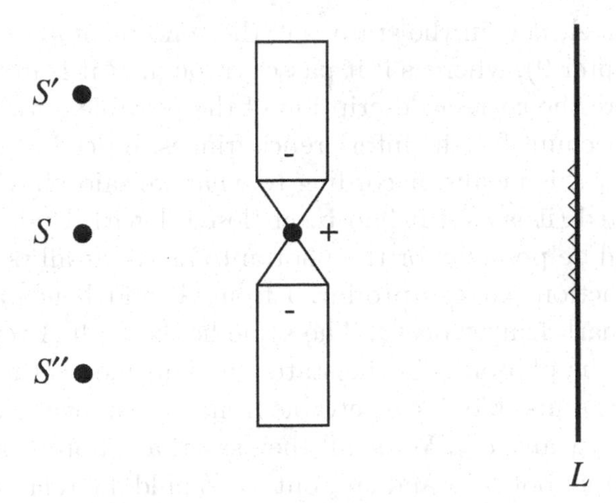

Fig. 1.8. Interference fringes with electrons, when they are deviated by an electric field.

Interestingly, the Fresnel biprism makes it possible to produce interference fringes with electrons. The source S is replaced by an electron gun and the biprism is replaced by a metallic panel where a slit has been produced. At the centre of the slit, a wire of silver-plated quartz is maintained at a potential slightly greater than the potential of the screen. The electrons are deviated by the electric field of the slit, and they reach the screen as if they were coming from two different sources (see figure 1.8). For simplicity, we can consider the Fresnel biprism and talk about photons, but of course this discussion can be repeated in precisely the same way for electrons.

How can one interpret the interference experiment in terms of photons? It is clear that bright fringes result from the arrival of several photons, whereas no photons arrive where dark fringes are observed. It therefore seems that the various photons interact with each other so as to give rise, on plate L, to an irregular distribution of photons, and hence bright as well as dark fringes are observed. If this is the case, what is going to happen if we reduce the intensity of the light emitted by S until only one photon at a time travels from the source S to the plate L? The answer is that we have then to increase the exposure time of the plate L, but eventually we will find the same interference fringes as discussed previously. Thus, the interpretation based upon the interaction among photons is incorrect: photons do not interfere with each other, but the only possible conclusion is that the interference involves the single photon, just as in the case of the superposition for polarization. However, according to a particle picture, a photon (or an electron) starting from S and arriving at L, *either* passes through A *or* passes through B. We shall say that, if it passes through A, it is in the state ψ_A (the concept of state will be fully defined in chapter 9), whereas if it passes through B it is in the state ψ_B. But if this were the correct description of the possible options, we would be unable to account for the interference fringes. Indeed, if the photon is in the state ψ_A this means, according to what we said above, that slit B can be neglected (it is as if it had been closed down). Under such conditions, it should be possible for the photon to arrive at all points on plate L of the diffraction pattern produced from A, and hence also at those points where dark fringes occur. The same holds, with A replaced by B, if we say that the photon is in the state ψ_B. This means that a third option should be admissible, inconceivable from the classical viewpoint, and different from ψ_A and ψ_B. We shall then say that photons are in a state ψ_C, different from both ψ_A and ψ_B, but ψ_C should 'be related', somehow, with both ψ_A and ψ_B. In other words, it is incorrect to say that photons pass through A or through B, but it is as if each of them were passing, at the same time, through both A and B. This conclusion is suggested by the wave-like interpretation of the interference phenomenon: if only slit A is open, there exists a wave $A(x, y, z, t)$ in between the screen and L, whereas, if only slit B is open, there exists a wave $B(x, y, z, t)$ in between the screen and L. If now both slits are opened up, the wave involved is neither $A(\vec{r}, t)$ nor $B(\vec{r}, t)$, but $C(\vec{r}, t) = A(\vec{r}, t) + B(\vec{r}, t)$.

But then, if the photon is passing 'partly through A and partly through B', what should we expect if we place two photomultipliers F_1 and F_2 in front of A and B, respectively, and a photon is emitted from S (see figure 1.9)? Should we expect that F_1 and F_2 register, at the same time, the passage of the photon? If this were the case, we would have achieved, in the laboratory, the 'division' of a photon! What happens, however, is that only one of the two photomultipliers registers the passage of the photon, and upon repeating the experiment several times one finds that, *on average*, half of the events can be ascribed to F_1 and half of the events can be ascribed to F_2. Does this mean that the existence of the state ψ_C is an incorrect assumption? Note, however, that the presence of photomultipliers has made it impossible to observe the interference fringes, since the photons are completely absorbed by such devices.

At this stage, one might think that, with the help of a more sophisticated experiment, one could still detect which path has been followed by photons, while maintaining the ability to observe interference fringes. For this purpose, one might think of placing a mirror S_1 behind slit A, and another mirror S_2 behind slit B (see figure 1.10). Such mirrors can be freely moved by hypothesis, so that, by observing their recoil, one

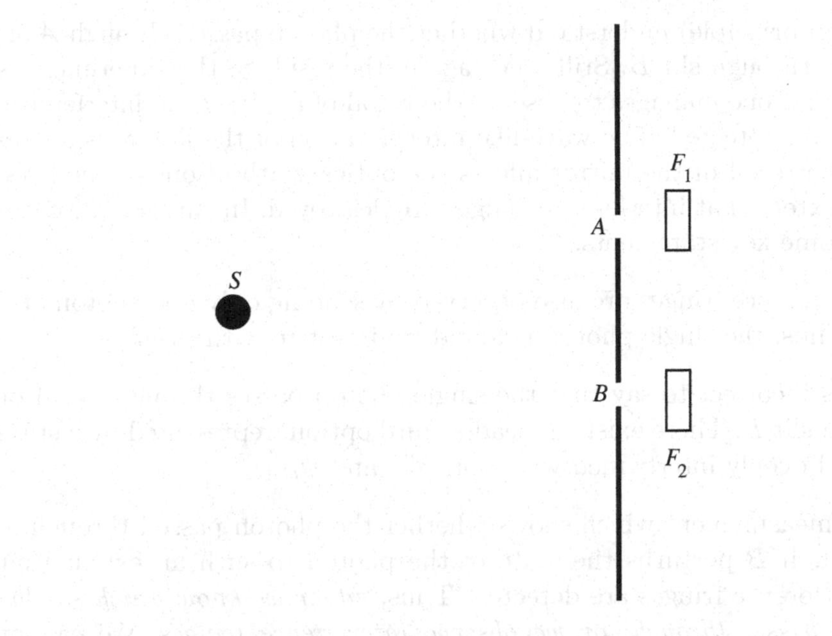

Fig. 1.9 Double-slit experiment with a photon.

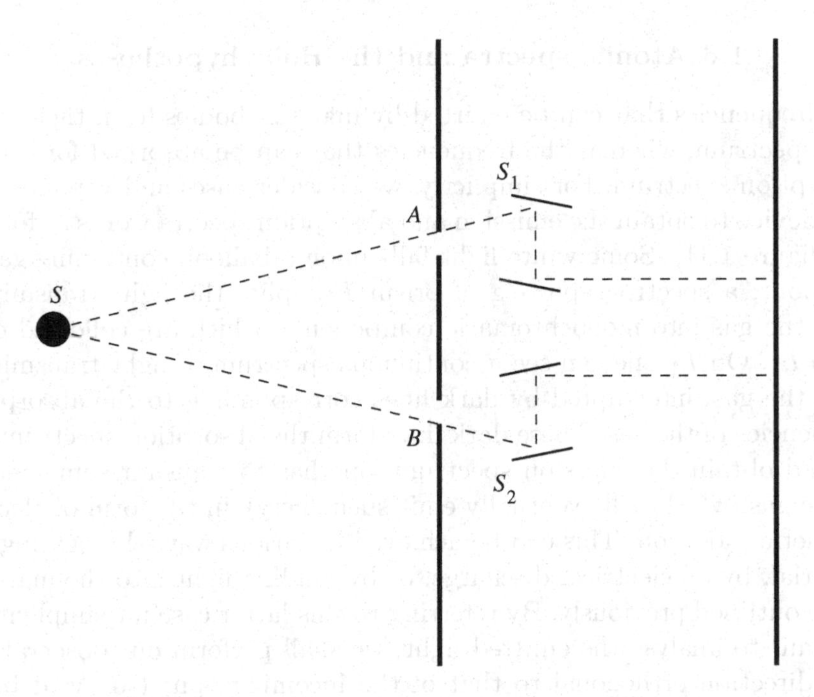

Fig. 1.10 Double-slit experiment supplemented by mirrors.

could (in principle) understand whether the photon passed through A or, instead, through slit B. Still, once again, the result of the experiment is negative: if one manages to observe the recoil of a mirror, no interference fringes are detected. The wave-like interpretation of the failure is as follows: the recoil of the mirror affects the optical path of one of the rays, to the extent that interference fringes are destroyed. In summary, we can make some key statements.

(i) Interference fringes are also observed by sending only one photon at a time. Thus, the single photon is found 'to interfere with itself'.

(ii) It is incorrect to say that the single photon passes through slit A or through slit B. There exists instead a third option, represented by a state ψ_C, and deeply intertwined with both ψ_A and ψ_B.

(iii) A measurement which shows whether the photon passed through A or through B perturbs the state of the photon to such an extent that no interference fringes are detected. Thus, *either we know which slit the photon passed through, or we observe interference fringes.* We cannot achieve both goals: the two possibilities are incompatible.

1.6 Atomic spectra and the Bohr hypotheses

The frequencies that can be emitted by material bodies form their emission spectrum, whereas the frequencies that can be absorbed form their absorption spectrum. For simplicity, we consider gases and vapours.

A device to obtain the emission and absorption spectra works as follows (see figure 1.11). Some white light falls upon a balloon containing gas or a vapour; a spectrograph, e.g. a prism P_1, splits the light transmitted from the gas into monochromatic components, which are collected on a plate L_1. On L_1 one can see a continuous spectrum of light transmitted from the gas, interrupted by dark lines corresponding to the absorption frequencies of the gas. These dark lines form the absorption spectrum. To instead obtain the emission spectrum, one has to transmit some energy to the gas, which will eventually emit such energy in the form of electromagnetic radiation. This can be achieved in various ways: by heating the material, by an electrical discharge, or by sending light into the material as we outlined previously. By referring to this latter case for simplicity, if we want to analyse the emitted light, we shall perform our observations in a direction orthogonal to that of the incoming light (to avoid being disturbed by such light). A second prism P_2 is inserted to decompose the radiation emitted from the gas, and this is collected on plate L_2. On L_2 one can see, on a dark background, some bright lines corresponding

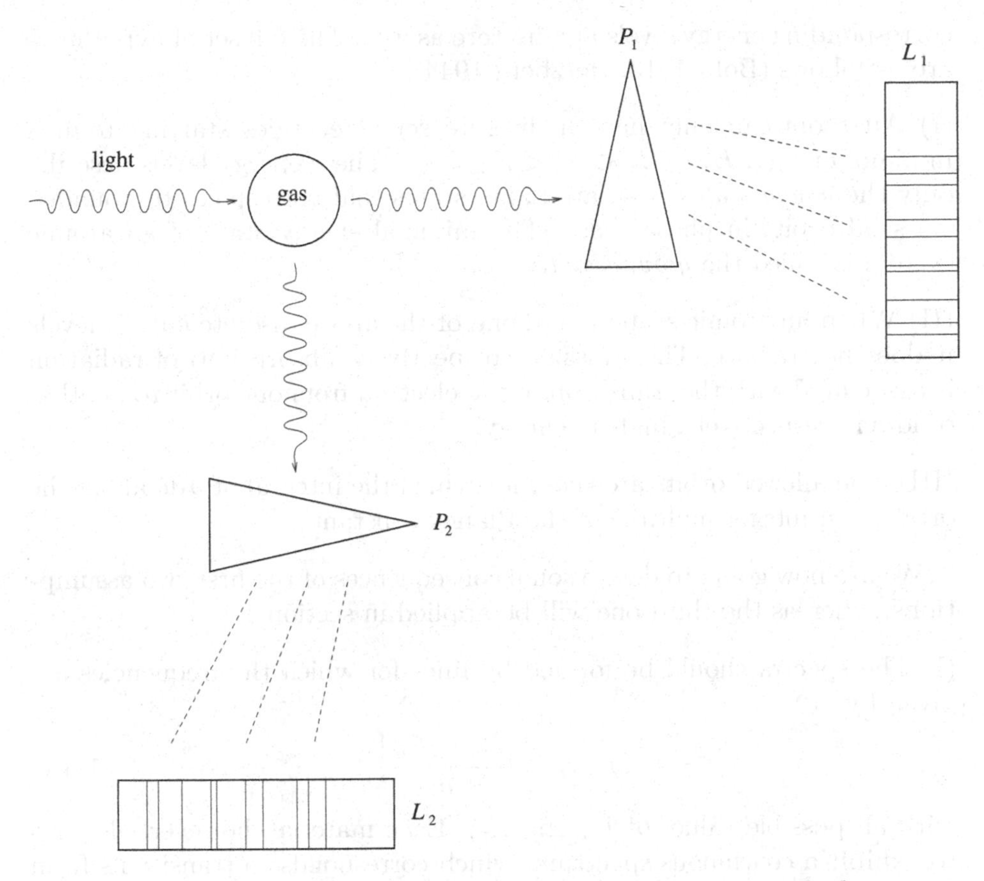

Fig. 1.11. Experimental setup used to obtain the emission and absorption spectra.

to the frequencies emitted from the gas. These lines form the emission spectrum.

First, the observations show that the emission and absorption spectra are quite different: the emission spectrum contains far more lines than the absorption spectrum, and one can find, within it, all lines of the absorption spectrum. Moreover, if the incoming radiation has a spectrum of frequencies ν greater than a certain value ν_1, it is also possible to observe, in the emission spectrum, lines corresponding to frequencies smaller than ν_1. To account for the emission and absorption spectra, Bohr made some assumptions (Bohr 1913) that, as in the case of Einstein's hypothesis, disagree with classical physics, which was indeed unable to account for the properties of the spectra. The basic idea was that privileged orbits for atoms exist that are stable. If the electrons in the atom lie on one of these orbits, they do not radiate. Such orbits are discrete, and hence the corresponding energy levels are discrete as well. The full set of hypotheses are as follows (Bohr 1913, Herzberg 1944).

(I) An atom can only have a discrete set of energies starting from a minimal energy: $E_1 < E_2 < \cdots < E_n < \cdots$. These energy levels describe only the *bound states* of an atom, i.e. states that correspond to bounded classical orbits in phase space. The minimal energy state of an atomic system is called the *ground state.*

(II) When an atomic system is in one of the above discrete energy levels it does not radiate. The emission (respectively, absorption) of radiation is associated with the transition of the electron from one orbit to another of lower (respectively, higher) energy.

(III) The allowed orbits are those for which the integral of $p\,dq$ along the orbit is an integer multiple of the Planck constant.

We are now going to derive some consequences of the first two assumptions, whereas the third one will be applied in section 1.8.

(i) The spectra should be formed by lines for which the frequencies are given by

$$\nu_{n,m} = \frac{|E_n - E_m|}{h}, \tag{1.6.1}$$

with all possible values of E_n and E_m. Each material, however, also has to exhibit a continuous spectrum, which corresponds to transitions from a bound state to ionization states (also called 'continuum states', because Bohr's hypothesis of discrete energies does not hold for them).

(ii) Bohr's assumptions are compatible with Einstein's hypothesis. Indeed, if an atom radiates energy in the form of discrete quanta, when the atom emits (or absorbs) a photon of frequency ν, its energy changes by an amount $h\nu$.

(iii) It is then clear why the emission spectra are richer than the absorption spectra. Indeed, at room temperature, the vast majority of atoms are in the ground state, and hence, in absorption, only the frequencies

$$\nu_{1,n} = \frac{E_n - E_1}{h} \tag{1.6.2}$$

are observed, which correspond to transitions from the ground state E_1 to the generic level E_n. Over a very short time period (of the order of 10^{-8} or 10^{-9} s), radiation is re-emitted in one or more transitions to lower levels, until the ground state is reached. Thus, during the emission stage, the whole spectrum given by the previous formula may be observed.

(iv) From what we have said it follows that all frequencies of the emission spectrum are obtained by taking differences of the frequencies of the absorption spectrum:

$$|\nu_n - \nu_m| = |(E_n - E_1)/h - (E_m - E_1)/h| = |(E_n - E_m)/h| = \nu_{n,m}. \quad (1.6.3)$$

This property, which was already well known to spectroscopists prior to Bohr's work, was known as the Ritz combination principle. More precisely, for a complex atom the lines of the spectrum can be classified into series, each of them being of the form

$$\frac{1}{\lambda} = R\left(\frac{1}{m^2} - \frac{1}{n^2}\right), \quad (1.6.4)$$

where n and m are integers, with m fixed and R being the Rydberg constant. From this experimental discovery one finds that, on the one hand, the frequency $\nu = \frac{c}{\lambda}$ is a 'more natural' parameter than the wavelength λ for indexing the lines of the spectrum and, on the other hand, *the spectrum is a set of differences of frequencies* (or spectral terms), i.e. there exists a set I of frequencies such that the spectrum is the set of differences

$$\nu_{ij} \equiv \nu_i - \nu_j \quad (1.6.5)$$

of arbitrary pairs of elements of I. Thus, one can combine two frequencies ν_{ij} and ν_{jk} to obtain a third one, i.e.

$$\nu_{ik} = \nu_{ij} + \nu_{jk}. \quad (1.6.6)$$

Such a corollary is the precise statement of the Ritz combination principle: the spectrum is endowed with a composition law, according to which the sum of the frequencies ν_{ij} and ν_{lk} is again a frequency of the spectrum only when $l = j$. Their combination is then expressed by Eq. (1.6.6).

(v) From the knowledge of the absorption spectrum one can derive the energies E_n, because, once the constant h is known, the absorption spectrum makes it possible to determine $E_2 - E_1, E_3 - E_1, \ldots, E_n - E_1$ and so on, up to $E_\infty - E_1$. Moreover, if one sets to zero the energy corresponding to the ionization threshold, i.e. to the limit level E_∞, one obtains $E_1 = -h\nu$, where ν is the limit frequency of the spectrum (for frequencies greater than ν one obtains a continuous spectrum and the atom is ionized).

(vi) Spectroscopists had been able to group together the lines of a (emission) spectrum, in such a way that the frequencies, or, more precisely, the wave numbers $\frac{1}{\lambda} = \frac{\nu}{c}$ corresponding to the lines of a spectrum could be expressed as differences between 'spectroscopic terms' (Balmer 1885):

$$\frac{1}{\lambda} = \frac{\nu}{c} = T(n) - T(m), \quad (1.6.7)$$

where n and m are positive integers. Each series is picked out by a particular value of n, and by all values of m greater than n. Thus, for example, the first series (which is the absorption series) corresponds to the wave numbers

$$\frac{1}{\lambda} = T(1) - T(m), \quad m > 1. \tag{1.6.8}$$

Now according to Bohr the spectroscopic terms $T(n)$ are nothing but the energy levels divided by hc:

$$T(n) = -\frac{E_n}{hc}, \tag{1.6.9}$$

and the various series correspond to transitions which share the same final level.

Property (iii) makes it possible to determine the energy levels of a system. It is indeed possible to perform the analysis for the hydrogen atom and hydrogen-like atoms, i.e. those systems where only one electron is affected by the field of a nucleus of charge Ze, where Z is the atomic number.

1.7 The experiment of Franck and Hertz

The experiment of Franck and Hertz, performed for the first time in 1914 (Franck and Hertz 1914), was intended to test directly the fundamental postulate of Bohr, according to which an atom can only have a discrete series of energy levels $E_0, E_1, E_2, \ldots$, corresponding to the frequencies $\nu_i = -\frac{E_i}{h}$, with $i = 0, 1, 2, \ldots$. The phenomenon under investigation is the collision of an electron with a monatomic substance. The atoms of such a substance are, to a large extent, in the ground state E_0. If an electron with a given kinetic energy collides with such an atom, which can be taken to be at rest both before the collision (by virtue of the small magnitude of its velocity, due to thermal agitation) and after this process (by virtue of its large mass), the collision is necessarily elastic if $T < E_1 - E_0$, where E_1 is the energy of the closest excited state. Thus, the atom remains in its ground state, and the conservation of energy implies that the electron is scattered in an arbitrary direction with the same kinetic energy, T. In contrast, if $T \geq E_1 - E_0$, inelastic collisions may occur that excite the atom to a level with energy E_1, while the electron is scattered with kinetic energy

$$T' = T - (E_1 - E_0). \tag{1.7.1}$$

The experiment is performed using a glass tube filled with monatomic vapour. On one side of the tube there is a metal filament F, which is heated by an auxiliary electric current. Electrons are emitted from F via the thermionic effect. On the other side of the tube there is a grid G

and a plate P. On taking the average potential of the filament as zero, one 'inserts' in between F and P an electromotive force $V - \varepsilon$ and a weak electromotive force ε in between G and P. A galvanometer, which is inserted in the circuit of P, makes it possible to measure the current at P and to study its variation as V is increased. Such a current is due to the electrons which, emitted from the filament, are attracted towards the grid, where they arrive with a kinetic energy $T = eV$, unless inelastic collisions occur. The electrons pass through the holes of the grid (overcoming the presence of the 'counterfield') and a large number of them reach the plate (despite the collisions occurring in between G and P). This occurs because the kinetic energy of the electrons is much larger than ε.

Since, for $eV < E_1 - E_0$, only elastic collisions may occur in between F and G, we have to expect that the higher the kinetic energy of the electrons, the larger the number of electrons reaching the plate will be. The experiment indeed shows that, for V increasing between 0 and the first excitation potential $\frac{E_1 - E_0}{e}$, the current detected at the plate increases continuously (see figure 1.12). However, as soon as V takes on larger values, inelastic collisions may occur in the neighbourhood of the grid, and if the density of the vapour is sufficiently high, a large number of electrons lose almost all of their kinetic energy in such collisions. Hence they are no longer able to reach the plate, because they do not have enough energy to overcome the 'counterfield' between the grid and the plate. This leads to a substantial reduction of the current registered at P. One now repeats the process: a further increase of the potential enhances the current at P provided that V remains smaller than the second excitation potential, $\frac{E_2 - E_0}{e}$, and so on. It is thus clear that one is measuring the excitation potentials, and the experimental data are in good agreement with the theoretical model.

It should be stressed that, when V increases so as to become larger than integer multiples of $\frac{E_1 - E_0}{e}$, the electron may undergo multiple collisions instead of a single inelastic collision. One then finds that the current at P starts decreasing for energies slightly larger than $(E_1 - E_0), 2(E_1 - E_0), 3(E_1 - E_0), \ldots$.

1.8 Wave-like behaviour and the Bragg experiment

In the light of a number of experimental results, one is led to formulate some key assumptions:

(i) The existence of photons (Einstein 1905, 1917; sections 1.3 and 1.4);

(ii) Bohr's assumption on the selection of classical orbits (Bohr 1913);

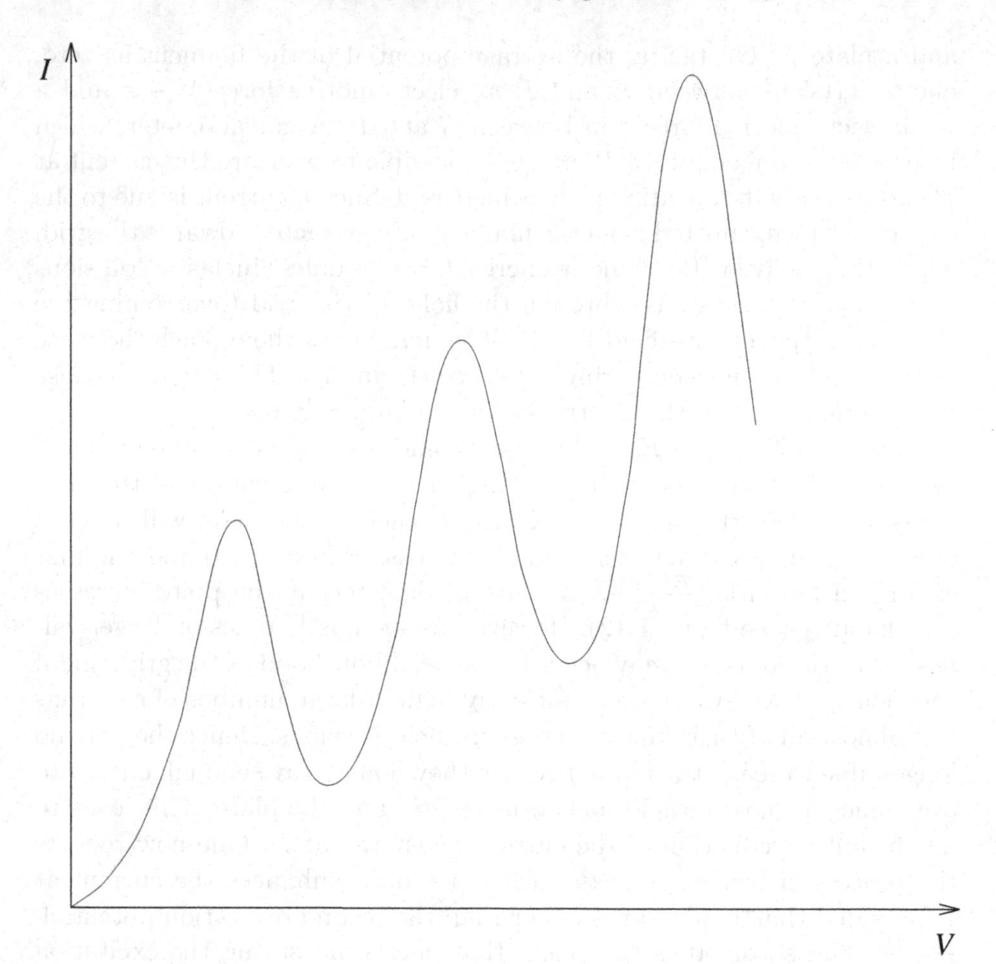

Fig. 1.12. Variation of the current I detected at the plate as the potential V is increased in the experiment of Franck and Hertz. For V increasing between 0 and the first excitation potential, the current I increases continuously. The subsequent decrease results from inelastic collisions occurring in the neighbourhood of the grid. All excitation potentials can be measured in this way.

(iii) Bohr's formula for the frequency of the radiation emitted (see Eq. (1.6.1)) and for the difference of the allowed values of energy.

These *ad hoc* assumptions make it possible to account successfully for a large number of experimental results, and A. Haas was indeed able to compute the Rydberg constant from elementary atomic parameters (Haas 1910a,b, 1925). (The Balmer formula contained a new constant, the Rydberg constant, in terms of which the energy differences of atomic levels was expressed. To the extent that the atomic levels depend on the energy of the electron subject to the Coulomb-like potential of the nucleus and the kinetic energy of the electron, it should be possible to relate the Rydberg constant to the other fundamental constants. This relation was deduced by Arthur Haas.) The resulting theoretical picture, however, is not satisfactory, in that classical physics is amended just when we strictly need to do so, but otherwise remains unaffected.

In 1923 L. de Broglie studied the following problem: the electromagnetic radiation, which has always been considered (quite legitimately) to be of a wave-like nature, also has a particle-like nature by virtue of the existence of photons. The link between the wave and particle aspects is obtained with the help of the Planck constant. The Planck constant, however, also plays a role, via the Bohr–Sommerfeld quantization conditions (see below, Eq. (1.8.1)), in problems where one deals with particles in the first place. Moreover, integer numbers occur in the quantization condition, and it is well known to physicists that their occurrence is associated with wave-like phenomena (e.g. interference, stationary waves). Louis de Broglie was then led to study whether it was possible to follow the opposite path, i.e. *to look for a wave-like aspect in what we always considered from a purely particle-like point of view* (Holland 1993).

Let us consider, for example, an electron in a circular orbit in the hydrogen atom, for which the Bohr–Sommerfeld condition

$$\oint \vec{p} \cdot {\rm d}\vec{q} = nh \tag{1.8.1}$$

implies that (with $p = |\vec{p}|$)

$$pL = nh, \tag{1.8.2}$$

where L is the length of the orbit. (The concept of an orbit becomes inappropriate when the full formalism of wave mechanics is developed, but remains useful for our introductory remarks.)

Equation (1.8.1) is a particular case of a method which can be considered for all dynamical systems subject to forces that can be derived from a potential and are independent of time. Let the given dynamical system have N degrees of freedom, described by position coordinates $q^1, \ldots, q^N$, with corresponding momenta $p_1, \ldots, p_N$. On choosing the position coordinates so as to achieve separation of variables, in that $\int \sum_{j=1}^{N} p_j \, {\rm d}q^j$ is a sum of functions each of which depends on one variable only, one can write that the integral of each term $p_j \, {\rm d}q^j$ taken over a complete cycle of the variable q^j should be equal to an integer multiple of the Planck constant:

$$\oint p_j \, {\rm d}q^j = n_j h, \quad \forall j = 1, \ldots, N. \tag{1.8.3}$$

These are the Bohr–Sommerfeld conditions. They can be stated in a form independent of the choice of variables, as was shown by Einstein in 1917. For this purpose one considers the Maupertuis action, for which $dS = \sum_{j=1}^{N} p_j \, dq^j$ (here there is some abuse of notation, since the right-hand side is not an exact differential), and one remarks that such an expression is invariant under point transformations. For any closed curve C in the region R where the motion takes place, the desired invariant principle can be stated in the form

$$\int_C \sum_{j=1}^{N} p_j \, dq^j = nh. \tag{1.8.4}$$

If the integration is performed in a system of coordinates where the separation of variables is obtained, one can write

$$\sum_{j=1}^{N} K_j \oint p_j \, dq^j = \sum_{j=1}^{N} K_j n_j h.$$

If the motions are bounded in space, the integral can be performed on any cycle of an invariant torus, and the quantum numbers correspond to the various independent cycles.

If we re-express Eq. (1.8.2) in the form $L = \frac{nh}{p}$, and bear in mind that, for a photon, $\frac{h}{p} = \lambda$, we obtain the following interpretation, known as the de Broglie hypothesis: *to every particle one can associate a wave. The relation between wavelength* λ *and momentum* p *is given, as in the case of photons, by*

$$\lambda = \frac{h}{p}. \tag{1.8.5}$$

The allowed orbits are those which contain an integer number of wavelengths.

The Bohr quantization condition is therefore studied under a completely different point of view. So far it is not yet clear whether we have only introduced a new terminology, or whether a more substantial step has been made. In particular, we do not know what sort of wave we are dealing with, and we are unable to identify it with a well-defined physical concept. We can, however, consider some peculiar properties of wave-like phenomena. In particular, it is well known that waves interfere, and hence, if de Broglie waves exist, it should be possible to detect these wave-like aspects by means of interference experiments. Before thinking of an experiment that provides evidence of the wave-like behaviour of particles, it is appropriate to gain an idea of the wavelengths we are dealing with.

For a free particle, the formulae (1.8.5) and $p = \sqrt{2mE}$ yield

$$\lambda = \frac{h}{\sqrt{2mE}}, \tag{1.8.6}$$

which is the de Broglie wavelength. On expressing λ in nanometres and E in electronvolts, one finds for an electron

$$\lambda = \frac{1.24}{\sqrt{E}} \text{ nm}, \tag{1.8.7}$$

and hence, for an electron with energy of 1 eV, λ equals 1.24 nm. Note that, while for photons λ is inversely proportional to E, for massive particles λ is inversely proportional to the square root of E. Thus, for electrons with an energy of the order of 100 eV, λ is of the order of a few Angstrom, i.e. the same order of magnitude as the wavelength of X-rays'.

A way to provide evidence in favour of X-rays being (electromagnetic) waves consists in analysing their reflection from a crystal. This phenomenon is known as Bragg reflection. In a crystal, the regular distribution of atoms (or ions) determines some grating planes for which the mutual separation d is called the *grating step*. The grating step is usually of the order of an Angstrom. If we let a beam of monochromatic X-rays fall upon a crystal, at an angle θ with respect to the surface of the crystal (see figure 1.13), and if we study the radiation emerging with the same inclination θ, we notice that the radiation is reflected only for particular values of θ, here denoted by $\theta_1, \theta_2, \ldots$. More precisely, one either finds some sharp maxima in the intensity of the reflected radiation, corresponding to the angles $\theta_1, \theta_2, \ldots$, or some minima, the intensity of which is very small and, indeed, virtually zero. To account for this behaviour, one has to think of the X-rays as being reflected by the various grating planes (see figure 1.14). The waves reflected from two adjacent planes differ by a phase shift, since they follow optical paths which differ by an amount $2d \sin\theta$.

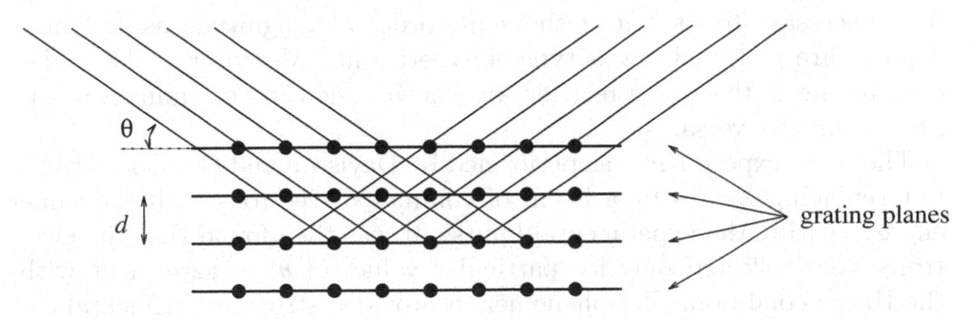

Fig. 1.13 A beam of monochromatic X-rays interacts with a grating.

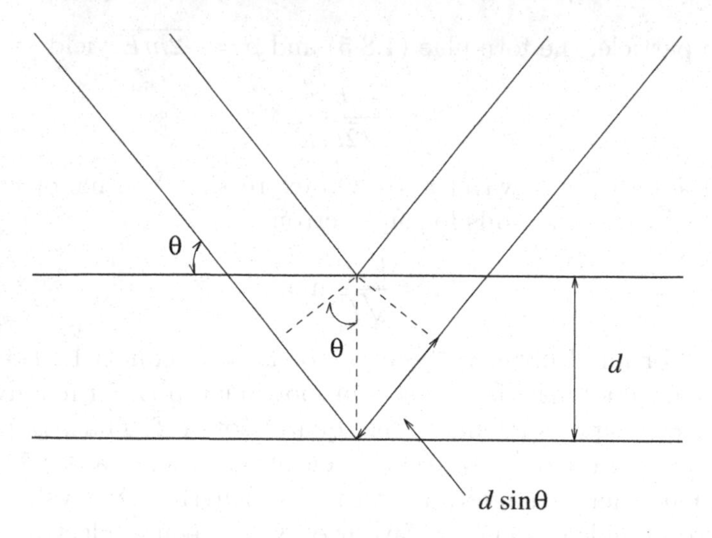

Fig. 1.14 Reflection of X-rays by grating planes.

When $2d \sin\theta$ is an integer multiple of the wavelength λ one has constructive interference of the reflected waves, i.e. a maximum of the intensity. If the crystal is slowly rotated, diffracted beams flash out momentarily every time the crystal satisfies the Bragg condition (see below). In contrast, if $2d \sin\theta$ is an odd multiple of $\frac{\lambda}{2}$, this leads to destructive interference of waves reflected from two adjacent planes, and hence no reflected radiation is observed. In the intermediate cases, the waves reflected from several grating planes interfere with each other, and this leads to an almost vanishing intensity. In summary, the maxima in the intensity of the reflected radiation are obtained for θ such that

$$2d \sin\theta = n\lambda, \tag{1.8.8}$$

which is called the *Bragg condition.* The number of maxima observed is the maximum integer contained in $\frac{2d}{\lambda}$. It is hence clear why, to observe easily the phenomenon in the case of X-rays (for which $\lambda \cong 0.1$ nm), it is necessary to have d of the same order of magnitude as λ. Thus, crystals are preferred in this type of experiments. Moreover, if the angles θ occurring in the Bragg relation are known, one can determine λ if d is known, or vice versa.

The same experiment was performed by Davisson and Germer in 1927, but replacing X-rays by a beam of collimated electrons with the same energy (within the experimental limits). It was then found that the electrons were reflected only for particular values of θ, in agreement with the Bragg condition. This phenomenon provided strong experimental evidence in favour of the electrons having a wave-like behaviour. Moreover, the experiment of Davisson and Germer makes it possible to determine λ from the Bragg relation, so that Eq. (1.8.6) is verified. A complete description is given in the following section.

1.9 The experiment of Davisson and Germer

Shortly after the appearance of de Broglie's original papers on wave mechanics, Elsasser (1925) had predicted that evidence for the wave nature of particle mechanics would be found in the interaction between a beam of electrons and a single crystal. He believed that evidence of this sort was already provided by experimental curves found by Davisson and Kunsman, showing the angular distribution of electrons scattered by a target of polycrystalline platinum. However, this was not quite the case, because the maxima in the scattering curves for platinum are unrelated to crystal structure.

At about the same time when Elsasser made his prediction, Davisson and Germer were continuing their investigation of the angular distribution of electrons scattered by a target of polycrystalline nickel. In April 1925, an unpredictable accident occurred, i.e. a liquid-air bottle exploded at a time when the target was at high temperature. Thus, the experimental tube was broken, and the target was heavily oxidized by the inrushing air. Interestingly, when the experiments were continued (after reducing the oxide and removing a layer of the target), Davisson and Germer found that the angular distribution of the scattered electrons had been completely changed. In particular, they were expecting that strong beams would be found issuing from the crystal along its *transparent directions*, i.e. the directions in which the atoms in the lattice are arranged along the smallest number of lines per unit area. In contrast, strong beams were found issuing from the crystal only when the speed of bombardment lies near one or another of a series of critical values, and in directions quite unrelated to crystal transparency (see figure 1.15).

Another peculiar property was a one-to-one correspondence between the strongest beams and the Laue beams that would be found issuing from the same crystal if the incident beam were a beam of X-rays. Certain other beams were instead the analogues of optical diffraction beams from plane reflection gratings.

By virtue of the similarities between the scattering of *electrons* by the crystal and the scattering of *waves* by three- and two-dimensional gratings, a description of the occurrence and behaviour of the electron diffraction beams in terms of the scattering of an *equivalent wave radiation* by the atoms of the crystal, and its subsequent interference, was 'not only possible, but most simple and natural', as Davisson and Germer pointed

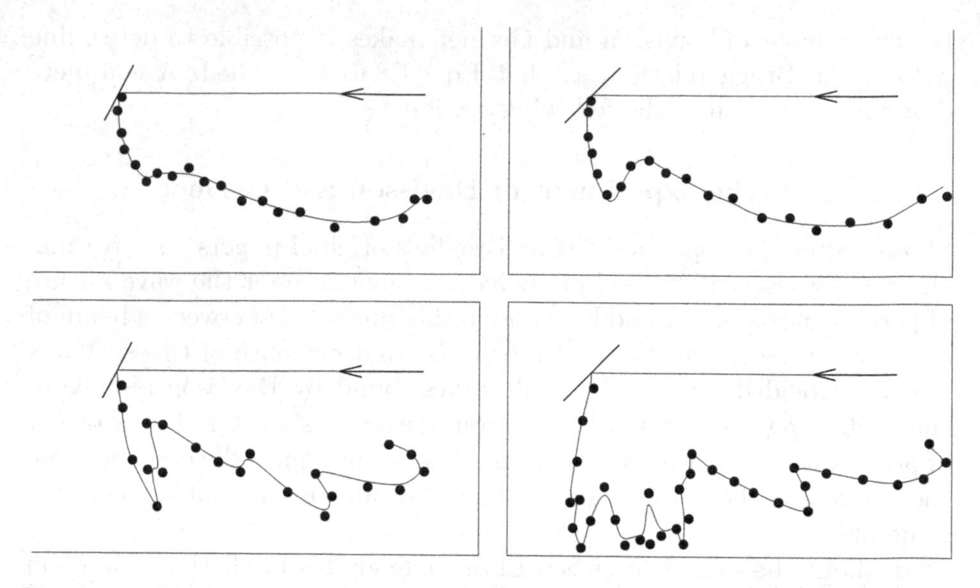

Fig. 1.15. The top two figures describe scattering of 75 V electrons from a block of nickel (many small crystals). The bottom two figures describe scattering of 75 V electrons from several large nickel crystals (Davisson and Germer 1927, © the American Physical Society).

out. One then associates a wavelength with the incident electron beam, which turns out to be in good agreement with the value $\frac{h}{mv}$ of wave mechanics (see also chapter 4).

The experimental apparatus used by Davisson and Germer consists of an electron gun G, a target T and a double Faraday box collector C. The electrons of the primary beam are emitted thermally from a tungsten ribbon F, and are projected from the gun into a field-free enclosure containing the target and collector; the outer walls of the gun, the target, the outer box of the collector and the box enclosing these parts are always held at the same potential. The beam of electrons is orthogonal to the target. High-speed electrons, scattered within the small solid angle defined by the collector opening, enter the inner box of the collector, and eventually pass through a sensitive galvanometer. In contrast, electrons leaving the target with speeds appreciably less than the speed of the incident electrons, are excluded from the collector by a retarding potential between the inner and outer boxes. The current of full speed electrons entering the collector is proportional to the current incident upon the target, and is otherwise a function of the bombarding potential and of the latitude and azimuth angles of the collector. One can thus perform three types of measurement, in each of which two of the independent variables are held constant, while the third is varied (Davisson and Germer 1927).

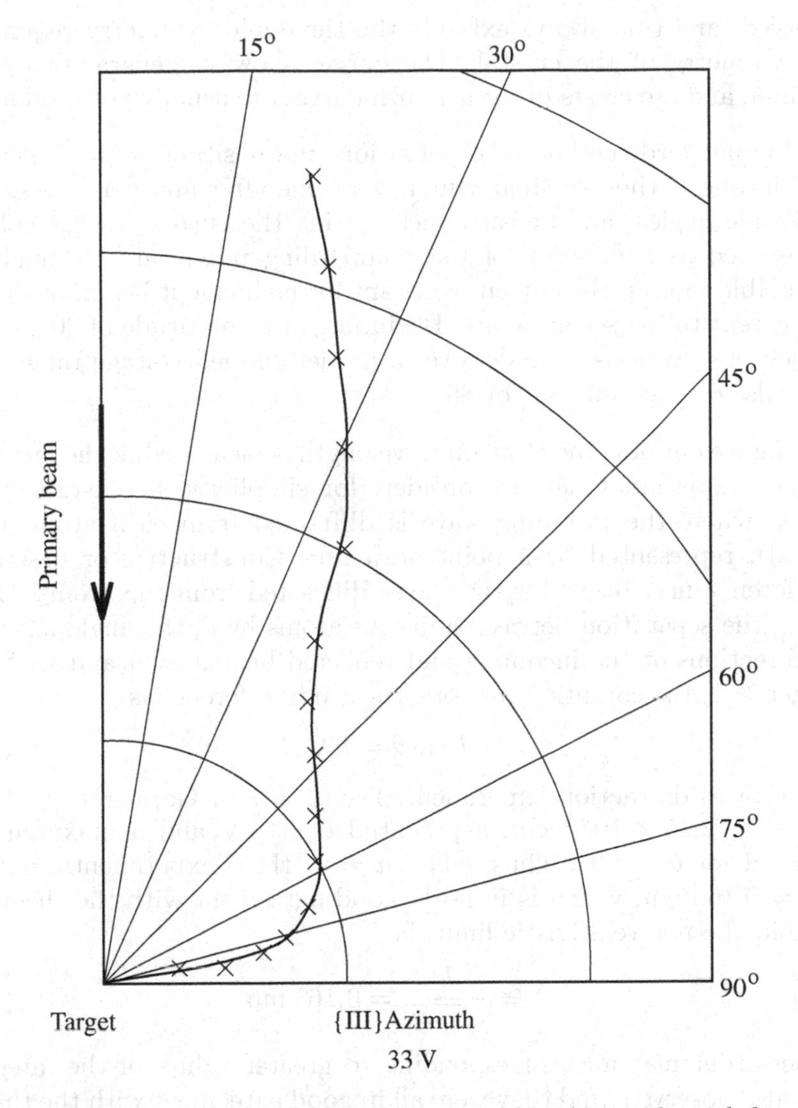

Fig. 1.16. Typical colatitude scattering curve for the single nickel crystal (Davisson and Germer 1927, © the American Physical Society).

(i) When the bombarding potential and azimuth are fixed and exploration is made in latitude a dependence of current upon angle is observed, which is of the form shown in figure 1.16. The current of scattered electrons vanishes in the plane of the target and increases regularly to a highest value at the limit of observations; colatitude 20°.

(ii) When the bombarding potential and latitude angle are fixed, and exploration is made in azimuth, a variation of collector current is always observed, and this always exhibits the three-fold symmetry required by the symmetry of the crystal. The curves show in general two sets of maxima, and the crests in the azimuth curves are usually not pronounced.

(iii) In the third method of observation, the position of the collector is fixed in one of the principal azimuths at one after another of a series of colatitude angles, and at each such setting the current to the collector is observed as a function of the bombarding potential. Although it is impossible to keep the current constant in the incident beam, one can fix the current to the second plate. Beginning at a colatitude of 20°, a series of such observations is made, over a predetermined voltage range, at 5° intervals to colatitude 80° or 85°.

To figure out how the electron wavelength is measured in the Davisson–Germer experiment, let us consider, for simplicity, a one-dimensional model, where the incoming wave is diffracted from each atom (of the crystal), represented by a point on a line. Constructive or destructive interference may occur for the waves diffracted from the atoms. On denoting the separation between adjacent atoms by d, the angle formed by the directions of the incoming and reflected beams by θ and with n an integer ≥ 1, the condition for constructive interference is

$$d \, \sin\theta = n\lambda. \tag{1.9.1}$$

In a typical diffraction experiment, Davisson and Germer were dealing with $d = 2.15 \times 10^{-8}$ cm, a potential of 54 eV and a maximum was observed for $\theta = 50°$. Thus, when $n = 1$, these experimental data led to $\lambda = 0.165$ nm, which is in fairly good agreement with the theoretical value in the non-relativistic limit, i.e.

$$\lambda \cong \frac{h}{\sqrt{2mT}} = 0.167 \text{ nm}. \tag{1.9.2}$$

Higher-order maxima, corresponding to greater values of the integer n, were also observed, and they were all in good agreement with the theoretical predictions. It is also clear, from Eq. (1.9.2), why a beam of electrons was actually chosen: since they have a very small mass, the corresponding wavelength is expected to be sufficiently large.

We conclude this section with a historical remark, which relies on the Nobel Laureate speech delivered by Davisson in 1937. The experiment performed by him and Germer in 1925 was not, *at first*, a proof of the validity of wave mechanics. It was only in the summer of 1926 that the physics community came to appreciate the relevance of such an investigation for the wave-like picture of physical phenomena, after a number of discussions between Davisson, Richardson, Born, Franck and other distinguished scientists. This kind of experiment has been repeated with protons, neutrons, helium atoms, ions, and in all these cases the Bragg relation for material particles has been verified.

1.10 Position and velocity of an electron

Suppose we deal with an electron for which the velocity is known. We are aiming to determine its position. For this purpose, one has to observe the electron. Monochromatic light of wavelength λ is shone upon the electron, and with the help of a microscope the light scattered from the electron is collected. Such light is focused at a point P' (the image point), and from the knowledge of the position of P' one can derive the location P of the electron (see figure 1.17). This is the basic idea used to measure the position of the particle (in the so-called Heisenberg microscope). Let us now assess more carefully how such a device can work. First, recall that each optical instrument has a finite resolution power. This means that, given two points P_1 and P_2 separated by a distance δ (see figure 1.18), there exists a δ_{min} such that, if $\delta < \delta_{\text{min}}$, the device is no longer able to distinguish them. Let us see how this can happen. A lens gives rise to diffraction, just as does a slit in a screen. Thus, the images of the points are not points but diffraction patterns produced by the lens, i.e. they are extended spots. Since each ray falling upon the lens gives rise to a cone of rays with opening $\sin(\alpha) \cong \frac{\lambda}{d}$, where d is the diameter of the lens, it

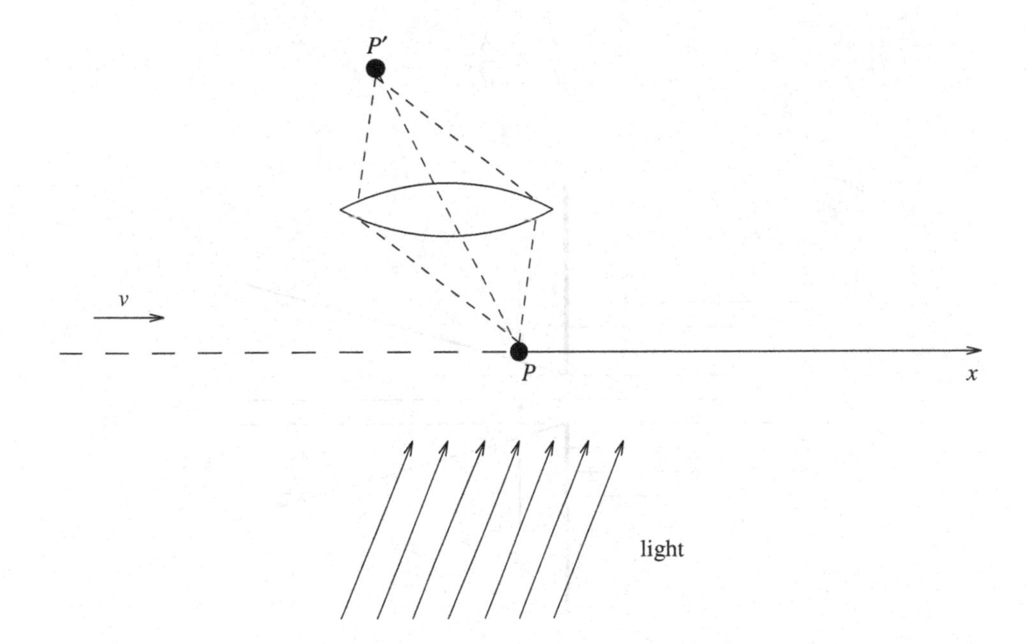

Fig. 1.17 The Heisenberg microscope.

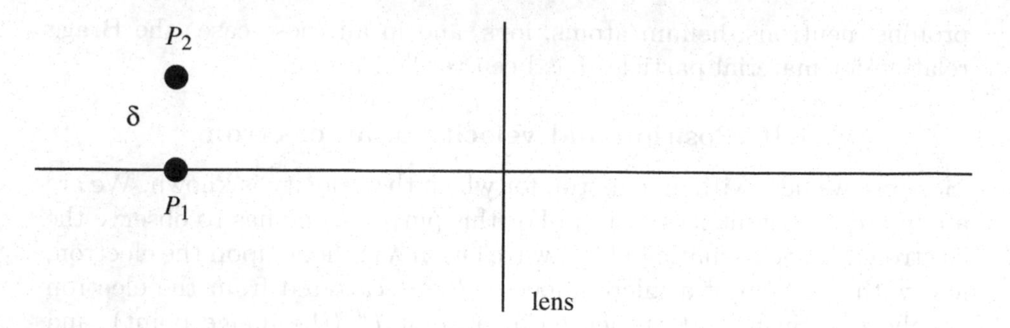

Fig. 1.18 A lens in vacuum.

follows that the image of a point is a spot of dimension (see figures 1.19 and 1.20)

$$l_2 \sin(\alpha) = l_2 \frac{\lambda}{d}. \tag{1.10.1}$$

The images of the points P_1 and P_2 are distinguishable if the two spots do not overlap. This occurs if

$$l_2 \frac{\lambda}{d} < \delta \frac{l_2}{l_1}, \tag{1.10.2a}$$

i.e. if

$$\frac{\delta}{l_1} > \frac{\lambda}{d}. \tag{1.10.2b}$$

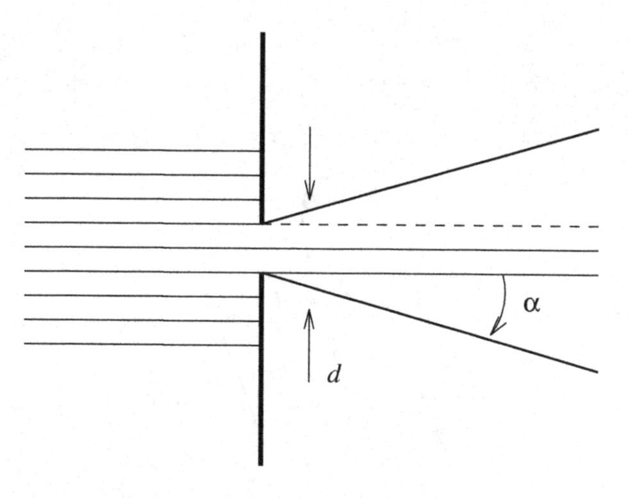

Fig. 1.19 Diffraction cone from a slit.

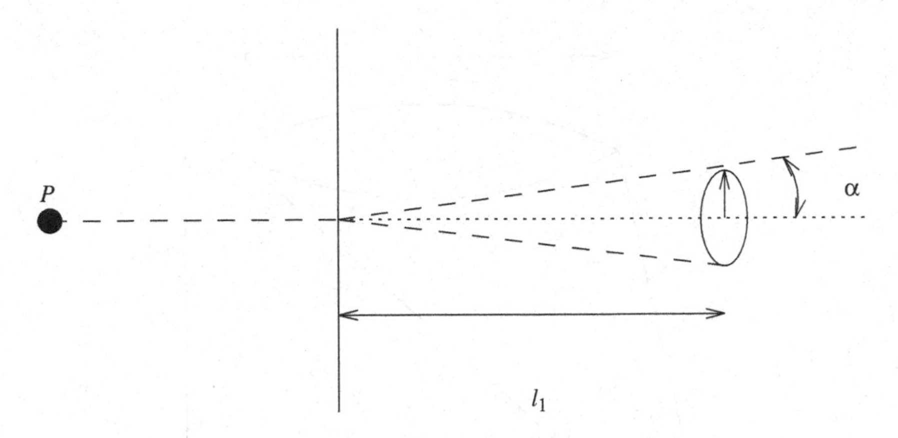

Fig. 1.20 The image of a point is a spot.

This means that, if one knows the location of the image point, one can derive the position of the source with an uncertainty equal to

$$\delta_{\text{min}} = l_1 \frac{\lambda}{d}. \tag{1.10.3}$$

Thus, to improve the resolution power of an optical device, one has to decrease λ, increase d, or both.

If we now consider the Heisenberg microscope, we realize that the position of the electron is known with an uncertainty (cf. section 4.2)

$$\Delta x \geq l_1 \frac{\lambda}{d}. \tag{1.10.4a}$$

On the other hand, denoting by 2φ the angle under which the electron 'sees' the lens, one has

$$\frac{l_1}{d} = \tan\left(\frac{\pi}{2} - 2\varphi\right) = \frac{\cos 2\varphi}{\sin 2\varphi},$$

which, at small φ, reduces to

$$\frac{l_1}{d} \cong \frac{1}{\sin 2\varphi} \cong \frac{1}{2\sin\varphi},$$

and hence

$$\Delta x \geq \frac{\lambda}{2\sin\varphi}. \tag{1.10.4b}$$

Δx can be as large or as small as we like by varying the parameters l_1, λ and d appropriately (see figure 1.21).

Now that we have studied a position measurement, let us try to understand what can we say about the velocity of the electron. If we have observed the electron with the help of a microscope, at least one photon

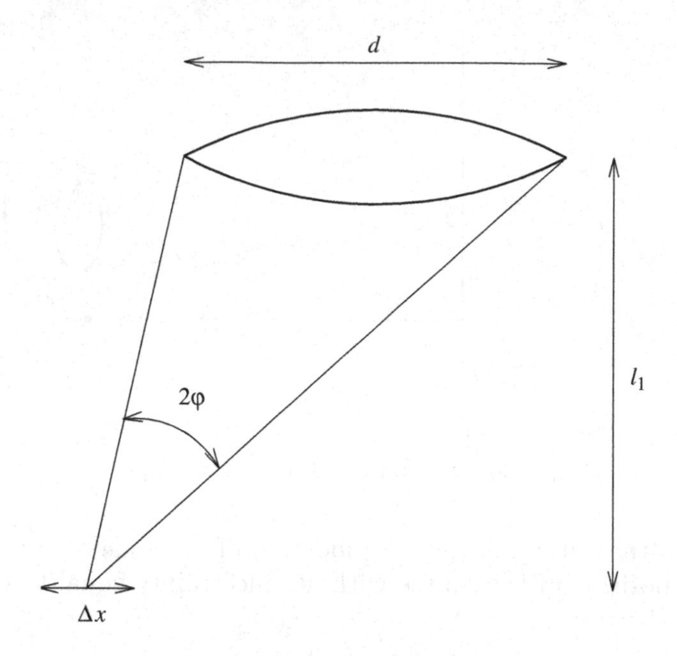

Fig. 1.21 Angle under which an electron 'sees' the lens.

has been scattered by the electron and has been able to pass through the lens: we register its arrival at P' but we do not know *where* it passed through the lens, i.e. its scattering direction. This implies that we cannot determine the momentum of the scattered photon exactly, and hence the amount of momentum exchanged by the electron and the photon. The uncertainty in the x-component of the momentum of the scattered photon is given by

$$\Delta p_x = 2\frac{h}{\lambda}\sin\varphi, \tag{1.10.5}$$

which, by virtue of the conservation of momentum, is also the uncertainty on the momentum of the electron, after its position has been measured. More precisely, the Heisenberg microscope does not show that the electron has an uncertainty $\Delta x \, \Delta p \cong h$, but *only shows that such a relation for photons only gives such an uncertain determination for the electron.* In other words, the electron has position and momentum but the tool with which we observe, i.e. the photon, has both wave and particle properties, and we are unable to specify both the location and the wavelength for light. At best the Heisenberg microscope demonstrates how the wave–particle duality is transmitted from the photon to the electron.

The uncertainty Δp_x can be made small or large by varying l, λ and d, but this affects severely the uncertainty of a position measurement, by virtue of the Heisenberg relation (see section 4.2)

$$\Delta x \cdot \Delta p_x \geq h.$$

Note that, while it is possible to obtain a small value of Δx, by decreasing λ, at a classical level we might think of decreasing the intensity of the radiation used to light up the electron as much as we like. By doing so, we might try to decrease the amount of momentum given to the electron arbitrarily. Of course, this argument is incorrect, since it does not take into account the fact that, once λ has been fixed, the intensity cannot be decreased by an arbitrary amount, by virtue of the fundamental equation $\lambda p = h$.

Chapters 4 and 5 are devoted to the systematic development of wave mechanics. Further material on the foundations of quantum theory can be found in the work by Ter Haar (1967), van der Waerden (1968), Mehra and Rechenberg (1982a–f, 1987a,b).

In this chapter we have seen that a quantum theory requires a kind of unification of the particle and wave viewpoints. In the following two chapters we shall summarize both aspects so that in chapter 4 we can tackle their unification.

1.11 Problems

1.P1. The stopping voltage for an experiment leading to photoelectric effect with monochromatic light falling upon a surface of Na is 1.85 V if the wavelength $\lambda = 300$ nm, and 0.82 V if $\lambda = 400$ nm. Find:

(i) the Planck constant;

(ii) the work necessary to extract photoelectrons;

(iii) the photoelectric threshold.

1.P2. In a Compton scattering, the photon is deviated by an angle $\theta = \frac{\pi}{2}$ (cf. figure 1.5) and the momentum of the recoiling electron is 100 MeV/c. Find the wavelength of the incoming radiation.

Appendix 1.A
The phase 1-form

Our discussion of the Einstein–de Broglie relations in section 1.2 assumes that the reader is already familiar with the phase 1-form in the classical theory of wave-like phenomena, but this is not necessarily the case. We therefore find it appropriate to recall the following basic properties of the Fourier expansion of a function.

(i) The expansion of the function $f : \mathbf{R}^4 \to \mathbf{C}$ in *monochromatic waves* is described by the formulae

$$f_\omega(\vec{x}) \equiv \frac{1}{\sqrt{2\pi}} \int_{-\infty}^{\infty} f(\vec{x}, t) e^{i\omega t} \, dt, \tag{1.A.1}$$

$$f(\vec{x}, t) = \frac{1}{\sqrt{2\pi}} \int_{-\infty}^{\infty} f_\omega(\vec{x}) e^{-i\omega t} \, d\omega. \tag{1.A.2}$$

(ii) The expansion of f in *plane waves* is instead given by

$$f_{\vec{k}}(t) \equiv \frac{1}{(2\pi)^{3/2}} \int_{\mathbf{R}^3} f(\vec{x}, t) e^{-i\vec{k}\cdot\vec{x}} \, d^3x, \tag{1.A.3}$$

$$f(\vec{x}, t) = \frac{1}{(2\pi)^{3/2}} \int_{\mathbf{R}^3} f_{\vec{k}}(t) e^{i\vec{k}\cdot\vec{x}} \, d^3k. \tag{1.A.4}$$

(iii) Many applications deal with the expansion in *monochromatic plane waves*, for which

$$f_{\vec{k}\omega} \equiv \frac{1}{(2\pi)^2} \int_{\mathbf{R}^4} f(\vec{x}, t) e^{-i(\vec{k}\cdot\vec{x} - \omega t)} \, d^3x \, dt, \tag{1.A.5}$$

$$f(\vec{x}, t) = \frac{1}{(2\pi)^2} \int_{\mathbf{R}^4} f_{\vec{k}\omega} e^{i(\vec{k}\cdot\vec{x} - \omega t)} \, d^3k \, d\omega. \tag{1.A.6}$$

In a similar way, a Fourier decomposition of differential forms can be considered, according to which

$$A_\mu(x, t) \, dx^\mu = dx^\mu \int_{\mathbf{R}^4} A_\mu(k, \omega) e^{ik_\rho x^\rho} \, d^3k \, d\omega, \tag{1.A.7}$$

and

$$\begin{aligned} dA &= \int_{\mathbf{R}^4} A_\mu(\vec{k}, \omega)(ik_\nu \, dx^\nu) \wedge dx^\mu \; e^{ik_\rho x^\rho} \, d^3k \, d\omega \\ &= \frac{i}{2} dx^\nu \wedge dx^\mu \int_{\mathbf{R}^4} (A_\mu k_\nu - A_\nu k_\mu) e^{ik_\rho x^\rho} \, d^3k \, d\omega. \end{aligned} \tag{1.A.8}$$

The 1-form $k_\nu \, dx^\nu$ is called the phase 1-form.

In analogy with phase space, the carrier space for Hamiltonian dynamics, parametrized by (p, x), momentum and position, one may introduce the 'optical phase space' parametrized by (k, x), wave vector and position. Therefore the optical phase 1-form plays the same role as the Liouville 1-form for phase space.

2

Classical dynamics

The aim of this chapter is to consider various formalisms of classical dynamics and their equivalence or lack of it. These considerations are very important when studying the 'classical limit' of quantum mechanics, i.e. which particular formulation of quantum mechanics will give rise, in the limiting process, to a particular formulation of classical mechanics?

Our review of basic concepts and tools in classical mechanics begins with the definition of Poisson brackets on functions on a manifold. The Poisson bracket is any map which is antisymmetric, bilinear, satisfies the Jacobi identity and obeys a fourth property (derivation) that relates the Poisson bracket with the commutative associative product. Symplectic geometry is then outlined, and an intrinsic definition of the Poisson bracket is given within that framework. The maps which preserve the Poisson-bracket structure are canonical transformations. They are presented in an implicit form in terms of the generating functions. The expressions for the new canonical variables in terms of the old canonical variables are non-linear in general, and they can be made explicit only locally. In the case of linear canonical transformations it can be extended to global definitions. Once a symplectic potential is selected one may identify four classes of generating functions of canonical transformations, and they are all presented. This makes it possible to cast the equations of motion in the simplest possible form after performing one set of canonical transformations. The problem of solving the Hamilton equations is then replaced by the analysis of a partial differential equation known as the Hamilton–Jacobi equation. In the time-dependent formulation, the solution is given by the Hamilton principal function, which leads, in turn, to the solution of the original problem for given initial conditions. If the Hamiltonian does not depend explicitly on time, however, it is more appropriate to re-express the Hamilton–Jacobi equation in a form that is solved by the Hamilton characteristic function. Some simple but very useful applications of the Hamilton–Jacobi method are also studied, i.e. the harmonic oscillator and motion in a central field. The chapter ends with a brief introduction to geometrical optics.

2.1 Poisson brackets

In our geometrical presentation of classical dynamics we rely upon some basic structures defined in appendices 2.A and 2.B, while the passage from Newtonian to Lagrangian dynamics is summarized in appendix 2.E. Here we are concerned with the Hamiltonian formalism, which is indeed usually presented starting with the Lagrangian formalism, while Poisson brackets are introduced afterwards. However, if classical mechanics is thought of as a suitable limit of quantum mechanics, it is convenient to follow a different route, i.e. we first consider a space endowed with Poisson brackets, then we use the symplectic formalism and eventually we try to understand whether it can result from a Lagrangian.

Given any (smooth) manifold M of dimension n, where n can be (for the time being) either even or odd, a Poisson bracket is any map ($\mathcal{F}(M)$ being the set of functions on M)

$$\{\,,\,\} : \mathcal{F}(M) \times \mathcal{F}(M) \rightarrow \mathcal{F}(M)$$

having the following properties:

$$\{f_1, f_2\} = -\{f_2, f_1\}, \tag{2.1.1}$$

$$\{f_1, \lambda f_2 + \mu f_3\} = \lambda \{f_1, f_2\} + \mu \{f_1, f_3\} \quad \lambda \in \mathbf{R}, \quad \mu \in \mathbf{R}, \tag{2.1.2}$$

$$\{f_1, \{f_2, f_3\}\} = \{\{f_1, f_2\}, f_3\} + \{f_2, \{f_1, f_3\}\}, \tag{2.1.3}$$

$$\{f_1, f_2 f_3\} = \{f_1, f_2\} f_3 + f_2 \{f_1, f_3\}, \tag{2.1.4}$$

for all $f_1, f_2, f_3 \in \mathcal{F}(M)$. The manifold M, endowed with a Poisson bracket, is said to be a Poisson manifold. Equations (2.1.1)–(2.1.3) express the antisymmetry, bilinearity and Jacobi identity, respectively, and are the properties which define a Lie-algebra structure on any vector space. The Jacobi identity may also be expressed as

$$\{f_1, \{f_2, f_3\}\} + \{f_2, \{f_3, f_1\}\} + \{f_3, \{f_1, f_2\}\} = 0. \tag{2.1.5}$$

This has a more algorithmic nature, but (2.1.3) has the advantage of clarifying the link with the Leibniz rule for derivations. Property (2.1.4) connects two different structures on $\mathcal{F}(M)$, i.e. Poisson brackets and the commutative associative product $f_1 \cdot f_2$. Note that antisymmetric maps can be defined that obey (2.1.1)–(2.1.3) but not (2.1.4). For example, on a vector space with $\partial_\mu \equiv \frac{\partial}{\partial \xi^\mu}$, the rule

$$[f_1, f_2] \equiv f_1 \partial_\mu f_2 - f_2 \partial_\mu f_1 \tag{2.1.6}$$

defines a Lie-algebra structure but does not satisfy (2.1.4) and hence is not a Poisson bracket. Our definition makes it clear that *the existence of Poisson brackets does not put restrictions on the dimension of the manifold.* In this respect, they are more fundamental than symplectic structures.

Properties (2.1.1)–(2.1.5) imply that every Poisson bracket determines uniquely a contravariant skew-symmetric 2-tensor field (Jost 1964). If $\xi_1, \xi_2, \ldots, \xi_n$ are coordinate functions one obtains a tensor field

$$\Lambda \equiv \{\xi_j, \xi_k\} \frac{\partial}{\partial \xi_j} \wedge \frac{\partial}{\partial \xi_k}.$$

In particular, all properties, being tensorial, are independent of the particular coordinate system used to describe them.

As an example of a Poisson bracket on $\mathbf{R}^3$ one can consider (here Latin indices run from 1 to 3)

$$\{x_i, x_j\} \equiv \varepsilon_{ijk} x_k. \tag{2.1.7}$$

On using the Hamiltonian given by

$$H \equiv \frac{1}{2}\left(I_1 x_1^2 + I_2 x_2^2 + I_3 x_3^2\right), \tag{2.1.8}$$

where I_1, I_2 and I_3 are the moments of inertia of a rigid rotator, one then finds equations of motion of the form

$$\dot{x}_1 = \{H, x_1\} = (I_3 - I_2) x_2 x_3, \tag{2.1.9}$$

$$\dot{x}_2 = \{H, x_2\} = (I_1 - I_3) x_1 x_3, \tag{2.1.10}$$

$$\dot{x}_3 = \{H, x_3\} = (I_2 - I_1) x_1 x_2. \tag{2.1.11}$$

It is now appropriate to introduce symplectic mechanics, so that the general reader may appreciate the difference between the two schemes.

2.2 Symplectic geometry

Let M be a manifold of dimension n. If we consider a non-degenerate Poisson bracket, i.e. such that

$$\left\{\xi^i, \xi^j\right\} \equiv \omega^{ij}$$

is an invertible matrix, we may define the inverse ω_{ij} by requiring

$$\omega_{ij}\omega^{jk} = \delta_i^{\ k}. \tag{2.2.1}$$

We define a tensorial quantity

$$\omega \equiv \frac{1}{2}\omega_{ij}\, \mathrm{d}\xi^i \wedge \mathrm{d}\xi^j, \tag{2.2.2}$$

which turns out to be a non-degenerate 2-form. This implies that the dimension of the manifold M is necessarily even. Furthermore, the form ω is closed: $\mathrm{d}\omega = 0$. Indeed, it can be shown that

$$\left\{\xi^i, \left\{\xi^j, \xi^k\right\}\right\} = \left\{\left\{\xi^i, \xi^j\right\}, \xi^k\right\} + \left\{\xi^j, \left\{\xi^i, \xi^k\right\}\right\}$$

is equivalent to

$$\frac{\partial}{\partial \xi^k}\omega_{ij} + \frac{\partial}{\partial \xi^i}\omega_{jk} + \frac{\partial}{\partial \xi^j}\omega_{ki} = 0,$$

i.e. $\mathrm{d}\omega = 0$ (Jost 1964).

The pair (M, ω) is then called a *symplectic manifold* and ω is a *symplectic form*. Note that the closure property does not imply that ω is exact. This only holds *locally*, in a way which is made precise by the Poincaré lemma.

Lemma. Let ω be such that $\mathrm{d}\omega = 0$ on an open set $U \subset M$, which is diffeomorphic to the open ball

$$\left\{x \in \mathbf{R}^n : |x|^2 < 1\right\},$$

then $\omega = \mathrm{d}\theta$ for some symplectic potential θ defined on U.

A symplectic manifold such that $\omega = \mathrm{d}\theta$ globally is said to be an *exact symplectic manifold*. The existence of a symplectic potential has strong implications on the topology of the manifold M. Under additional mild conditions (De Filippo *et al.* 1989) it can be shown that M is diffeomorphic to the cotangent bundle of some configuration manifold Q, i.e. $M \cong T^*Q$. Under such conditions one can define a dual bundle TQ that is the carrier space for a Lagrangian description of dynamics. From here it should be clear that a Poisson description is more general than a symplectic description, which in turn is more general than a Lagrangian description (see appendix 2.E).

For any symplectic manifold, in the neighbourhood of each point it is possible to define 'canonical coordinates' $(q^1, \ldots, q^n; p_1, \ldots, p_n)$ for which the symplectic form reads

$$\omega = \mathrm{d}q^i \wedge \mathrm{d}p_i.$$

A choice of canonical coordinates amounts to choosing a symplectic potential $\theta = p_i\,\mathrm{d}q^i$, which is defined up to an exact differential $\mathrm{d}f$. As for the Poisson bracket among canonical coordinates, one finds

$$\left\{q^i, p_j\right\} = \delta^i{}_j, \tag{2.2.3}$$

$$\{p_i, p_j\} = \left\{q^i, q^j\right\} = 0. \tag{2.2.4}$$

By virtue of the derivation property (2.1.4), the Poisson bracket of any two functions F, G therefore takes the form

$$\{F, G\} = \frac{\partial F}{\partial q^j}\frac{\partial G}{\partial p_j} - \frac{\partial F}{\partial p_j}\frac{\partial G}{\partial q^j}. \tag{2.2.5}$$

For any function f which does not depend explicitly on time, one has the time derivative along trajectories of some Hamiltonian dynamical vector field

$$\frac{\mathrm{d}}{\mathrm{d}t} f = \{f, H\}, \tag{2.2.6}$$

while the Hamilton equations read

$$\frac{\mathrm{d}}{\mathrm{d}t} q^j = \left\{q^j, H\right\}, \tag{2.2.7a}$$

$$\frac{\mathrm{d}}{\mathrm{d}t} p_j = \{p_j, H\}. \tag{2.2.7b}$$

Thus, the general expression of the dynamical vector field is

$$\frac{\mathrm{d}}{\mathrm{d}t} = \left\{q^j, H\right\}\frac{\partial}{\partial q^j} + \{p_j, H\}\frac{\partial}{\partial p_j}. \tag{2.2.8}$$

If H depends on time, it is appropriate to use a time-dependent formalism, and this will be developed in section 2.3.

If we start with a symplectic manifold (M, ω), it is possible to define a Poisson bracket on it. More precisely, the map

$$\{\,,\,\} : \mathcal{F}(M) \times \mathcal{F}(M) \rightarrow \mathcal{F}(M)$$

defined by

$$\{f, h\} \equiv \omega(X_f, X_h) \equiv L_{X_h} f, \tag{2.2.9}$$

is a *Poisson bracket*, where X_f and X_h are given by

$$i_{X_f}\omega \equiv \left(X_f^i\, \omega_{ij}\right)\mathrm{d}\xi^j = \frac{\partial f}{\partial \xi^j}\,\mathrm{d}\xi^j = \mathrm{d}f, \tag{2.2.10}$$

$$i_{X_h}\omega \equiv \left(X_h^i\, \omega_{ij}\right)\mathrm{d}\xi^j = \frac{\partial h}{\partial \xi^k}\,\mathrm{d}\xi^k = \mathrm{d}h. \tag{2.2.11}$$

Equation (2.2.10) (one can follow an analogous procedure for (2.2.11)) may be written in the form

$$\left(X_f^i\, \omega_{ij} - \frac{\partial f}{\partial \xi^j}\right)\mathrm{d}\xi^j = 0, \tag{2.2.12}$$

which implies (using the inverse of ω_{ij} defined in (2.2.1))

$$X_f^i = \frac{\partial f}{\partial \xi^j} \omega^{ji}, \tag{2.2.13}$$

and hence X_f reads

$$X_f = \frac{\partial f}{\partial \xi^j} \, \omega^{jk} \, \frac{\partial}{\partial \xi^k}. \tag{2.2.14}$$

The local form of the associated Poisson brackets is therefore

$$\{f, h\} = \frac{\partial f}{\partial \xi^j} \, \omega^{jk} \, \frac{\partial h}{\partial \xi^k}. \tag{2.2.15}$$

If a symplectic manifold (M, ω) or a Poisson manifold $(M, \{ \ , \ \})$ is given, a map $\phi : M \rightarrow M$ is called a *canonical transformation* if and only if

$$\phi^* \{f, h\} = \{\phi^* f, \phi^* h\}, \quad \forall f, h \in \mathcal{F}(M). \tag{2.2.16}$$

This means that ϕ preserves the Poisson-bracket structure. Using canonical coordinates for ω and the transformed symplectic structure $\phi^* \omega$, one finds that ϕ is canonical if

$$\mathrm{d}q^i \wedge \mathrm{d}p_i = \mathrm{d}Q^i \wedge \mathrm{d}P_i, \tag{2.2.17}$$

where

$$Q^i = Q^i(q, p),$$

$$P_i = P_i(q, p),$$

represents the transformation ϕ.

Remark. The position of indices means that if we perform transformations among the q, the induced transformations on the P, deduced by imposing the invariance of $p_i \, \mathrm{d}q^i$, have the same property that a transformation on vectors induces on 'covectors' (i.e. a transformation on a vector space induces one on the dual). Therefore, in canonical coordinates, a neighbourhood of a point of the manifold is represented as an open subset of $\mathbf{R}^n \times (\mathbf{R}^n)^*$.

If we identify $\mathbf{R}^n$ with $((\mathbf{R}^n)^*)^*$, we may also consider the invariance of $q^i \, \mathrm{d}p_i$, where now p_i are viewed as 'independent' variables. From now on, all our considerations will be mainly restricted to an exact symplectic manifold, which in addition is assumed to be a vector space $T^* \mathbf{R}^n \cong \mathbf{R}^n \times (\mathbf{R}^n)^*$.

2.3 Generating functions of canonical transformations

Let $W : \mathbf{R}^n \times \mathbf{R}^n \rightarrow \mathbf{R}$ be a smooth real-valued function such that, at all points of $\mathbf{R}^n \times \mathbf{R}^n$, the Hessian matrix

$$\mathcal{H}_W \equiv \left\| \frac{\partial^2 W}{\partial x^k \partial y^h} \right\|$$

is invertible. We define two invertible maps

$$\mathbf{R}^n \times \mathbf{R}^n \rightarrow \mathbf{R}^n \times (\mathbf{R}^n)^*,$$

i.e.

$$(x, y) \rightarrow \left(x, \frac{\partial W}{\partial x} \right), \tag{2.3.1}$$

and

$$(x, y) \rightarrow \left(y, -\frac{\partial W}{\partial y} \right), \tag{2.3.2}$$

respectively. By putting them together we have a map

$$\mathbf{R}^n \times \mathbf{R}^n \rightarrow \mathbf{R}^n \times (\mathbf{R}^n)^* \times \left[\mathbf{R}^n \times (\mathbf{R}^n)^* \right],$$

for which the action reads

$$(x, y) \rightarrow \left(x, \frac{\partial W}{\partial x}; y, -\frac{\partial W}{\partial y} \right). \tag{2.3.3}$$

On each 'factor' $\mathbf{R}^n \times (\mathbf{R}^n)^*$ we have a canonical 1-form $p_k \, \mathrm{d}x^k$ and $P_k \, \mathrm{d}y^k$, therefore on the 'graph' of our previous map we find

$$P_k \, \mathrm{d}y^k - p_k \, \mathrm{d}x^k = -\mathrm{d}W(x, y). \tag{2.3.4}$$

The previous construction can be interpreted as follows: on some $2n$-dimensional sub-manifold Σ of $\left[\mathbf{R}^n \times (\mathbf{R}^n)^* \right] \times \left[\mathbf{R}^n \times (\mathbf{R}^n)^* \right]$ for which the restriction of

$$\mathrm{d}p_k \wedge \mathrm{d}x^k - \mathrm{d}P_k \wedge \mathrm{d}y^k$$

vanishes identically, it is possible to find (at least locally) a function $W : \mathbf{R}^n \times \mathbf{R}^n \rightarrow \mathbf{R}$ such that Eq. (2.3.4) holds, i.e.

$$\mathrm{d}\left(P_k \, \mathrm{d}y^k - p_k \, \mathrm{d}x^k \right) = 0$$

implies, for a sub-space $U \subset \Sigma$ where the Poincaré lemma applies, that

$$\left(P_k \, \mathrm{d}y^k - p_k \, \mathrm{d}x^k \right)|_U = \mathrm{d}W$$

for a function W defined on U. Thus, canonical transformations

$$\phi : \mathbf{R}^n \times (\mathbf{R}^n)^* \rightarrow \mathbf{R}^n \times (\mathbf{R}^n)^*$$

may be associated with 'generating functions' via the previous construction.

Remark. On any $2n$-dimensional symplectic manifold (M, ω), any n-dimensional sub-manifold Σ on which the restriction of ω identically vanishes is called a Lagrangian sub-manifold (Marmo *et al.* 1985). Our previous construction shows that canonical transformations $\varphi : M \rightarrow M$ may be identified with Lagrangian sub-manifolds of $M \times M$, with the symplectic form being given by the difference of that on the first factor with that on the second factor. Moreover, at least locally, every such sub-manifold may be described as the graph of a map associated with a 'generating function'. Since the dual of $(\mathbf{R}^n)^*$ is $\mathbf{R}^n$, it is clear that we may identify vectors and corresponding 'covectors' in many different ways.

2.3.1 Time-dependent Hamiltonian formalism

In a time-dependent formalism, the configuration space Q is replaced by a factorizable 'extended configuration space' $\widetilde{Q} \equiv Q \times \mathbf{R}$. As long as we do not use this factorizability property, whatever we have said in the general case also holds in this 'extended setting'. It should be mentioned that in a non-relativistic framework one usually considers a given factorization and this is called an *Aristotelian setting.* When the projection $\widetilde{Q} \rightarrow \mathbf{R}$ is preserved but no specific factorization is considered, one is in a Galilean setting; if no factorization is used but $\widetilde{Q}$ is endowed with a Lorentzian metric, one is dealing with an Einstein setting. In what follows we shall limit ourselves to the specific case of $\widetilde{Q} \equiv \mathbf{R}^4$. By considering the phase space $\mathbf{R}^4 \times (\mathbf{R}^4)^*$ we have a symplectic potential, or canonical 1-form,

$$\theta = p_0 \, \mathrm{d}x^0 + p_k \, \mathrm{d}x^k. \tag{2.3.5}$$

If we use the Hamiltonian function $H = H(x, p)$ we have the equations of motion

$$\frac{\mathrm{d}x^0}{\mathrm{d}s} = \frac{\partial H}{\partial p_0}, \quad \frac{\mathrm{d}x^k}{\mathrm{d}s} = \frac{\partial H}{\partial p_k}, \tag{2.3.6}$$

$$\frac{\mathrm{d}p_0}{\mathrm{d}s} = -\frac{\partial H}{\partial x^0}, \quad \frac{\mathrm{d}p_k}{\mathrm{d}s} = -\frac{\partial H}{\partial x^k}. \tag{2.3.7}$$

Note that s is an external parameter. When it is possible to identify a coordinate function with the evolution parameter s, i.e. $\frac{\partial H}{\partial p_0}$ is nowhere vanishing, we may set

$$\frac{\mathrm{d}x^0}{\mathrm{d}s} = 1, \tag{2.3.8}$$

which is equivalent to requiring the following special form for the Hamiltonian:

$$H = p_0 + h(p_1, p_2, p_3, x^k). \tag{2.3.9}$$

Remark. When $\frac{\partial H}{\partial p_0}$ is nowhere vanishing one could think of reparametrizing the dynamical vector field by multiplying by the inverse of $\frac{\partial H}{\partial p_0}$. One should be aware, however, that the resulting vector field need no longer be Hamiltonian. Still, we may define a surface Σ by fixing a value of H (e.g. $H = 0$). On this sub-manifold the dynamics will be defined by the kernel of $\mathrm{d}\theta$ restricted to Σ, i.e.

$$\left(\mathrm{d}p_k \wedge \mathrm{d}x^k + \mathrm{d}p_0 \wedge \mathrm{d}s\right)\Big|_\Sigma .$$

Since the parametrization is arbitrary we can now fix it by imposing the condition (2.3.8).

If this identification is required to be preserved in going from one reference frame to another, transformations should preserve $\mathrm{d}x^0$. Indeed, as $\mathrm{d}s$ is an 'external parameter', it should not be affected by point transformations. Therefore all reference frames consistent with this identification will have a common notion of simultaneity (see appendix 2.D).

If, instead of considering the equations of motion on $\mathbf{R}^4 \times \mathbf{R}^4$, we restrict them to a particular level set of H (see the previous remark), we identify what is usually understood as a time-dependent formalism. By selecting a specific level set of H, e.g. $H = 0$, we find on this sub-manifold Σ_0,

$$p_0 = -h(p_1, p_2, p_3, x^1, x^2, x^3; x^0). \tag{2.3.10}$$

The restriction of θ_0 to this sub-manifold gives

$$\theta = p_k \,\mathrm{d}x^k - h(p, x; x^0)\,\mathrm{d}x^0 . \tag{2.3.11}$$

If Eq. (2.3.9) is globally valid, Σ is isomorphic with $T^*\mathbf{R}^3 \times \mathbf{R}$. Of course Σ, being odd-dimensional, can no longer be a symplectic manifold, nevertheless we may still consider generating functions on $(T^*\mathbf{R}^3) \times (T^*\mathbf{R}^3) \times \mathbf{R}$ for which

$$\mathrm{d}W = \left(p_k \,\mathrm{d}x^k - h\,\mathrm{d}s\right) - \left(P_k \,\mathrm{d}y^k - \widetilde{h}\,\mathrm{d}s\right), \tag{2.3.12}$$

where we have used $\mathrm{d}x^0 = \mathrm{d}s$. Equation (2.3.12) implies that

$$\begin{aligned} p_k \,\mathrm{d}x^k - h\,\mathrm{d}s &- \left(P_k \,\mathrm{d}y^k - \widetilde{h}\,\mathrm{d}s\right) \\ &= \frac{\partial W}{\partial x^k}\mathrm{d}x^k + \frac{\partial W}{\partial y^k}\mathrm{d}y^k + \frac{\partial W}{\partial s}\mathrm{d}s. \end{aligned} \tag{2.3.13}$$

Thus, we find the transformation described by

$$p_k = \frac{\partial W}{\partial x^k}, \tag{2.3.14}$$

$$P_k = -\frac{\partial W}{\partial y^k}, \tag{2.3.15}$$

$$\widetilde{h} - h = \frac{\partial W}{\partial s}. \tag{2.3.16}$$

The variable p_0 has disappeared from this theory, and we only have variables $(x, p; s)$. By virtue of Eq. (2.3.16) we can ask for a transformation defined by a function W such that we have the result $\widetilde{h} = 0$, i.e. $H = \widetilde{p}_0$, or we have to solve for W the following partial differential equation:

$$h\left(x, \frac{\partial W}{\partial x^k}, s\right) + \frac{\partial W}{\partial s} = 0. \tag{2.3.17}$$

This equation is usually called the Hamilton–Jacobi equation for the Hamilton principal function W. We notice that in the (P_k, y^k) variables $\widetilde{h}$ vanishes and hence the associated Hamilton equations of motion are

$$\frac{\mathrm{d}P_k}{\mathrm{d}s} = 0, \quad \frac{\mathrm{d}y^k}{\mathrm{d}s} = 0, \quad \frac{\mathrm{d}x_0}{\mathrm{d}s} = 1.$$

Thus, if we find a globally defined principal function W solving Eq. (2.3.17) and depending on additional parameters Y^j such that

$$\det \left\| \frac{\partial^2 W}{\partial x^k \partial Y^j} \right\| \neq 0,$$

we can straighten out the flow completely. This rather strong result also indicates that in most cases we will not be able to find a complete solution of Eq. (2.3.17) globally defined on $T^*\mathbf{R}^3 \times \mathbf{R}$.

In the Poisson-bracket formalism one now has

$$\frac{\mathrm{d}}{\mathrm{d}s} f = \{f, H\} + \frac{\partial f}{\partial s}, \tag{2.3.18}$$

with the general expression for a Hamiltonian vector field

$$\frac{\mathrm{d}}{\mathrm{d}s} = \frac{\partial}{\partial s} + \left\{x^j, H\right\} \frac{\partial}{\partial x^j} + \{p_j, H\} \frac{\partial}{\partial p_j}. \tag{2.3.19}$$

As far as Poisson brackets are concerned we use an 'equal-time' Poisson bracket, i.e. for f, g also depending on time we have

$$\{f, g\} = \frac{\partial f}{\partial x^j} \frac{\partial g}{\partial p_j} - \frac{\partial f}{\partial p_j} \frac{\partial g}{\partial x^j}, \tag{2.3.20}$$

with time behaving as a parameter. In other words, the Poisson bracket is such that the coordinate function $x^0 = s$ commutes with any function f.

2.3.2 Dynamical time

In general, on any phase space, for any specific dynamical system, a dynamical time $\tau_{\rm D}$ is any function on phase space satisfying

$$\frac{\rm d}{{\rm d}t}\tau_{\rm D} = 1, \tag{2.3.21}$$

or, for Hamiltonian systems,

$$\{\tau_{\rm D}, H\} = 1, \tag{2.3.22}$$

i.e. any function canonically conjugate to H. Note that $\tau_{\rm D}$ is defined up to the addition of constants of motion. In this subsection we consider dynamical time for a Galilean invariant two-particle system. This $\tau_{\rm D}$ is a function on the cotangent bundle of $\mathbf{R}^6$ satisfying the above equation $\frac{\rm d}{{\rm d}t}\tau_{\rm D} = 1$ along the trajectory of the dynamical vector field. This dynamical time is associated with the 'free' dynamics defined by H (readers not familiar with the Galilei group are referred to appendix 2.B, where the Galilei group is defined jointly with other groups).

Since true dynamical time must not depend on the choice of coordinates origin, we take a system of two free particles with canonical coordinates $\vec{q}_1, \vec{p}_1, \vec{q}_2, \vec{p}_2$. Then the Galilean generators are

$$\vec{P} = \vec{p}_1 + \vec{p}_2, \tag{2.3.23}$$

$$\vec{J} = \vec{q}_1 \wedge \vec{p}_1 + \vec{q}_2 \wedge \vec{p}_2, \tag{2.3.24}$$

$$H = \frac{p_1^2}{2m_1} + \frac{p_2^2}{2m_2}, \tag{2.3.25}$$

$$\vec{G} = m_1\vec{q}_1 + m_2\vec{q}_2. \tag{2.3.26}$$

Then $\vec{q}_1 - \vec{q}_2, \vec{p}_1, \vec{p}_2$ are translation invariant, while $\vec{q}_1, \vec{q}_2, \frac{\vec{p}_1}{m_1} - \frac{\vec{p}_2}{m_2}$ are boost invariant. Thus, a dynamical time $\tau_{\rm D}$ is defined by

$$\tau_{\rm D} \equiv \frac{(\vec{q}_1 - \vec{q}_2)\cdot\left(\frac{\vec{p}_1}{m_1} - \frac{\vec{p}_2}{m_2}\right)}{\left(\frac{\vec{p}_1}{m_1} - \frac{\vec{p}_2}{m_2}\right)^2}. \tag{2.3.27}$$

This is invariant under $\vec{P}, \vec{G}, \vec{J}$ but it is displaced by H, because

$$\{\tau_{\rm D}, H\} = 1.$$

In the Galilean case we can also write down 'cyclic time' or other forms of time when there is a Galilean invariant interaction. In particular, if

$$H_{\mathrm{int}} = \frac{1}{4}(\vec{q}_1 - \vec{q}_2)^2, \tag{2.3.28}$$

the corresponding dynamical time is periodic and given by

$$\tau = \tan^{-1}\tau_{\mathrm{D}}, \tag{2.3.29}$$

where τ_{D} is given in (2.3.27).

2.3.3 Various generating functions

We have obtained a scheme where the transformation

$$(q, t; p, H) \rightarrow (Q, t; P, K)$$

is *canonical* if a generating function W exists such that

$$p_i \, \mathrm{d}q^i - H \, \mathrm{d}t = P_i \, \mathrm{d}Q^i - K \, \mathrm{d}t + \mathrm{d}W, \tag{2.3.30}$$

where K is the new 'Hamiltonian', obtained from the original Hamiltonian H by the equation (see problem 2.P6)

$$K(Q, P; t) = H(q, p; t) + \frac{\partial W}{\partial t}. \tag{2.3.31}$$

Remark. As we have seen in sections 2.2 and 2.3, this W depends on the specific choice of canonical coordinate system, i.e. on the choice of symplectic potential.

Independently of the choice of Hamiltonian function H, if one considers the solutions of the Hamilton equations (for $i = 1, \ldots, n$)

$$\frac{\mathrm{d}}{\mathrm{d}t} q^i = \frac{\partial H}{\partial p_i}, \tag{2.3.32}$$

$$\frac{\mathrm{d}}{\mathrm{d}t} p_i = -\frac{\partial H}{\partial q^i}, \tag{2.3.33}$$

the new coordinates Q^i and momenta P_i depend on the time in such a way that, for all $i = 1, \ldots, n$,

$$\frac{\mathrm{d}}{\mathrm{d}t} Q^i = \frac{\partial K}{\partial P_i}, \tag{2.3.34}$$

$$\frac{\mathrm{d}}{\mathrm{d}t} P_i = -\frac{\partial K}{\partial Q^i}. \tag{2.3.35}$$

The time-dependent canonical transformations obtained in such a way are often called *completely canonical*.

For a given Hamiltonian H, one looks for a time-dependent generating function W on phase space such that K takes a particularly simple form. This leads, in turn, to a set of Hamilton equations (2.3.34) and (2.3.35), for which integration is straightforward, and a solution for the original coordinates q^i and momenta p_i is eventually obtained (see the following sections). In particular, it may happen that K vanishes, and hence the right-hand sides of Eqs. (2.3.34) and (2.3.35) vanish as well. When this is achieved, one says that the flow has been completely straightened out. This is the case for Eq. (2.3.17).

We now summarize the schemes resulting from the various possibilities (Whittaker 1937, Sudarshan and Mukunda 1974, Arnol'd 1978, Goldstein 1980, Abraham and Marsden 1985, José and Saletan 1998). For convenience of notation, we shall distinguish four generating functions, although, strictly, one actually deals with a generating function obtained by means of four different procedures. These are associated with various choices of symplectic potential in the form $p\,\mathrm{d}q, -q\,\mathrm{d}p, P\,\mathrm{d}Q, -Q\,\mathrm{d}P$. It is clear that with many degrees of freedom one could select various combinations on different sub-spaces.

(i) The generating function W_1

In such a case one has

$$\sum_i p_i\,\mathrm{d}q^i - H(q,p;t)\,\mathrm{d}t = \sum_i P_i\,\mathrm{d}Q^i - K\,\mathrm{d}t + \mathrm{d}W_1(q,Q;t). \tag{2.3.36}$$

Furthermore,

$$\mathrm{d}W_1(q,Q;t) = \sum_i \frac{\partial W_1}{\partial q^i}\mathrm{d}q^i + \sum_i \frac{\partial W_1}{\partial Q^i}\mathrm{d}Q^i + \frac{\partial W_1}{\partial t}\mathrm{d}t. \tag{2.3.37}$$

One therefore finds by comparison of the left- and right-hand sides of (2.3.36) the equations

$$p_i = \frac{\partial W_1}{\partial q^i}, \tag{2.3.38}$$

$$P_i = -\frac{\partial W_1}{\partial Q^i}, \tag{2.3.39}$$

$$K_1 = H + \frac{\partial W_1}{\partial t}. \tag{2.3.40}$$

(ii) The generating function W_2

Equation (2.3.39) suggests obtaining W_2 from W_1 via a Legendre transform, defined as

$$W_2(q,P;t) \equiv W_1(q,Q;t) + \sum_i P_i Q^i. \tag{2.3.41}$$

We now re-express W_1 from (2.3.41) and insert it into (2.3.36). Collecting together the terms involving dP_i and dq^i one finds the equations

$$p_i = \frac{\partial W_2}{\partial q^i}, \tag{2.3.42}$$

$$Q^i = \frac{\partial W_2}{\partial P_i}, \tag{2.3.43}$$

$$K_2 = H + \frac{\partial W_2}{\partial t}. \tag{2.3.44}$$

(iii) The generating function W_3

Equation (2.3.38) suggests obtaining W_1 from W_3 by means of the Legendre transform

$$W_1(q,Q;t) \equiv \sum_i p_i q^i + W_3(Q,p;t). \tag{2.3.45}$$

This relation is inserted into (2.3.36), and after taking the differential of W_3 and collecting similar terms one finds

$$P_i = -\frac{\partial W_3}{\partial Q^i}, \tag{2.3.46}$$

$$q^i = -\frac{\partial W_3}{\partial p_i}, \tag{2.3.47}$$

$$K_3 = H + \frac{\partial W_3}{\partial t}. \tag{2.3.48}$$

(iv) The W_4 choice

Lastly, one may consider the generating function defined by

$$W_4(p,P;t) \equiv W_1(q,Q;t) + \sum_i P_i Q^i - \sum_i p_i q^i. \tag{2.3.49}$$

The insertion of (2.3.49) into (2.3.36) yields

$$\sum_i p_i \, dq^i - H = \sum_i P_i \, dQ^i - K + \sum_i \frac{\partial W_4}{\partial p_i} dp_i + \sum_i \frac{\partial W_4}{\partial P_i} dP_i$$

$$+\frac{\partial W_4}{\partial t}\mathrm{d}t - \sum_i (\mathrm{d}P_i)Q^i - \sum_i P_i\,\mathrm{d}Q^i$$
$$+\sum_i p_i\,\mathrm{d}q^i + \sum_i (\mathrm{d}p_i)q^i, \tag{2.3.50}$$

which implies

$$q^i = -\frac{\partial W_4}{\partial p_i}, \tag{2.3.51}$$

$$Q^i = \frac{\partial W_4}{\partial P_i}, \tag{2.3.52}$$

$$K_4 = H + \frac{\partial W_4}{\partial t}. \tag{2.3.53}$$

In all of these cases we assume that at least locally we can solve for the old variables in terms of the new, from (2.3.38) and (2.3.39). But these may be degenerate and do not allow for explicit solution. These degenerate cases can occur for all four generating functions, and can be treated (Sudarshan and Mukunda 1974). This happens when the defining equations cannot give the expressions for the new variables, but instead there are auxiliary functional relations among a subset of them.

2.3.4 An example: particle in a repulsive potential

Consider the generating function (the $2m$ factor is absorbed into our notation at some places, for simplicity)

$$W_2(q,P) \equiv \int_{x_0}^{q} \sqrt{P^2 - V(x)}\,\mathrm{d}x, \tag{2.3.54}$$

with the associated equations

$$p = \frac{\partial W_2}{\partial q} = \sqrt{P^2 - V(q)}, \tag{2.3.55}$$

$$q = \frac{\partial W_2}{\partial P} = \int_{x_0}^{q} \frac{P}{\sqrt{P^2 - V(x)}}\,\mathrm{d}x, \tag{2.3.56}$$

or

$$P = \sqrt{p^2 + V(q)}, \tag{2.3.57}$$

$$q = \int_{x_0}^{q} \frac{P}{\sqrt{P^2 - V(x)}}\,\mathrm{d}x = \int_{x_0}^{q} \frac{\sqrt{p^2 + V(x)}}{p}\,\mathrm{d}x, \tag{2.3.58}$$

$$K_2 = H + \frac{\partial W_2}{\partial t} = \frac{P^2}{2m}. \tag{2.3.59}$$

This transformation is defined for all q, p.

2.3.5 The harmonic oscillator

Another example of local canonical transformation is given by the one-dimensional harmonic oscillator with the generating function W_1:

$$W_1 = \frac{m}{2}\omega q^2 \cot Q. \tag{2.3.60}$$

According to (2.3.38) and (2.3.39) we obtain the old and new momenta as

$$p = \frac{\partial W_1}{\partial q} = m\omega q \, \cot Q, \tag{2.3.61}$$

$$P = -\frac{\partial W_1}{\partial Q} = \frac{m\omega q^2}{2\sin^2 Q}. \tag{2.3.62}$$

Equation (2.3.62) may be solved for q:

$$q = \sqrt{\frac{2P}{m\omega}} \sin Q, \tag{2.3.63}$$

where Q is defined only modulo 2π, and the result may be inserted into (2.3.61) to give

$$p = \sqrt{2m\omega P} \cos Q. \tag{2.3.64}$$

By virtue of (2.3.63) and (2.3.64), the Hamiltonian $H = \frac{p^2}{2m} + \frac{m\omega^2}{2}q^2$ becomes linear in P:

$$H = \omega \, P, \tag{2.3.65}$$

and hence is cyclic with respect to Q. The equation of motion for Q is $\dot{Q} = \frac{\partial H}{\partial P} = \omega$, which implies

$$Q = \omega t + \alpha. \tag{2.3.66}$$

This leads to (see Eq. (2.3.63))

$$q = \sqrt{\frac{2P}{m\omega}} \, \sin(\omega t + \alpha). \tag{2.3.67}$$

We must note, however, that this transformation is not global and the topology of the coordinate manifold $-\infty < q < +\infty$ is changed into the periodic compact phase space $0 \leq Q < 2\pi$ (mod 2π). In fact all the so-called 'canonical transformations' to action-angle variables have this effect; and so does the discussion in section 2.5.2 (cf. Eq. (2.5.20)). But in the case of the free particle discussed in section 2.5.1 one can make the definition global.

2.4 Hamilton and Hamilton–Jacobi equations

Starting with the Hamilton equations (2.3.6) and (2.3.7), solutions are given by

$$q(s) = q(q(0), p(0), s^0; s), \tag{2.4.1}$$

$$p(s) = p(q(0), p(0), s^0; s). \tag{2.4.2}$$

This can be thought of as a transformation from 'initial' position and momenta to 'final' position and momenta (q, p). Differentiability of the pair $(q(s), p(s))$ with respect to the initial conditions and the invertibility of the map are ruled by the standard theorems on solutions of ordinary differential equations (Yosida 1991).

This transformation is known to be canonical. The existence of solutions for $-\infty \leq s \leq \infty$ and for any initial condition gives rise to a one-parameter group of canonical transformations. The vector field

$$\Gamma \equiv \frac{\partial}{\partial s} + \frac{\partial H}{\partial p_k}\frac{\partial}{\partial x^k} - \frac{\partial H}{\partial x^k}\frac{\partial}{\partial p_k} \tag{2.4.3}$$

is referred to as the infinitesimal generator of this one-parameter group of canonical transformations.

Let us now consider the function W defined by

$$W(x_0, s_0; x_f, s) \equiv \int_{x_0, s_0}^{x_f, s_f} (p\, \mathrm{d}x - h\, \mathrm{d}s), \tag{2.4.4}$$

where the integral is evaluated along a solution of the Hamilton equations passing through (x_0, s_0) and (x_f, s). We are considering neighbourhoods of (x_0, s_0) and (x_f, s) such that for any pair of points, one in the first neighbourhood and one in the second, there is a unique solution. When this is not the case, W is multi-valued. On fixing the initial value s_0 of s and letting (x_0, x, s) vary, and by taking the differential of both sides of (2.4.4) we find

$$\mathrm{d}W = (p\, \mathrm{d}x - h\, \mathrm{d}s)|_{(x_f, s_f)} - (p\, \mathrm{d}x - h\, \mathrm{d}s)|_{(x_0, s_0)}, \tag{2.4.5}$$

i.e.

$$\begin{aligned} &\frac{\partial W}{\partial x_0^k}\, \mathrm{d}x_0^k + \frac{\partial W}{\partial x_f^k}\, \mathrm{d}x_f^k + \frac{\partial W}{\partial s_f}\, \mathrm{d}s_f \\ &= p_{f,k}\, \mathrm{d}x_f^k - h_f\, \mathrm{d}s_f - p_{0,k}\, \mathrm{d}x_0^k, \end{aligned} \tag{2.4.6}$$

which implies

$$p_{0,k} = -\frac{\partial W}{\partial x_0^k}, \tag{2.4.7}$$

$$p_{f,k} = \frac{\partial W}{\partial x_f^k}, \tag{2.4.8}$$

$$\frac{\partial W}{\partial s_f} + h_f\left(x_f, \frac{\partial W}{\partial x_f}, s_f\right) = 0. \tag{2.4.9}$$

To sum up, we have found the interesting result that the integral of the action along a solution, when the initial instant of time is fixed but the initial position and the final position are considered to be variable, provides a solution of the Hamilton–Jacobi equation. Conversely, if W is a solution of the Hamilton–Jacobi equation, from

$$p_{0,k} = -\frac{\partial W}{\partial x_0^k}(x_0, x_f, s; s_0), \tag{2.4.10}$$

on solving with respect to x_f via the implicit function theorem to obtain

$$x_f(s) = x_f(x_0, s_0, s),$$

and setting

$$p_{f,k} = \frac{\partial W}{\partial x_f}(x_0, x_f(s), s, s_0), \tag{2.4.11}$$

we obtain a solution of the Hamilton equations. We therefore conclude that a one-to-one correspondence exists between families of solutions of the Hamilton–Jacobi equation and families of solutions of the Hamilton equations. According to the discussion following Eq. (2.3.17), this implies that with respect to the 'initial conditions' thought of as (P_k, y^k) variables, one has

$$\frac{\mathrm{d}P_{0,k}}{\mathrm{d}s} = 0, \quad \frac{\mathrm{d}x_{0,k}}{\mathrm{d}s} = 0.$$

As a last remark we note that, in the course of evaluating (see Eq. (2.4.4))

$$W = \int_{x_0,s_0}^{x,s} (p\,\mathrm{d}x' - h\,\mathrm{d}s), \tag{2.4.12}$$

when we integrate along a solution of the Hamilton equations, if the Hamiltonian is time-independent the function h is a constant of motion and hence we have

$$W = \int_{x_0,s_0}^{x,s} p\,\mathrm{d}x' - h(x_0, x)(s - s_0). \tag{2.4.13}$$

This means that, on solving the Hamilton–Jacobi equation for time-independent Hamiltonian functions it is possible to use separation of variables, requiring linearity in the evolution parameter.

2.5 The Hamilton principal function

Now we are going to investigate the Hamilton principal function in three cases of physical interest, i.e. the free particle, the one-dimensional harmonic oscillator and a system having a time-dependent Hamiltonian. The detailed steps are as follows.

2.5.1 Free particle on a line

For a free particle on a line, the Hamiltonian is

$$H = \frac{p^2}{2m}, \tag{2.5.1}$$

and we consider two complete integrals of the Hamilton–Jacobi equation of the form

$$W_1(q, Q_1, t) = \frac{m(q - Q_1)^2}{2t}, \quad t > 0, \tag{2.5.2}$$

and

$$W_2(q, Q_2, t) = qQ_2 - \frac{Q_2^2}{2m}t. \tag{2.5.3}$$

From W_1 one finds

$$p = m\frac{(q - Q_1)}{t}, \tag{2.5.4}$$

with P_1 and p constant, so that the equations of motion are solved by

$$q(t) = \frac{p}{m}t + Q_1, \tag{2.5.5}$$

where $Q_1 = q_0 = q(0)$ and $P_1 = p_0 = p(0)$. The principal function W_2 leads to

$$p = Q_2 = \text{constant}, \tag{2.5.6}$$

$$P_2 = -q + \frac{Q_2}{m}t, \tag{2.5.7}$$

i.e. $Q_2 = p_0$ and $P_2 = -q_0$.

While W_1 is singular for fixed q as $t \to 0^+$, W_2 is a complete integral corresponding to the Cauchy data

$$W_0(q) = W(q, t = 0) = Qq. \tag{2.5.8}$$

If the mass is set to 1 for simplicity of notation, the initial condition

$$W_0(q) = \frac{q^2}{2} \tag{2.5.9}$$

provides a solution with

$$W(q,t) = \frac{q^2}{(1+t)}, \quad t > -1. \tag{2.5.10}$$

This is not a complete integral because there are no free constants. In general we may say that complete integrals, when viewed as functions on phase space, are related by

$$\frac{\mathrm{d}W}{\mathrm{d}t} \equiv L_\Gamma W = \frac{\partial W}{\partial q^i}\dot{q}^i + \frac{\partial W}{\partial t} = p_i \dot{q}^i - H \equiv \mathcal{L}. \tag{2.5.11}$$

Thus, if W_1 and W_2 are two complete integrals of the Hamilton–Jacobi equation, one has

$$\frac{\mathrm{d}}{\mathrm{d}t}(W_1 - W_2) = 0, \tag{2.5.12}$$

i.e. any two complete integrals may differ by a constant of motion. Hence one can write

$$W(q,t;q_0,t_0) = \int_\gamma \mathcal{L}\,\mathrm{d}t + W_0. \tag{2.5.13}$$

2.5.2 One-dimensional harmonic oscillator

For a one-dimensional harmonic oscillator, the Hamiltonian reads

$$H = \frac{p^2}{2m} + \frac{k}{2}q^2, \tag{2.5.14}$$

and hence the Hamilton–Jacobi equation takes the form

$$\frac{1}{2m}\Big(\partial W/\partial q\Big)^2 + \frac{k}{2}q^2 + \frac{\partial W}{\partial t} = 0. \tag{2.5.15}$$

This suggests using separation of variables and looking for a solution in the form (see Eq. (2.4.13))

$$W(q,\alpha,t) = S(q,\alpha) - \alpha t, \tag{2.5.16}$$

which implies

$$\Big(\partial S/\partial q\Big)^2 = mk\Big(\frac{2\alpha}{k} - q^2\Big). \tag{2.5.17}$$

One can thus express the unknown function W as

$$W = \sqrt{mk}\int\sqrt{\frac{2\alpha}{k} - q^2}\,\mathrm{d}q - \alpha t. \tag{2.5.18}$$

The general formula (2.3.43) shows that the coordinate Q is

$$Q = \frac{\partial W}{\partial \alpha} = \sqrt{\frac{m}{k}}\int\frac{\mathrm{d}q}{\sqrt{2\alpha/k - q^2}} - t. \tag{2.5.19}$$

The integral (2.5.19) can be evaluated exactly to give

$$q = \sqrt{\frac{2\alpha}{k}} \cos \omega(t + Q). \tag{2.5.20}$$

For example, if we choose the initial conditions

$$p_0 = p(t = 0) = 0, \tag{2.5.21}$$

$$q_0 = q(t = 0) \neq 0, \tag{2.5.22}$$

we find that $\alpha = \frac{kq_0^2}{2} = \frac{m\omega^2 q_0^2}{2}$, and Q has to vanish.

2.5.3 *Time-dependent Hamiltonian*

We here consider a system for which the Hamiltonian depends explicitly on time and has the form

$$H = \frac{p^2}{2m} - mktq. \tag{2.5.23}$$

The resulting equations of motion are solved by

$$q(t) = q_0 + \frac{p_0}{m} t + \frac{1}{6} kt^3, \tag{2.5.24}$$

$$p(t) = m\dot{q}(t) = p_0 + \frac{1}{2} mkt^2. \tag{2.5.25}$$

The Lagrangian associated to the Hamiltonian (2.5.23) is

$$\mathcal{L} = \frac{p^2}{2m} + mktq. \tag{2.5.26}$$

Along a trajectory parametrized by the initial conditions (q_0, p_0) it reads, by virtue of (2.5.24) and (2.5.25),

$$\mathcal{L}(t) = \frac{p_0^2}{2m} + mkq_0 t + \frac{3}{2} kp_0 t^2 + \frac{7}{24} mk^2 t^4. \tag{2.5.27}$$

The integration with respect to time yields therefore (see Eqs. (2.4.4) and (2.5.13))

$$\begin{aligned} W(q, t; q_0, 0) &= \int_0^t \mathcal{L}(t')\, \mathrm{d}t' \\ &= \frac{p_0^2}{2m} t + \frac{mkq_0}{2} t^2 + \frac{kp_0}{2} t^3 + \frac{7mk^2}{120} t^5. \end{aligned} \tag{2.5.28}$$

Note now that Eq. (2.5.24) makes it possible to express the initial condition p_0 in the form

$$p_0 = m \frac{(q - q_0)}{t} - \frac{1}{6} mkt^2, \tag{2.5.29}$$

and its insertion into (2.5.28) yields the desired formula for the principal function in terms of q, the initial condition q_0 and the time, i.e.

$$W(q, q_0; t) = \frac{m}{2} \left\{ \frac{(q - q_0)^2}{t} + \frac{k}{3}(2q + q_0)t^2 - \frac{1}{45}k^2 t^5 \right\}. \tag{2.5.30}$$

It can be checked that such a principal function satisfies the Hamilton–Jacobi equation

$$\frac{\partial W}{\partial t} + \frac{1}{2m}\left(\frac{\partial W}{\partial q}\right)^2 - mktq = 0. \tag{2.5.31}$$

2.6 The characteristic function

Subsection 2.5.2 studied the case when the Hamiltonian does not depend explicitly on time. It suggests that, in all such cases, one can set

$$W(q^i, \alpha_i, t) = S(q^i, \alpha_i) - \alpha t, \tag{2.6.1}$$

and hence the Hamilton–Jacobi equation reduces to

$$H\left(q^i, \frac{\partial S}{\partial q^i}\right) = \alpha. \tag{2.6.2}$$

The unknown function S is called the Hamilton *characteristic function.* The resulting scheme differs from that obtained by applying the principal function, and is as follows.

The Hamilton–Jacobi equation takes the form (2.6.2), and the first-order equations of motion read as

$$\frac{\mathrm{d}}{\mathrm{d}t}Q^i = \frac{\partial K}{\partial P_i} = \nu^i, \tag{2.6.3}$$

$$\frac{\mathrm{d}}{\mathrm{d}t}P_i = -\frac{\partial K}{\partial Q^i} = 0. \tag{2.6.4}$$

Their integration yields (where β^i are some constants)

$$Q^i = \nu^i t + \beta^i, \tag{2.6.5}$$

$$P_i = \gamma_i(\alpha_1, \ldots, \alpha_n). \tag{2.6.6}$$

The integration of (2.6.5) and (2.6.6) involves $n - 1$ constants which, jointly with α_1, form a set of n independent constants. One can then solve locally for

$$q^i = q^i(\beta^k, \gamma_k, t), \tag{2.6.7}$$

$$\beta^i = \beta^i\left(q^k_{(0)}, p^{(0)}_k\right), \tag{2.6.8}$$

$$\gamma_i = \gamma_i\left(q_{(0)}^k, p_k^{(0)}\right). \tag{2.6.9}$$

As a further application, let us consider the motion in a central potential. The Hamiltonian of a particle of mass m can then be written as

$$H = \frac{1}{2m}\left(p_r^2 + \frac{p_\phi^2}{r^2}\right) + V(r). \tag{2.6.10}$$

The Hamiltonian is cyclic with respect to ϕ, and hence we are led to write the characteristic function S as the sum

$$S(r,\phi) = S_1(r) + \alpha_\phi \phi, \tag{2.6.11}$$

where α_ϕ is the momentum conjugate to ϕ. The Hamilton–Jacobi equation therefore reads as

$$\frac{1}{2m}\left[\left(\partial S_1/\partial r\right)^2 + \frac{\alpha_\phi^2}{r^2}\right] + V(r) = \alpha \tag{2.6.12}$$

and hence S is given by

$$S = \int \sqrt{2m(\alpha - V(r)) - \frac{\alpha_\phi^2}{r^2}}\, \mathrm{d}r + \alpha_\phi \phi. \tag{2.6.13}$$

Differentiation of S with respect to α and α_ϕ enables one to derive the time evolution of r, and the equation of the orbit, respectively. The application of the method leads to the following integrals (Landau and Lifshitz 1960, Goldstein 1980):

$$t + \beta_1 = \frac{\partial S}{\partial \alpha} = \int \frac{m\, \mathrm{d}r}{\sqrt{2m[\alpha - V(r)] - \alpha_\phi^2/r^2}}, \tag{2.6.14}$$

$$\beta_2 = \frac{\partial S}{\partial \alpha_\phi} = -\int \frac{\alpha_\phi\, \mathrm{d}r}{r^2\sqrt{2m[\alpha - V(r)] - \alpha_\phi^2/r^2}} + \phi. \tag{2.6.15}$$

To evaluate these integrals it is convenient to set $u \equiv \frac{1}{r}$. The orbit may or may not be closed, depending on the form of $V(r)$. In many applications, $V(r)$ contains finitely many negative powers of r. The signs of the various coefficients, e.g.

$$V(r) = \frac{\sigma}{r} + \frac{\tau}{r^3}, \tag{2.6.16}$$

play a crucial role and deserve careful investigation (see problem 2.P5).

2.6.1 *Principal versus characteristic function*

To appreciate the difference between the principal and the characteristic function, we now give two examples. First, a free particle in $\mathbf{R}^3$ has, for $t > t_0$, the principal function

$$W(q,t) = \frac{m(q-q_0)^2}{2(t-t_0)}, \tag{2.6.17}$$

while the characteristic function reads as

$$S(E,q) = \sqrt{2mE}|q-q_0|. \tag{2.6.18}$$

These formulae correspond to 'spherical wave fronts' (section 3.3), whereas plane-wave fronts are associated with the principal function

$$W(q,t) = p \cdot q - \frac{p^2}{2m}t, \tag{2.6.19}$$

with the characteristic function

$$S(E,q) = p \cdot q, \tag{2.6.20}$$

since $E = \frac{p^2}{2m}$.

If $0 < t - t_0 < \frac{\pi}{\omega}$, the one-dimensional harmonic oscillator has the principal function

$$W(q,t) = \frac{m\omega}{2\sin\omega(t-t_0)} \left[(q^2+q_0^2)\cos\omega(t-t_0) - 2qq_0 \right], \tag{2.6.21}$$

while the characteristic function reads as

$$S(E,q) = \frac{E}{\omega}\arcsin\left(\sqrt{\frac{m}{2E}}\omega q\right) + \sqrt{\frac{mE}{2}}q\sqrt{1-\frac{m\omega^2q^2}{2E}}. \tag{2.6.22}$$

2.7 Hamilton equations associated with metric tensors

Consider again the Hamilton equations (2.3.6) and (2.3.7), and assume we have a solution of the Hamilton–Jacobi equation (2.3.17). Using the replacement $p_k = \frac{\partial W}{\partial x^k}$ we may turn the Hamilton equation for x^k into an equation on the configuration space only:

$$\frac{{\rm d}x^k}{{\rm d}s} = \frac{\partial H}{\partial p_k}\left(x, \frac{\partial W}{\partial x}, s\right). \tag{2.7.1}$$

Solutions of this ordinary differential equation define a congruence of lines on the configuration space usually called 'rays' in analogy with geometrical optics (see section 2.8). If we use a solution W on $Q \times \mathbf{R}$ we determine a sub-space on the extended configuration space by setting $W = c$, e.g. $W = 0$. By intersecting this sub-space with the sub-manifold $s =$ constant, we obtain a codimension-2 surface for each value of the constant, therefore if we let s increase, for instance, we find a 'moving' codimension-1 surface on the sub-space $W = 0$. The congruence defined by Eq. (2.7.1) when considered with respect to these 'wave fronts' behaves like a family of 'rays'. Instead of showing this construction in general, we shall perform it in the particular case in which the starting relation is quadratic in the momenta, i.e. it is associated with a metric tensor on space–time (see appendix 2.D).

We therefore consider the Hamiltonian function

$$H \equiv \frac{1}{2} g^{ab} p_a p_b \tag{2.7.2}$$

on $\mathbf{R}^4 \times (\mathbf{R}^4)^*$, where $g_{ab}(x)\, \mathrm{d}x^a \otimes \mathrm{d}x^b$ is the metric on $\mathbf{R}^4$, and the 'contravariant' form of the metric tensor is defined by

$$g^{ab} g_{bc} = \delta^a_{\ c}. \tag{2.7.3}$$

The associated Hamilton–Jacobi equation is given by

$$g^{ab} \frac{\partial W}{\partial x^a} \frac{\partial W}{\partial x^b} = 0. \tag{2.7.4}$$

The differential equation for the congruence of trajectories on $\mathbf{R}^4$ is given by

$$\frac{\mathrm{d}x^a}{\mathrm{d}s} = g^{ab} \frac{\partial W}{\partial x^b}, \tag{2.7.5}$$

where W is a solution of the Hamilton–Jacobi equation. Note that

$$\frac{\mathrm{d}W}{\mathrm{d}s} = \frac{\partial W}{\partial x^a} \frac{\mathrm{d}x^a}{\mathrm{d}s} = \frac{\partial W}{\partial x^a} g^{ab} \frac{\partial W}{\partial x^b} = 0,$$

i.e. W is a constant of motion for the dynamics associated with Eq. (2.7.5). In particular, the associated vector field is tangent to each level set, i.e. trajectories of Eq. (2.7.5) with initial conditions on $W =$ constant will remain on the surface. Moreover,

$$\frac{\mathrm{d}x^0}{\mathrm{d}s} = g^{0b} \frac{\partial W}{\partial x^b} \neq 0$$

shows that the congruence of trajectories is transverse to the sub-manifold $s =$ constant.

The level surfaces of W, for which $W(x) =$ constant, define 3-surfaces with null normal vector (hence they are called null 3-surfaces). By also setting s equal to a constant, one finds for various values of s a 'moving surface' in $\mathbf{R}^3$, i.e. a 'wave front' (Synge 1954, Guckenheimer 1973). The solutions of the equation (cf. Eq. (2.7.5))

$$\frac{\mathrm{d}x^i}{\mathrm{d}s} = g^{ai} \frac{\partial W}{\partial x^a} \tag{2.7.6}$$

are transversel to the 'wave fronts'. Our general considerations show that either we start with trajectories and, by integration of $p_k \, \mathrm{d}x^k - h \, \mathrm{d}s$ along them, we find W, or we start with a solution W and find the trajectories. We have checked that the two viewpoints are completely equivalent.

So far we have been associating equations of motion on $\mathbf{R}^4 \times (\mathbf{R}^4)^*$ with a given function H. Our previous example of a Hamiltonian function shows that, in the case of the flat Minkowski metric $\eta_{ab} \, \mathrm{d}x^a \otimes \mathrm{d}x^b$, the expression

$$\eta^{ab} p_a p_b = -p_0^2 + p_1^2 + p_2^2 + p_3^2 = -m_0^2 c^2 \tag{2.7.7}$$

is the square of the energy–momentum of a particle with rest mass m_0. This defines a codimension-1 surface, here denoted by Σ_{m_0}. Possible motions of a particle of mass m_0 are associated only with initial conditions on the 'mass shell', therefore the motion should take place not on $\mathbf{R}^4 \times (\mathbf{R}^4)^*$ but only on Σ_{m_0}, the 'mass shell'. The associated equations are

$$\frac{\mathrm{d}x^a}{\mathrm{d}s} = \eta^{ab} p_b, \tag{2.7.8}$$

$$\frac{\mathrm{d}p_a}{\mathrm{d}s} = 0, \tag{2.7.9}$$

subject to the restriction (2.7.7), and the associated Hamilton–Jacobi equation is given by (cf. Eq. (2.7.4))

$$\eta^{ab} \frac{\partial W}{\partial x^a} \frac{\partial W}{\partial x^b} = -m_0^2 c^2. \tag{2.7.10}$$

This shows that the particular choice of sub-manifold Σ, which we introduced in the time-dependent formalism has, in this case, a physical interpretation connected with the mass of the particle. Therefore the Hamiltonian function $H = \eta^{ab} p_a p_b$ on $\mathbf{R}^4 \times (\mathbf{R}^4)^*$ describes all relativistic particles with all possible values of the mass.

2.8 Introduction to geometrical optics

Geometrical optics is a mathematical theory of light rays. It is not concerned with the properties of light rays as waves (i.e. wave propagation, wavelength and frequency), but studies their properties as *pencils of rays.* It relies on three fundamental laws.

(i) The law of rectilinear propagation in a homogeneous medium.

(ii) The law of reflection, i.e. the angles of incidence and reflection on a smooth plane are equal.

(iii) The law of refraction: if θ and θ' are the angles of incidence and refraction of a light ray refracted from a uniform medium to a second uniform medium, and if n and n' are the refractive indices of the first and second medium, respectively, then

$$n \sin\theta = n' \sin\theta'. \tag{2.8.1}$$

These three laws follow from the Fermat principle, which states that the path of a light ray travelling from a point A' to A in a medium with refractive index $n(P)$ at the point P is such that the integral, called the optical distance from A' to A:

$$\int_{A'}^{A} n(P) \, ds,$$

attains its extremal value, where ds is the line element along the path. Therefore, the Fermat principle may be taken as a foundation of geometrical optics and is the analogue of the variational principle in particle dynamics, i.e. the Maupertuis principle according to which

$$\delta \int \sqrt{2m[h - U(P)]} \, ds = 0. \tag{2.8.2}$$

This principle is satisfied by the path of a particle of mass m having constant total energy h and passing through a field of potential $U(P)$. The quantity $\sqrt{2m[h - U(P)]}$ corresponds to the refractive index n. A geometrical view of the world lines, that is, parametrized objects in relativistic mechanics, is given in Mukunda and Sudarshan (1981) and Balachandran *et al.* (1982a, 1984).

An optical system works as follows. A ray starts at a point $P^i \equiv (\vec{r}^{\,i})$ and traces a direction determined by the unit vector $\vec{v}^{\,i}$. After interacting with the optical apparatus the ray reaches a point $P^f \equiv (\vec{r}^{\,f})$ in the direction determined by $\vec{v}^{\,f}$. The fundamental problem of geometrical optics requires, once the initial data $(\vec{r}^{\,i}, \vec{v}^{\,i})$ and a specification of the optical apparatus are given, to determine the final quantities $(\vec{r}^{\,f}, \vec{v}^{\,f})$. The most relevant result of the method developed by Hamilton is expressed by the statement: *any map associated with an optic apparatus is a canonical transformation, and hence it has a generating function called the characteristic function* (or eikonal).

Optical rays travel in a three-dimensional medium parametrized by Cartesian orthogonal coordinates (x, y, z), characterized by a medium function $n(\vec{r}) = n(x, y, z)$. A homogeneous medium corresponds to $n =$ constant. In vacuum $n = 1$, whereas any other medium has $n > 1$. Usually the z coordinate is taken to play the same role as time in mechanics. We assume that the optical path intersects each plane $z =$ constant at one point. It is therefore possible to introduce a map

$$\gamma : \mathbf{R} \rightarrow \mathbf{R}^2 : z \rightarrow (q_1(z), q_2(z)).$$

We denote by Π^i and Π^f the initial and final plane, which are also called the plane of the object and the plane of the image, respectively. When the optical path does not possess discontinuities in the 'velocities' it is possible to use the standard 'indirect method' to find solutions of the variational problems. To stress the analogy with motion of material particles we introduce an optical Lagrangian given by

$$\mathcal{L} \equiv n\sqrt{1 + \dot{q}_1^2 + \dot{q}_2^2}, \tag{2.8.3}$$

with

$$\dot{q}_1 = \frac{\mathrm{d}x}{\mathrm{d}z}, \quad \dot{q}_2 = \frac{\mathrm{d}y}{\mathrm{d}z}. \tag{2.8.4}$$

The optical distance is defined using the expression for the line element:

$$(\mathrm{d}s)^2 = (\mathrm{d}x)^2 + (\mathrm{d}y)^2 + (\mathrm{d}z)^2 \tag{2.8.5}$$

and setting

$$\int_{P^i}^{P^f} n \, \mathrm{d}s = \int_{z^i}^{z^f} n(q_1, q_2; z)\sqrt{1 + \dot{q}_1^2 + \dot{q}_2^2} \, \mathrm{d}z. \tag{2.8.6}$$

The integrand on the right-hand side of Eq. (2.8.6) is called the optical Lagrangian $\mathcal{L}$. As usual, the equations for the rays are given by the Euler–Lagrange equations (here $k = 1, 2$)

$$\frac{\mathrm{d}}{\mathrm{d}z}\frac{\partial \mathcal{L}}{\partial \dot{q}_k} - \frac{\partial \mathcal{L}}{\partial q_k} = 0. \tag{2.8.7}$$

Using the Legendre transform we introduce the 'optical phase space'. The Hamiltonian formulation follows by introducing the optical direction cosines

$$p_k \equiv \frac{\partial \mathcal{L}}{\partial \dot{q}_k}, \tag{2.8.8}$$

and setting, as usual,

$$H \equiv \dot{q}_1 p_1 + \dot{q}_2 p_2 - \mathcal{L}. \tag{2.8.9}$$

More specifically, the definitions (2.8.8) lead to

$$p_k = n(q_1, q_2; z)\frac{\dot{q}_k}{\sqrt{1 + \dot{q}_1^2 + \dot{q}_2^2}}. \tag{2.8.10}$$

Note that $\vec{p} \equiv (p_1, p_2)$ and $\vec{q} \equiv (q_1, q_2)$ satisfy the condition

$$p_1^2 + p_2^2 = n^2 \frac{\dot{q}_1^2 + \dot{q}_2^2}{(1 + \dot{q}_1^2 + \dot{q}_2^2)} = (n \sin\theta)^2 \leq n^2, \tag{2.8.11}$$

where

$$|\dot{\vec{q}}| = \left| \frac{\mathrm{d}\vec{q}}{\mathrm{d}z} \right| = \tan\theta, \tag{2.8.12}$$

with θ being the angle between the ray and the optical axis. In particular,

$$(\sin\theta)^2 = \frac{(\mathrm{d}q_1)^2 + (\mathrm{d}q_2)^2}{[(\mathrm{d}z)^2 + (\mathrm{d}q_1)^2 + (\mathrm{d}q_2)^2]} = \frac{\dot{q}_1^2 + \dot{q}_2^2}{(1 + \dot{q}_1^2 + \dot{q}_2^2)}. \tag{2.8.13}$$

Rays parallel to the optical axis have $p = 0$. The optical phase space is $\mathbf{R}^2 \times D_2$, i.e. one deals with 'momenta' on a disc

$$D_2 \equiv \left\{ \vec{p} \in \mathbf{R}^2 : p^2 \leq n^2 \right\}. \tag{2.8.14}$$

Using the linear form (cf. Eq. (2.3.11))

$$\theta = p_1 \, \mathrm{d}q_1 + p_2 \, \mathrm{d}q_2 - H \, \mathrm{d}z, \tag{2.8.15}$$

the variation of the integral along the light path may be written as (cf. Eq. (2.4.5))

$$\theta(A) - \theta(A') = \mathrm{d}S(A', A), \tag{2.8.16}$$

because the 'evolution' along light paths defines a canonical transformation, by virtue of their property of being solutions of a variational problem. Thus, the optical direction cosines and the Hamiltonians of the system at A' and A are given by

$$\frac{\partial S}{\partial q_k^i} = p_k^i, \tag{2.8.17}$$

$$\frac{\partial S}{\partial z^i} = -H^i, \frac{\partial S}{\partial z^f} = H^f, \tag{2.8.18}$$

$$\frac{\partial S}{\partial q_k^f} = -p_k^f, \tag{2.8.19}$$

and hence

$$\left(\frac{\partial S}{\partial q_1^i} \right)^2 + \left(\frac{\partial S}{\partial q_2^i} \right)^2 + \left(\frac{\partial S}{\partial z^i} \right)^2 = (n^i)^2, \tag{2.8.20}$$

$$\left(\frac{\partial S}{\partial q_1^f} \right)^2 + \left(\frac{\partial S}{\partial q_2^f} \right)^2 + \left(\frac{\partial S}{\partial z^f} \right)^2 = (n^f)^2. \tag{2.8.21}$$

As a corollary of these relations one obtains the Malus theorem (Born and Wolf 1959), which states that a pencil of light rays perpendicular to a given surface at a given moment remains perpendicular to the surface after an arbitrary number of reflections and refractions.

2.8.1 Variational principles

The Maupertuis principle mentioned earlier is a formulation of the variational problem appropriate for the case in which one has only to determine the path of a mechanical system, without reference to time. On assuming that the energy is conserved:

$$H(q,p) = E = \text{constant}, \tag{2.8.22}$$

and that the final time t varies, while the initial and final coordinates $(q^i_{\rm in}, q^i_{\rm fin})$ and initial time t_0 are fixed, the action functional

$$I \equiv \int \left(\sum_i p_i \, {\rm d}q^i - H \, {\rm d}t \right) \tag{2.8.23}$$

has to fulfil the variational condition

$$\delta I + E \delta t = 0. \tag{2.8.24}$$

By virtue of these assumptions, Eq. (2.8.24) reduces to finding the extremals of the *reduced action*

$$I_0 \equiv \int \sum_i p_i \, {\rm d}q^i, \tag{2.8.25}$$

with respect to all paths that satisfy the law (2.8.22) of conservation of energy and pass through the final point at any instant t. The action I is defined on all paths joining the initial point and the final point. Only when we evaluate it along a solution of the Hamilton equations does its value coincide with W obtained in Eq. (2.4.4).

In particular, for a particle of unit mass one has (${\rm d}s$ being the line element along the path)

$$E - U(q) = \frac{1}{2} \left(\frac{{\rm d}s}{{\rm d}t} \right)^2, \tag{2.8.26}$$

and hence

$$p_i \, {\rm d}q^i = \left(\frac{{\rm d}s}{{\rm d}t} \right)^2 dt = \sqrt{2[E - U(q)]} \, {\rm d}s. \tag{2.8.27}$$

One therefore deals with the variational problem

$$\delta \int_{A'}^{A} \sqrt{2[E - U(q)]} \, {\rm d}s = 0, \tag{2.8.28}$$

subject to the conservation of energy

$$\frac{p^2}{2} + U(q) = E. \tag{2.8.29}$$

On making the identification

$$p(E,x,y,z) = \frac{c}{\lambda(\omega,x,y,z)}, \tag{2.8.30}$$

the two variational principles, for particle dynamics at fixed energy, and light rays at fixed frequency, are therefore found to coincide.

2.9 Problems

2.P1. The following equations of motion are given on $\mathbf{R}^3$:

$$\frac{\mathrm{d}}{\mathrm{d}t}x = \frac{(z-y)}{1+(x+y+z)^2}, \tag{2.9.1}$$

$$\frac{\mathrm{d}}{\mathrm{d}t}y = \frac{(x-z)}{1+(x+y+z)^2}, \tag{2.9.2}$$

$$\frac{\mathrm{d}}{\mathrm{d}t}z = \frac{(y-x)}{1+(x+y+z)^2}. \tag{2.9.3}$$

Setting $x_1 = x, x_2 = y, x_3 = z$, find a definition of Poisson bracket for $\{x_i, x_j\}$, and among any two functions F and G, such that the above equations can be written as Hamilton equations

$$\frac{\mathrm{d}}{\mathrm{d}t}x_i = \{x_i, H\}. \tag{2.9.4}$$

Find the corresponding Hamiltonian function H.

2.P2. Recall the Poisson brackets

$$\left\{q^i, q^j\right\} = 0, \tag{2.9.5}$$

$$\{p_i, p_j\} = 0, \tag{2.9.6}$$

$$\left\{q^i, p_j\right\} = \delta^i_{\ j}. \tag{2.9.7}$$

For a charged particle moving under the influence of an electromagnetic field, derive the Poisson brackets $\{v_i, q^j\}$ and $\{v_i, v_j\}$, bearing in mind that $p_i = mv_i + \frac{e}{c}A_i$, where e is the charge of the particle.

Long hint: the condition (2.9.6) implies that

$$\{v_i, v_j\} = \frac{e}{mc}\left[\{A_j, v_i\} + \{v_j, A_i\}\right]. \tag{2.9.8}$$

Now for any pair of functions F and G of the (q, v) coordinates one has

$$\begin{aligned}\{F,G\} = {} & \frac{\partial F}{\partial q^k}\left\{q^k, q^l\right\}\frac{\partial G}{\partial q^l} + \frac{\partial F}{\partial v^k}\left\{v^k, v^l\right\}\frac{\partial G}{\partial v^l} \\ & + \frac{\partial F}{\partial q^k}\left\{q^k, v^l\right\}\frac{\partial G}{\partial v^l} + \frac{\partial F}{\partial v^k}\left\{v^k, q^l\right\}\frac{\partial G}{\partial q^l},\end{aligned} \tag{2.9.9}$$

and hence

$$\{p_i, A_j\} = -\frac{\partial A_j}{\partial q^i}, \tag{2.9.10}$$

because

$$\left\{v_i, q^j\right\} = \frac{1}{m}\left\{p_i, q^j\right\} = -\frac{1}{m}\delta_i^{\ j}. \tag{2.9.11}$$

Eventually, these equations lead to

$$\{v_i, v_j\} = \frac{e}{mc}\left[-\frac{1}{m}\frac{\partial A_i}{\partial q^j} + \frac{1}{m}\frac{\partial A_j}{\partial q^i}\right] = \frac{e}{m^2 c}\varepsilon_{ij}^{\ \ k}B_k. \tag{2.9.12}$$

2.P3. Can one find a Newtonian description of the equations of motion of a charged particle, when the radiation reaction is taken into account? Write an essay on this topic, after reading sections 17.1, 17.2 and 17.6 of Jackson (1975).

2.P4. Try to obtain the various generating functions of canonical transformations by replacing in a suitable way the $P\,dQ$-like 1-forms. For example, which generating function is obtained if $P_i\,dQ^i$ is replaced by $-Q^i\,dP_i$?

2.P5. Study under which conditions closed orbits occur in the central potential (2.6.16), after reducing the problem to a one-dimensional case. Hint: Equation (2.6.10) suggests defining the *effective potential*

$$\widetilde{V}(r) \equiv \frac{p_\phi^2}{2mr^2} + V(r). \tag{2.9.13}$$

2.P6. Prove that the 'new Hamiltonian' K is related to the original Hamiltonian H by Eq. (2.3.31). Hint: study the variational problems for which the functionals

$$\int_{t_1}^{t_2} \left[P_i \dot{Q}^i - K(Q,P;t)\right] dt$$

and

$$\int_{t_1}^{t_2} \left[p_i \dot{q}^i - H(q,p;t)\right] dt$$

should be stationary, and hence derive Eq. (2.3.31). You may instead follow the derivation in section 5.3.3 of José and Saletan (1998).

2.P7. Solve by exponentiation the system

$$\frac{d}{dt}x = V_x, \quad \frac{d}{dt}y = V_y, \quad \frac{d}{dt}z = V_z, \tag{2.9.14}$$

$$\frac{d}{dt}V_x = 2V_y + \frac{\partial U}{\partial x}, \quad \frac{d}{dt}V_y = -2V_x + \frac{\partial U}{\partial y}, \quad \frac{d}{dt}V_z = \frac{\partial U}{\partial z}, \tag{2.9.15}$$

where U is a smooth function of x, y and z. Apply the general formula to the particular case when

$$U(x,y,z) = \frac{1}{2}(x^2+y^2) + \frac{(1-\mu)}{r_1} + \frac{\mu}{r_2}, \tag{2.9.16}$$

where μ is a real parameter, while r_1 and r_2 are defined by

$$(r_1)^2 \equiv (\mu + x)^2 + y^2 + z^2, \tag{2.9.17}$$

$$(r_2)^2 \equiv (1 - \mu - x)^2 + y^2 + z^2. \tag{2.9.18}$$

These equations describe the restricted three-body problem (Pars 1965, Szebehely 1967). Can you find a constant of motion?

Appendix 2.A

Vector fields

When we first learn the intrinsic definition of tangent vectors, we think of them as an equivalence class of curves at a point p of some manifold M. The equivalence relation is that two curves are tangent at the point p. The equivalence class of a particular curve σ is denoted by $[\sigma]$. A *vector field* X on a C^∞ manifold M is a *smooth* assignment of a tangent vector $X_p \in T_p(M)$ for each point $p \in M$. By smooth we mean that, for all $f \in C^\infty(M)$, the function

$$p \in M \rightarrow (Xf)(p) \equiv X_p(f)$$

is C^∞. Thus, a vector field may be viewed as a map

$$X : C^\infty(M) \rightarrow C^\infty(M)$$

in which $f \rightarrow X(f)$ is defined as above. The function $X(f)$ is called the Lie derivative (see appendix 2.C) of the function f along the vector field X, and is usually denoted by $L_X f$.

A vector field can also be defined as a smooth *cross-section* of the tangent bundle TM. In other words, TM is a bundle over M with projection map π. A vector field is instead a map from M to TM, such that its composition with the projection yields the identity map on M. Such a map from M to TM is C^∞.

The map $f \rightarrow X(f)$ has the following properties (Marmo *et al.* 1985, Isham 1989, José and Saletan 1998):

$$X(f+h) = X(f) + X(h) \quad \forall f, h \in C^\infty(M), \tag{2.A.1}$$

$$X(rf) = rX(f) \quad \forall f \in C^\infty(M) \text{ and } r \in \mathbf{R}, \tag{2.A.2}$$

$$X(fh) = fX(h) + hX(f) \quad \forall f, h \in C^\infty(M). \tag{2.A.3}$$

These formulae should be compared with those defining a *derivation* at a point $p \in M$. These are maps $v : C^\infty(M) \rightarrow \mathbf{R}$ such that

$$v(f+h) = v(f) + v(h) \quad \forall f, h \in C^\infty(M), \tag{2.A.4}$$

$$v(rf) = rv(f) \quad \forall f \in C^\infty(M) \text{ and } r \in \mathbf{R}, \tag{2.A.5}$$

$$v(fh) = f(p)v(h) + h(p)v(f) \quad \forall f, h \in C^\infty(M). \tag{2.A.6}$$

The set of all derivations at $p \in M$ is denoted by $D_p M$.

Equations (2.A.1) and (2.A.2) show that X is a linear map from the vector space $C^\infty(M)$ into itself, and (2.A.3) shows that X is a derivation of the set $C^\infty(M)$. It is therefore possible to *define* a vector field as a derivation of the set of functions $C^\infty(M)$ satisfying (2.A.1)–(2.A.3). Such a map assigns a derivation, in the sense of (2.A.4)–(2.A.6), to each point $p \in M$, denoted X_p and defined by

$$X_p(f) \equiv [X(f)](p), \tag{2.A.7}$$

for each $f \in C^\infty(M)$. By construction, X_p satisfies Eqs. (2.A.4)–(2.A.6). But then the map $p \rightarrow X_p$ assigns a field of tangent vectors, and this assignment is smooth. Thus, a completely equivalent way of defining a vector field in the first place is to regard it as a derivation on the set $C^\infty(M)$. This alternative approach, where one goes from a vector field to a derivation at a point $p \in M$, is useful in many applications. If (U, ϕ) is a local coordinate chart on the manifold M, the derivations X_p associated with a vector field X defined on U make it possible to express $(Xf)(p)$ as

$$(Xf)(p) = X_p(f) = \sum_{\mu=1}^{m} X_p(x^\mu) \left(\frac{\partial}{\partial x^\mu} \right)_p f, \tag{2.A.8}$$

which implies that

$$X = \sum_{\mu=1}^{m} X(x^\mu) \frac{\partial}{\partial x^\mu}. \tag{2.A.9}$$

This formula shows the precise sense in which a vector field may be viewed as a first-order differential operator on the functions on a manifold. The functions $X(x^\mu)$ are defined on the open set U that defines the coordinate chart with the coordinate functions x^μ, and are the components of the vector field X with respect to this coordinate system.

A crucial question is whether or not one can 'multiply' two vector fields X and Y to obtain a third field. Indeed, we know that X and Y can be viewed as linear maps of $C^\infty(M)$ into itself, and hence we can define the composite map $X \cdot Y : C^\infty(M) \rightarrow C^\infty(M)$ as

$$X \cdot Y(f) \equiv X(Yf). \tag{2.A.10}$$

By construction, this is a linear map, in that it satisfies (2.A.1) and (2.A.2). However, (2.A.10) is not a vector field since it fails to satisfy (2.A.3). In contrast, one finds

$$\begin{aligned} X \cdot Y(fh) &= X(Y(fh)) = X(hY(f) + fY(h)) \\ &= X(h)Y(f) + hX(Y(f)) \\ &\quad + X(f)Y(h) + fX(Y(h)), \end{aligned} \tag{2.A.11}$$

which does not equal $hX(Y(f)) + fX(Y(h))$. However, one also has

$$Y \cdot X(fh) = Y(h)X(f) + hY(X(f)) + Y(f)X(h) + fY(X(h)). \tag{2.A.12}$$

One can now subtract (2.A.11) and (2.A.12) to find

$$\Big(X \cdot Y - Y \cdot X\Big)(fh) = h(X \cdot Y - Y \cdot X)(f) + f\Big(X \cdot Y - Y \cdot X\Big)(h). \tag{2.A.13}$$

We have thus found that $X \cdot Y - Y \cdot X$ is a vector field, even though the individual pieces $X \cdot Y$ and $Y \cdot X$ are not. The new vector field just obtained is called the commutator of X and Y, and is denoted by $[X, Y]$. If X, Y and Z are any three vector fields on M, their commutators satisfy the Jacobi identity:

$$[X,[Y,Z]] + [Y,[Z,X]] + [Z,[X,Y]] = 0. \tag{2.A.14}$$

The consideration of such structures becomes natural within the framework of Lie algebras, as shown in appendix 2.B.

In our book we are interested in the interpretation of vector fields suggested by systems of first-order differential equations. More precisely, we know that, given any differential equation

$$\dot{x} = F(x,y), \tag{2.A.15}$$

$$\dot{y} = G(x,y), \tag{2.A.16}$$

one defines the derivative of functions along solutions by setting

$$\frac{\mathrm{d}}{\mathrm{d}t} f = \frac{\partial f}{\partial x}\frac{\mathrm{d}x}{\mathrm{d}t} + \frac{\partial f}{\partial y}\frac{\mathrm{d}y}{\mathrm{d}t} = \left(F\frac{\partial}{\partial x} + G\frac{\partial}{\partial y}\right) f. \tag{2.A.17}$$

The homogeneous first-order differential operator

$$X \equiv F\frac{\partial}{\partial x} + G\frac{\partial}{\partial y} \tag{2.A.18}$$

is called a vector field. If our differential equation admits solutions for any initial condition from $-\infty$ to $+\infty$ in time, it defines the action of a one-parameter group of transformations. The vector field associated with the differential equation is called the *infinitesimal* generator of the one-parameter group ϕ_t and is said to be a *complete* vector field.

Appendix 2.B

Lie algebras and basic group theory

The notion of Lie algebra is introduced to deal with several ordinary differential equations at the same time. Their integration gives rise to the notion of the action of a many-parameter group of transformations, instead of a one-parameter group of transformations.

A vector space E over the real or complex numbers is said to be a Lie algebra if a bilinear map $E \times E \rightarrow E$ exists, often denoted by $[\ ,\ ]$, such that

$$[u,v] = -[v,u], \tag{2.B.1}$$

$$[u,[v,w]] = [[u,v],w] + [v,[u,w]]. \tag{2.B.2}$$

Equation (2.B.2) describes the fundamental Jacobi identity, and is written in a way that emphasizes the link with the Leibniz rule for derivations.

Examples of Lie algebras are provided by $n \times n$ matrices with real or complex coefficients, for which the bilinear map is defined by

$$[A,B] \equiv A \cdot B - B \cdot A, \tag{2.B.3}$$

where the *dot* denotes the row by column product. Yet another familiar example is provided by the vector product on three-dimensional vector spaces:

$$[\vec{u},\vec{v}] \equiv \vec{u} \times \vec{v}. \tag{2.B.4}$$

The reader may check that (2.B.1) and (2.B.2) are then satisfied. A realization of E is a map ρ from E to the linear transformations on some vector space $\mathcal{F}$, with the property $\rho([X,Y]) = \rho(X)\rho(Y) - \rho(Y)\rho(X)$.

Algebras

In general, an associative *algebra* is meant to be a vector space endowed with an associative and bilinear multiplication map, for which therefore

$$(xy)z = x(yz), \tag{2.B.5}$$

$$(\alpha x + \beta y)z = \alpha(xz) + \beta(yz), \tag{2.B.6}$$

$$x(\beta y + \gamma z) = \beta(xy) + \gamma(xz), \tag{2.B.7}$$

for all vectors x, y, z and all scalars α, β, γ. The associativity can be replaced by other requirements, and indeed Lie algebras are non-associative. An algebra $\mathcal{A}$ is said to have an identity $\mathbb{I}$ if

$$\mathbb{I}a = a\mathbb{I} = a \ \forall a \quad \in \mathcal{A}. \tag{2.B.8}$$

Given $a \in \mathcal{A}$, its inverse a^{-1}, if it exists, is an element of $\mathcal{A}$ such that

$$a^{-1}a = aa^{-1} = \mathbb{I}. \tag{2.B.9}$$

Commutative algebras are those particular algebras for which $ab = ba \ \forall a, b \in \mathcal{A}$. The *centre* is a subset given by the elements $a \in \mathcal{A}$ such that

$$ax = xa \quad \forall x \in \mathcal{A}. \tag{2.B.10}$$

Groups and sub-groups

A *group* is a set G such that:

(i) a binary operation is defined, usually called a product and denoted by $\cdot$, for which

$$g_1 \cdot g_2 = g_3 \in G, \quad \forall g_1, g_2 \in G; \tag{2.B.11}$$

(ii) the product is associative, i.e.

$$g_1 \cdot (g_2 \cdot g_3) = (g_1 \cdot g_2) \cdot g_3; \tag{2.B.12}$$

(iii) the identity exists, i.e. an element $\mathbb{I}$ such that

$$\mathbb{I}g = g\mathbb{I} = g, \quad \forall g \in G; \tag{2.B.13}$$

(iv) any element g of G has an inverse $\gamma \equiv g^{-1}$, which satisfies the condition

$$g\gamma = \gamma g = \mathbb{I}. \tag{2.B.14}$$

Note that the product need not be commutative, i.e. $g_1 \cdot g_2 \neq g_2 \cdot g_1$ in general.

In part II we will need the concept of *topological group.* Now G is endowed with a topology compatible with the group structure, i.e. all group operations are continuous in the given topology. When the topology is actually that resulting from a differentiable structure on G and the group operations are smooth (differentiable), then the group becomes a *Lie group.* In many physical situations the group is actually defined as a group of transformations, therefore it arises from infinitesimal generators, i.e. by integrating simultaneously a set of vector fields. Since the integration may give rise to various problems, it is common to introduce Lie groups on their own and then show that the infinitesimal generators of a Lie group define a Lie algebra.

Sub-groups of a given group G are sub-sets of G closed under the binary operation defining the group. When the group G is a topological group, the sub-group should be a topological group with the induced topology. When the sub-set is a differentiable sub-manifold compatible with the composition law, the sub-group is a Lie sub-group.

The general linear group $GL(n, \mathbf{R})$

An important example of a Lie group, considered directly as a group of linear transformations, is the set of linear transformations represented as invertible matrices from $\mathbf{R}^n$ into itself. This set is clearly a group under the usual multiplication of matrices, the unit matrix acts like the identity. The invertibility requirement means that the set we are considering is the inverse image of $(\mathbf{R} - \{0\})$ under $\det : \mathrm{Mat}(n, \mathbf{R}) \to \mathbf{R}$. Since the determinant is a polynomial map, it is clear that the inverse image of $(\mathbf{R} - \{0\})$ is an open sub-manifold of $\mathbf{R}^{n^2}$. The binary operation defined by the matrix product is clearly differentiable, and hence the group $GL(n, \mathbf{R})$ is a Lie group. Since the set is the inverse image of a disconnected set, $(\mathbf{R} - \{0\})$, $GL(n, \mathbf{R})$ is not connected.

Examples of groups of interest for physical applications are the Euclidean and rotation groups, the Galilei group and the Poincaré group (see below). In physics they arise again as transformation groups acting on space–time (or their higher-dimensional generalizations).

Euclidean and rotation group

A first important example of group is provided here by the Euclidean group on $\mathbf{R}^3$, i.e. the group of affine transformations that preserve the length of vectors (the length of vectors being evaluated with a Euclidean metric g). It contains translations and linear homogeneous transformations that satisfy the condition

$$g(Tx, Ty) = g(x, y). \tag{2.B.15}$$

Using a basis for $\mathbf{R}^3$, e.g. orthonormal vectors represented by row or column vectors of the form $(1,0,0), (0,1,0), (0,0,1)$, a generic matrix

$$R \equiv \begin{pmatrix} \alpha_1 & \alpha_2 & \alpha_3 \\ \beta_1 & \beta_2 & \beta_3 \\ \gamma_1 & \gamma_2 & \gamma_3 \end{pmatrix} \tag{2.B.16}$$

represents a rotation if

$$g(\vec{\alpha}, \vec{\alpha}) = g(\vec{\beta}, \vec{\beta}) = g(\vec{\gamma}, \vec{\gamma}) = 1, \tag{2.B.17}$$

$$g(\vec{\alpha}, \vec{\beta}) = g(\vec{\alpha}, \vec{\gamma}) = g(\vec{\beta}, \vec{\gamma}) = 0. \tag{2.B.18}$$

It is easy to show that these six conditions imply that R is a rotation matrix if and only if its transpose equals its inverse. Now we see that

$$(R_1 R_2)^{\,\mathrm{t}} (R_1 R_2) = R_1 R_2 \, {}^{\mathrm{t}}R_2 \, {}^{\mathrm{t}}R_1 = \mathbb{1}.$$

Thus, the product of two rotations is again a rotation, and the identity matrix is of course a rotation. By virtue of all these properties, the set of rotations is a sub-group of the group of invertible matrices. The rotation group is denoted by $O(3)$ (the rotation matrices having a $+1$ determinant are the group $SO(3)$). It is three-dimensional because one has to subtract from the dimension of $\mathbf{R}^9$ the number of relations in (2.B.17) and (2.B.18). Note that the defining relations of the sub-group $O(3)$ of $GL(3, \mathbf{R})$ are differentiable and satisfy the condition for the application of the implicit function theorem. Thus, the set $O(3)$ is a closed differentiable sub-manifold of $GL(3, \mathbf{R})$, compatible with the composition law, and hence is a Lie (sub)group.

For the affine part of the Euclidean group, one has to introduce it as a sub-group of $GL(4, \mathbf{R})$ of the form

$$\begin{pmatrix} 1 & 0 & 0 & a \\ 0 & 1 & 0 & b \\ 0 & 0 & 1 & c \\ 0 & 0 & 0 & 1 \end{pmatrix},$$

acting on elements of the form $\begin{pmatrix} x \\ y \\ z \\ 1 \end{pmatrix}$.

Galilei group

The Galilei group expresses the geometrical invariance properties of the equations of motion of a non-relativistic classical dynamical system when the system is isolated from external influences. The general Galilei transformation $G(R, \vec{v}, \vec{\xi}, \tau)$ takes a point of space–time with coordinates x_1, x_2, x_3, t to another point with coordinates x_1', x_2', x_3', t' given by

$$\vec{x}' = R\vec{x} + \vec{v}t + \vec{\xi}, \tag{2.B.19}$$

$$t' = t + \tau, \tag{2.B.20}$$

where $R \in SO(3)$, $\vec{\xi}$ and $\vec{v}$ are fixed vectors in $\mathbf{R}^3$, and τ is a real constant. The resulting group multiplication law is

$$G(R_2, \vec{v}_2, \vec{\xi}_2, \tau_2) G(R_1, \vec{v}_1, \vec{\xi}_1, \tau_1)$$
$$= G(R_2 R_1, R_2 \vec{v}_1 + \vec{v}_2, R_2 \vec{\xi}_1 + \vec{\xi}_2 + \vec{v}_2 \tau_1, \tau_2 + \tau_1). \tag{2.B.21}$$

The one-parameter sub-groups are rotations about a fixed axis, transformations to frames moving in a fixed direction, displacements of the origin in a fixed direction, and time displacements. These correspond to fixed axes of rotation for R (three sub-groups), fixed directions of $\vec{v}$ (three sub-groups), fixed directions of $\vec{\xi}$ (three sub-groups) and time displacements (one sub-group). The Galilei group is hence 10-dimensional. The commutation relations for its Lie algebra, when realized on phase-space (e.g. the cotangent bundle of $\mathbf{R}^3$), are

$$\{M_\alpha, M_\beta\} = \varepsilon_{\alpha\beta\gamma} M_\gamma, \tag{2.B.22}$$

$$\{M_\alpha, P_\beta\} = \varepsilon_{\alpha\beta\gamma} P_\gamma, \tag{2.B.23}$$

$$\{M_\alpha, G_\beta\} = \varepsilon_{\alpha\beta\gamma} G_\gamma, \tag{2.B.24}$$

$$\{H, G_\alpha\} = -P_\alpha, \tag{2.B.25}$$

$$\{M_\alpha, H\} = \{P_\alpha, G_\beta\} = \{G_\alpha, G_\beta\} = \{P_\alpha, P_\beta\} = \{P_\alpha, H\} = 0, \tag{2.B.26}$$

where M_α is the infinitesimal generator of rotations about the α-axis, G_α of Galilei transformations to frames moving in the fixed α-direction, P_α of displacements of the origin in the α-direction and H of time displacements.

Lorentz and Poincaré groups

The Lorentz group is defined with the Euclidean metric g being replaced by the Minkowski metric η in (2.B.15). One still has $\eta(Tx,Ty)=\eta(x,y)$, and by virtue of the signature of the metric, if one writes

$$T \equiv \begin{pmatrix} \alpha_0 & \alpha_1 & \alpha_2 & \alpha_3 \\ \beta_0 & \beta_1 & \beta_2 & \beta_3 \\ \gamma_0 & \gamma_1 & \gamma_2 & \gamma_3 \\ \delta_0 & \delta_1 & \delta_2 & \delta_3 \end{pmatrix} \tag{2.B.27}$$

one finds

$$\eta(\alpha,\alpha)=-1, \quad \eta(\beta,\beta)=\eta(\gamma,\gamma)=\eta(\delta,\delta)=1, \tag{2.B.28}$$

$$\eta(\rho,\sigma)=0 \quad \forall \rho\neq\sigma, \tag{2.B.29}$$

for all $\rho,\sigma=\alpha,\beta,\gamma,\delta$. These 10 independent conditions determine a six-parameter group in the 16-dimensional space of 4×4 matrices. On adding to this the linear homogeneous transformations known as space–time translations one obtains the Poincaré group. More precisely, this is the abstract group isomorphic to the geometric group of transformations of a 'world point' $(\vec{r},t)$ as follows.

(i) Space displacements: $\vec{r}\,'=\vec{r}+\vec{a},\ t'=t$.

(ii) Time displacements: $\vec{r}\,'=\vec{r},\ t'=t+b$.

(iii) Moving frames:

$$\vec{r}\,'=\frac{(\vec{v}\wedge\vec{r})\wedge\vec{v}}{v^2}+\frac{\vec{v}}{v^2}\frac{\vec{v}\cdot\vec{r}-v^2t}{\sqrt{1-v^2}}, \quad |\vec{v}|<1, \tag{2.B.30}$$

$$t'=\frac{t-\vec{v}\cdot\vec{r}}{\sqrt{1-v^2}}. \tag{2.B.31}$$

(iv) Space rotations: $\vec{r}\,'=R\,\vec{r},\ t'=t$.

If we represent the general transformation by $T(\vec{a},b,\vec{v},R)$, with the convention

$$T(\vec{a},b,\vec{v},R)=T(\vec{a},0,\vec{0},\mathbb{1})T(\vec{0},b,\vec{0},\mathbb{1})T(\vec{0},0,\vec{v},\mathbb{1})T(\vec{0},0,\vec{0},R), \tag{2.B.32}$$

the effect of a general transformation on $(\vec{r},t)$ is

$$\vec{r}\,'=\vec{a}+\frac{(\vec{v}\wedge R\vec{r})\wedge\vec{v}}{v^2}+\frac{\vec{v}}{v^2}\frac{\vec{v}\cdot R\vec{r}-v^2t}{\sqrt{1-v^2}}, \tag{2.B.33}$$

$$t'=b+\frac{t-\vec{v}\cdot R\vec{r}}{\sqrt{1-v^2}}. \tag{2.B.34}$$

Actually the group that we have so far defined is the connected Lie group, since we have not considered the inversion operations

$$R_p:\ \vec{r}\,'=-\vec{r},\ t'=t,$$

$$R_T:\ \vec{r}\,'=\vec{r},\ t'=-t,$$

$$R_TR_p=R_S:\ \vec{r}\,'=-\vec{r},\ t'=-t.$$

Thus, the group we have defined may be more properly called the connected proper inhomogeneous orthochronous Lorentz group. We refer to this briefly as the Poincaré group.

One can exhibit an inverse for every element $T(\vec{a},b,\vec{v},R)$ and show that a multiplication exists which is associative so that one has a group; moreover the composition functions $z_r\equiv f_r(x,y)$, with $x=(\vec{a},b,\vec{v},R)$, etc. are differentiable functions of x and y, so that the system thus defined is a Lie group with 10 parameters, which is clearly non-commutative. The 10 generators in the Lie algebra are defined with the help of a Poisson bracket on $T^*\mathbf{R}^4$ as:

$$\vec{P}=\{P_j\},\ j=1,2,3;\ \text{space displacement},$$

$$H=P_0,\ \text{time displacement},$$

$$\vec{K}=\{K_j\},\ j=1,2,3;\ \text{moving frames},$$

$$\vec{J}=\{J_j\},\ j=1,2,3;\ \text{space rotations}.$$

The composition law for the Poincaré group is then equivalent to the following bracket relations for the generators:

$$\{P_j, P_k\} = 0,\ \{P_j, H\} = 0,\ \{K_j, P_k\} = \delta_{jk} H,\ \{J_j, P_k\} = \varepsilon_{jkl} P_l, \tag{2.B.35}$$

$$\{K_j, H\} = P_j,\ \{J_j, H\} = 0, \tag{2.B.36}$$

$$\{K_j, K_k\} = -\varepsilon_{jkl} J_l,\ \{J_j, K_k\} = \varepsilon_{jkl} K_l, \tag{2.B.37}$$

$$\{J_j, J_k\} = \varepsilon_{jkl} J_l. \tag{2.B.38}$$

From our experience with the rotation, Euclidean and Galilei groups, we know that, as applied to dynamical systems, generally the infinitesimal generators (i.e. the elements of the Lie algebra) have an immediate physical interpretation. We then expect a similar situation to also arise for the Poincaré group. By the close correspondence with the Galilei group, we identify $\vec{P}$ with linear momentum, H with the energy and $\vec{J}$ with the angular momentum. We may also consider $\vec{K}$ to be a relativistic 'moment'. Bearing this in mind we may now look at the Poisson-bracket relations; the generators now play a dual role. On the one hand, they represent generators of infinitesimal transformations and on the other hand, they represent physical quantities. Thus, the Poisson-bracket relations

$$\{J_j, P_k\} = \varepsilon_{jkl} P_l$$

may be interpreted either as stating that the linear momentum transforms like a vector under a rotation (i.e. the increment in $\vec{P}$ in an infinitesimal rotation is at right angles to itself and the axis of rotation) or as stating that the angular momentum increases by a quantity proportional to the normal component of momentum under a displacement. Similarly, the first of the relations (2.B.36) may be taken to mean that the energy changes on transforming to a moving frame by a quantity proportional to the component of linear momentum along the direction of relative velocity. Equally well it may be taken to mean that the relativistic moment $\vec{K}$ is not a constant but changes linearly with respect to time by a quantity proportional to the linear momentum.

Unitary group

The unitary matrices A of degree n form the elements of the n^2-parameter unitary group $U(n)$, which leaves the Hermitian form $\sum_{i=1}^{n} z_i z_i^*$ invariant.

Since the unitarity of the matrices A requires that $AA^\dagger = A^\dagger A = \mathbb{1}$, the range of the matrix elements a_{ij} is restricted by the requirement that

$$\sum_l a_{il} a_{lj}^* = \delta_{ij}, \tag{2.B.41}$$

and hence $|a_{ij}|^2 \leq 1$. Thus, in this case the domain of the n^2 parameters is bounded, and $U(n)$ is found to be a compact group.

Appendix 2.C
Some basic geometrical operations

In this appendix we also use the following geometrical concepts.

(i) *Exterior product of forms.* Given two 1-forms $A = a_\mu \, dx^\mu$ and $B = b_\nu \, dx^\nu$, their exterior product, denoted by the wedge symbol, is defined as

$$A \wedge B \equiv a_\mu b_\nu \, dx^\mu \wedge dx^\nu = \frac{1}{2}(a_\mu b_\nu - a_\nu b_\mu) \, dx^\mu \wedge dx^\nu, \tag{2.C.1}$$

where

$$dx^\mu \wedge dx^\nu = \det \begin{pmatrix} dx^\mu & 0 \\ 0 & dx^\nu \end{pmatrix}. \tag{2.C.2}$$

The exterior product generalizes the vector product to any number of dimensions and moreover has the advantage of being associative, unlike the vector product. One has

$$(A \wedge B) \wedge C = A \wedge (B \wedge C) = a_\mu b_\nu c_\rho \, dx^\mu \wedge dx^\nu \wedge dx^\rho. \tag{2.C.3}$$

Furthermore, for forms of any degree one finds

$$\alpha \wedge \beta = (-1)^{|\alpha||\beta|} \beta \wedge \alpha, \tag{2.C.4}$$

where $|\alpha|$ and $|\beta|$ are the degrees of the forms α and β, respectively.

(ii) *Exterior derivative of forms.* For any 1-form $g\,\mathrm{d}f$, the differential operator d, called the exterior derivative, is extended by setting

$$\mathrm{d}(g\,\mathrm{d}f) = \mathrm{d}g \wedge \mathrm{d}f, \tag{2.C.5}$$

and, in general,

$$\mathrm{d}(\alpha \wedge \beta) = \mathrm{d}\alpha \wedge \beta + (-1)^{|\alpha|} \alpha \wedge \mathrm{d}\beta. \tag{2.C.6}$$

(iii) *Contraction (or inner multiplication) of forms.* Given a 1-form $\mathrm{d}f$ and a vector field X, they can be contracted to yield a function h, which is written as

$$h \equiv i_X\,\mathrm{d}f \equiv \mathrm{d}f(X) = X^k \frac{\partial f}{\partial x^k}, \tag{2.C.7}$$

where the last equality holds in a local chart. If α and β are 1-forms, their wedge product previously defined is the 2-form $\alpha \wedge \beta$ such that

$$i_X(\alpha \wedge \beta) = (i_X\alpha)\beta - (i_X\beta)\alpha. \tag{2.C.8}$$

The operation of contraction can be further extended to forms of higher degree. For example, if α is an n-form and β a p-form, then

$$i_X(\alpha \wedge \beta) = (i_X\alpha) \wedge \beta + (-1)^n \alpha \wedge (i_X\beta). \tag{2.C.9}$$

Note that $i_X(\alpha \wedge \beta)$ is itself a form, of degree $n+p-1$, so that contraction (or inner multiplication) is a degree-lowering operation on forms.

(iv) *Lie derivatives.* The action of a first-order differential operator on functions is called Lie derivative. For example, if X is the vector field given in Eq. (2.A.18), one defines

$$L_X f \equiv F\frac{\partial f}{\partial x} + G\frac{\partial f}{\partial y}. \tag{2.C.10}$$

The notation L_X is used when the definition of a Lie derivative is extended to tensor fields. The action of L_X is indeed extended to forms and to tensors by setting

$$L_X\,\mathrm{d}f = \mathrm{d}L_X f, \tag{2.C.11}$$

and requiring that the Leibniz rule with respect to tensor product should be satisfied, i.e.

$$L_X\left(a\,\mathrm{d}f \otimes \mathrm{d}g\right) = \left(L_X a\right)\mathrm{d}f \otimes \mathrm{d}g + a\left(\mathrm{d}L_X f\right) \otimes \mathrm{d}g + a\,\mathrm{d}f \otimes \left(\mathrm{d}L_X g\right). \tag{2.C.12}$$

One then finds

$$L_X\left(\mathrm{d}f \wedge \mathrm{d}g\right) = \left(\mathrm{d}L_X f\right) \wedge \mathrm{d}g + \mathrm{d}f \wedge \left(\mathrm{d}L_X g\right). \tag{2.C.13}$$

It is possible to extend the Lie derivative to vector fields by setting (see appendix 2.A)

$$(L_X Y)(f) \equiv L_X(L_Y f) - L_Y(L_X f) \equiv L_{[X,Y]} f. \tag{2.C.14}$$

Using the Leibniz rule we can extend the Lie derivative to any tensor field (being formed with tensor products of forms and vector fields). A fundamental identity is the Cartan identity, which is known to provide a rule for the evaluation of the Lie derivative along a vector field X by composing the action of i_X with the exterior differentiation as follows:

$$L_X = i_X\,\mathrm{d} + \mathrm{d}\,i_X. \tag{2.C.15}$$

The two derivations coincide on functions and both commute with the exterior derivative. Both are extended to forms using the Leibniz rule, therefore they coincide on forms. It should be noticed that L_X is defined on all tensor fields, while $i_X\,\mathrm{d} + \mathrm{d}\,i_X$ is defined *on forms only.*

As an example of the application of these structures, consider rotations in the plane. They are expressed by

$$\begin{pmatrix} x(\theta) \\ y(\theta) \end{pmatrix} = \begin{pmatrix} \cos\theta & \sin\theta \\ -\sin\theta & \cos\theta \end{pmatrix} \begin{pmatrix} x \\ y \end{pmatrix}. \tag{2.C.16}$$

The orbit is a circle, because

$$x^2(\theta_1) + y^2(\theta_1) = x^2(\theta_2) + y^2(\theta_2), \quad \forall\, \theta_1, \theta_2. \tag{2.C.17}$$

If Eq. (2.C.16) is viewed as expressing a flow, we can ask which is the differential equation having $x(\theta)$ and $y(\theta)$ as the solution, i.e.

$$\left[\frac{\mathrm{d}}{\mathrm{d}\theta}\begin{pmatrix} x(\theta) \\ y(\theta) \end{pmatrix}\right]_{\theta=0} = \begin{pmatrix} 0 & 1 \\ -1 & 0 \end{pmatrix}\begin{pmatrix} x \\ y \end{pmatrix}. \tag{2.C.18}$$

This equation is solved by exponentiation, bearing in mind that

$$\exp\left[\theta\begin{pmatrix} 0 & 1 \\ -1 & 0 \end{pmatrix}\right] = \cos\theta\begin{pmatrix} 1 & 0 \\ 0 & 1 \end{pmatrix} + \sin\theta\begin{pmatrix} 0 & 1 \\ -1 & 0 \end{pmatrix}. \tag{2.C.19}$$

Eq. (2.C.18) can also be expressed by means of a vector field

$$\Gamma \equiv a\frac{\partial}{\partial x} + b\frac{\partial}{\partial y} = \frac{\mathrm{d}}{\mathrm{d}\theta}, \tag{2.C.20}$$

and one finds the following Lie derivatives:

$$L_\Gamma x = \frac{\mathrm{d}}{\mathrm{d}\theta}x(\theta)\bigg|_{\theta=0} = y, \tag{2.C.21}$$

$$L_\Gamma y = \frac{\mathrm{d}}{\mathrm{d}\theta}y(\theta)\bigg|_{\theta=0} = -x. \tag{2.C.22}$$

Thus, the vector field

$$\Gamma \equiv y\frac{\partial}{\partial x} - x\frac{\partial}{\partial y} \tag{2.C.23}$$

is the infinitesimal generator of rotations in the plane. It is a tensor quantity, and hence is independent of any choice of coordinates.

Digression. The transition from the algebra to the Lie group is achieved by solving a simultaneous set of first-order differential equations. Each equation gives rise to a one-parameter group of transformations, and the different parameters of the equations determine the parameters of the Lie group of transformations. When the differential equations can be written as linear differential equations, the solutions are simply found by exponentiation, i.e. $\dot{x} = Ax$ is solved by $x(t) = \mathrm{e}^{tA}x_0$, where x_0 is the initial condition. If we replace A with the associated vector field $X_A \equiv A_j^{\ k}\,\xi^j\frac{\partial}{\partial \xi^k}$, when considering the Taylor expansion of the exponential we need to make sense of expressions such as $X_A \cdot X_A \cdot \cdots X_A$ appearing in the series expansion. When the full Lie group is considered we encounter expressions such as $X_a \cdot X_b \cdot \cdots X_c$. When these expressions are considered for a Lie algebra, independent of the particular way we realize its elements (i.e. matrices, vector fields or other structures), the previous expressions give rise to the notion of *enveloping algebra*. Given a Lie algebra g, we introduce its tensor algebra $T(g) \equiv \sum_r T^r(g)$, with $T^1(g) \equiv g, T^2(g) \equiv g \otimes g$, and so on. Any realization ρ of g extends to a realization of $T(g)$ by setting

$$\rho(x_1 \otimes \cdots \otimes x_k) \equiv \rho(x_1)\cdots\rho(x_k).$$

Since ρ is a realization, the induced ρ is identically vanishing on the ideal $J(g)$ of $T(g)$ (i.e. a sub-set of $T(g)$ closed under multiplication by all the elements of $T(g)$) generated by elements of the form

$$X \otimes Y - Y \otimes X - [X, Y].$$

Therefore it is effective on the quotient $T(g)/J(g)$. This quotient is called the *universal enveloping algebra* of g and is denoted by $U(g)$.

Casimir elements of the Lie algebra g are elements of $U(g)$ and belong to the centre of $U(g)$. It is an important property that every unitary representation of a Lie group $\mathcal{G}$ gives rise to a self-adjoint representation of its Lie algebra g. The representatives of the Casimir elements, called *Casimir operators*, commute with all representatives of $\mathcal{G}$ and generate an Abelian algebra. Thus, by virtue of the Schur lemma (Wybourne 1974), in an irreducible representation every Casimir operator is a multiple of the identity operator. Such numbers make it possible to label the irreducible representation. In particular, the universal enveloping algebra of the Poincaré group has two independent Casimir elements. They are interpreted in terms of mass and a particular kind of angular momentum (the latter being called *spin* in the language of chapter 6).

Appendix 2.D
Space–time

The unification of space and time into a four-dimensional continuum seems to reflect all we know about the structure of our world up to scales of order 10^{-13} cm (Hawking and Ellis 1973).

Aristotelian space–time, where both space and time are absolute, is simply the Cartesian product $\mathbf{R}^4 = \mathbf{R}^3 \times \mathbf{R}$, where $\mathbf{R}$ stands for time and $\mathbf{R}^3$ for space. Usually it is taken to be equipped with a Euclidean structure. Linear transformations preserving the Euclidean structure constitute the group of Euclidean motions.

Galilean and Newtonian space–times differ from Aristotelean space–time in that time remains absolute but space becomes relativistic. It is not a Cartesian product but there is a projection $t : \mathbf{R}^4 \rightarrow \mathbf{R}$, which associates with any event $m \in \mathbf{R}^4$ the corresponding instant of time $t = t(m)$. The inverse image $t^{-1}(a)$ is the space of simultaneous events in $\mathbf{R}^4$. The group of motions of Galilean space that preserves the structure defined by the projection and the Euclidean structure on every $t^{-1}(a), a \in \mathbf{R}$, is the 10-parameter Galilean group.

The space–time of the Einstein theory of special relativity is known as Minkowski space–time. Here the space of events is endowed with a Minkowski metric, and the group of motions preserving this metric is the inhomogeneous Lorentz group (also called the Poincaré group). Eventually, the Einstein theory of gravitation, known as general relativity, is characterized by a metric of Lorentzian signature $(-,+,+,+)$ (also called of hyperbolic type), and is associated with the group of smooth coordinate transformations, i.e. space–time diffeomorphisms.

Appendix 2.E
From Newton to Euler–Lagrange

In Newtonian mechanics, the equations of motion are written as second-order differential equations (Newton 1999)

$$m \frac{\mathrm{d}^2 x^i}{\mathrm{d}t^2} = F^i \left(x, \frac{\mathrm{d}x}{\mathrm{d}t} \right). \tag{2.E.1}$$

With these equations we associate a set of first-order differential equations of the form

$$\frac{\mathrm{d}}{\mathrm{d}t} x^i = v^i, \tag{2.E.2}$$

$$\frac{\mathrm{d}}{\mathrm{d}t} (m v^i) = F^i. \tag{2.E.3}$$

On the other hand, with any set of first-order differential equations we can associate a first-order differential operator, for which the action on functions is given by

$$\frac{\mathrm{d}}{\mathrm{d}t} f = \frac{\partial f}{\partial x^i} \frac{\mathrm{d}x^i}{\mathrm{d}t} + \frac{\partial f}{\partial v^i} \frac{\mathrm{d}v^i}{\mathrm{d}t}. \tag{2.E.4}$$

If $\frac{\mathrm{d}}{\mathrm{d}t} m$ vanishes, one obtains from Eqs. (2.E.2)–(2.E.4) that

$$\frac{\mathrm{d}}{\mathrm{d}t} f = \left(v^i \frac{\partial}{\partial x^i} + \frac{F^i}{m} \frac{\partial}{\partial v^i} \right) f, \tag{2.E.5}$$

and hence $\frac{\mathrm{d}f}{\mathrm{d}t}$ is nothing but the Lie derivative of f along the vector field (see appendices 2.A and 2.C)

$$\Gamma \equiv v^i \frac{\partial}{\partial x^i} + \frac{F^i}{m} \frac{\partial}{\partial v^i}. \tag{2.E.6}$$

Equations (2.E.2) and (2.E.3) show that all (time-independent) mechanical systems of Newtonian type can be described by systems of equations of the form

$$\dot{y} = f(y),$$

where $(y) \equiv (x, v)$. The set of solutions of such equation defines the *flow* of the dynamical system associated with the map

$$y \in \mathbf{R}^{2n} \rightarrow f(y) \in \mathbf{R}^{2n}.$$

An important theorem guarantees that if, for some $y_0 \in \mathbf{R}^{2n}$, one has $f(y_0) \neq 0$, the flow in a neighbourhood of y_0 can be described as a 'uniform rectilinear motion', provided a suitable choice of coordinate system is made. One then achieves what is called the *straightening of the flow* (this idea is exploited in section 2.3). The required change of coordinates will 'mix', in general, positions and velocities.

In a Lagrangian formulation, the variational problem, that requires the stationarity of the action functional under suitable boundary conditions, leads to the Euler–Lagrange equations (Lagrange 1788)

$$\frac{\mathrm{d}}{\mathrm{d}t}\frac{\partial \mathcal{L}}{\partial v^i} - \frac{\partial \mathcal{L}}{\partial x^i} = 0, \tag{2.E.7}$$

where $\mathcal{L}$ is the Lagrangian, and the relation between positions x^i and velocities v^i is that given in Eq. (2.E.2). With our notation, the differential of $\mathcal{L}$, assumed to be a function of positions and velocities only, reads as

$$\mathrm{d}\mathcal{L} = \frac{\partial \mathcal{L}}{\partial x^i}\mathrm{d}x^i + \frac{\partial \mathcal{L}}{\partial v^i}\mathrm{d}v^i. \tag{2.E.8}$$

Thus, if we multiply Eq. (2.E.7) by $\mathrm{d}x^i$, use the Leibniz rule and bear in mind that

$$\frac{\mathrm{d}}{\mathrm{d}t}\mathrm{d}x^i = \mathrm{d}\frac{\mathrm{d}}{\mathrm{d}t}x^i = \mathrm{d}v^i, \tag{2.E.9}$$

we obtain the *intrinsic* (i.e. coordinate-independent) form of the Euler–Lagrange equations, for a time-independent Lagrangian, as

$$\frac{\mathrm{d}}{\mathrm{d}t}\left(\frac{\partial \mathcal{L}}{\partial v^i}\mathrm{d}x^i\right) - \mathrm{d}\mathcal{L} = 0. \tag{2.E.10}$$

This property shows that Euler–Lagrange equations are naturally expressed in terms of differential forms, while Newton equations involve vector fields in the first place.

In general, the Euler–Lagrange equations cannot be given a Newtonian form. This crucial point can be appreciated by writing in explicit form the action of $\frac{\mathrm{d}}{\mathrm{d}t}$ on $\frac{\partial \mathcal{L}}{\partial v^i}$ in (2.E.7), which leads to

$$\frac{\partial^2 \mathcal{L}}{\partial v^i \partial v^j}\frac{\mathrm{d}v^j}{\mathrm{d}t} + \frac{\partial^2 \mathcal{L}}{\partial v^i \partial x^j}\frac{\mathrm{d}x^j}{\mathrm{d}t} - \frac{\partial \mathcal{L}}{\partial x^i} = 0. \tag{2.E.11}$$

It is now clear that, if

$$\det\left(\frac{\partial^2 \mathcal{L}}{\partial v^i \partial v^j}\right) \neq 0, \tag{2.E.12}$$

it is possible to express $\frac{\mathrm{d}}{\mathrm{d}t}v^j$ from (2.E.11) and hence cast the Euler–Lagrange equations in Newtonian form. However, in many relevant physical applications the condition (2.E.12) is not satisfied (e.g. the relativistic free particle, Abelian and non-Abelian gauge theories, Einstein's theory of gravitation). The converse problem, i.e. providing a Lagrangian description for a given set of Newtonian equations of motion, is highly non-trivial and is known as the inverse problem in the calculus of variations (Morandi *et al.* 1990).

A physical system S is said to have a symmetry property if it is invariant under an invertible map, which is called a symmetry (transformation) of S. The identity, the inverse and the composition of symmetry transformations are symmetries, and hence the set of all symmetries of S forms a group called the symmetry group of S.

Symmetries are very often connected with constants of motion (conservation laws) via a Lagrangian description. To exhibit this link, we multiply (2.E.7) by some functions A^i depending on the x coordinates. We denote by $\delta_A \mathcal{L}$ the following expression:

$$\delta_A \mathcal{L} \equiv A^i \frac{\partial \mathcal{L}}{\partial x^i} + \frac{\mathrm{d}A^i}{\mathrm{d}t}\frac{\partial \mathcal{L}}{\partial v^i} = A^i \frac{\partial \mathcal{L}}{\partial x^i} + \frac{\partial A^i}{\partial x^j}v^j \frac{\partial \mathcal{L}}{\partial v^i}. \tag{2.E.13}$$

The above procedure, jointly with the Leibniz rule, leads to

$$\frac{\mathrm{d}}{\mathrm{d}t}\left(A^i \frac{\partial \mathcal{L}}{\partial v^i}\right) - \delta_A \mathcal{L} = 0. \tag{2.E.14}$$

When $\delta_A \mathcal{L} = \frac{\mathrm{d}f}{\mathrm{d}t}$, we find a version of Noether's theorem, i.e.

$$\frac{\mathrm{d}}{\mathrm{d}t}\left(A^i \frac{\partial \mathcal{L}}{\partial v^i} - f\right) = 0. \tag{2.E.15}$$

Thus, $A^i \frac{\partial \mathcal{L}}{\partial v^i} - f$ is a constant of motion. It should be stressed that in general, in Eq. (2.E.15), A^i is not required to be a function only of positions x^i. When this is the case we say we are dealing with infinitesimal *point transformations*. For example, if

$$A^i = \dot{x}^i = \frac{\mathrm{d}x^i}{\mathrm{d}t}, \tag{2.E.16}$$

then Eq. (2.E.13) leads to

$$\delta_A \mathcal{L} = \dot{x}^i \frac{\partial \mathcal{L}}{\partial x^i} + \frac{\mathrm{d}\dot{x}^i}{\mathrm{d}t}\frac{\partial \mathcal{L}}{\partial \dot{x}^i} = \frac{\mathrm{d}}{\mathrm{d}t}\mathcal{L}, \tag{2.E.17}$$

and one finds

$$\frac{\mathrm{d}}{\mathrm{d}t}\left(\dot{x}^i \frac{\partial \mathcal{L}}{\partial \dot{x}^i} - \mathcal{L}\right) = 0, \tag{2.E.18}$$

i.e. a conservation of 'energy' for time-independent Lagrangians.

3
Wave equations

A physics-oriented review is first given of wave equations: examples, the Cauchy problem, solutions in various coordinates, their symmetries, how to build a wave packet. Fourier analysis and dispersion relations are then introduced, and here the key tool is the symbol of differential operators, which makes it possible to study them in terms of algebraic polynomials involving cotangent-bundle variables. Further basic material deals with geometrical optics from the wave equation, phase and group velocity (and their dual relationship in momentum space), the Helmholtz equation and the eikonal approximation for the scalar wave equation.

The Schrödinger equation is then derived with emphasis on the wave packet and its relation to classical behaviour in the light of the Einstein–de Broglie relation. For this purpose, it is shown that it is possible to build a wave packet for which the wave-like properties manifest themselves for distances of the order of atomic dimensions. The Fourier transform with respect to time of the wave packet obeys the stationary Schrödinger equation, while the wave function is found to obey a partial differential equation which is of first order in time.

3.1 The wave equation

The wave equation occurs in several branches of classical physics, e.g. the theory of sound, electromagnetic phenomena in vacuum and in material media, and elastic vibrations of material bodies. We consider first the simplest situation, i.e. the propagation of waves in a homogeneous, isotropic and stationary medium. Propagation is characterized by the refractive index n, a quantity that is independent of the point in space-time. If c represents a fundamental velocity (c is the velocity of sound in acoustics, and the velocity of light in electrodynamics) one can write

$$\Delta u = \left(\frac{\partial^2}{\partial x^2} + \frac{\partial^2}{\partial y^2} + \frac{\partial^2}{\partial z^2} \right) u = \frac{n^2}{c^2} \frac{\partial^2 u}{\partial t^2}, \tag{3.1.1}$$

or

$$\Box u = 0, \tag{3.1.2}$$

where a fourth coordinate $x_0 \equiv ct$ is introduced to write

$$\Box u = \left(\frac{\partial^2}{\partial x^2} + \frac{\partial^2}{\partial y^2} + \frac{\partial^2}{\partial z^2} - n^2 \frac{\partial^2}{\partial x_0^2} \right) u = 0. \tag{3.1.3}$$

Many other partial differential equations arise in the description of particular physical phenomena. For example, one can study the Laplace equation

$$\Delta u = 0 \tag{3.1.4}$$

or the Poisson equation

$$\Delta u = -\rho. \tag{3.1.5}$$

The latter is known, in the theory of gravitation, as the expression of the field-theoretical approach, as opposed to the action-at-a-distance approach of Newton. The operator Δ is known as the Laplace operator. In Cartesian coordinates in $\mathbf{R}^3$ it reads as

$$\Delta \equiv \frac{\partial^2}{\partial x^2} + \frac{\partial^2}{\partial y^2} + \frac{\partial^2}{\partial z^2}, \tag{3.1.6}$$

while its coordinate-free definition is

$$\Delta \equiv \text{div grad}. \tag{3.1.7}$$

These same equations are fundamental for electrostatic and magnetic fields.

In the general theory of elasticity one has, as a special case, the equation for the transverse vibrations of a thin disc:

$$\Delta^2 u = -\frac{1}{c^2} \frac{\partial^2 u}{\partial t^2}, \tag{3.1.8}$$

with

$$\Delta^2 \equiv \Delta\Delta = \frac{\partial^4}{\partial x^4} + 2 \frac{\partial^4}{\partial x^2 \partial y^2} + \frac{\partial^4}{\partial y^4}, \tag{3.1.9}$$

where, for dimensional reasons, c does not stand for the velocity of sound in the elastic material, as it does in acoustics. Analogously, the equation for an oscillating elastic rod is

$$\frac{\partial^4 u}{\partial x^4} = -\frac{1}{c^2} \frac{\partial^2 u}{\partial t^2}. \tag{3.1.10}$$

Another type of equations arise in equalization processes. As a chief representative one may mention the heat equation (equalization of energy differences). One should point out that diffusion (equalization of differences of material densities), elastic conduction (equalization of potential differences) all follow the same pattern. These equations are represented by

$$\triangle u = \frac{1}{k}\frac{\partial u}{\partial t}, \tag{3.1.11}$$

where k is the thermal conductivity.

All previous equations display similarities that result from the invariance under rotations and translations. Such invariance properties hold because we are dealing with isotropic and homogeneous media. The differential operator associated with this invariance is the Laplacian $\triangle$. In the case of the space-time invariance of relativity this role is played by the d'Alembert operator $\Box$. For the case of an anisotropic medium, the Laplacian is replaced by the operator

$$v_x^2\frac{\partial^2}{\partial x^2} + v_y^2\frac{\partial^2}{\partial y^2} + v_z^2\frac{\partial^2}{\partial z^2}.$$

The coefficients of second derivatives may also depend on position in inhomogeneous media. An example is provided by light propagation in the atmosphere.

3.2 Cauchy problem for the wave equation

In this section we consider the solutions of the homogeneous and isotropic wave equations by means of the Fourier transform. For this purpose, we consider the Cauchy problem

$$\left(\triangle - \frac{1}{v^2}\frac{\partial^2}{\partial t^2}\right) u(\vec{x}, t) = 0, \tag{3.2.1}$$

$$u(\vec{x}, 0) = \varphi_0(\vec{x}), \tag{3.2.2}$$

$$\frac{\partial u}{\partial t}(\vec{x}, 0) = \varphi_1(\vec{x}), \tag{3.2.3}$$

where $u(\vec{x}, t)$ is a wave function, $\triangle$ is the Laplacian in Cartesian coordinates on $\mathbf{R}^3$ and v (having dimension of velocity) is taken to be constant. The problem is well posed if the initial data φ_0 and φ_1 are sufficiently regular.

The crucial step in our analysis is the Fourier transform of Eq. (3.2.1) with respect to the spatial variable, i.e.

$$(2\pi)^{-3/2}\int_{\mathbf{R}^3} \mathrm{e}^{-\mathrm{i}\vec{k}\cdot\vec{x}} \triangle u(\vec{x}, t)\, \mathrm{d}^3x$$

$$= (2\pi)^{-3/2} \int_{\mathbf{R}^3} \frac{1}{v^2} \mathrm{e}^{-\mathrm{i}\vec{k}\cdot\vec{x}} \frac{\partial^2 u}{\partial t^2}(\vec{x},t)\, \mathrm{d}^3 x. \tag{3.2.4}$$

To ensure that all the operations we are performing are meaningful, we assume that $u(\vec{x},t)$ is a C^∞ function that decreases rapidly at infinity with respect to the $\vec{x}$ variable, for all values of t. Moreover, we assume that the absolute values of the first and second derivatives of u with respect to t remain smaller than some functions f_1 and f_2, which are integrable on $\mathbf{R}^3$, for all finite intervals $(0,T)$. The latter assumption is a sufficient condition for us to be able to interchange the operations of second derivative with respect to t and integration over $\mathbf{R}^3$.

If such conditions are satisfied, one finds, from the definition of the Fourier transform:

$$\widetilde{u}(\vec{k},t) \equiv (2\pi)^{-3/2} \int_{\mathbf{R}^3} \mathrm{e}^{-\mathrm{i}\vec{k}\cdot\vec{x}} u(\vec{x},t)\, \mathrm{d}^3 x, \tag{3.2.5}$$

the equation (see the left-hand side of (3.2.4))

$$(2\pi)^{-3/2} \int_{\mathbf{R}^3} \mathrm{e}^{-\mathrm{i}\vec{k}\cdot\vec{x}} \triangle\, u(\vec{x},t)\, \mathrm{d}^3 x = -k^2 \widetilde{u}(\vec{k},t), \tag{3.2.6}$$

and, eventually,

$$\left(\frac{1}{v^2} \frac{\partial^2}{\partial t^2} + k^2 \right) \widetilde{u}(\vec{k},t) = 0, \tag{3.2.7}$$

with v^2 independent of x, in that $\frac{\partial v^2}{\partial x} = 0$. The general solution of Eq. (3.2.7) is

$$\widetilde{u}(\vec{k},t) = A(\vec{k})\mathrm{e}^{-\mathrm{i}\omega t} + B(\vec{k})\mathrm{e}^{\mathrm{i}\omega t}, \tag{3.2.8}$$

where $\omega \equiv kv$.

We now have to take the Fourier transforms of our initial data, i.e.

$$\widetilde{u}(\vec{k},0) = \widetilde{\varphi}_0(\vec{k}), \tag{3.2.9}$$

$$\frac{\partial \widetilde{u}}{\partial t}(\vec{k},0) = \widetilde{\varphi}_1(\vec{k}). \tag{3.2.10}$$

On using (3.2.8), this leads to

$$\widetilde{\varphi}_0(\vec{k}) = A(\vec{k}) + B(\vec{k}), \tag{3.2.11}$$

$$\widetilde{\varphi}_1(\vec{k}) = \mathrm{i}\omega \Big[B(\vec{k}) - A(\vec{k}) \Big], \tag{3.2.12}$$

and hence

$$A(\vec{k}) = \frac{1}{2} \Big[\widetilde{\varphi}_0(\vec{k}) + \frac{\mathrm{i}}{\omega} \widetilde{\varphi}_1(\vec{k}) \Big], \tag{3.2.13}$$

$$B(\vec{k}) = \frac{1}{2} \Big[\widetilde{\varphi}_0(\vec{k}) - \frac{\mathrm{i}}{\omega} \widetilde{\varphi}_1(\vec{k}) \Big]. \tag{3.2.14}$$

Last, on anti-transforming, one finds

$$u(\vec{x},t) = (2\pi)^{-3/2}\left[\int_{\mathbf{R}^3} A(\vec{k})\mathrm{e}^{\mathrm{i}(\vec{k}\cdot\vec{x}-\omega t)}\,\mathrm{d}^3k + \int_{\mathbf{R}^3} B(\vec{k})\mathrm{e}^{\mathrm{i}(\vec{k}\cdot\vec{x}+\omega t)}\,\mathrm{d}^3k\right], \tag{3.2.15}$$

and it is now possible to prove that such a function solves the Cauchy problem given by (3.2.1)–(3.2.3).

3.3 Fundamental solutions

The previous section describes how to approach in rigorous terms the solution of the wave equation by a Fourier transform. The underlying idea is that the space of solutions is a vector space because the equation is linear, and therefore it makes sense to look for a basis of this vector space (we have used monochromatic plane waves). Of course, in infinite dimensions the use of bases in vector spaces is much more subtle than in finite dimensions (more relevant than the question of convergence, when we deal with infinite sums, is the question of completeness of the system of functions used as a basis). For instance, if the function u is meant to represent the electric and magnetic field, it is meaningful to require that it should be square integrable, but if it represents the vector potential this requirement is not justified. In this section we consider other solutions of the wave equations which may exhibit other features.

3.3.1 Wave equations in spherical polar coordinates

Using the change of variables from Cartesian coordinates to spherical polar coordinates $x = r\sin\theta\cos\varphi, y = r\sin\theta\sin\varphi, z = r\cos\theta$ one obtains

$$\Box u = \frac{1}{r^2}\frac{\partial}{\partial r}\left(r^2\frac{\partial u}{\partial r}\right) + \frac{1}{r^2\sin\theta}\frac{\partial}{\partial\theta}\left(\sin\theta\frac{\partial u}{\partial\theta}\right) + \frac{1}{r^2\sin^2\theta}\frac{\partial^2 u}{\partial\varphi^2} - \frac{n^2}{c^2}\frac{\partial^2 u}{\partial t^2} = 0. \tag{3.3.1}$$

It is also possible to use cylindrical coordinates

$$x = r\cos\varphi, \quad y = r\sin\varphi, \quad z = z$$

to obtain

$$\left(\frac{\partial^2}{\partial r^2} + \frac{1}{r}\frac{\partial}{\partial r} + \frac{1}{r^2}\frac{\partial^2}{\partial\varphi^2} + \frac{\partial^2}{\partial z^2} - \frac{n^2}{c^2}\frac{\partial^2}{\partial t^2}\right)u = 0. \tag{3.3.2}$$

For the spherically symmetric solutions one easily finds that

$$u(r,t) = \frac{1}{r} f\left(r - \frac{ct}{n}\right) + \frac{1}{r} g\left(r + \frac{ct}{n}\right). \tag{3.3.3}$$

Here the first term represents an outgoing wave and the second term an incoming wave. Similarly, one can find cylindrically symmetric solutions. Less useful solutions are provided by functions such as

$$u = \frac{1}{6}(x^2 + y^2 + z^2) - \frac{c^2}{n^2} t^2. \tag{3.3.4}$$

What we would like to stress in this section is that, because the same mathematical equation may describe different physical phenomena, depending on the identification of physical variables we will find that some mathematical solutions may be disposed of by virtue of the physical interpretation.

3.4 Symmetries of wave equations

It is quite usual (often justified by simplicity requirements) to find the discussion of the wave equation in one space and one time dimension, i.e.

$$\left(\frac{n^2}{c^2}\frac{\partial^2}{\partial t^2} - \frac{\partial^2}{\partial x^2}\right) u = 0. \tag{3.4.1}$$

This equation is usually rewritten in the form

$$\left(\frac{n}{c}\frac{\partial}{\partial t} - \frac{\partial}{\partial x}\right)\left(\frac{n}{c}\frac{\partial}{\partial t} + \frac{\partial}{\partial x}\right) u = 0, \tag{3.4.2}$$

which shows that

$$u = f\left(x - \frac{c}{n}t\right) + g\left(x + \frac{c}{n}t\right) \tag{3.4.3}$$

is a general solution.

Since any unit vector in $\mathbf{R}^3$ can be transformed into any other by performing a rotation, it is clear that, by virtue of the rotational invariance of the d'Alembert operator, if R represents the rotation

$$R \begin{pmatrix} 1 \\ 0 \\ 0 \end{pmatrix} = \begin{pmatrix} k_1 \\ k_2 \\ k_3 \end{pmatrix}, \tag{3.4.4}$$

one obtains

$$u_R = f\left(k_1 x + k_2 y + k_3 z - \frac{c}{n}t\right) + g\left(k_1 x + k_2 y + k_3 z + \frac{c}{n}t\right). \tag{3.4.5}$$

More generally, we may apply any symmetry of the wave equation to any one of the previous solutions to obtain new solutions. It is therefore quite evident that knowledge of the full symmetry group of our equation may be a powerful tool for generating solutions.

By exploiting the previous remark, we may transform the solution of the Cauchy problem for the wave equation (3.2.1)–(3.2.3) for one space dimension:

$$u(x,t) = \frac{\varphi_0(nx - ct) + \varphi_0(nx + ct)}{2} + \frac{n}{2c}\int_{nx-ct}^{nx+ct} \varphi_1(s)\,\mathrm{d}s, \tag{3.4.6}$$

into one in $\mathbf{R}^4$, provided that the initial conditions are also transformed.

3.5 Wave packets

Monochromatic plane waves (appendix 1.A), which have been used as a family of fundamental solutions to build general ones, represent an excessive idealization, since they are everywhere in space and exist forever. Actually, a wave occupies a bounded region of space at each given time, and always has a limited duration. Such a wave is what we call a wave packet. We use combinations

$$u_k(x,t) = A(k)\mathrm{e}^{\mathrm{i}k(x-vt)} + B(k)\mathrm{e}^{\mathrm{i}k(x+vt)} \tag{3.5.1}$$

to build wave packets

$$u(x,t) = \int_{\mathbf{R}^3} u_k(x,t)\,\mathrm{d}^3k. \tag{3.5.2}$$

The more sharply we want to 'localize' our solution in space, the larger will be the number of monochromatic waves we have to use in the process of building our packet. If $\delta k = k_2 - k_1$ is the width of the interval of monochromatic waves we use and $\delta x = x_2 - x_1$ is the width of the interval in space where our solution is localized, general results from Fourier analysis imply that

$$(\delta k)(\delta x) \geq 1. \tag{3.5.3}$$

Any phenomenon dealing with waves therefore leads to a form of indeterminacy.

3.6 Fourier analysis and dispersion relations

The solution method based on Fourier analysis relies on the use of monochromatic plane waves

$$\phi(x,t) = \mathrm{e}^{\mathrm{i}(kx \pm \omega t)}. \tag{3.6.1}$$

When they are used as a fundamental family of solutions one finds

$$\frac{\partial^2 \phi}{\partial t^2} = -\omega^2 \phi, \quad \frac{\partial^2 \phi}{\partial x^2} = -k^2 \phi, \tag{3.6.2}$$

and hence ϕ is an elementary solution of the wave equation (3.2.1) provided that

$$\omega^2 n^2 - k^2 c^2 = 0. \tag{3.6.3}$$

The optical phase-space 1-form $k\,\mathrm{d}x - \omega\,\mathrm{d}t$ that we have used in the Einstein–de Broglie identification (section 1.2)

$$\vec{p} \cdot \mathrm{d}\vec{x} - E\,\mathrm{d}t = \hbar(\vec{k} \cdot \mathrm{d}\vec{x} - \omega \mathrm{d}t), \tag{3.6.4}$$

evaluated along the solution ϕ, satisfies

$$-\mathrm{i}\phi^{-1}\,\mathrm{d}\phi = k\,\mathrm{d}x - \omega\,\mathrm{d}t. \tag{3.6.5}$$

The use of (k, ω) as additional variables with respect to (x, t) suggests interpreting $n^2 \frac{\omega^2}{c^2} - k^2$ as a function on $T^*\mathbf{R}^4$ when $\Box$ is the d'Alembert operator on $\mathbf{R}^4$. We can extend the map associating Eq. (3.6.3) with the wave equation to any other partial differential equation.

3.6.1 The symbol of differential operators

Given any differential operator D on $\mathbf{R}^4$ we define a map

$$\sigma : \mathrm{D} \rightarrow \sigma(\mathrm{D}) \equiv \mathrm{e}^{-\alpha k_\mu x^\mu}\, \mathrm{D}\, \mathrm{e}^{\alpha k_\mu x^\mu}, \tag{3.6.6}$$

with $\sigma(\mathrm{D}) \in \mathcal{F}(T^*\mathbf{R}^4)$, i.e. the set of functions on the cotangent bundle of $\mathbf{R}^4$. For $\mathrm{D} = \Box$ one obtains

$$\sigma(\Box) = \alpha^2 \left(\vec{k} \cdot \vec{k} - n^2 k_0^2\right). \tag{3.6.7}$$

For $\mathrm{D} = \frac{\partial^4}{\partial x^4} + \frac{n^2}{c^2}\frac{\partial^2}{\partial t^2}$ one finds

$$\sigma(\mathrm{D}) = \alpha^4 k_x^4 + \alpha^2 n^2 k_0^2. \tag{3.6.8}$$

If the d'Alembert operator acquires a mass term, one has

$$\sigma(\Box - m_0^2 c^2) = \alpha^2 \left(\vec{k} \cdot \vec{k} - n^2 k_0^2\right) - m_0^2 c^2. \tag{3.6.9}$$

By setting $\sigma(\mathrm{D}) = 0$ one finds necessary conditions for $\mathrm{e}^{\alpha k_\mu x^\mu}$ to be an elementary solution of the corresponding differential equation $\mathrm{D}u = 0$. The function $\sigma(\mathrm{D})$ is called the *symbol* of D and it represents a function on $T^*\mathbf{R}^4$ if D is a differential operator on $\mathbf{R}^4$. From the definition it follows that σ is a linear map. The order of the differential operator becomes the degree of the polynomial in k for the function on $T^*\mathbf{R}^4$. In general,

$$\mathrm{D} = \sum_m a_m(x)\left(\frac{\partial}{\partial x}\right)^m \tag{3.6.10}$$

is mapped onto

$$\sigma(\mathrm{D}) = \sum_m a_m(x)\alpha^m k^m. \tag{3.6.11}$$

First-order differential operators such as $\vec{a}\cdot\frac{\partial}{\partial \vec{x}}$ and $\frac{\partial}{\partial x_0}$ are mapped onto $\alpha\vec{a}\cdot\vec{k}$ and αk_0, respectively. If $\vec{a}$ is any constant vector, all operators generated by taking powers of $\vec{a}\cdot\frac{\partial}{\partial \vec{x}}$ and $\frac{\partial}{\partial x_0}$ with constant coefficients constitute an associative algebra. The map σ becomes a map between this algebra and the algebra of polynomials in k with constant coefficients. The map is a homomorphism of algebras and is invertible, i.e. with any constant-coefficient polynomial on $(\mathbf{R}^4)^* \subset T^*\mathbf{R}^4$ it associates a constant-coefficient differential operator.

It should be stressed that, since the algebra on the set of functions on the cotangent bundle of $\mathbf{R}^4$ is always commutative, the homomorphism is guaranteed whenever one considers the algebra of differential operators generated by using commuting generators, even if the coefficients are not constant (for instance, as provided by any non-linear realization of the translation group). We recall that $T^*\mathbf{R}^4 = \mathbf{R}^4 \times (\mathbf{R}^4)^*$, with local coordinates $(\vec{x}, t; \vec{k}, \omega)$ can also be considered as the cotangent bundle constructed on $(\mathbf{R}^4)^*$, i.e. using as 'independent variables' the momenta instead of positions. In this way we would consider differential operators with generators $\frac{\partial}{\partial k_\mu}$. The symbol then becomes a polynomial in the x variables, and the associated canonical 1-form is $x^\mu\, \mathrm{d}k_\mu - t\, \mathrm{d}\omega$. By exploiting this property, a Lagrangian sub-manifold of $T^*\mathbf{R}^4$ may be identified either as a graph of $\mathrm{d}W : \mathbf{R}^4 \to T^*\mathbf{R}^4$ or as a graph of

$$\mathrm{d}\widetilde{W} : (\mathbf{R}^4)^* \to T^*(\mathbf{R}^4)^*.$$

The transition from one representation to the other is achieved via the Fourier transform.

3.6.2 Dispersion relations

The use of complex-valued functions to describe elementary solutions and to make a full use of Fourier analysis requires that we also accept complex-valued functions on $T^*\mathbf{R}^4$, and therefore as symbols. In particular, it means that in Eq. (3.6.6) α can be a complex number, which is what we assume here. The equation $\sigma(\mathrm{D}) = 0$, associated with any operator on $\mathbf{R}^4$, is usually called the *dispersion relation*. It shows how the frequency ω

depends on the wave vector (and hence on the wavelength). For constant-coefficient equations for which the solution may be written as a wave packet

$$\psi(\vec{x},t) = \int_{\mathbf{R}^3} A(k) e^{\mathrm{i}[\vec{k}\cdot\vec{x}-\omega(k)t]} \, \mathrm{d}^3k, \tag{3.6.12}$$

with A being smooth and with compact support away from the origin: $A \in C_0^\infty(\mathbf{R}^3)$, one has

$$\frac{\partial\psi}{\partial t} = \int_{\mathbf{R}^3} -\mathrm{i}\omega(k) A(k) e^{\mathrm{i}[\vec{k}\cdot\vec{x}-\omega(k)t]} \, \mathrm{d}^3k, \tag{3.6.13}$$

$$\frac{\partial\psi}{\partial x^l} = \int_{\mathbf{R}^3} \mathrm{i}k_l A(k) e^{\mathrm{i}[\vec{k}\cdot\vec{x}-\omega(k)t]} \, \mathrm{d}^3k, \quad \forall l = 1,2,3. \tag{3.6.14}$$

This leads to the following prescription for deriving the dispersion relation associated to the equation $\mathrm{D}\psi = 0$ (having set $\alpha = \mathrm{i}$):

$$\frac{\partial}{\partial t} \rightarrow -\mathrm{i}\omega, \quad \frac{\partial}{\partial \vec{x}} \rightarrow \mathrm{i}\vec{k}.$$

In the above expressions for ψ and its partial derivatives and throughout the following presentation, we denote by $\mathrm{d}^3k = \mathrm{d}k_1\,\mathrm{d}k_2\,\mathrm{d}k_3$ the volume element in the three-dimensional wave-vector space (which by the above identification may be called the momentum space). We also have to stress that (3.6.13) is not a four-dimensional Fourier integral as might be used to give an integral representation of any function of the variables x_μ. The integral should be understood as being made on the sub-manifold of $\mathbf{R}^4$ defined by $\sigma(\mathrm{D}) = 0$, i.e. $\omega = \omega(k)$. When this sub-manifold is in one-to-one correspondence with $\mathbf{R}^3$ we may integrate over it and use the measure d^3k.

For the operator

$$\mathrm{D} \equiv \sum_{q=0}^{r} B_q \frac{\partial^q}{\partial x^q} + \sum_{j=0}^{s} C_j \frac{\partial^j}{\partial t^j} \tag{3.6.15}$$

the dispersion relation therefore reads as

$$\sum_{q=0}^{r} B_q (\mathrm{i}k)^q + \sum_{j=0}^{s} C_j (-\mathrm{i}\omega(k))^j = 0. \tag{3.6.16}$$

And vice versa, if the dispersion relation is given, one can recover the associated equation via the prescription

$$\omega \rightarrow \mathrm{i}\frac{\partial}{\partial t}, \quad \vec{k} \rightarrow -\mathrm{i}\frac{\partial}{\partial \vec{x}}.$$

With every form of the dispersion relation ($\sigma(\mathrm{D}; k,\omega)$ being our notation for the symbol D in the coordinates k,ω)

$$\sigma(\mathrm{D}; k,\omega) = 0 \tag{3.6.17}$$

one can therefore associate a wave equation, along with a Hamilton–Jacobi equation

$$\sigma\left(\mathrm{D};\left(\frac{\partial W}{\partial \vec{x}},\frac{\partial W}{\partial t}\right)\right)=0. \tag{3.6.18}$$

When the differential operators are no longer constant-coefficient operators, ordering problems arise in associating a differential operator with a function on $T^*\mathbf{R}^4$, but of course, the Hamilton–Jacobi equation remains well defined. Various examples of dispersion relation are now given (while the proper physical interpretation of some equations is postponed).

(a) The dispersion relation

$$\omega(k)=\frac{\hbar k^2}{2m} \tag{3.6.19}$$

leads to the non-relativistic wave equation

$$\left(\mathrm{i}\hbar\frac{\partial}{\partial t}+\frac{\hbar^2}{2m}\triangle\right)\psi=0. \tag{3.6.20}$$

(b) The dispersion relation

$$\omega=c\sqrt{k^2+\chi^2}, \tag{3.6.21}$$

leads to the relativistic wave equation

$$\left(\mathrm{i}\frac{\partial}{\partial t}-c\sqrt{-\triangle+\chi^2}\right)\psi=0. \tag{3.6.22}$$

The 'nice' property of this equation is its first-order nature in the time variable. However, it is non-linear in $\triangle$, and hence it is very difficult to extend it when external forces occur. This is a particular instance of the Salpeter equation (Bethe and Salpeter 1957).

(c) On squaring up the dispersion relation (3.6.22) one finds the Klein–Gordon equation (Klein 1926, Gordon 1926)

$$\left(\square-\chi^2\right)\psi=0. \tag{3.6.23}$$

(d) If one takes

$$\omega=\pm\sqrt{\alpha^2k^2+\beta^2} \tag{3.6.24}$$

one finds the equation

$$\left(\frac{\partial^2}{\partial t^2}-\alpha^2\frac{\partial^2}{\partial x^2}+\beta^2\right)\varphi=0. \tag{3.6.25}$$

(e) The choice

$$\omega = \pm\gamma k^2 \tag{3.6.26}$$

leads to

$$\left(\frac{\partial^2}{\partial t^2} + \gamma^2 \frac{\partial^4}{\partial x^4}\right)\varphi = 0. \tag{3.6.27}$$

(f) The dispersion relation

$$\omega = \alpha k - \beta k^3 \tag{3.6.28}$$

implies, for φ, the equation

$$\left(\frac{\partial}{\partial t} + \alpha\frac{\partial}{\partial x} + \beta\frac{\partial^3}{\partial x^3}\right)\varphi = 0. \tag{3.6.29}$$

(g) An integral equation for φ can also be obtained, e.g.

$$\frac{\partial\varphi}{\partial t} + \int_{-\infty}^{\infty} G(x-\xi)\frac{\partial\varphi}{\partial\xi}\,\mathrm{d}\xi = 0, \tag{3.6.30}$$

where G, the *kernel*, is a given function. Equation (3.6.30) admits elementary solutions with amplitude A, frequency ω and wave number k, i.e.

$$\varphi(x,t) = A\mathrm{e}^{\mathrm{i}(kx-\omega t)}, \tag{3.6.31}$$

provided that

$$-\mathrm{i}\omega\mathrm{e}^{\mathrm{i}kx} + \int_{-\infty}^{\infty} G(x-\xi)\mathrm{i}k\mathrm{e}^{\mathrm{i}k\xi}\,\mathrm{d}\xi = 0, \tag{3.6.32}$$

which implies

$$C(k) \equiv \frac{\omega}{k} = \int_{-\infty}^{\infty} G(\eta)\mathrm{e}^{-\mathrm{i}k\eta}\,\mathrm{d}\eta. \tag{3.6.33}$$

The right-hand side of Eq. (3.6.33) is the Fourier transform of the kernel, and hence from the inversion formula one finds

$$G(x-\xi) = \frac{1}{2\pi}\int_{-\infty}^{\infty} C(k)\mathrm{e}^{\mathrm{i}k(x-\xi)}\,\mathrm{d}k. \tag{3.6.34}$$

It is thus possible to obtain *any dispersion relation by choosing the kernel G as the Fourier transform of the desired phase velocity.* To go beyond the class of constant-coefficient equations let us point out that, if the elementary solutions are proportional to $\mathrm{e}^{\mathrm{i}[\vec{k}\cdot\vec{x}-\omega(k)t]}$, one can then write, at least formally,

$$\varphi(\vec{x},t) = \int_{\mathbf{R}^3} F(\vec{k})\mathrm{e}^{\mathrm{i}[\vec{k}\cdot\vec{x}-\omega(k)t]}\,\mathrm{d}^3k, \tag{3.6.35}$$

which is required to be a solution. The arbitrary function F can be chosen in such a way as to satisfy the given initial (or boundary) conditions, provided that such conditions are sufficiently regular to admit a Fourier transform. If there exist n modes, with n distinct roots of $\omega(k)$, there will be n distinct terms like the previous one, with n arbitrary functions $F(\vec{k})$. It will then be appropriate to assign n initial conditions to determine the solution. For example, if $\omega(k) = \pm W(k)$, one has in one dimension

$$\varphi(x,t) = \int_{-\infty}^{\infty} F_1(k) \mathrm{e}^{\mathrm{i}[kx - W(k)t]} \, \mathrm{d}k + \int_{-\infty}^{\infty} F_2(k) \mathrm{e}^{\mathrm{i}[kx + W(k)t]} \, \mathrm{d}k, \quad (3.6.36)$$

with initial conditions $\varphi(x,0) = \varphi_0(x)$ and $\frac{\partial \varphi}{\partial t}(x,0) = \varphi_1(x)$. Taking into account the initial conditions, we require that

$$\varphi_0(x) = \int_{-\infty}^{\infty} [F_1(k) + F_2(k)] \mathrm{e}^{ikx} \, \mathrm{d}k, \quad (3.6.37)$$

$$\varphi_1(x) = -\mathrm{i} \int_{-\infty}^{\infty} W(k)[F_1(k) - F_2(k)] \mathrm{e}^{ikx} \, \mathrm{d}k. \quad (3.6.38)$$

The inverse formulae yield

$$F_1(k) + F_2(k) = \phi_0(k) = \frac{1}{2\pi} \int_{-\infty}^{\infty} \varphi_0(x) \mathrm{e}^{-ikx} \, \mathrm{d}x, \quad (3.6.39)$$

$$-\mathrm{i}W(k)[F_1(k) - F_2(k)] = \phi_1(k) = \frac{1}{2\pi} \int_{-\infty}^{\infty} \varphi_1(x) \mathrm{e}^{-ikx} \, \mathrm{d}x, \quad (3.6.40)$$

and hence F_1 and F_2 are determined by

$$F_1(k) = \frac{1}{2} \left[\phi_0(k) + \mathrm{i} \frac{\phi_1(k)}{W(k)} \right], \quad (3.6.41)$$

$$F_2(k) = \frac{1}{2} \left[\phi_0(k) - \mathrm{i} \frac{\phi_1(k)}{W(k)} \right]. \quad (3.6.42)$$

Since $\varphi_0(x)$ and $\varphi_1(x)$ are real-valued, one has $\phi_0(-k) = \phi_0^*(k)$ and $\phi_1(-k) = \phi_1^*(k)$, where the star denotes complex conjugation. Hence it follows that, for $W(k)$ odd,

$$F_1(-k) = F_1^*(k), \quad F_2(-k) = F_2^*(k), \quad (3.6.43)$$

and for $W(k)$ even:

$$F_1(-k) = F_2^*(k), \quad F_2(-k) = F_1^*(k). \quad (3.6.44)$$

Note that real-valued initial conditions lead to real-valued solutions. A standard solution, from which the other can be derived, is obtained by imposing that

$$\varphi_0(x) = \delta(x), \quad \varphi_1(x) = 0. \quad (3.6.45)$$

This implies that $F_1(k) = F_2(k) = \frac{1}{4\pi}$, and hence

$$\varphi(x,t) = \frac{1}{\pi}\int_0^\infty (\cos kx)[\cos W(k)t]\, \mathrm{d}k. \tag{3.6.46}$$

Of course, this formal integral should be interpreted within the framework of generalized functions, i.e. distribution theory.

3.7 Geometrical optics from the wave equation

We may now use the transition from Hamilton equations to Lagrange equations to find the usual description of geometrical optics as discussed in section 2.8. Starting with the operator

$$\Box \equiv \triangle - \frac{n^2}{c^2}\frac{\partial^2}{\partial t^2}, \tag{3.7.1}$$

and using the symbol $\sigma : \Box \rightarrow \sigma(\Box)$ defined by

$$\sigma(\Box) \equiv \mathrm{e}^{-\mathrm{i}(\vec{k}\cdot\vec{x}-\omega t)} \; \Box \; \mathrm{e}^{\mathrm{i}(\vec{k}\cdot\vec{x}-\omega t)} \tag{3.7.2}$$

one finds

$$\sigma(\Box) = (\mathrm{i}\vec{k})^2 - \frac{n^2}{c^2}(-\mathrm{i}\omega)^2 = \frac{n^2\omega^2}{c^2} - (\vec{k})^2. \tag{3.7.3}$$

This function on $T^*\mathbf{R}^4$ can be used to define the Hamilton equations of motion associated with the symplectic structure

$$\mathrm{d}\vec{k} \wedge \mathrm{d}\vec{x} - \mathrm{d}\omega \wedge \mathrm{d}t$$

and the Hamilton–Jacobi equation associated with the zero-level set

$$\Sigma_0 \equiv \left\{(\omega, \vec{k}) : \sigma(\Box) = 0\right\} \subset T^*\mathbf{R}^4.$$

One then has

$$\frac{n^2}{c^2}\left(\frac{\partial W}{\partial t}\right)^2 - \left(\frac{\partial W}{\partial \vec{x}}\right)^2 = 0, \tag{3.7.4}$$

and

$$\frac{\mathrm{d}\vec{x}}{\mathrm{d}s} = \frac{\partial \sigma(\Box)}{\partial \vec{k}}, \quad \frac{\mathrm{d}x^0}{\mathrm{d}s} = \frac{\partial \sigma(\Box)}{\partial \omega}, \tag{3.7.5}$$

$$\frac{\mathrm{d}\vec{k}}{\mathrm{d}s} = -\frac{\partial \sigma(\Box)}{\partial \vec{x}}, \quad \frac{\mathrm{d}\omega}{\mathrm{d}s} = -\frac{\partial \sigma(\Box)}{\partial x^0}. \tag{3.7.6}$$

If the refractive index is independent of x_0 but depends on $\vec{x}$, one finds the equations of motion restricted to Σ_0. They are called the characteristic and bicharacteristic equations, respectively. If one writes $u(x)$ in the form

$$u(x_\mu) = a(x_\mu)\mathrm{e}^{\mathrm{i}W(x_\mu)\nu}$$

with W the phase, a real-valued function, and a the amplitude, ν the frequency, Eq. (3.7.4) provides us with a phase while the amplitude a is evaluated by integration along the solutions of (3.7.5) and (3.7.6).

It is convenient to use the time-dependent formalism directly as developed in section 2.3 to take into account the restriction to Σ_0. We set

$$h \equiv \frac{c^2}{2n^2}\left(k_1^2 + k_2^2 + k_3^2\right) \tag{3.7.7}$$

and find the associated Hamiltonian vector field (see Eq. (2.3.19))

$$\Gamma_h = \frac{\partial}{\partial t} + \frac{1}{n^2}\left(k_1 \frac{\partial}{\partial x} + k_2 \frac{\partial}{\partial y} + k_3 \frac{\partial}{\partial z}\right) + \frac{k_1^2 + k_2^2 + k_3^2}{n^2}\left(\frac{\partial n}{\partial x}\frac{\partial}{\partial k_1} + \frac{\partial n}{\partial y}\frac{\partial}{\partial k_2} + \frac{\partial n}{\partial z}\frac{\partial}{\partial k_3}\right). \tag{3.7.8}$$

If n is independent of time, one looks for solutions of the Hamilton–Jacobi equation of the form

$$W = S(x, y, z) - ct. \tag{3.7.9}$$

In this way one obtains the equation

$$\text{grad}\, S \cdot \text{grad}\, S = n^2, \tag{3.7.10}$$

known as the eikonal equation. For any such solution, the space part of the Hamiltonian vector field on $\mathbf{R}^3$ is given by

$$\Gamma = \frac{1}{n^2}\frac{\partial S}{\partial \vec{x}}\frac{\partial}{\partial \vec{x}}, \tag{3.7.11}$$

having replaced $\vec{k} = \frac{\partial S}{\partial \vec{x}}$, because Γ_h is tangent to the graph of $\mathrm{d}W$ in $\Sigma_0 \subset T^*\mathbf{R}^4$. The congruence of solutions defined by Γ represents the 'rays' associated with the 'wave fronts' in $\mathbf{R}^3$ defined by $S =$ constant.

Remark. Having the Hamiltonian h we can use it to associate an optical Lagrangian and repeat the analysis performed in section 2.8.

3.8 Phase and group velocity

For wave-like phenomena, the surface in space-time defined by the equation (with our notation, $k_0 S$ expresses a function which reduces to the familiar $\vec{k} \cdot \vec{x}$ when the refractive index n equals 1)

$$k_0 S - \omega t = 0 \tag{3.8.1}$$

defines for each t a surface in configuration space. When t varies we obtain a family of surfaces in configuration space parametrized by t, i.e. the wave front for which the location varies in time in agreement with $S = \frac{\omega t}{k_0}$. The constant-phase surfaces in space-time are therefore characterized by the equation

$$k_0 \, \mathrm{d}S - \omega \mathrm{d}t = 0, \tag{3.8.2}$$

i.e.

$$k_0 \frac{\partial S}{\partial \vec{x}} \, \mathrm{d}\vec{x} = \omega \, \mathrm{d}t, \tag{3.8.3}$$

and this implies

$$\frac{\partial S}{\partial \vec{x}} \frac{\mathrm{d}\vec{x}}{\mathrm{d}t} = \frac{\omega}{k_0}, \tag{3.8.4}$$

which yields the *phase velocity*

$$v_f \equiv |\mathrm{d}\vec{x}/\mathrm{d}t| = \frac{\omega}{k_0} \frac{1}{|\mathrm{grad}\, S|} = \frac{\omega}{n k_0}, \tag{3.8.5}$$

because the vectors $\frac{\mathrm{d}\vec{x}}{\mathrm{d}t}$ and $\frac{\partial S}{\partial \vec{x}}$ are parallel (from the Hamilton equations $\frac{\mathrm{d}\vec{x}}{\mathrm{d}t} = \vec{p} = \frac{\partial S}{\partial \vec{x}}$ when evaluated for the particular solution of the Hamilton–Jacobi equation). Such a property holds provided that the Hamiltonian has the form $\frac{p^2}{2m} + V(x)$. It should be stressed that the phase velocity is defined for any S, and its specific value depends on the particular equation by requiring that the graph of

$$\mathrm{d}S : M \rightarrow T^* M$$

should be contained in the sub-space defined by the dispersion relation.

Now consider two solutions Φ_1 and Φ_2 of the wave equation with slightly different frequencies and moving along the x-axis with cosine form, i.e.

$$\Phi_1 = A \cos(\omega_1 t - k_1 x), \tag{3.8.6}$$

$$\Phi_2 = A \cos(\omega_2 t - k_2 x), \tag{3.8.7}$$

where $k_1 \equiv \frac{2\pi}{\lambda_1}$ and $k_2 \equiv \frac{2\pi}{\lambda_2}$. The amplitudes of Φ_1 and Φ_2 are taken to coincide for simplicity. The superposition

$$\Phi \equiv \Phi_1 + \Phi_2 \tag{3.8.8}$$

exhibits some beats with total amplitude varying between 0 and $2A$ (correspondingly, an amplitude modulation is said to occur). On setting

$$\omega_1 \equiv \omega + \varepsilon, \quad \omega_2 \equiv \omega - \varepsilon, \tag{3.8.9}$$

$$k_1 \equiv k + \alpha, \quad k_2 \equiv k - \alpha, \tag{3.8.10}$$

it is possible to express Φ in the form

$$\Phi = 2A \cos(\varepsilon t - \alpha x) \cos(\omega t - k x). \tag{3.8.11}$$

Such a superposition of solutions may therefore be viewed as a plane wave with mean angular frequency ω and wavelength $\lambda = \frac{2\pi}{k}$, but with variable amplitude

$$A' = 2A\cos(\varepsilon t - \alpha x). \tag{3.8.12}$$

This modulated amplitude travels with velocity

$$V = \frac{\varepsilon}{\alpha} = \frac{\omega_1 - \omega_2}{k_1 - k_2} \cong \frac{d\omega}{dk} \equiv v_g, \tag{3.8.13}$$

which is called the *group velocity*.

For a thorough treatment, we should actually consider the wave packet (here $\nu \equiv \frac{\omega}{2\pi}$, $\vec{\kappa} \equiv \frac{\vec{k}}{2\pi}$, and $\vec{r}$ has components $r^1 = x, r^2 = y, r^3 = z$)

$$u(\vec{r},t) = \int_{\mathbf{R}^3} A(\vec{\kappa}) e^{-2\pi\nu i t + 2\pi i(\kappa_x x + \kappa_y y + \kappa_z z)} \, d^3\kappa, \tag{3.8.14}$$

which is a superposition of plane waves with propagation vectors and frequencies very close to each other, i.e. $A(\vec{\kappa})$ is significantly different from zero only around some given value (see below). By construction, at a given time t_0 the wave packet is non-vanishing only in a narrow spatial region around a point with coordinates (x_0, y_0, z_0). On denoting by $\vec{\kappa}^0$ the associated mean value of $\vec{\kappa}$ one can therefore set

$$\kappa_l = \kappa_l^0 + \delta\kappa_l, \quad \forall l = 1, 2, 3, \tag{3.8.15}$$

and the frequency ν can be expressed, to a first approximation, by

$$\nu = \nu^0 + \sum_{l=1}^{3} \frac{\partial \nu}{\partial \kappa_l} \delta\kappa_l, \tag{3.8.16}$$

where the partial derivatives of ν are evaluated at the mean value of κ. Upon making these approximations, the wave packet can be re-expressed in the form (Majorana 1938)

$$u(\vec{r},t) = e^{i\rho} \int_{\mathbf{R}^3} A(\vec{\kappa}) \exp\left[2\pi i \sum_{l=1}^{3} \delta\kappa_l \left(r^l - \frac{\partial \nu}{\partial \kappa_l} t\right)\right] d^3\kappa, \tag{3.8.17}$$

where

$$\rho \equiv -2\pi\nu^0 t + 2\pi \vec{\kappa}^0 \cdot \vec{r}. \tag{3.8.18}$$

The *group velocity* is the vector $\vec{v}$ in $\mathbf{R}^3$ with components

$$v^l \equiv \frac{\partial \nu}{\partial \kappa_l}. \tag{3.8.19}$$

Its meaning is that, up to the phase factor

$$e^{i\varphi} \equiv e^{-2\pi i \nu^0 t + 2\pi i \vec{\kappa}^0 \cdot \tau \vec{v}}, \tag{3.8.20}$$

the vibration u takes at time $t+\tau$ (at the point with position vector $\vec{r}+\vec{v}\tau$) the same value it had at $\vec{r}$ at time t, i.e.

$$u(\vec{r}+\vec{v}\tau, t+\tau) = e^{i\varphi}u(\vec{r}, t). \tag{3.8.21}$$

Since u is assumed to represent a wave packet at time t, which is therefore non-vanishing only as one gets closer and closer to $\vec{r}$, the *group velocity turns out to be the velocity at which the wave packet is effectively moving, in a first approximation, without being deformed.* For the modulus of the group velocity one has, upon taking $\nu = \nu(\kappa_x, \kappa_y, \kappa_z) = \nu(\kappa)$,

$$v = \frac{d\nu}{d\kappa} = \frac{d\omega}{dk}. \tag{3.8.22}$$

This differs from the phase velocity defined in Eq. (3.8.5), and the relation among the two is given by

$$\begin{aligned}\frac{1}{v_g} &= \frac{dk}{d\omega} = \frac{d(nk_0)}{d\omega} \\ &= \frac{d(\omega/v_f)}{d\omega} = \frac{1}{v_f} - \frac{\omega}{v_f^2}\frac{dv_f}{d\omega}.\end{aligned} \tag{3.8.23}$$

This derivation of the group velocity depends specifically on the dynamics, i.e. the surface Σ or the Hamiltonian H or the dispersion relation. The relation between the group velocity and the phase velocity relies on the fact that

$$\text{grad}(dS) \subset \Sigma$$

or that H is constant on the graph of dS.

Remark. We have seen that the phase velocity is associated with constant-phase surfaces in $\mathbf{R}^3$, i.e. (we now set $n=1$ for simplicity)

$$\vec{k}\cdot d\vec{x} - \omega\, dt = dW = 0,$$

implying $\vec{k}\cdot\frac{d\vec{x}}{dt} = \omega$. Here we have used the phase expressed in terms of independent variables $(\vec{x}, t)$, i.e. associated with

$$dW : \mathbf{R}^4 \to T^*\mathbf{R}^4.$$

If we use, instead, $(\vec{k}, \omega) \in (\mathbf{R}^4)^*$ as independent variables, the same sub-manifold in $T^*\mathbf{R}^4$ may be represented as a graph

$$d\widetilde{W} : (\mathbf{R}^4)^* \to T^*(\mathbf{R}^4)^*.$$

The associated constant-phase surface in 'momentum space' reads

$$d\widetilde{W} = \vec{x}\cdot d\vec{k} - t\, d\omega = 0, \tag{3.8.24}$$

which implies that

$$\vec{x} = t \frac{d\omega}{d\vec{k}}. \tag{3.8.25}$$

Thus, *the group velocity* $\frac{d\omega}{d\vec{k}}$ *may also be viewed as the phase velocity in momentum space.*

3.9 The Helmholtz equation

If a partial differential equation is invariant under time translations, it is possible to Fourier transform it with respect to time. Thus, if the wave function u solves the wave equation (3.2.1), its partial Fourier transform with respect to time is defined by

$$\varphi(\vec{x}, \omega) \equiv \frac{1}{\sqrt{2\pi}} \int_{-\infty}^{\infty} u(\vec{x}, t) e^{i\omega t} \, dt, \tag{3.9.1}$$

so that, using an inverse transformation, $u(\vec{x}, t)$ can be expressed as

$$u(\vec{x}, t) = \frac{1}{\sqrt{2\pi}} \int_{-\infty}^{\infty} \varphi(\vec{x}, \omega) e^{-i\omega t} \, d\omega. \tag{3.9.2}$$

The sign $+i\omega$ in Eq. (3.9.1) results from the convention $(-, +, +, +)$ for the Minkowski metric η^{ab}, once we bear in mind that $x_0 = ct, k_0 = \frac{\omega}{c}$, and

$$-ik_b x^b = -i\eta^{cd} k_c x_d = -i\eta^{00} k_0 x_0 - i\eta^{pr} k_p x_r,$$

with p and r ranging from 1 to 3.

Another fundamental equation is obtained by expressing $u(\vec{x}, t)$ according to Eq. (3.9.2) in the wave equation (3.2.1). One then finds

$$\int_{-\infty}^{\infty} \left[\Delta\varphi(\vec{x}, \omega) + \frac{\omega^2}{v^2} \varphi(\vec{x}, \omega) \right] e^{-i\omega t} \, d\omega = 0, \tag{3.9.3}$$

which implies, for the Fourier transform with respect to time,

$$\left(\Delta + \frac{\omega^2}{v^2} \right) \varphi(\vec{x}, \omega) = 0. \tag{3.9.4}$$

Equation (3.9.4) is known as the Helmholtz equation associated with the wave equation (3.2.1). In mathematical language one can say that to the wave equation, which is hyperbolic, one can associate an elliptic equation, i.e. the Helmholtz equation, via a Fourier transform with respect to the time.

If the parameter v depends on the frequency, for each fixed value of ω the plane-wave elementary solutions are still given by Eq. (3.6.1). However, it is no longer true that the resulting wave function is the sum of two functions of $\vec{x} - \vec{v}t$ and $\vec{x} + \vec{v}t$. In physical language, one can say that the wave described by $u(\vec{x},t)$ changes its shape during the propagation. The dispersion relation reads (as is clear from previous discussions, this is connected with the surface Σ in $T^*\mathbf{R}^4$)

$$\omega(k) = \pm k v(\omega(k)), \tag{3.9.5}$$

the crucial difference with respect to Eq. (3.6.3) now relies on the functional dependence of v on $\omega(k)$. More precisely, in such a case it is the single monochromatic component

$$\Phi(\vec{x},\omega,t) \equiv \varphi(\vec{x},\omega)\mathrm{e}^{-\mathrm{i}\omega t} \tag{3.9.6}$$

that obeys the wave equation

$$\left[\triangle - \frac{1}{v^2(\omega)}\frac{\partial^2}{\partial t^2}\right]\Phi(\vec{x},\omega,t) = 0. \tag{3.9.7}$$

This leads to the Helmholtz equation for $\varphi(\vec{x},\omega)$:

$$\left[\triangle + \frac{\omega^2}{v^2(\omega)}\right]\varphi(\vec{x},\omega) = 0. \tag{3.9.8}$$

3.10 Eikonal approximation for the scalar wave equation

If the velocity parameter v depends on both $\vec{x}$ and $\omega(k)$, also the form (3.9.5) of the dispersion relation is lost (in addition to the impossibility of expressing $u(\vec{x},t)$ as the combination of two functions of $\vec{x}-\vec{v}t$ and $\vec{x}+\vec{v}t$). We only know that every monochromatic component obeys the 'wave equation'

$$\left[\triangle - \frac{1}{v^2(\vec{x},\omega)}\frac{\partial^2}{\partial t^2}\right]\Phi(\vec{x},\omega,t) = 0, \tag{3.10.1}$$

with a corresponding Helmholtz equation

$$\left[\triangle + \frac{\omega^2}{v^2(\vec{x},\omega)}\right]\varphi(\vec{x},\omega) = 0. \tag{3.10.2}$$

This analysis will find an important application in section 3.11, where the arguments leading to the fundamental equation of wave mechanics will be presented in detail.

It is of course easier to work with constant-coefficient partial differential operators, and for example the existence of Green functions is guaranteed by a theorem of Malgrange (1955) and Ehrenpreis (1954) (see section IX.5 of Reed and Simon (1975)). For operators with variable coefficients the corresponding analysis is, in general, more difficult (Garabedian 1964). However, if the relative variation of $v(\vec{x},\omega)$ is much smaller than 1 over distances of the order of the wavelength, we may look for approximate solutions of Eq. (3.10.1) in the form

$$\Phi(\vec{x},\omega,t) = A(\vec{x},\omega)\mathrm{e}^{\mathrm{i}F(\vec{x},t;\vec{k},\omega)}, \tag{3.10.3}$$

where F is a real-valued function of its arguments $\vec{x}, t, \vec{k}$ and ω. Of course, Eq. (3.10.3) is the familiar Gauss expression of complex-valued quantities, but the nature of the approximations will be made clear below. By virtue of this ansatz, we can already say that the equation for F will involve the coordinates of the cotangent bundle of $\mathbf{R}^4$. Moreover, we require that the elementary solutions of our partial differential equation with variable coefficients should reduce to

$$A(\omega)\mathrm{e}^{\mathrm{i}[\vec{k}\cdot\vec{x}-\omega(k)t]}$$

when v is a constant. The equation obeyed by $u(\vec{x},t)$ is in general integro-differential (see problem 3.P2). In building the wave packet we look for a fundamental system of solutions so that every solution can be expressed through it. The insertion of the ansatz (3.10.3) into the wave equation (3.10.1) leads to (with C.C. denoting complex conjugation)

$$\begin{aligned}&\left\{A\left[\frac{(\triangle A)}{A} - (\text{grad } F)^2 + \frac{1}{v^2}\left(\frac{\partial F}{\partial t}\right)^2\right]\right.\\ &\quad\left. + \mathrm{i}A\left[2\frac{(\text{grad } F)\cdot(\text{grad } A)}{A} + \triangle F - \frac{1}{v^2}\frac{\partial^2 F}{\partial t^2}\right]\right\}\mathrm{e}^{\mathrm{i}F}\\ &\quad + \text{C.C.} = 0.\end{aligned} \tag{3.10.4}$$

Hereafter we shall assume that the amplitude A is a slowly varying function, in that

$$\frac{|\text{grad } A|}{|A|} \ll \frac{1}{\lambda}, \tag{3.10.5}$$

$$\frac{|\triangle A|}{|A|} \ll \frac{1}{\lambda^2}, \tag{3.10.6}$$

where the spatial variation of the wavelength is taken to obey the restrictions

$$|\text{grad } \lambda| \ll 1, \tag{3.10.7}$$

$$|\lambda(\triangle\lambda)| \ll 1. \tag{3.10.8}$$

One has now to split A into its real and imaginary parts. In particular, if A is assumed to be real-valued one finds eventually the equations

$$\frac{(\triangle A)}{A} - (\text{grad } F)^2 + \frac{1}{v^2}\left(\frac{\partial F}{\partial t}\right)^2 = 0, \tag{3.10.9}$$

$$2\frac{(\text{grad } F)\cdot(\text{grad } A)}{A} + \triangle F - \frac{1}{v^2}\frac{\partial^2 F}{\partial t^2} = 0. \tag{3.10.10}$$

Equations (3.10.9) and (3.10.10) 'replace' the dispersion relation when v also depends on $\vec{x}$ but has a small relative variation in the sense specified above. If $\frac{(\triangle A)}{A}$ satisfies the inequality (3.10.6), these equations reduce to

$$(\text{grad } F)^2 = \frac{1}{v^2(\vec{x},\omega)}\left(\frac{\partial F}{\partial t}\right)^2. \tag{3.10.11}$$

Moreover, if F is a linear function of t, one finds

$$(\text{grad } F)\cdot(\text{grad } A) = -\frac{1}{2}A\,\triangle F. \tag{3.10.12}$$

The form of Eq. (3.10.12) suggests multiplying both sides by A, so that it reads as

$$\frac{1}{2}(\text{grad } F)\cdot(\text{grad } A^2) = -\frac{1}{2}A^2\,\triangle F, \tag{3.10.13}$$

which implies

$$\text{div}\Big(A^2 \text{grad } F\Big) = 0. \tag{3.10.14}$$

Now on taking

$$F = k_0 S(\vec{x},\omega) - \omega(k)t, \tag{3.10.15}$$

Eq. (3.10.11) leads to the *eikonal equation* for S:

$$(\text{grad } S)^2 = \frac{\omega^2}{v^2(\vec{x},\omega)}\frac{1}{k_0^2} = \frac{c^2}{v^2(\vec{x},\omega)} \equiv n^2(\vec{x},\omega). \tag{3.10.16}$$

Furthermore, if A is independent of $\vec{x}$ so that we may choose an F of the form $\vec{k}\cdot\vec{x} - \omega(k)t$, Eq. (3.10.9) reduces to

$$-k^2 + \frac{\omega^2}{v^2} = 0,$$

whereas Eq. (3.10.10) reduces to an identity. Note also that, by virtue of (3.10.15), one finds

$$\text{grad } F = k_0 \text{ grad } S, \tag{3.10.17}$$

and Eq. (3.10.14) may be re-expressed in the form

$$\text{div}\Big(A^2 \text{ grad } S\Big) = 0, \tag{3.10.18}$$

since k_0 is a constant. The quantity $A^2 \,\text{grad}\, S$ can be interpreted as the analogue of the Poynting vector (Jackson 1975) for the scalar field $\psi \equiv A\mathrm{e}^{-ik_0 S}$.

3.10.1 Application: from the eikonal to the Schrödinger equation

We now recall from sections 2.3 the Hamilton–Jacobi equation of classical mechanics:

$$H\left(x, \frac{\partial W}{\partial x}\right) + \frac{\partial W}{\partial t} = 0, \tag{3.10.19}$$

and we focus on the Hamilton principal function $W = S(x, \alpha) - Et$ appropriate for the case when the Hamiltonian is independent of time: $H = \frac{p^2}{2m} + V(x)$. Of course, α is the notation for the set of parameters on which the complete integral of the Hamilton–Jacobi equation depends in this case. Equation (3.10.19) can also be written, in Cartesian coordinates, in the form

$$\frac{1}{2m}\left[\left(\frac{\partial S}{\partial x}\right)^2 + \left(\frac{\partial S}{\partial y}\right)^2 + \left(\frac{\partial S}{\partial z}\right)^2\right] + V = E, \tag{3.10.20}$$

or

$$(\text{grad } S)^2 = 2m(E - V). \tag{3.10.21}$$

On the other hand, we know from the analysis of the scalar 'wave equation' with the velocity parameter depending on both position and frequency that if the amplitude A satisfies the inequality (3.10.6) one finds the eikonal equation (3.10.16), where S is defined by the ansatz (see Eqs. (3.10.3) and (3.10.15))

$$\Phi(\vec{x}, \omega, t) = A(\vec{x}, \omega) e^{i[k_0 S(\vec{x}, \omega) - \omega t]}. \tag{3.10.22}$$

The formal analogy between the eikonal equation and the Hamilton–Jacobi equation for a particle in a potential is one of the properties that suggests a wave mechanics might exist from which classical mechanics can be obtained as a limit, in the same way that geometrical optics can be obtained as a limit from wave optics (see figure 3.1). We are now going to investigate in detail how a corresponding wave mechanics can be constructed. We will see that the construction *is, by no means, unique*, but nevertheless consistent. The key elements of our analysis will be as follows.

(i) It is possible to build a wave packet such that the wave-like properties manifest themselves for distances of the order of atomic dimensions.

(ii) The velocity of the wave packet is the group velocity, and it reduces, in the classical limit, to the velocity of a classical particle.

(iii) The Fourier transform with respect to time of the wave packet obeys a second-order differential equation, if one proceeds by analogy with the properties of geometrical optics. Such an equation yields, in the eikonal approximation, the Hamilton–Jacobi equation.

Wave optics	Wave mechanics
Geometrical optics	Point particle mechanics
Wave packet	Point mass
Pencilsof rays	Collection of trajectories
Group velocity	Velocity
Phase velocity	No simple analogue
Refractive index	Potential
Frequency	Energy

Fig. 3.1 Optics–mechanics analogy.

(iv) The equation for the 'wave function' $\psi(\vec{x},t)$ *can be assumed to be of first order in time.* At this point the analogy with classical wave equation ceases.

The naturally occurring question is what should be the starting point in the attempt to build a theory that describes wave-like properties. As a first step, we consider waves, i.e. oscillating functions of time, by writing $e^{if(x,t)}$, where $f(x,t)$ is expected to be related to Hamilton's principal function W, because $W(x,t)$ is the analogue of the function which characterizes the eikonal approximation. The $f(x,t)$ we are looking for, however, cannot be simply equal to $W(x,t)$, because, by definition, $W = S - Et$, which has dimensions of energy multiplied by time. For f to be dimensionless we have thus to introduce a physical quantity that also has dimensions of energy multiplied by time. The experiments described in chapter 1 suggest identifying this with the Planck constant h. Thus, the oscillating functions of time that we are going to consider are the exponentials (cf. Eq. (2.6.1))

$$e^{i\frac{[S(x,\alpha)-Et]}{h}},$$

which correspond to $e^{i(k_0 S-\omega t)}$ (note that, instead of $k_0 S$ (cf. Eq. (3.10.15)), we now consider only S). With our notation, α denotes the triple of parameters $\alpha_1, \alpha_2, \alpha_3$, with $\alpha_3 \equiv E$, and corresponding integration measure $d^3\alpha \equiv d\alpha_1 \, d\alpha_2 \, d\alpha_3 = d\alpha_1 \, d\alpha_2 \, dE$ (see below). Since we are aiming to build a wave-like theory, we use the functions written above to build the wave packet

$$\psi(x,t) \equiv \int A(\alpha) e^{i \frac{[S(x,\alpha) - Et]}{\hbar}} \, d^3\alpha. \tag{3.10.23a}$$

A fundamental remark is now in order: we are considering as fundamental quantities those of wave-like type, and hence we are going to build a purely wave-like theory, where for the time being the concept of energy does not occur. What is meaningful is the ratio $\frac{E}{h}$, which, for dimensional reasons, is identified with a frequency ν. Only after showing that, in the 'classical limit', Eq. (3.10.23a) describes the motion of a particle, can the quantity E be interpreted as the energy of the particle. Thus, *we are not studying a system with a definite energy, but rather a system with a given set of frequencies.* In other words, what is meaningful for us are wave-like quantities, which, in a suitable classical limit, are reduced to classical quantities. The fact that, in this classical limit, we want to recover the Hamilton–Jacobi equation *does not mean* that E admits immediately to a classical interpretation. Our limit in principle will give equations on the optical phase space. Only from the Einstein–de Broglie relation can the identification with 'particle attributes' be made.

Having stressed this crucial point we remark that, by virtue of our notation for the α parameters, the wave packet can be written in the form (cf. Schrödinger in Ludwig (1968))

$$\psi(x,t) = \int A(\alpha_1, \alpha_2, E) e^{i \frac{[S(x,\alpha_1,\alpha_2,E) - Et]}{\hbar}} \, d\alpha_1 \, d\alpha_2 \, dE, \tag{3.10.23b}$$

where S is a solution of the Hamilton–Jacobi equation for fixed values of E, α_1 and α_2. Hereafter, whenever we mention the mean value of E and of the wavelength, we mean what follows. The original wave packet is obtained by integration over a continuous set of values of the parameter E, where the integrand contains a function of E for which the values differ substantially from 0 only in the neighbourhood of a certain value E'. The corresponding *mean frequency* of the wave packet is

$$\langle \nu \rangle \equiv \frac{E'}{h} \equiv \frac{\langle E \rangle}{h},$$

where the concept of mean frequency is actually already familiar from the definition of group velocity in section 3.8. We thus want a certain mean value $\langle E \rangle$ to equal the product of h with a certain mean frequency, but some checks of the orders of magnitude are in order, before we can claim that, in the classical limit, such a relation holds. The order of magnitude of h is such that, in the theory we are going to construct, the wave-like properties manifest themselves at a microscopic level, but disappear at a macroscopic level, where the laws of classical mechanics provide the appropriate description of physical phenomena. This is in full agreement with what we expect from requiring that the Einstein–de Broglie relation should hold.

For the time being, we have considered the integral (3.10.23b), where S solves the equation (3.10.21). From these assumptions one finds that the W = constant surfaces (which are defined on space-time) move with velocity $\frac{E}{\sqrt{2m(E-V)}}$, which is the *phase velocity* v_f. This occurs because on such surfaces one has

$$\mathrm{d}S - E\,\mathrm{d}t = 0,$$

which implies

$$\operatorname{grad} S \cdot \frac{\mathrm{d}\vec{x}}{\mathrm{d}t} = E.$$

If $\frac{\mathrm{d}\vec{x}}{\mathrm{d}t}$ is orthogonal to the surface σ, it is parallel to the gradient of S, and hence

$$v_f \equiv \left|\frac{\mathrm{d}\vec{x}}{\mathrm{d}t}\right| = \frac{E}{|\operatorname{grad} S|} = \frac{E}{\sqrt{2m(E-V)}}. \tag{3.10.24}$$

This makes it possible to define the de Broglie wavelength

$$\lambda \equiv \frac{v_f}{\nu} = \frac{h}{\sqrt{2m(E-V)}}. \tag{3.10.25}$$

Such a λ is a wave-like quantity that is independent of the lowest value of the parameter E, since it depends on $(E-V)$. In the classical limit, one thus finds the mean value of the wavelength

$$\langle \lambda(x) \rangle = \frac{h}{\sqrt{2m(\langle E \rangle - V)}}. \tag{3.10.26}$$

After this, a second check is in order, i.e. whether the group velocity $v_{\rm g}$ turns out to be equal, in the classical limit, to the velocity of the particle. For this purpose, recall from section 3.8 that

$$\begin{aligned} v_{\rm g} &= \frac{\mathrm{d}\omega}{\mathrm{d}k} = \frac{\mathrm{d}2\pi\nu}{\mathrm{d}(2\pi/\lambda)} = \frac{\mathrm{d}\nu}{\mathrm{d}(1/\lambda)} = \frac{\mathrm{d}\nu}{\mathrm{d}(\nu/v_f)} \\ &= h\frac{\mathrm{d}\nu}{\mathrm{d}\sqrt{2m(h\nu-V)}} = h\frac{1}{\frac{\mathrm{d}}{\mathrm{d}\nu}\sqrt{2m(h\nu-V)}} \\ &= \frac{\sqrt{2m(h\nu-V)}}{m} = \frac{\sqrt{2m(E-V)}}{m}. \end{aligned} \tag{3.10.27}$$

In the classical limit, the right-hand side of Eq. (3.10.27) becomes the ratio $\frac{\sqrt{2m(\langle E \rangle - V)}}{m}$.

Now, instead of considering any value of E over which we integrate in Eq. (3.10.23*b*), we consider the maximum of the wave packet, and hence we compute all quantities for the value $E = E_{\rm max} = \langle E \rangle$; the velocity of the wave packet is the group velocity, and is related to the average value $\langle E \rangle$ in the same way in which the velocity of the classical particle is related to the energy of such a particle. This is the precise meaning of Eq. (3.10.27). In the classical limit, Eq. (3.10.27) therefore shows that the group velocity reduces to the velocity of a classical particle, because

$$\langle E \rangle - V = \frac{m}{2} v^2.$$

We have thus obtained a wave-like perturbation localized in a region, and which varies in time according to a classical trajectory. It is already a non-trivial achievement having shown that one can 'mimic' the motion of a classical particle by means of the motion of a wave (it is as if the classical particle were being turned into a wave; only our inability to understand the nature of the phenomena led us to think of particles instead of waves). On the other hand, although we have shown that a wave group exists which follows the trajectory of classical particles, we have not yet derived the wave equation we were looking for (we have, indeed, only the 'would-be solutions' of such an equation). For this purpose, let us now recall that, for the scalar wave equation, the following Helmholtz equation holds for the Fourier transform with respect to time (cf. Eq. (3.9.4)):

$$\left(\triangle + k_0^2 n^2 \right) \varphi(\vec{x}, \omega) = 0, \tag{3.10.28}$$

which can be re-expressed as

$$\left(\triangle + \frac{4\pi^2 \nu^2}{v_f^2} \right) \varphi(\vec{x}, \omega) = 0. \tag{3.10.29}$$

But we know, from the previous analysis, that

$$v_f = \frac{h\nu}{\sqrt{2m(E - V)}}. \tag{3.10.30}$$

Thus, the wave equation we are looking for is (cf. Fermi 1965)

$$\left[\triangle + \frac{8\pi^2 m(E - V)}{h^2} \right] \varphi(\vec{x}, \omega) = 0. \tag{3.10.31}$$

This is, by construction, an equation for the Fourier transform φ. On setting $\hbar \equiv \frac{h}{2\pi}$, it reads, eventually, as

$$\left(-\frac{\hbar^2}{2m} \triangle + V \right) \varphi(\vec{x}, \omega) = E \varphi(\vec{x}, \omega). \tag{3.10.32}$$

It should be stressed, however, that *this is not the only differential equation which, in the eikonal approximation, yields the Hamilton–Jacobi equation.*

The problem now arises of finding an equation for $\psi(\vec{x}, t)$. Indeed, the Helmholtz equation (3.9.4), which is a second-order partial differential equation for the Fourier transform with respect to time of the wave function, leads, in the eikonal approximation, to Eq. (3.10.16), which is formally analogous to the Hamilton–Jacobi equation (3.10.21). Thus, if a wave mechanics exists, leading to Eq. (3.10.21) in the eikonal approximation, we find for the Fourier transform with respect to time, defined as

$$\varphi(\vec{x}, \omega) \equiv \frac{1}{\sqrt{2\pi}} \int_{-\infty}^{\infty} \psi(\vec{x}, t) \mathrm{e}^{\mathrm{i}\omega t} \, \mathrm{d}t, \tag{3.10.33}$$

the second-order equation (3.10.32). The very existence of the transform (3.10.33) is a non-trivial assumption that we need to derive a time-dependent equation for $\psi(\vec{x}, t)$. At this stage we exploit the identity

$$\int_{-\infty}^{\infty} \frac{\partial \psi}{\partial t} \mathrm{e}^{\mathrm{i}\omega t} \, \mathrm{d}t = -\mathrm{i}\omega \int_{-\infty}^{\infty} \psi(\vec{x}, t) \mathrm{e}^{\mathrm{i}\omega t} \, \mathrm{d}t. \tag{3.10.34}$$

By virtue of (3.10.33) and (3.10.34), and of the identification $E = \hbar\omega$ from the Einstein–de Broglie relation, Eq. (3.10.32) leads to

$$\int_{-\infty}^{\infty} \left(-\frac{\hbar^2}{2m} \triangle + V \right) \psi(\vec{x}, t) \mathrm{e}^{\mathrm{i}\omega t} \, \mathrm{d}t = \int_{-\infty}^{\infty} \mathrm{i}\hbar \frac{\partial \psi}{\partial t} \mathrm{e}^{\mathrm{i}\omega t} \, \mathrm{d}t, \tag{3.10.35}$$

and hence 'one can' assume that the desired 'wave equation' has the form

$$\mathrm{i}\hbar \frac{\partial}{\partial t} \psi(\vec{x}, t) = H\psi(\vec{x}, t). \tag{3.10.36}$$

The first-order derivative with respect to t results from the linearity in E of Eq. (3.10.32). At the same time one can derive another equation for the complex conjugate wave function:

$$-\mathrm{i}\hbar \frac{\partial}{\partial t} \psi^*(\vec{x}, t) = H\psi^*(\vec{x}, t). \tag{3.10.37}$$

Since we have been led by formal analogy (and physical considerations), starting from a second-order equation for the Fourier transform with respect to time, no uniqueness theorem for the wave equation solved by $\psi(\vec{x}, t)$ can be proved on purely mathematical grounds (see problem 4.P1). Note also that Eq. (3.10.36) is an equation *which is of first order in the time variable and has a complex coefficient.* These properties were not familiar when Schrödinger tried to formulate the quantization as an eigenvalue problem (motivated by his familiarity with the theory of vibrating bodies and partial differential equations), and indeed he first proposed a 'wave equation' with real coefficients (see problem 4.P1). The price to be paid, however, was the occurrence of fourth-order derivatives with respect to the spatial variables (cf. the theory of elastic solids), and this property was undesirable from the point of view of the initial-value problem (Schrödinger 1977).

3.11 Problems

3.P1. Suppose one studies the wave equation for $u(x,t)$ in one spatial dimension with vanishing Cauchy data at $t=0$ for $x>0$, and knowing that $u(0,t)=\phi(t)$ for $t>0$. Find the form of $u(x,t)$ when both x and t are positive.

3.P2. If the single monochromatic component obeys the 'wave equation' (3.10.1), prove that $u(\vec{x},t)$ obeys the integro-differential equation

$$\Delta u(\vec{x},t) + \frac{1}{2\pi}\int_{-\infty}^{\infty} \mathrm{d}\omega \, \frac{\omega^2}{v^2(\vec{x},\omega)} \int_{-\infty}^{\infty} \mathrm{d}\tau \, u(\vec{x},\tau) \mathrm{e}^{\mathrm{i}\omega(\tau-t)} = 0. \tag{3.11.1}$$

Thus, at a deeper level, wave-like phenomena in material media are associated to the analysis of integro-differential operators (Maslov and Fedoriuk 1981).

3.P3. A candle emits light that is received by a screen of area 1 cm^2 located at a distance of 3 m. Find the distribution of energy on the surface of the screen.

4
Wave mechanics

This chapter begins by exploiting the Einstein–de Broglie relation and the notion of symbol, as defined in Eq. (3.6.6). The following steps are the local and global conservation laws associated to the Schrödinger equation, the probabilistic interpretation of the wave function, the spreading of wave packets and transformation properties of wave functions.

The solution of the Schrödinger equation for a given initial condition is then studied. For this purpose, one first expands the initial condition in terms of eigenfunctions of the Hamiltonian operator. This is then 'propagated' to determine the evolution of the wave function. Thus, one is led to consider the Green kernel of the Schrödinger equation, and the boundary conditions on the wave function that lead to complete knowledge of stationary states. A dynamics involving an isometric non-unitary operator is also studied. The chapter ends with an elementary introduction to harmonic oscillators and to the JWKB method in wave mechanics, plus examples on the Bohr–Sommerfeld quantization.

4.1 From classical to wave mechanics

The correspondence between differential operators and polynomials in k relies on the notion of the symbol defined in Eq. (3.6.6). Now we are aiming to pass from a function on phase space, i.e. the classical Hamiltonian, to an operator which, by analogy, is called the Hamiltonian operator. Such a transition is non-trivial because the map from symbols to operators contains some ambiguity. Such a transition is always possible if polynomials in p have constant coefficients. When this is not the case, i.e. if the coefficients are functions on the configuration space, there is an obvious ambiguity in the association of an operator with xp or px, for instance. A first way out would be to consider ordered polynomials, i.e. all p should occur to the right before a differential operator is associated with it. In this way we would obtain an isomorphism of vector spaces between ordered polynomials of a given degree and differential operators of a given order. However the correspondence, even though one-to-one, will not be an algebra isomorphism. The pointwise product of polynomials is commutative while the operator product is not commutative (indeed this property is responsible for the uncertainty relations as treated in section 4.2).

If we use the Einstein–de Broglie identification in the form

$$\vec{p} \cdot {\rm d}\vec{x} - H \, {\rm d}t = \hbar(\vec{k} \cdot {\rm d}\vec{x} - \omega \, {\rm d}t),$$

the symbol of an operator reads, in general, as

$$\sigma({\rm D}) = {\rm e}^{-\frac{{\rm i}}{\hbar}(\vec{p}\cdot\vec{x}-Et)} \, {\rm D} \, {\rm e}^{\frac{{\rm i}}{\hbar}(\vec{p}\cdot\vec{x}-Et)}, \tag{4.1.1}$$

which yields

$$\sigma\left(\frac{\partial}{\partial t}\right) = -\frac{{\rm i}}{\hbar}E, \tag{4.1.2}$$

and

$$\sigma\left(\frac{\partial}{\partial \vec{x}}\right) = \frac{{\rm i}}{\hbar}\vec{p}. \tag{4.1.3}$$

These properties justify the associations

$$E \rightarrow {\rm i}\hbar\frac{\partial}{\partial t}, \quad \vec{p} \rightarrow -{\rm i}\hbar\frac{\partial}{\partial \vec{x}},$$

already encountered in section 3.6. Moreover, by virtue of the definition (3.6.6), the symbol of a multiplication operator is the operator itself, viewed as a function. We therefore write, using a hat to denote operators,

$$V(\vec{x}) \rightarrow \hat{V}(\vec{x}), \quad x^i \rightarrow \hat{x}^i,$$

and we associate to the classical Hamiltonian

$$H = \frac{p^2}{2m} + V(\vec{x})$$

the Hamiltonian operator

$$\frac{1}{2m}\left(-{\rm i}\hbar\frac{\partial}{\partial \vec{x}}\right)^2 + \hat{V}(\vec{x}).$$

Thus, since classically $\frac{p^2}{2m} + V(x) = E$, we eventually obtain the partial differential equation

$${\rm i}\hbar\frac{\partial \psi}{\partial t} = -\frac{\hbar^2}{2m}\frac{\partial^2 \psi}{\partial \vec{x}^2} + \hat{V}(\vec{x})\psi, \tag{4.1.4}$$

bearing in mind that operator equations like ours can be turned into partial differential equations, once the operators therein are viewed as acting on suitable 'functions'. The passage from the classical Hamiltonian to the Hamiltonian operator is called 'quantization', and is the counterpart of the quantization rules of the old quantum theory mentioned in chapter 1. The symbol map, however, is severely affected by the frame of reference. For example, the symbol map is invertible if the coefficients of the differential operator D are constant, but on changing coordinates, in general one obtains a variable-coefficient operator, so that inversion of the symbol map is no longer possible. We are therefore *forced to quantize in Cartesian coordinates*, although nothing prevents us from changing coordinates once we are dealing with differential operators.

Remark. The Schrödinger equation (4.1.4) just obtained belongs formally to the same scheme as the heat equation. However, since the real constant of thermal conductivity occurring therein is replaced here by the imaginary constant $\frac{i\hbar}{2m}$, the Schrödinger equation describes oscillations rather than equalization processes. One can see this more clearly on passing to the Helmholtz equation. Then the 'stationary' Schrödinger equation becomes, in the free case,

$$\left(\triangle + \frac{2mE}{\hbar^2}\right)\varphi = 0,$$

which has the same form as what we would obtain from the wave equation on letting $\frac{2mE}{\hbar^2}$ be replaced by $\frac{\omega^2}{c^2}$ (cf. Eq. (3.10.2)).

The ground is now ready for the investigation of all properties of a quantum theory based on a wave equation, and this is the object of the present sections. Here we are going to show that, in wave mechanics, the wave function obeys a local and a corresponding, global conservation law. This result leads to the consideration of square-integrable functions on $\mathbf{R}^3$, and to the physical interpretation of the wave function. Some key properties of wave packets are eventually studied.

4.1.1 Continuity equation

Now we consider the Schrödinger equation (4.1.4) and its complex conjugate equation which is automatically satisfied:

$$-i\hbar\frac{\partial \psi^*}{\partial t} = \left(-\frac{\hbar^2}{2m}\triangle + V\right)\psi^*(\vec{x},t), \tag{4.1.5}$$

and we multiply (4.1.4) by $\psi^*(\vec{x},t)$, and (4.1.5) by $\psi(\vec{x},t)$. On subtracting one equation from the other one then finds

$$-\frac{\hbar^2}{2m}\left(\psi^*\triangle\psi - \psi\triangle\psi^*\right) = i\hbar\left(\psi^*\frac{\partial\psi}{\partial t} + \psi\frac{\partial\psi^*}{\partial t}\right). \tag{4.1.6}$$

Interestingly, on defining

$$\vec{j} \equiv \frac{\hbar}{2mi}\Big[\psi^* \text{grad}(\psi) - \psi \,\text{grad}(\psi^*)\Big], \tag{4.1.7}$$

and $\rho \equiv \psi^* \psi$, Eq. (4.1.6) takes the form of the *continuity equation* for a current:

$$\frac{\partial \rho}{\partial t} + \text{div}(\vec{j}) = 0. \tag{4.1.8a}$$

This is a *local conservation law.* If one integrates the continuity equation on a volume $\mathcal{V}$, the divergence theorem yields

$$\begin{aligned}\int_{\mathcal{V}} \frac{\partial}{\partial t}\psi^* \psi \, \mathrm{d}^3x &= \frac{\partial}{\partial t}\int_{\mathcal{V}} \psi^* \psi \, \mathrm{d}^3x \\ &= \frac{\mathrm{d}}{\mathrm{d}t}\int_{\mathcal{V}} \psi^* \psi \, \mathrm{d}^3x \\ &= -\frac{\hbar}{2mi}\int_{\sigma}\left(\psi^* \frac{\partial \psi}{\partial n} - \psi \frac{\partial \psi^*}{\partial n}\right) \mathrm{d}\sigma, \end{aligned} \tag{4.1.8b}$$

where σ is the boundary surface of $\mathcal{V}$ and $\frac{\partial}{\partial n}$ denotes differentiation along the direction normal to σ. Thus, if ψ vanishes in a sufficiently rapid way as $|\vec{x}| \rightarrow \infty$, and if its first derivatives remain bounded in that limit, the integral on the right-hand side of (4.1.8*b*) vanishes when the surface σ is pushed off to infinity. The volume $\mathcal{V}$ extends then to the whole of $\mathbf{R}^3$, and one finds the *global conservation property*

$$\frac{\mathrm{d}}{\mathrm{d}t}\int_{\mathbf{R}^3} \psi^* \psi \, \mathrm{d}^3x = 0,$$

which shows that the integral of $|\psi|^2$ over the whole space is independent of t. By virtue of the linearity of the Schrödinger equation, if ψ is square-integrable, one can then rescale the wave function so that the integral is set to 1:

$$\int_{\mathbf{R}^3} \psi^* \psi \, \mathrm{d}^3x = 1.$$

It is therefore clear that we are considering a theory for which the wave functions belong to the space of square-integrable functions on $\mathbf{R}^3$. With the notation of appendix 4.A, this is a Hilbert space. The probabilistic interpretation of ψ, presented in subsection 4.1.2, is then tenable.

For a wave function that is square-integrable on $\mathbf{R}^3$, one can write

$$\psi(\vec{x}, t) = (2\pi\hbar)^{-3/2}\int_{\mathbf{R}^3} A(\vec{p}) \mathrm{e}^{\mathrm{i}(\vec{p}\cdot\vec{x} - Et)/\hbar} \, \mathrm{d}^3p, \tag{4.1.9a}$$

and its complex conjugate is

$$\psi^*(\vec{x}, t) = (2\pi\hbar)^{-3/2}\int_{\mathbf{R}^3} A^*(\vec{p}) \mathrm{e}^{-\mathrm{i}(\vec{p}\cdot\vec{x} - Et)/\hbar} \, \mathrm{d}^3p, \tag{4.1.10a}$$

where $E = E(p)$. Now by setting

$$\varphi(\vec{p},t) \equiv A(\vec{p})\mathrm{e}^{-\mathrm{i}Et/\hbar} \tag{4.1.11}$$

one can re-express the above formulae as

$$\psi(\vec{x},t) = (2\pi\hbar)^{-3/2} \int_{\mathbf{R}^3} \varphi(\vec{p},t)\mathrm{e}^{\mathrm{i}\vec{p}\cdot\vec{x}/\hbar}\,\mathrm{d}^3p, \tag{4.1.9b}$$

$$\psi^*(\vec{x},t) = (2\pi\hbar)^{-3/2} \int_{\mathbf{R}^3} \varphi^*(\vec{p},t)\mathrm{e}^{-\mathrm{i}\vec{p}\cdot\vec{x}/\hbar}\,\mathrm{d}^3p. \tag{4.1.10b}$$

These equations imply that the spatial Fourier transform of the wave function can be expressed as

$$\varphi(\vec{p},t) = (2\pi\hbar)^{-3/2} \int_{\mathbf{R}^3} \psi(\vec{x},t)\mathrm{e}^{-\mathrm{i}\vec{p}\cdot\vec{x}/\hbar}\,\mathrm{d}^3x, \tag{4.1.12}$$

and hence, from (4.1.11),

$$A(\vec{p}) = (2\pi\hbar)^{-3/2} \int_{\mathbf{R}^3} \psi(\vec{x},t)\mathrm{e}^{-\mathrm{i}(\vec{p}\cdot\vec{x}-Et)/\hbar}\,\mathrm{d}^3x. \tag{4.1.13}$$

Moreover, the following relation, known as the Parseval lemma (Hörmander 1983), holds:

$$\int_{\mathbf{R}^3} \psi^*\psi\,\mathrm{d}^3x = \int_{\mathbf{R}^3} \varphi^*\varphi\,\mathrm{d}^3p = \int_{\mathbf{R}^3} A^*A\,\mathrm{d}^3p. \tag{4.1.14}$$

It should be emphasized that, on passing from the x- to the p-integration via a Fourier transform, the position operator becomes a first-order differential operator:

$$x : \varphi \rightarrow \mathrm{i}\hbar\frac{\partial\varphi}{\partial p},$$

whereas the momentum operator acts in a multiplicative way:

$$p : \varphi \rightarrow p\,\varphi.$$

These equations define the *momentum representation* (see figure 4.1), which is the wave-mechanical counterpart of the well-known transformation in classical mechanics:

$$(p,q) \rightarrow (P = -q, Q = p),$$

i.e. a canonical transformation because of the identity

$$p\,\mathrm{d}q = -q\,\mathrm{d}p + \mathrm{d}(pq).$$

The momentum representation amounts to considering as 'independent variables', to express our operators, the momenta p instead of the positions q. This alternative option looks rather natural if one bears in mind our subsection 3.6.1, which implies that symbols can be polynomials in the q or p variables depending on which basis is chosen for the cotangent bundle and hence for square-integrable functions on the dual of $\mathbf{R}^4$ or on $\mathbf{R}^4$ itself.

4.1.2 Physical interpretation of the wave function

Let us now consider the physical meaning which can be attributed to $\psi(\vec{x}, t)$. Born's proposal was to regard

$$\rho(\vec{x}, t)\, \mathrm{d}^3x = \psi^*(\vec{x}, t)\psi(\vec{x}, t)\, \mathrm{d}^3x \tag{4.1.15}$$

as a probability measure. In other words, this quantity should represent the probability of observing a particle at time t within the volume element

Classical dynamical variable	Coordinate repr.	Momentum repr.
E	$\mathrm{i}\hbar\frac{\partial}{\partial t}$	E
x	x	$\mathrm{i}\hbar\frac{\partial}{\partial p}$
p	$-\mathrm{i}\hbar\frac{\partial}{\partial x}$	p
x^n	x^n	$(\mathrm{i}\hbar)^n\frac{\partial^n}{\partial p^n}$
p^n	$(-\mathrm{i}\hbar)^n\frac{\partial^n}{\partial x^n}$	p^n
$\frac{p^2}{2m}$	$-\frac{\hbar^2}{2m}\triangle$	$\frac{p^2}{2m}$

Fig. 4.1 Coordinate vs. momentum representation.

d^3x around the point for which the position vector is $\vec{x}$. Similarly, the quantity

$$\rho(\vec{p}, t)\, d^3p = \varphi^*(\vec{p}, t)\varphi(\vec{p}, t)\, d^3p \tag{4.1.16}$$

represents the probability of observing a particle at time t within the 'volume element' d^3p around the point with momentum $\vec{p}$. This interpretation is possible if ψ is square-integrable and normalized. Born was led to his postulate by comparing how the scattering of a particle by a potential is described in classical and quantum mechanics. For example, if an electron is allowed to interact with a short-range potential, and if a screen is placed at a distance from the interaction region of a few metres, the electron is always detected at a fixed point on the screen. On repeating the experiment, the electron is detected at a different point, and so on. After several experiments, the fraction of the number of times the electron is detected at $\vec{x}$ at time t is proportional to $|\psi(\vec{x}, t)|^2$, which is therefore the probability density of such an event. Remarkably, the electron exhibits both a corpuscular and a wave-like nature by virtue of the probabilistic interpretation of its wave function. This means that the 'quantum electron' does not coincide with the notion of a 'classical electron' we may have developed in a classical framework.

We will entertain the idea that, in quantum mechanics, *no elementary phenomenon is a phenomenon unless it is registered* (Wheeler and Zurek 1983). For example, it would be meaningless to say that a particle passed through a hole in a wall unless one has a measuring device which provides evidence that it did so. Ultimately, we have to give up the attempt to describe the particle motion as if we could use what we are familiar with from everyday experience. We have instead to limit ourselves to an abstract mathematical description, which makes it possible to extract information *from the experiments that we are able to perform.* This crucial point (which is still receiving careful consideration in the current literature) is well emphasized by the interference from a double slit. Now, in the light of the previous paragraphs on the properties and meaning of the wave function, we are in a position to perform a more careful investigation, which should be compared with section 1.5.

Suppose that a beam of electrons emitted from the source S, and having a sufficiently well-defined energy, hits a screen Σ_1 with two slits F_1 and F_2. A second screen Σ_2 is placed thereafter (see figure 4.2). To each electron of the beam one can associate a wave which is partially reflected from Σ_1 and partially diffracted through the two slits. Now let ψ_1 and ψ_2 represent the two diffracted waves. The full wave in the neighbourhood

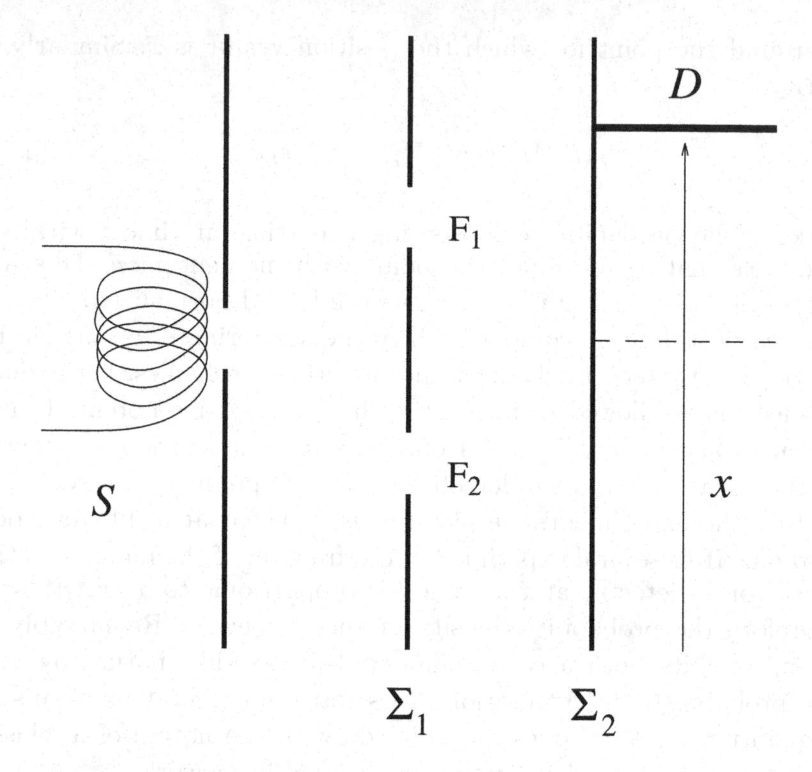

Fig. 4.2 Double-slit experiment with two screens.

of Σ_2 is given by

$$\psi(\vec{x},t) = \psi_1(\vec{x},t) + \psi_2(\vec{x},t). \tag{4.1.17}$$

Thus, bearing in mind what one knows from the analysis of the continuity equation, one can say that the probability that a detector D placed on Σ_2 detects, in a given time interval, the particle, is given by

$$\begin{aligned} \vec{j}\cdot\vec{n}\,\mathrm{d}\sigma &= \frac{\hbar}{2mi}\Big[\psi^*\mathrm{grad}(\psi) - \psi\,\mathrm{grad}(\psi^*)\Big]\cdot\vec{n}\,\mathrm{d}\sigma \\ &= \frac{\hbar}{2mi}\Big\{\Big[\psi_1^* + \psi_2^*\Big]\mathrm{grad}\Big[\psi_1 + \psi_2\Big] \\ &\quad -\Big[\psi_1 + \psi_2\Big]\mathrm{grad}\Big[\psi_1^* + \psi_2^*\Big]\Big\}\cdot\vec{n}\,\mathrm{d}\sigma, \end{aligned} \tag{4.1.18}$$

where, according to a standard notation, $\vec{n}$ is the normal to Σ_2 and $\mathrm{d}\sigma$ denotes the section of the detector, which is viewed as being 'infinitesimal'. Since all electrons are (essentially) under the same conditions, the temporal average of Eq. (4.1.18) represents the effective flux of electrons through D per unit time. As the position of the detector on the screen varies, the observed flux is expected to exhibit the sequence of maxima and minima, which is typical of an interference process.

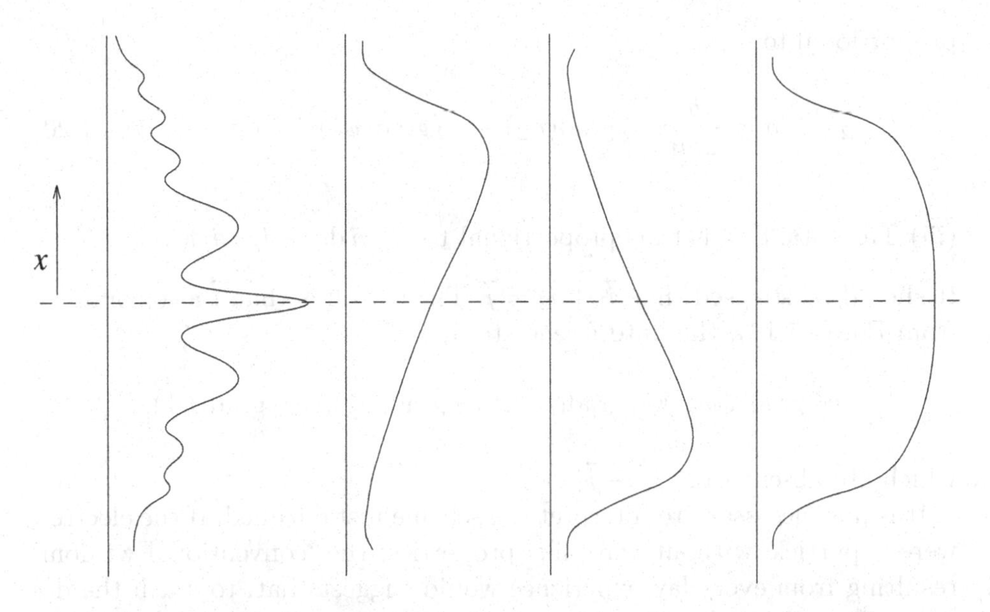

Fig. 4.3. Probability distribution in a Young experiment in the four cases: (i) no detector is placed; (ii) a detector in front of slit 1; (iii) a detector in front of slit 2; (iv) detectors in front of both slits.

However, one may want to supplement the experimental apparatus, by inserting yet more detectors, C_1 and C_2, in front of the slits F_1 and F_2, respectively. This is a highly non-trivial point: taking into account the dimensions of the slits F_1 and F_2, one sees that it is extremely difficult to *build* detectors that are able to distinguish particles passing through F_1 from particles passing through F_2. Thus, from now on, we are describing what is, strictly, a *gedanken experiment*, i.e. a conceptual construction which is consistent with all the rules of the theory, but whose actual implementation is very difficult (see, however, the encouraging progress described in Scully *et al.* (1991)). With this understanding, we can now distinguish the following cases (see figure 4.3).

(i) Some particles interact with the detector C_1, and hence we can say that they passed through the slit F_1. Their flux is proportional to

$$\vec{j}_1 \cdot \vec{n}\, d\sigma = \frac{\hbar}{2mi}\Big[\psi_1^* \, \mathrm{grad}(\psi_1) - \psi_1 \, \mathrm{grad}(\psi_1^*)\Big] \cdot \vec{n}\, d\sigma. \qquad (4.1.19)$$

(ii) The other particles interact with the detector C_2, and hence it is legitimate to say that they passed through the slit F_2. Their flux is proportional to

$$\vec{j}_2 \cdot \vec{n}\, d\sigma = \frac{\hbar}{2mi}\Big[\psi_2^* \,\mathrm{grad}(\psi_2) - \psi_2 \,\mathrm{grad}(\psi_2^*)\Big] \cdot \vec{n}\, d\sigma. \qquad (4.1.20)$$

(iii) The total flux is thus proportional to $\vec{j}_1 \cdot \vec{n}\, d\sigma + \vec{j}_2 \cdot \vec{n}\, d\sigma$.

It should be stressed that $\vec{j}_1 + \vec{j}_2 \neq \vec{j}$. The reason is that $\vec{j}$ also contains, from Eq. (4.1.18), the 'interference terms'

$$\psi_1^* \,\mathrm{grad}(\psi_2), \quad \psi_2^* \,\mathrm{grad}(\psi_1), \quad \psi_1 \,\mathrm{grad}(\psi_2^*), \quad \psi_2 \,\mathrm{grad}(\psi_1^*),$$

which are absent from $\vec{j}_1 + \vec{j}_2$.

It is now necessary to interpret the scheme just outlined. If the electron were a particle without wave-like properties, the 'conventional wisdom' resulting from everyday experience would suggest that, to reach the detector D on Σ_2, it should always pass through F_1 or F_2. This would imply, in turn, that the total flux is always proportional to $\left(\vec{j}_1 \cdot \vec{n} + \vec{j}_2 \cdot \vec{n}\right) d\sigma$, *whether or not* the detectors C_1 and C_2 are placed in front of the two slits. However, this conclusion contradicts all that is known from diffraction and interference experiments on electrons. The crucial point is, as we stated before, that one can only say that the electron passed through F_1 or F_2 upon placing the detectors C_1 and C_2. Thus, the particle picture, without any reference to wave-like properties, is inconsistent (for a related discussion, see Sudarshan and Rothman (1991)).

On the other hand, the purely wave-like description is also inappropriate, because it disagrees with the property of the detectors of being able to register the passage of only one electron at a time. In other words, when the detectors C_1 and C_2 are placed in front of F_1 and F_2, *only one of them* (*either* C_1 *or* C_2) provides evidence of the passage of the electron. Moreover, if the single detector D on the screen Σ_2 is replaced by a set of detectors distributed all over the surface of Σ_2, only one detector at a time is able to indicate the passage of electrons, and this phenomenon occurs in a random way. The statistical distribution of countings, however, agrees with the predictions obtained from the Schrödinger equation. Furthermore, the current of electrons can be reduced so that no more than one electron is present in the apparatus at any time, thus showing that an electron interferes with itself.

For recent theoretical developments on interference experiments, we refer the reader to the work in Bimonte and Musto (2003a,b) and references therein.

4.1.3 Mean values

The probabilistic interpretation of the wave function ψ makes it possible to define mean values, and hence the formalism developed so far leads to a number of simple properties, which can be tested against observation. For example, one can define the mean values of position and momentum operators,

$$\langle x_l \rangle \equiv \int_{\mathbf{R}^3} \psi^* x_l \psi \, \mathrm{d}^3x, \tag{4.1.21}$$

$$\langle p_l \rangle \equiv \int_{\mathbf{R}^3} \varphi^* p_l \varphi \, \mathrm{d}^3p. \tag{4.1.22}$$

The corresponding standard deviations can also be defined, i.e.

$$\langle (\triangle x_l)^2 \rangle \equiv \int_{\mathbf{R}^3} \psi^* (x_l - \langle x_l \rangle)^2 \psi \, \mathrm{d}^3x, \tag{4.1.23}$$

$$\langle (\triangle p_l)^2 \rangle \equiv \int_{\mathbf{R}^3} \varphi^* (p_l - \langle p_l \rangle)^2 \varphi \, \mathrm{d}^3p. \tag{4.1.24}$$

The formulae described so far can be generalized to *entire rational functions* of x_l and p_l:

$$\langle F(x_l) \rangle = \int_{\mathbf{R}^3} \psi^* F(x_l) \psi \, \mathrm{d}^3x = \int_{\mathbf{R}^3} \varphi^* F\left(\mathrm{i}\hbar \frac{\partial}{\partial p_l}\right) \varphi \, \mathrm{d}^3p, \tag{4.1.25}$$

$$\langle F(p_l) \rangle = \int_{\mathbf{R}^3} \psi^* F\left(\frac{\hbar}{\mathrm{i}} \frac{\partial}{\partial x_l}\right) \psi \, \mathrm{d}^3x = \int_{\mathbf{R}^3} \varphi^* F(p_l) \varphi \, \mathrm{d}^3p. \tag{4.1.26}$$

Indeed, weaker conditions on F can also be considered, e.g. C^2 functions with a rapid fall off. Moreover, elementary differentiation, e.g.

$$\psi^* \frac{\partial^2 \psi}{\partial x_l^2} = \frac{\partial}{\partial x_l}\left(\psi^* \frac{\partial \psi}{\partial x_l}\right) - \frac{\partial \psi^*}{\partial x_l} \frac{\partial \psi}{\partial x_l},$$

can be used to compute the following mean values:

$$\begin{aligned}\langle p_l^2 \rangle &= \int_{\mathbf{R}^3} \varphi^* p_l^2 \varphi \, \mathrm{d}^3p = \int_{\mathbf{R}^3} \psi^* \left(-\hbar^2 \frac{\partial^2}{\partial x_l^2}\right) \psi \, \mathrm{d}^3x \\ &= \hbar^2 \int_{\mathbf{R}^3} \frac{\partial \psi^*}{\partial x_l} \frac{\partial \psi}{\partial x_l} \, \mathrm{d}^3x,\end{aligned} \tag{4.1.27}$$

$$\begin{aligned}\langle x_l^2 \rangle &= \int_{\mathbf{R}^3} \psi^* x_l^2 \psi \, \mathrm{d}^3x = \int_{\mathbf{R}^3} \varphi^* \left(-\hbar^2 \frac{\partial^2}{\partial p_l^2}\right) \varphi \, \mathrm{d}^3p \\ &= \hbar^2 \int_{\mathbf{R}^3} \frac{\partial \varphi^*}{\partial p_l} \frac{\partial \varphi}{\partial p_l} \, \mathrm{d}^3p.\end{aligned} \tag{4.1.28}$$

In the course of deriving the results (4.1.27) and (4.1.28) we have imposed, once more, suitable fall-off conditions at infinity on the wave function, so that all total derivatives give a vanishing contribution.

To understand how the mean value $\langle x_l \rangle$ evolves in time with a dynamics ruled by $H = \frac{p^2}{2m} + V(x)$, we now perform some formal manipulations. By virtue of Eq. (4.1.9*b*), one finds

$$\begin{aligned}\langle x_l \rangle &= (2\pi\hbar)^{-3/2} \int_{\mathbf{R}^3} x_l \psi^* \, \mathrm{d}^3x \int_{\mathbf{R}^3} \varphi(p) \mathrm{e}^{\mathrm{i}\vec{p}\cdot\vec{x}/\hbar} \, \mathrm{d}^3p \\ &= (2\pi\hbar)^{-3/2} \int_{\mathbf{R}^3} \psi^* \, \mathrm{d}^3x \int_{\mathbf{R}^3} \varphi(p) \frac{\hbar}{\mathrm{i}} \frac{\partial}{\partial p_l} \left(\mathrm{e}^{\mathrm{i}\vec{p}\cdot\vec{x}/\hbar} \right) \mathrm{d}^3p. \qquad (4.1.29)\end{aligned}$$

Now we use the Leibniz rule in the integrand to express

$$\begin{aligned}\int_{\mathbf{R}^3} \frac{\hbar}{\mathrm{i}} \varphi(p) \frac{\partial}{\partial p_l} \left(\mathrm{e}^{\mathrm{i}\vec{p}\cdot\vec{x}/\hbar} \right) \mathrm{d}^3p &= \int_{\mathbf{R}^3} \frac{\hbar}{\mathrm{i}} \frac{\partial}{\partial p_l} \left[\varphi(p) \mathrm{e}^{i\vec{p}\cdot\vec{x}/\hbar} \right] \mathrm{d}^3p \\ &\quad - \int_{\mathbf{R}^3} \frac{\hbar}{\mathrm{i}} \frac{\partial \varphi}{\partial p_l} \mathrm{e}^{\mathrm{i}\vec{p}\cdot\vec{x}/\hbar} \, \mathrm{d}^3p. \qquad (4.1.30)\end{aligned}$$

The total derivative on the right-hand side of the first line of (4.1.30) can be studied for each component of the momentum. For example, when $l = 1$, one finds (Σ being here a domain in momentum space)

$$\begin{aligned}\int_{\Sigma} \frac{\hbar}{\mathrm{i}} \frac{\partial}{\partial p_1} \left[\varphi(p) \mathrm{e}^{\mathrm{i}\vec{p}\cdot\vec{x}/\hbar} \right] \mathrm{d}p_1 \, \mathrm{d}p_2 \, \mathrm{d}p_3 &= \int_{\Sigma} \frac{\hbar}{\mathrm{i}} \mathrm{d} \left(\varphi \mathrm{e}^{\mathrm{i}\vec{p}\cdot\vec{x}/\hbar} \, \mathrm{d}p_2 \, \mathrm{d}p_3 \right) \\ &= \int_{\partial\Sigma} \frac{\hbar}{\mathrm{i}} \varphi \mathrm{e}^{\mathrm{i}\vec{p}\cdot\vec{x}/\hbar} \, \mathrm{d}p_2 \, \mathrm{d}p_3 = 0, \qquad (4.1.31)\end{aligned}$$

because φ vanishes on $\partial\Sigma$, and an analogous procedure can be applied when $l = 2, 3$. Thus, the total derivative yields a vanishing contribution upon integration (on requiring the usual fall-off conditions at infinity), and Eq. (4.1.29) leads to

$$\begin{aligned}\langle x_l \rangle &= (2\pi\hbar)^{-3/2} \int_{\mathbf{R}^3} \psi^* \, \mathrm{d}^3x \int_{\mathbf{R}^3} \mathrm{i}\hbar \frac{\partial \varphi}{\partial p_l} \mathrm{e}^{i\vec{p}\cdot\vec{x}/\hbar} \, \mathrm{d}^3p \\ &= (2\pi\hbar)^{-3/2} \int_{\mathbf{R}^3} \mathrm{i}\hbar \frac{\partial \varphi}{\partial p_l} \, \mathrm{d}^3p \int_{\mathbf{R}^3} \psi^* \mathrm{e}^{i\vec{p}\cdot\vec{x}/\hbar} \, \mathrm{d}^3x \\ &= \int_{\mathbf{R}^3} \varphi^* \mathrm{i}\hbar \frac{\partial \varphi}{\partial p_l} \, \mathrm{d}^3p, \qquad (4.1.32)\end{aligned}$$

where we have used Eq. (4.1.12) after interchanging the order of integrations. Now we express $\varphi(\vec{p}, t)$ by means of (4.1.11), to find

$$\begin{aligned}\langle x_l \rangle &= \int_{\mathbf{R}^3} A^*(\vec{p}) \mathrm{e}^{\mathrm{i}Et/\hbar} \mathrm{i}\hbar \frac{\partial}{\partial p_l} \left[A(\vec{p}) \mathrm{e}^{-\mathrm{i}Et/\hbar} \right] \mathrm{d}^3p \\ &= \int_{\mathbf{R}^3} A^*(\vec{p}) \mathrm{i}\hbar \frac{\partial A}{\partial p_l} \mathrm{d}^3p + t \int_{\mathbf{R}^3} \frac{\partial E}{\partial p_l} A^* A \, \mathrm{d}^3p, \qquad (4.1.33)\end{aligned}$$

because $\frac{\partial}{\partial p_l} e^{-iEt} = -i \frac{\partial E}{\partial p_l} e^{-iEt}$. Thus, differentiation with respect to time yields (bearing in mind that $E = \frac{p^2}{2m} + V(x)$)

$$\frac{d}{dt} \langle x_l \rangle = \int_{\mathbf{R}^3} \frac{\partial E}{\partial p_l} A^* A \, d^3 p = \langle \partial E / \partial p_l \rangle = \langle v_l \rangle = \langle p_l / m \rangle. \tag{4.1.34}$$

Similarly, for a system of N particles, for which the Hamiltonian reads

$$\hat{H} = \sum_{j=1}^{N} \frac{\hat{p}_j^2}{2m_j} + V(\hat{x}_1, \ldots, \hat{x}_N), \tag{4.1.35}$$

one finds

$$\frac{d}{dt} \langle x_j \rangle = \frac{1}{m_j} \langle p_j \rangle, \tag{4.1.36}$$

jointly with

$$\frac{d}{dt} \langle p_j \rangle = - \left\langle \frac{\partial V}{\partial x_j} \right\rangle. \tag{4.1.37}$$

The result expressed by Eqs. (4.1.36) and (4.1.37) is known as the Ehrenfest theorem (Ehrenfest 1927). It is crucial to compute first the partial derivatives with respect to x_j and then to take the mean value in Eq. (4.1.37). If these operations are performed in the opposite order, one no longer obtains the time derivative of the mean value of p_j if V is an arbitrary function of q and p. The theorem can be extended to functions of q and p which are at most quadratic in p, provided the Hamiltonians are of the form $\frac{p^2}{2m} + V(x)$. The modifications necessary for generic dynamical variables were discovered in Moyal (1949). To appreciate the points raised here, consider a one-dimensional model where the first derivative of the potential can be expanded in a power series:

$$\begin{aligned} V'(x) &= V'(\langle x \rangle_t) + (x - \langle x \rangle_t) V''(\langle x \rangle_t) \\ &\quad + \frac{1}{2!} (x - \langle x \rangle_t)(x - \langle x \rangle_t) V'''(\langle x \rangle_t) + \cdots . \end{aligned} \tag{4.1.38}$$

The resulting mean value of $V'(x)$ is then equal to

$$\langle V'(x) \rangle_t = V'(\langle x \rangle_t) + \frac{(\triangle x)_t^2}{2} V'''(\langle x \rangle_t) + \cdots . \tag{4.1.39}$$

The first correction to the formula expressing complete analogy between equations of motion for mean values and classical equations of motion for dynamical variables is therefore found to involve the third derivative of the potential, evaluated at a point equal to the mean value of position at time t.

Of course, if the mean values of x and p vanish, the previous analysis provides the time rate of change of the 'localization':

$$\langle(\triangle x_l)^2\rangle \quad \text{and} \quad \langle(\triangle p_l)^2\rangle.$$

When a wave packet is considered which, at $t = 0$, is localized in a certain interval, what is going to happen during the time evolution? It is clear that, if the wave packet were to remain localized in an interval of the same dimensions, we could then think of such a packet as describing a 'particle' (see the analysis in the end of the following section).

4.1.4 Eigenstates and eigenvalues

In our introductory presentation of wave mechanics we can now call 'eigenstates' of any quantity A depending on the q and p variables those particular wave functions for which the standard deviation of A vanishes. The mean value of A in such a state can therefore be viewed as its 'eigenvalue'. This would imply that the probability distribution is concentrated on a specific value. Moreover, if the mean value of A is constant in time on all states ψ for which the Hamiltonian has vanishing dispersion, A is said to be a constant of motion.

4.2 Uncertainty relations for position and momentum

We have seen that wave functions in $\mathbf{R}^3$ can be written in the form (4.1.9a). Moreover, on using the factorization (4.1.11) for the spatial Fourier transform of the wave function, one can also express such a transform as in (4.1.12). We have also learned that the quantities $\psi^*\psi$ and $\varphi^*\varphi$ are probability densities that make it possible to define mean values of functions of x and p.

The uncertainty relations result directly from Fourier analysis and hence are not an exclusive property of quantum mechanics. A possible formulation is as follows: *a non-vanishing function and its Fourier transform cannot be both localized with precision.* Indeed, in the framework of classical physics, if $f(t)$ represents the amplitude of a signal (e.g. an acoustic wave or an electromagnetic wave) at time t, its Fourier transform $\widetilde{f}$ shows how f is constructed from sine waves of various frequencies. The uncertainty relation expresses a restriction with respect to the measurement in which the signal can be bounded in time and in the frequency band.

The previous statement can be made more explicit by considering the mean values (4.1.21) and (4.1.22) with quadratic deviations (4.1.23) and

(4.1.24), respectively. If one then takes the positive-definite quantity

$$K \equiv \left| \frac{x - \langle x \rangle}{2(\triangle x)^2} \psi + \frac{\partial \psi}{\partial x} \right|^2 \geq 0, \tag{4.2.1}$$

the integration over $\mathbf{R}^3$ yields

$$\int_{\mathbf{R}^3} K \, \mathrm{d}^3 x = \frac{1}{\hbar^2} (\triangle p)_\psi^2 - \frac{1}{4} \frac{1}{(\triangle x)_\psi^2} \geq 0, \tag{4.2.2}$$

which implies

$$(\triangle p)_\psi^2 (\triangle x)_\psi^2 \geq \frac{\hbar^2}{4}. \tag{4.2.3}$$

This inequality is invariant under scale changes and displacements in p and x. But it is not canonically invariant, even though the commutation relations among x and p can be shown to be invariant under a larger group including a rotation in the x, p variables. Therefore there must be a stronger relation between the uncertainties. This was discovered in Robertson (1930) and in Schrödinger (1930), and is mentioned in chapter 9 (dealing with more advanced aspects). To prove the inequality (4.2.2) we can simplify the calculation by taking a $\psi(x, t)$ in which the mean values (4.1.21) and (4.1.22) vanish. These assumptions lead to

$$\begin{aligned} K &= \frac{x^2}{4\langle x^2 \rangle^2} \psi \psi^* + \frac{\partial \psi}{\partial x} \frac{\partial \psi^*}{\partial x} + \frac{x}{2\langle x^2 \rangle} \left(\psi \frac{\partial \psi^*}{\partial x} + \psi^* \frac{\partial \psi}{\partial x} \right) \\ &= \frac{1}{4\langle x^2 \rangle^2} \left(x^2 - 2\langle x^2 \rangle \right) \psi^* \psi + \frac{\partial \psi}{\partial x} \frac{\partial \psi^*}{\partial x} \\ &\quad + \frac{1}{2} \frac{\partial}{\partial x} \left(\frac{x}{\langle x^2 \rangle} \psi \psi^* \right). \end{aligned} \tag{4.2.4}$$

By virtue of (4.2.4) one finds

$$\int_{\mathbf{R}^3} K \, \mathrm{d}^3 x = -\frac{1}{4} \frac{\langle x^2 \rangle_\psi}{\langle x^2 \rangle_\psi^2} + \frac{1}{\hbar^2} \langle p^2 \rangle_\psi \geq 0, \tag{4.2.5}$$

where we have used the fall off at infinity of the wave function and the general property (4.2.1). This completes the proof of (4.2.2) in the particular case considered here.

Note that the inequality (4.2.5) reduces to an equality only when the integrand vanishes. This condition leads to the first-order equation

$$\frac{x}{2\langle x^2 \rangle} \psi + \frac{\partial \psi}{\partial x} = 0, \tag{4.2.6}$$

which is solved by a Gaussian curve, i.e.

$$\psi = C \, \mathrm{e}^{-\frac{1}{4} \frac{x^2}{\langle x^2 \rangle}}, \tag{4.2.7}$$

where C is a normalization constant. More precisely, this is a one-parameter family of Gaussian curves that will be encountered again in section 10.3.

4.2.1 Uncertainty relations in relativistic systems

The Heisenberg uncertainty relation for position and momentum operators, and its generalizations to polynomials of higher degree than quadratic, is usually considered for non-relativistic particle position and momentum. Any measurement of position tends to concentrate the wave function in the neighbourhood of a suitable point. This process of a wave function 'shrinking' in coordinate space is taken as if it is instantaneous without any further discussion. As Schrödinger has emphasized any measurement is a process that takes place over a small interval of time. The instantaneous collapse of the wave function into the neighbourhood of a given point is both unnecessary and unphysical. Nevertheless we would like to see how this is seen in another moving inertial frame.

For this purpose we may choose the simplest system of a free particle of mass m. The canonical variables q and p transform as follows:

$$p^{\|} \rightarrow p^{\|} \cosh(\nu) + \omega \sinh(\nu), \tag{4.2.8}$$

$$p^{\perp} \rightarrow p^{\perp}, \tag{4.2.9}$$

$$q^{\|} \rightarrow \left[\cosh(\nu) + \frac{p^{\|}}{\omega} \sinh(\nu) \right]^{-1} q^{\|}, \tag{4.2.10}$$

with a more complicated transformation law for $q^{\perp}$. We can easily verify that this is a canonical transformation, and hence the measurements of q and p will yield the familiar Heisenberg relation. The non-linear transformations given above signify highly non-local behaviour of the coordinate $q^{\|}$ in so far as it depends non-trivially on functions of the momentum. So while the classical phase points go into phase points, at the quantum level this is not so.

The wave function of a particle of mass m localized at q_0 is $\psi(p) = e^{ipq_0}\omega^{-1/2}$, where $\omega \equiv \sqrt{p^2 + m^2}$. Under a Lorentz transformation this wave function is no longer localized at *any* point but spread out all over the configuration space (Newton and Wigner 1949, Wightman and Schweber 1955). Hence the appearance in the new frame is not shrinking to the neighbourhood of some point, but is seen as being non-zero over a very extended region in space. Of course in the new frame there are states localized in some region, but the inverse image in the original frame is not of a localized state.

We therefore conclude that there are no paradoxes in the context of the Heisenberg uncertainty relation in relativistic systems.

4.3 Transformation properties of wave functions

To understand how the standard deviation evolves in time one can use the mean values of x^2 and p^2 provided that the mean values of x and p vanish. We shall perform this calculation by transforming our description from a reference frame to another one by using the Galilei transformations. To achieve this it is necessary to first discuss the transformation properties of the Schrödinger equation, as we do hereafter.

When the transformation properties of the Schrödinger equation and of the wave function are analysed, it can be useful to assess the role played by the action $S \equiv \int \mathcal{L}\, \mathrm{d}t = \int (p\, \mathrm{d}q - H\, \mathrm{d}t)$ (see sub-section 3.10.1). Our analysis begins with classical considerations. For this purpose, let us consider the map (f being a smooth function depending on x and t)

$$(x, p) \rightarrow \left(x, p + \frac{\partial f}{\partial x}\right), \tag{4.3.1a}$$

which implies, from the definition $\mathcal{L}\, \mathrm{d}t \equiv p\, \mathrm{d}x - H\, \mathrm{d}t$, the transformation property

$$\mathcal{L}' = \mathcal{L} + \frac{\mathrm{d}f}{\mathrm{d}t}, \tag{4.3.1b}$$

provided that the velocity v is identified with $\frac{\mathrm{d}x}{\mathrm{d}t}$, because then

$$p' = \frac{\partial \mathcal{L}'}{\partial v} = p + \frac{\partial}{\partial v}\left(\frac{\mathrm{d}x}{\mathrm{d}t}\frac{\partial f}{\partial x}\right) = p + \frac{\partial f}{\partial x}.$$

Thus, the momentum variable, p, is *gauge-dependent* (see Eq. (4.3.1a)), whereas the resulting Lagrangians lead to the same equations of motion. A physically significant transformation of this type is generated by the map $\vec{p} \rightarrow \vec{p} + \vec{a}$, $\vec{x} \rightarrow \vec{x}$. This amounts to using $(\vec{p} + \vec{a}) \cdot \mathrm{d}\vec{x}$ as the phase 1-form on $T^*\mathbf{R}^3$.

In wave mechanics, if one evaluates the mean values of the quantum-mechanical operators, which correspond to the left- and right-hand sides of (4.3.1a), one finds

$$\langle p \rangle = \int_{\mathbf{R}^3} \mathrm{d}^3x\, \psi^* \frac{\hbar}{\mathrm{i}} \frac{\partial}{\partial x} \psi, \tag{4.3.2}$$

$$\langle p' \rangle = \int_{\mathbf{R}^3} \mathrm{d}^3x\, \psi^* \left(\frac{\hbar}{\mathrm{i}} \frac{\partial}{\partial x} \psi + \frac{\partial f}{\partial x} \psi\right), \tag{4.3.3}$$

and hence the mean values of p and p' do not coincide. This is a clear indication that the wave function cannot remain unaffected, but has to change if we want to make sure that the mean value of $\langle p \rangle$ remains the same. The desired transformation law is written here in the form

$$\psi' = \mathrm{e}^{-\mathrm{i}f/\hbar}\, \psi. \tag{4.3.4}$$

Note that it is possible to arrive at the same conclusion, i.e. the form of Eq. (4.3.4), using the formula (3.10.23*b*) for the wave function, jointly with Eq. (4.3.1*b*) and the definition of the action $S' \equiv \int \mathcal{L}'\, \mathrm{d}t$. By explicit calculation one then finds

$$\begin{aligned} \langle p' \rangle &= \int_{\mathbf{R}^3} \mathrm{d}^3x\, (\psi')^* \left(\frac{\hbar}{\mathrm{i}} \frac{\partial}{\partial x} \psi' \right) + \int_{\mathbf{R}^3} \mathrm{d}^3x\, \psi^* \frac{\partial f}{\partial x} \psi \\ &= \int_{\mathbf{R}^3} \mathrm{d}^3x\, \psi^* \frac{\hbar}{\mathrm{i}} \frac{\partial \psi}{\partial x} = \langle p \rangle. \end{aligned} \tag{4.3.5}$$

Remark. In higher-dimensional configuration spaces, the operators associated with $p_k + \frac{\partial f}{\partial x_k}$, i.e. $\frac{\hbar}{\mathrm{i}} \frac{\partial}{\partial x_k} + \frac{\partial f}{\partial x_k}$, constitute a commuting set of differential operators and therefore can be used in setting the correspondence between polynomials and differential operators. This amounts to defining the symbol with the exponential of $\alpha k_\mu x^\mu + f(x)$, then (cf. (3.6.6))

$$\mathrm{e}^{-\alpha k_\mu x^\mu - f(x)} \frac{\partial}{\partial x^\nu} \mathrm{e}^{\alpha k_\mu x^\mu + f(x)} = \alpha k_\nu + \frac{\partial f}{\partial x^\nu}.$$

One sees here very clearly the occurrence of a 'gauge transformation', in the way it is usually understood.

Let us now study point transformations that change the Lagrangian by a total time derivative, as in (4.3.1*b*). In particular, bearing in mind that we are interested in transformations such that the mean value of p vanishes, we consider the Galilei transformations. For this purpose, let us consider two frames Σ and Σ' with coordinates (x, t) and (x', t'), respectively. We assume that the coordinate transformation is that resulting from the Galilei group:

$$x' = x - x_0 - vt, \quad t' = t, \tag{4.3.6}$$

which implies that $\dot{x}' = \dot{x} - v$. One thus finds that the Lagrangian $\mathcal{L} = \frac{1}{2} m \dot{x}^2$ transforms as follows:

$$\begin{aligned} \mathcal{L}' &= \frac{1}{2} m (\dot{x}')^2 = \frac{1}{2} m \dot{x}^2 + \frac{1}{2} m v^2 - m \dot{x} v \\ &= \mathcal{L} - \frac{\mathrm{d}}{\mathrm{d}t} \Big(m x v - \frac{1}{2} m v^2 t \Big). \end{aligned} \tag{4.3.7}$$

The comparison with Eq. (4.3.4) and the consideration of physical dimensions therefore shows that the wave function undergoes the transformation

$$\psi' = \mathrm{e}^{\frac{\mathrm{i}}{\hbar}(m x v - \frac{1}{2} m v^2 t)}\, \psi. \tag{4.3.8}$$

4.3.1 Direct approach to the transformation properties of the Schrödinger equation

Let us now consider the effect of the transformations (4.3.6) on the differential operator occurring in the Schrödinger equation. Indeed, the elementary rules of differentiation of composite functions yield

$$\frac{\partial}{\partial t'} = \frac{\partial t}{\partial t'}\frac{\partial}{\partial t} + \frac{\partial \vec{x}}{\partial t'}\frac{\partial}{\partial \vec{x}} = \frac{\partial}{\partial t} + \vec{v}\frac{\partial}{\partial \vec{x}}, \tag{4.3.9}$$

$$\frac{\partial}{\partial \vec{x}} = \frac{\partial t'}{\partial \vec{x}}\frac{\partial}{\partial t'} + \frac{\partial \vec{x}'}{\partial \vec{x}}\frac{\partial}{\partial \vec{x}'} = \frac{\partial}{\partial \vec{x}'}, \tag{4.3.10}$$

which implies

$$\mathrm{i}\hbar\frac{\partial}{\partial t'} + \frac{\hbar^2}{2m}\triangle' = \mathrm{i}\hbar\left(\frac{\partial}{\partial t} + \vec{v}\frac{\partial}{\partial \vec{x}}\right) + \frac{\hbar^2}{2m}\triangle. \tag{4.3.11}$$

By virtue of Eqs. (4.3.4) and (4.3.11) one has

$$\psi'(x',t') = \mathrm{e}^{\frac{\mathrm{i}}{\hbar}\left(\frac{1}{2}mv^2 t - m\vec{v}\cdot\vec{x}\right)}\,\psi(x,t). \tag{4.3.12}$$

Therefore, in the frame Σ', one finds that $\psi'(x',t')$ solves the equation

$$\left(\mathrm{i}\hbar\frac{\partial}{\partial t'} + \frac{\hbar^2}{2m}\triangle'\right)\psi' = 0, \tag{4.3.13}$$

if ψ solves the equation

$$\left(\mathrm{i}\hbar\frac{\partial}{\partial t} + \frac{\hbar^2}{2m}\triangle\right)\psi = 0 \tag{4.3.14}$$

in the frame Σ. This happens because the left-hand sides of Eqs. (4.3.13) and (4.3.14) differ by the function

$$\mathrm{e}^{\mathrm{i}f/\hbar}\left\{-\frac{\partial f}{\partial t}\psi - \vec{v}\frac{\partial f}{\partial \vec{x}}\psi + \mathrm{i}\hbar\vec{v}\frac{\partial \psi}{\partial \vec{x}} + \frac{\hbar^2}{2m}\sum_{l=1}^{3}\left[\frac{2\mathrm{i}}{\hbar}\frac{\partial f}{\partial x_l}\frac{\partial \psi}{\partial x_l} + \frac{\mathrm{i}}{\hbar}\frac{\partial^2 f}{\partial x_l^2}\psi - \frac{1}{\hbar^2}\left(\frac{\partial f}{\partial x_l}\right)^2\psi\right]\right\},$$

which is found to vanish if $f = \frac{1}{2}mv^2 t - m\vec{v}\cdot\vec{x}$.

Remark. Note that the phase of a plane wave: $\frac{\mathrm{i}}{\hbar}\left(\vec{p}\cdot\vec{x} - Et\right)$ is not invariant under Galilei transformations, because

$$p'x' - E't' = (px - Et) + \frac{1}{2}mv^2 t - m\vec{v}\cdot\vec{x}. \tag{4.3.15}$$

From the relation $\vec{p}\cdot \mathrm{d}\vec{x} - H\,\mathrm{d}t \equiv \mathcal{L}\,\mathrm{d}t$, it follows that the phase changes exactly by the quantity occurring in the variation of the Lagrangian. The transformation (4.3.4) is called a *gauge transformation*, and the physical quantities that remain invariant under such transformations are called *gauge-invariant*. The analysis performed in this section shows therefore that the wave function of quantum mechanics does not change as a (scalar) function under transformations of reference frames. Its deeper meaning can only become clear after a thorough investigation of the geometrical formulation of modern physical theories, but this task goes beyond the limits of an introductory course.

4.3.2 Width of the wave packet

As an application of the transformation properties that we have just derived, let us now consider how the mean quadratic deviation of position and momentum evolves in time for a free particle. This calculation is quite important to understanding whether a wave packet 'localized' in a certain interval $[x-\delta x, x+\delta x]$ remains localized and hence may be identified with a sort of particle. Of course, if the mean values of p_l and x_l vanish, the mean quadratic deviations coincide with $\langle p_l^2 \rangle$ and $\langle x_l^2 \rangle$, respectively. We can thus think of choosing a reference frame where $\langle p_l \rangle = 0$, and then perform a translation of coordinates so that $\langle x_l \rangle = 0$ as well. For this purpose (as we anticipated after Eq. (4.3.5)) we define a suitable change of coordinates consisting of the Galilei transformations (4.3.6). We then consider

$$\begin{aligned} x_l' &\equiv x_l - \langle x_l \rangle_\psi \\ &= x_l - \int_{\mathbf{R}^3} A^* \mathrm{i}\hbar \frac{\partial A}{\partial p_l} \, \mathrm{d}^3 p - t \int_{\mathbf{R}^3} \frac{\partial E}{\partial p_l} A^* A \, \mathrm{d}^3 p, \end{aligned} \tag{4.3.16}$$

and we perform the identifications

$$x_0 \equiv \int_{\mathbf{R}^3} A^* \mathrm{i}\hbar \frac{\partial A}{\partial p_l} \, \mathrm{d}^3 p, \tag{4.3.17}$$

$$v_l \equiv \int_{\mathbf{R}^3} \frac{\partial E}{\partial p_l} A^* A \, \mathrm{d}^3 p. \tag{4.3.18}$$

We now take into account that we are dealing with a free particle:

$$E = \frac{1}{2m} p^2 \Longrightarrow \frac{\partial E}{\partial \vec{p}} = \frac{\vec{p}}{m} = \vec{v}.$$

The Galilei transformation induces the following transformation on the momenta:

$$\begin{aligned} p_l' &= p_l - m v_l = p_l - m \int_{\mathbf{R}^3} \frac{\partial E}{\partial p_l} A^* A \, \mathrm{d}^3 p \\ &= p_l - \int_{\mathbf{R}^3} p_l A^* A \, \mathrm{d}^3 p = p_l - \langle p_l \rangle_\psi. \end{aligned} \tag{4.3.19}$$

In the frame Σ' we hence obtain

$$\langle x_l' \rangle_\psi = 0, \quad \langle p_l' \rangle_\psi = 0. \tag{4.3.20}$$

We now evaluate the mean values of $p_l'^2$ and $x_l'^2$, omitting hereafter, for simplicity of notation, the 'prime'. For the former, one finds

$$\begin{aligned} \langle p_l^2 \rangle &= \int_{\mathbf{R}^3} p_l^2 \varphi^* \varphi \, \mathrm{d}^3 p = \int_{\mathbf{R}^3} 2mE\varphi^* \varphi \, \mathrm{d}^3 p \\ &= \int_{\mathbf{R}^3} 2mEA^* A \, \mathrm{d}^3 p, \end{aligned} \tag{4.3.21}$$

which is a constant in time. Moreover, by virtue of Eq. (4.1.11), one finds

$$\begin{aligned} \langle x_l^2 \rangle &= \hbar^2 \int_{\mathbf{R}^3} \frac{\partial \varphi^*}{\partial p_l} \frac{\partial \varphi}{\partial p_l} \mathrm{d}^3 p = \hbar^2 \int_{\mathbf{R}^3} \frac{\partial A^*}{\partial p_l} \frac{\partial A}{\partial p_l} \mathrm{d}^3 p \\ &+ i\hbar t \int_{\mathbf{R}^3} \frac{\partial E}{\partial p_l} \left(A^* \frac{\partial A}{\partial p_l} - A \frac{\partial A^*}{\partial p_l} \right) \mathrm{d}^3 p \\ &+ t^2 \int_{\mathbf{R}^3} \left(\frac{\partial E}{\partial p_l} \right)^2 A^* A \, \mathrm{d}^3 p. \end{aligned} \tag{4.3.22}$$

On using the well-known property $\vec{x} = \frac{\partial E}{\partial \vec{p}} = \frac{\vec{p}}{m}$, one can re-express the result (4.3.22) in the form

$$\begin{aligned} \langle (\vec{x})^2 \rangle_\psi &= \hbar^2 \int_{\mathbf{R}^3} \frac{\partial A^*}{\partial \vec{p}} \cdot \frac{\partial A}{\partial \vec{p}} \mathrm{d}^3 p + \mathrm{i} \frac{\hbar t}{m} \int_{\mathbf{R}^3} \vec{p} \cdot \left(A^* \frac{\partial A}{\partial \vec{p}} - A \frac{\partial A^*}{\partial \vec{p}} \right) \mathrm{d}^3 p \\ &+ \frac{t^2}{m^2} \langle (\vec{p})^2 \rangle_\psi. \end{aligned} \tag{4.3.23}$$

One thus finds that the wave packet has a 'width' that grows rapidly after the passage through a minimum; it also grows rapidly for earlier times.

The result (4.3.23) can also be expressed in terms of the wave function $\psi(\vec{x}, t)$. Let us denote by ψ_0 the initial value of the wave function: $\psi_0 \equiv \psi(\vec{x}, 0)$. The mean value of x^2 at the time $t = 0$ is given by

$$\langle x^2(0) \rangle_\psi = \int_{\mathbf{R}^3} \psi_0^* x^2 \psi_0 \, \mathrm{d}^3 x, \tag{4.3.24}$$

while

$$\vec{j}(0) = \frac{\hbar}{2mi} \left(\psi_0^* \operatorname{grad} \psi_0 - \psi_0 \operatorname{grad} \psi_0^* \right) \tag{4.3.25}$$

is the value of the current $\vec{j}(t)$ at $t = 0$. One thus finds

$$\langle x^2(t) \rangle_\psi = \langle x^2(0) \rangle_\psi + 2t \int_{\mathbf{R}^3} \vec{x} \cdot \vec{j}(0) \, \mathrm{d}^3 x + \frac{t^2}{m^2} \langle p^2 \rangle_\psi. \tag{4.3.26}$$

It is also possible to re-express the result in the original frame Σ:

$$\langle(\triangle x)^2\rangle_\psi = \langle(\triangle x(0))^2\rangle_\psi + 2t\int_{\mathbf{R}^3}\Big(\vec{x} - \langle\vec{x}\rangle\Big)\cdot\Big[\vec{j}(0) - \frac{\rho(0)}{m}\vec{p}\Big]\,\mathrm{d}^3x + \frac{t^2}{m^2}\langle[\triangle p(0)]^2\rangle_\psi. \tag{4.3.27}$$

This relation holds for both positive and negative times.

Having established the physical interpretation of the Schrödinger equation and its solutions, our next task is an outline of the basic steps which are necessary to solve the Schrödinger equation.

4.4 Green kernel of the Schrödinger equation

The main technical problem of wave mechanics is the solution of the Schrödinger equation once an initial condition is given. We are going to see that two key steps should be undertaken for this purpose, i.e.

(i) to find (at least implicitly) eigenvalues and eigenfunctions of the Hamiltonian operator H (taken to be independent of t), once a domain of essential self-adjointness for H (see appendix 4.A) has been determined;

(ii) to evaluate the Green kernel, which makes it possible to 'propagate' the initial condition and hence leads to complete knowledge of the wave function at all times.

To begin our analysis it is helpful to consider a simpler problem, i.e. how to solve the linear equation

$$\frac{\mathrm{d}\varphi}{\mathrm{d}t} = A\varphi \tag{4.4.1}$$

on a finite-dimensional vector space V. If the matrix A is diagonalizable and all eigenvalues are simple (i.e. without degeneracy), the space V has a basis of right eigenvectors $\{v_j\}$:

$$Av_j = \varepsilon_j v_j, \tag{4.4.2}$$

and the initial condition can be expanded in the form

$$\varphi(0) = \sum_{k=1}^{N} B_k v_k, \tag{4.4.3}$$

where the coefficients of linear combination are obtained by using the left eigenvectors v^j satisfying

$$v^j A = \varepsilon_j v^j$$

in the form

$$B_j = (v^j, \varphi(0)), \tag{4.4.4}$$

since left and right eigenvectors belonging to different eigenvalues are orthogonal, i.e.

$$(v^j, v_k) = \delta^j_{\ k}. \tag{4.4.5}$$

The solution of our first-order equation is thus found to be

$$\varphi(t) = \mathrm{e}^{tA}\varphi(0) = \sum_{k=1}^{N} B_k v_k \, \mathrm{e}^{\varepsilon_k t}, \tag{4.4.6}$$

because any power of the matrix A acts as a multiplication operator on the right eigenvectors v_k:

$$A^r v_k = (\varepsilon_k)^r v_k \quad \forall r = 0, 1, \ldots \tag{4.4.7}$$

by virtue of the eigenvalue equation (4.4.2). The left and right eigenvectors are conjugates of each other if the matrix A is Hermitian: $(A^*)^{\mathrm{T}} = A$.

Similarly, in quantum mechanics if we assume that H possesses a complete set of orthonormal eigenvectors $\{u_j\}$:

$$H u_j(\vec{x}) = E_j u_j(\vec{x}). \tag{4.4.8}$$

Then the initial condition for the Schrödinger equation can be expanded as

$$\psi(\vec{x}, 0) = \sum_{j=1}^{\infty} C_j u_j(\vec{x}), \tag{4.4.9}$$

where the Fourier coefficients C_j can be computed using the formula

$$C_j = \int_{\mathbf{R}^3} \mathrm{d}^3x' \; u_j^*(\vec{x}')\psi(\vec{x}', 0), \tag{4.4.10}$$

since eigenvectors belonging to different eigenvalues are orthogonal:

$$\int_{\mathbf{R}^3} \mathrm{d}^3x \; u_j^*(\vec{x}) u_l(\vec{x}) = \delta_{jl}.$$

The solution of the initial-value problem for the Schrödinger equation is thus found to be (bearing in mind that our equation is of first order in the time variable)

$$\psi(\vec{x}, t) = \mathrm{e}^{-\mathrm{i}tH/\hbar}\psi(\vec{x}, 0) = \sum_{j=1}^{\infty} C_j \sum_{r=0}^{\infty} \frac{(-\mathrm{i}t/\hbar)^r}{r!} H^r u_j(\vec{x})$$

$$= \sum_{j=1}^{\infty} C_j u_j(\vec{x}) \mathrm{e}^{-\mathrm{i}E_j t/\hbar}, \tag{4.4.11}$$

where we have used the 'formal' Taylor series for $\mathrm{e}^{-\mathrm{i}tH/\hbar}$, jointly with the eigenvalue equation (4.4.8) and the purely discrete nature of the spectrum of H (for a generalization, see below). In other words, the general solution is expressed as an infinite sum of elementary solutions $u_j(\vec{x})e^{-iE_jt/\hbar}$, and it is now clear why, to use this approach, one first needs to solve the eigenvalue problem for the stationary Schrödinger equation (4.4.8).

Another useful expression of the solution is obtained after inserting the result (4.4.10) for the coefficients C_j into Eq. (4.4.11), which leads to

$$\psi(\vec{x},t) = \sum_{j=1}^{\infty} \int d^3x' \, u_j^*(\vec{x}')\psi(\vec{x}',0)u_j(\vec{x})e^{-iE_jt/\hbar}$$
$$= \int d^3x' \, G(\vec{x},\vec{x}';t)\psi(\vec{x}',0), \tag{4.4.12}$$

where $G(\vec{x},\vec{x}';t)$ is the standard notation for the Green function (see the comments below):

$$G(\vec{x},\vec{x}';t) \equiv \sum_{n=1}^{\infty} u_n^*(\vec{x}')u_n(\vec{x})e^{-iE_nt/\hbar}. \tag{4.4.13}$$

In other words, once the initial condition $\psi(\vec{x},0)$ is known, the solution at a time $t \neq 0$ is obtained by means of Eq. (4.4.12), where $G(\vec{x},\vec{x}';t)$ is the Green kernel of the operator $e^{-itH/\hbar}$. This is, by definition, a solution for $t \neq 0$ of the equation

$$\left(i\hbar\frac{\partial}{\partial t} - H_{(x)}\right) G(\vec{x},\vec{x}';t) = 0, \tag{4.4.14}$$

subject to the initial condition (where ρ is a suitably smooth function)

$$\lim_{t\to 0} \int d^3x' \, G(\vec{x},\vec{x}';t)\rho(\vec{x}') = \rho(\vec{x}). \tag{4.4.15}$$

This is a more careful way to express the distributional behaviour of the Green kernel. In the physics literature, Eqs. (4.4.14) and (4.4.15) are more frequently re-expressed as follows:

$$G(\vec{x},\vec{x}',t) = \theta(t) \sum_{n=1}^{\infty} u_n^*(\vec{x}')u_n(\vec{x})e^{-iE_nt/\hbar}, \tag{4.4.16a}$$

$$\left(i\hbar\frac{\partial}{\partial t} - H_{(x)}\right) G(\vec{x},\vec{x}';t) = \delta(\vec{x},\vec{x}')\delta(t), \tag{4.4.16b}$$

$$G(\vec{x},\vec{x}';0) = \delta(\vec{x},\vec{x}'), \tag{4.4.16c}$$

where we have multiplied the right-hand side of the definition (4.4.13) by $\theta(t)$ (θ being the step function) to recover the effect of $\delta(t)$ in Eq. (4.4.16b) (see the details given in section 15.1). On considering the equations (4.4.11) and (4.4.12) one therefore says that $G(\vec{x},\vec{x}';t)$ is the Schrödinger kernel for the one-parameter strongly continuous unitary group $e^{-itH/\hbar}$ (cf. chapter 9). It propagates both forward and backward in time. Some authors prefer to say that the kernel (4.4.13) is the *propagator*, while the kernel (4.4.16*a*), which incorporates the step function, is called the *Green function.*

Note that in Eq. (4.4.12) the initial condition $\psi(\vec{x}',0)$ can be any vector in the infinite-dimensional Hilbert space of the problem, since one deals with an integral. On the other hand, the formal exponentiation used in (4.4.11) would require, in general, a C^∞ initial condition. However, if the Hilbert space can be decomposed into a direct sum of finite-dimensional sub-spaces invariant under H, the operator H becomes a Hermitian matrix on every single sub-space, and hence the exponentiation reduces to the well-defined operation used in the finite-dimensional case.

We should now stress that the hypothesis of a purely discrete spectrum for H is very restrictive. For example, the stationary Schrödinger equation for a free particle

$$-\frac{\hbar^2}{2m}\,\Delta u = Eu, \tag{4.4.17}$$

has a continuous spectrum $E = \frac{\hbar^2 k^2}{2m}$, since $E = \hbar\omega$, and the dispersion relation is then $\omega = \frac{\hbar k^2}{2m}$. The corresponding solutions (also called improper eigenfunctions)

$$u_{\vec{k}}(\vec{x}) = e^{i\vec{k}\cdot\vec{x}}, \tag{4.4.18}$$

are not, by themselves, normalizable. They are, however, of algebraic growth, in that, as $|\vec{x}| \to \infty$, one can find a polynomial p such that (see section 4.6)

$$|u_{\vec{k}}(\vec{x})| \leq p(x). \tag{4.4.19}$$

One can thus form a meaningful wave packet for the general solution by means of a Fourier transform:

$$\psi(\vec{x},t) = (2\pi)^{-3/2} \int_{\mathbf{R}^3} d^3k\; e^{i\vec{k}\cdot\vec{x}}\; e^{-i\hbar k^2 t/2m}\; \hat{u}(k), \tag{4.4.20}$$

where

$$\hat{u}(k) = (2\pi)^{-3/2} \int_{\mathbf{R}^3} d^3x\; e^{-i\vec{k}\cdot\vec{x}}\; \psi(\vec{x},0). \tag{4.4.21}$$

Remark. In general, the spectrum of H, assumed to be self-adjoint, may consist of a discrete part $\sigma_{\rm d}$ and a continuous part $\sigma_{\rm c}$, and hence the general solution of the time-dependent Schrödinger equation reads as

$$\psi(\vec{x},t) = \int_{\sigma(H)} d\mu(E) \sum_{\alpha} C_\alpha(E)\psi_{E,\alpha}(\vec{x})e^{-iEt/\hbar}. \tag{4.4.22}$$

With our notation, the symbol

$$\int_{\sigma(H)} \mathrm{d}\mu(E)$$

is a condensed notation for the summation over discrete eigenvalues and integration over the continuous spectrum. The corresponding spectral representation of the Hamiltonian operator reads as (see items (iv) and (v) in appendix 4.A)

$$H = \int \lambda \, \mathrm{d}\hat{E}_\lambda = \sum_{\lambda \in \sigma_{\mathrm{d}}} \lambda \hat{P}_\lambda + \int_{\lambda \in \sigma_{\mathrm{c}}} \lambda \, \mathrm{d}\hat{E}_\lambda. \tag{4.4.23}$$

This holds by virtue of the Lebesgue decomposition of a measure on $\mathbf{R}$: any measure is the sum of a part $\mathrm{d}\mu_{\mathrm{ac}} = f(\alpha)\,\mathrm{d}\alpha$, with $f \geq 0$ and locally integrable, which is absolutely continuous with respect to the Lebesgue measure $\mathrm{d}\alpha$; a part $\mathrm{d}\mu_{\mathrm{p}}$ concentrated on some separate points

$$\mathrm{d}\mu_{\mathrm{p}} = \mathrm{d}\alpha \sum_n c_n \delta(\alpha - \alpha_n), \quad \alpha_n \in \mathbf{R},$$

and a remainder $\mathrm{d}\mu_{\mathrm{s}}$, the singular spectrum. This last part is 'pathological' and will not occur in the problems considered in our book (although there exist one-electron band models with a non-empty singular spectrum). Each of the three pieces of the measure is concentrated on null sets with respect to the others, and $\mathcal{L}^2(\mathbf{R}, \mathrm{d}\mu)$ admits an orthogonal decomposition as

$$\mathcal{L}^2(\mathbf{R}, \mathrm{d}\mu) = \mathcal{L}^2(\mathbf{R}, \mathrm{d}\mu_{\mathrm{p}}) \oplus \mathcal{L}^2(\mathbf{R}, \mathrm{d}\mu_{\mathrm{ac}}) \oplus \mathcal{L}^2(\mathbf{R}, \mathrm{d}\mu_{\mathrm{s}}).$$

4.4.1 Free particle

A Green-function approach to the Schrödinger equation for a free particle is rather instructive. For this purpose, we look for a 'fundamental solution' of Eq. (4.4.14) which is singular at $t - t' = 0$ in such a way that, for every finite region of integration, one has

$$\lim_{t \to t'} \int_V G(\vec{x} - \vec{x}'; t - t') \, \mathrm{d}^3x' = 1 \quad \text{if } \vec{x} - \vec{x}' \text{ is in } V, \tag{4.4.24}$$

and 0 otherwise. Since the Schrödinger equation is linear, the desired solution is obtained from the integral (cf. Eq. (4.4.12))

$$\psi(\vec{x}, t) = \int G(\vec{x} - \vec{x}'; t - t') \psi(\vec{x}', t') \, \mathrm{d}^3x', \tag{4.4.25}$$

where G solves the equation (cf. Eq. (4.4.14))

$$\left(\mathrm{i}\hbar \frac{\partial}{\partial t} + \frac{\hbar^2}{2m} \triangle \right) G(\vec{x} - \vec{x}'; t - t') = 0 \quad \text{for } t \neq t', \tag{4.4.26}$$

and satisfies the initial condition (cf. Eq. (4.4.15))

$$G(\vec{x}-\vec{x}';0)=\delta(\vec{x}-\vec{x}'). \tag{4.4.27}$$

For a free particle in one dimension, the Green function is given by (we set $x'=0$ for simplicity)

$$G(x,0,t)=\frac{C}{\sqrt{t}}\mathrm{e}^{\frac{im}{2\hbar}\frac{x^2}{t}}. \tag{4.4.28}$$

The calculation shows indeed that, $\forall t\neq 0$, such a $G(x,0,t)$ solves the equation (see appendix 4.B)

$$\frac{\partial G}{\partial t}=\frac{\mathrm{i}\hbar}{2m}\frac{\partial^2 G}{\partial x^2}. \tag{4.4.29}$$

Moreover, motivated by (4.4.24), we consider the integral

$$\int_{x_1}^{x_2} G(x,0,t)\,\mathrm{d}x=C\sqrt{\frac{2\hbar}{m}}\int_{\sqrt{\frac{m}{2\hbar}}\frac{x_1}{\sqrt{t}}}^{\sqrt{\frac{m}{2\hbar}}\frac{x_2}{\sqrt{t}}}\mathrm{e}^{\mathrm{i}\xi^2}\,\mathrm{d}\xi. \tag{4.4.30}$$

Now we recall that, if $a\rightarrow\infty$ and $b\rightarrow\infty$, one has

$$\lim_{a,b\rightarrow\infty}\int_a^b \mathrm{e}^{\mathrm{i}\xi^2}\,\mathrm{d}\xi=0, \tag{4.4.31}$$

whereas, if $a\rightarrow-\infty$ and $b\rightarrow\infty$, one finds

$$\int_a^b \mathrm{e}^{\mathrm{i}\xi^2}\,\mathrm{d}\xi\rightarrow\int_{-\infty}^{\infty}\mathrm{e}^{\mathrm{i}\xi^2}\,\mathrm{d}\xi=\sqrt{\pi}\mathrm{e}^{\mathrm{i}\frac{\pi}{4}}. \tag{4.4.32}$$

By virtue of (4.4.30)–(4.4.32), the property (4.4.24) is indeed satisfied, the origin corresponding to $x=0$, and the region V to the interval $[x_1,x_2]$, with

$$C=\mathrm{e}^{-\mathrm{i}\frac{\pi}{4}}\sqrt{\frac{m}{2\pi\hbar}}. \tag{4.4.33}$$

In $\mathbf{R}^3$, one can thus write (see Eq. (4.B.20))

$$\begin{aligned}G(\vec{x},\vec{0},t)&=G(x_1,0,t)G(x_2,0,t)G(x_3,0,t)\\&=\mathrm{e}^{-\mathrm{i}\frac{3\pi}{4}}\left(\frac{m}{2\pi\hbar}\right)^{3/2}t^{-3/2}\mathrm{e}^{\frac{im}{2\hbar}\frac{(x_1^2+x_2^2+x_3^2)}{t}},\end{aligned} \tag{4.4.34}$$

and this formula should be used to evaluate the right-hand side of Eq. (4.4.25). As an example, let us consider a one-dimensional wave packet which, at $t=0$, is given by

$$\psi(x',0)=\widetilde{C}\,\mathrm{e}^{-(x'/2\triangle_0 x)^2+\frac{imvx'}{\hbar}}, \tag{4.4.35}$$

where $\widetilde{C}$ is a constant, and $\Delta_0 x$ is the mean quadratic deviation at $t = 0$. By virtue of Eq. (4.4.25), adapted to our one-dimensional problem, one finds (use (4.4.28) with x replaced by $x - x'$, and (4.4.33))

$$\psi(x,t) = \gamma(m/\hbar t)^{1/2} \mathrm{e}^{-\mathrm{i}\frac{\pi}{4}} \mathrm{e}^{\frac{\mathrm{i}mx^2}{2\hbar t}} \int_{-\infty}^{\infty} \exp\left\{ \left[\frac{\mathrm{i}m}{2\hbar t} - \frac{1}{(2\,\Delta_0\,x)^2} \right] x'^2 + \frac{\mathrm{i}m}{\hbar t}(vt - x)x' \right\} \mathrm{d}x'. \quad (4.4.36)$$

In this equation, $\gamma \equiv \frac{\widetilde{C}}{\sqrt{2\pi}}$, and one deals with an integral of the kind

$$\int_{-\infty}^{\infty} \mathrm{e}^{-ax^2+2\mathrm{i}bx}\, \mathrm{d}x = \sqrt{\frac{\pi}{a}}\, \mathrm{e}^{-b^2/a}. \quad (4.4.37)$$

Thus, on defining

$$(2\,\Delta\,x)^2 \equiv (2\,\Delta_0\,x)^2 \left\{ 1 + \left[\frac{2\hbar t}{m(2\,\Delta_0\,x)^2} \right]^2 \right\}, \quad (4.4.38)$$

one finds

$$|\psi(x,t)|^2 = \left|\widetilde{C}\right|^2 \frac{(\Delta_0 x)}{(\Delta x)} \mathrm{e}^{-\frac{(x-vt)^2}{2(\Delta x)^2}}. \quad (4.4.39)$$

The physical interpretation is that the centre of the wave packet moves with velocity v, and its spreading in time is described by Eq. (4.4.38). In particular, for Δx to become twice as large as $\Delta_0 x$, one has to wait for a time

$$t = 2\sqrt{3}\frac{m}{\hbar}(\Delta_0 x)^2. \quad (4.4.40)$$

Thus, if $m = 1.7 \times 10^{-24}$ g (as for the hydrogen atom), with $\Delta_0 x = 10^{-8}$ cm, one finds $t = 5.5 \times 10^{-13}$ s. In contrast, if $m = 10^{-3}$ g, with $\Delta_0 x = 10^{-3}$ cm, one finds $t \cong 3.3 \times 10^{18}$ s, i.e. a time of the order of 10^{11} years! Finally, the evolution is time symmetric, so for $t < 0$ there is also a greater spread.

4.5 Example of isometric non-unitary operator

Although one normally studies unitary operators in quantum dynamics, it may be now instructive to study a problem where the dynamics involves a non-unitary operator. For this purpose, we consider a massive particle moving on the positive half-line and reflected totally and elastically at the origin. The mathematical formulation is obtained in terms of a complex-valued function defined on $\mathbf{R}^+ \times \mathbf{R}$, with $x \in \mathbf{R}^+$ and $t \in \mathbf{R}$:

$$\psi : \mathbf{R}^+ \times \mathbf{R} \rightarrow \mathbf{C} : (x,t) \rightarrow \psi(x,t),$$

which obeys the equation (v being a constant with the dimension of velocity)

$$i\hbar \frac{\partial \psi}{\partial t} = v \frac{\hbar}{i} \frac{\partial \psi}{\partial x}, \tag{4.5.1a}$$

and the conditions

$$\int_0^\infty |\psi(x,t)|^2 \, dx < \infty, \tag{4.5.2}$$

$$\psi(0,t) = 0 \quad \forall t. \tag{4.5.3}$$

First, note that the operator

$$A \equiv v \frac{\hbar}{i} \frac{\partial}{\partial x}, \tag{4.5.4}$$

which plays the role of the 'Hamiltonian' operator in Eq. (4.5.1a), is symmetric on the domain $\mathcal{D}$ specified by the above conditions, because, on defining for any f and $g \in \mathcal{D}$,

$$(f,g) \equiv \int_0^\infty f^*(x,0) g(x,0) \, dx, \tag{4.5.5}$$

one finds

$$(g, Af) - (Ag, f) = \frac{\hbar}{i} \Big[g^*(x,0) f(x,0) \Big]_{x=0}^{x=\infty} = 0. \tag{4.5.6}$$

The evaluation of the left-hand side of (4.5.6) involves an integration by parts, which is performed with the help of the identity

$$\frac{\hbar}{i} \frac{\partial}{\partial x} (g^* f) = - \left(\frac{\hbar}{i} \frac{\partial g}{\partial x} \right)^* f + \frac{\hbar}{i} g^* \frac{\partial f}{\partial x}. \tag{4.5.7}$$

On the other hand, the boundary condition (4.5.3) is so 'strong' that the functions in the domain of the adjoint of A are not forced to obey it as well (see also the example of section 5.4, where we study the operator $i\frac{d}{dx}$ on the space of square-integrable functions on the closed interval $[0,1]$). Thus, the operator A is not 'diagonalizable'. This does not prevent us, however, from being able to solve Eq. (4.5.1a), which can be cast in the more convenient form

$$\left(\frac{\partial}{\partial t} + v \frac{\partial}{\partial x} \right) \psi(x,t) = 0, \tag{4.5.1b}$$

solved by

$$\psi(x,t) = f(x - vt). \tag{4.5.8}$$

In general, such solutions may have a distributional nature, but we are not concerned with such properties. For our purposes, it is more important to remark that the vector field in Eq. (4.5.1b), i.e.

$$\Gamma \equiv \frac{\partial}{\partial t} + v \frac{\partial}{\partial x}, \tag{4.5.9}$$

maps the initial condition $\psi(x,0)$ into $\psi(x,t)$ and preserves the length of vectors. The associated flow, however, is not unitary. This is clear if we bear in mind that the solution (4.5.8) is, by construction, constant on the lines of the equation

$$x = vt + c, \tag{4.5.10}$$

where c is a constant. Consider now, in particular, the line passing through the origin, for which c vanishes. As t varies in the closed interval $[0,\tau]$, the boundary condition (4.5.3) implies that only functions vanishing as $x \in [0, v\tau]$ are obtained by the application of $U \equiv \mathrm{e}^{-\mathrm{i}\frac{At}{\hbar}}$ to the initial condition, because then

$$\psi(x,t) = f(x - vt) = f(0) = \psi(0,0) = 0.$$

Thus, the inverse of U is not defined on the functions which, at $t = \tau$, are non-vanishing as $x \in [0, v\tau]$, and hence the operator A cannot be diagonalized on the set of complex-valued functions satisfying (4.5.1)–(4.5.3).

In other words, if an equation of the type

$$\mathrm{i}\hbar \frac{\partial \psi}{\partial t} = A\psi \tag{4.5.11}$$

is studied with a non-self-adjoint operator A, a temporal evolution may still exist, but is no longer described by a unitary operator.

Another example of an isometric operator is the map of the free particle continuum wave functions on to the continuum states of particles in an attractive potential with one or more bound states, for example the Hulthén potential (Hulthén 1942, 1943): $V(r) \equiv -\frac{AB\mathrm{e}^{-Br}}{(1-\mathrm{e}^{-Br})}$.

4.6 Boundary conditions

The eigenvalue problem for the Hamiltonian operator cannot be solved unless we specify the class of potentials we are interested in, with the associated boundary conditions and self-adjointness domain for H. A great variety of potentials can indeed occur in the investigation of physical phenomena, e.g. central potentials behaving as r^{-n} $(n \geq 1)$, Yukawa terms $\frac{\mathrm{e}^{-\mu r}}{r}$, isotropic and anisotropic harmonic oscillators, polynomials of suitable degree, Laurent series, logarithmic terms and periodic potentials.

Here we study the Schrödinger equation for stationary states (cf. Eq. (4.4.8))

$$\left(-\frac{\hbar^2}{2m} \triangle + V(x) \right) u(x) = Eu(x), \tag{4.6.1}$$

where the potential is taken to be of the form

$$V(x) = \sum_{j=1}^{l} \frac{g_j}{|\vec{x} - \vec{x}_j|} + V_1(x), \tag{4.6.2}$$

with g_j being real constants and V_1 being such that: (i) it is continuous with the exception of a finite number of surfaces $\sigma_1, \sigma_2, \ldots, \sigma_r$, where it has finite discontinuities; (ii) it is bounded from below; (iii) it can diverge at infinity, but not faster than a polynomial. The points $x_1, \ldots, x_l$ and the surfaces $\sigma_1, \ldots, \sigma_r$ are said to be the singular points and singular surfaces of the potential V, respectively. Hereafter, we shall denote by H the Hamiltonian operator in round brackets on the left-hand side of Eq. (4.6.1), by $D_0(H)$ the set

$$D_0(H) \equiv \Biggl\{ f \in \mathcal{L}^2(\mathbf{R}^3) : f \in C^2(\mathbf{R}^3), \\ \left[-\frac{\hbar^2}{2m} \triangle + V(x) \right] f \in \mathcal{L}^2(\mathbf{R}^3) \Biggr\}, \tag{4.6.3}$$

and by $D(H)$ the (as yet unknown) domain where H is self-adjoint. In general, it is not easy to know explicitly *a priori* the self-adjointness domain $D(H)$, but this technical difficulty will not affect what we are going to do. The set $D_0(H)$ represents a possible characterization of the domain of H, and for $u \in D_0(H)$ Eq. (4.6.1) may be viewed as an ordinary differential equation in $C^2(\mathbf{R}^3)$. However, if the potential has singularities, Eq. (4.6.1) does not have solutions in $C^2(\mathbf{R}^3)$, and hence the (formal) eigenvalue equation does not have solutions belonging to $D_0(H)$. In contrast, for a singularity-free potential, it can be proved that the eigenfunctions of H in the self-adjointness domain $D(H)$ coincide with the eigenfunctions in $D_0(H)$.

First, note that, if q is any element of $C_0^\infty(\mathbf{R}^3)$ (the consideration of this space is suggested by the property of H being essentially self-adjoint on it), one can study, for any locally integrable function u, the equation

$$(u, Hq) = E(u, q), \tag{4.6.4a}$$

or, explicitly (our scalar product being anti-linear in the first argument),

$$\int_{\mathbf{R}^3} d^3x \, u^*(x) \left(-\frac{\hbar^2}{2m} \triangle + V(x) \right) q(x) = E \int_{\mathbf{R}^3} d^3x \, u^*(x) q(x). \tag{4.6.4b}$$

It is therefore meaningful to consider solutions of Eq. (4.6.4*b*), *whether or not* they belong to the Hilbert space $\mathcal{L}^2(\mathbf{R}^3)$. Such solutions are said to be *generalized* (or *weak*) *solutions* of Eq. (4.6.1). The *eigenfunctions* of H in $D(H)$ are then those particular generalized solutions of Eq. (4.6.1) that belong to $\mathcal{L}^2(\mathbf{R}^3)$.

The generalized solutions of Eq. (4.6.1) possess four fundamental properties.

(i) They are of class C^2 outside the singularity set $\mathcal{S}$ of the potential V.

(ii) They are ordinary solutions of Eq. (4.6.1) outside $\mathcal{S}$.

(iii) On the singular surfaces, they are continuous jointly with their normal derivatives:

$$u_+(x) = u_-(x) \quad x \in \sigma_i, \tag{4.6.5}$$

$$\frac{\partial u_+(x)}{\partial n} = \frac{\partial u_-(x)}{\partial n} \quad x \in \sigma_i, \tag{4.6.6}$$

where the $\pm$ labels are used for the limiting values of such functions on the two sides of the singular surfaces. These equations express the boundary conditions in the eigenvalue problem for the Hamiltonian operator.

(iv) At the singular points they take a finite value, in that

$$\lim_{x \to x_j} u(x) = \text{finite quantity}. \tag{4.6.7}$$

Property (i) is a theorem due to Weyl (Weyl 1940, Hellwig 1964), and will not be proved here. Property (ii) follows because, if (i) holds, and if q vanishes in a neighbourhood of the singularities, one finds

$$\int_{\mathbf{R}^3} d^3x \, u^*(x) \, \triangle \, q(x) - \int_{\mathbf{R}^3} d^3x \, [\Delta u^*(x)] q(x)$$
$$= \int_{\mathbf{R}^3} d^3x \, \text{div} \Big[u^*(\text{grad } q) - (\text{grad } u^*) q \Big] = 0, \tag{4.6.8}$$

and hence Eq. (4.6.4b) becomes (since we can bring u^* to the right of the Laplace operator)

$$\int_{\mathbf{R}^3} d^3x \left[\left(-\frac{\hbar^2}{2m} \triangle + V(x) \right) u^*(x) \right] q(x) = E \int_{\mathbf{R}^3} d^3x \, u^*(x) q(x). \tag{4.6.9}$$

By virtue of the arbitrariness in the choice of q, Eq. (4.6.9) leads to Eq. (4.6.1) outside the singularity set $\mathcal{S}$.

Lastly, to prove (iii) and (iv), which play a key role in all the applications of quantum mechanics, we surround the singular surfaces by narrow tubes, and the singular points by spheres ω of radius r_j. This is necessary to be able to apply the divergence theorem. On taking a function q of arbitrary form, one then finds (cf. Eq. (4.6.8))

$$\int_{\mathbf{R}^3} d^3x \, u^*(x) \, \triangle \, q(x) - \int_{\mathbf{R}^3} d^3x \, [\Delta u^*(x)] q(x)$$
$$= -\sum_i \int_{\sigma_i} d\sigma \left[\left(u_+^* \frac{\partial q}{\partial n} - \frac{\partial u_+^*}{\partial n} q \right) - \left(u_-^* \frac{\partial q}{\partial n} - \frac{\partial u_-^*}{\partial n} q \right) \right]$$
$$- \sum_j \lim_{r_j \to 0} \int_{\omega(x_j, r_j)} d\sigma_j \left(u^* \frac{\partial q}{\partial n_j} - \frac{\partial u^*}{\partial n_j} q \right)$$

$$= -\sum_i \int_{\sigma_i} \mathrm{d}\sigma \left[(u_+ - u_-)^* \frac{\partial q}{\partial n} - \left(\frac{\partial u_+}{\partial n} - \frac{\partial u_-}{\partial n} \right)^* q \right]$$

$$- \sum_j \lim_{r_j \to 0} r_j^2 \int \mathrm{d}\Omega \left(u^* \frac{\partial q}{\partial r_j} - \frac{\partial u^*}{\partial r_j} q \right). \qquad (4.6.10)$$

If Eq. (4.6.4*b*) is satisfied, and if Eq. (4.6.1) holds up to a zero-measure set, the boundary terms on the right-hand side of Eq. (4.6.10) have to vanish, since the left-hand side is zero by virtue of

$$\triangle u^* = -\frac{2m}{\hbar^2}(E - V)u^*$$

and

$$\int_{\mathbf{R}^3} \mathrm{d}^3x \, u^* \triangle q = -\frac{2m}{\hbar^2} \int_{\mathbf{R}^3} \mathrm{d}^3x \, u^*(E - V)q.$$

Bearing in mind that q and its normal derivative can be fixed in an arbitrary way on the singular surfaces, and also the arbitrariness in choosing the values of q at the singular points, one finds that the boundary conditions (4.6.5) and (4.6.6) are indeed satisfied. Lastly, to prove the result (4.6.7), one should bear in mind that, since the potential is taken to be of the form (4.6.2), the solutions of Eq. (4.6.1) behave, in a neighbourhood of the singular points, as r_j^ν, with $\nu = 0, \pm 1, \pm 2, \ldots$. Hence one finds the qualitative behaviour

$$r_j^2 \frac{\partial u}{\partial r_j} \sim \nu r_j^{\nu+1} \sim \nu r_j u,$$

which implies (remember that $\mathrm{d}\sigma_j = r_j^2 \, \mathrm{d}\Omega$)

$$\lim_{r_j \to 0} \int \mathrm{d}\Omega \, r_j^2 \left(u^* \frac{\partial q}{\partial r_j} - \frac{\partial u^*}{\partial r_j} q \right) = -\nu \lim_{r_j \to 0} \int \mathrm{d}\Omega \, r_j u^* q. \qquad (4.6.11)$$

Since such a limit should vanish *for any* q, ν cannot take negative values, which leads in turn to the condition (4.6.7). Conversely, it is obvious that a function u satisfying the properties (i)–(iv) is a generalized solution of Eq. (4.6.1).

In summary, if the potential V is everywhere continuous, the ordinary solutions of Eq. (4.6.1) are also generalized solutions. If V has a finite number of singularities, however, the space is divided by the discontinuity surfaces into a number of disconnected regions $\mathcal{D}_i$. After finding the ordinary solutions of Eq. (4.6.1) within each $\mathcal{D}_i$, the boundary conditions (4.6.5) and (4.6.6) make it possible to match such ordinary solutions and hence to extend the solution to the whole space, with the exception of the singular points x_j. Among the (infinitely many) solutions built in this way, those satisfying the finiteness property (4.6.7) at the singular points are the desired generalized solutions of Eq. (4.6.1). Moreover, to pick out the *eigenfunctions* of H, one has to choose, among the generalized solutions obtained with the above procedure, those belonging to the Hilbert space $\mathcal{L}^2(\mathbf{R}^3)$. This is only possible for some real and discrete values of E.

In the applications of wave mechanics we shall use the boundary conditions (4.6.5) and (4.6.6), but the reader should be aware that it is not strictly necessary that these equations be satisfied (indeed, they are not satisfied for an infinite potential step!). It is sufficient to require continuity of the density $\psi^*\psi$ and the current $\frac{i}{2}\Big(\psi^* \text{grad}\, \psi - \psi \, \text{grad}\, \psi^*\Big)$. This leaves an arbitrary phase in ψ between two regions. Such a phase is irrelevant and can be set to zero for convenience.

4.6.1 Particle confined by a potential

If the potential has an infinite jump, this may be viewed as the limiting case of a problem where the discontinuity of V is finite (as one takes the limit for the 'jump' tending to infinity), and hence the above rules for the boundary conditions can be applied to solve all quantum-mechanical problems. For example, if a particle of energy $E \in]0, V_0[$ and mass m is subject to the potential

$$\widetilde{V}(x) = V_0 \quad x \in]-\infty, -a[, \tag{4.6.12a}$$

$$\widetilde{V}(x) = 0 \quad x \in]-a, a[, \tag{4.6.12b}$$

$$\widetilde{V}(x) = V_0 \quad x \in]a, \infty[, \tag{4.6.12c}$$

the solutions of the Schrödinger equation for stationary states read, on setting $\Gamma \equiv \frac{\sqrt{2m(V_0-E)}}{\hbar}, \kappa \equiv \frac{\sqrt{2mE}}{\hbar}$,

$$u_1(x) = A_1 e^{\Gamma x} + A_2 e^{-\Gamma x} = A_1 e^{\Gamma x} \quad x \in]-\infty, -a[, \tag{4.6.13}$$

$$u_2(x) = A_3 \cos \kappa x + A_4 \sin \kappa x \quad x \in]-a, a[, \tag{4.6.14}$$

$$u_3(x) = A_5 e^{\Gamma x} + A_6 e^{-\Gamma x} = A_6 e^{-\Gamma x} \quad x \in]a, \infty[, \tag{4.6.15}$$

where we have set $A_2 = A_5 = 0$ to obtain a wave function $\in \mathcal{L}^2(\mathbf{R}, dx)$. The continuity conditions at $x = -a$ and at $x = a$:

$$\lim_{x \to -a^-} u_1(x) = \lim_{x \to -a^+} u_2(x), \tag{4.6.16}$$

$$\lim_{x \to -a^-} u_1'(x) = \lim_{x \to -a^+} u_2'(x), \tag{4.6.17}$$

$$\lim_{x \to a^-} u_2(x) = \lim_{x \to a^+} u_3(x), \tag{4.6.18}$$

$$\lim_{x \to a^-} u_2'(x) = \lim_{x \to a^+} u_3'(x), \tag{4.6.19}$$

lead to the homogeneous system

$$A_1 e^{-\Gamma a} - A_3 \cos \kappa a + A_4 \sin \kappa a = 0, \tag{4.6.20}$$

$$A_1 e^{-\Gamma a} - \frac{\kappa}{\Gamma} A_3 \sin \kappa a - \frac{\kappa}{\Gamma} A_4 \cos \kappa a = 0, \tag{4.6.21}$$

$$A_3 \cos \kappa a + A_4 \sin \kappa a - A_6 e^{-\Gamma a} = 0, \tag{4.6.22}$$

$$-A_3 \sin \kappa a + A_4 \cos \kappa a + \frac{\Gamma}{\kappa} A_6 e^{-\Gamma a} = 0. \tag{4.6.23}$$

To find a non-trivial solution for A_1, A_3, A_4, A_6, the determinant of the matrix of coefficients should vanish, and this leads to the eigenvalue condition

$$2\kappa\Gamma \cos(2\kappa a) + \left(\Gamma^2 - \kappa^2\right) \sin(2\kappa a) = 0. \tag{4.6.24}$$

When $V_0 \to \infty$, Eq. (4.6.24) reduces to $\Gamma^2 \sin(2\kappa a) = 0$, which implies

$$\sin(2\kappa a) = 0, \tag{4.6.25}$$

an equation solved by $2\kappa a = n\pi$, and hence one recovers the well-known spectrum

$$E_n = n^2 \frac{\hbar^2 \pi^2}{8ma^2} \quad \forall\, n = 1, 2, \ldots \tag{4.6.26}$$

for the particle confined within the closed interval $[-a, a]$ by the infinite-wall potential

$$V(x) = \infty \quad x \in \,]-\infty, -a[, \tag{4.6.27a}$$

$$V(x) = 0 \quad x \in \,]-a, a[, \tag{4.6.27b}$$

$$V(x) = \infty \quad x \in \,]a, \infty[. \tag{4.6.27c}$$

Note also that, when $V_0 \to \infty$, $u_1(x)$ in (4.6.13) and $u_3(x)$ in (4.6.15) tend to zero for any value of x. This is why one says that, in the potential (4.6.27), the wave function vanishes outside the interval $]-a, a[$. Furthermore, (4.6.17) and (4.6.19) no longer hold, but uu' is continuous at these points.

4.6.2 Improper eigenfunctions

Another little bit of notation is now necessary to characterize an important class of 'eigenfunctions' that frequently occur. For this purpose, we say that $\{u_W(x)\}$ is a family of *improper eigenfunctions* of the Hamiltonian operator if the following conditions hold:

(i) $\{u_W\}$ is continuous, or at least locally integrable, with respect to the parameter W in a subset σ of the real line, with positive measure;

(ii) each function u_W is a generalized solution of Eq. (4.6.1) for the specific value of W (and hence satisfies the regularity properties and the boundary conditions previously derived);

(iii) for any interval $(W, W+\delta W)$ having intersection of positive measure with σ, the *eigen-differential*

$$F_{(W,W+\delta W)}u(x) \equiv \int_W^{W+\delta W} d\gamma \; u_\gamma(x) \tag{4.6.28}$$

belongs to $\mathcal{L}^2(\mathbf{R})$ and is non-vanishing.

The improper eigenfunctions are, in general, *tempered distributions.* The class of ordinary functions, which correspond to tempered distributions, can be essentially identified with the set of locally integrable functions of *algebraic growth.* By definition, the latter are functions bounded at infinity by a polynomial: for each u there exists $r \in N$ such that, as $|x| \rightarrow \infty$, one finds

$$|u(x)| \leq \sum_{s=0}^{r} a_s x^s. \tag{4.6.29}$$

The property of having algebraic growth can be used, for differential operators, as a requisite to select solutions of the (formal) eigenvalue equation among which one should look for eigenfunctions. For example, according to this criterion, one should rule out solutions of the form (see Eqs. (4.6.13) and (4.6.15))

$$e^{\sqrt{W}|x|} \quad W > 0,$$

whereas, in $\mathcal{L}^2(\mathbf{R})$, one can build wave functions $\psi(x,t)$ from the improper eigenfunctions

$$e^{i\sqrt{W}x}, \; e^{-i\sqrt{W}x} \quad W > 0.$$

In $\mathcal{L}^2(\mathbf{R}^p)$, with $p > 1$, some extra care is necessary to select suitable improper eigenfunctions.

4.7 Harmonic oscillator

One of the few systems for which the Schrödinger equation can be solved is the harmonic oscillator (for more examples, see chapters 5 and 12). From a physical point of view, all cases when 'equilibrium states' are approached can be studied using harmonic oscillators.

4.7.1 One-dimensional oscillator

Our analysis begins with the Schrödinger equation for stationary states of a one-dimensional harmonic oscillator, i.e.

$$\left(-\frac{\hbar^2}{2m}\frac{\mathrm{d}^2}{\mathrm{d}x^2}+\frac{m}{2}\omega^2x^2\right)\psi_n=E_n\psi_n. \tag{4.7.1}$$

Rather than using a direct approach for the solution of this equation, we rely upon some 'ad hoc' methods, which nevertheless can also be exploited for other problems. For this purpose, we notice that Eq. (4.7.1) can be factorized in the form

$$-\frac{\hbar^2}{2m}\left(\frac{\mathrm{d}}{\mathrm{d}x}-\frac{m\omega}{\hbar}x\right)\left(\frac{\mathrm{d}}{\mathrm{d}x}+\frac{m\omega}{\hbar}x\right)\psi_n=\left(E_n-\frac{\hbar\omega}{2}\right)\psi_n, \tag{4.7.2}$$

because the Leibniz rule implies that

$$\frac{\mathrm{d}}{\mathrm{d}x}(x\psi_n)=x\frac{\mathrm{d}}{\mathrm{d}x}\psi_n+\psi_n. \tag{4.7.3}$$

The factorized form (4.7.2) shows that the operator $\frac{\mathrm{d}}{\mathrm{d}x}-\frac{m\omega}{\hbar}x$ makes it possible to generate a new solution from a given one. Thus, the function ψ_{n+1} defined by

$$\psi_{n+1}\equiv\left(\frac{\mathrm{d}}{\mathrm{d}x}-\frac{m\omega}{\hbar}x\right)\psi_n, \tag{4.7.4}$$

is a solution of Eq. (4.7.1) with eigenvalue $E_n+\hbar\omega$, and the function

$$\psi_{n-1}\equiv\left(\frac{\mathrm{d}}{\mathrm{d}x}+\frac{m\omega}{\hbar}x\right)\psi_n \tag{4.7.5}$$

solves Eq. (4.7.1) with eigenvalue $E_n-\hbar\omega$. The latter property follows easily from the factorization

$$-\frac{\hbar^2}{2m}\left(\frac{\mathrm{d}}{\mathrm{d}x}+\frac{m\omega}{\hbar}x\right)\left(\frac{\mathrm{d}}{\mathrm{d}x}-\frac{m\omega}{\hbar}x\right)\psi_n=\left(E_n+\frac{\hbar\omega}{2}\right)\psi_n \tag{4.7.6}$$

and from the relation

$$\hat{H}\left(\frac{\mathrm{d}}{\mathrm{d}x}+\frac{m\omega}{\hbar}x\right)\psi_n=(E_n-\hbar\omega)\left(\frac{\mathrm{d}}{\mathrm{d}x}+\frac{m\omega}{\hbar}x\right)\psi_n. \tag{4.7.7}$$

Repeated application shows that $\hat{H}\left(\frac{d}{dx}+\frac{m\omega}{\hbar}x\right)^r \psi_n$ is equal to $(E_n - r\hbar\omega)\left(\frac{d}{dx}+\frac{m\omega}{\hbar}x\right)^r \psi_n$. Such properties suggest calling the operators $\frac{d}{dx}-\frac{m\omega}{\hbar}x$ and $\frac{d}{dx}+\frac{m\omega}{\hbar}x$ the creation and annihilation operators, respectively.

At this stage, a naturally occurring question is whether one can obtain negative energy eigenvalues by repeated application of the annihilation operator. Indeed, if ψ_0 is a normalized eigenfunction belonging to the eigenvalue E_0 such that $E_0 > 0$, $E_0 - \hbar\omega < 0$, one can define $\psi_{-1} \equiv \left(\frac{d}{dx}+\frac{m\omega}{\hbar}x\right)\psi_0$, and integration by parts, jointly with Eq. (4.7.2), yields

$$\int_{-\infty}^{\infty} \psi_{-1}^* \psi_{-1} \, dx = \int_{-\infty}^{\infty} \psi_0^* \left[\left(-\frac{d}{dx}+\frac{m\omega}{\hbar}x\right)\left(\frac{d}{dx}+\frac{m\omega}{\hbar}x\right)\psi_0\right] dx$$
$$= \frac{2m}{\hbar^2}\left(E_0 - \frac{\hbar\omega}{2}\right). \tag{4.7.8}$$

If E_0 were smaller than $\frac{\hbar\omega}{2}$ we would obtain a negative norm of ψ_{-1}, which is impossible. If E_0 were larger than $\frac{\hbar\omega}{2}$ (but smaller than $\hbar\omega$), we might iterate the procedure and consider

$$\psi_{-2} \equiv \left(\frac{d}{dx}+\frac{m\omega}{\hbar}x\right)\psi_{-1}, \tag{4.7.9}$$

the 'norm' of which would then be negative by construction. To avoid having such inconsistencies we can only accept that $E_0 = \frac{\hbar\omega}{2}$, which leads to

$$\left(\frac{d}{dx}+\frac{m\omega}{\hbar}x\right)\psi_0 = 0. \tag{4.7.10}$$

This is a first-order equation, which is solved by separation of variables to find

$$\psi_0 = K_0 \, e^{-\frac{m\omega x^2}{2\hbar}}, \tag{4.7.11}$$

where K_0 is a normalization constant obtained by requiring that

$$K_0^2 \int_{-\infty}^{\infty} e^{-\frac{m\omega x^2}{\hbar}} \, dx = 1, \tag{4.7.12}$$

which implies

$$K_0 = \left(\frac{m\omega}{\pi\hbar}\right)^{1/4}. \tag{4.7.13}$$

All the eigenfunctions of the harmonic oscillator can be built starting from the ground state ψ_0 and repeatedly applying the creation operator $\frac{d}{dx}-\frac{m\omega}{\hbar}x$. In such a way one obtains

$$\psi_n(x) = K_n \left(\frac{d}{dx}-\frac{m\omega}{\hbar}x\right)^n \psi_0, \quad n \in \mathcal{N}, \tag{4.7.14}$$

where the normalization constant K_n is such that

$$K_n^2 = \left(\frac{\hbar}{m\omega}\right)^n \sqrt{\frac{m\omega}{\pi\hbar}} \frac{1}{2^n n!}. \tag{4.7.15}$$

4.7.2 Hermite polynomials

Note now that

$$\left(\frac{\mathrm{d}}{\mathrm{d}x} + \frac{m\omega}{\hbar}x\right)\psi = \mathrm{e}^{-\frac{m\omega x^2}{2\hbar}} \frac{\mathrm{d}}{\mathrm{d}x}\left(\mathrm{e}^{\frac{m\omega x^2}{2\hbar}}\psi\right). \tag{4.7.16}$$

The generalization to arbitrary powers of the annihilation and creation operators is straightforward:

$$\left(\frac{\mathrm{d}}{\mathrm{d}x} + \frac{m\omega}{\hbar}x\right)^n \psi = \mathrm{e}^{-\frac{m\omega x^2}{2\hbar}} \frac{\mathrm{d}^n}{\mathrm{d}x^n}\left(\mathrm{e}^{\frac{m\omega x^2}{2\hbar}}\psi\right) \tag{4.7.17}$$

and

$$\left(\frac{\mathrm{d}}{\mathrm{d}x} - \frac{m\omega}{\hbar}x\right)^n \psi = \mathrm{e}^{\frac{m\omega x^2}{2\hbar}} \frac{\mathrm{d}^n}{\mathrm{d}x^n}\left(\mathrm{e}^{-\frac{m\omega x^2}{2\hbar}}\psi\right). \tag{4.7.18}$$

This remark leads to the introduction of the Hermite polynomials. This means that, on defining the variable

$$\xi \equiv \sqrt{\frac{m\omega}{\hbar}}\, x, \tag{4.7.19}$$

we deal with

$$H_n(\xi) \equiv (-1)^n \mathrm{e}^{\xi^2} \frac{\mathrm{d}^n}{\mathrm{d}\xi^n} \mathrm{e}^{-\xi^2}. \tag{4.7.20}$$

By virtue of (4.7.18), the term on the right-hand side of (4.7.20) can be re-expressed in the form

$$\mathrm{e}^{\xi^2/2}\left(\frac{\mathrm{d}}{\mathrm{d}\xi} - \xi\right)^n \mathrm{e}^{-\xi^2/2} = \mathrm{e}^{\xi^2} \frac{\mathrm{d}^n}{\mathrm{d}\xi^n} \mathrm{e}^{-\xi^2}. \tag{4.7.21}$$

The Hermite polynomials therefore make it possible to write, for the harmonic oscillator eigenfunctions,

$$\psi_n(\xi) = \widetilde{K}_n H_n(\xi) \mathrm{e}^{-\xi^2/2}, \tag{4.7.22}$$

having set

$$\widetilde{K}_n \equiv K_n \left(\frac{m\omega}{\hbar}\right)^{n/2} (-1)^n. \tag{4.7.23}$$

For example, for the lowest values of the integer n, the Hermite polynomials read as

$$H_0(\xi) = 1, \quad H_1(\xi) = 2\xi, \tag{4.7.24}$$

$$H_2(\xi) = 4\xi^2 - 2, \quad H_3(\xi) = 8\xi^3 - 12\xi. \tag{4.7.25}$$

Thus, the eigenfunctions are even or odd depending on whether n is even or odd, respectively:

$$\psi_{2k}(x) = \psi_{2k}(-x), \tag{4.7.26}$$

$$\psi_{2k+1}(x) = -\psi_{2k+1}(-x). \tag{4.7.27}$$

The previous results can be used to discuss the harmonic oscillator in three dimensions. In this case, the potential reads as

$$U(x, y, z) = \frac{m}{2}\left(\omega_1^2 x^2 + \omega_2^2 y^2 + \omega_3^2 z^2\right), \tag{4.7.28}$$

and the corresponding eigenvalue equation for the Hamiltonian operator is

$$\left[-\frac{\hbar^2}{2m}\left(\frac{\partial^2}{\partial x^2} + \frac{\partial^2}{\partial y^2} + \frac{\partial^2}{\partial z^2}\right) + U(x, y, z)\right]\psi_E = E\psi_E. \tag{4.7.29}$$

Such an equation can be solved by separation of variables, i.e. by writing

$$\psi_E(x, y, z) = \psi_{n_1}(x)\psi_{n_2}(y)\psi_{n_3}(z), \tag{4.7.30}$$

where ψ_{n_1}, ψ_{n_2} and ψ_{n_3} are eigenfunctions of the one-dimensional harmonic oscillator. One thus obtains

$$E = \hbar\omega_1\left(n_1 + \frac{1}{2}\right) + \hbar\omega_2\left(n_2 + \frac{1}{2}\right) + \hbar\omega_3\left(n_3 + \frac{1}{2}\right). \tag{4.7.31}$$

In particular, if two frequencies coincide, e.g. $\omega = \omega_1 = \omega_2 \neq \omega_3$, one finds

$$E = \hbar\omega(n_1 + n_2 + 1) + \hbar\omega_3\left(n_3 + \frac{1}{2}\right). \tag{4.7.32}$$

If all frequencies are equal: $\omega = \omega_1 = \omega_2 = \omega_3$, one has

$$E = \hbar\omega\left(n_1 + n_2 + n_3 + \frac{3}{2}\right). \tag{4.7.33}$$

If the frequencies coincide the resulting energy levels are degenerate. On defining

$$n \equiv n_1 + n_2 + n_3, \tag{4.7.34}$$

we remark that all eigenfunctions obtained as a product of one-dimensional eigenfunctions for which the indices add up to n belong to the same eigenvalue E. The number of such eigenfunctions can be easily evaluated by pointing out that it is given by the number of ways in which we can choose the integers n_1 and n_2 subject to the condition

$$0 \leq n_1 + n_2 \leq n.$$

Since for $n_1 + n_2 = j$, with $j \leq n$, there exist $j+1$ ways of choosing n_1 and n_2 so that their sum equals j, one finds that the desired degeneracy is given by (in the isotropic case)

$$\sum_{j=0}^{n}(j+1) = \frac{(n+1)(n+2)}{2}. \tag{4.7.35}$$

Moreover, some degeneracy may survive if the frequencies are rational, i.e. there exist some integers j_1, j_2 and j_3 for which

$$j_1\omega_1 = j_2\omega_2 = j_3\omega_3. \tag{4.7.36}$$

4.8 JWKB solutions of the Schrödinger equation

The constant of nature that is a peculiar property of quantum mechanics with respect to classical mechanics is the Planck constant h. Thus, if quantum mechanics is more fundamental, because it leads to a good understanding of atomic physics unlike classical theory, it is of particular interest to find a method that expresses in a quantitative and predictive way by 'how much' quantum theory differs from classical theory, and relates this to the Planck constant. This is made possible by the JWKB method (Jeffreys–Wentzel–Kramers–Brillouin), which evaluates the wave function by means of an asymptotic expansion (see appendix 4.C) in powers of $\hbar$ (Froman and Froman 1965). To zeroth order in $\hbar$ one has to recover the Hamilton–Jacobi equation, plus corrections that we are going to compute. The method turns out to be a valuable tool in obtaining approximate solutions of the Schrödinger equation in cases where an exact analytic solution is very difficult or impossible (e.g. with long-range potentials, or phenomenological potentials that approximate a very involved configuration in nuclear structure). For this purpose, in the Schrödinger equation we now look for solutions in the form

$$\psi(\vec{x}, t) = A(\vec{x}, t)\mathrm{e}^{\mathrm{i}W(\vec{x},t)/\hbar}, \tag{4.8.1}$$

where A and W are real-valued functions (if A were complex, one might absorb its phase into the definition of W). This means that a purely oscillatory behaviour is modulated by some function of $\vec{x}$ and t, i.e. the prefactor A. On separating the real and imaginary parts of the resulting equation, one finds (cf. Merzbacher (1961), where the prefactor A is instead absorbed into the definition of W)

$$\frac{\partial A}{\partial t} = -\frac{1}{2m}\Big[A \triangle W + 2(\mathrm{grad}\, A)\cdot(\mathrm{grad}\, W)\Big], \tag{4.8.2}$$

$$\frac{\partial W}{\partial t} = -\left[\frac{(\mathrm{grad}\, W)^2}{2m} + V(\vec{x}) - \frac{\hbar^2}{2m}\frac{(\triangle A)}{A}\right]. \tag{4.8.3}$$

On defining

$$\rho(\vec{x},t) \equiv A^2(\vec{x},t) = |\psi|^2, \tag{4.8.4}$$

$$\vec{v} \equiv \frac{1}{m} \text{grad}\, W, \tag{4.8.5}$$

equation (4.8.2) leads to the continuity equation (cf. Eq. (4.1.8a))

$$\frac{\partial \rho}{\partial t} + \text{div}(\rho \vec{v}) = 0. \tag{4.8.6}$$

Moreover, if one neglects in Eq. (4.8.3) the term proportional to $\hbar^2$, one obtains the Hamilton–Jacobi equation associated with the Hamiltonian $H = \frac{p^2}{2m} + V(\vec{x})$. Recall, in fact, that this can be obtained from the relation

$$H(\vec{p}, \vec{x}) = E, \tag{4.8.7}$$

bearing in mind that the equation

$$\text{d}W = \vec{p} \cdot \text{d}\vec{x} - E\, \text{d}t \tag{4.8.8}$$

leads to

$$\vec{p} = \frac{\partial W}{\partial \vec{x}}, \quad E = -\frac{\partial W}{\partial t}. \tag{4.8.9}$$

Under such conditions, W reduces to a complete integral of the Hamilton–Jacobi equation

$$\frac{(\text{grad}\, W)^2}{2m} + V(x) = E,$$

and it is possible to show that (appendix 4.B)

$$\rho = \left| \det \left(\frac{\partial^2 W}{\partial x_i \partial \alpha_k} \right) \right|$$

solves the continuity equation (4.8.6). One then finds an approximate solution of the Schrödinger equation of the form $\psi = \sqrt{\rho}\, \text{e}^{\frac{\text{i}W}{\hbar}}$.

In general, the term proportional to $\hbar^2$ in Eq. (4.8.3) may be viewed as a quantum-mechanical 'correction' to the classical potential, so that the JWKB Lagrangian reads

$$\mathcal{L} = \frac{m}{2} v^2 - \left[V(\vec{x}) - \frac{\hbar^2}{2m} \frac{(\triangle A)}{A} \right], \tag{4.8.10}$$

with

$$m \frac{\text{d}\vec{v}}{\text{d}t} = -\text{grad} \left[V(\vec{x}) - \frac{\hbar^2}{2m} \frac{(\triangle A)}{A} \right]. \tag{4.8.11}$$

An equivalent (and useful) expression of Eq. (4.8.3) is obtained in terms of $\rho(\vec{x},t)$ defined in Eq. (4.8.4):

$$\frac{\partial W}{\partial t} = -\left\{\frac{(\mathrm{grad}\, W)^2}{2m} + V - \frac{\hbar^2}{4m}\left[\frac{(\triangle\rho)}{\rho} - \frac{1}{2}\frac{(\mathrm{grad}\,\rho)^2}{\rho^2}\right]\right\}. \tag{4.8.12}$$

It is now instructive to consider the eikonal approximation for the stationary Schrödinger equation in one spatial dimension, i.e. we study

$$\left[-\frac{\hbar^2}{2m}\frac{\mathrm{d}^2}{\mathrm{d}x^2} + V(x)\right]\varphi_E = E\varphi_E, \tag{4.8.13}$$

and we look for solutions in the form $A(x)\mathrm{e}^{\mathrm{i}S(x)/\hbar}$. We then obtain

$$\frac{\mathrm{d}^2\varphi}{\mathrm{d}x^2} = \left[\frac{\mathrm{d}^2 A}{\mathrm{d}x^2} + \frac{2\mathrm{i}}{\hbar}\frac{\mathrm{d}A}{\mathrm{d}x}\frac{\mathrm{d}S}{\mathrm{d}x} + \frac{\mathrm{i}}{\hbar}A\frac{\mathrm{d}^2 S}{\mathrm{d}x^2} - \frac{A}{\hbar^2}\left(\frac{\mathrm{d}S}{\mathrm{d}x}\right)^2\right]\mathrm{e}^{\mathrm{i}S/\hbar}. \tag{4.8.14}$$

The resulting non-linear equation is

$$A\left[\frac{1}{2m}\left(\frac{\mathrm{d}S}{\mathrm{d}x}\right)^2 + V(x) - E\right] - \frac{\mathrm{i}\hbar}{2m}\left(2\frac{\mathrm{d}A}{\mathrm{d}x}\frac{\mathrm{d}S}{\mathrm{d}x} + A\frac{\mathrm{d}^2 S}{\mathrm{d}x^2}\right) - \frac{\hbar^2}{2m}\frac{\mathrm{d}^2 A}{\mathrm{d}x^2} = 0. \tag{4.8.15}$$

On neglecting second-order terms in $\hbar$, one finds

$$\frac{\mathrm{d}S}{\mathrm{d}x} = \pm\sqrt{2m[E - V(x)]} = \pm p(x), \tag{4.8.16}$$

$$S(x) = S(x_0) \pm \int_{x_0}^{x} p(y)\,\mathrm{d}y. \tag{4.8.17}$$

The second large bracketed term in (4.8.15), with imaginary coefficient, can be written in the form

$$2\frac{\mathrm{d}A}{\mathrm{d}x}p + A\frac{\mathrm{d}p}{\mathrm{d}x} = \frac{1}{A}\frac{\mathrm{d}}{\mathrm{d}x}(A^2 p) = 0. \tag{4.8.18}$$

Now $\varphi_E^*\varphi_E = A^2$ yields

$$\varphi_E^*\varphi_E\frac{p}{m} = A^2\frac{p}{m}, \tag{4.8.19}$$

which is invariant under translations by virtue of (4.8.18). One thus obtains

$$A = \frac{C}{\sqrt{p}}. \tag{4.8.20}$$

This implies that approximate solutions take the form

$$\varphi_E^{\pm}(x) = \frac{C_\pm}{\sqrt{p}}\mathrm{e}^{\pm\frac{i}{\hbar}\int p\,\mathrm{d}x}, \tag{4.8.21}$$

where

$$p = \sqrt{2m[E - V(x)]}. \tag{4.8.22}$$

We are here assuming that $E - V(x) > 0$, but in general $E - V(x)$ may have zeros that separate regions where $E - V(x) > 0$ from regions where $E - V(x) < 0$. The problem then arises of deriving junction conditions for stationary states (Merzbacher 1961, Maslov and Fedoriuk 1981).

The result (4.8.21) may be re-expressed in terms of the initial conditions, and hence one has

$$\varphi_E^{\pm}(x) = \varphi_E^{\pm}(x_0)\sqrt{\frac{p(x_0)}{p(x)}}\ e^{\pm\frac{i}{\hbar}\int_{x_0}^{x} p(y)\, dy}. \tag{4.8.23}$$

More precisely, using

$$E = \frac{p^2(x_0)}{2m} + V(x_0)$$

one finds

$$\frac{dS}{dx}(x, p_0(x_0)) = \pm\sqrt{p_0^2(x_0) + 2m[V(x_0) - V(x)]} = \pm p(x).$$

Hence one obtains

$$\frac{\partial^2 S}{\partial x \partial p_0} = \mp \frac{p_0(x_0)}{p(x)},$$

which leads, in turn, to $\rho = \left|\frac{p(x_0)}{p(x)}\right|$ and to the result expressed by (4.8.23).

If the term of order $\hbar^2$ is not neglected in Eq. (4.8.15), one obtains the equation

$$A\left[\frac{1}{2m}\left(\frac{dS}{dx}\right)^2 + V(x) - E\right] = \frac{\hbar^2}{2m}\frac{d^2A}{dx^2}, \tag{4.8.24}$$

which leads to (see Eq. (4.8.22))

$$\frac{dS}{dx} = \pm\sqrt{p^2(x) + \frac{\hbar^2}{A}\frac{d^2A}{dx^2}} = \pm p\sqrt{1 + \frac{\hbar^2}{p^2}\frac{1}{A}\frac{d^2A}{dx^2}}. \tag{4.8.25}$$

For sufficiently large values of p, we consider a series expansion of Eq. (4.8.25) in the form

$$\frac{dS}{dx} = \pm\left[p(x) + \frac{\hbar^2}{2pA}\frac{d^2A}{dx^2} + \cdots\right]. \tag{4.8.26}$$

To this order, A is proportional to $\frac{1}{\sqrt{p}}$, and one obtains

$$\frac{dS}{dx} \cong \left[p(x) + \frac{1}{2}\hbar^2 p(x)^{-1/2}\frac{d^2}{dx^2}p(x)^{-1/2}\right]. \tag{4.8.27}$$

The integration of Eq. (4.8.27) yields

$$S(x) = S(x_0) \pm \left(\int_{x_0}^{x} p(y)\, dy + \delta S \right), \tag{4.8.28}$$

with

$$\delta S = \frac{\hbar^2}{2} \int_{x_0}^{x} p(y)^{-1/2} \frac{d^2}{dy^2} p(y)^{-1/2}\, dy. \tag{4.8.29}$$

Thus, a better approximation for φ_E is given by

$$\varphi_E^{\pm}(x) = \varphi_E^{\pm}(x_0) \sqrt{\frac{p(x_0)}{p(x)}}\, e^{\pm \frac{i}{\hbar} \left[\int_{x_0}^{x} p(y)\, dy + \delta S \right]}. \tag{4.8.30}$$

The transition from quantum to classical mechanics is formally analogous to the transition from wave optics to geometrical optics. It is thus possible to use this formal analogy to understand under which conditions one can use classical mechanics instead of quantum mechanics. For this purpose, one has to require that (cf. Eq. (3.10.7))

$$\frac{\lambda}{2\pi} \frac{|\text{grad}\, \lambda|}{\lambda} = |\text{grad}(\lambda/2\pi)| \ll 1. \tag{4.8.31}$$

Thus, the equation (3.10.25) for the wavelength in the Schrödinger theory makes it possible to re-express the inequality (4.8.31) in the form

$$\frac{m\hbar}{[2m(E-V)]^{3/2}} |\text{grad}\, V| \ll 1. \tag{4.8.32}$$

A more convenient form of the inequality is obtained by considering the modulus of the velocity and the modulus of the acceleration of the particle:

$$[2m(E-V)]^{1/2} = mv, \tag{4.8.33}$$

$$|\text{grad}\, V| = ma. \tag{4.8.34}$$

Hence one finds the condition (Fock 1978)

$$\frac{\hbar a}{mv^3} \ll 1. \tag{4.8.35}$$

Moreover, denoting by r the curvature radius of the orbit of the particle, one finds

$$a^2 = \left(\frac{v^2}{r}\right)^2 + \left(\frac{dv}{dt}\right)^2, \tag{4.8.36}$$

which implies

$$a \geq \frac{v^2}{r}, \tag{4.8.37}$$

and hence, from (4.8.35),

$$\frac{\hbar}{mvr} \ll 1. \tag{4.8.38}$$

Interestingly, on denoting by λ the de Broglie wavelength, one thus finds the desired condition in the form

$$\frac{\lambda}{2\pi r} \ll 1. \tag{4.8.39}$$

4.8.1 On the meaning of semi-classical

In a more mathematical presentation, one starts from the time-dependent Schrödinger equation for which a Cauchy problem with rapidly oscillating initial data is studied. The initial condition is written in the form

$$\psi(\vec{x},0) = \psi_0(\vec{x}) e^{\frac{i}{\hbar} S_0(\vec{x})}, \tag{4.8.40}$$

where the potential V and S_0 are real-valued and infinitely differentiable, and ψ_0 is taken to be infinitely differentiable and with compact support. One then looks for an asymptotic solution with initial condition (4.8.40) as $\hbar \rightarrow 0^+$ and $\vec{x} \in \mathbf{R}^3$, within an arbitrary finite time t. The corresponding asymptotic formulae are said to be a *semi-classical approximation* or semi-classical asymptotics whenever they are expressed only in terms of functions that characterize the underlying classical mechanics (Maslov and Fedoriuk 1981). Note that, when the limit $\hbar \rightarrow 0$ is used, this is not exactly a limit in the usual sense one has learned in analysis, since $\hbar$ is not a dimensionless quantity.

4.8.2 Example: α-decay

Two kinds of spontaneous emission are known to exist in nature: α- and β-decay. Experiments in which a sufficiently large number of α-particles enter a chamber with a thin mica window and are collected show that α-rays correspond to positively charged particles with a $\frac{q}{m}$ ratio that is the same as that of a doubly ionized helium atom: He^{++}. Their identification with He^{++} is made possible because, when a gas of α-particles produces light, it indeed reveals the spectroscopic lines of He^{++}. An important quantitative law of α-decay was discovered, at the experimental level, by Geiger and Nuttall in 1911. It states that the velocity v of the α-particle, and the mean life T of the emitting nucleus, are related by the empirical equation

$$\log(T) = A + B \log(v), \tag{4.8.41}$$

where the constants A and B can be determined in the laboratory. We are now aiming to develop an elementary theory of α-decay, following Born

(1969). For further details, we refer the reader to the work in Gamow (1928) and Fermi (1950).

Let $V(r)$ be the potential for a particle emitted by a nucleus of atomic number Z. By virtue of the shielding effect, it is as if the particle were 'feeling' the effect of a field with residual atomic number $Z-2$. At large distances, i.e. for $r > r_0$, the resulting potential is Coulombian, while for $r < r_0$ the precise form of $V(r)$ is not known, but for phenomenological reasons we expect it should have the shape of a *crater* (the α-particle being 'trapped' in the field of the nucleus). In other words, we consider a potential such that

$$V(r) = -V_0 < 0 \quad \text{if } r \in]0, r_0[, \tag{4.8.42}$$

$$V(r) = \frac{2(Z-2)e^2}{r} \quad \text{if } r > r_0, \tag{4.8.43}$$

and a positive value of the energy E such that $E < V$ if $r \in]r_0, r_1[$, $E = V$ at $r = r_1$ and $E > V$ if $r > r_1$. The emission frequency, ν, can be expressed as the product of the number n of times that the α-particle hits against the crater with probability p that it can 'tunnel' through the barrier given by the crater: $\nu = np$. To find p, one has to solve the stationary Schrödinger equation with the lowest possible value of angular momentum (see section 5.4). The part of the stationary state depending on spatial variables is written as $\varphi(r) = \frac{\psi(r)}{r}$, where $\psi(r)$ satisfies the equation

$$\left\{ \frac{d^2}{dr^2} + \frac{2m}{\hbar^2}[E - V(r)] \right\} \psi = 0. \tag{4.8.44}$$

In between r_0 and r_1, the wave function is exponentially damped, and hence one has, approximately,

$$p = |\psi(r_1)/\psi(r_0)|^2. \tag{4.8.45}$$

In such an intermediate region the potential is well approximated by the Coulomb part (4.8.43). For large values of the atomic number Z, one thus looks for solutions of Eq. (4.8.44) in the form (cf. Eq. (4.8.1))

$$\psi = \exp\left[\frac{y(r)}{\hbar}\right], \tag{4.8.46}$$

which leads to a non-linear equation for y:

$$\hbar y'' + y'^2 - F(r) = 0, \tag{4.8.47}$$

having defined

$$F(r) \equiv 2m \left[-E + \frac{2(Z-2)e^2}{r} \right]. \tag{4.8.48}$$

If the term $\hbar y''$ can be neglected in Eq. (4.8.47), one finds

$$y(r) = \int_a^r \sqrt{F(x)}\, dx, \tag{4.8.49}$$

$$\frac{\psi(r_1)}{\psi(r_0)} = \exp\left[\frac{y(r_1) - y(r_0)}{\hbar}\right] = \exp\left[\frac{1}{\hbar}\int_{r_0}^{r} \sqrt{F(x)}\, dx\right]. \tag{4.8.50}$$

The integral occurring in Eq. (4.8.50) is non-trivial but only involves standard techniques (e.g. Gradshteyn and Ryzhik (1965)), so that one can eventually express the probability as

$$p = \exp\left[-(2n_0 - \sin 2n_0)\frac{8\pi e^2}{h}\frac{(Z-2)}{v}\right], \tag{4.8.51}$$

where n_0 is a root of the equation

$$\cos^2 n_0 = \frac{r_0 E}{2(Z-2)e^2}. \tag{4.8.52}$$

For small values of the energy E of α-particles, and for a deep crater, the probability can be approximated as follows:

$$p = \exp\left[-8\pi^2 e^2 \frac{(Z-2)}{hv} + \frac{16\pi e}{h}\sqrt{m(Z-2)r_0}\right]. \tag{4.8.53}$$

The decay constant, ν_D, is thus given by

$$\log(\nu_D) = \log(h/4mr_0^2) - 8\pi^2 e^2 \frac{(Z-2)}{hv} + \frac{16\pi e}{h}\sqrt{m(Z-2)r_0}. \tag{4.8.54}$$

Indeed, Eq. (4.8.54) differs from the empirical law (4.8.41), because it predicts that $\log(\nu_D)$ depends linearly on $\frac{1}{v}$. Moreover, it introduces a dependence on Z and r_0. However, the discrepancy is not very severe from the point of view of the actual values taken by the physical parameters: the velocity varies in between 1.4×10^9 and 2×10^9 cm s^{-1}, and the variation of Z and r_0 is negligible for the elements of radioactive series. Note also that the range of values of the decay constant is quite large, since $\frac{8\pi^2 e^2}{h}$ is very large.

4.9 From wave mechanics to Bohr–Sommerfeld

Suppose we can solve the Schrödinger equation by making the geometrical-optics approximation, according to which

$$\psi_{E,\alpha}(x, y, z, t) = A\, e^{\frac{2\pi i}{h}[Et - S(x,y,z,\alpha)]}, \tag{4.9.1}$$

where S is a complete integral of the Hamilton–Jacobi equation. For the wave function to be single-valued at each space-time point, it is necessary that along any closed curve C_k where the propagation takes place one can write

$$\int_{C_k} p_j \, dq^j = h \, n_k(E, \alpha). \tag{4.9.2}$$

Interestingly, one therefore recovers the Einstein formulation of the Bohr–Sommerfeld quantization conditions.

Thus, if we can obtain a solution of the Schrödinger equation with the help of a complete integral of the Hamilton–Jacobi equation as in (4.9.1), wave mechanics yields immediately an interpretation of the previously 'ad hoc' Bohr–Sommerfeld conditions. It is now instructive to consider in detail some applications of the Bohr–Sommerfeld method, first outlined in section 1.8. They are as follows.

4.9.1 Quantization of Keplerian motion

Recall that, in classical mechanics, for a particle of mass m in a central potential, the kinetic energy in polar coordinates reads as

$$T = \frac{m}{2}\left(\dot{r}^2 + r^2\dot{\theta}^2\right), \tag{4.9.3}$$

with associated momenta

$$p_r = \frac{\partial T}{\partial \dot{r}} = m\dot{r}, \quad p_\theta = \frac{\partial T}{\partial \dot{\theta}} = mr^2\dot{\theta}. \tag{4.9.4}$$

The conservation of angular momentum (area law) implies that p_θ is a constant of motion. The Coulomb law makes it possible to write

$$H = E = \frac{m}{2}\left(\dot{r}^2 + r^2\dot{\theta}^2\right) - \frac{k}{r}. \tag{4.9.5}$$

On solving with respect to $m\dot{r}$, and replacing p_θ with γ, one finds

$$p_r = m\dot{r} = \pm\sqrt{2mE - \frac{\gamma^2}{r^2} + \frac{2mk}{r}} = \pm\sqrt{\varphi(r)}. \tag{4.9.6}$$

If E is negative, $\varphi(r)$ has two positive roots r_1 and r_2, and the modulus of $\vec{r}$ lies in between such roots. This is therefore a periodic motion.

The Bohr–Sommerfeld condition for the azimuth variable is

$$\int_{C_\theta} p_j \, dq^j = \oint p_\theta \, d\theta = 2\pi\gamma = n_1 h, \tag{4.9.7}$$

which implies that $\gamma = n_1 \frac{h}{2\pi}$, where n_1 is called the azimuth quantum number. The second quantization condition reads as

$$\int_{C_r} p_j \, dq^j = \oint p_r \, dr = n_2 h, \tag{4.9.8}$$

where n_2 is called the radial quantum number. Now the integral (4.9.8) can be expressed in the form

$$\int_{r_1}^{r_2} \sqrt{\varphi(r)}\,\mathrm{d}r + \int_{r_2}^{r_1} -\sqrt{\varphi(r)}\,\mathrm{d}r = 2\int_{r_1}^{r_2} \sqrt{\varphi(r)}\,\mathrm{d}r. \tag{4.9.9}$$

Note that it is possible to express p in terms of q, which is necessary to evaluate the Bohr–Sommerfeld integrals, starting from the Hamilton–Jacobi equation

$$\frac{1}{2m}\left[\left(\frac{\partial S}{\partial r}\right)^2 + \frac{1}{r^2}\left(\frac{\partial S}{\partial \theta}\right)^2\right] - \frac{k}{r} = E. \tag{4.9.10}$$

On setting $\frac{\partial S}{\partial \theta} = \gamma$ one finds for $p_r = \frac{\partial S}{\partial r}$ the result (4.9.6). The Jacobi function

$$S = \gamma\theta + \oint p_r\,\mathrm{d}r \tag{4.9.11}$$

is a complete integral because it contains two arbitrary constants, i.e. the pair (γ, E). The admissible values of these constants are determined from the quantization conditions (4.9.7) and (4.9.8). Sommerfeld was able to evaluate, with the help of the residue method, the integral in Eq. (4.9.8). On equating the result to $n_2 h$, and replacing γ with $n_1 \frac{h}{2\pi}$, he found, for the energy of the orbit characterized by the quantum numbers n_1 and n_2, the formula

$$E_{n_1,n_2} = -\frac{2\pi m e^4}{h^2(n_1+n_2)^2}. \tag{4.9.12}$$

Note that the energy of bound states depends only on the sum $n_1 + n_2$, and this feature leads to degeneracy of the eigenvalues.

4.9.2 Harmonic oscillator

The one-dimensional harmonic oscillator has a classical Hamiltonian

$$H = \frac{p^2}{2m} + \frac{k}{2}x^2, \tag{4.9.13}$$

and hence the Newton equation is solved by

$$x(t) = A\sin\left(\sqrt{\frac{k}{m}}t + \alpha\right). \tag{4.9.14}$$

Over a period, one has

$$S(x) = \int_{x_0}^{x} p_{x'}\,\mathrm{d}x' = m\int_{x_0}^{x} v\,\mathrm{d}x'. \tag{4.9.15}$$

Thus, on defining $\nu \equiv \frac{1}{2\pi}\sqrt{\frac{k}{m}}$, one finds

$$J = \oint \mathrm{d}S = \oint mv \, \mathrm{d}x = m \int_0^{1/\nu} v^2 \, \mathrm{d}t$$
$$= 4\pi^2 \nu^2 A^2 m \int_0^{1/\nu} \cos^2 \left(\sqrt{\frac{k}{m}} t + \alpha \right) \mathrm{d}t$$
$$= 2\pi^2 \nu m A^2. \tag{4.9.16}$$

The energy E of the particle equals the kinetic energy that it has on passing through the equilibrium position, because at such a point the potential energy vanishes. One can thus write

$$\frac{E}{\omega} = \frac{J}{2\pi} = n\hbar. \tag{4.9.17}$$

The Bohr–Sommerfeld method leads therefore to energy eigenvalues which are integer multiples of $\hbar\omega$:

$$E = 0, \hbar\omega, 2\hbar\omega, \ldots, n\hbar\omega, \ldots \tag{4.9.18}$$

unlike the experimentally verified result of section 4.7, according to which all of the above values should be corrected by adding $\frac{1}{2}\hbar\omega$.

4.9.3 Rotator in a plane

A particle of mass m moving on a circle of radius R has kinetic energy

$$T = \frac{m}{2} R^2 \dot{\theta}^2, \tag{4.9.19}$$

with azimuthal momentum

$$p_\theta = \frac{\partial T}{\partial \dot{\theta}} = mR^2 \dot{\theta}. \tag{4.9.20}$$

The Bohr–Sommerfeld quantization condition yields

$$\int_0^{2\pi} mR^2 \dot{\theta} \, \mathrm{d}\theta = \int_0^{2\pi} mvR \, \mathrm{d}\theta = nh, \tag{4.9.21}$$

which implies

$$mvR = \frac{nh}{2\pi}, \tag{4.9.22}$$

and hence

$$E_n = T = \frac{m}{2} v^2 = \frac{n^2 h^2}{8\pi^2 m R^2} = \frac{n^2 h^2}{8\pi^2 I}, \tag{4.9.23}$$

where $I = mR^2$ is the moment of inertia of the rotator.

On the other hand, the Schrödinger equation for stationary states reads, in our case, as

$$\frac{1}{R^2}\frac{d^2a}{d\theta^2} + \frac{8\pi^2 m}{h^2}Ea = 0, \tag{4.9.24}$$

which is solved by

$$a(\theta) = A\sin\left[\frac{2\pi}{h}\sqrt{2mE}R(\theta - \theta_0)\right], \tag{4.9.25}$$

where A and θ_0 are integration constants. For a to be a well-defined function of θ, the following periodicity condition should hold:

$$a(\theta) = a(\theta + 2\pi), \tag{4.9.26}$$

which leads to

$$2\pi\frac{\sqrt{2mE}}{h}R = n, \tag{4.9.27}$$

and hence one finds again the formula (4.9.23) for the energy eigenvalues. The eigenfunctions belonging to these eigenvalues are

$$a_n(\theta) = A\sin n(\theta - \theta_0), \tag{4.9.28}$$

and the resulting wave function reads

$$\psi_n(\theta, t) = A\sin n(\theta - \theta_0)\, e^{\frac{2\pi i}{h}(E_n t + \alpha)}. \tag{4.9.29}$$

Note that the above eigenfunctions are an orthogonal system, in that

$$\int_0^{2\pi} a_r a_l R\, d\theta = 0 \quad \text{if } r \neq l. \tag{4.9.30}$$

The requirement of normalization implies that $A = \frac{1}{\sqrt{\pi R}}$, because

$$A^2 \int_0^{2\pi} \sin^2[n(\theta - \theta_0)]\, d\theta = \pi R. \tag{4.9.31}$$

By virtue of (4.9.29), the probability density takes the form

$$\psi_n \psi_n^* = A^2 \sin^2[n(\theta - \theta_0)], \tag{4.9.32}$$

and hence does not evolve in time. The corresponding current (see the definition (4.1.7)) is found to vanish.

4.10 Problems

4.P1. Consider the equation originally studied by Schrödinger to construct wave mechanics:

$$\frac{\partial^2}{\partial t^2}\psi(\vec{x},t) = -\frac{H^2}{\hbar^2}\psi(\vec{x},t). \tag{4.10.1}$$

(i) Which initial conditions are necessary to solve Eq. (4.10.1)?

(ii) How many conditions on partial derivatives with respect to $\vec{x}$ are necessary to specify the Cauchy problem?

(iii) Prove that Eq. (4.10.1) is compatible with Eq. (3.10.32) for the Fourier transform of ψ.

(iv) Derive how $\psi(\vec{x},t)$ depends on the initial conditions and on the Green kernel of the Hamiltonian operator.

(v) Re-express Eq. (4.10.1) in first-order form. Is it possible to associate to such a first-order partial differential equation a local conservation law?

4.P2. What would be the effect in the Schrödinger equation

$$\mathrm{i}\hbar\frac{\partial}{\partial t}\psi(\vec{x},t) = \left[-\frac{\hbar^2}{2m}\triangle + V(\vec{x})\right]\psi(\vec{x},t) \tag{4.10.2}$$

of a complex-valued potential?

4.P3. The Schrödinger equation in a central potential $U(r)$ leads to the analysis, in n spatial dimensions, of the following equation for stationary states:

$$\left[\frac{\mathrm{d}^2}{\mathrm{d}r^2} + \frac{(n-1)}{r}\frac{\mathrm{d}}{\mathrm{d}r} - \frac{l(l+n-2)}{r^2} + \kappa\right]\psi(r) = V(r)\psi(r), \tag{4.10.3}$$

where $\kappa \equiv \frac{2mE}{\hbar^2}$, $V(r) \equiv \frac{2m}{\hbar^2}U(r)$.

(i) Look for JWKB solutions: $\psi(r) = A(r)\mathrm{e}^{\frac{\mathrm{i}}{\hbar}S(r)}$, deriving the differential equations for the functions A and S.

(ii) Compute, to first order in $\hbar$, the prefactor $A(r)$ and the phase $S(r)$.

(iii) Evaluate the corrections resulting from $O(\hbar^2)$ terms.

Hint: use the method of section 4.8, and then look at section 4 of Esposito (1998a).

Appendix 4.A

Glossary of functional analysis

This appendix describes, without proofs, some basic notions and results of functional analysis that are needed in the present chapter and in the rest of our book. By doing so, we hope that our manuscript will be eventually self-contained. For more extensive treatments, the reader is referred to the work in von Neumann (1955), Gel'fand and Vilenkin (1964), Gel'fand and Shilov (1967), Reed and Simon (1980).

(i) *Inner product spaces and Hilbert spaces.* A complex vector space V is called an *inner product space* if there exists a complex-valued function $(\cdot,\cdot)$ on $V \times V$ that satisfies the following conditions $\forall x, y, z \in V$ and $\forall \alpha \in \mathbb{C}$:

$$(x,x) \geq 0, \text{ and } (x,x) = 0 \Longleftrightarrow x = 0, \tag{4.A.1}$$

$$(x, y+z) = (x,y) + (x,z), \tag{4.A.2}$$

$$(x, \alpha y) = \alpha (x,y), \tag{4.A.3}$$

$$(x,y) = (y,x)^*, \tag{4.A.4}$$

where * is the operation of complex conjugation. For example, on denoting by $C[a,b]$ the complex-valued continuous functions on the closed interval $[a,b]$, the *inner product* of f and $g \in C[a,b]$ may be defined by

$$(f,g) \equiv \int_a^b f^*(x) g(x) \, dx. \tag{4.A.5}$$

If x and y are vectors of an inner product space, the Schwarz inequality holds:

$$|(x,y)| \leq \|x\| \, \|y\|. \tag{4.A.6}$$

Every inner product space V is a normed linear space with the norm

$$\|x\| \equiv \sqrt{(x,x)}. \tag{4.A.7}$$

Thus, one has a natural metric

$$d(x,y) \equiv \sqrt{(x-y, x-y)} \tag{4.A.8}$$

in V, and the notion of *completeness*: V is complete if all Cauchy sequences converge.

A complete inner product space is, by definition, a *Hilbert space*, whereas inner product spaces (without the completeness property) are also called pre-Hilbert spaces. In the applications, separability (i.e. existence of a countable basis) of Hilbert spaces will be assumed. As a first example, we consider $\mathcal{L}^2[a,b]$, i.e. the set of complex-valued measurable functions on the closed interval $[a,b]$ such that

$$\int_a^b |f(x)|^2 \, dx < \infty, \tag{4.A.9}$$

with inner product given by Eq. (4.A.5). The set $\mathcal{L}^2[a,b]$ is complete and is hence a Hilbert space.

(ii) *Maps between Hilbert spaces.* Two Hilbert spaces $\mathcal{H}_1$ and $\mathcal{H}_2$ are said to be *isomorphic* if there exist invertible maps U from $\mathcal{H}_1$ onto $\mathcal{H}_2$ such that

$$(Ux, Uy)_{\mathcal{H}_2} = (x,y)_{\mathcal{H}_1}, \quad \forall x, y \in \mathcal{H}_1. \tag{4.A.10}$$

In particular, an isomorphism of a Hilbert space $\mathcal{H}$ onto itself is called an automorphism of a Hilbert space. The set of all automorphisms of $\mathcal{H}$ is a group, and is called the *automorphism group* of $\mathcal{H}$, denoted by $\text{Aut}(\mathcal{H})$.

The set of pairs $\langle x, y \rangle$ with $x \in \mathcal{H}_1$, $y \in \mathcal{H}_2$ is a Hilbert space with inner product

$$(\langle x_1, y_1 \rangle, \langle x_2, y_2 \rangle) \equiv (x_1, x_2)_{\mathcal{H}_1} + (y_1, y_2)_{\mathcal{H}_2}. \tag{4.A.11}$$

The resulting space is the direct sum of Hilbert spaces $\mathcal{H}_1$ and $\mathcal{H}_2$, and is denoted by $\mathcal{H}_1 \oplus \mathcal{H}_2$.

Now for all $\varphi_1 \in \mathcal{H}_1$ and $\varphi_2 \in \mathcal{H}_2$, let us denote by $\varphi_1 \otimes \varphi_2$ the actions of bilinear forms on $\mathcal{H}_1 \times \mathcal{H}_2$ given by

$$\Big(\varphi_1 \otimes \varphi_2 \Big) \langle \psi_1, \psi_2 \rangle \equiv (\psi_1, \varphi_1) \, (\psi_2, \varphi_2). \tag{4.A.12}$$

Let Λ be the set of finite linear combinations of such conjugate linear forms endowed with an inner product defined by

$$\Big(\varphi \times \psi, \eta \times \mu \Big) \equiv (\varphi, \eta) \, (\psi, \mu). \tag{4.A.13}$$

The tensor product of $\mathcal{H}_1$ and $\mathcal{H}_2$, denoted by $\mathcal{H}_1 \otimes \mathcal{H}_2$, is defined as the completion of Λ under the inner product given by Eq. (4.A.13).

(iii) *Operators on Hilbert spaces.* The Hellinger–Toeplitz theorem (1910) states that, if A is an everywhere defined linear operator on a Hilbert space $\mathcal{H}$, with

$$(Ax, y) = (x, Ay), \quad \forall x, y \in \mathcal{H},$$

then A is bounded, i.e. for some $C \geq 0$ one has $\|Ax\| \leq C\|x\|$, $\forall x \in \mathcal{H}$. Thus, if B is an unbounded operator (e.g. differential operators are, in general, unbounded), it cannot be everywhere defined, but is only defined on a sub-space $D(B)$, called the *domain* of B, of the Hilbert space $\mathcal{H}$. For example, defining what one means by $B+C$ or BC, if B and C are unbounded, may be difficult. In particular, the sum $B+C$ is only defined on $D(B) \cap D(C)$, which might be the empty set!

(1) By definition, an *operator*

$$A : D(A) \rightarrow \mathcal{H}$$

on a Hilbert space $\mathcal{H}$ is a linear map from its domain, a linear sub-space of $\mathcal{H}$, into $\mathcal{H}$. Unless otherwise stated, we will assume that the domain, denoted by $D(A)$, is dense. The *graph* $\Gamma(A)$ of A is the set of pairs

$$\{(\varphi, A\varphi) | \varphi \in D(A)\}.$$

The operator A is *closed* if its graph is a closed subset of $\mathcal{H} \times \mathcal{H}$.

(2) If A and B are operators on $\mathcal{H}$, and if the graph of B includes the graph of A, B is said to be an *extension* of A, and one writes $A \subset B$. This implies that $D(A) \subset D(B)$ and $B\varphi = A\varphi$, $\forall \varphi \in D(A)$.

(3) An operator A is *closable* if it has a closed extension. Every closable operator has a smallest closed extension, called its *closure* and denoted by $\overline{A}$.

(4) Let A be a densely defined linear operator on a Hilbert space $\mathcal{H}$, and let $D(A^\dagger)$ be the set of $\varphi \in \mathcal{H}$ for which there exists an $\eta \in \mathcal{H}$ with

$$(A\psi, \varphi) = (\psi, \eta), \quad \forall \psi \in D(A). \tag{4.A.14}$$

For each such $\varphi \in D(A^\dagger)$, one defines

$$A^\dagger \varphi \equiv \eta. \tag{4.A.15}$$

The operator $A^\dagger$ is, by definition, the *adjoint* of A.

(5) A densely defined operator A on a Hilbert space is called symmetric, or Hermitian, if $D(A) \subset D(A^\dagger)$ and $A\varphi = A^\dagger \varphi$, $\forall \varphi \in D(A)$. This implies that A is symmetric if and only if

$$(A\varphi, \psi) = (\varphi, A\psi), \quad \forall \varphi, \psi \in D(A). \tag{4.A.16}$$

(6) A densely defined operator A is called *self-adjoint* if and only if it is symmetric and its domain coincides with the domain of its adjoint: $D(A) = D(A^\dagger)$. This condition on the domains is crucial. In sections 4.5 and 5.4 we give examples of symmetric operators that are not self-adjoint.

(7) A symmetric operator A is called *essentially self-adjoint* if its closure, $\overline{A}$, is self-adjoint. One can then prove that a necessary and sufficient condition for A to be essentially self-adjoint is that A has one and only one self-adjoint extension.

(8) Given a symmetric operator A on a Hilbert space $\mathcal{H}$, its self-adjointness can be proved by studying the equations

$$A^\dagger \varphi = \pm \mathrm{i} \varphi. \tag{4.A.17}$$

On defining the *deficiency sub-spaces*

$$\mathcal{H}^+(A) \equiv \left\{ \varphi \in D(A^\dagger) : A^\dagger \varphi = \mathrm{i}\varphi \right\}, \tag{4.A.18}$$

$$\mathcal{H}^-(A) \equiv \left\{ \varphi \in D(A^\dagger) : A^\dagger \varphi = -\mathrm{i}\varphi \right\}, \tag{4.A.19}$$

the operator A turns out to be self-adjoint if its *deficiency indices*, defined by

$$n_\pm(A) \equiv \dim \mathcal{H}^\pm(A), \tag{4.A.20}$$

satisfy the condition

$$n_+(A) = n_-(A) = 0. \tag{4.A.21}$$

In other words, one looks for vectors in the domain of the operator which are also eigenfunctions of its adjoint belonging to the eigenvalue $+\mathrm{i}$ or $-\mathrm{i}$. The dimensions of the spaces of solutions of such equations are the deficiency indices of the operator A, and they should both vanish if A is self-adjoint. At first sight, this condition might seem a bit odd, but is almost straightforward if one remarks that, by virtue of self-adjointness, if the first equation in (4.A.17) holds (i.e. with $+\mathrm{i}$), then also

$$A\varphi = \mathrm{i}\varphi, \tag{4.A.22}$$

because $D(A) = D(A^\dagger)$. This leads to (the inner product being anti-linear in the first argument, according to (4.A.5))

$$-\mathrm{i}(\varphi,\varphi) = (\mathrm{i}\varphi,\varphi) = (A\varphi,\varphi) = (\varphi, A^\dagger\varphi) \\ = (\varphi, A\varphi) = (\varphi, \mathrm{i}\varphi) = \mathrm{i}(\varphi,\varphi), \tag{4.A.23}$$

which implies that $\varphi = 0$. Similarly, one proves that the second equation in (4.A.17) has no solutions if A is self-adjoint:

$$\mathrm{i}(\varphi,\varphi) = (-\mathrm{i}\varphi,\varphi) = (A\varphi,\varphi) = (\varphi, A^\dagger\varphi) \\ = (\varphi, A\varphi) = (\varphi, -\mathrm{i}\varphi) = -\mathrm{i}(\varphi,\varphi). \tag{4.A.24}$$

If the deficiency indices are equal but non-vanishing, then A has self-adjoint extensions. An example with $n_+ = n_- = 1$ is the operator $\mathrm{i}\frac{\mathrm{d}}{\mathrm{d}x}$ on square-integrable functions on $[a, b]$ with periodic boundary conditions. For $\mathcal{L}^2(-\infty, a)$ this operator is not self-adjoint since $n_- = 1, n_+ = 0$, while for $\mathcal{L}^2(\mathbf{R})$ it has $n_+ = n_- = 0$ and hence is self-adjoint.

(iv) *Spectral resolution of operators.* The proper treatment of Eq. (4.4.23) requires some care. For this purpose, let us first consider a bounded self-adjoint operator A defined on a separable Hilbert space $\mathcal{H}$. It can be proved that, for each interval $I = (\alpha, \beta)$ on the real axis, there exists a maximal invariant subspace $\mathcal{H}_I \subset \mathcal{H}$ for which the quadratic form (Ax, x) takes values in the interval (α, β) for $\|x\| = 1$. Thus, the restriction of the operator A to the subspace $\mathcal{H}_I$ realizes a map of that subspace into itself, which differs in norm from multiplication by the number α by no more than $\beta - \alpha$ (Gel'fand and Shilov 1967). The projection operator of the subspace $\mathcal{H}_I$ is denoted here by $E(I)$. Moreover, if $I_{-\infty}^{\lambda}$ denotes the interval $(-\infty, \lambda)$, the projector $E(I_{-\infty}^{\lambda})$ will be abbreviated as I_λ. The set of all projectors $E(I)$ is called the *spectral family* of the operator A. If A is a *compact operator*, i.e. if it maps any bounded set into a compact set, the spectral family $E(I)$ consists of the projectors on the subspaces spanned by the eigenvectors for which the eigenvalues lie in the interval I. In the general case, one can decompose the interval $\Big[-\|A\|, \|A\|\Big]$ of the real axis, $\forall \varepsilon > 0$, into a sum of intervals I_j $(j = 1, 2, \ldots, N; N \leq \frac{2}{\varepsilon}\|A\|)$ of length smaller than ε, such that the Hilbert space $\mathcal{H}$ is decomposed into the orthogonal sum of subspaces $\mathcal{H}_I$. In each of these subspaces, the operator A acts as a map which differs in norm by less than ε from multiplication with the eigenvalue λ_j. Hence the operator A differs, in norm, by less than ε from the operator

$$\sum_{j=1}^{n} \lambda_j E(I_j),$$

throughout the whole Hilbert space. For $\varepsilon \to 0$ this sum converges to a limit which is often denoted as

$$A = \int_{-\infty}^{\infty} \lambda \, \mathrm{d}E_\lambda. \tag{4.A.25}$$

The integration in (4.A.25) is actually a Stieltjes integral over the interval

$$\Big[-\|A\|, \|A\|\Big],$$

and we need to be more precise. The correct way to interpret Eq. (4.A.25) is expressed by the formula

$$(Af, g) = \int_{-\infty}^{\infty} \lambda \, \mathrm{d}(E_\lambda f, g), \quad \forall f, g \in \mathcal{H}. \tag{4.A.26}$$

This is the *spectral resolution* of the operator A, which replaces the eigenvector expansion valid for finite-dimensional or compact operators.

The work in von Neumann (1955) extended these results to the case of an unbounded self-adjoint operator A, which is defined on a dense subset $\mathcal{H}_A$ of the Hilbert space $\mathcal{H}$. For self-adjoint operators, one can construct invariant subspaces $\mathcal{H}_I$ with the properties described above, but the interval $\Big[-\|A\|, \|A\|\Big]$ is replaced by the whole real axis, since an unbounded operator has no finite norm. The theorem on the spectral decomposition of a self-adjoint operator can be stated after defining the concept of a resolution of unity. For this purpose, suppose that to every interval $I = [a, b)$ on the real line there corresponds a bounded self-adjoint operator $E(I)$ in a Hilbert space $\mathcal{H}$, and that the following three properties are satisfied.

(1) For any two intervals I_1 and I_2

$$E(I_1)E(I_2) = E(I_1 \cap I_2), \tag{4.A.27}$$

(2) The asymptotic conditions hold according to which

$$\lim_{x \to +\infty} (E(x)f, g) = (f, g), \quad \forall f, g \in \mathcal{H}, \tag{4.A.28}$$

$$\lim_{x \to -\infty} (E(x)f, g) = 0, \quad \forall f, g \in \mathcal{H}, \tag{4.A.29}$$

where

$$E(x) \equiv E(I_x), \tag{4.A.30}$$

and I_x is the interval $(-\infty, x)$.

(3) If the interval I is a countable union of non-intersecting intervals, i.e. $I = \cup_{n=1}^{\infty} I_n$, then

$$E(I) = \sum_{n=1}^{\infty} E(I_n). \tag{4.A.31}$$

A system of operators $E(I)$ with the above properties is called a *resolution of unity*. From Eq. (4.A.27) it follows that for any interval I one has $E^2(I) = E(I)$. This implies that $E(I)$ is a projection operator, projecting the space $\mathcal{H}$ onto the subspace $\mathcal{H}_I \equiv E(I)\mathcal{H}$. The operators $E(I)$ are positive-definite for any $f \in \mathcal{H}$, because

$$(E(I)f, f) = (E^2(I)f, f) = (E(I)f, E(I)f) \geq 0. \tag{4.A.32}$$

For any interval I and any element f of the Hilbert space $\mathcal{H}$ one defines

$$\mu_f(I) \equiv (E(I)f, f). \tag{4.A.33}$$

By construction, $\mu_f(I)$ is a countably additive positive measure defined on the intervals I, and is called the *spectral measure* corresponding, by the resolution of unity $E(I)$, to the element f.

The theorem on the spectral decomposition of a self-adjoint operator can be now stated as follows (Gel'fand and Vilenkin 1964).

T.1 Let A be a self-adjoint operator on a Hilbert space $\mathcal{H}$. Then there exists a resolution of unity $E(I)$ such that the operator A is defined on the set Ω_A of those elements $f \in \mathcal{H}$ for which the integral

$$\int_{-\infty}^{\infty} x^2 \, d\mu_f(x)$$

converges. The operator A is given, for these elements f, by the formula

$$Af = \int_{-\infty}^{\infty} x \, d(E(x)f), \tag{4.A.34}$$

i.e. for all $f, g \in \Omega_A$ one has

$$(Af, g) = \int_{-\infty}^{\infty} x \, d(E(x)f, g). \tag{4.A.35}$$

(v) *Cyclic operators.* A self-adjoint operator A is called a *cyclic operator* if there exists a vector f such that the linear combinations of the vectors $E(I)f$ are everywhere dense in $\mathcal{H}$. The vector f is called a *cyclic vector.* If A is a cyclic operator, the Hilbert space $\mathcal{H}$ can be realized as a space $\mathcal{L}_f^2$ of square-integrable functions φ with respect to the measure $\mu_f(I)$, whereby to the operator A corresponds the operator of multiplication of the functions φ by x. The domain of definition of A by this realization consists of those functions $\varphi \in \mathcal{L}_f^2$ for which the integral

$$\int_{-\infty}^{\infty} x^2 |\varphi(x)|^2 \, d\mu_f(x)$$

converges.

This realization can be achieved as follows. With each vector of the form $E(I)f$ one associates the characteristic function of the interval I:

$$E(I)f \to \chi_I(x).$$

In particular, one associates with f the function identically equal to 1 on the real line:

$$f \to 1.$$

Such a correspondence is isometric, i.e.

$$\|E(I)f\|^2 = \int_{-\infty}^{\infty} |\chi_I(x)|^2 \, d\mu_f(x). \tag{4.A.36}$$

In fact, one has (Gel'fand and Vilenkin 1964)

$$\begin{aligned} \|E(I)f\|^2 &= (E(I)f, E(I)f) = (E(I)f, f) \\ &= \mu_f(I) = \int_I d\mu_f(x) = \int_{-\infty}^{\infty} |\chi_I(x)|^2 \, d\mu_f(x). \end{aligned} \tag{4.A.37}$$

The isometric correspondence $E(I)f \rightarrow \chi_I(x)$ is then extended using linear combinations and passing to the limit. Since, on the one hand, the linear combinations of the vectors $E(I)f$ are everywhere dense in $\mathcal{H}$, and, on the other hand, the linear combinations of the characteristic functions χ_I are everywhere dense in $\mathcal{L}_f^2$, the isometry between $\mathcal{H}$ and $\mathcal{L}_f^2$ is proved (Gel'fand and Vilenkin 1964).

Note now that, from Eqs. (4.A.27) and (4.A.35), one finds for any interval I the equality

$$(AE(I)f, g) = \int_{-\infty}^{\infty} x \, \mathrm{d}(E(x)E(I)f, g) = \int_I x \, \mathrm{d}(E(x)f, g). \tag{4.A.38}$$

In particular, on taking $g = f$, this leads to

$$\begin{aligned}(AE(I)f, f) &= \int_I x \, \mathrm{d}(E(x)f, f) = \int_I x \, \mathrm{d}\mu_f(x) \\ &= \int_{-\infty}^{\infty} x \, \chi_I(x) \, \mathrm{d}\mu_f(x).\end{aligned} \tag{4.A.39}$$

Equation (4.A.39) shows that to the operator A there corresponds in $\mathcal{L}_f^2$ the operator mapping the characteristic functions χ_I into the functions $x\chi_I(x)$. Since the functions χ_I form an everywhere dense set in $\mathcal{L}_f^2$, we have proved that, for the realization considered here, there corresponds to A the operator of multiplication by x of the functions $\varphi \in \mathcal{L}_f^2$ (Gel'fand and Vilenkin 1964).

(vi) If φ is an eigenfunction of the operator A, it satisfies the eigenvalue equation $A\varphi = \lambda\varphi$, and hence φ belongs to the kernel of $(A - \lambda\mathbb{1})$, in that $(A - \lambda\mathbb{1})\varphi = 0$. The operator $(A - \mu\mathbb{1})$ may therefore fail to be invertible. The inverse

$$R_\mu(A) \equiv (A - \mu\mathbb{1})^{-1}, \tag{4.A.40}$$

when it exists, is called the *resolvent*, and μ is then said to belong to the *resolvent set*. The spectrum of the operator A can thus be identified as the set of singularities of the resolvent $R_\mu(A)$. The spectra may be discrete, continuous or singular (see comments following Eq. (4.4.23)).

(vii) *The Dirac notation.* In parts II and III of our monograph, we shall use the Dirac notation for *ket* and *bra* vectors. Originally, Dirac wrote the scalar product between a pair of vectors as $\langle\psi|\phi\rangle$, and called it a 'bracket'. He then split the bracket in two, and thought of it as a combination of the 'bra' $\langle\psi|$ and the 'ket' $|\phi\rangle$, which, at this stage, were viewed as entities in their own right. Of course, the ket $|\phi\rangle$ is just an alternative notation for the vector ϕ in an abstract Hilbert space $\mathcal{H}$. However, a 'bra' really belongs to the dual space of $\mathcal{H}$, here denoted by $\mathcal{H}^*$, and hence is a linear map from $\mathcal{H}$ into the set of complex numbers. In the case of a Hilbert space, it can be shown that a one-to-one correspondence exists between bras and kets, and $\langle\psi|\phi\rangle$ can be thought of as being either the scalar product between vectors $|\psi\rangle$ and $|\phi\rangle$ or the value of the dual vector $\langle\psi|$ on the vector $|\phi\rangle$. In the applications, kets are often written down using the quantum numbers relevant for the problem that is being studied. With the Dirac notation, the tensor product of two states $|\psi\rangle$ and $|\phi\rangle$ is written as $|\psi\rangle|\phi\rangle$, i.e. $|\psi\rangle \otimes |\phi\rangle \equiv |\psi\rangle|\phi\rangle$. For a recent and detailed assessment of the Dirac notation, we refer the reader to the work in Gieres (2000).

Appendix 4.B
JWKB approximation

If the function W solves the time-dependent Hamilton–Jacobi equation, it is possible to prove that the continuity equation

$$\mathrm{div}\left(\frac{\rho}{m}\mathrm{grad}\, W\right) + \frac{\partial \rho}{\partial t} = 0 \tag{4.B.1}$$

is solved by

$$\rho = \left|\det\left(\frac{\partial^2 W}{\partial x_i \partial x_j^0}\right)\right|, \tag{4.B.2}$$

and hence

$$\psi = \sqrt{\rho}\, \mathrm{e}^{\frac{\mathrm{i}W}{\hbar}} \tag{4.B.3}$$

is an approximate solution of the Schrödinger equation determined from the solution of the Hamilton–Jacobi equation. To see in detail how all of this works, let us consider a complete integral of the Hamilton–Jacobi equation (hence depending on as many parameters as the number of variables) and let us define (with α_k being basically the initial data)

$$\frac{\partial W}{\partial \alpha_k} \equiv \beta_k. \tag{4.B.4}$$

Hereafter we deal with

$$(x_j, p_j) \in T^*\mathbf{R}^3, \quad (\alpha_k, \beta_k) \in T^*\mathbf{R}^3, \quad t \in \mathbf{R}.$$

β_k are constants of motion because

$$\frac{\partial}{\partial \alpha_k}\frac{\partial W}{\partial t} = -\frac{\partial}{\partial \alpha_k}\left[\frac{1}{2m}(\text{grad } W)^2 + V(x)\right]$$

implies that

$$\frac{\partial}{\partial t}\frac{\partial W}{\partial \alpha_k} = -\frac{\text{grad } W}{m}\frac{\partial}{\partial \vec{x}}\frac{\partial W}{\partial \alpha_k},$$

and eventually

$$L_X \beta_k \equiv \left(\frac{\partial}{\partial t} + \frac{\text{grad } W}{m}\frac{\partial}{\partial \vec{x}}\right)\beta_k = 0.$$

One then studies a vector field

$$Y \equiv \rho\left(\frac{\partial}{\partial t} + v_i \frac{\partial}{\partial x_i}\right)$$

chosen in such a way that

$$\begin{aligned} d\beta_1 \wedge d\beta_2 \wedge d\beta_3 = & \left(\frac{\partial^2 W}{\partial \alpha_1 \partial x_1}\mathrm{d}x_1 + \frac{\partial^2 W}{\partial \alpha_1 \partial x_2}\mathrm{d}x_2 \right. \\ & \left. + \frac{\partial^2 W}{\partial \alpha_1 \partial x_3}\mathrm{d}x_3 + \frac{\partial^2 W}{\partial \alpha_1 \partial t}\mathrm{d}t\right) \\ & \wedge \left(\frac{\partial^2 W}{\partial \alpha_2 \partial x_1}\mathrm{d}x_1 + \frac{\partial^2 W}{\partial \alpha_2 \partial x_2}\mathrm{d}x_2 + \frac{\partial^2 W}{\partial \alpha_2 \partial x_3}\mathrm{d}x_3 + \frac{\partial^2 W}{\partial \alpha_2 \partial t}\mathrm{d}t\right) \\ & \wedge \left(\frac{\partial^2 W}{\partial \alpha_3 \partial x_1}\mathrm{d}x_1 + \frac{\partial^2 W}{\partial \alpha_3 \partial x_2}\mathrm{d}x_2 + \frac{\partial^2 W}{\partial \alpha_3 \partial x_3}\mathrm{d}x_3 + \frac{\partial^2 W}{\partial \alpha_3 \partial t}\mathrm{d}t\right) \\ = & \, \rho \, \mathrm{d}x_1 \wedge \mathrm{d}x_2 \wedge \mathrm{d}x_3 - \rho v_1 \, \mathrm{d}t \wedge \mathrm{d}x_2 \wedge \mathrm{d}x_3 \\ & - \rho v_2 \, \mathrm{d}x_1 \wedge \mathrm{d}t \wedge \mathrm{d}x_3 - \rho v_3 \, \mathrm{d}x_1 \wedge \mathrm{d}x_2 \wedge \mathrm{d}t \\ = & \, i_Y(\mathrm{d}t \wedge \mathrm{d}x_1 \wedge \mathrm{d}x_2 \wedge \mathrm{d}x_3). \end{aligned} \tag{4.B.5}$$

By application of i_Y to both sides of Eq. (4.B.5) we find that β_k are also constants of motion for the vector field Y, since $(i_Y)^2 = 0$ and hence $i_Y(\mathrm{d}\beta_1 \wedge \mathrm{d}\beta_2 \wedge \mathrm{d}\beta_3) = 0$. Thus, the vector field Y is proportional to

$$\frac{\partial}{\partial t} + \frac{\text{grad } W}{m}\frac{\partial}{\partial \vec{x}}$$

because $\beta_1, \beta_2, \beta_3$ are independent constants of motion for the two vector fields X and Y. The explicit computation shows that

$$\rho = \left|\det\left(\frac{\partial^2 W}{\partial x_j \partial \alpha_k}\right)\right|. \tag{4.B.6}$$

The vector field Y is now found to have vanishing divergence by virtue of the identity

$$\mathrm{d}(i_Y(\mathrm{d}t \wedge \mathrm{d}x_1 \wedge \mathrm{d}x_2 \wedge \mathrm{d}x_3)) = (\text{div } Y)(\mathrm{d}t \wedge \mathrm{d}x_1 \wedge \mathrm{d}x_2 \wedge \mathrm{d}x_3) \tag{4.B.7}$$

and bearing in mind that $\mathrm{d}^2 = 0$:

$$\mathrm{d}(\mathrm{d}\beta_1 \wedge \mathrm{d}\beta_2 \wedge \mathrm{d}\beta_3) = 0, \tag{4.B.8}$$

because the left-hand side of (4.B.7) is equal to the differential of our volume form by virtue of (4.B.5), which vanishes according to (4.B.8). One therefore obtains, eventually,

$$\text{div}(\rho\vec{v}) + \frac{\partial \rho}{\partial t} = 0, \tag{4.B.9}$$

with

$$\vec{v} = \frac{\text{grad}\, W}{m}, \tag{4.B.10}$$

and ρ given by (4.B.2).

A simple but important application of (4.B.2) and (4.B.3) is provided by the quantum-mechanical analysis of a free particle of mass m, for which the classical solution of the equations of motion is

$$x = x_0 + v_x t, \quad y = y_0 + v_y t, \quad z = z_0 + v_z t. \tag{4.B.11}$$

Thus, upon integration of the Lagrangian along the motion one finds

$$W = \int_0^t \frac{mv^2}{2} \mathrm{d}t' = \frac{mv^2}{2} t, \tag{4.B.12}$$

i.e.

$$W = \frac{m}{2t} \left[(x - x_0)^2 + (y - y_0)^2 + (z - z_0)^2 \right]. \tag{4.B.13}$$

The second derivatives of W read

$$\frac{\partial^2 W}{\partial x \partial x_0} = \frac{\partial^2 W}{\partial y \partial y_0} = \frac{\partial^2 W}{\partial z \partial z_0} = -\frac{m}{t}, \tag{4.B.14}$$

and hence the formula (4.B.2) leads to $\rho = (\frac{m}{t})^3$. By virtue of (4.B.3), one finds eventually

$$\psi = C \left(\frac{m}{t} \right)^{3/2} \mathrm{e}^{\frac{im}{2\hbar t}[(x-x_0)^2+(y-y_0)^2+(z-z_0)^2]}, \tag{4.B.15}$$

where C is a dimensionful constant. The insertion of (4.B.15) into the Schrödinger equation shows that we have obtained (in this particular case) an *exact* solution for all $t \neq 0$, i.e. the Schrödinger kernel (also called the Green function) for the free particle (see section 4.4). It is possible to show that this situation prevails for all quadratic Hamiltonians.

Appendix 4.C
Asymptotic expansions

In our book we need a few key definitions concerning asymptotic expansions and their properties (the concept was first introduced in Poincaré (1886), which is a seminal paper on the irregular integrals of linear equations).

Let f be a function defined in an unbounded domain Ω. A power series $\sum_{n=0}^{\infty} a_n z^{-n}$, *convergent* or *divergent*, is said to be the *asymptotic expansion* of f if, *for all* fixed values of the integer N, one has

$$f(z) = \sum_{n=0}^{N} a_n z^{-n} + \mathrm{O}(z^{-(N+1)}), \tag{4.C.1}$$

as z tends to ∞. Hence one finds

$$\lim_{z \to \infty} z^N |f(z) - S_N(z)| = 0, \tag{4.C.2}$$

where S_N is the sum of the first $N+1$ terms of the series, and one writes

$$f(z) \sim \sum_{n=0}^{\infty} a_n z^{-n} \quad \text{as } z \to \infty. \tag{4.C.3}$$

The asymptotic expansions (hereafter denoted by A.E.) have some basic properties, which are standard (by now) but very useful. They are as follows.

(i) The A.E. of f, *if it exists*, is unique.

(ii) A.E. may be summed, i.e. if $f(z) \sim \sum_{n=0}^{\infty} a_n z^{-n}$ and $g(z) \sim \sum_{n=0}^{\infty} b_n z^{-n}$, then

$$\rho f(z) + \sigma g(z) \sim \sum_{n=0}^{\infty} \left(\rho a_n + \sigma b_n \right) z^{-n}, \tag{4.C.4}$$

as z tends to ∞ in Ω.

(iii) A.E. can be multiplied, i.e.

$$f(z)g(z) \sim \sum_{n=0}^{\infty} c_n z^{-n}, \tag{4.C.5}$$

where $c_n \equiv \sum_{s=0}^{n} a_s b_{n-s}$.

(iv) If f is continuous in the domain Ω defined by $|z| > a$, $\arg(z) \in [\theta_0, \theta_1]$, and if (4.C.3) holds, then

$$\int_z^{\infty} \left[f(t) - a_0 - \frac{a_1}{t} \right] \mathrm{d}t \sim \sum_{n=1}^{\infty} \frac{a_{n+1}}{n} z^{-n}, \tag{4.C.6}$$

as $z \to \infty$ in Ω, where the integration is taken along a line $z \to \infty$ with fixed argument. This is what one means by term-by-term integration of A.E.

(v) Term-by-term differentiation can also be performed, provided that, in the domain Ω defined by the conditions $|z| > R$, $\arg(z) \in]\theta_0, \theta_1[$, the function f satisfying (4.C.3) has a continuous derivative f', and f' has an A.E. as $z \to \infty$ in Ω. One then finds

$$f'(z) \sim - \sum_{n=1}^{\infty} n a_n z^{-(n+1)} \quad \text{as } z \to \infty \text{ in } \Omega. \tag{4.C.7}$$

It should be stressed that, if f admits an asymptotic expansion, and one writes

$$f(z) = \sum_{n=0}^{N} a_n z^{-n} + R_N(z), \tag{4.C.8}$$

the 'rest' does not tend to 0 as $N \to \infty$, but the series approximates very well the value of f as $z \to \infty$, *once that N has been fixed.* If this cannot be repeated for all values of N, one has the weaker concept of a series that is asymptotic to a function up to some *finite order* as $z \to \infty$. An example of how (4.C.1) works is provided by

$$f(z) \equiv \int_z^{\infty} \frac{\mathrm{e}^{z-y}}{y} \mathrm{d}y = \frac{1}{z} \sum_{n=0}^{N} \frac{(-1)^n n!}{z^n} + R_{N+1}(z). \tag{4.C.9}$$

The result (4.C.9) is obtained after repeated integration by parts. The series on the right-hand side of (4.C.9) diverges for all values of z. Nevertheless, if one keeps N *fixed*, the rest tends to 0 as $z \to \infty$:

$$|R_N(z)| = N! \int_z^{\infty} \frac{\mathrm{e}^{z-y}}{y^{N+1}} \mathrm{d}y < \frac{N!}{z^{N+1}}. \tag{4.C.10}$$

5
Applications of wave mechanics

One-dimensional problems are studied as an introduction to quantum dynamics, i.e. reflection and transmission of a wave packet, the step-like potential and the tunnelling effect. Stationary states of a particle in a linear potential are then considered and the basic properties of Bessel functions are introduced. After this, the Schrödinger equation in a central potential is analysed in detail. As an application, the Balmer formula for bound states of the hydrogen atom is derived. At this stage, a self-consistent treatment of angular momentum is presented, and the homomorphism between the rotation group in three dimensions and the $SU(2)$ group is constructed. The chapter ends with a derivation of energy bands in the presence of periodic potentials.

5.1 Reflection and transmission

The investigation of reflection and transmission of wave packets elucidates many peculiar properties of the quantum-mechanical analysis of the motion of particles or beams of particles. For example, let a particle of mass m and positive energy, $W > 0$, move in the one-dimensional potential

$$V(x) = -|V_0| \quad \text{if } x \in]-b, b[, \quad 0 \text{ if } |x| > b. \tag{5.1.1}$$

Since also $W + |V_0|$ is positive, one finds the improper eigenfunctions

$$u_{\rm I}(x) = A{\rm e}^{{\rm i}kx} + B{\rm e}^{-{\rm i}kx}, \tag{5.1.2}$$

$$u_{\rm II}(x) = C{\rm e}^{{\rm i}\overline{k}x} + D{\rm e}^{-{\rm i}\overline{k}x}, \tag{5.1.3}$$

$$u_{\rm III}(x) = E{\rm e}^{{\rm i}kx} + F{\rm e}^{-{\rm i}kx}, \tag{5.1.4}$$

where $k \equiv \sqrt{2mW}/\hbar$, $\overline{k} \equiv \sqrt{2m(W + |V_0|)}/\hbar$, and the labels I, II, III refer to the intervals $x \in]-\infty, -b[, x \in]-b, b[, x \in]b, \infty[$, respectively. Recall from subsection 4.6.2 that these stationary 'states' are not in Hilbert space, but are used to build a normalizable wave function. Thus, in this problem, the eigendifferential

$$\int_k^{k+\delta k} \mathrm{d}\gamma \ u_\gamma(x)$$

belongs to $\mathcal{L}^2(\mathbf{R})$ for any k and δk and for any choice of the two arbitrary constants in Eqs. (5.1.2)–(5.1.4). This means that the whole interval $W > 0$ belongs to the continuous spectrum of H, and a twofold degeneracy occurs. In the problems studied here, the choice of a particular formal solution of the eigenvalue equation is made in such a way that it is then possible to build a wave function satisfying certain asymptotic conditions. For potentials vanishing at infinity more rapidly than $\frac{1}{|\vec{x}|}$ one requires that it should be possible to build a wave function $\psi(\vec{x}, t)$ which, for $t \rightarrow -\infty$ or $t \rightarrow +\infty$ behaves as a free particle, before interacting (or after interacting) with the potential for $t \approx 0$. In one-dimensional problems one can also make statements on the motion of the particle in the presence of potentials that remain non-vanishing at infinity. In such a case an asymptotic condition as $t \rightarrow -\infty$ is obtained by imposing that the particle should be located at $x \rightarrow +\infty$ or at $x \rightarrow -\infty$, and that it should move towards the origin, hence specifying how the linear momentum is directed as $t \rightarrow -\infty$. Similarly, as $t \rightarrow +\infty$, one can impose that the particle is located at $x \rightarrow +\infty$ or at $x \rightarrow -\infty$ and that it is moving away from the origin.

We may exploit the arbitrariness in the value of two constants by choosing

$$F = 0, \ A = \frac{1}{\sqrt{2\pi\hbar}}. \tag{5.1.5}$$

Of course, we might have chosen, instead of $F = 0$, the alternative condition $E = 0$. The value for A is chosen to make it easier to compare with the case of the free particle (cf. section 4.4), although the crucial role will be played by ratios of coefficients involving A. We set (bearing in mind that $p \equiv \hbar k$)

$$\beta(p) \equiv \frac{B}{A}, \quad \varepsilon(p) \equiv \frac{E}{A}. \tag{5.1.6}$$

The improper eigenfunctions resulting from our choice of parameters are thus

$$u_p^{(+)}(x) = \frac{1}{\sqrt{2\pi\hbar}} \left[\mathrm{e}^{\frac{\mathrm{i}}{\hbar}px} + \beta(p)\mathrm{e}^{-\frac{\mathrm{i}}{\hbar}px} \right] \quad \text{if } x < -b, \tag{5.1.7}$$

$$u_p^{(+)}(x) = \frac{1}{\sqrt{2\pi\hbar}} \varepsilon(p)\mathrm{e}^{\frac{\mathrm{i}}{\hbar}px} \quad \text{if } x > b. \tag{5.1.8}$$

Of course, nothing prevents us from writing the form of $u_p^+(x)$ if $x \in]-b, b[$, but this calculation is not relevant for the analysis of reflection and transmission coefficients, and is hence omitted.

One is eventually interested in a solution $\psi(x,t)$ of the time-dependent Schrödinger equation, which consists of a wave packet built from the above improper eigenfunctions, according to the well-established rule (cf. section 4.4)

$$\psi(x,t) = \int_{-\infty}^{\infty} \mathrm{d}p \; c(p) u_p^{(+)}(x) \mathrm{e}^{-\frac{\mathrm{i}}{\hbar}\frac{p^2}{2m}t}. \tag{5.1.9}$$

By virtue of Eqs. (5.1.7) and (5.1.8), Eq. (5.1.9) leads to a first non-trivial result:

$$\psi(x,t) = \psi_{\text{in}}(x,t) + \psi_{\text{rifl}}(x,t) = \frac{1}{\sqrt{2\pi\hbar}} \int_{-\infty}^{\infty} \mathrm{d}p \; c(p) \mathrm{e}^{\frac{\mathrm{i}}{\hbar}\left(px - \frac{p^2}{2m}t\right)}$$
$$+ \frac{1}{\sqrt{2\pi\hbar}} \int_{-\infty}^{\infty} \mathrm{d}p \; c(p)\beta(p) \mathrm{e}^{-\frac{\mathrm{i}}{\hbar}\left(px + \frac{p^2}{2m}t\right)} \quad \text{if } x < -b, \tag{5.1.10}$$

$$\psi(x,t) = \psi_{\text{tr}}(x,t) = \frac{1}{\sqrt{2\pi\hbar}} \int_{-\infty}^{\infty} \mathrm{d}p \; c(p)\varepsilon(p) \mathrm{e}^{\frac{\mathrm{i}}{\hbar}\left(px - \frac{p^2}{2m}t\right)} \quad \text{if } x > b. \tag{5.1.11}$$

If $c(p)$ is sufficiently smooth and takes non-vanishing values only in a small neighbourhood of a particular value p_0 of p, such formulae describe wave packets moving with speed $\frac{p_0}{m}, -\frac{p_0}{m}, \frac{p_0}{m}$, respectively, without a sensible spreading effect. The first and third wave packet move from the left to the right, while the second wave packet moves from the right to the left (appendix 5.A).

In other words, at the beginning of the motion, $\psi(x,t)$ coincides with a solution of the Schrödinger equation for a free particle, and represents a wave packet moving from the left to the right. When such a packet reaches the region where the potential takes a non-vanishing value it splits into two new packets, one of which is transmitted, while the other is reflected. Both packets have eventually the same speed, in magnitude, as the initial packet. From the physical point of view, the solution (5.1.7)–(5.1.9), jointly with the hypothesis that $c(p)$ differs from zero only for p in the neighbourhood of a value p_0, describes a particle which has a certain probability τ of being transmitted, and a probability ρ of being reflected. Unlike classical mechanics, τ is, in general, smaller than 1, and ρ does not vanish.

To complete the calculations of general nature note first that, as $t \rightarrow -\infty$, one has

$$\int_{-\infty}^{\infty} \mathrm{d}x \, |\psi(x,t)|^2 \longrightarrow \int_{-\infty}^{\infty} \mathrm{d}x \, |\psi_{\text{in}}(x,t)|^2 = \int_{-\infty}^{\infty} \mathrm{d}p \, |c(p)|^2. \tag{5.1.12}$$

On the other hand, the terms on both sides of Eq. (5.1.12) are essentially independent of time, and hence should also coincide for finite values of the time variable. The normalizability condition of $\psi(x,t)$ therefore implies

$$\int_{-\infty}^{\infty} dp \, |c(p)|^2 = 1. \tag{5.1.13}$$

Now the probability that, at large t, the particle has been transmitted, is given by ($c(p)$ being of compact support)

$$\begin{aligned}\tau \equiv \lim_{t\to\infty} \int_b^{\infty} dx \, |\psi(x,t)|^2 &= \int_{-\infty}^{\infty} dx \, |\psi_{\rm tr}(x,t)|^2 \\ &= \int_{-\infty}^{\infty} dp \, |c(p)|^2 |\varepsilon(p)|^2 \\ &\sim |\varepsilon(p_0)|^2 \int_{-\infty}^{\infty} dp \, |c(p)|^2 = |\varepsilon(p_0)|^2,\end{aligned} \tag{5.1.14}$$

where we have used the Parseval lemma to go from the x- to the p-integration, and, in the last equality, we have used Eq. (5.1.13). Moreover, the reflection probability can be evaluated as follows:

$$\begin{aligned}\rho \equiv \lim_{t\to\infty} \int_{-\infty}^{-b} dx \, |\psi(x,t)|^2 &= \int_{-\infty}^{\infty} dx \, |\psi_{\rm rifl}(x,t)|^2 \\ &= \int_{-\infty}^{\infty} dp \, |c(p)|^2 |\beta(p)|^2 \sim |\beta(p_0)|^2,\end{aligned} \tag{5.1.15}$$

where we have again used the Parseval lemma and the property of c of being a function with compact support, jointly with Eq. (5.1.13). The explicit calculation in the potential (5.1.1), bearing in mind the definitions (5.1.6), shows that

$$\rho + \tau = 1. \tag{5.1.16}$$

Indeed, the continuity conditions at the points $-b$ and b lead to the system

$$A\Big(e^{-ikb} + \beta e^{ikb}\Big) = Ce^{-i\overline{k}b} + De^{i\overline{k}b}, \tag{5.1.17}$$

$$kA\Big(e^{-ikb} - \beta e^{ikb}\Big) = \overline{k}\Big(Ce^{-i\overline{k}b} - De^{i\overline{k}b}\Big), \tag{5.1.18}$$

$$Ce^{i\overline{k}b} + De^{-i\overline{k}b} = A\varepsilon e^{ikb}, \tag{5.1.19}$$

$$\overline{k}\Big(Ce^{i\overline{k}b} - De^{-i\overline{k}b}\Big) = A\varepsilon k \, e^{ikb}, \tag{5.1.20}$$

which, on setting

$$\gamma \equiv \frac{\overline{k}}{k}, \tag{5.1.21}$$

is solved by

$$\beta = \frac{2\mathrm{i}\sin(2\bar{k}b)\mathrm{e}^{-2\mathrm{i}(k+\bar{k})b}(1-\gamma)/(1+\gamma)}{\left[\frac{(1-\gamma)^2}{(1+\gamma)^2} - \mathrm{e}^{-4\mathrm{i}\bar{k}b}\right]}, \tag{5.1.22}$$

$$\varepsilon = -\frac{4\gamma \mathrm{e}^{-2\mathrm{i}(k+\bar{k})b}}{(1+\gamma)^2\left[\frac{(1-\gamma)^2}{(1+\gamma)^2} - \mathrm{e}^{-4\mathrm{i}\bar{k}b}\right]}, \tag{5.1.23}$$

and a patient check shows that $|\beta|^2 + |\varepsilon|^2 = 1$, because $|\beta|^2 + |\varepsilon|^2$ is found to coincide with the ratio $\frac{F(\gamma,k)}{F(\gamma,k)}$, where

$$F(\gamma, k) \equiv \frac{16\gamma^2}{(1+\gamma)^4} + 4\frac{(1-\gamma)^2}{(1+\gamma)^2}\sin^2(2k\gamma b). \tag{5.1.24}$$

The functions ρ and τ are called the *reflection* and *transmission coefficient*, respectively. The underlying reason for such names is that, if a sufficiently large number N of particles, all with the same initial value of the momentum, move in the potential (5.1.1), then by virtue of the 'law of big numbers' the number of reflected particles is ρN, and the number of transmitted particles is τN. Indeed, if one considers a *beam of particles*, all of them under the same initial conditions, and in sufficiently large number so that statistical fluctuations can be neglected, then $|\psi|^2 \, \mathrm{d}^3x$ can describe the *percentage of particles observed within the volume* d^3x. Similarly, if a beam of particles of sufficiently high intensity is available, all of them being in the same initial conditions, it is possible to interpret the quantity (cf. Eq. (4.1.18))

$$\vec{j} \cdot \vec{n} \, \mathrm{d}\sigma \, \mathrm{d}t$$

as the percentage of particles that cross the surface $\mathrm{d}\sigma$ in the time interval $\mathrm{d}t$. This interpretation holds both for charged and neutral particles. What we say remains true if the relative phase of the wave function between the two regions is arbitrary (see the remarks prior to subsection 4.6.1).

5.2 Step-like potential: tunnelling effect

We know from section 4.6 that, whenever the potential in the Schrödinger equation has finite discontinuities at some points, one can impose, correspondingly, continuity conditions on the wave function and its first derivative. These boundary conditions ensure (essential) self-adjointness of the Hamiltonian operator, under suitable assumptions on the potential (Reed and Simon 1975). Consider, for simplicity, the stationary Schrödinger equation in one-dimensional problems. According to this scheme, in each open interval of the real line where the potential is continuous, one has to solve the second-order equation

$$\left\{ \frac{d^2}{dx^2} + \frac{2m}{\hbar^2}[E - V_i(x)] \right\} u = 0, \tag{5.2.1}$$

where V_i is the potential in the interval I_i, subject to the continuity conditions

$$\lim_{x \to x_i^-} u(x) = \lim_{x \to x_i^+} u(x), \tag{5.2.2}$$

$$\lim_{x \to x_i^-} \frac{du}{dx} = \lim_{x \to x_i^+} \frac{du}{dx}, \tag{5.2.3}$$

where x_i denote the various discontinuity points of the potential $V(x)$.

5.2.1 Step-like potential

A first non-trivial consequence of these properties is that, if a beam of particles with the same mass m and energy $E > W$ enters a region where the potential is step-like and given by

$$V(x) = 0 \quad \text{if } x < 0, \; W \; \text{if } x > 0, \tag{5.2.4}$$

one has a solution of the kind (Squires 1995)

$$u(x) = e^{i\kappa x} + R e^{-i\kappa x} \quad \text{if } x < 0, \tag{5.2.5}$$

$$u(x) = T e^{i\gamma x} \quad \text{if } x > 0, \tag{5.2.6}$$

where $\kappa^2 \equiv \frac{2mE}{\hbar^2}, \gamma^2 \equiv \frac{2m(E-W)}{\hbar^2}$, and the application of (5.2.2) and (5.2.3) leads to the system

$$1 + R = T, \tag{5.2.7}$$

$$i\kappa(1 - R) = iT\gamma, \tag{5.2.8}$$

which implies

$$R = \frac{1 - \gamma/\kappa}{1 + \gamma/\kappa}, \quad T = \frac{2}{1 + \gamma/\kappa}. \tag{5.2.9}$$

Note that the irrelevance of the phase of the wave function at interfaces would make no difference to the calculation of the R and T for a step-like potential. The solution of the time-dependent Schrödinger equation is then found to be (hereafter $p \equiv \hbar\kappa, \overline{p} \equiv \hbar\gamma$)

$$\psi(x,t) = \frac{1}{\sqrt{2\pi\hbar}} \int_{-\infty}^{\infty} c(p) \left[e^{\frac{i}{\hbar}\left(px - \frac{p^2}{2m}t\right)} + R(p) e^{-\frac{i}{\hbar}\left(px + \frac{p^2}{2m}t\right)} \right] dp, \tag{5.2.10}$$

if $x < 0$, and

$$\psi(x,t) = \frac{1}{\sqrt{2\pi\hbar}} \int_{-\infty}^{\infty} c(p)T(p)e^{\frac{i}{\hbar}[\overline{p}x - E(p)t]} \, dp \tag{5.2.11}$$

if $x > 0$. The integration variable is p for both wave packets. However, the latter reads (factors of $\sqrt{2\pi\hbar}$ are here omitted for simplicity)

$$\psi_{II}(x,t) = \int_{-\infty}^{\infty} f(p,t)e^{\frac{i}{\hbar}\overline{p}x} \, dp, \tag{5.2.12}$$

where

$$f(p,t) \equiv c(p)T(p)e^{-\frac{i}{\hbar}E(p)t}, \tag{5.2.13}$$

whereas, to be able to apply the Parseval lemma, we would like it to be able to express ψ_{II} in the form

$$\psi_{II}(x,t) = \int_{-\infty}^{\infty} \widetilde{\psi}_{II}(\overline{p},t)e^{\frac{i}{\hbar}\overline{p}x} \, d\overline{p}. \tag{5.2.14}$$

Since the integrals (5.2.12) and (5.2.14) represent the same wave packet, we have to change variables in Eq. (5.2.12), writing

$$\psi_{II}(x,t) = \int_{-\infty}^{\infty} f(p(\overline{p}),t)\frac{dp}{d\overline{p}}e^{\frac{i}{\hbar}\overline{p}x} \, d\overline{p}, \tag{5.2.15}$$

and hence the spatial Fourier transform of the wave packet ψ_{II} reads

$$\widetilde{\psi}_{II}(\overline{p},t) = f(p(\overline{p}),t)\frac{dp}{d\overline{p}} = \frac{\overline{p}}{p}c(p)T(p)e^{-\frac{i}{\hbar}E(p)t}, \tag{5.2.16}$$

having exploited the property

$$\frac{dp}{d\overline{p}} = \frac{d}{d\overline{p}}\left(\overline{p}^2 + 2mW\right)^{1/2} = \frac{\overline{p}}{p}. \tag{5.2.17}$$

Now the Parseval lemma and the mean-value theorem yield (since $c(p)$ has compact support)

$$\begin{aligned}
(\psi_{II},\psi_{II}) &= \left(\widetilde{\psi}_{II},\widetilde{\psi}_{II}\right) \\
&= \int_{-\infty}^{\infty} |c(p(\overline{p}))|^2 |T(p(\overline{p}))|^2 \frac{\overline{p}}{p}\frac{\overline{p}}{p} \, d\overline{p} \\
&= \frac{\overline{p}}{p}|T|^2 \int_{-\infty}^{\infty} |c(p(\overline{p}))|^2 \frac{\overline{p}}{p} \, d\overline{p} \\
&= \frac{\overline{p}}{p}|T|^2 \int_{-\infty}^{\infty} |c(p)|^2 \, dp = \frac{\overline{p}}{p}|T|^2 \\
&= \frac{\gamma}{\kappa}|T|^2.
\end{aligned} \tag{5.2.18}$$

This is the transmission coefficient

$$\tau \equiv \lim_{t\to\infty} \int_0^\infty |\psi(x,t)|^2 \, dx = \frac{\gamma}{\kappa} |T|^2, \tag{5.2.19}$$

under the assumption that the apparatuses used for preparation and detection refer to the same state.

The evaluation of the reflection coefficient is easier, since no change of measure is necessary when $V(x) = 0$ for $x \in (-\infty, 0)$. One then finds

$$\rho \equiv \lim_{t\to\infty} \int_{-\infty}^0 |\psi(x,t)|^2 \, dx = |R|^2. \tag{5.2.20}$$

Thus, although all particles have energy greater than W, a non-vanishing fraction is reflected, and is expressed by $|R|^2$. A consistency check shows that

$$1 = |R|^2 + \frac{\gamma}{\kappa}|T|^2 = |R|^2 + \frac{\overline{p}}{p}|T|^2. \tag{5.2.21}$$

The stationary phase method (see appendix 5.A) shows that, as $t \to \infty$, the first of the two terms in (5.2.10) does not contribute at all, because it corresponds to a wave packet located at $x = +\infty$, i.e. outside the domain of validity of (5.2.10). Thus, as $t \to \infty$, Eq. (5.2.10) reduces to a reflected wave packet located at $x = -\infty$. Moreover, in the limit $t \to \infty$, Eq. (5.2.11) indeed describes a wave packet located at $x = +\infty$, which is an acceptable asymptotic solution, because it agrees with the domain of definition of Eq. (5.2.11).

In contrast, if all particles have energy $E < W$, the stationary state again takes the form (5.2.5) if $x < 0$, but for $x > 0$ is a decreasing exponential:

$$u(x) = Q e^{-\Gamma x} \quad \text{if } x > 0, \tag{5.2.22}$$

where $\Gamma^2 \equiv \frac{2m(W-E)}{\hbar^2}$. Thus, the application of (5.2.2) and (5.2.3) leads instead to the system

$$1 + R = Q, \tag{5.2.23}$$

$$i\kappa(1 - R) = -\Gamma Q, \tag{5.2.24}$$

the solution of which is (Squires 1995)

$$R = \frac{1 - i\Gamma/\kappa}{1 + i\Gamma/\kappa}, \quad Q = \frac{2}{1 + i\Gamma/\kappa}. \tag{5.2.25}$$

Interestingly, although all particles have energy smaller than W, a non-vanishing solution of the Schrödinger equation *for stationary states* exists if $x > 0$. However, if one defines $\rho \equiv \frac{\Gamma}{\kappa} \equiv \tan\theta$, one finds from (5.2.25) that

$$R = \mathrm{e}^{-2\mathrm{i}\theta}, \tag{5.2.26}$$

while the stationary state for $x > 0$ reads (see Eq. (5.2.22))

$$u(x) = 2\mathrm{e}^{-\mathrm{i}\theta} \cos\theta \, \mathrm{e}^{-\Gamma x}. \tag{5.2.27}$$

It should be stressed that, from (5.2.26), the definition of reflection coefficient used so far, jointly with the stationary phase method, implies that

$$\rho = |R|^2 = 1, \tag{5.2.28}$$

i.e. *all particles are reflected.* This is in agreement with Eq. (5.2.27), because the exponential fall-off of u implies that $\psi(x,t)$ tends to 0 as $t \rightarrow \infty$ when $x \rightarrow \infty$. Thus, it would be incorrect to claim that the transmission coefficient is $|Q|^2$, with Q expressed by the second equation in (5.2.25). Waves falling off exponentially like (5.2.27) are called 'evanescent waves', and carry no current or momentum. But the density does not vanish abruptly; the waves are continuous across the boundary and gradually tend to zero far from the boundary.

5.2.2 *Tunnelling effect*

The above potential can be modified to show that quantum mechanics makes it possible to build a theoretical model for penetration through a potential barrier. This is called the tunnelling effect. The complete scheme, in one-dimensional problems, requires the introduction of a potential V such that

$$V(x) = 0 \text{ if } x < 0 \text{ or } x > a, \; W \text{ if } x \in]0,a[, \tag{5.2.29}$$

and a beam of particles of mass m and energy $E < W$ is affected by $V(x)$. By virtue of (5.2.29), one deals with three regions. The corresponding forms of the stationary states are:

$$u(x) = u_1(x) = \mathrm{e}^{\mathrm{i}\kappa x} + R\mathrm{e}^{-\mathrm{i}\kappa x} \quad \text{if } x < 0, \tag{5.2.30}$$

$$u(x) = u_2(x) = A\mathrm{e}^{\Gamma x} + B\mathrm{e}^{-\Gamma x} \quad \text{if } x \in]0,a[, \tag{5.2.31}$$

$$u(x) = u_3(x) = T\mathrm{e}^{\mathrm{i}\kappa x} \quad \text{if } x > a. \tag{5.2.32}$$

Of course, since we deal with three second-order differential equations with four boundary conditions, resulting from the application of (5.2.2) and (5.2.3) at $x = 0$ and a, we expect to have in general $6 - 4 = 2$ unknown coefficients. At this stage, physical considerations concerning reflected and transmitted particles, suggested by the simpler examples discussed at the beginning of this section and in section 5.1, lead to (5.2.30)–(5.2.32), where the number of coefficients, i.e. A, B, R and T, is equal to the number of boundary conditions. Indeed, from (5.2.2) and (5.2.3) one has

$$u_1(0) = u_2(0), \quad u_1'(0) = u_2'(0), \quad u_2(a) = u_3(a), \quad u_2'(a) = u_3'(a). \tag{5.2.33}$$

The insertion of (5.2.30)–(5.2.32) into (5.2.33) leads to

$$1 + R = A + B, \tag{5.2.34}$$

$$\mathrm{i}\kappa(1 - R) = \Gamma(A - B), \tag{5.2.35}$$

$$A\mathrm{e}^{\Gamma a} + B\mathrm{e}^{-\Gamma a} = T\mathrm{e}^{\mathrm{i}\kappa a}, \tag{5.2.36}$$

$$\Gamma\left(A\mathrm{e}^{\Gamma a} - B\mathrm{e}^{-\Gamma a}\right) = \mathrm{i}T\kappa\mathrm{e}^{\mathrm{i}\kappa a}. \tag{5.2.37}$$

It is now convenient to define new parameters:

$$\rho \equiv \frac{\Gamma}{\kappa}, \quad \theta \equiv \Gamma a, \quad \delta \equiv \kappa a. \tag{5.2.38}$$

In terms of them, one finds (from Eqs. (5.2.36) and (5.2.37))

$$A = \frac{1}{2}\left(1 + \frac{\mathrm{i}}{\rho}\right) T\mathrm{e}^{\mathrm{i}\delta - \theta}, \tag{5.2.39}$$

$$B = \frac{1}{2}\left(1 - \frac{\mathrm{i}}{\rho}\right) T\mathrm{e}^{\mathrm{i}\delta + \theta}. \tag{5.2.40}$$

One now inserts (5.2.39) and (5.2.40) into (5.2.34) and (5.2.35). On adding term by term, this yields

$$4\mathrm{e}^{-\mathrm{i}\delta} = T\left[\left(2 + \frac{\mathrm{i}}{\rho} - \mathrm{i}\rho\right)\mathrm{e}^{-\theta} + \left(2 + \mathrm{i}\rho - \frac{\mathrm{i}}{\rho}\right)\mathrm{e}^{\theta}\right]. \tag{5.2.41}$$

The exponentials on the right-hand side of Eq. (5.2.41) are now re-expressed in terms of the well-known hyperbolic functions, so that the formula for the coefficient T reads

$$T = \frac{\mathrm{e}^{-\mathrm{i}\delta}}{\left[\cosh\theta + \frac{\mathrm{i}}{2}\left(\rho - \frac{1}{\rho}\right)\sinh\theta\right]}. \tag{5.2.42}$$

Thus, the fraction of transmitted particles is

$$\begin{aligned} \mathcal{F} = |T|^2 &= \left[1 + \frac{(\rho^2+1)^2}{4\rho^2}\sinh^2\theta\right]^{-1} \\ &= \left[1 + \frac{1}{4}\frac{W^2}{E(W-E)}\sinh^2\Gamma a\right]^{-1}, \end{aligned} \tag{5.2.43}$$

because $\rho^2 = \frac{W-E}{E}$. In particular, in the limit as $W \rightarrow \infty$ and $a \rightarrow 0$, while $Wa = b =$ constant (this corresponds to the *delta-like* potential), the result (5.2.43) reduces to (Squires 1995)

$$\mathcal{F} = \left(1 + \frac{mb^2}{2\hbar^2 E}\right)^{-1}. \tag{5.2.44}$$

This calculation shows once again that, when the potential has an infinite discontinuity, one can first consider a problem where it takes finite values, and then evaluate the limit on the solution obtained with finite values of V (cf. Eqs. (4.6.12)–(4.6.27)). However, one can instead impose boundary conditions on the density and the current as we mentioned prior to subsection 4.6.1.

5.3 Linear potential

The quantum-mechanical analysis of a particle of mass m moving in the homogeneous gravitational field over the Earth's surface is a non-trivial problem that deserves careful consideration. In our investigation, the Earth is assumed to reflect the particle elastically. Denoting by x the height over the Earth's surface, the gravitational potential is $V(x) = mgx$, and one has to solve the stationary Schrödinger equation for a one-dimensional problem in a linear potential:

$$\left[\frac{\mathrm{d}^2}{\mathrm{d}x^2} + \frac{2m}{\hbar^2}(E - mgx)\right] u(x) = 0, \tag{5.3.1}$$

with $x \geq 0$. Classically, the region $x > \frac{E}{mg}$ is forbidden, whereas only $x \in \left[0, \frac{E}{mg}\right]$ is allowed, because $E = \frac{p^2}{2m} + mgx \geq mgx$. The boundary conditions are elastic reflection at $x = 0$:

$$u(0) = 0, \tag{5.3.2}$$

which may be obtained by considering $V(x) = \infty$ for $x < 0$, jointly with fall-off condition at infinity (to ensure normalizability of our solution):

$$\lim_{x \to \infty} u(x) = 0. \tag{5.3.3}$$

It is now convenient to introduce the dimensionless parameter λ, and the parameter l having dimension of length, defined by the equations (Flügge 1971)

$$\frac{2mE}{\hbar^2} \equiv \frac{\lambda}{l^2}, \quad \frac{2m^2 g}{\hbar^2} \equiv \frac{1}{l^3}. \tag{5.3.4}$$

In terms of these parameters, Eq. (5.3.1) reads

$$\left(\frac{\mathrm{d}^2}{\mathrm{d}x^2} + \frac{\lambda}{l^2} - \frac{x}{l^3}\right) u = 0. \tag{5.3.5}$$

Equation (5.3.5) suggests defining a new independent variable, i.e.

$$\xi \equiv \frac{x}{l} - \lambda, \tag{5.3.6}$$

so that the Schrödinger equation takes the form

$$\left(\frac{d^2}{d\xi^2} - \xi \right) u = 0. \tag{5.3.7}$$

The boundary conditions (5.3.2) and (5.3.3) now read as

$$u(-\lambda) = 0, \quad \lim_{\xi \to \infty} u(\xi) = 0. \tag{5.3.8}$$

Following what we know from appendix 5.B, we want to re-express Eq. (5.3.7) in terms of the Bessel equation. This means that we want to transform Eq. (5.3.7) into an equation with a Fuchsian singularity at the origin. Indeed, if one defines

$$u(\xi) \equiv \xi^p \, y(\xi), \tag{5.3.9}$$

one finds the equation

$$\left[\frac{d^2}{d\xi^2} + \frac{2p}{\xi} \frac{d}{d\xi} + \frac{p(p-1)}{\xi^2} - \xi \right] y = 0. \tag{5.3.10}$$

Equation (5.3.10) has a Fuchsian singularity at $\xi = 0$ (corresponding to $x = l\lambda$), but is not yet of the Bessel type, because the term linear in ξ survives in the square brackets. However, if we introduce yet another parameter, here denoted by α, and define instead

$$u(\xi) \equiv \xi^p \, y\left(\frac{1}{\alpha} \xi^\alpha \right), \tag{5.3.11}$$

we find, setting

$$\chi(\xi) \equiv \frac{1}{\alpha} \xi^\alpha, \tag{5.3.12}$$

the second-order equation

$$\left[\frac{d^2}{d\chi^2} + (\alpha + 2p - 1) \frac{1}{\xi^\alpha} \frac{d}{d\chi} + \frac{p(p-1)}{\xi^{2\alpha}} - \xi^{3-2\alpha} \right] y = 0. \tag{5.3.13}$$

At this stage, no term linear in ξ occurs provided that one takes $\alpha = \frac{3}{2}$. Moreover, if $p = \frac{1}{2}$, Eq. (5.3.13) becomes

$$\left[\frac{d^2}{d\chi^2} + \frac{1}{\chi} \frac{d}{d\chi} - \left(\frac{\frac{1}{4\alpha^2}}{\chi^2} + 1 \right) \right] y = 0. \tag{5.3.14}$$

The regular solution of Eq. (5.3.14) is (up to a multiplicative constant)

$$y = I_\rho\left(\frac{1}{\alpha}\xi^\alpha\right), \tag{5.3.15}$$

where

$$\rho = \sqrt{\frac{1}{4\alpha^2}} = \pm\frac{1}{3} = \rho_\pm. \tag{5.3.16}$$

Denoting by C a constant, one thus finds

$$\begin{aligned} u(\xi) &= C\xi^p\left[I_{\rho_+}\left(\frac{1}{\alpha}\xi^\alpha\right) + I_{\rho_-}\left(\frac{1}{\alpha}\xi^\alpha\right)\right] \\ &= C\sqrt{\xi}\left[I_{1/3}\left(\frac{2}{3}\xi^{3/2}\right) + I_{-1/3}\left(\frac{2}{3}\xi^{3/2}\right)\right], \end{aligned} \tag{5.3.17}$$

where we have set equal to 1 the relative weight of $I_{1/3}$ and $I_{-1/3}$. More precisely, if ξ is *negative* (i.e. $x \leq l\lambda$), the general solution is obtained from ordinary Bessel functions in the form

$$u(\xi) = \widetilde{C}\sqrt{|\xi|}\left[J_{1/3}\left(\frac{2}{3}|\xi|^{3/2}\right) + J_{-1/3}\left(\frac{2}{3}|\xi|^{3/2}\right)\right], \tag{5.3.18}$$

which is obtained by analytic continuation of (5.3.17). The first boundary condition in (5.3.8) implies therefore the following equation for the eigenvalues:

$$J_{1/3}\left(\frac{2}{3}\lambda^{3/2}\right) + J_{-1/3}\left(\frac{2}{3}\lambda^{3/2}\right) = 0. \tag{5.3.19}$$

Equation (5.3.19) provides a good example of a general feature: with a few exceptions, the equation obeyed by the eigenvalues by virtue of the boundary conditions yields them only *implicitly*, i.e. as the zeros of some function (cf. Eq. (4.6.24)). Such zeros are known, numerically, to a high accuracy, but one cannot generate analytic formulae for their exact evaluation. Nevertheless, one can obtain some useful asymptotic estimates. In our case, if λ is much greater than 1, one can use the asymptotic formulae valid at large z:

$$J_{1/3}(z) \sim \sqrt{\frac{2}{\pi z}}\cos\left(z - \frac{5\pi}{12}\right) + \mathrm{O}\left(z^{-3/2}\right), \tag{5.3.20}$$

$$J_{-1/3}(z) \sim \sqrt{\frac{2}{\pi z}}\cos\left(z - \frac{\pi}{12}\right) + \mathrm{O}\left(z^{-3/2}\right). \tag{5.3.21}$$

These properties, jointly with (5.3.19) and the identity

$$\cos\gamma + \cos\delta = 2\cos\left(\frac{\gamma+\delta}{2}\right)\cos\left(\frac{\gamma-\delta}{2}\right), \tag{5.3.22}$$

lead to

$$\cos\left(\frac{2}{3}\lambda^{3/2} - \frac{\pi}{4}\right) = 0, \tag{5.3.23}$$

which implies (n being an integer)

$$\frac{2}{3}\lambda^{3/2} - \frac{\pi}{4} = -\frac{\pi}{2} + n\pi, \tag{5.3.24}$$

i.e. (Flügge 1971)

$$\frac{2}{3}\lambda^{3/2} = \left(2n - \frac{1}{2}\right)\frac{\pi}{2}, \tag{5.3.25}$$

and hence, from (5.3.4),

$$E_n = \frac{\hbar^2}{2ml^2}\left[\frac{3\pi}{4}\left(2n - \frac{1}{2}\right)\right]^{2/3}. \tag{5.3.26}$$

A plot shows that the energy eigenvalues tend to get closer and closer.

Unlike the classical analysis of the problem, which is confined to the interval $\xi \in [-\lambda, 0]$, in quantum mechanics one also finds a non-trivial (generalized) solution at positive values of ξ. In particular, as $\xi \to \infty$, the general solution of Eq. (5.3.14), i.e.

$$y(\xi) = C_1 I_\rho\left(\frac{1}{\alpha}\xi^\alpha\right) + C_2 K_\rho\left(\frac{1}{\alpha}\xi^\alpha\right), \tag{5.3.27}$$

reduces to (see Eq. (5.3.8))

$$y(\xi) = C_2 K_{1/3}\left(\frac{2}{3}\xi^{3/2}\right), \tag{5.3.28}$$

because $I_{1/3}$ increases exponentially at infinity, and hence cannot represent a wave function. In contrast, $K_{1/3}$ decreases exponentially at infinity, and one finds, from

$$K_\nu(z) \sim \sqrt{\frac{\pi}{2z}}\,\mathrm{e}^{-z} \quad \text{as } z \to \infty, \tag{5.3.29}$$

the result (see Eq. (5.3.11))

$$u(\xi) \sim \frac{C_2}{2\sqrt{\pi}}\,\xi^{-1/4}\,\mathrm{e}^{-2/3\xi^{3/2}} \quad \text{as } \xi \to \infty. \tag{5.3.30}$$

This is therefore another example of an evanescent wave. Note also that no eigenvalue condition is obtained from the asymptotic behaviour.

In the literature, the solution $u(\xi)$ is known to be the Airy function. Its integral representation can be obtained by solving our problem in the *momentum representation*. In other words, we know that the Fourier transform makes it possible to associate the symbol with a differential operator. If $\phi(p)$ denotes the Fourier transform of $u(x)$, the operator $\hat{p}$

acts as multiplication, and the operator $\hat{x}$ is a first-order operator (see chapter 4):

$$\hat{p}\,\phi(p) = p\,\phi(p), \tag{5.3.31}$$

$$\hat{x}\,\phi(p) = \mathrm{i}\hbar \frac{\partial \phi}{\partial p}. \tag{5.3.32}$$

The resulting form of the stationary Schrödinger equation is

$$\frac{p^2}{2m}\phi(p) + img\hbar\phi'(p) = E\,\phi(p), \tag{5.3.33}$$

which implies

$$\frac{\phi'}{\phi} = \frac{(2mE - p^2)}{2\mathrm{i}\hbar m^2 g}. \tag{5.3.34}$$

The general solution of this equation is

$$\phi(p) = N \exp\left[-\frac{iE}{mg\hbar}p + \frac{ip^3}{6m^2 g\hbar}\right], \tag{5.3.35}$$

where N is an integration constant. Thus, defining

$$\mu \equiv \frac{p}{(2m^2 g\hbar)^{1/3}}, \tag{5.3.36}$$

one finds

$$u(x) = N \int_{-\infty}^{\infty} \exp\left[\frac{\mathrm{i}}{3}\mu^3 + \mathrm{i}\mu \frac{(mgx - E)}{(mg^2\hbar^2/2)^{1/3}}\right] d\mu. \tag{5.3.37}$$

The link with the integral representation of the Airy function, denoted by Φ, is now clear, because (cf. Airy 1838)

$$\Phi(x) \equiv \frac{1}{2\pi} \int_{-\infty}^{\infty} \exp\left(\mathrm{i}x\xi + \frac{\mathrm{i}}{3}\xi^3\right) d\xi. \tag{5.3.38}$$

Note that, in the momentum representation, the differential equation is solved more easily. One pays, however, a non-trivial price, because the boundary condition (5.3.2) is expressed by the vanishing of an integral:

$$u(0) = 0 = N \int_{-\infty}^{\infty} \exp\left[\frac{\mathrm{i}}{3}\mu^3 - \frac{\mathrm{i}\mu E}{(mg^2\hbar^2/2)^{1/3}}\right] d\mu. \tag{5.3.39}$$

On using the identity $\mathrm{e}^{\mathrm{i}z} = \cos(z) + \mathrm{i}\sin(z)$, and bearing in mind that cos is an even function, Eq. (5.3.39) may be re-expressed as an integral equation where only the cos function occurs in the integrand (Mavromatis 1987). This yields, implicitly, the eigenvalues E. Agreement with Eq. (5.3.19) is proved upon inserting therein the integral representation of Bessel functions.

5.4 The Schrödinger equation in a central potential

A central potential in $\mathbf{R}^3$ is a real-valued function that only depends on the magnitude $r \equiv \sqrt{x^2+y^2+z^2}$ of the position vector, where x, y, z are Cartesian coordinates in $\mathbf{R}^3$. Our aim is to build a general formalism to analyse the Hamiltonian

$$H = \frac{1}{2m}\left(p_x^2 + p_y^2 + p_z^2\right) + V, \tag{5.4.1a}$$

where p_x, p_y, p_z are Cartesian components of the linear momentum, and V is a central potential. By virtue of the hypothesis on the potential, the symmetry group of our problem is the rotation group in three dimensions (see section 5.6), since the rotations preserve the length of vectors in $\mathbf{R}^3$ and their real nature. In classical mechanics it is therefore convenient to pass from Cartesian to spherical coordinates in $\mathbf{R}^3$. On doing so, p^2 can be re-expressed by considering the orbital angular momentum $\vec{L} \equiv \vec{r} \wedge \vec{p}$ and the 'radial component' $p_r \equiv \frac{\vec{r}}{r} \cdot \vec{p}$ of the linear momentum. By virtue of the identity

$$L^2 = r^2 p^2 - (\vec{r} \cdot \vec{p})^2$$

one then finds

$$H = \frac{1}{2m}\left(p_r^2 + \frac{L^2}{r^2}\right) + V(r). \tag{5.4.1b}$$

In quantum mechanics, we have to find an operator realization of the counterpart of the classical Hamiltonian (5.4.1b). For this purpose, our first problem is the construction of the operator $\hat{p}_r$ in a central potential. Indeed, one has the classical formula

$$p_r \equiv \frac{\vec{r}}{r} \cdot \vec{p} = \frac{1}{2}\left(\frac{\vec{r}}{r} \cdot \vec{p} + \vec{p} \cdot \frac{\vec{r}}{r}\right). \tag{5.4.2}$$

However, in quantum mechanics, if we were to consider, naively, the operator

$$\hat{D}_r \equiv \frac{\vec{r}}{r} \cdot \vec{p} = \frac{\hbar}{\mathrm{i}}\left(\frac{x}{r}\frac{\partial}{\partial x} + \frac{y}{r}\frac{\partial}{\partial y} + \frac{z}{r}\frac{\partial}{\partial z}\right) = \frac{\hbar}{\mathrm{i}}\frac{\partial}{\partial r}, \tag{5.4.3}$$

we would realize that this is not the appropriate choice. What happens is that $\hat{D}_r$ is not Hermitian, because

$$\begin{aligned}
\hat{D}_r^\dagger &\equiv \frac{\vec{p} \cdot \vec{r}}{r} = \frac{\hbar}{\mathrm{i}}\left(\frac{\partial}{\partial x}\frac{x}{r} + \frac{\partial}{\partial y}\frac{y}{r} + \frac{\partial}{\partial z}\frac{z}{r}\right) \\
&= \frac{\hbar}{\mathrm{i}}\left(\frac{3}{r} + x\frac{\partial}{\partial x}r^{-1} + y\frac{\partial}{\partial y}r^{-1} + z\frac{\partial}{\partial z}r^{-1} + \frac{x}{r}\frac{\partial}{\partial x} + \frac{y}{r}\frac{\partial}{\partial y} + \frac{z}{r}\frac{\partial}{\partial z}\right) \\
&= \frac{\hbar}{\mathrm{i}}\left[\frac{3}{r} - \frac{(x^2+y^2+z^2)}{r^3} + \frac{\partial}{\partial r}\right] = \frac{\hbar}{\mathrm{i}}\left(\frac{\partial}{\partial r} + \frac{2}{r}\right).
\end{aligned} \tag{5.4.4}$$

The calculations (5.4.3) and (5.4.4) suggest defining (Dirac 1958)

$$\hat{p}_r \equiv \frac{1}{2}\left(\hat{D}_r + \hat{D}_r^\dagger\right) = \frac{\hbar}{\mathrm{i}}\left(\frac{\partial}{\partial r} + \frac{1}{r}\right). \tag{5.4.5}$$

Of course, the formula (5.4.5) can also be derived from the scalar product

$$(f, g) \equiv \int_0^\infty f^*(r) g(r) r^2 \, \mathrm{d}r.$$

This operator is Hermitian by construction, and obeys the desired form of the commutation rules:

$$\left[\hat{r}, \hat{p}_r\right] \equiv \hat{r}\, \hat{p}_r - \hat{p}_r\, \hat{r} = \mathrm{i}\hbar. \tag{5.4.6}$$

Moreover, it leads to the following form of the operator $\hat{p}_r^2$:

$$\hat{p}_r^2 = -\hbar^2\left(\frac{\partial^2}{\partial r^2} + \frac{2}{r}\frac{\partial}{\partial r}\right) = -\frac{\hbar^2}{r^2}\frac{\partial}{\partial r}\left(r^2 \frac{\partial}{\partial r}\right)$$
$$= -\frac{\hbar^2}{r}\frac{\partial^2}{\partial r^2}(r\cdot). \tag{5.4.7}$$

Note that the conditions of regularity at the origin, and square integrability on $(0, \infty)$, imply that $\hat{p}_r$ is symmetric on

$$C_0^\infty\left(\mathbf{R}^3 - \{0\}\right),$$

whereas $\hat{p}_r^2$ is self-adjoint. Without giving too many details, we can, however, make contact with what one knows in one-dimensional problems concerning these sorts of issues.

Let T be the operator $\mathrm{i}\frac{\mathrm{d}}{\mathrm{d}x}$ on $\mathcal{L}^2(0,1)$, with domain (for simplicity, we consider a closed interval)

$$D(T) = \{\varphi : \varphi \in AC[0,1], \varphi(0) = 0 = \varphi(1)\}. \tag{5.4.8}$$

In other words, the domain of T consists of absolutely continuous functions in $[0,1]$ that vanish at the end points of such interval. The adjoint of T, here denoted by $T^\dagger$, is again the operator $\mathrm{i}\frac{\mathrm{d}}{\mathrm{d}x}$, and for $\varphi \in D(T)$ and $\psi \in D(T^\dagger)$ one finds

$$\left(T\varphi, \psi\right) - \left(\varphi, T^\dagger \psi\right) = -\mathrm{i}[\varphi^*(1)\psi(1) - \varphi^*(0)\psi(0)] = 0. \tag{5.4.9}$$

It is hence clear that the boundary conditions on the functions $\in D(T)$ are so 'strong' that no boundary conditions whatsoever are necessary for the functions $\in D(T^\dagger)$. This is why T is not self-adjoint (although one can obtain a family of self-adjoint extensions by requiring proportionality of the boundary values at 0 and 1). However, if one studies the operator $A \equiv \frac{d^2}{dx^2}$ on $\mathcal{L}^2(0,1)$, the condition

$$\Big(Au, v\Big) - \Big(u, A^\dagger v\Big) = 0 \tag{5.4.10}$$

is fulfilled if and only if both the functions $\in D(A)$ and the functions $\in D(A^\dagger)$ obey the *same* boundary conditions, because integration by parts leads to the condition

$$\left(\frac{du^*}{dx} v - u^* \frac{dv}{dx} \right)_0^1 = 0, \tag{5.4.11}$$

which is satisfied if both $u \in D(A)$ and $v \in D(A^\dagger)$ vanish at 0 and at 1.

Now we can come back to our original problem, and we remark that the symmetry of the problem makes it possible to look for solutions of the stationary Schrödinger equation in the form

$$\psi(r, \theta, \varphi) = R(r) Y(\theta, \varphi), \tag{5.4.12}$$

where θ and φ are the angular variables occurring in the change of coordinates ($\theta \in [0, \pi], \varphi \in [0, 2\pi[$)

$$x = r \sin\theta \, \cos\varphi, \tag{5.4.13}$$

$$y = r \sin\theta \, \sin\varphi, \tag{5.4.14}$$

$$z = r \cos\theta. \tag{5.4.15}$$

Separation of variables in our eigenvalue equation therefore leads, for some real parameter λ, to the equations

$$\frac{d}{dr}\left(r^2 \frac{dR}{dr}\right) + \frac{2mr^2}{\hbar^2}\Big[E - V(r)\Big] R = \lambda R(r), \tag{5.4.16}$$

$$\frac{1}{\sin\theta} \frac{\partial}{\partial\theta}\left(\sin\theta \frac{\partial Y}{\partial\theta}\right) + \frac{1}{\sin^2\theta} \frac{\partial^2 Y}{\partial\varphi^2} = -\lambda Y(\theta, \varphi). \tag{5.4.17}$$

At this stage, we make a further separation of variables, and look for $Y(\theta, \varphi)$ in the form of a product

$$Y(\theta, \varphi) = \widetilde{Y}(\theta) \sigma(\varphi). \tag{5.4.18}$$

This leads to ordinary differential equations for σ and $\widetilde{Y}$, respectively:

$$\left(\frac{d^2}{d\varphi^2} + \mu \right) \sigma = 0, \tag{5.4.19}$$

$$\left[\frac{1}{\sin\theta} \frac{d}{d\theta}\left(\sin\theta \frac{d}{d\theta}\right) + \left(\lambda - \frac{\mu}{\sin^2\theta}\right) \right] \widetilde{Y} = 0. \tag{5.4.20}$$

If one imposes the periodicity conditions

$$\sigma(0) = \sigma(2\pi), \quad \sigma'(0) = \sigma'(2\pi), \tag{5.4.21}$$

these imply that, in the general form of the solution:

$$\sigma(\varphi) = A + B\varphi \quad \text{if } \mu = 0, \tag{5.4.22}$$

$$\sigma(\varphi) = A\mathrm{e}^{\mathrm{i}\sqrt{\mu}\varphi} + B\mathrm{e}^{-\mathrm{i}\sqrt{\mu}\varphi} \quad \text{if } \mu \neq 0, \tag{5.4.23}$$

one has to set $B = 0$, while $\sqrt{\mu}$ is an integer. Thus, one has

$$\sigma_m(\varphi) = A_m \mathrm{e}^{im\varphi}, \quad m = 0, \pm 1, \pm 2 \ldots . \tag{5.4.24}$$

The constant A_m is equal to $\frac{1}{\sqrt{2\pi}}$, from the condition

$$\int_0^{2\pi} |\sigma_m|^2 \, \mathrm{d}\varphi = 1. \tag{5.4.25}$$

At a deeper level, the integer values of m, including 0, are the spectrum of the z-component of the orbital angular momentum (cf. section 11.1 for the general theory of angular momentum operators).

The change of variable

$$\eta \equiv \cos\theta \tag{5.4.26}$$

therefore turns Eq. (5.4.20) into

$$\left[\frac{\mathrm{d}^2}{\mathrm{d}\eta^2} - \frac{2\eta}{(1-\eta^2)} \frac{\mathrm{d}}{\mathrm{d}\eta} + \frac{\lambda}{(1-\eta^2)} - \frac{m^2}{(1-\eta)^2(1+\eta)^2} \right] \widetilde{Y} = 0. \tag{5.4.27}$$

This equation has Fuchsian singularities at $\eta = 1$ and -1. In the (left) neighbourhood of $\eta = 1$, it has two linearly independent integrals

$$Y_1(\eta) = (1-\eta)^{|m|/2} h_1(\eta), \tag{5.4.28}$$

$$Y_2(\eta) = (1-\eta)^{|m|/2} h_1(\eta) \log(1-\eta) + (1-\eta)^{-|m|/2} h_2(\eta), \tag{5.4.29}$$

where the functions h_1 and h_2 are regular at $\eta = 1$. Moreover, in the (right) neighbourhood of $\eta = -1$, its linearly independent integrals are

$$\widetilde{Y}_1(\eta) = (1+\eta)^{|m|/2} \widetilde{h}_1(\eta), \tag{5.4.30}$$

$$\widetilde{Y}_2(\eta) = (1+\eta)^{|m|/2} \widetilde{h}_1(\eta) \log(1+\eta) + (1+\eta)^{-|m|/2} \widetilde{h}_2(\eta), \tag{5.4.31}$$

where the functions $\widetilde{h}_1$ and $\widetilde{h}_2$ are regular at $\eta = -1$. Thus, regularity of $\widetilde{Y}$ for $\eta \in [-1, 1]$ picks out a solution that can be written in the form

$$\widetilde{Y}(\eta) = (1-\eta^2)^{|m|/2} f(\eta), \tag{5.4.32}$$

where the function f is regular at $\eta = 1$ and -1. The power series for $f(\eta)$ involves numerical coefficients C_s, which turn out to obey the recurrence relations (s being an integer)

$$C_{s+2} = \frac{\left[(s+|m|)(s+|m|+1) - \lambda\right]}{(s+2)(s+1)} C_s, \tag{5.4.33}$$

by virtue of (5.4.27) and (5.4.32). One now remarks that, if these coefficients were all non-vanishing, one would find

$$\lim_{s \to \infty} |C_s / C_{s+2}| = 1, \tag{5.4.34}$$

and hence the radius of convergence of the series for f would not be greater than 1, leading in turn to infinitely many coefficients and to a solution that is not square-integrable on S^2. This means that, in contrast, an integer p exists such that

$$C_p \neq 0, \quad C_{p+2} = 0, \tag{5.4.35}$$

and the corresponding value of λ is

$$\lambda_p = (p+|m|)(p+|m|+1). \tag{5.4.36}$$

At a deeper level, Eq. (5.4.36) reflects the general properties of harmonic polynomials and of the operator $L^2 \equiv L_x^2 + L_y^2 + L_z^2$ (see section 5.6).

In Eq. (5.4.16) it is now convenient to set

$$R(r) \equiv \frac{u(r)}{r}. \tag{5.4.37}$$

The function R is square integrable with respect to the measure $r^2 \, dr$, whereas the function u is square integrable with respect to the measure dr. In our analysis of the eigenvalue problem in a central potential, we are interested in a subspace of $\mathcal{L}^2(\mathbf{R}^3)$, which is invariant under rotations and is spanned by vectors of the form (5.4.12), where, after considering (5.4.37), one studies the eigenvalue equation (where E denotes the desired eigenvalues)

$$\left\{ \frac{d^2}{dr^2} + \frac{2m}{\hbar^2} \left[E - V(r) - \frac{l(l+1)\hbar^2}{2mr^2} \right] \right\} u = 0, \tag{5.4.38}$$

for a particle of mass m. Equation (5.4.38) should be supplemented by the $\mathcal{L}^2$ condition

$$\int_0^\infty |u(r)|^2 \, dr < \infty, \tag{5.4.39}$$

and the boundary condition ensuring essential self-adjointness of the Hamiltonian operator:

$$u(0) = 0. \tag{5.4.40a}$$

In other words, Eq. (5.4.38) holds in the sense of differential operators on $\mathcal{L}^2(0,\infty)$ with the boundary condition (5.4.40a), rather than in the sense of classical differential equations. If the potential V is C^∞ with compact support, a theorem ensures that the solutions of (5.4.38) are C^∞ (this goes under the name of the *elliptic regularity* theorem). In the applications of section 5.5, we will see that regularity of u, for given values of l and E, is specified, more precisely, by the conditions

$$u_l(0,E) = 0, \tag{5.4.40b}$$

$$\lim_{r\to 0} r^{-l-1} u_l(r,E) = 1. \tag{5.4.41}$$

5.5 Hydrogen atom

The investigation of the hydrogen atom is a two-body problem in quantum mechanics. Before quantization, one has a classical Hamiltonian

$$H = \frac{p_{\rm n}^2}{2m_{\rm n}} + \frac{p_{\rm e}^2}{2m_{\rm e}} - \frac{e^2}{|\vec{x}_{\rm n} - \vec{x}_{\rm e}|}, \tag{5.5.1}$$

where the subscripts 'n' and 'e' refer to the nucleus and the electron, respectively. What we are really interested in is the relative motion of the electron and the nucleus. For this purpose, we pass to new variables according to the standard formulae for two-body problems:

$$\vec{P} \equiv \vec{p}_{\rm n} + \vec{p}_{\rm e}, \quad \vec{X} \equiv \frac{\left(m_{\rm e}\vec{x}_{\rm e} + m_{\rm n}\vec{x}_{\rm n}\right)}{(m_{\rm e} + m_{\rm n})}, \tag{5.5.2}$$

$$\vec{p}_{\rm en} \equiv \frac{\left(m_{\rm n}\vec{p}_{\rm e} - m_{\rm e}\vec{p}_{\rm n}\right)}{(m_{\rm e} + m_{\rm n})}, \quad \vec{r} \equiv \vec{x}_{\rm e} - \vec{x}_{\rm n}. \tag{5.5.3}$$

These formulae are obtained by requiring that a canonical transformation is performed on going from $(\vec{x}_{\rm e}, \vec{x}_{\rm n}, \vec{p}_{\rm e}, \vec{p}_{\rm n})$ to $(\vec{r}, \vec{X}, \vec{p}_{\rm en}, \vec{P})$. The resulting form of the Hamiltonian reads as

$$H = \frac{P^2}{2M} + \frac{p_{\rm en}^2}{2\mu} - \frac{e^2}{r}, \tag{5.5.4}$$

where μ is the reduced mass:

$$\mu \equiv \frac{m_{\rm n} m_{\rm e}}{(m_{\rm n} + m_{\rm e})}. \tag{5.5.5}$$

At this stage, we quantize in the coordinate representation, and factorize the stationary state as

$$\psi(\vec{X}, r) = \Omega(\vec{X})\psi(\vec{r}). \tag{5.5.6}$$

The centre-of-mass degrees of freedom represent a free particle and are encoded into $\Omega(\vec{X})$. The motion of the electron with respect to the nucleus is instead associated to $\psi(\vec{r})$. In the frame where the centre of mass is at rest, $\psi(\vec{r})$ obeys a stationary Schrödinger equation in a central field, and hence can be written as in (5.4.12). Bearing in mind (5.4.37) and (5.4.38), and defining

$$a \equiv \frac{\hbar^2}{\mu e^2}, \quad \nu \equiv \frac{\mu e^2}{\hbar\sqrt{-2\mu E}}, \quad \varrho \equiv r(\nu a)^{-1}, \tag{5.5.7}$$

our form of the stationary Schrödinger equation becomes

$$\left\{ \frac{d^2}{d\varrho^2} + \left[-1 + \frac{2\nu}{\varrho} - \frac{l(l+1)}{\varrho^2} \right] \right\} u(\varrho) = 0. \tag{5.5.8}$$

For the time being, ν is only a real-valued parameter, and this is emphasized by the choice of a Greek letter. This equation has a Fuchsian singularity at $\varrho = 0$, and is a particular case of a more general class of differential equations, i.e.

$$Gy \equiv \left[\frac{d^2}{dx^2} + \left(A + \frac{B}{x} \right) \frac{d}{dx} + \left(C + \frac{D}{x} + \frac{F}{x^2} \right) \right] y(x) = 0. \tag{5.5.9}$$

If one sets

$$y(x) \equiv e^{\alpha x} x^{\beta} f(x), \tag{5.5.10}$$

one finds that

$$Gy = e^{\alpha x} x^{\beta} \widetilde{G} f, \tag{5.5.11}$$

where

$$\widetilde{G} \equiv \frac{d^2}{dx^2} + \left(\widetilde{A} + \frac{\widetilde{B}}{x} \right) \frac{d}{dx} + \left(\widetilde{C} + \frac{\widetilde{D}}{x} + \frac{\widetilde{F}}{x^2} \right), \tag{5.5.12}$$

with

$$\widetilde{A} \equiv A + 2\alpha, \tag{5.5.13}$$

$$\widetilde{B} \equiv B + 2\beta, \tag{5.5.14}$$

$$\widetilde{C} \equiv C + \alpha A + \alpha^2, \tag{5.5.15}$$

$$\widetilde{D} \equiv D + \alpha B + (A + 2\alpha)\beta, \tag{5.5.16}$$

$$\widetilde{F} \equiv F + (B - 1)\beta + \beta^2. \tag{5.5.17}$$

Thus, the transformations (5.5.10) preserve Eq. (5.5.9) in that

$$\widetilde{G} f = 0. \tag{5.5.18}$$

For some choices of α and ρ (see below) it is then clear that one can set $\widetilde{C} = \widetilde{F} = 0$. In this case, defining

$$\zeta \equiv -\widetilde{A}\, x, \quad b \equiv \frac{\widetilde{D}}{\widetilde{A}}, \quad c \equiv \widetilde{B}, \tag{5.5.19}$$

Eq. (5.5.18) reduces to the confluent hypergeometric equation

$$\left[\frac{\mathrm{d}^2}{\mathrm{d}\zeta^2} + \left(\frac{c}{\zeta} - 1\right)\frac{\mathrm{d}}{\mathrm{d}\zeta} - \frac{b}{\zeta}\right] f = 0. \tag{5.5.20}$$

This is a second-order differential equation with a Fuchsian singularity at the origin where the coefficients of first- and zeroth-order derivative both have a first-order pole at $\zeta = 0$. Its general solution reads as

$$f(\zeta) = Q\ {}_1F_1(b, c; \zeta) + \widetilde{Q}\zeta^{1-c}{}_1F_1(b - c + 1, 2 - c; \zeta), \tag{5.5.21}$$

where Q and $\widetilde{Q}$ are some constants, and ${}_1F_1$ can be expressed by the series

$${}_1F_1(b, c; x) = \sum_{k=0}^{\infty} \frac{(b)_k}{(c)_k} \frac{x^k}{k!} = 1 + \frac{b}{c}x + \frac{b(b+1)}{c(c+1)}\frac{x^2}{2!} + \cdots. \tag{5.5.22}$$

Note now that the second-order operator in Eq. (5.5.8) is obtained from that in Eq. (5.5.9) by setting $A = B = 0, C = -1, D = 2\nu, F = -l(l+1)$. Thus, imposing $\widetilde{C} = \widetilde{F} = 0$, one finds, from (5.5.15) and (5.5.17), the algebraic equations

$$-1 + \alpha^2 = 0, \tag{5.5.23}$$

$$-l(l+1) - \beta + \beta^2 = 0, \tag{5.5.24}$$

which are solved by $\alpha_+ = 1, \alpha_- = -1, \beta_+ = l + 1, \beta_- = -l$. Regularity of the solution at $\varrho = 0$ picks out $\beta_+ = l + 1$, and square integrability on $(0, \infty)$ selects $\alpha_- = -1$. By virtue of (5.5.13), (5.5.14) and (5.5.16) one thus finds $\widetilde{A} = -2, \widetilde{B} = 2(l+1) = c, \widetilde{D} = 2(\nu - l - 1)$, and hence $b = l + 1 - \nu$. In the general formula (5.5.21) one has therefore to set $\widetilde{Q} = 0$ to ensure regularity of f (and u) at $\varrho = 0$, which implies

$$u(\varrho) = Q\varrho^{l+1}\mathrm{e}^{-\varrho}\ {}_1F_1(l + 1 - \nu, 2(l+1); 2\varrho). \tag{5.5.25}$$

At this stage, we only need to use a simple but important property of confluent hypergeometric functions according to which, as $x \to \infty$, one has the asymptotic expansion

$${}_1F_1(b, c; x) \sim \frac{\Gamma(c)}{\Gamma(b)}\mathrm{e}^{x}x^{b-c}. \tag{5.5.26}$$

In our problem, $\Gamma(c) = \Gamma(2(l+1))$ with $l \geq 0$, and hence $\Gamma(c)$ does not vanish. However, since the Γ-function is a meromorphic function with simple *poles* at $0, -1, -2, \ldots$, the term

$$\frac{1}{\Gamma(b)} = \frac{1}{\Gamma(l+1-\nu)}$$

has simple *zeros* at $0, -1, -2, \ldots$, for which

$$l+1-\nu = 0, \quad l+1-\nu = -1, \quad l+1-\nu = -2, \ldots .$$

These are the only admissible values of ν, since otherwise our solution $u(\varrho)$ would fail to be square integrable on $(0, \infty)$. This analysis proves that ν can only take the *integer* values $n = l+1, l+2, l+3, \ldots$. Such a property leads, in turn, to the Balmer formula for the energy spectrum of bound states (see the second definition in (5.5.7)):

$$E_n = -\frac{\mu e^4}{2\hbar^2} \frac{1}{n^2}. \tag{5.5.27}$$

The number n is called the principal quantum number, and is ≥ 1. For a given value of n, there exist n^2 linearly independent states with the *same* energy. The quantum number l takes all integer values from 0 to $n-1$, while m ranges over all values from $-l$ to $+l$, including 0 (see Eq. (5.4.24)). For example, the values $l = 0, 1, 2$ correspond to the so-called s-, p- and d-wave sectors, respectively (sharp, principal, diffuse). From the allowed ranges, it is clear that there exist one $1s$ state ($n = 1, l = m = 0$), one $2s$ state ($n = 2, l = m = 0$), three $2p$ states ($n = 2, l = 1, m = -1, 0, 1$), one $3s$ state ($n = 3, l = m = 0$), three $3p$ states ($n = 3, l = 1, m = -1, 0, 1$), five $3d$ states ($n = 3, l = 2, m = -2, -1, 0, 1, 2$), and so on.

For the first few values of the quantum numbers n, l, one has for example, for hydrogen-like atoms (see Eq. (5.4.37), bearing in mind that Z is the atomic number),

$$R_{10}(r) = (Z/a_0)^{3/2} 2 \mathrm{e}^{-\frac{1}{2}\rho_1}, \tag{5.5.28}$$

$$R_{20}(r) = (Z/a_0)^{3/2} \frac{1}{2\sqrt{2}} (2 - \rho_2) \mathrm{e}^{-\frac{1}{2}\rho_2}, \tag{5.5.29}$$

$$R_{21}(r) = (Z/a_0)^{3/2} \frac{1}{2\sqrt{6}} \rho_2 \mathrm{e}^{-\frac{1}{2}\rho_2}, \tag{5.5.30}$$

$$R_{30}(r) = (Z/a_0)^{3/2} \frac{1}{9\sqrt{3}} \left(6 - 6\rho_3 + \rho_3^2\right) \mathrm{e}^{-\frac{1}{2}\rho_3}, \tag{5.5.31}$$

$$R_{31}(r) = (Z/a_0)^{3/2} \frac{1}{9\sqrt{6}} (4 - \rho_3) \rho_3 \mathrm{e}^{-\frac{1}{2}\rho_3}, \tag{5.5.32}$$

$$R_{32}(r) = (Z/a_0)^{3/2} \frac{1}{9\sqrt{30}} \rho_3^2 \mathrm{e}^{-\frac{1}{2}\rho_3}, \tag{5.5.33}$$

where

$$\rho_n(r) \equiv \frac{2Zr}{na_0}. \tag{5.5.34}$$

The occurrence of factors which decrease exponentially (which are, of course, expected from (5.5.25)), implies that the probability of finding the electron at large distances becomes very small indeed. Here, very large means a few times the Bohr radius a_0. Nevertheless, it is a non-trivial property of quantum mechanics that such a probability can be considered for all (positive) values of r.

5.5.1 A simpler derivation of the Balmer formula

We have just derived the Balmer formula from the zeros of the Γ-function occurring in the asymptotic expansion of the hypergeometric function. However, if the reader does not remember the basic properties of special functions, he can still derive the Balmer formula. For this purpose, one can start from the eigenvalue equation (5.4.38), with m replaced by μ, and look for solutions in the form

$$u(r) = \mathrm{e}^{-wr} f(r), \tag{5.5.35}$$

where $w^2 \equiv -\varepsilon \equiv \frac{2\mu|E|}{\hbar^2}$. Of course, one only takes into account a decreasing exponential, to ensure that $u \in \mathcal{L}^2(0,\infty)$. At this stage, one writes (see appendix 5.B)

$$f(r) = r^s \sum_{k=0}^{\infty} a_k r^k, \tag{5.5.36}$$

where the parameter s will be derived by imposing regularity at the origin. Differentiation of u with respect to r yields

$$u'' = \Big(f'' - 2wf' + w^2 f\Big)\mathrm{e}^{-wr}, \tag{5.5.37}$$

and hence equation (5.5.36), jointly with (5.5.35), leads to the equation

$$\begin{aligned} & a_0\Big[s(s-1) - l(l+1)\Big]r^{s-2} \\ & + \sum_{k=0}^{\infty} a_{k+1}\Big[(k+s+1)(k+s) - l(l+1)\Big]r^{k+s-1} \\ & + \sum_{k=0}^{\infty} a_k \left[\frac{2\mu e^2}{\hbar^2} - 2w(k+s)\right] r^{k+s-1} = 0. \end{aligned} \tag{5.5.38}$$

The idea is, of course, to use the property according to which, if

$$\sum_{k=0}^{\infty} b_k r^k = 0 \quad \forall r,$$

then $b_k = 0 \ \forall k$. The coefficient a_0 should not vanish, and hence one has to set to zero what is multiplied by a_0, i.e.

$$s(s-1) - l(l+1) = 0. \tag{5.5.39}$$

Among the two roots of Eq. (5.5.39): $s_1 = l+1, s_2 = -l$, only s_1 is compatible with having a regular solution at $r = 0$.

From the remaining part of Eq. (5.5.38), one finds recurrence relations among a_{k+1} and a_k. If the algorithm could be implemented for all integer values of k, nothing would guarantee that the series expressing $f(r)$ through (5.5.36) is convergent. This would lead, in turn, to a solution $u(r)$ that is not square integrable on $(0, \infty)$. One thus finds that a value k^* of k exists such that

$$a_{k^*} \neq 0, \tag{5.5.40}$$

but

$$a_{k^*+1} = 0. \tag{5.5.41}$$

In other words, the series in (5.5.36) actually reduces to a polynomial, and one finds (recall that $s_1 = l + 1$)

$$\frac{\mu e^2}{\hbar^2} - w\Big(k^* + s_1\Big) = 0. \tag{5.5.42}$$

The comparison with the definition of w leads to

$$\varepsilon = -\frac{\mu e^4}{2\hbar^2}\frac{1}{n^2}, \tag{5.5.43}$$

which is indeed the Balmer formula for bound states of the hydrogen atom (we have set $n \equiv k^* + s_1 \equiv k^* + l + 1$).

5.6 Introduction to angular momentum

The results of sections 5.4 and 5.5 are best understood if one uses the quantum theory of angular momentum. However, we are not yet ready to study a rigorous mathematical formulation of quantum mechanics (see, in particular, chapters 9 and 11), and hence we limit ourselves to an introductory presentation. The key steps of our analysis are as follows.

5.6.1 Lie algebra of $O(3)$ and associated vector fields

We begin with classical considerations concerning rotations, first defined in appendix 2.B, by pointing out that if we associate to any vector $\vec{u}$ in $\mathbf{R}^3$ the linear transformation defined by

$$R_{\vec{u}}(\vec{v}) \equiv \vec{u} \times \vec{v}, \tag{5.6.1}$$

in a given orthonormal basis we can associate a 3×3 matrix with $R_{\vec{u}}$. One thus finds, associated with the unit vectors $\vec{u}_x, \vec{u}_y, \vec{u}_z$,

$$\mathcal{R}_x \begin{pmatrix} 0 \\ 1 \\ 0 \end{pmatrix} = \begin{pmatrix} 0 \\ 0 \\ 1 \end{pmatrix}, \quad \mathcal{R}_x \begin{pmatrix} 1 \\ 0 \\ 0 \end{pmatrix} = \begin{pmatrix} 0 \\ 0 \\ 0 \end{pmatrix}, \quad \mathcal{R}_x \begin{pmatrix} 0 \\ 0 \\ 1 \end{pmatrix} = \begin{pmatrix} 0 \\ -1 \\ 0 \end{pmatrix}, \tag{5.6.2}$$

which implies

$$\mathcal{R}_x = \begin{pmatrix} 0 & 0 & 0 \\ 0 & 0 & -1 \\ 0 & 1 & 0 \end{pmatrix}, \tag{5.6.3}$$

while a calculation along the same lines yields

$$\mathcal{R}_y = \begin{pmatrix} 0 & 0 & 1 \\ 0 & 0 & 0 \\ -1 & 0 & 0 \end{pmatrix}, \tag{5.6.4}$$

$$\mathcal{R}_z = \begin{pmatrix} 0 & -1 & 0 \\ 1 & 0 & 0 \\ 0 & 0 & 0 \end{pmatrix}. \tag{5.6.5}$$

The Lie algebra of the connected component $SO(3)$ of $O(3)$ is generated by the matrices $\mathcal{R}_x, \mathcal{R}_y, \mathcal{R}_z$. Note now that to any matrix A we can associate the vector field

$$X_A \equiv x^j A^i{}_j \frac{\partial}{\partial x^i}. \tag{5.6.6}$$

Although the matrix A and the vector field X_A contain the same information, the analysis in terms of X_A makes it possible to consider the effect of non-linear transformations upon it. For our purposes, the use of vector fields also has the advantage of making it very clear that they correspond to angular momentum (see below).

By virtue of (5.6.3)–(5.6.6) we obtain the following form of infinitesimal generators of rotations viewed as vector fields:

$$R_x = y\frac{\partial}{\partial z} - z\frac{\partial}{\partial y}, \tag{5.6.7}$$

$$R_y = z\frac{\partial}{\partial x} - x\frac{\partial}{\partial z}, \tag{5.6.8}$$

$$R_z = x\frac{\partial}{\partial y} - y\frac{\partial}{\partial x}. \tag{5.6.9}$$

Since we have replaced the matrices (5.6.3)–(5.6.5) with vector fields according to (5.6.7)–(5.6.9), it is now possible to perform generic coordinate transformations. For example, in the spherical coordinates defined by (5.4.13)–(5.4.15), the formulae (5.6.7)–(5.6.9), which imply (here $R^1 \equiv R_x, R^2 \equiv R_y, R^3 \equiv R_z$)

$$R^i x^j = -\varepsilon^{ij}{}_k \, x^k, \tag{5.6.10}$$

lead to

$$R_x = -\sin\varphi\frac{\partial}{\partial\theta} - \cos\varphi\cot\theta\frac{\partial}{\partial\varphi}, \tag{5.6.11}$$

$$R_y = \cos\varphi\frac{\partial}{\partial\theta} - \sin\varphi\cot\theta\frac{\partial}{\partial\varphi}, \tag{5.6.12}$$

$$R_z = \frac{\partial}{\partial\varphi}. \tag{5.6.13}$$

Note also that, by exploiting the invariance of $p_i \, \mathrm{d}q^i$, the momenta p_r, p_θ and p_φ can be derived by requiring that

$$p_x \, \mathrm{d}x + p_y \, \mathrm{d}y + p_z \, \mathrm{d}z = p_r \, \mathrm{d}r + p_\theta \, \mathrm{d}\theta + p_\varphi \, \mathrm{d}\varphi. \tag{5.6.14}$$

For this purpose one has to evaluate the differentials $\mathrm{d}x, \mathrm{d}y$ and $\mathrm{d}z$ from (5.4.13)–(5.4.15), i.e.

$$\begin{pmatrix} \mathrm{d}x \\ \mathrm{d}y \\ \mathrm{d}z \end{pmatrix} = \begin{pmatrix} \sin\theta\cos\varphi & r\cos\theta\cos\varphi & -r\sin\theta\sin\varphi \\ \sin\theta\sin\varphi & r\cos\theta\sin\varphi & r\sin\theta\cos\varphi \\ \cos\theta & -r\sin\theta & 0 \end{pmatrix} \begin{pmatrix} \mathrm{d}r \\ \mathrm{d}\theta \\ \mathrm{d}\varphi \end{pmatrix}, \tag{5.6.15}$$

and the insertion of the formulae (5.6.15) into Eq. (5.6.14) leads to

$$\begin{pmatrix} p_r \\ p_\theta \\ p_\varphi \end{pmatrix} = \begin{pmatrix} \sin\theta\cos\varphi & \sin\theta\sin\varphi & \cos\theta \\ r\cos\theta\cos\varphi & r\cos\theta\sin\varphi & -r\sin\theta \\ -r\sin\theta\sin\varphi & r\sin\theta\cos\varphi & 0 \end{pmatrix} \begin{pmatrix} p_x \\ p_y \\ p_z \end{pmatrix}. \tag{5.6.16}$$

5.6.2 Quantum definition of angular momentum

Using the association between symbols and operators, i.e.

$$\hat{x} = x, \quad \hat{y} = y, \quad \hat{z} = z, \tag{5.6.17a}$$

$$\hat{p}_x \equiv \frac{\hbar}{\mathrm{i}}\frac{\partial}{\partial x}, \quad \hat{p}_y \equiv \frac{\hbar}{\mathrm{i}}\frac{\partial}{\partial y}, \quad \hat{p}_z \equiv \frac{\hbar}{\mathrm{i}}\frac{\partial}{\partial z}, \tag{5.6.17b}$$

we are led to associate an operator with the orbital angular momentum according to the prescription

$$\hat{L}_x \equiv \frac{\hbar}{\mathrm{i}} R_x = y \frac{\hbar}{\mathrm{i}} \frac{\partial}{\partial z} - z \frac{\hbar}{\mathrm{i}} \frac{\partial}{\partial y}$$
$$= \hat{y}\hat{p}_z - \hat{z}\hat{p}_y, \tag{5.6.18}$$

along with the other components

$$\hat{L}_y \equiv \frac{\hbar}{\mathrm{i}} R_y = \hat{z}\hat{p}_x - \hat{x}\hat{p}_z, \tag{5.6.19}$$

$$\hat{L}_z \equiv \frac{\hbar}{\mathrm{i}} R_z = \hat{x}\hat{p}_y - \hat{y}\hat{p}_x. \tag{5.6.20}$$

These operators obey the commutation relations

$$\hat{L}_j \hat{L}_k - \hat{L}_k \hat{L}_j = \mathrm{i}\hbar \varepsilon_{jks} \hat{L}_s, \tag{5.6.21}$$

In spherical coordinates, having obtained the formulae (5.6.11)–(5.6.13) for the classical operators, the quantum operators read as

$$\hat{L}_x \equiv \frac{\hbar}{\mathrm{i}} R_x = -\frac{\hbar}{\mathrm{i}} \left(\sin\varphi \frac{\partial}{\partial \theta} + \cos\varphi \cot\theta \frac{\partial}{\partial \varphi} \right), \tag{5.6.22}$$

$$\hat{L}_y \equiv \frac{\hbar}{\mathrm{i}} R_y = \frac{\hbar}{\mathrm{i}} \left(\cos\varphi \frac{\partial}{\partial \theta} - \sin\varphi \cot\theta \frac{\partial}{\partial \varphi} \right), \tag{5.6.23}$$

$$\hat{L}_z \equiv \frac{\hbar}{\mathrm{i}} R_z = \frac{\hbar}{\mathrm{i}} \frac{\partial}{\partial \varphi}. \tag{5.6.24}$$

We recall from section 4.1 that we are forced to quantize in Cartesian coordinates (cf. DeWitt 1952), while changes of coordinates can be performed only after we have obtained the operator. The reader should also remember what we have found in section 5.4 about the operator $\hat{p}_r$ to obtain a well-defined (formally) self-adjoint operator.

Interestingly, the quantum operator defined in (5.6.18)–(5.6.20) turns out to be essentially self-adjoint on the set of linear combinations of vectors of the kind (Thirring 1981)

$$\psi_k \equiv \mathrm{e}^{-r^2/2} \, x^{k_1} \, y^{k_2} \, z^{k_3}, \tag{5.6.25}$$

where $k_i \in \{0, 1, 2, \ldots\}$. The sub-space

$$D_k \equiv \{\psi_k : k_1 + k_2 + k_3 \leq k\} \tag{5.6.26}$$

is finite-dimensional, and hence the operator $\vec{L}$ is represented by finite-dimensional matrices.

5.6.3 Harmonic polynomials and spherical harmonics

On considering the transition from Cartesian coordinates to spherical polar coordinates, we pass from $\mathbf{R}^3$ to $S^2 \times \mathbf{R}^+$. It is reasonable to expect that, for problems with rotational symmetry, $S^2 \times \mathbf{R}^+$ is more convenient than $\mathbf{R}^3$. On $\mathbf{R}^3$, we know that $\frac{\partial}{\partial x}, \frac{\partial}{\partial y}$ and $\frac{\partial}{\partial z}$ generate translations and commute. When polar coordinates are used, one looks for operators compatible with the split $S^2 \times \mathbf{R}^+$. A possible choice is given by the operators

$$x\frac{\partial}{\partial y} - y\frac{\partial}{\partial x}, \quad y\frac{\partial}{\partial z} - z\frac{\partial}{\partial y}, \quad z\frac{\partial}{\partial x} - x\frac{\partial}{\partial z},$$

$$x\frac{\partial}{\partial x} + y\frac{\partial}{\partial y} + z\frac{\partial}{\partial z}.$$

We now look for the base of a Hilbert space on S^2, which should be completed by specifying its dependence on $r \in \mathbf{R}^+$. Indeed, starting from the above first-order operators, one can see that polynomials of degree l in the variables x, y, z, generated by the monomials $x^m y^n z^{l-m-n}$, form an invariant space of dimension $\frac{(l+1)(l+2)}{2}$, since the generators of rotations are homogeneous of degree 0. Note that an invariant sub-space exists, defined by polynomials of the form $\left(x^2 + y^2 + z^2\right)P_{l-2}$, where P_{l-2} are the homogeneous polynomials of degree $l-2$ (their space has dimension $\frac{l(l-1)}{2}$). Another invariant sub-space H_l can be built as follows. Let $\triangle$ be the second-order operator defined by

$$\triangle \equiv \frac{\partial^2}{\partial x^2} + \frac{\partial^2}{\partial y^2} + \frac{\partial^2}{\partial z^2}. \tag{5.6.27}$$

By construction, $\triangle$ is a map $\triangle : P_l \rightarrow P_{l-2}$. Since the algebra of the rotation group commutes with $\triangle$, one has that

$$\text{Ker}\,\triangle \equiv H_l \tag{5.6.28}$$

is an invariant sub-space. By definition, the elements of H_l are the *harmonic polynomials*.

The solutions of the corresponding Laplace equation: $\triangle P_l = 0$, can be found by introducing new variables

$$u \equiv \frac{1}{2}(x + \mathrm{i}y), \quad v \equiv \frac{1}{2}(x - \mathrm{i}y), \tag{5.6.29}$$

in terms of which $\triangle$ reads as

$$\triangle = \frac{\partial^2}{\partial u \partial v} + \frac{\partial^2}{\partial z^2}. \tag{5.6.30}$$

If P_l is a homogeneous polynomial of degree l in u, v, z, then it is also homogeneous in x, y, z. Some solutions of the Laplace equation can be immediately written: $u^l, v^l, u^{l-1}z, v^{l-1}z$. One can now look for solutions of the form $u^{l-m}v^m$, $m = 0, 1, \ldots, l$. These are not, by themselves, solutions of our second-order equation, because

$$\triangle(u^{l-m}v^m) = (l-m)mu^{l-m-1}v^{m-1}. \tag{5.6.31}$$

As a first step, one tries to 'compensate' the term on the right-hand side of (5.6.31) by adding the term

$$(-1)(l-m)mu^{l-m-1}v^{m-1}\frac{z^2}{2}.$$

One then finds

$$\triangle\left(u^{l-m}v^m - \frac{(l-m)mu^{l-m-1}v^{m-1}z^2}{2}\right)$$
$$= -\frac{(l-m)(l-m-1)m(m-1)u^{l-m-2}v^{m-2}z^2}{2}. \tag{5.6.32}$$

This result makes it necessary to introduce yet another 'compensating' term, now of the form

$$\frac{(l-m)(l-m-1)m(m-1)u^{l-m-2}v^{m-2}z^4}{4!}.$$

The iteration of this algorithm leads, eventually, to a harmonic polynomial of degree l, which can be written in the form

$$f_{l,m}(u,v,z) = \sum_{p=0}^{a_{l,m}} \frac{(-1)^p(l-m)!m!}{(l-m-p)!(m-p)!(2p)!} u^{l-m-p}v^{m-p}z^{2p}, \tag{5.6.33}$$

where $a_{l,m} \equiv \min(l-m, m)$. Note that, in all terms of the sum defining $f_{l,m}$, the exponent of u minus the exponent of v equals $l - 2m$. There exist $l + 1$ functions of the form (5.6.33).

Similarly, one can start from a term of the kind $u^{l-m}v^{m-1}z$, where $m = 1, 2, \ldots, l$. This gives rise to solutions of the Laplace equation of the form

$$h_{l,m}(u,v,z) = \sum_{p=0}^{b_{l,m}} \frac{(-1)^p(l-m)!(m-1)!u^{l-m-p}v^{m-p-1}z^{2p+1}}{(l-m-p)!(m-p-1)!(2p+1)!}, \tag{5.6.34}$$

where $b_{l,m} \equiv \min(l-m, m-1)$. In all terms of the sum defining $h_{l,m}$, the exponent of u minus the exponent of v equals $l - 2m + 1$. There exist l solutions of the form (5.6.34).

Thus, there exist $2l+1$ harmonic polynomials of degree l. They can be labelled by the number s, which, by definition, is the difference between the exponent of u and the exponent of v. On dividing by r^l the harmonic polynomials of degree l, here denoted by $K_{l,s}$, one obtains the *spherical harmonics*. Using the fact that R_x, R_y, R_z commute with the Laplacian, we may derive the same result by iterating the action of these operators on the starting solutions.

The rotation group acts on the space of functions $K_{l,s}$ and defines the representation $\triangle^l(R)$. If R represents a rotation about the z-axis, one has the transformation property

$$x' = x\cos\phi + y\sin\phi, \tag{5.6.35}$$

$$y' = -x\sin\phi + y\cos\phi, \tag{5.6.36}$$

$$z' = z, \tag{5.6.37}$$

which implies

$$2u' = x' + \mathrm{i}y' = \mathrm{e}^{-\mathrm{i}\phi}x + \mathrm{i}\mathrm{e}^{-\mathrm{i}\phi}y = 2\mathrm{e}^{-\mathrm{i}\phi}u. \tag{5.6.38}$$

Hence one finds that v, defined in (5.6.29), transforms as $v' = \mathrm{e}^{\mathrm{i}\phi}v$. These basic rules imply the following transformation property for monomials:

$$u^\alpha\, v^\beta\, z^\gamma \rightarrow \mathrm{e}^{\mathrm{i}(\beta-\alpha)\phi}\, u^\alpha\, v^\beta\, z^\gamma, \tag{5.6.39}$$

and hence

$$K_{l,s}(R^{-1}\vec{x}) = \mathrm{e}^{-\mathrm{i}s\phi}\, K_{l,s}(\vec{x}). \tag{5.6.40}$$

The matrix which represents the rotation about the z-axis is therefore

$$\triangle^l(R)_{st} = \mathrm{e}^{-\mathrm{i}s\phi}\, \delta_{st}. \tag{5.6.41}$$

The results of section 5.4 are easily re-interpreted in terms of spherical harmonics and orbital angular momentum. What happens is that, in $\hbar = 1$ units, the Laplacian in spherical coordinates reads as

$$\triangle_{\mathrm{LP}} = -\left(\frac{\partial^2}{\partial r^2} + \frac{2}{r}\frac{\partial}{\partial r}\right) - \frac{1}{r^2}L^2, \tag{5.6.42}$$

where

$$L^2 = \frac{1}{\sin\theta}\frac{\partial}{\partial\theta}\left(\sin\theta\frac{\partial}{\partial\theta}\right) + \frac{1}{\sin^2\theta}\frac{\partial^2}{\partial\varphi^2}. \tag{5.6.43}$$

On the other hand, the harmonic polynomials obey the Laplace equation

$$\triangle_{\mathrm{LP}}\Big(r^l Y_l(\theta,\varphi)\Big) = 0, \tag{5.6.44}$$

which implies

$$L^2 Y_l(\theta, \varphi) = -l(l+1) Y_l(\theta, \varphi), \tag{5.6.45}$$

by virtue of (5.6.42). The quantum operator $\hat{L}^2$ is then equal to $-\hbar^2 L^2$, and hence has eigenvalues $l(l+1)\hbar^2$.

In spherical coordinates, the components of the orbital angular momentum take the form (5.6.22)–(5.6.24). In particular, Eq. (5.6.24) leads immediately to the spectrum in (5.4.24), if the first boundary condition in (5.4.21) is imposed. Other useful formulae are those for the *raising* and *lowering* operators:

$$\hat{L}_+ \equiv \hat{L}_x + \mathrm{i}\hat{L}_y = \hbar \mathrm{e}^{\mathrm{i}\varphi} \left(\frac{\partial}{\partial \theta} + \mathrm{i} \cot \theta \frac{\partial}{\partial \varphi} \right), \tag{5.6.46}$$

$$\hat{L}_- \equiv \hat{L}_x - \mathrm{i}\hat{L}_y = \hbar \mathrm{e}^{-\mathrm{i}\varphi} \left(-\frac{\partial}{\partial \theta} + \mathrm{i} \cot \theta \frac{\partial}{\partial \varphi} \right). \tag{5.6.47}$$

They are called in this way because the operator L_- maps the eigenfunction of L_z with eigenvalue m into the eigenfunction of L_z with eigenvalue $m-1$, and in general one has (for some constant N_{km})

$$Y_{km} = N_{km} [(\hat{L}_x - \mathrm{i}\hat{L}_y)/\hbar]^{k-m} Y_{kk}. \tag{5.6.48}$$

Some formulae for spherical harmonics that are frequently used in the applications are as follows:

$$Y_{0,0} = \frac{1}{\sqrt{4\pi}}, \tag{5.6.49}$$

$$Y_{1,0} = \sqrt{\frac{3}{4\pi}} \cos \theta, \tag{5.6.50}$$

$$Y_{1,\pm 1} = \mp \sqrt{\frac{3}{8\pi}} \sin \theta \, \mathrm{e}^{\pm \mathrm{i}\varphi}, \tag{5.6.51}$$

$$Y_{2,0} = \sqrt{\frac{5}{16\pi}} \left(3 \cos^2 \theta - 1 \right), \tag{5.6.52}$$

$$Y_{2,\pm 1} = \mp \sqrt{\frac{15}{8\pi}} (\cos \theta)(\sin \theta) \mathrm{e}^{\pm \mathrm{i}\varphi}, \tag{5.6.53}$$

$$Y_{2,\pm 2} = \sqrt{\frac{15}{32\pi}} \sin^2 \theta \, \mathrm{e}^{\pm 2\mathrm{i}\varphi}. \tag{5.6.54}$$

One can prove that the spherical harmonics $Y_{km}(\theta,\varphi)$ form an orthonormal system, i.e.

$$\int_0^{2\pi} d\varphi \int_0^{\pi} d\theta \ \sin\theta \, Y_{km}^*(\theta,\varphi) Y_{k'm'}(\theta,\varphi) = \delta_{kk'}\delta_{mm'}. \qquad (5.6.55)$$

Moreover, such functions also form a complete system, because every square-integrable function on S^2 can be expanded according to the relation

$$f(\theta,\varphi) = \sum_{k=0}^{\infty} \sum_{m=-k}^{k} C_{km} Y_{km}(\theta,\varphi), \qquad (5.6.56)$$

where the Fourier coefficients C_{km} are given by

$$C_{km} = \int_0^{2\pi} d\varphi \int_0^{\pi} d\theta \ \sin\theta \, Y_{km}^*(\theta,\varphi) f(\theta,\varphi), \qquad (5.6.57)$$

and satisfy the condition

$$\sum_{k=0}^{\infty} \sum_{m=-k}^{k} |C_{km}|^2 = 1. \qquad (5.6.58)$$

The property expressed by (5.6.58) suggests interpreting $|C_{km}|^2$ as the probability of finding the eigenvalues $k(k+1)\hbar^2$ and $m\hbar$ on performing a measurement of $\hat{L}^2$ and $\hat{L}_z$ in the state described by the wave function $f(\theta,\varphi)$. For example, such an interpretation can be applied to the state

$$f(\theta,\varphi) = \sqrt{\frac{3}{8\pi}} \Big(\sin\theta \sin\varphi + \mathrm{i}\cos\theta \Big), \qquad (5.6.59)$$

after re-expressing it as a linear combination of spherical harmonics with the help of (5.6.50) and (5.6.51), i.e.

$$f(\theta,\varphi) = \mathrm{i} \left(\frac{1}{\sqrt{2}} Y_{1,0} + \frac{1}{2} Y_{1,1} + \frac{1}{2} Y_{1,-1} \right). \qquad (5.6.60)$$

5.6.4 Back to central potentials in $\mathbf{R}^3$

We can now gain a better understanding of the Schrödinger equation in a central potential in $\mathbf{R}^3$, first studied in section 5.4. Indeed, if E is an eigenvalue of the Hamiltonian operator $\hat{H} = \frac{\hat{p}^2}{2m} + \hat{V}$, then the set

$$\Big\{ \psi : \hat{H}\psi = E\psi \Big\}$$

is a rotationally invariant sub-space of the general set of square-integrable functions on $\mathbf{R}^3$, and hence is spanned by vectors of the form

$$\psi(r,\theta,\varphi) = r^{-1} u(r) Y_{lm}(\theta,\varphi), \qquad (5.6.61)$$

where Y_{lm} are the spherical harmonics and

$$\int_0^\infty |u(r)|^2 \, \mathrm{d}r < \infty. \tag{5.6.62}$$

The number l is called the angular momentum quantum number of ψ, and $u(r)$ obeys the differential equation (5.4.38), with boundary condition $u(0) = 0$. The form of Eq. (5.4.38) is obtained upon using the classical formula $p^2 = p_r^2 + \frac{L^2}{r^2}$ (see Eq. (5.4.1b)), the relation $\hat{L}^2 = -\hbar^2 L^2$, and the eigenvalue equation (5.6.43). Moreover, the operators $\hat{H}, \hat{L}^2, \hat{L}_z$ commute with one another:

$$(\hat{H}\hat{L}^2 - \hat{L}^2\hat{H}) = 0, \quad (\hat{H}\hat{L}_z - \hat{L}_z\hat{H}) = 0, \quad (\hat{L}^2\hat{L}_z - \hat{L}_z\hat{L}^2) = 0, \tag{5.6.63}$$

and have common eigenvectors

$$\hat{H}\psi(r,\theta,\varphi) = E\psi(r,\theta,\varphi), \tag{5.6.64}$$

$$\hat{L}^2\psi(r,\theta,\varphi) = l(l+1)\hbar^2\psi(r,\theta,\varphi), \tag{5.6.65}$$

$$\hat{L}_z\psi(r,\theta,\varphi) = m\hbar\psi(r,\theta,\varphi), \tag{5.6.66}$$

where the integer m ranges from $-l$ to $+l$, including 0.

As a further example, we now consider the isotropic harmonic oscillator in three dimensions. The angular part of the stationary states is given by the spherical harmonics as we just mentioned, while Eq. (5.4.38) reads, in this case, as

$$\left[\frac{\mathrm{d}^2}{\mathrm{d}r^2} + \frac{2mE}{\hbar^2} - \frac{m^2\omega^2}{\hbar^2}r^2 - \frac{l(l+1)}{r^2}\right] u(r) = 0. \tag{5.6.67}$$

We now define (Flügge 1971)

$$\frac{2mE}{\hbar^2} \equiv k^2, \quad \frac{m\omega}{\hbar} \equiv \lambda, \quad \frac{k^2}{2\lambda} \equiv \frac{E}{\hbar\omega} \equiv \mu, \tag{5.6.68}$$

and point out that, as $r \rightarrow 0$, the regular solution of Eq. (5.6.67) is proportional to r^{l+1}, while, as $r \rightarrow \infty$, the square-integrable solution behaves as $\mathrm{e}^{-\frac{\lambda}{2}r^2}$. These properties suggest looking for an exact solution in the form

$$u(r) = r^{l+1} \mathrm{e}^{-\frac{\lambda}{2}r^2} v(r). \tag{5.6.69}$$

Moreover, it is convenient to define the new independent variable

$$t \equiv \lambda r^2, \tag{5.6.70}$$

so that the Schrödinger equation for stationary states takes the form

$$\left\{ t\frac{\mathrm{d}^2}{\mathrm{d}t^2} + \left[\left(l + \frac{3}{2}\right) - t\right]\frac{\mathrm{d}}{\mathrm{d}t} - \left[\frac{1}{2}\left(l + \frac{3}{2}\right) - \frac{\mu}{2}\right] \right\} v(r) = 0. \tag{5.6.71}$$

This is a Kummer equation (Kummer 1836), for which the general integral is

$$v(r) = C_a \, {}_1F_1\left(\frac{1}{2}\left(l+\frac{3}{2}-\mu\right), l+\frac{3}{2}; \lambda r^2\right) + C_b r^{-(2l+1)} \, {}_1F_1\left(\frac{1}{2}\left(-l+\frac{1}{2}-\mu\right), -l+\frac{1}{2}; \lambda r^2\right). \quad (5.6.72)$$

However, to obtain a square-integrable solution $\forall l \neq 0$, we have to set $C_b = 0$. From now on the procedure is very similar to the case of the hydrogen atom (cf. section 5.5). Since, as $x \to \infty$, a confluent hypergeometric function behaves as in (5.5.26), this implies that $u(r)$ would diverge exponentially at large r:

$$u(r) \sim r^{l+1} e^{\frac{\lambda}{2}r^2} r^{-(l+\frac{3}{2}+\mu)}.$$

We then exploit the fact that the inverse of the Γ-function is a function with simple zeros, and hence the series can reduce to a polynomial provided that

$$b = \frac{1}{2}\left(l+\frac{3}{2}-\mu\right) = -n_r, \quad (5.6.73)$$

which leads to the spectrum

$$E = \left(2n_r + l + \frac{3}{2}\right)\hbar\omega, \quad (5.6.74)$$

where the radial quantum number n_r takes all values $0, 1, 2, \ldots$. The complete $\mathcal{L}^2$ solution in spherical coordinates therefore reads as

$$\psi(r,\theta,\varphi) = C r^l e^{-\frac{\lambda}{2}r^2} \, {}_1F_1\left(-n_r, l+\frac{3}{2}; \lambda r^2\right) Y_{lm}(\theta,\varphi), \quad (5.6.75)$$

where C is a normalization constant, and the degeneracy is that evaluated in section 4.7, with $n \equiv 2n_r + l$.

5.7 Homomorphism between $SU(2)$ and $SO(3)$

In the previous section we have studied the rotation group in $\mathbf{R}^3$ more thoroughly. As a manifold, this can be viewed as a sub-manifold of $\mathbf{R}^9$ defined by equations (2.B.19) and (2.B.20). To investigate its topological properties, it is more convenient to consider its double cover, i.e. the group $SU(2)$ of 2×2 unitary matrices with unit determinant. The present section is therefore devoted to the relation between these two groups. We shall see that the two groups are related by a homomorphism. In general, for any two groups G_1 and G_2, a *homomorphism* is a map $\phi : G_1 \to G_2$ such that

$$\phi(g_1 \cdot g_2) = \phi(g_1)\phi(g_2),$$

$$\phi(g^{-1}) = (\phi(g))^{-1}.$$

The work in this section follows closely the presentation in Wigner (1959), which relies in turn on a method suggested by H. Weyl. We begin with some elementary results in the theory of matrices, which turn out to be very useful for our purposes.

(i) A matrix which transforms every real vector into a real vector is itself real, i.e. all its elements are real. If this matrix is applied to the kth unit vector (which has kth component $= 1$, with all others vanishing), the result is the vector which forms the kth row of the matrix. Thus, this row must be real. But this argument can be applied to all k, and hence all the rows of the matrix must be real.

(ii) It is also well known that a matrix $\mathcal{O}$ is orthogonal if it preserves the scalar product of two arbitrary vectors, i.e. if (see appendix 2.B)

$$(\vec{a}, \vec{b}) = (\mathcal{O}\vec{a}, \mathcal{O}\vec{b}). \tag{5.7.1}$$

An equivalent condition can be stated in terms of one arbitrary vector: a matrix $\mathcal{O}$ is orthogonal if the length of every single arbitrary vector $\vec{v}$ is left unchanged under transformation by $\mathcal{O}$. Consider now two arbitrary vectors $\vec{a}$ and $\vec{b}$, and write $\vec{v} = \vec{a} + \vec{b}$. Then our condition for the orthogonality of $\mathcal{O}$ is

$$(\vec{v}, \vec{v}) = (\mathcal{O}\vec{v}, \mathcal{O}\vec{v}). \tag{5.7.2}$$

By virtue of the symmetry of the scalar product: $(\vec{a}, \vec{b}) = (\vec{b}, \vec{a})$, this yields

$$\begin{aligned}(\vec{a} + \vec{b}, \vec{a} + \vec{b}) &= (\vec{a}, \vec{a}) + (\vec{b}, \vec{b}) + 2(\vec{a}, \vec{b}) \\ &= (\mathcal{O}\vec{a}, \mathcal{O}\vec{a}) + (\mathcal{O}\vec{b}, \mathcal{O}\vec{b}) + 2(\mathcal{O}\vec{a}, \mathcal{O}\vec{b}).\end{aligned} \tag{5.7.3}$$

However, orthogonality also implies that $(\vec{a}, \vec{a})$ equals $(\mathcal{O}\vec{a}, \mathcal{O}\vec{a})$, and the same with $\vec{a}$ replaced by $\vec{b}$. It then follows from (5.7.3) that (appendix 2.B)

$$(\vec{a}, \vec{b}) = (\mathcal{O}\vec{a}, \mathcal{O}\vec{b}), \tag{5.7.4}$$

which implies that $\mathcal{O}$ is orthogonal. It can be shown, in a similar way, that $\mathcal{U}$ is unitary only if $(\vec{v}, \vec{v}) = (\mathcal{U}\vec{v}, \mathcal{U}\vec{v})$ holds for every vector.

By definition, a matrix that leaves each real vector real, and leaves the length of every vector unchanged and preserves the origin, is a *rotation* (see appendix 2.B). Indeed, when all lengths are equal in the original and transformed figures, the angles must also be equal; hence the transformation is merely a rotation.

(iii) We now want to determine the general form of a two-dimensional unitary matrix

$$\mathbf{u} = \begin{pmatrix} a & b \\ c & d \end{pmatrix} \tag{5.7.5}$$

of determinant +1 by considering the elements of the product

$$\mathbf{u}\mathbf{u}^\dagger = \mathbb{I}. \tag{5.7.6}$$

Recall that the † operation means taking the complex conjugate and then the transpose of the original matrix. Thus, the condition (5.7.6) implies that

$$a^* c + b^* d = 0, \tag{5.7.7}$$

which leads to $c = -b^* d/a^*$. The insertion of this result into the condition of unit determinant:

$$ad - bc = 1, \tag{5.7.8}$$

yields $\left(aa^* + bb^*\right)d/a^* = 1$. Moreover, since $aa^* + bb^* = 1$ from (5.7.6), it follows that $d = a^*$ and $c = -b^*$. The general two-dimensional unitary matrix with unit determinant is hence

$$\mathbf{u} = \begin{pmatrix} a & b \\ -b^* & a^* \end{pmatrix}, \tag{5.7.9}$$

where, of course, we still have to require that $aa^* + bb^* = 1$. Note that, if one writes $a = y_0 + \mathrm{i}y_3$ and $b = y_1 + \mathrm{i}y_2$, one finds

$$\det \mathbf{u} = y_0^2 + y_1^2 + y_2^2 + y_3^2 = 1.$$

This is the equation of a unit 3-sphere centred at the origin, which means that $SU(2)$ has 3-sphere topology and is hence simply connected (the n-sphere is simply connected for all $n > 1$). More precisely, $SU(2)$ is homeomorphic to $S^3 \subset \mathbf{R}^4$, where we remark that S^3 can be obtained as the quotient space $\left(\mathbf{R}^4 - \{0\}\right)/\mathbf{R}^+$.

Consider now the so-called Pauli matrices:

$$\sigma_x \equiv \begin{pmatrix} 0 & 1 \\ 1 & 0 \end{pmatrix}, \tag{5.7.10}$$

$$\sigma_y \equiv \begin{pmatrix} 0 & -\mathrm{i} \\ \mathrm{i} & 0 \end{pmatrix}, \tag{5.7.11}$$

$$\sigma_z \equiv \begin{pmatrix} 1 & 0 \\ 0 & -1 \end{pmatrix}. \tag{5.7.12}$$

Every traceless two-dimensional matrix, here denoted by $\mathbf{h}$, can be expressed as a linear combination of these matrices:

$$\mathbf{h} = x\sigma_x + y\sigma_y + z\sigma_z = (r, \sigma). \tag{5.7.13}$$

Explicitly, one has

$$\mathbf{h} = \begin{pmatrix} z & x - \mathrm{i}y \\ x + \mathrm{i}y & -z \end{pmatrix}. \tag{5.7.14}$$

In particular, if x, y, and z are real, then $\mathbf{h}$ is Hermitian.

If one transforms $\mathbf{h}$ by an arbitrary unitary matrix $\mathbf{u}$ with unit determinant, one again obtains a matrix with zero trace, $\overline{\mathbf{h}} = \mathbf{u}\mathbf{h}\mathbf{u}^\dagger$. Thus, $\overline{\mathbf{h}}$ can also be written as a linear combination of $\sigma_x, \sigma_y, \sigma_z$:

$$\overline{\mathbf{h}} = \mathbf{u}\mathbf{h}\mathbf{u}^\dagger = \mathbf{u}(r, \sigma)\mathbf{u}^\dagger = x'\sigma_x + y'\sigma_y + z'\sigma_z = (r', \sigma), \tag{5.7.15}$$

and in matrix form we write

$$\begin{pmatrix} a & b \\ -b^* & a^* \end{pmatrix} \begin{pmatrix} z & x - \mathrm{i}y \\ x + \mathrm{i}y & -z \end{pmatrix} \begin{pmatrix} a^* & -b \\ b^* & a \end{pmatrix} = \begin{pmatrix} z' & x' - \mathrm{i}y' \\ x' + \mathrm{i}y' & -z' \end{pmatrix}. \tag{5.7.16}$$

Equation (5.7.16) determines x', y', z' as linear functions of x, y, z. The transformation R_u, which carries $r = (x, y, z)$ into $R_u r = r' = (x', y', z')$, can be found from Eq. (5.7.16). It is

$$x' = \frac{1}{2}\Big(a^2 + a^{*2} - b^2 - b^{*2}\Big)x + \frac{\mathrm{i}}{2}\Big(a^{*2} - a^2 + b^{*2} - b^2\Big)y - \Big(a^*b^* + ab\Big)z, \tag{5.7.17}$$

$$y' = \frac{\mathrm{i}}{2}\Big(a^2 - a^{*2} + b^{*2} - b^2\Big)x + \frac{1}{2}\Big(a^2 + a^{*2} + b^2 + b^{*2}\Big)y + \mathrm{i}\Big(a^*b^* - ab\Big)z, \tag{5.7.18}$$

$$z' = (a^*b + ab^*)x + \mathrm{i}(a^*b - ab^*)y + (aa^* - bb^*)z. \tag{5.7.19}$$

The particular form of the matrix R_u does not matter; it is important only that

$$x'^2 + y'^2 + z'^2 = x^2 + y^2 + z^2, \tag{5.7.20}$$

since

$$\det \overline{\mathbf{h}} = \det \mathbf{h},$$

$\mathbf{u}$ being an element of $SU(2)$. According to the analysis in (ii), this implies that the transformation R_u must be orthogonal. Such a property can also be seen directly from (5.7.17)–(5.7.19).

Moreover, $\overline{\mathbf{h}}$ is Hermitian if $\mathbf{h}$ is. In other words, $r' = (x', y', z')$ is real if $r = (x, y, z)$ is real. This implies, by virtue of (i), that R_u is pure real, as can also be seen directly from (5.7.17)–(5.7.19). Thus, R_u is a rotation: every two-dimensional unitary matrix $\mathbf{u}$ of unit determinant corresponds to a three-dimensional rotation R_u; the correspondence is given by (5.7.15) or (5.7.16).

It should be stressed that the determinant of R_u is +1, since as $\mathbf{u}$ is changed continuously into a unit matrix, R_u goes continuously into the three-dimensional unit matrix. If its determinant were −1 at the beginning of this process, it would have to make the jump to +1. This is impossible, since the function 'det' is continuous, and hence the matrices with negative determinant cannot be connected to the identity of the group (appendix 2.B). As a corollary of these properties, R_u is a *pure rotation* for all $\mathbf{u}$.

The above correspondence is such that the product $\mathbf{qu}$ of two unitary matrices $\mathbf{q}$ and $\mathbf{u}$ corresponds to the product $R_{qu} = R_q \cdot R_u$ of the corresponding rotations. According to (5.7.15), applied to $\mathbf{q}$ instead of $\mathbf{u}$,

$$\mathbf{q}(r, \sigma)\mathbf{q}^\dagger = \Big(R_q r, \sigma\Big), \tag{5.7.21}$$

and upon transformation with $\mathbf{u}$ this yields

$$\mathbf{uq}(r, \sigma)\mathbf{q}^\dagger\mathbf{u}^\dagger = \mathbf{u}(R_q r, \sigma)\mathbf{u}^\dagger = (R_u R_q r, \sigma) = (R_{uq} r, \sigma), \tag{5.7.22}$$

using (5.7.15) again, with $R_q r$ replacing r and $\mathbf{uq}$ replacing $\mathbf{u}$. Thus, a homomorphism exists between the group of two-dimensional unitary matrices of determinant +1 (the group $SU(2)$) and three-dimensional rotations; the correspondence is given by (5.7.15) or (5.7.17)–(5.7.19). However, we note that so far we have not shown that the homomorphism exists between the $SU(2)$ group and the *whole* connected component of the rotation group (called *pure* rotation group by Wigner). This would imply that R_u ranges over all rotations as $\mathbf{u}$ covers the whole unitary group. This will be proved shortly. It should also be noticed that the homomorphism *is not an isomorphism*, since more than one unitary matrix corresponds to the same rotation (see below).

We first assume that $\mathbf{u}$ is a diagonal matrix, here denoted by $\mathbf{u}_1(\alpha)$ (i.e. we set $b = 0$, and, for convenience, we write $a = \mathrm{e}^{-\frac{\mathrm{i}}{2}\alpha}$). Then $|a^2| = 1$ and α is real:

$$\mathbf{u}_1(\alpha) = \begin{pmatrix} \mathrm{e}^{-\frac{\mathrm{i}}{2}\alpha} & 0 \\ 0 & \mathrm{e}^{\frac{\mathrm{i}}{2}\alpha} \end{pmatrix}. \tag{5.7.23}$$

From (5.7.17)–(5.7.19) one can see that the corresponding rotation:

$$R_{u_1} = \begin{pmatrix} \cos\alpha & \sin\alpha & 0 \\ -\sin\alpha & \cos\alpha & 0 \\ 0 & 0 & 1 \end{pmatrix} \tag{5.7.24}$$

is a rotation about Z through an angle α. Next, we assume that $\mathbf{u}$ is real:

$$\mathbf{u}_2(\beta) = \begin{pmatrix} \cos\frac{\beta}{2} & -\sin\frac{\beta}{2} \\ \sin\frac{\beta}{2} & \cos\frac{\beta}{2} \end{pmatrix}. \tag{5.7.25}$$

From (5.7.17)–(5.7.19) the corresponding rotation is found to be

$$R_{u_2} = \begin{pmatrix} \cos\beta & 0 & -\sin\beta \\ 0 & 1 & 0 \\ \sin\beta & 0 & \cos\beta \end{pmatrix}, \tag{5.7.26}$$

i.e. a rotation about Y through an angle β. The product of the three unitary matrices $\mathbf{u}_1(\alpha)\mathbf{u}_2(\beta)\mathbf{u}_1(\gamma)$, with $\alpha, \beta, \gamma \in S^1$, corresponds to the product of a rotation about Z through an angle γ, about Y through β, and about Z through α, in other words, to a rotation with Euler angles α, β, γ. It follows from this that the correspondence defined in (5.7.15) not only specified a three-dimensional rotation for every two-dimensional unitary matrix, but also *at least one unitary matrix* for every pure rotation. Specifically, the matrix

$$\begin{pmatrix} e^{-\frac{i}{2}\alpha} & 0 \\ 0 & e^{\frac{i}{2}\alpha} \end{pmatrix} \begin{pmatrix} \cos\frac{\beta}{2} & -\sin\frac{\beta}{2} \\ \sin\frac{\beta}{2} & \cos\frac{\beta}{2} \end{pmatrix} \begin{pmatrix} e^{-\frac{i}{2}\gamma} & 0 \\ 0 & e^{\frac{i}{2}\gamma} \end{pmatrix}$$

$$= \begin{pmatrix} e^{-\frac{i}{2}\alpha}\cos\frac{\beta}{2}e^{-\frac{i}{2}\gamma} & -e^{-\frac{i}{2}\alpha}\sin\frac{\beta}{2}e^{\frac{i}{2}\gamma} \\ e^{\frac{i}{2}\alpha}\sin\frac{\beta}{2}e^{-\frac{i}{2}\gamma} & e^{\frac{i}{2}\alpha}\cos\frac{\beta}{2}e^{\frac{i}{2}\gamma} \end{pmatrix} \tag{5.7.27}$$

corresponds to the rotation $\{\alpha\beta\gamma\}$. Thus, the homomorphism is, in fact, a homomorphism of the group $SU(2)$ onto the *whole* three-dimensional connected component of the rotation group.

The question remains of the *multiplicity* of the homomorphism, i.e. how many unitary matrices $\mathbf{u}$ correspond to the same rotation. For this purpose, it is sufficient to check how many unitary matrices $\mathbf{u}_0$ correspond to the identity of the rotation group, i.e. to the transformation $x' = x, y' = y, z' = z$. For these particular $\mathbf{u}_0$s, the identity $\mathbf{u}_0\mathbf{h}\mathbf{u}_0^\dagger = \mathbf{h}$ should hold for all $\mathbf{h}$; this can only be the case when $\mathbf{u}_0$ is a constant matrix: $b = 0$ and $a = a^*$, $\mathbf{u}_0 = \pm 1\!\!1$ (since $|a|^2 + |b|^2 = 1$). Thus, the two unitary matrices $+1\!\!1$ and $-1\!\!1$, and *only these*, correspond to the identity of the rotation group. These two elements form an invariant sub-group of the group $SU(2)$, and the elements $\mathbf{u}$ and $-\mathbf{u}$, and only those, correspond to the same rotation. Indeed, on defining

$$\chi \equiv \begin{pmatrix} -1 & 0 \\ 0 & -1 \end{pmatrix},$$

one can express the 2×2 identity matrix as

$$\begin{pmatrix} 1 & 0 \\ 0 & 1 \end{pmatrix} = \chi^2,$$

and hence from the identity

$$\mathbf{u}(\vec{r} \cdot \vec{\sigma})\mathbf{u}^{\dagger} = \mathbf{u}\chi(\vec{r} \cdot \vec{\sigma})\chi\mathbf{u}^{\dagger}$$

it follows that, if $\mathbf{u} \rightarrow R_u$, then also $\mathbf{u}\chi \rightarrow R_u$, so that our homomorphism is two-to-one.

Alternatively, one can simply note that only the half-Euler-angles occur in (5.7.27). The Euler angles are determined by a rotation only up to a multiple of 2π; the half-angles, only up to a multiple of π. This implies that the trigonometric functions in (5.7.27) are determined only up to a sign.

A very important result has been thus obtained: there exists a two-to-one homomorphism of the group of two-dimensional unitary matrices with determinant 1 *onto* the three-dimensional connected component of the rotation group: there is a one-to-one correspondence between *pairs* of unitary matrices $\mathbf{u}$ and $-\mathbf{u}$ and rotations R_u in such a way that, from $\mathbf{uq} = \mathbf{t}$ it also follows that $R_u R_q = R_t$; conversely, from $R_u R_q = R_t$, one has that $\mathbf{uq} = \pm\mathbf{t}$. If the unitary matrix $\mathbf{u}$ is known, the corresponding rotation is best obtained from (5.7.17)–(5.7.19). Conversely, the unitary matrix for a rotation $\{\alpha\beta\gamma\}$ is best found from (5.7.27), showing that $SU(2)$ covers the whole connected component of the rotation group.

5.8 Energy bands with periodic potentials

In one-dimensional problems, a periodic potential for which

$$V(x + l) = V(x) \tag{5.8.1}$$

can be applied to obtain an approximate description of quantum behaviour of electrons in a metal, and is therefore of high physical relevance. The resulting energy spectrum turns out to be purely continuous and consists of a set of intervals called *energy bands*. The allowed energies range from a minimum value through to $+\infty$, and in general the larger the energy the larger the width of the band.

By virtue of (5.8.1), if $\varphi(x)$ solves the stationary Schrödinger equation then so does $\varphi(x+l)$. On denoting by φ_1 and φ_2 two linearly independent solutions one can therefore write

$$\begin{pmatrix} \varphi_1(x+l) \\ \varphi_2(x+l) \end{pmatrix} = \begin{pmatrix} C_{11} & C_{12} \\ C_{21} & C_{22} \end{pmatrix} \begin{pmatrix} \varphi_1(x) \\ \varphi_2(x) \end{pmatrix}, \tag{5.8.2}$$

where C_{ij} are some coefficients. Note now that, for any linear combination u of φ_1 and φ_2, i.e.

$$u(x) = A\varphi_1(x) + B\varphi_2(x), \tag{5.8.3}$$

if one requires

$$u(x + l) = \lambda u(x), \tag{5.8.4}$$

one finds from (5.8.2) the equation

$$\Big(AC_{11}+BC_{21}\Big)\varphi_1(x)+\Big(AC_{12}+BC_{22}\Big)\varphi_2(x) = \lambda\Big[A\varphi_1(x)+B\varphi_2(x)\Big]. \tag{5.8.5}$$

This leads to the linear homogeneous system

$$\Big(C_{11}-\lambda\Big)A+C_{21}B=0, \tag{5.8.6}$$

$$C_{12}A+\Big(C_{22}-\lambda\Big)B=0, \tag{5.8.7}$$

which has non-trivial solutions if and only if

$$\Big(C_{11}-\lambda\Big)\Big(C_{22}-\lambda\Big)-C_{12}C_{21}=0. \tag{5.8.8}$$

The corresponding roots λ_1 and λ_2 are such that

$$u_1(x+l)=\lambda_1 u_1(x), \tag{5.8.9}$$

$$u_2(x+l)=\lambda_2 u_2(x), \tag{5.8.10}$$

and hence the Wronskian, defined by

$$W(x)\equiv(u_1u_2'-u_1'u_2)(x), \tag{5.8.11}$$

satisfies the condition

$$W(x+l)=\lambda_1\lambda_2 W(x). \tag{5.8.12}$$

On the other hand, the Wronskian of the stationary Schrödinger equation is independent of x, and hence

$$\lambda_1\lambda_2=1. \tag{5.8.13}$$

Equally well, with $C_{11}C_{22}-C_{12}C_{21}=1$, the product of the roots of

$$\lambda^2-(C_{11}+C_{22})\lambda+\det C=0$$

must be $\lambda_1\lambda_2=\det C=1$. Repeated application of translations by l yields the equations

$$u_1(x+nl)=(\lambda_1)^n u_1(x), \tag{5.8.14}$$

$$u_2(x+nl)=(\lambda_2)^n u_2(x). \tag{5.8.15}$$

Now if it were possible to obtain $|\lambda_1|>1$ and hence $|\lambda_2|<1$, u_1 would blow up exponentially fast as $x\rightarrow+\infty$, and u_2 would diverge exponentially as $x\rightarrow-\infty$. None of them is of algebraic growth and no acceptable improper eigenfunction exists under such circumstances. We are therefore left with the case

$$|\lambda_1|=|\lambda_2|=1. \tag{5.8.16}$$

A real-valued k therefore exists for which $\lambda_1 = e^{ikl} = \lambda_2^{-1}$. Values of k are determined up to multiples of $\frac{\pi}{l}$, and if one assumes

$$-\frac{\pi}{l} \leq k \leq \frac{\pi}{l}, \tag{5.8.17}$$

equations (5.8.9) and (5.8.10) can be described by a single equation, i.e.

$$u(x+l) = u_k(x+l) = e^{ikl} u_k(x). \tag{5.8.18}$$

This implies that $u_k(x)$ takes the form

$$u_k(x) = e^{ikx} v_k(x), \tag{5.8.19}$$

where v_k is a function with period l: $v_k(x+l) = v_k(x)$. Note that the eigendifferential built from $u_k(x)$ is always square-integrable on $\mathbf{R}$, because

$$\int_k^{k+\delta k} u_\gamma(x)\, d\gamma = 2v_k(x) \frac{\sin \frac{\delta k}{2} x}{x} e^{i\left(k+\frac{\delta k}{2}\right)x}. \tag{5.8.20}$$

Equation (5.8.18) expresses periodicity of improper eigenfunctions in the presence of a periodic potential, and this property is the Bloch theorem (Bloch 1928). When the linear combination (5.8.3) is inserted into (5.8.18) one obtains the equation

$$A\varphi_1(x+l) + B\varphi_2(x+l) = e^{ikl}\Big[A\varphi_1(x) + B\varphi_2(x)\Big]. \tag{5.8.21}$$

Both sides of this equation solve the same second-order equation, and they agree for all x if and only if they and their first derivatives agree at $x = 0$, i.e.

$$A\varphi_1(l) + B\varphi_2(l) = e^{ikl}\Big[A\varphi_1(0) + B\varphi_2(0)\Big], \tag{5.8.22}$$

$$A\varphi_1'(l) + B\varphi_2'(l) = e^{ikl}\Big[A\varphi_1'(0) + B\varphi_2'(0)\Big]. \tag{5.8.23}$$

This linear homogeneous system has non-trivial solutions for A and B if and only if the determinant of the matrix of coefficients vanishes. Bearing in mind that the Wronskian is independent of x, and hence $W(l) = W(0)$ (see Eq. (5.8.11)), one obtains the equation

$$e^{i2kl} + \frac{\Gamma}{W(0)} e^{ikl} + 1 = 0, \tag{5.8.24}$$

where

$$\Gamma \equiv \varphi_2(0)\varphi_1'(l) + \varphi_2(l)\varphi_1'(0) - \varphi_1(0)\varphi_2'(l) - \varphi_1(l)\varphi_2'(0). \tag{5.8.25}$$

Equation (5.8.24) can be re-expressed in the form

$$e^{ikl}\left[2\cos kl + \frac{\Gamma}{W(0)}\right] = 0, \tag{5.8.26}$$

which eventually yields

$$\cos kl = -\frac{\Gamma}{2W(0)}. \tag{5.8.27}$$

Since $-1 \leq \cos kl \leq 1$, the energy bands correspond to those energy values for which the right-hand side of Eq. (5.8.27) takes values in the closed interval $[-1, 1]$.

5.9 Problems

5.P1. For a one-dimensional potential with a finite step, the wave function and its derivative are continuous across the step. Compute the wave function for $V \to \infty$ and comment on the result: are the continuity conditions satisfied? Moreover, re-express the boundary conditions in terms of density and current as is first suggested prior to subsection 4.6.1.

5.P2. Consider the motion of a particle of mass m and energy E in the one-dimensional potential

$$V(x) = 0 \text{ if } x \in]-a, a[, \quad \infty \text{ if } |x| > a. \tag{5.9.1}$$

(i) Find the normalized solutions of the Schrödinger equation for stationary states, with the corresponding energies.

(ii) Plot the wave functions for the four lowest energy values.

(iii) Analyse the parity properties of the wave functions.

(iv) Consider, eventually, the potential

$$\widetilde{V}(x) = 0 \text{ if } x \in]-a, a[, \quad W > E \text{ if } |x| > a, \tag{5.9.2}$$

and solve the problem with the potential (5.9.1) as a limiting case of the problem with the potential (5.9.2), when $W \to \infty$.

(v) Prove that the Hamiltonian operator is self-adjoint with the boundary conditions describing the case of an infinite potential wall.

5.P3. A particle of mass m and energy $E > 0$ interacts with the one-dimensional potential

$$V(x) = 0 \text{ if } x < -b, \; -|U_0| \text{ if } x \in]-b, b[, \; 0 \text{ if } x > b. \tag{5.9.3}$$

(i) Solve the Schrödinger equation for stationary states in the three regions.

(ii) Compute the reflection and transmission coefficients.

(iii) Compute the particular values of the energy for which the reflection coefficient vanishes.

(iv) What happens if, instead, $-|U_0| < E < 0$ or $E \leq -|U_0|$?

5.P4. A particle of energy E interacts with the potential

$$V(x) = V_1 \text{ if } x \in]-\infty, 0[, \tag{5.9.4a}$$

$$V(x) = V_2 \text{ if } x \in]0, \infty[, \tag{5.9.4b}$$

where $V_1 > 0, V_2 > V_1$.

(i) Solve the Schrödinger equation for stationary states in the two intervals, in the three cases $E < V_1$, $E \in]V_1, V_2[$, $E > V_2$, and interpret the result.

(ii) If $E > V_2$, solve the time-dependent Schrödinger equation, and analyse the behaviour of the solution as $t \rightarrow \infty$.

(iii) Compute the reflection and transmission coefficients if E is greater than V_2.

5.P5. A beam of particles with energy $E \in]0, W[$ interacts with the following potential in one spatial dimension:

$$V(x) = 0 \text{ if } x < 0 \text{ or } x > a, \tag{5.9.5a}$$

$$V(x) = W \text{ if } x \in]0, a[. \tag{5.9.5b}$$

(i) Solve the Schrödinger equation for stationary states in the three intervals.

(ii) Compute the fraction of particles which enter the region where $x > a$, and try to describe the behaviour of the wave function $\psi(x,t)$ as $t \rightarrow \infty$ in such a region.

(iii) When $W \rightarrow \infty$ and $a \rightarrow 0$ in such a way that their product remains constant, prove that the fraction of particles entering the region with $x > a$ remains non-vanishing.

5.P6. A current of electrons, with initial energy W, is emitted by a metallic surface, in the positive x-direction, and a homogeneous electric field, $\vec{E}$, is applied in the same direction. One wants to study the corresponding one-dimensional Schrödinger equation. For this purpose, consider the characteristic length l and the dimensionless energy parameter, λ, defined by the equations

$$\frac{2meE}{\hbar^2} \equiv \frac{1}{l^3}, \quad \frac{2mW}{\hbar^2} \equiv \frac{\lambda}{l^2}. \tag{5.9.6}$$

Find the approximate form of the solution when x is much larger than l.

As a next step, denoting by $u(x)$ the spatial part of the wave function, compute when $x \gg l$ the *flux* of particles, defined by the equation

$$\varphi \equiv \frac{\hbar}{2mi}\left(u^* \frac{du}{dx} - u\frac{du^*}{dx}\right), \tag{5.9.7}$$

and the density of particles: $\rho \equiv u^* u$. Find, from the ratio $\frac{\varphi}{\rho}$, the velocity of the electrons.

5.P7. Consider the motion of a particle of mass m in a central potential. As a first step, discuss in detail the construction of the Hamiltonian operator. After this, consider the potential having the form

$$V(r) = V_0 \text{ if } r \leq a, \ 0 \text{ if } r > a, \tag{5.9.8}$$

and study the existence of bound states when the quantum number l takes the values $0, 1, 2$, with positive and negative values of V_0.

A valuable collection of problems in quantum mechanics can be found in Flügge (1971), Ter Haar (1975), Grechko *et al.* (1977), Mavromatis (1987), Squires (1995), Lim (1998) and Lizzi *et al.* (1999).

Appendix 5.A

Stationary phase method

The stationary phase method makes it possible to study integrals of the type

$$I(t) \equiv \int_{-\infty}^{\infty} \varphi(x) e^{itF(x)} \, dx, \tag{5.A.1}$$

where φ is a function with compact support. At large values of $|t|$, the exponential $e^{itF(x)}$ oscillates rapidly and hence its contributions to the integral tend to cancel each other, with the exception of points where $F'(x)$ vanishes, which correspond to a slow variation of F. Any point x_0 such that

$$F'(x_0) = 0 \tag{5.A.2}$$

is called a *stationary phase point.* What is crucial is to understand whether or not x_0 belongs to the support of φ. In the affirmative case, the significant contribution to $I(t)$, as $|t| \rightarrow +\infty$, is given by those points x in the neighbourhood of x_0. In contrast, if x_0 does not belong to the support of φ, $I(t)$ tends rapidly to zero as $|t| \rightarrow +\infty$. More precisely, many relevant applications of the method rely on the following theorem.

Theorem 5.1 Let $\varphi \in C_0^\infty(\mathbf{R})$, and let $F \in C_0^\infty(\mathbf{R})$ be a real-valued function, such that the equation $F'(x) = 0$ has a unique solution x_0 belonging to the support of φ, where $F''(x) \neq 0$. The integral (5.A.1) then has the asymptotic expansion

$$I(t) \sim \mathrm{e}^{\mathrm{i}tF(x_0)} \sum_{j=0}^{n} a_j(\varphi, F) t^{-j-\frac{1}{2}} + \mathrm{O}\left(t^{-n-\frac{3}{2}}\right). \tag{5.A.3}$$

In sections 5.1 and 5.2, the stationary phase method has been applied to wave packets of the form

$$\psi_{\mathrm{I}}(x,t) = \int_{-\infty}^{\infty} C_{\mathrm{I}}(p) \left[\mathrm{e}^{\frac{\mathrm{i}}{\hbar}\left(px - \frac{p^2}{2m}t\right)} + R(p) \mathrm{e}^{-\frac{\mathrm{i}}{\hbar}\left(px + \frac{p^2}{2m}t\right)} \right] \mathrm{d}p, \tag{5.A.4}$$

$$\psi_{\mathrm{II}}(x,t) = \int_{-\infty}^{\infty} C_{\mathrm{II}}(p) T(p) \mathrm{e}^{\frac{\mathrm{i}}{\hbar}\left(\tilde{p}x - \frac{p^2}{2m}t\right)} \mathrm{d}p. \tag{5.A.5}$$

With our notation, C_{I} and C_{II} are functions with compact support, and all functions

$$R, T, C_{\mathrm{I}}, C_{\mathrm{II}}$$

have, in general, a phase depending on p:

$$C_{\mathrm{I}}(p) = \chi_1(p)\, \mathrm{e}^{\mathrm{i}\varphi_1(p)}, \tag{5.A.6}$$

$$R(p) = \chi_2(p)\, \mathrm{e}^{\mathrm{i}\varphi_2(p)}, \tag{5.A.7}$$

$$C_{\mathrm{II}}(p) = \chi_3(p)\, \mathrm{e}^{\mathrm{i}\varphi_3(p)}, \tag{5.A.8}$$

$$T(p) = \chi_4(p)\, \mathrm{e}^{\mathrm{i}\varphi_4(p)}. \tag{5.A.9}$$

Moreover, ψ_{I} is defined for $x \in]-\infty, -a[$, ψ_{II} is defined for $x \in]a, \infty[$, with $a \geq 0$, and $\tilde{p}$ either coincides with p or is a more complicated function of p: $\tilde{p} = \tilde{p}(p)$. Thus, stationarity of the phase implies, for the two parts of ψ_{I}, the conditions

$$\left(\frac{\partial \varphi_1}{\partial p} + \frac{x}{\hbar} - \frac{p}{m}\frac{t}{\hbar} \right)_{p=p_1} = 0, \tag{5.A.10a}$$

$$\left(\frac{\partial \varphi_1}{\partial p} + \frac{\partial \varphi_2}{\partial p} - \frac{x}{\hbar} - \frac{p}{m}\frac{t}{\hbar} \right)_{p=p_2} = 0, \tag{5.A.11a}$$

whereas, for ψ_{II}, it leads to the equation

$$\left(\frac{\partial \varphi_3}{\partial p} + \frac{\partial \varphi_4}{\partial p} + \frac{x}{\hbar}\frac{\partial \tilde{p}}{\partial p} - \frac{p}{m}\frac{t}{\hbar} \right)_{p=p_3} = 0. \tag{5.A.12a}$$

These equations may be re-expressed in the form

$$x = \frac{p_1}{m} t - \hbar \left. \frac{\partial \varphi_1}{\partial p} \right|_{p_1}, \tag{5.A.10b}$$

$$x = -\frac{p_2}{m} t + \hbar \left. \frac{\partial(\varphi_1 + \varphi_2)}{\partial p} \right|_{p_2}, \tag{5.A.11b}$$

$$x = \frac{p_3}{m} t - \hbar \left. \frac{\partial(\varphi_3 + \varphi_4)}{\partial p} \right|_{p_3}. \tag{5.A.12b}$$

The following comments are now in order.

(i) As $t \rightarrow \infty$, Eq. (5.A.10b) implies that $x \rightarrow \infty$, which is incompatible with the domain of definition of ψ_{I}. This implies that no incident packet survives after the interaction with a potential of compact support, in agreement with what is expected on physical grounds.

(ii) As $t \to \infty$, Eq. (5.A.11b) implies that $x \to -\infty$, which is compatible with the domain of definition of $\psi_{\rm I}$. Thus, after the interaction with the potential, the only asymptotic state in the negative x region is a reflected wave packet.

(iii) As $t \to \infty$, Eq. (5.A.12b) implies that $x \to \infty$, which is compatible with the domain of definition of $\psi_{\rm II}$. Thus, after the interaction with the potential, the only asymptotic state in the positive x region is a transmitted wave packet.

(iv) All the above conclusions hold for a state which, as $t \to -\infty$, describes a free particle located at $x = -\infty$ and moving from the left to the right with velocity $\frac{p_1}{m}$. However, one might equally well require that, as $t \to -\infty$, the initial wave packet is located at $x = +\infty$, and evaluate the probability of detecting an asymptotic state (i.e. as $t \to \infty$) at $x = -\infty$ after the interaction with a short-range potential.

(v) The localization of the wave packet at large times does not contradict the spreading of the wave packet evaluated in section 4.3, because the former results from an asymptotic calculation, whereas the latter refers to finite time intervals.

Appendix 5.B

Bessel functions

In the applications of quantum mechanics one quite often deals with second-order differential equations for which the coefficients are, themselves, some function of the independent variable x:

$$\left[\frac{{\rm d}^2}{{\rm d}x^2} + p(x)\frac{\rm d}{{\rm d}x} + q(x)\right] y(x) = 0. \tag{5.B.1}$$

Although the functions p and q may be singular at some points, the solution of Eq. (5.B.1) may remain regular therein. A well-known theorem on this problem is due to Fuchs (1866), and can be formulated for differential equations of arbitrary order, although we focus on the second-order case. The Fuchs theorem states that a *necessary and sufficient* condition for the singular point x_0 to be regular (i.e. the solution is regular at x_0) is that p and q take the form (cf. Eq. (5.5.9))

$$p(x) = \frac{\alpha(x)}{(x - x_0)}, \tag{5.B.2}$$

$$q(x) = \frac{\beta(x)}{(x - x_0)^2}, \tag{5.B.3}$$

for some holomorphic functions α and β which are regular at x_0. In other words, p should have a pole at x_0 of order not greater than 1, and q should have a pole at x_0 of order not greater than 2. If Eqs. (5.B.2) and (5.B.3) hold, the point x_0 is said to be a Fuchsian singularity (or a *regular* singular point) for Eq. (5.B.1). In the neighbourhood of a Fuchsian singularity, one can look for a solution in the form

$$y(x) = (x - x_0)^r \sum_{k=0}^{\infty} a_k (x - x_0)^k, \tag{5.B.4}$$

where the parameter r is found by solving an algebraic equation of degree 2. In particular, if $x_0 = 0$, and considering the series expansions of α and β:

$$\alpha(x) = \sum_{k=0}^{\infty} \alpha_k x^k, \tag{5.B.5}$$

$$\beta(x) = \sum_{k=0}^{\infty} \beta_k x^k, \tag{5.B.6}$$

the algebraic equation for r reduces to

$$r^2 + (\alpha_0 - 1)r + \beta_0 = 0. \tag{5.B.7}$$

Interestingly, the operator in Eq. (5.B.1) can then be mapped into a constant-coefficient operator upon defining the independent variable $X \equiv \log(x)$, and provided that

$$p(x) = \frac{\alpha_0}{x}, \quad q(x) = \frac{\beta_0}{x^2}.$$

Such a transformation property is peculiar to the Fuchsian case. The roots r_1 and r_2 of Eq. (5.B.7) are called *characteristic exponents* of Eq. (5.B.1). The difference $r_1 - r_2$ may or may not be an integer. In both cases, an integral of Eq. (5.B.1) is

$$y_1(x) = x^{r_1} \sum_{k=0}^{\infty} a_k x^k. \tag{5.B.8}$$

Moreover, if $r_1 - r_2$ is not an integer, a second, linearly independent integral of Eq. (5.B.1) is of the form

$$y_2(x) = x^{r_2} \sum_{k=0}^{\infty} b_k x^k. \tag{5.B.9}$$

However, if $r_1 - r_2$ is an integer, one can prove that the second independent solution of Eq. (5.B.1) reads (cf. Eqs. (5.4.29) and (5.4.31))

$$\tilde{y}_2(x) = y_1(x) \log(x) + x^{r_2 - r_1} \psi(x) y_1(x), \tag{5.B.10}$$

where ψ is a holomorphic function, regular at the origin.

The Bessel equation is a particular case of a second-order differential equation with a Fuchsian singularity at the origin:

$$\left[\frac{d^2}{dx^2} + \frac{1}{x}\frac{d}{dx} + \left(1 - \frac{\lambda^2}{x^2}\right) \right] y(x) = 0, \tag{5.B.11}$$

where λ may take, in general, complex values. The corresponding integral is called a Bessel function of the first kind and can be written in the form (Watson 1966)

$$J_\lambda(x) = \left(\frac{x}{2}\right)^\lambda \sum_{m=0}^{\infty} \frac{(-1)^m}{m!\Gamma(\lambda + m + 1)} \left(\frac{x}{2}\right)^{2m}, \tag{5.B.12}$$

where Γ is the meromorphic function known as the Γ-function, defined by

$$\begin{aligned} \Gamma(z) &\equiv \int_0^\infty e^{-y} y^{z-1} \, dy \\ &= \sum_{l=0}^{\infty} \frac{(-1)^l}{l!(z+l)} + \int_1^\infty e^{-y} y^{z-1} \, dy. \end{aligned} \tag{5.B.13}$$

Such a notation means that the second line of Eq. (5.B.13) is the analytic extension of the Γ-function, originally defined on the half-plane $\mathrm{Re}(z) > 0$. This yields a meromorphic function, with poles at $0, -1, -2, \ldots, -\infty$.

The function $x^{-\lambda} J_\lambda(x)$ is an entire function (i.e. analytic on the whole complex plane) which, at $x = 0$, takes the value $\frac{1}{2^\lambda \Gamma(\lambda+1)}$. If 2λ is not an integer, the general solution of Eq. (5.B.11) is (cf. Eq. (5.3.18))

$$y(x) = C_1 J_\lambda(x) + C_2 J_{-\lambda}(x), \tag{5.B.14}$$

for some constants C_1 and C_2.

Bessel functions of the second kind are instead given by

$$Y_\lambda(x) \equiv \frac{J_\lambda(x)\cos(\lambda\pi) - J_{-\lambda}(x)}{\sin(\lambda\pi)}, \tag{5.B.15}$$

with the property that, for integer values of n,

$$Y_n(x) \equiv \lim_{\lambda \to n} Y_\lambda(x). \tag{5.B.16}$$

The general solution of the Bessel equation (5.B.11) is therefore

$$y(x) = B_1 J_\lambda(x) + B_2 Y_\lambda(x), \tag{5.B.17}$$

for some constants B_1 and B_2.

The second-order equation

$$\left[\frac{d^2}{dx^2} + \frac{1}{x}\frac{d}{dx} - \left(1 + \frac{\lambda^2}{x^2}\right) \right] y(x) = 0 \tag{5.B.18}$$

is instead solved by the so-called modified Bessel functions:

$$I_\lambda(x) \equiv \mathrm{i}^{-\lambda} J_\lambda(\mathrm{i}x), \tag{5.B.19}$$

$$K_\lambda(x) \equiv \frac{I_{-\lambda}(x) - I_\lambda(x)}{\sin(\lambda\pi)}, \tag{5.B.20}$$

$$K_n(x) \equiv \lim_{\lambda \to n} K_\lambda(x), \tag{5.B.21}$$

and the general solution of Eq. (5.B.18) is

$$y(x) = A_1 I_\lambda(x) + A_2 K_\lambda(x), \tag{5.B.22}$$

for some constants A_1 and A_2. The function I_λ has a finite limit as $x \to 0$ and increases exponentially at infinity, whereas the converse holds for K_λ.

Spherical Bessel functions are, by definition, the solutions of the equation (l being an integer ≥ 0)

$$\left\{ \frac{\mathrm{d}^2}{\mathrm{d}x^2} + \frac{2}{x}\frac{\mathrm{d}}{\mathrm{d}x} + \left[1 - \frac{l(l+1)}{x^2}\right] \right\} y(x) = 0. \tag{5.B.23}$$

The two linearly independent integrals of Eq. (5.B.23) are

$$j_l(x) \equiv (-1)^l x^l \left(\frac{1}{x}\frac{\mathrm{d}}{\mathrm{d}x}\right)^l \left[\frac{(\sin x)}{x}\right], \tag{5.B.24}$$

$$n_l(x) \equiv (-1)^{l+1} x^l \left(\frac{1}{x}\frac{\mathrm{d}}{\mathrm{d}x}\right)^l \left[\frac{(\cos x)}{x}\right]. \tag{5.B.25}$$

For example, the explicit calculation shows that

$$j_0(x) = \frac{(\sin x)}{x}, \tag{5.B.26}$$

$$j_1(x) = \frac{(\sin x)}{x^2} - \frac{(\cos x)}{x}, \tag{5.B.27}$$

$$n_0(x) = -\frac{(\cos x)}{x}, \tag{5.B.28}$$

$$n_1(x) = -\frac{(\cos x)}{x^2} - \frac{(\sin x)}{x}. \tag{5.B.29}$$

A monograph would not be enough to derive all properties of Bessel functions, with their applications to theoretical physics and mathematics. Our appendix ends instead with the asymptotic expansion of Bessel functions of integer order as $x \to \infty$ (see applications in (5.3.20) and (5.3.21)):

$$J_n(x) \sim \sqrt{\frac{2}{\pi x}} \cos\left(x - \frac{n\pi}{2} - \frac{\pi}{4}\right) + \mathrm{O}\left(x^{-3/2}\right), \tag{5.B.30}$$

which is obtained from the integral representation of Bessel functions:

$$J_n(x) = \frac{1}{2\pi} \int_{-\pi}^{\pi} \mathrm{e}^{\mathrm{i}(x \sin\theta - n\theta)} \mathrm{d}\theta. \tag{5.B.31}$$

6
Introduction to spin

First, the experimental foundations for the existence of a new dynamical variable, the spin of particles, are presented. The Pauli equation is then derived in detail, two applications are given and the energy levels of a particle with spin in a constant magnetic field are studied. This is the analysis of Landau levels, which can be performed using the known results on the spectrum of harmonic oscillators.

6.1 Stern–Gerlach experiment and electron spin

The hypothesis that the electron has a magnetic moment and an angular momentum, in short a 'spin', was first suggested in Uhlenbeck and Goudsmit (1926). They noticed, even before the discovery of quantum mechanics, that a complete description of spectra was not possible unless a magnetic moment and a mechanical moment were ascribed to the electron, and hence the concept of an electron as a point charge was insufficient. First, let us therefore try to understand how can one associate a magnetic moment to an atomic system. For this purpose, consider for simplicity the Bohr model of an hydrogen atom, where the electron moves along a circular orbit and rotates with orbital angular momentum $\vec{L}$. A moving charge is equivalent to an electric current, hence an electron along a circular orbit can be treated as a loop along which an electric current flows, and such a loop has a magnetic moment. Starting from the magnetic moment associated to each individual electron, a magnetic moment for the whole atom can be derived. Indeed, a current i along a loop enclosing a small area S_δ gives rise to a dipole magnetic moment

$$\vec{M} = \vec{n}\frac{i}{c}S_\delta, \tag{6.1.1}$$

where $\vec{n}$ is the normal to the plane containing the loop. If the loop has a uniform charge density ρ_e with magnitude $\frac{e_0}{2\pi r}$, the modulus of $\vec{M}$

therefore reads as ($e = -e_0$ being the charge of the electron)

$$M = \frac{e_0 v}{2\pi r}\frac{\pi r^2}{c} = \frac{e_0 v r}{2c} = \frac{e_0 L}{2mc}. \tag{6.1.2}$$

Since the rotation of the electron is opposite to the current, the corresponding vectors are related by

$$\vec{M} = -\frac{e_0}{2mc}\vec{L}. \tag{6.1.3}$$

The Bohr quantization rules suggest replacing the orbital angular momentum by $\frac{\vec{L}}{\hbar}$ in Eq. (6.1.3) (see Eqs. (5.6.28)–(5.6.30)), and hence we write eventually

$$\vec{M} = -\frac{e_0\hbar}{2mc}\frac{\vec{L}}{\hbar} = -\mu_{\rm B}\frac{\vec{L}}{\hbar}. \tag{6.1.4}$$

The quantity

$$\mu_{\rm B} \equiv \frac{e_0\hbar}{2mc}$$

has, of course, the dimensions of a magnetic moment and is called the Bohr magneton (equal to 0.9274×10^{-20} erg gauss^{-1}. In general, a system of electrons with total angular momentum $\vec{J}$ has a magnetic moment $\vec{M}$ antiparallel to $\vec{J}$ and one usually writes

$$\vec{M} = -g\mu_{\rm B}\frac{\vec{J}}{\hbar}, \tag{6.1.5}$$

where g is called the gyromagnetic ratio.

If an atom with magnetic moment $\vec{M}$ is affected by a magnetic field $\vec{B}$, the interaction potential is

$$V_{\rm I} = -\vec{M}\cdot\vec{B} \equiv H_I, \tag{6.1.6}$$

so that the resulting Hamilton equations read as

$$\frac{{\rm d}\vec{x}}{{\rm d}t} = \frac{\vec{p}}{m}, \tag{6.1.7}$$

$$\frac{{\rm d}\vec{p}}{{\rm d}t} = -\frac{\partial H_{\rm I}}{\partial\vec{x}} = \vec{M}\cdot\frac{\partial\vec{B}}{\partial\vec{x}}. \tag{6.1.8}$$

These formulae make it clear that it is the gradient of the magnetic field that really plays the key role in the equations of motion. In particular, if the magnetic field is uniform, the total force on the magnetic dipole is vanishing.

We now describe in some detail the key steps in the experimental detection of spin. First, in Gerlach and Stern (1922) the authors measured the possible values of the *magnetic dipole moment* for silver atoms by sending a beam of these atoms throughout a non-uniform magnetic field. In this experiment, a beam of neutral atoms is formed by evaporating silver from an oven. The beam is collimated by a diaphragm and it enters a magnet (see figure 6.1). The cross-sectional view of the magnet shows that it produces a field that increases in intensity in the z-direction, which is also the direction of the magnetic field itself in the region of the beam. Since the atoms are neutral overall, the only net force acting on them is a force proportional to M_z and the gradient of the external magnetic field. Each atom is hence deflected, in passing through the non-uniform magnetic field, by an amount that is proportional to M_z. This means that the beam is analysed into components, depending on the various values of M_z. Lastly, deflected atoms strike a metallic plate, upon which they condense and leave a visible trace. According to the general properties of angular momentum operators, M_z can only take discrete values (see section 5.6)

$$M_z = -g_l \, \mu_{\rm B} \, m_l, \tag{6.1.9}$$

where m_l ranges from $-l$ to $+l$, including 0. Thus, according to quantum mechanics, the deflected beam should be split into several discrete components, and for all orientations of the analysing magnet. In other words, the experimental setting consists of an oven, a collimator, a magnet and eventually a detector plate. The magnet acts essentially as a measuring device, which investigates the quantization of the component of the magnetic dipole moment along a z-axis. Such an axis is defined by the direction in which its field increases in intensity.

Stern and Gerlach found that the beam of silver atoms is split into *two* discrete components, one component being bent in the positive z-direction and the other bent in the negative z-direction. Moreover, they

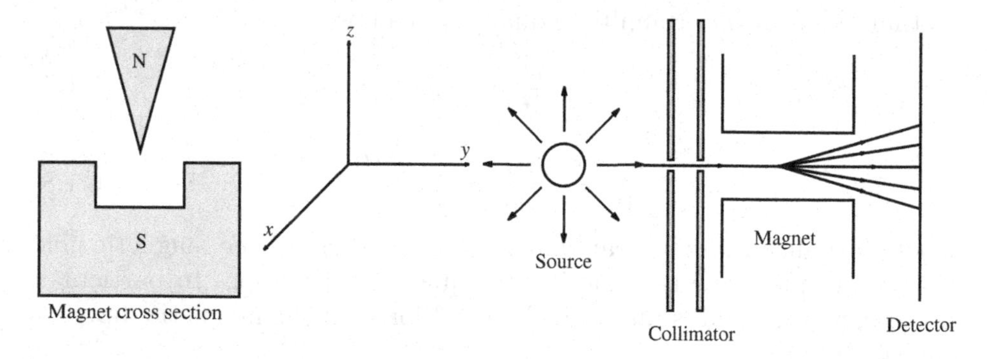

Fig. 6.1. Experimental apparatus used in the Stern–Gerlach experiment (Gottfried 1966, by kind permission of Professor Gottfried (© 1966, reprinted by permission of Perseus Books Publishers, a member of Perseus Books, L.L.C)).

found that these results hold independently of the choice of the z-direction. The experiment was repeated using several other species of atoms, and in all cases it was found that the deflected beam is split into two, or more, discrete components. However, these experimental data were puzzling for a simple but fundamental reason: since l is an integer, the number of possible values of M_z is always odd. This is incompatible with the beam of silver atoms being split into *only two* components, both of which are deflected.

Besides all this in 1927 Phipps and Taylor used the Stern–Gerlach technique on a beam of hydrogen atoms. This was a crucial test, since these atoms contain a single electron. The atoms in the beam were kept in their ground state by virtue of the relatively low temperature of the oven, and hence the quantum mechanical theory predicts that the quantum number l can only take the 0 value, and correspondingly m_l vanishes as well. Thus, if only the orbital angular momentum were involved, one would expect that the beam should be unaffected by the non-uniform magnetic field. Interestingly, however, Phipps and Taylor found that the beam was split into two symmetrically deflected components. This implies in turn that a sort of magnetic dipole moment exists in the atom, though not quite of the sort suggested by the theory of orbital angular momentum. Note that we are treating the translational degree of freedom classically, and using the quantum spin in the external magnetic field.

The data obtained by Phipps and Taylor can indeed be understood if the formalism described in section 5.6 holds for the *spin angular momentum*. In other words, one is led to assume that a new (quantum) number m_s exists, which implies the existence of a new degree of freedom, the possible values of which range from $-s$ to $+s$, as is true of the quantum numbers m_l and l for orbital angular momentum. To agree with the results by Phipps and Taylor, the two possible values of m_s are

$$m_s = -\frac{1}{2}, \ +\frac{1}{2}, \tag{6.1.10}$$

and hence s can only take the value

$$s = \frac{1}{2}. \tag{6.1.11}$$

As far as the very idea of *electron spin* is concerned, credit is given, appropriately, to Goudsmit and Uhlenbeck, as was stated at the beginning of this section. In 1925 they were young graduate students who were trying to understand why certain lines of the optical spectra of hydrogen atom and alkali atoms consist of a closely spaced pair of lines. They were thus dealing with the *fine structure* of these atoms, and proposed that the electron has an intrinsic angular momentum and magnetic dipole moment with the properties outlined above. The non-trivial step was to assign a fourth quantum number to the electron, rather than the three that would be obtained from the (usual) Schrödinger theory. They tried to understand the electron spin in terms of a model where the electron is rotating. However, Lorentz studied the electromagnetic properties of rotating electrons, and was able to show that serious inconsistencies would result from such a model. In particular, the magnetic energy would be so large that, by the equivalence of mass and energy, the electron would have a larger mass than the proton, and would be bigger than the whole atom! None of the people concerned, including Wolfgang Pauli, were aware of Elie Cartan's discovery of spinors (Cartan 1938) and their properties. Thus, the understanding of the underlying reasons for the existence of electron spin was completely lacking when Goudsmit and Uhlenbeck first brought their idea to the attention of the scientific community (thanks to the enthusiastic support of Ehrenfest).

In 'classical physics', the Hamiltonian description of a non-relativistic particle with spin can be obtained by considering the vectors $\vec{x}, \vec{p}$ and $\vec{s}$, and requiring the fundamental Poisson brackets

$$\{x_i, x_j\} = \{p_i, p_j\} = 0, \tag{6.1.12}$$

$$\{x_i, p_j\} = \delta_{ij}, \tag{6.1.13}$$

$$\{s_j, x_k\} = \{s_j, p_k\} = 0, \quad \{s_i, s_j\} = \varepsilon_{ijk} s_k. \tag{6.1.14}$$

On using the Hamiltonian $H \equiv H_0 + \mu \vec{s} \cdot \vec{B}$, the resulting Hamilton equations of motion read as

$$\dot{x}_i = \frac{p_i}{m}, \tag{6.1.15}$$

$$\dot{p}_i = -\mu s_j \partial_i B_j, \tag{6.1.16}$$

$$\dot{s}_i = \mu \varepsilon_{ijk} B_j s_k. \tag{6.1.17}$$

6.2 Wave functions with spin

Wave functions for particles with spins are represented by column vectors of square-integrable functions on $\mathbf{R}^3$, with norm given by (in this section we study the stationary theory)

$$\begin{aligned} \|\psi\|^2 = (\psi, \psi) &= \int d^3x \, \psi^\dagger(\vec{x}) \psi(\vec{x}) \\ &= \int d^3x \left(\psi_1^* \psi_1 + \psi_2^* \psi_2 + \cdots + \psi_{2s+1}^* \psi_{2s+1} \right). \end{aligned} \tag{6.2.1}$$

Let us now consider in more detail this non-relativistic formalism for the electron, which is a particle of spin $\frac{1}{2}$. Its wave function reads as

$$\psi(\vec{x}) = \begin{pmatrix} \psi_1(\vec{x}) \\ \psi_2(\vec{x}) \end{pmatrix} = \psi_1(\vec{x}) \begin{pmatrix} 1 \\ 0 \end{pmatrix} + \psi_2(\vec{x}) \begin{pmatrix} 0 \\ 1 \end{pmatrix}. \tag{6.2.2}$$

Thus, on defining

$$\chi_+ \equiv \begin{pmatrix} 1 \\ 0 \end{pmatrix}, \tag{6.2.3}$$

$$\chi_- \equiv \begin{pmatrix} 0 \\ 1 \end{pmatrix}, \tag{6.2.4}$$

one can write

$$\psi(\vec{x}) = \psi_1(\vec{x})\chi_+ + \psi_2(\vec{x})\chi_-. \tag{6.2.5}$$

The column vectors χ_+ and χ_- are a basis in the space $\mathbf{C}^2$ of spin-states for a spin-$\frac{1}{2}$ particle. They are orthonormal, because

$$(\chi_+, \chi_+) = (\chi_-, \chi_-) = 1, \tag{6.2.6}$$

$$(\chi_+, \chi_-) = 0. \tag{6.2.7}$$

Moreover, every element χ of $\mathbf{C}^2$ can be written as a linear combination of χ_+ and χ_-, i.e.

$$\chi = C_+\chi_+ + C_-\chi_-, \tag{6.2.8}$$

where

$$C_+ = (\chi_+, \chi), \tag{6.2.9}$$

$$C_- = (\chi_-, \chi). \tag{6.2.10}$$

Our basis vectors χ_+ and χ_- form therefore a complete orthonormal system in $\mathbf{C}^2$.

Given now the spin operators

$$\hat{S}_x \equiv \frac{\hbar}{2}\sigma_x, \quad \hat{S}_y \equiv \frac{\hbar}{2}\sigma_y, \quad \hat{S}_z \equiv \frac{\hbar}{2}\sigma_z, \tag{6.2.11}$$

where σ_x, σ_y and σ_z can be represented by 2×2 matrices coinciding with the Pauli matrices, one finds that, by virtue of (5.7.10)–(5.7.12), these operators obey the commutation relations of angular momentum:

$$\hat{S}_k\hat{S}_l - \hat{S}_l\hat{S}_k = \mathrm{i}\hbar\varepsilon_{klm}\hat{S}_m. \tag{6.2.12}$$

Moreover, one has

$$\hat{S}^2\chi_\pm = \frac{3}{4}\hbar^2\chi_\pm, \tag{6.2.13}$$

$$\hat{S}_z \chi_\pm = \pm \frac{\hbar}{2} \chi_\pm. \tag{6.2.14}$$

The 'raising' and 'lowering' operators can also be defined:

$$\hat{S}_+ \equiv \hat{S}_x + \mathrm{i}\hat{S}_y, \tag{6.2.15}$$

$$\hat{S}_- \equiv \hat{S}_x - \mathrm{i}\hat{S}_y, \tag{6.2.16}$$

for which

$$\hat{S}_+ \chi_+ = 0, \quad \hat{S}_+ \chi_- = \hbar \chi_+, \quad \hat{S}_- \chi_+ = \hbar \chi_-, \quad \hat{S}_- \chi_- = 0. \tag{6.2.17}$$

As a corollary, $\hat{S}_+$ and $\hat{S}_-$ are nilpotent, in that $\hat{S}_+^2 = \hat{S}_-^2 = 0$.

Furthermore, bearing in mind the split (6.2.2) one finds

$$\hat{S}_x \psi(\vec{x}) = \frac{\hbar}{2} \Big[\psi_2(\vec{x}) \chi_+ + \psi_1(\vec{x}) \chi_- \Big], \tag{6.2.18}$$

$$\hat{S}_y \psi(\vec{x}) = -\frac{\mathrm{i}\hbar}{2} \Big[\psi_2(\vec{x}) \chi_+ - \psi_1(\vec{x}) \chi_- \Big], \tag{6.2.19}$$

$$\hat{S}_z \psi(\vec{x}) = \frac{\hbar}{2} \Big[\psi_1(\vec{x}) \chi_+ - \psi_2(\vec{x}) \chi_- \Big], \tag{6.2.20}$$

$$\hat{S}^2 \psi(\vec{x}) = \frac{3}{4} \hbar^2 \psi(\vec{x}). \tag{6.2.21}$$

Several examples concerning spin in quantum mechanics can be found in Cohen-Tannoudji *et al.* (1977a,b).

6.2.1 Addition of orbital and spin angular momentum

On considering the quantum theory of the electron in Cartesian coordinates in $\mathbf{R}^3$, and bearing in mind the definition (5.6.18)–(5.6.20) of its orbital angular momentum, we are led to define its total angular momentum $\hat{J}$ by the equations

$$\hat{J}_x \equiv \hat{L}_x + \hat{S}_x, \quad \hat{J}_y \equiv \hat{L}_y + \hat{S}_y, \quad \hat{J}_z \equiv \hat{L}_z + \hat{S}_z. \tag{6.2.22}$$

By virtue of the relations (5.6.21) and (6.2.12), and assuming that the orbital and spin angular momentum commute:

$$\hat{L}_k \hat{S}_m - \hat{S}_m \hat{L}_k = 0, \quad \forall k, m = 1, 2, 3, \tag{6.2.23}$$

the operators $\hat{J}_x, \hat{J}_y$ and $\hat{J}_z$ are indeed found to obey the commutation relations of angular momentum. In other words, one obtains

$$\begin{aligned} \hat{J}_k \hat{J}_l - \hat{J}_l \hat{J}_k &= \Big(\hat{L}_k \hat{L}_l - \hat{L}_l \hat{L}_k \Big) + \Big(\hat{S}_k \hat{S}_l - \hat{S}_l \hat{S}_k \Big) \\ &= \mathrm{i}\hbar \varepsilon_{klm} \Big(\hat{L}_m + \hat{S}_m \Big) = \mathrm{i}\hbar \varepsilon_{klm} \hat{J}_m. \end{aligned} \tag{6.2.24}$$

The use of Cartesian coordinates is not mandatory. For example, the operators of total angular momentum can be written in spherical coordinates using the standard transformation formulae (Fock 1978). In this chapter we limit ourselves to the above properties, while the deeper nature of our operations is described in sections 11.1 and 11.4.

In addition to infinitesimal rotations one can calculate the effect of finite rotations, including rotations through 2π around any axis. For spin $\frac{1}{2}$ this rotation, which is no rotation at all, is represented by $-\mathbb{1}$ rather than by the identity. These representations of rotations are sometimes called 'double valued'.

6.3 The Pauli equation

The analysis of an electron in an external electromagnetic field is an important problem in quantum mechanics. As we mentioned in appendix 1.A, one can describe the electromagnetic field in terms of the potential A_μ, with components (ϕ, A_x, A_y, A_z). With our notation, ϕ corresponds to the temporal component (the so-called 'scalar potential'), and A_x, A_y and A_z are the components of the vector potential in Cartesian coordinates. When the relativistic effects are negligible, the Lagrangian is thus found to be (see appendix 6.A)

$$\mathcal{L} = \frac{m}{2}\left(\dot{x}^2 + \dot{y}^2 + \dot{z}^2\right) - \frac{e_0}{c}\left(\dot{x}A_x + \dot{y}A_y + \dot{z}A_z\right) + e_0\phi. \tag{6.3.1}$$

Interestingly, the canonical momenta $p_i \equiv \frac{\partial \mathcal{L}}{\partial \dot{q}^i}$ do not coincide with the kinematic momenta (i.e. the components of linear momentum) $P_x \equiv m\dot{x}, P_y \equiv m\dot{y}, P_z \equiv m\dot{z}$, but one finds

$$p_k = m\dot{x}_k - \frac{e_0}{c}A_k = P_k - \frac{e_0}{c}A_k. \tag{6.3.2}$$

The energy of the particle is a quadratic function of $\dot{x}, \dot{y}$ and $\dot{z}$, without linear terms, because one has

$$E = \dot{x}p_x + \dot{y}p_y + \dot{z}p_z - \mathcal{L} = \frac{m}{2}\left(\dot{x}^2 + \dot{y}^2 + \dot{z}^2\right) - e_0\phi. \tag{6.3.3}$$

Thus, on using (again) Eq. (6.3.2), one can eventually express the Hamiltonian in terms of the canonical momenta as

$$H = \frac{1}{2m}\left[\left(p_x + \frac{e_0}{c}A_x\right)^2 + \left(p_y + \frac{e_0}{c}A_y\right)^2 + \left(p_z + \frac{e_0}{c}A_z\right)^2\right] - e_0\phi. \tag{6.3.4}$$

Note that the terms linear in the velocities disappear in the Hamiltonian, by virtue of the Euler theorem on homogeneous functions. If the external magnetic field vanishes, a gauge choice makes it possible to write

$$H = \frac{1}{2m}\left(p_x^2 + p_y^2 + p_z^2\right) - e_0\phi. \tag{6.3.5}$$

Because of the existence of spin, the quantum Hamiltonian operator has to act on wave functions

$$\psi : \mathbf{R}^3 \rightarrow \mathbf{C}^2.$$

Thus, our three momentum operators should be combined into a 2×2 matrix-valued operator. Our experience with the matrix (5.7.14) suggests considering the operator

$$\vec{\sigma} \cdot \vec{p} = \frac{\hbar}{\mathrm{i}} \begin{pmatrix} \frac{\partial}{\partial z} & \frac{\partial}{\partial x} - \mathrm{i}\frac{\partial}{\partial y} \\ \frac{\partial}{\partial x} + \mathrm{i}\frac{\partial}{\partial y} & -\frac{\partial}{\partial z} \end{pmatrix}. \tag{6.3.6}$$

Thus, on taking into account the electron spin, the Hamiltonian operator is taken to be

$$H_0 = \frac{1}{2m}\left(\vec{\sigma} \cdot \vec{p}\right)^2 + U(x, y, z)\sigma_0. \tag{6.3.7}$$

If the external magnetic field vanishes, however, no new effect can be appreciated, and the operator H_0 acts in a diagonal way on the wave function.

In contrast, when an external magnetic field $\vec{B}$ is switched on, it is convenient to use the gauge-invariant kinematic momenta P_x, P_y and P_z. By virtue of Eq. (6.3.2) one obtains

$$\frac{\mathrm{i}}{\hbar}(P_y P_z - P_z P_y) = \frac{e_0}{c}\left(\frac{\partial A_z}{\partial y} - \frac{\partial A_y}{\partial z}\right) = \frac{e_0}{c} B_x, \tag{6.3.8}$$

$$\frac{\mathrm{i}}{\hbar}(P_z P_x - P_x P_z) = \frac{e_0}{c}\left(\frac{\partial A_x}{\partial z} - \frac{\partial A_z}{\partial x}\right) = \frac{e_0}{c} B_y, \tag{6.3.9}$$

$$\frac{\mathrm{i}}{\hbar}(P_x P_y - P_y P_x) = \frac{e_0}{c}\left(\frac{\partial A_y}{\partial x} - \frac{\partial A_x}{\partial y}\right) = \frac{e_0}{c} B_z. \tag{6.3.10}$$

The crucial step in building the Hamiltonian operator with spin is now to use the gauge-invariant kinematic momenta by writing (cf. Eq. (6.3.4))

$$H = \frac{1}{2m}\left(\vec{\sigma} \cdot \vec{P}\right)^2 + U(x, y, z)\sigma_0. \tag{6.3.11}$$

On using the labels 1, 2 and 3 for the components along the x-, y- and z-axes, respectively, this leads to the lengthy but useful formula

$$\begin{aligned} H = \frac{1}{2m}\Big(&\sigma_1^2 P_1^2 + \sigma_2^2 P_2^2 + \sigma_3^2 P_3^2 + \sigma_1\sigma_2 P_1 P_2 + \sigma_2\sigma_1 P_2 P_1 \\ &+ \sigma_1\sigma_3 P_1 P_3 + \sigma_3\sigma_1 P_3 P_1 + \sigma_2\sigma_3 P_2 P_3 + \sigma_3\sigma_2 P_3 P_2\Big) \\ &+ U(x, y, z)\sigma_0. \end{aligned} \tag{6.3.12}$$

The cross-terms in Eq. (6.3.12) are conveniently rearranged with the help of the identity

$$\sigma_j \sigma_k = \delta_{jk} + \mathrm{i}\varepsilon_{jkl}\sigma_l,$$

and hence one finds

$$\begin{aligned} H &= \widetilde{H}_0 + \frac{1}{2m}\Big[\sigma_1\sigma_2(P_1P_2 - P_2P_1) + \sigma_2\sigma_3(P_2P_3 - P_3P_2) \\ &\quad + \sigma_3\sigma_1(P_3P_1 - P_1P_3)\Big] \\ &= \widetilde{H}_0 + \mu_{\mathrm{B}}\vec{\sigma}\cdot\vec{B}, \end{aligned} \tag{6.3.13}$$

where

$$\widetilde{H}_0 \equiv \frac{1}{2m}\left(\sigma_1^2 P_1^2 + \sigma_2^2 P_2^2 + \sigma_3^2 P_3^2\right) + U(x,y,z)\sigma_0. \tag{6.3.14}$$

The Pauli equation for a particle with spin in an external magnetic field is therefore

$$\mathrm{i}\hbar\frac{\partial\psi}{\partial t} = H\psi, \tag{6.3.15}$$

with the Hamiltonian H given by Eq. (6.3.13). The coupling term $\vec{\sigma}\cdot\vec{B}$ was derived by Pauli in a highly original way, at a time when the relativistic quantum theory of the electron was not yet developed. It should be stressed, however, that such a theory (see section 16.1), due to Dirac (1928, 1958), can be used to put on firmer ground the *ad hoc* non-relativistic derivation of Pauli, in which 'spin degrees of freedom' are added 'by hand'.

6.4 Solutions of the Pauli equation

In the Pauli equation (6.3.15) only the term $\mu_{\mathrm{B}}\vec{\sigma}\cdot\vec{B}$ in the Hamiltonian depends on the spin operators. Thus, on solving Eq. (6.3.15) by separation of variables, with $\hat{X}, \hat{Y}, \hat{Z}, \hat{S}^2$ and $\hat{S}_z$ as fundamental operators, the time evolution of that part of the wave function depending on spin variables is ruled by the equation (here μ is the magnetic moment)

$$\mathrm{i}\hbar\frac{\partial\chi}{\partial t} = -\frac{2\mu}{\hbar}\vec{B}\cdot\vec{S}\chi, \tag{6.4.1}$$

where

$$\chi(\vec{S},t) = \begin{pmatrix} \chi_1(\vec{S},t) \\ \chi_2(\vec{S},t) \end{pmatrix} = \chi_1(\vec{S},t)\chi_+ + \chi_2(\vec{S},t)\chi_-. \tag{6.4.2}$$

Two important features are hence found to emerge:

(i) the Pauli equation leads to a coupled system of first-order differential equations, to be solved for given initial conditions;

(ii) the operator on the right-hand side of Eq. (6.4.1) may be time-dependent. Thus, some care is necessary to map the problem into one for which the well-known exponentiation of time-independent matrices can be applied (see Eqs. (4.4.1)–(4.4.7)).

Following Landau and Lifshitz (1958), we first consider the Pauli equation for a neutral particle of spin $\frac{1}{2}$ in a magnetic field $\vec{B}$ for which the only non-vanishing component is directed along the z-axis: $B_z = B(t)$. Its explicit form is not specified. Equations (6.4.1) and (6.4.2) therefore lead to a first-order system which, in this particular case, is decoupled:

$$\frac{\partial \chi_1}{\partial t} = \frac{\mathrm{i}\mu B}{\hbar}\chi_1, \tag{6.4.3}$$

$$\frac{\partial \chi_2}{\partial t} = -\frac{\mathrm{i}\mu B}{\hbar}\chi_2. \tag{6.4.4}$$

The solution is hence given by

$$\chi_1(\vec{S},t) = C_1 \mathrm{e}^{\frac{\mathrm{i}\mu}{\hbar}\int_{t_0}^{t} B(t')\,\mathrm{d}t'}, \tag{6.4.5}$$

$$\chi_2(\vec{S},t) = C_2\, \mathrm{e}^{-\frac{\mathrm{i}\mu}{\hbar}\int_{t_0}^{t} B(t')\,\mathrm{d}t'}, \tag{6.4.6}$$

where the constants C_1 and C_2 can be determined once the initial conditions are given.

In a second example, we consider instead a magnetic field $\vec{B}$ with components (Landau and Lifshitz 1958)

$$B_x = B \sin\theta \cos\omega t, \tag{6.4.7}$$

$$B_y = B \sin\theta \sin\omega t, \tag{6.4.8}$$

$$B_z = B \cos\theta. \tag{6.4.9}$$

In this case, the modulus of $\vec{B}$ is constant in time, whereas all its components change smoothly according to (6.4.7)–(6.4.9). Thus, the Pauli equation can be expressed by the coupled system

$$\mathrm{i}\hbar\frac{\partial}{\partial t}\begin{pmatrix}\chi_1\\ \chi_2\end{pmatrix} = -\mu B\begin{pmatrix}\cos\theta & \sin\theta\, \mathrm{e}^{-\mathrm{i}\omega t}\\ \sin\theta\, \mathrm{e}^{\mathrm{i}\omega t} & -\cos\theta\end{pmatrix}\begin{pmatrix}\chi_1\\ \chi_2\end{pmatrix}. \tag{6.4.10}$$

The matrix on the right-hand side of Eq. (6.4.10) depends explicitly on time. We can map our equation into an equivalent equation where the time dependence of the matrix disappears on setting

$$\phi_1 \equiv \mathrm{e}^{\frac{\mathrm{i}}{2}\omega t}\,\chi_1, \tag{6.4.11}$$

$$\phi_2 \equiv \mathrm{e}^{-\frac{\mathrm{i}}{2}\omega t}\,\chi_2. \tag{6.4.12}$$

On using (6.4.10)–(6.4.12) and the Leibniz rule one then finds the first-order equation

$$\frac{\partial}{\partial t}\begin{pmatrix}\phi_1\\ \phi_2\end{pmatrix} = \frac{\mathrm{i}}{2}\begin{pmatrix}(\omega + 2\omega_B\cos\theta) & 2\omega_B\sin\theta\\ 2\omega_B\sin\theta & -(\omega + 2\omega_B\cos\theta)\end{pmatrix}\begin{pmatrix}\phi_1\\ \phi_2\end{pmatrix}, \qquad (6.4.13)$$

where $\omega_B \equiv \frac{\mu B}{\hbar}$. The system (6.4.13) can be decoupled by acting with $\frac{\partial}{\partial t}$ on both sides and then using the original first-order equations again. One then finds

$$\phi_1(\vec{S}, t) = A_1 \mathrm{e}^{\frac{\mathrm{i}}{2}\Omega t} + A_2 \mathrm{e}^{-\frac{\mathrm{i}}{2}\Omega t}, \qquad (6.4.14)$$

$$\phi_2(\vec{S}, t) = C_1 \mathrm{e}^{\frac{\mathrm{i}}{2}\Omega t} + C_2 \mathrm{e}^{-\frac{\mathrm{i}}{2}\Omega t}, \qquad (6.4.15)$$

where

$$\Omega \equiv \sqrt{(\omega + 2\omega_B\cos\theta)^2 + 4\omega_B^2\sin^2\theta}, \qquad (6.4.16)$$

$$C_1 \equiv \frac{2\omega_B\sin\theta}{(\Omega + \omega + 2\omega_B\cos\theta)} A_1, \qquad (6.4.17)$$

$$C_2 \equiv \frac{-2\omega_B\sin\theta}{(\Omega - \omega - 2\omega_B\cos\theta)} A_2. \qquad (6.4.18)$$

At this stage, a naturally occurring problem is how to solve the Pauli equation in an external magnetic field of a more general nature. For this purpose, we assume that the components B_x, B_y and B_z of the magnetic field are at least of class C^2 in the time variable, but are otherwise arbitrary. The Pauli equation (6.4.1) now leads to the system

$$\frac{\partial}{\partial t}\begin{pmatrix}\chi_1\\ \chi_2\end{pmatrix} = \frac{\mathrm{i}\mu}{\hbar}\begin{pmatrix}B_z & B_x - \mathrm{i}B_y\\ B_x + \mathrm{i}B_y & -B_z\end{pmatrix}\begin{pmatrix}\chi_1\\ \chi_2\end{pmatrix}, \qquad (6.4.19)$$

which implies

$$\left(\frac{\partial}{\partial t} - \frac{\mathrm{i}\mu}{\hbar}B_z\right)\chi_1 = \frac{\mathrm{i}\mu}{\hbar}(B_x - \mathrm{i}B_y)\chi_2, \qquad (6.4.20)$$

$$\left(\frac{\partial}{\partial t} + \frac{\mathrm{i}\mu}{\hbar}B_z\right)\chi_2 = \frac{\mathrm{i}\mu}{\hbar}(B_x + \mathrm{i}B_y)\chi_1. \qquad (6.4.21)$$

We can now express χ_1 from Eq. (6.4.21) in the form

$$\chi_1 = (\mathrm{i}\mu/\hbar)^{-1}(B_x + \mathrm{i}B_y)^{-1}\left(\frac{\partial}{\partial t} + \frac{\mathrm{i}\mu}{\hbar}B_z\right)\chi_2. \qquad (6.4.22)$$

Its insertion into Eq. (6.4.20) leads to the second-order equation

$$\left[\frac{\partial^2}{\partial t^2} + P(t)\frac{\partial}{\partial t} + Q(t)\right]\chi_2 = 0, \qquad (6.4.23)$$

where, on setting (ρ and θ being real-valued functions)

$$B_x(t) + \mathrm{i}B_y(t) \equiv \rho(t)\mathrm{e}^{\mathrm{i}\theta(t)}, \tag{6.4.24}$$

$$B_x(t) - \mathrm{i}B_y(t) \equiv \rho(t)\mathrm{e}^{-\mathrm{i}\theta(t)}, \tag{6.4.25}$$

$$B_z(t) \equiv b(t), \tag{6.4.26}$$

one finds

$$P(t) \equiv -\left(\frac{\dot{\rho}}{\rho} + \mathrm{i}\dot{\theta}\right), \tag{6.4.27}$$

$$Q(t) \equiv \frac{\mathrm{i}\mu}{\hbar}\left(\dot{b} - \frac{\dot{\rho}}{\rho}b - \mathrm{i}\dot{\theta}b\right) + \frac{\mu^2}{\hbar^2}(b^2 + \rho^2). \tag{6.4.28}$$

The equation for χ_2 can be mapped into a simpler equation upon defining

$$K(t) \equiv \chi_2(t)\exp\left[\int \frac{1}{2}P(t)\,\mathrm{d}t\right]. \tag{6.4.29}$$

The function K is then found to obey the differential equation

$$\left[\frac{\partial^2}{\partial t^2} + I(t)\right]K(t) = 0, \tag{6.4.30}$$

where

$$I(t) \equiv Q(t) - \frac{1}{4}P^2(t) - \frac{1}{2}\frac{\partial P}{\partial t}. \tag{6.4.31}$$

If one is able to find K from Eq. (6.4.30), one obtains χ_2 from the definition (6.4.29) and eventually χ_1 from (6.4.22), i.e.

$$\chi_1(t) = \frac{\mathrm{e}^{-\mathrm{i}\theta}}{\rho}\left(-\frac{\mathrm{i}\hbar}{\mu}\frac{\partial}{\partial t} + b\right)\chi_2(t), \tag{6.4.32}$$

bearing in mind that

$$\rho(t) = \sqrt{B_x^2(t) + B_y^2(t)}, \tag{6.4.33}$$

$$\theta(t) = \tan^{-1}\left(\frac{B_y(t)}{B_x(t)}\right), \tag{6.4.34}$$

as is clear from the definitions (6.4.24) and (6.4.25).

6.5 Landau levels

Here we are interested in the energy levels of a particle with spin in a time-independent constant magnetic field. We will see that a deep link exists between this problem and the spectrum of harmonic oscillators (Landau 1930).

It is convenient to take the vector potential in the form

$$A_x = -By, \tag{6.5.1}$$

$$A_y = A_z = 0, \tag{6.5.2}$$

if the magnetic field is directed along the z-axis. Denoting by

$$\hat{\mu} = \frac{\mu}{s}\hat{s} \tag{6.5.3}$$

the intrinsic magnetic moment of the particle of mass m and charge e, where $\hat{s}$ is the spin operator of the particle, the Hamiltonian operator reads (Landau and Lifshitz 1958)

$$\hat{H} = \frac{1}{2m}\Big(\hat{p}_x + \frac{eB}{c}\hat{y}\Big)^2 + \frac{\hat{p}_y^2}{2m} + \frac{\hat{p}_z^2}{2m} - \frac{\mu}{s}B\hat{s}_z, \tag{6.5.4}$$

where $\hat{p}_i \equiv m\hat{v}_i + \frac{e}{c}\hat{A}_i$, $\forall i = x, y, z$. Since $\hat{s}_x$ and $\hat{s}_y$ do not contribute to the Hamiltonian operator, the projection of the spin operator along the z-axis is a conserved quantity. Hereafter, we shall thus replace $\hat{s}_z$ by its eigenvalue $s_z = \sigma$. At this stage, since $\frac{\mu}{s}$ is a constant, the stationary state in the Schrödinger equation may be expressed as the product of a function φ depending on spatial coordinates only, with a function $\widetilde{\psi}$, depending on s_x, s_y and s_z. The equation obeyed by φ is

$$\frac{1}{2m}\left[\Big(\hat{p}_x + \frac{eB}{c}\hat{y}\Big)^2 + \hat{p}_y^2 + \hat{p}_z^2\right]\varphi - \frac{\mu}{s}\sigma B\varphi = E\varphi. \tag{6.5.5}$$

The Hamiltonian of this form of the stationary Schrödinger equation does not depend explicitly on x and z. This implies that p_x and p_z are conserved quantities of our problem, and φ may be expressed in the form

$$\varphi(x, y, z) = e^{\frac{i}{\hbar}(p_x x + p_z z)}\, \chi(y). \tag{6.5.6}$$

Note that the eigenvalues p_x and p_z take all values from $-\infty$ to $+\infty$. Moreover, since $A_z = 0$, the p_z component of the canonical momentum coincides with the z component of the linear momentum, and hence the component v_z of the velocity of the particle may take arbitrary values. This peculiar property is expressed by saying that the motion along the field *is not quantized.*

On requiring that

$$\hat{p}_x \equiv \frac{\hbar}{\mathrm{i}}\frac{\partial}{\partial x}, \quad \hat{p}_y \equiv \frac{\hbar}{\mathrm{i}}\frac{\partial}{\partial y}, \quad \hat{p}_z \equiv \frac{\hbar}{\mathrm{i}}\frac{\partial}{\partial z},$$

as is always the case for canonical momenta in Cartesian coordinates, the insertion of (6.5.6) into (6.5.5) leads to

$$\chi'' + \frac{2m}{\hbar^2}\left[\left(E + \frac{\mu\sigma}{s}B - \frac{p_z^2}{2m}\right) - \frac{m}{2}\omega_B^2(y - y_0)^2\right]\chi = 0, \tag{6.5.7}$$

where we have defined

$$y_0 \equiv -\frac{cp_x}{eB}, \tag{6.5.8}$$

$$\omega_B \equiv \frac{|e|B}{mc}. \tag{6.5.9}$$

Note that Eq. (6.5.7) is the stationary Schrödinger equation for a one-dimensional harmonic oscillator of frequency ω_B, and we know from chapter 4 that the energy eigenvalues are $\left(n + \frac{1}{2}\right)\hbar\omega_B$. In other words, one has

$$E_n = \left(n + \frac{1}{2}\right)\hbar\omega_B + \frac{p_z^2}{2m} - \frac{\mu\sigma}{s}B. \tag{6.5.10}$$

For an electron, one has $\frac{\mu}{s} = -\frac{|e|\hbar}{mc}$, so that the energy spectrum becomes

$$E_n = \left(n + \frac{1}{2} + \sigma\right)\hbar\omega_B + \frac{p_z^2}{2m}. \tag{6.5.11}$$

The corresponding eigenfunctions are expressed through the Hermite polynomials in the standard way

$$\chi_n(y) = \frac{1}{\pi^{1/4}a_B^{1/2}\sqrt{2^n n!}}\exp\left[-\frac{(y - y_0)^2}{2a_B^2}\right]H_n\left(\frac{y - y_0}{a_B}\right), \tag{6.5.12}$$

where $a_B \equiv \sqrt{\frac{\hbar}{m\omega_B}}$.

A striking consequence of (6.5.10) is that, since p_x does not contribute to the energy spectrum, while it ranges continuously over all values from $-\infty$ to $+\infty$, the energy levels have an *infinite degeneracy*. It is usually suggested that such an infinite degeneracy may be removed by confining the motion in the xy plane to a large but finite area. The problem remains, however, to understand what is the deeper underlying reason for the occurrence of such an infinite degeneracy, from the point of view of the general formalism of quantum mechanics. For this purpose one needs a more advanced treatment, which is presented in section 10.2.

6.6 Problems

6.P1. A particle with spin is affected by a homogeneous magnetic field. Find the eigenvalues of the Hamiltonian when the vector potential takes the forms

$$A_x = -By, \quad A_y = A_z = 0, \tag{6.6.1}$$

$$A_x = 0, \quad A_y = -Bz, \quad A_z = 0, \tag{6.6.2}$$

$$A_x = A_y = 0, \quad A_z = -Bx. \tag{6.6.3}$$

Discuss, in all three cases, how the eigenvalues depend on the conserved quantities and on the parameters of the particle with spin. Analyse the degeneracy of the eigenvalues. How do the above choices for the vector potential differ from the choices

$$A_x = -\frac{B}{2}y, \quad A_y = \frac{B}{2}x, \quad A_z = 0, \tag{6.6.4}$$

$$A_x = 0, \quad A_y = -\frac{B}{2}z, \quad A_z = \frac{B}{2}y, \tag{6.6.5}$$

$$A_x = \frac{B}{2}z, \quad A_y = 0, \quad A_z = -\frac{B}{2}x. \tag{6.6.6}$$

Show that the particle motion is unaffected. Such freedom to change the vector potential is a case of *gauge invariance*.

6.P2. Derive in a neater way the result (6.3.13), after writing the commutation relations (6.3.8)–(6.3.10) in the concise form

$$\frac{\mathrm{i}}{\hbar}\Bigl(P_k P_l - P_l P_k\Bigr) = \frac{e_0}{c}\varepsilon_{klm}B_m \tag{6.6.7}$$

and using the identity

$$\varepsilon_{ijk}\varepsilon_{klm} = \Bigl(\delta_{il}\delta_{jm} - \delta_{im}\delta_{jl}\Bigr). \tag{6.6.8}$$

6.P3. Find the JWKB solution of the Pauli equation for a particle of spin $\frac{1}{2}$. If in doubt, compare your result with section 6 of the work in Bolte and Keppeler (1999).

6.P4. Let W be a smooth function such that $|W(x)| \rightarrow \infty$ when $|x| \rightarrow \infty$, and consider the self-adjoint operators

$$Q_1 \equiv \frac{1}{2}\Bigl[\sigma_1 p + \sigma_2 W(x)\Bigr], \tag{6.6.9}$$

$$Q_2 \equiv \frac{1}{2}\Bigl[\sigma_2 p - \sigma_1 W(x)\Bigr]. \tag{6.6.10}$$

One can then build the Hamiltonian operator

$$H \equiv Q_1^2 + Q_2^2 = \frac{1}{2}\left\{\mathbb{1}\left[-\hbar^2\frac{\mathrm{d}^2}{\mathrm{d}x^2} + W^2(x)\right] + \hbar\sigma_3\frac{\mathrm{d}W}{\mathrm{d}x}\right\}. \tag{6.6.11}$$

(i) Prove that, for any state vector ψ, one has

$$(\psi, H\psi) \geq 0. \tag{6.6.12}$$

(ii) The Hamiltonian has an eigenstate with vanishing energy if $W(x)$ has an odd number of zeros. Find such a state. (Hint: for the underlying ideas, see the work in Witten (1981).)

Appendix 6.A

Lagrangian of a charged particle

In classical electrodynamics, a particle of charge q and mass m is described by the Lagrangian

$$\mathcal{L} = \frac{m}{2}\delta_{ij}v^i v^j - q\left[\phi(\vec{x},t) - \frac{1}{c}v^j A_j(\vec{x},t)\right], \tag{6.A.1}$$

where ϕ is the scalar potential and A_j are the components of the vector potential. The Lorentz force is indeed easily obtained from such a Lagrangian because (recall that $\frac{\partial v^j}{\partial v^i} = \delta_i^{\ j}$)

$$\frac{\partial \mathcal{L}}{\partial v^i} = mv_i + \frac{q}{c}A_i, \tag{6.A.2}$$

$$\begin{aligned}\frac{\mathrm{d}}{\mathrm{d}t}\frac{\partial \mathcal{L}}{\partial v^i} &= \frac{\mathrm{d}}{\mathrm{d}t}(mv_i) + \frac{q}{c}\left(\frac{\partial A_i}{\partial t} + \frac{\partial A_i}{\partial x^j}\frac{\mathrm{d}x^j}{\mathrm{d}t}\right)\\ &= \frac{\mathrm{d}}{\mathrm{d}t}(mv_i) + \frac{q}{c}\left(\frac{\partial A_i}{\partial t} + v^j\frac{\partial A_i}{\partial x^j}\right),\end{aligned} \tag{6.A.3}$$

$$\frac{\partial \mathcal{L}}{\partial x^i} = -q\left(\frac{\partial \phi}{\partial x^i} - \frac{1}{c}v^j\frac{\partial A_j}{\partial x^i}\right), \tag{6.A.4}$$

and hence the Euler–Lagrange equations read as

$$\begin{aligned}\frac{\mathrm{d}}{\mathrm{d}t}(mv_i) &= q\left(-\frac{\partial \phi}{\partial x^i} - \frac{1}{c}\frac{\partial A_i}{\partial t}\right) + \frac{q}{c}v^j\left(\frac{\partial A_j}{\partial x^i} - \frac{\partial A_i}{\partial x^j}\right)\\ &= q\left[E_i + \frac{1}{c}\left(\vec{v}\wedge\vec{B}\right)_i\right],\end{aligned} \tag{6.A.5}$$

which are the three components of the Lorentz force, but expressed in a form involving covariant vectors (or covectors), in agreement with what we know from chapter 2 concerning differential forms and Euler–Lagrange equations. Note that, to obtain the last line of Eq. (6.A.5), we have used the simple but non-trivial identity

$$\begin{aligned}\left(\vec{v}\wedge\vec{B}\right)_i &= \varepsilon_i^{\ jk}v_j B_k = \varepsilon_{ij}^{\ \ k}v^j\varepsilon_k^{\ lp}\frac{\partial}{\partial x^l}A_p\\ &= \varepsilon_{ij}^{\ \ k}\varepsilon_k^{\ lp}v^j\frac{\partial}{\partial x^l}A_p = \left(\delta_i^{\ l}\delta_j^{\ p} - \delta_i^{\ p}\delta_j^{\ l}\right)v^j\frac{\partial A_p}{\partial x^l}\\ &= v^j\left(\frac{\partial A_j}{\partial x^i} - \frac{\partial A_i}{\partial x^j}\right).\end{aligned} \tag{6.A.6}$$

In section 6.3, we have $q = -e_0$ for the electron.

Appendix 6.B

Charged particle in a monopole field

The electric charge has an electric field isotropically and radially distributed about it. To describe the motion of a charged particle in an 'external' field we have to use the coupling of the charge with the scalar potential, and the momentum of the particle is gauge invariantly coupled with the replacement $\vec{p} \rightarrow \vec{p} - e\vec{A}$. The magnetic monopole has a radial isotropic magnetic field and may be treated in terms of a vector potential $\vec{A}(\vec{r})$, the curl of which produces this radial magnetic field. For this purpose we need to solve the equation

$$\text{curl}\,\vec{A}(\vec{r}) = \vec{B}(\vec{r}) = \frac{g\vec{r}}{r^3}. \tag{6.B.1}$$

A particular solution is

$$A_x(\vec{r}) = \frac{gy}{r(r-z)}, \quad A_y(\vec{r}) = -\frac{gx}{r(r-z)}, \quad A_z(\vec{r}) = 0. \tag{6.B.2}$$

In polar coordinates this reads as

$$A_x = \frac{g}{r}\frac{\sin\theta\ \sin\varphi}{(1-\cos\theta)}, \quad A_y = -\frac{g}{r}\frac{\sin\theta\ \cos\varphi}{(1-\cos\theta)}, \quad A_z = 0. \tag{6.B.3}$$

This vector potential is defined everywhere except along the positive z-axis, $\theta = 0$. There is no singularity in the magnetic field (except at the origin) and it is spherically symmetric. So there must exist other solutions for $\vec{A}(\vec{r})$ that have the singularity elsewhere. In particular, we may choose

$$A_x = -\frac{g}{r}\frac{\sin\theta\ \sin\varphi}{(1+\cos\theta)}, \quad A_y = \frac{g}{r}\frac{\sin\theta\ \cos\varphi}{(1+\cos\theta)}, \quad A_z = 0, \tag{6.B.4}$$

which has a singularity along the negative z-axis ($\theta = \pi$). The difference between them is a pure gauge term for which the curl vanishes, and the gauge transformation is singular along the whole z-axis.

If we now demand, along with Dirac, that the charged particle wave function be single-valued (Dirac 1931), there is some restriction:

$$e\oint \vec{A}\cdot d\vec{r} = 2n\pi, \tag{6.B.5}$$

since this is the phase change around a closed loop. This integral is best evaluated along the equator $\theta = \frac{\pi}{2}$. Then $d\vec{r} = r\,d\hat{\varphi}$ and hence

$$eg = \frac{2n\pi}{\oint r\vec{a}\cdot d\hat{\varphi}}, \tag{6.B.6}$$

where $\vec{a}$ is the vector potential divided by g. The resulting magnetic flux crossing the hemisphere limited by the equator is then quantized as well.

The wave function of a charged particle in a generic monopole field cannot be defined by a single function everywhere on the sphere of any radius around the monopole. Instead we have two wave functions, one which is valid everywhere except in a small solid angle along the positive z-axis, and the other everywhere except in a small solid angle along the negative z-axis. Such wave functions are discussed later in section 11.7. An extensive analysis of monopole fields can be found in Balachandran *et al.* (1983, 1991) and Marmo and Rubano (1988).

7

Perturbation theory

The subject of perturbation theory in non-relativistic quantum mechanics is introduced. First, perturbation theory for stationary states in the absence of degeneracy is studied. The case of nearby levels, perturbations of a one-dimensional harmonic oscillator, the occurrence of degeneracy, Stark and Zeeman effects are then described in detail. After a brief outline of the variational method, the Dyson series for time-dependent perturbation theory is derived, with application to harmonic perturbations. The Fermi golden rule is also presented. The chapter ends with an assessment of four branches of the subject: regular perturbation theory, asymptotic perturbation theory, and summability methods, spectral concentration and singular perturbations.

7.1 Approximate methods for stationary states

It is frequently the case, in physical problems, that the full Hamiltonian operator H consists of an operator H_0 whose eigenvalues and eigenfunctions are known exactly, and a second term V resulting from a variety of sources (e.g. the interaction with an electric field, or a magnetic field, or the relativistic corrections to the kinetic energy, or the effects of spin). The problems and aims of perturbation theory for stationary states can be therefore summarized as follows.

(i) Once the domain of (essential) self-adjointness of the 'unperturbed' Hamiltonian H_0 is determined, find on which domain the 'sum' $H = H_0 + V$ represents an (essentially) self-adjoint operator (the reader should remember from appendix 4. A that in general, once some operators A and B are given, the intersection of their domains, on which their sum is defined, might be the empty set).

(ii) Use the explicit knowledge of eigenvalues and eigenfunctions of H_0 to compute eigenvalues and eigenfunctions of H as a series of 'corrections' resulting from V.

(iii) To pick out in V a suitable (dimensionless) parameter λ that makes it possible to use perturbation techniques.

(iv) To understand in what sense the smallness of λ makes it possible to evaluate 'small corrections' to the spectrum of H_0. For example, one might find that any non-vanishing value of λ may even change the nature of the spectrum, in that an unperturbed Hamiltonian with discrete spectrum is mapped into a full Hamiltonian with a continuous spectrum. This is indeed the case in some relevant applications, and the task of the physicist is then to understand the meaning of the algorithm he has developed to deal with such cases.

(v) To understand whether the perturbation series obtained for eigenvalues and eigenfunctions are convergent or, instead, are asymptotic expansions. In the latter case, one will eventually face the task of performing a more detailed analysis of the asymptotic series, so as to obtain a proper understanding of why they encode all relevant information about the 'correction' one is interested in.

For the time being, we shall focus on applications of the formalism to a number of problems, while the reader more interested in the mathematical foundations of our operations is referred to section 7.10 and appendix 7.A. At this stage, we assume that H can be split into the sum of two terms

$$H = H_0 + \lambda W, \tag{7.1.1}$$

with H_0 such that its eigenvalues and eigenfunctions can be determined exactly. The operators H_0 and λW are taken to be self-adjoint on the same domain. It is hence possible to choose as a 'reference frame' the base of eigenvectors of H_0, and regard λW as a 'small' time-independent perturbation. Let

$$H_0 \varphi_n = E_n^{(0)} \varphi_n \tag{7.1.2}$$

be the eigenvalue equation for H_0, which is solved (by hypothesis) with φ_n normalized and non-degenerate. The eigenvalue equation for the full Hamiltonian, with eigenvector ψ, can be re-expressed in the form

$$H_0 \psi = (E\mathbb{I} - \lambda W)\psi. \tag{7.1.3}$$

For $\varepsilon \in \mathbf{R}$ one therefore has

$$(\varepsilon \mathbb{I} - H_0)\psi = [(\varepsilon - E)\mathbb{I} + \lambda W]\psi, \tag{7.1.4}$$

and, formally,

$$\psi = (\varepsilon \mathbb{I} - H_0)^{-1}[(\varepsilon - E)\mathbb{I} + \lambda W]\psi. \tag{7.1.5}$$

This expression has only a formal value because ψ also occurs on the right-hand side, and $(\varepsilon \mathbb{I} - H_0)^{-1}$ becomes meaningless as soon as ε happens to coincide with an eigenvalue of H_0. It is therefore convenient to introduce the projection operator

$$Q \equiv \mathbb{I} - \varphi_n(\varphi_n, \cdot), \tag{7.1.6}$$

which, by construction, rules out the eigenfunction φ_n in that the latter is annihilated by Q. On taking for ψ the condition

$$(\varphi_n, \psi) = 1, \tag{7.1.7}$$

one finds

$$\psi = \varphi_n + Q\psi, \tag{7.1.8}$$

which implies, from (7.1.5),

$$\psi = \varphi_n + Q(\varepsilon \mathbb{I} - H_0)^{-1}[(\varepsilon - E)\mathbb{I} + \lambda W]\psi. \tag{7.1.9}$$

If the state $[(\varepsilon - E)\mathbb{I} + \lambda W]\psi$ turns out to be proportional to φ_n when $\varepsilon = E_n^{(0)}$, the operator Q is able to get rid of the resulting singularity, because

$$Q(\varepsilon \mathbb{I} - H_0)^{-1}\varphi_n = 0, \tag{7.1.10}$$

and hence the expression of ψ becomes regular. Note now that, by virtue of Eq. (7.1.9), one obtains

$$\begin{aligned}\psi &= \varphi_n \\ &\quad + Q(\varepsilon \mathbb{I} - H_0)^{-1}[(\varepsilon - E)\mathbb{I} + \lambda W] \\ &\quad \cdot \left\{\varphi_n + Q(\varepsilon \mathbb{I} - H_0)^{-1}[(\varepsilon - E)\mathbb{I} + \lambda W]\psi\right\},\end{aligned} \tag{7.1.11}$$

and, by repeated application of (7.1.9),

$$\psi = \sum_{r=0}^{\infty} \left\{Q(\varepsilon \mathbb{I} - H_0)^{-1}[(\varepsilon - E)\mathbb{I} + \lambda W]\right\}^r \varphi_n. \tag{7.1.12}$$

It is rather hard to study the convergence of this series expansion, and when the assumptions concerning the decomposition (7.1.1) of the Hamiltonian are not satisfied, it only yields an asymptotic expansion of the eigenvector (see section 7.10). Note also what follows.

(1) The state φ_n is an arbitrary eigenfunction of H_0, and hence the method makes it possible to evaluate any eigenfunction of H starting from the eigenfunctions of H_0.

(2) The state ψ is not normalized.

To find the corresponding energy eigenvalue E it is enough to take the scalar product of Eq. (7.1.3) with φ_n. One then obtains

$$(\varphi_n, H_0\psi) = E(\varphi_n, \psi) - \lambda(\varphi_n, W\psi), \tag{7.1.13}$$

which implies, by virtue of (7.1.7),

$$E - E_n^{(0)} = \lambda(\varphi_n, W\psi). \tag{7.1.14}$$

It is now possible to obtain a series expansion for the eigenvalue by virtue of (7.1.12) and (7.1.14), i.e.

$$\lambda\left(\varphi_n, W\sum_{r=0}^{\infty}\left\{Q(\varepsilon \mathbb{I} - H_0)^{-1}[(\varepsilon - E)\mathbb{I} + \lambda W]\right\}^r \varphi_n\right)$$
$$= E - E_n^{(0)}. \tag{7.1.15}$$

It is possible to choose ε in different ways so as to simplify the resulting formulae, and this procedure is described in the following subsections.

7.1.1 Rayleigh–Schrödinger expansion

If ε is set equal to $E_n^{(0)}$, Eq. (7.1.12) leads to

$$\psi = \sum_{r=0}^{\infty}\left\{Q(E_n^{(0)}\mathbb{I} - H_0)^{-1}\Big[(E_n^{(0)} - E)\mathbb{I} + \lambda W\Big]\right\}^r \varphi_n$$
$$= \lambda^0\psi^{(0)} + \lambda^1\psi^{(1)} + \lambda^2\psi^{(2)} + \cdots, \tag{7.1.16}$$

where $\psi^{(0)} \equiv \varphi_n$. The first-order term in the eigenvector is therefore

$$\begin{aligned}\lambda^1\psi^{(1)} &= Q(E_n^{(0)}\mathbb{I} - H_0)^{-1}\Big[(E_n^{(0)} - E)\mathbb{I} + \lambda W\Big]\varphi_n \\ &= \Big(E_n^{(0)} - E\Big)Q(E_n^{(0)}\mathbb{I} - H_0)^{-1}\varphi_n \\ &\quad + Q(E_n^{(0)}\mathbb{I} - H_0)^{-1}\lambda W\varphi_n \\ &= Q(E_n^{(0)}\mathbb{I} - H_0)^{-1}\lambda W\varphi_n \\ &= \sum_m Q(E_n^{(0)}\mathbb{I} - H_0)^{-1}\lambda\varphi_m(\varphi_m, W\varphi_n) \\ &= \sum_{m\neq n}\frac{\lambda(\varphi_m, W\varphi_n)}{(E_n^{(0)} - E_m^{(0)})}\varphi_m,\end{aligned} \tag{7.1.17}$$

where we have inserted a resolution of the identity and then used the property expressed by Eq. (7.1.10). From Eq. (7.1.17) it is clear that a necessary (but not sufficient!) condition for the convergence of the series is

$$\left| \frac{(\varphi_m, W\varphi_n)}{(E_n^{(0)} - E_m^{(0)})} \right| \ll 1. \tag{7.1.18}$$

As far as the energy is concerned, the expansion (7.1.15) becomes

$$\begin{aligned} E &= E_n^{(0)} \\ &+ \left(\varphi_n, \lambda W \sum_{r=0}^{\infty} \left\{ Q(E_n^{(0)} \mathbb{I} - H_0)^{-1} \Big[(E_n^{(0)} - E)\mathbb{I} + \lambda W \Big] \right\}^r \varphi_n \right) \\ &= E_n^{(0)} + E_n^{(1)} + E_n^{(2)} + \cdots . \end{aligned} \tag{7.1.19}$$

Thus, to first order in λ, one finds

$$E_n^{(1)} = (\varphi_n, \lambda W \varphi_n). \tag{7.1.20}$$

Moreover, to second order in λ,

$$\begin{aligned} E_n^{(2)} &= \left\{ \varphi_n, \lambda W Q(E_n^{(0)} \mathbb{I} - H_0)^{-1} \Big[(E_n^{(0)} - E)\mathbb{I} + \lambda W \Big] \varphi_n \right\} \\ &= \sum_{k \neq n} \frac{|(\varphi_n, \lambda W \varphi_k)|^2}{(E_n^{(0)} - E_k^{(0)})}, \end{aligned} \tag{7.1.21}$$

where we have again made use of a resolution of the identity. Note that, for the lowest eigenvalue: $E_n^{(0)} = E_0^{(0)}$, the second-order correction (7.1.21) is always negative.

The calculation of ψ can also be pushed to second order:

$$\begin{aligned} \psi^{(2)} &= Q(E_n^{(0)} \mathbb{I} - H_0)^{-1} \Big[(E_n^{(0)} - E)\mathbb{I} + \lambda W \Big] \\ &\quad \cdot Q(E_n^{(0)} \mathbb{I} - H_0)^{-1} \Big[(E_n^{(0)} - E)\mathbb{I} + \lambda W \Big] \varphi_n \\ &= \sum_{k \neq n} \sum_{l \neq n} \frac{\varphi_k}{(E_n^{(0)} - E_k^{(0)})} \left(\varphi_k, \Big[(E_n^{(0)} - E)\mathbb{I} + \lambda W \Big] \varphi_l \right) \\ &\quad \cdot \frac{1}{(E_n^{(0)} - E_l^{(0)})} \left(\varphi_l, \Big[(E_n^{(0)} - E)\mathbb{I} + \lambda W \Big] \varphi_n \right) \\ &= \sum_{k \neq n, l \neq n} \delta_{kl} (E_n^{(0)} - E) \frac{(\varphi_k, \lambda W \varphi_n)}{(E_n^{(0)} - E_k^{(0)})^2} \varphi_k \\ &\quad + \sum_{k \neq n, l \neq n} \frac{(\varphi_k, \lambda W \varphi_l)(\varphi_l, \lambda W \varphi_n)}{(E_n^{(0)} - E_k^{(0)})(E_n^{(0)} - E_l^{(0)})} \varphi_k . \end{aligned} \tag{7.1.22}$$

This formula is incomplete because it contains the unknown eigenvalue E. On the other hand, $\psi^{(2)}$ should be of second order in λ, and hence one can replace E with its first-order approximation given in (7.1.14).

7.1.2 Brillouin–Wigner expansion

This algorithm consists in setting $\varepsilon = E$ in the formulae (7.1.12) and (7.1.15). This leads to

$$\psi = \sum_{r=0}^{\infty} \left\{ Q(E\mathbb{1} - H_0)^{-1} \lambda W \right\}^r \varphi_n, \tag{7.1.23}$$

$$E = E_n^{(0)} + \left(\varphi_n, \lambda W \sum_{r=0}^{\infty} \left\{ Q(E\mathbb{1} - H_0)^{-1} \lambda W \right\}^r \varphi_n \right). \tag{7.1.24}$$

The resulting terms in the expansion of the eigenvector are, for the first few orders,

$$\psi^{(0)} = \varphi_n, \tag{7.1.25}$$

$$\psi^{(1)} = Q(E\mathbb{1} - H_0)^{-1} \lambda W \varphi_n, \tag{7.1.26}$$

$$\psi^{(2)} = Q(E\mathbb{1} - H_0)^{-1} \lambda W Q(E\mathbb{1} - H_0)^{-1} \lambda W \varphi_n, \tag{7.1.27}$$

whereas for the energy eigenvalue one finds that $E_n^{(1)}$ is again given by (7.1.20), while

$$E_n^{(2)} = \sum_{m \neq n} \frac{|(\varphi_n, \lambda W \varphi_m)|^2}{(E - E_m^{(0)})}. \tag{7.1.28}$$

This scheme is well suited for a preliminary investigation of systems for which the first-order contribution to the energy vanishes, and the unperturbed situation exhibits degenerate or almost degenerate levels (see the next section).

7.1.3 Remark on quasi-stationary states

On taking the limit as the parameter $\lambda \to 0$, eigenfunctions and eigenvalues of the full Hamiltonian H are expected to be turned in a continuous way into eigenfunctions and eigenvalues of the original, unperturbed Hamiltonian H_0. In some relevant cases, however, the perturbation may turn a discrete spectrum into a continuous one. For example, in a one-dimensional problem with potential energy

$$U(x) = \frac{1}{2} m\omega^2 x^2 + \lambda x^3, \tag{7.1.29}$$

the spectrum is discrete when $\lambda = 0$ and coincides with the spectrum of a harmonic oscillator. However, if λ does not vanish, the resulting Hamiltonian operator acquires a continuous spectrum. This Hamiltonian is unbounded from below and hence is physically unacceptable. In this case, perturbation theory describes non-stationary states, in that the particle can enter the region of negative values of x and then escape to infinity. Nevertheless, as $\lambda \rightarrow 0$, the probability of such processes is negligibly small, and the resulting perturbative states are called *quasi-stationary*.

7.2 Very close levels

It is clear from the previous analysis that, if $(E_l^0 - E_n^0)$ is very small for some n, the perturbative formulae become unreliable, since the 'correction' is actually very large. Nevertheless, if there are only a few eigenvalues in a small neighbourhood of E_l^0, it is possible to modify the procedure so as to avoid inconsistent results. To prove this property, let us consider the case when the unperturbed Hamiltonian, H_0, has two nearby eigenvalues E_1^0 and E_2^0 with eigenfunctions φ_1 and φ_2, respectively:

$$H_0\varphi_1 = E_1^0\varphi_1, \tag{7.2.1}$$

$$H_0\varphi_2 = E_2^0\varphi_2. \tag{7.2.2}$$

Let us now look for solutions of Eq. (7.1.3) in the form

$$\psi = a\varphi_1 + b\varphi_2. \tag{7.2.3}$$

This means that we study the equation (Davydov 1981)

$$(H - E)(a\varphi_1 + b\varphi_2) = 0, \tag{7.2.4}$$

which leads, by taking scalar products with φ_1 and φ_2, to the equations

$$\Big(\varphi_1, (H - E)(a\varphi_1 + b\varphi_2)\Big) = 0, \tag{7.2.5}$$

$$\Big(\varphi_2, (H - E)(a\varphi_1 + b\varphi_2)\Big) = 0. \tag{7.2.6}$$

Thus, defining $H_{ij} \equiv \Big(\varphi_i, H\varphi_j\Big)$ $\forall i, j = 1, 2$, and bearing in mind that $\Big(\varphi_i, \varphi_j\Big) = \delta_{ij}$, one obtains the homogeneous linear system

$$(H_{11} - E)a + H_{12}b = 0, \tag{7.2.7}$$

$$H_{21}a + (H_{22} - E)b = 0. \tag{7.2.8}$$

Non-trivial solutions for a and b only exist if the determinant of the matrix of coefficients vanishes, which implies

$$E^2 - (H_{11} + H_{22})E + H_{11}H_{22} - H_{12}H_{21} = 0. \tag{7.2.9}$$

This equation admits two roots:

$$E_{1,2} = \frac{1}{2}(H_{11} + H_{22}) \pm \frac{1}{2}\sqrt{(H_{11} - H_{22})^2 + 4|H_{12}|^2}. \tag{7.2.10}$$

We now distinguish two limiting cases:

$$|H_{11} - H_{22}| \gg |H_{12}|, \tag{7.2.11}$$

or

$$|H_{11} - H_{22}| \ll |H_{12}|. \tag{7.2.12}$$

If the inequality (7.2.11) is satisfied, one finds

$$\begin{aligned} E_1 &= \frac{1}{2}(H_{11} + H_{22}) + \frac{1}{2}(H_{11} - H_{22})\sqrt{1 + \frac{4|H_{12}|^2}{(H_{11} - H_{22})^2}} \\ &\cong \frac{1}{2}(H_{11} + H_{22}) + \frac{1}{2}(H_{11} - H_{22})\left[1 + \frac{2|H_{12}|^2}{(H_{11} - H_{22})^2}\right] \\ &= H_{11} + \frac{|H_{12}|^2}{(H_{11} - H_{22})} \\ &= E_1^0 + V_{11} + \frac{|V_{12}|^2}{(E_1^0 + V_{11} - E_2^0 - V_{22})}, \end{aligned} \tag{7.2.13}$$

and, similarly,

$$\begin{aligned} E_2 &\cong H_{22} - \frac{|H_{12}|^2}{(H_{11} - H_{22})} \\ &= E_2^0 + V_{22} + \frac{|V_{12}|^2}{(E_2^0 + V_{22} - E_1^0 - V_{11})}. \end{aligned} \tag{7.2.14}$$

In contrast, if the inequality (7.2.12) holds, the energy eigenvalues are given by

$$\begin{aligned} E &= \frac{1}{2}(H_{11} + H_{22}) \pm |H_{12}|\sqrt{1 + \frac{(H_{11} - H_{22})^2}{4|H_{12}|^2}} \\ &\cong \frac{1}{2}(H_{11} + H_{22}) \pm \left[|H_{12}| + \frac{(H_{11} - H_{22})^2}{8|H_{12}|}\right]. \end{aligned} \tag{7.2.15}$$

A useful parametrization of eigenvalues and eigenfunctions is obtained by using the 'exact' formula (7.2.10) and then defining

$$\tan\beta \equiv \frac{2H_{12}}{(H_{11} - H_{22})}. \tag{7.2.16}$$

If $E = E_1$, the ratio $\frac{a}{b}$ is then found to be (see Eq. (7.2.7))

$$\begin{aligned} (a/b)_1 &= \frac{H_{12}}{(E_1 - H_{11})} = \frac{\tan\beta}{-1 + \sqrt{1 + \tan^2\beta}} \\ &= \frac{\sin\beta}{(1 - \cos\beta)} = \cot\frac{\beta}{2}. \end{aligned} \tag{7.2.17}$$

At this stage, on imposing the normalization condition for the linear combination (7.2.3) one finds $b = \sin\frac{\beta}{2}$, and hence

$$\psi_1 = \varphi_1 \cos\frac{\beta}{2} + \varphi_2 \sin\frac{\beta}{2}. \tag{7.2.18}$$

An entirely analogous procedure yields, when $E = E_2$, the result

$$\psi_2 = -\varphi_1 \sin\frac{\beta}{2} + \varphi_2 \cos\frac{\beta}{2}. \tag{7.2.19}$$

7.3 Anharmonic oscillator

Given a one-dimensional harmonic oscillator, here we consider a perturbed potential (Grechko *et al.* 1977)

$$U(\hat{x}) = \frac{1}{2}m\omega^2\hat{x}^2 + \varepsilon_1\hat{x}^3 + \varepsilon_2\hat{x}^4, \tag{7.3.1}$$

where ε_1 and ε_2 are two 'small' parameters. Our aim is to evaluate how the energy spectrum and the eigenfunctions are affected by the addition of $\varepsilon_1\hat{x}^3 + \varepsilon_2\hat{x}^4$ to the unperturbed potential $\frac{1}{2}m\omega^2\hat{x}^2$.

Our problem provides a non-trivial application of time-independent perturbation theory in the non-degenerate case, and we can apply the standard formulae of section 7.1 (see, however, section 7.10):

$$E_n \sim E_n^0 + (\psi_n^0, \hat{W}\psi_n^0) + \sum_{k\neq n} \frac{\left|(\psi_n^0, \hat{W}\psi_k^0)\right|^2}{(E_n^0 - E_k^0)}, \tag{7.3.2}$$

$$\psi_n \sim \psi_n^0 + \sum_{k\neq n} \frac{(\psi_k^0, \hat{W}\psi_n^0)}{(E_n^0 - E_k^0)}\psi_k^0, \tag{7.3.3}$$

where, defining $\xi \equiv x\sqrt{\frac{m\omega}{\hbar}}$, one has

$$\psi_n^0 = C_n e^{-\xi^2/2} H_n(\xi), \tag{7.3.4}$$

$$(\psi_k^0, \hat{W}\psi_n^0) = (\psi_k^0, \hat{W}_1\psi_n^0) + (\psi_k^0, \hat{W}_2\psi_n^0), \tag{7.3.5}$$

$$\hat{W}_1 \equiv \varepsilon_1\hat{x}^3, \tag{7.3.6}$$

$$\hat{W}_2 \equiv \varepsilon_2\hat{x}^4. \tag{7.3.7}$$

The building blocks of our calculation for the anharmonic oscillator are the matrix elements

$$(\psi_n^0, \hat{x}\psi_k^0) = \sqrt{\frac{\hbar}{2m\omega}}\left(\sqrt{n}\,\delta_{n,k+1} + \sqrt{n+1}\;\delta_{n,k-1}\right). \tag{7.3.8}$$

Consider first the effect of $\hat{W}_1$. Its matrix elements are

$$(\psi_n^0, \hat{W}_1 \psi_k^0) = \varepsilon_1 (\psi_n^0, \hat{x}^3 \psi_k^0) = \varepsilon_1 \sum_l (\psi_n^0, \hat{x}^2 \psi_l^0)(\psi_l^0, \hat{x}\psi_k^0), \tag{7.3.9}$$

where we have inserted a resolution of the identity. In an analogous way, the matrix elements $(\psi_n^0, \hat{x}^2 \psi_l^0)$ can be evaluated as follows:

$$\begin{aligned}
(\psi_n^0, \hat{x}^2 \psi_l^0) &= \sum_p (\psi_n^0, \hat{x}\psi_p^0)(\psi_p^0, \hat{x}\psi_l^0) \\
&= \frac{\hbar}{2m\omega} \sum_p \Big(\sqrt{n}\delta_{n,p+1} + \sqrt{n+1}\, \delta_{n+1,p} \Big) \\
&\quad \times \Big(\sqrt{p}\delta_{p,l+1} + \sqrt{p+1}\, \delta_{l,p+1} \Big) \\
&= \frac{\hbar}{2m\omega} \Big[\sqrt{n(l+1)} \sum_p \delta_{n-1,p}\delta_{p,l+1} \\
&\quad + \sqrt{(n+1)(l+1)} \sum_p \delta_{n+1,p}\delta_{p,l+1} \\
&\quad + \sqrt{nl} \sum_p \delta_{n-1,p}\delta_{p,l-1} + \sqrt{(n+1)l} \sum_p \delta_{n+1,p}\delta_{p,l-1} \Big].
\end{aligned} \tag{7.3.10}$$

At this stage, we remark that

$$\begin{aligned}
\sqrt{n(l+1)} \sum_p \delta_{n-1,p}\delta_{p,l+1} &= \sqrt{n(l+1)}\delta_{n-1,l+1} \\
&= \sqrt{n(n-1)}\delta_{n,l+2},
\end{aligned} \tag{7.3.11}$$

and, similarly,

$$\sqrt{(n+1)(l+1)} \sum_p \delta_{n+1,p}\delta_{p,l+1} = (n+1)\delta_{n,l}, \tag{7.3.12}$$

$$\sqrt{nl} \sum_p \delta_{n-1,p}\delta_{p,l-1} = n\delta_{n,l}, \tag{7.3.13}$$

$$\sqrt{(n+1)l} \sum_p \delta_{n+1,p}\delta_{p,l-1} = \sqrt{(n+1)(n+2)}\delta_{n,l-2}. \tag{7.3.14}$$

This leads to

$$\begin{aligned}
(\psi_n^0, \hat{x}^2 \psi_l^0) = \frac{\hbar}{2m\omega} \Big[&\sqrt{n(n-1)}\delta_{n,l+2} + (2n+1)\delta_{n,l} \\
&+ \sqrt{(n+1)(n+2)}\delta_{n,l-2} \Big],
\end{aligned} \tag{7.3.15}$$

which implies (see Eqs. (7.3.8) and (7.3.9))

$$(\psi_n^0, \hat{x}^3 \psi_k^0) = \left(\frac{\hbar}{2m\omega}\right)^{3/2} \Big[\sqrt{n(n-1)(n-2)}\,\delta_{n,k+3} + 3n^{3/2}\delta_{n,k+1} + 3(n+1)^{3/2}\delta_{n,k-1} + \sqrt{(n+1)(n+2)(n+3)}\,\delta_{n,k-3}\Big]. \tag{7.3.16}$$

This implies that $(\psi_n^0, \hat{W}_1 \psi_n^0) = 0$, and hence $\hat{W}_1$ only contributes to second-order corrections of the energy spectrum, when k equals $n \pm 1$ and $n \pm 3$. The complete formula for first-order corrections to the eigenfunctions is therefore (see Eq. (7.3.3))

$$\psi_n \sim \psi_n^0 + \varepsilon_1 \Big[\psi_{n-3}^0 \frac{\sqrt{n(n-1)(n-2)}}{3\hbar\omega} + \psi_{n-1}^0 \frac{3n^{3/2}}{\hbar\omega} + \psi_{n+1}^0 \frac{3(n+1)^{3/2}}{-\hbar\omega} + \psi_{n+3}^0 \frac{\sqrt{(n+1)(n+2)(n+3)}}{-3\hbar\omega}\Big] \left(\frac{\hbar}{2m\omega}\right)^{3/2}. \tag{7.3.17}$$

It should be stressed that $\hat{W}_2$ does not contribute to (7.3.17), since ε_2 is of higher order with respect to ε_1, in that

$$\frac{\varepsilon_2}{(\varepsilon_1)^2} = \mathrm{O}(1). \tag{7.3.18}$$

The term $\hat{W}_2$, however, affects the energy spectrum (see the expansion (7.3.2)). Indeed, one has

$$(\psi_n^0, \hat{W}_2 \psi_n^0) = \varepsilon_2 (\psi_n^0, \hat{x}^4 \psi_n^0), \tag{7.3.19}$$

where

$$\begin{aligned}(\psi_n^0, \hat{x}^4 \psi_n^0) &= \sum_k (\psi_n^0, \hat{x}^3 \psi_k^0)(\psi_k^0, \hat{x} \psi_n^0) \\ &= \left(\frac{\hbar}{2m\omega}\right)^2 \Big[3n^{3/2}\sqrt{n}\,\delta_{n-1,n-1} + 3(n+1)^{3/2}\sqrt{n+1}\,\delta_{n+1,n+1}\Big] \\ &= 3\Big(2n^2 + 2n + 1\Big)\left(\frac{\hbar}{2m\omega}\right)^2,\end{aligned} \tag{7.3.20}$$

which implies

$$E_n \sim \left(n + \frac{1}{2}\right)\hbar\omega + \varepsilon_2 (\psi_n^0, \hat{x}^4 \psi_n^0) + (\varepsilon_1)^2 \Big[\frac{|(\psi_n^0, \hat{x}^3 \psi_{n-3}^0)|^2}{3\hbar\omega} + \frac{|(\psi_n^0, \hat{x}^3 \psi_{n-1}^0)|^2}{\hbar\omega}$$

$$+ \frac{|(\psi_n^0, \hat{x}^3 \psi_{n+1}^0)|^2}{-\hbar\omega} + \frac{|(\psi_n^0, \hat{x}^3 \psi_{n+3}^0)|^2}{-3\hbar\omega}\Bigg]$$

$$= \left(n + \frac{1}{2}\right)\hbar\omega + 3\varepsilon_2 \left(\frac{\hbar}{2m\omega}\right)^2 \left(2n^2 + 2n + 1\right)$$

$$- \frac{(\varepsilon_1)^2}{\hbar\omega}\left(\frac{\hbar}{2m\omega}\right)^3 \left(30n^2 + 30n + 11\right). \tag{7.3.21}$$

On defining

$$\sigma \equiv \sqrt{\frac{\hbar}{2m\omega}}, \tag{7.3.22}$$

$$A \equiv 6\varepsilon_2 - 15\frac{\varepsilon_1^2}{m\omega^2}, \quad B \equiv 3\varepsilon_2 - \frac{11}{2}\frac{\varepsilon_1^2}{m\omega^2}, \tag{7.3.23}$$

the result (7.3.21) may be cast in the very convenient form

$$E_n \sim \left(n + \frac{1}{2}\right)\hbar\omega + \sigma^4\Big[An(n+1) + B\Big], \tag{7.3.24}$$

which implies that, to avoid level-crossing (i.e. a sign change of perturbed eigenvalues for sufficiently large values of n), one should impose the following inequality:

$$\varepsilon_2 > \frac{5}{2}\frac{\varepsilon_1^2}{m\omega^2}. \tag{7.3.25}$$

7.4 Occurrence of degeneracy

Let us assume that the level with energy E_l^0 has a degeneracy of multiplicity f. This is indeed what happens in the majority of physical problems (e.g. the hydrogen atom or the harmonic oscillator in dimension greater than 1). As a zeroth-order approximation one may consider the linear combination

$$\psi_l = \sum_{k=1}^{f} a_k \varphi_{lk}, \tag{7.4.1}$$

where φ_{lk} are taken to be solutions of the equation

$$(H_0 - E_l^0 \mathbb{1})\varphi_{lk} = 0. \tag{7.4.2}$$

The insertion of the combination (7.4.1) into the eigenvalue equation (7.1.3), jointly with the scalar product with the unperturbed eigenfunctions (cf. Eqs. (7.2.5)–(7.2.8)), leads to a homogeneous linear system of f equations:

$$\sum_{k=1}^{f} \Big(H_{mk} - E_l \delta_{mk}\Big) a_k = 0 \quad \forall m = 1, 2, \ldots, f. \tag{7.4.3}$$

The condition for finding non-trivial solutions is, of course, that the determinant of the matrix of coefficients should vanish:

$$\det\Big(H_{mk} - E_l \delta_{mk}\Big) = 0. \tag{7.4.4}$$

This yields an algebraic equation of degree f in the unknown E_l. Such an equation is called the *secular* equation. In particular, it may happen that all roots are distinct. If this turns out to be the case, the effect of the perturbation is to remove the degeneracy completely, and the level with energy E_l^0 splits into f distinct levels, for each of which a different eigenfunction exists:

$$\psi_{lk} = \sum_m a_{mk} \varphi_m. \tag{7.4.5}$$

In Eq. (7.4.5), the coefficients a_{mk} are evaluated by replacing E_l with E_{lk} in the system (7.4.3).

If some roots of the secular equation (7.4.4) coincide, however, the perturbation does not completely remove the degeneracy, and in the following section one can find a first concrete example of how this method works. At a deeper level, the occurrence of degenerate eigenvalues can be dealt with as follows. Let $\varepsilon_0 \equiv E_a^{(0)}$ be a degenerate eigenvalue of the unperturbed Hamiltonian H_0, with eigenspace $\mathcal{H}^{(0)}$ having dimension $d > 1$. Let i be a degeneracy label ranging from 1 to d, and let us assume that the full Hamiltonian (7.1.1) has eigenvalues $E_i^{(\lambda)}$ and eigenfunctions φ_λ^i:

$$H\varphi_\lambda^i = E_i^{(\lambda)} \varphi_\lambda^i \tag{7.4.6}$$

given by the (formal) power series

$$E_i^{(\lambda)} = \varepsilon_0 + \sum_{n=1}^{\infty} \lambda^n \varepsilon_{n,i}, \tag{7.4.7}$$

$$\varphi_\lambda^i = \chi_0^i + \sum_{n=1}^{\infty} \lambda^n \chi_n^i. \tag{7.4.8}$$

The projector onto the eigenspace $\mathcal{H}^{(0)}$ reads as

$$P^{(0)} \equiv \sum_{k=1}^{d} \chi_0^k \Big(\chi_0^k, \cdot\Big), \tag{7.4.9}$$

because the unperturbed eigenfunctions are taken to be mutually orthogonal:

$$\Big(\chi_0^i, \chi_0^j\Big) = \delta_{ij}. \tag{7.4.10}$$

One therefore finds

$$H_0 P^{(0)} = P^{(0)} H_0 = \varepsilon_0 P^{(0)}, \tag{7.4.11}$$

$$\chi_0^i = P^{(0)} \chi_0^i. \tag{7.4.12}$$

Moreover, on denoting by β the degeneracy label for the unperturbed eigenfunctions Φ_b^β belonging to the eigenvalue $E_b^{(0)}$, with $b \neq a$, the projector onto the orthogonal complement $\mathcal{H}_\perp^{(0)}$ can be expressed in the form

$$Q^{(0)} = \sum_{b \neq a} \sum_{\beta} \Phi_b^\beta \left(\Phi_b^\beta, \cdot \right). \tag{7.4.13}$$

Of course, the sum of the complementary projectors $P^{(0)}$ and $Q^{(0)}$ yields the identity:

$$P^{(0)} + Q^{(0)} = \mathbb{I}. \tag{7.4.14}$$

We now assume the validity of the condition

$$\left(\chi_0^i, \varphi_\lambda^i \right) = 1 \quad \forall i = 1, \ldots, d, \tag{7.4.15}$$

which implies, by virtue of (7.4.8) and (7.4.10), that

$$\left(\chi_0^i, \chi_p^i \right) = \delta_{p,0}. \tag{7.4.16}$$

Equations (7.4.6)–(7.4.8) therefore yield

$$\lambda \Big[(H_0 - \varepsilon \mathbb{I}) \chi_1^i + W \chi_0^i \Big] + \lambda^2 \Big[(H_0 - \varepsilon \mathbb{I}) \chi_2^i + W \chi_1^i \Big] + \mathrm{O}(\lambda^3)$$
$$= \lambda \varepsilon_{1,i} \chi_0^i + \lambda^2 (\varepsilon_{2,i} \chi_0^i + \varepsilon_{1,i} \chi_1^i) + \mathrm{O}(\lambda^3), \tag{7.4.17a}$$

i.e. the fundamental equation

$$(H_0 - \varepsilon_0 \mathbb{I}) \chi_p^i + W \chi_{p-1}^i - \sum_{n=1}^{p} \varepsilon_{n,i} \chi_{p-n}^i = 0, \tag{7.4.17b}$$

for all $p = 1, 2, \ldots, \infty$ and for all $i = 1, \ldots, d$. First, we project Eq. (7.4.17b) onto the eigenspace $\mathcal{H}^{(0)}$ by applying the projector $P^{(0)}$ defined in (7.4.9). This yields

$$P^{(0)} W \chi_{p-1}^i = \sum_{n=1}^{p} \varepsilon_{n,i} P^{(0)} \chi_{p-n}^i. \tag{7.4.18}$$

Secondly, we project Eq. (7.4.17b) onto the base vectors of $\mathcal{H}_\perp^{(0)}$ and find

$$\left(\Phi_b^\beta, \chi_p^i \right) = - \frac{\left(\Phi_b^\beta, W \chi_{p-1}^i \right)}{(E_b^{(0)} - \varepsilon_0)} + \sum_{n=1}^{p} \varepsilon_{n,i} \frac{\left(\Phi_b^\beta, \chi_{p-n}^i \right)}{(E_b^{(0)} - \varepsilon_0)}. \tag{7.4.19}$$

When $p = 1$, which corresponds to first-order corrections, Eqs. (7.4.12) and (7.4.18) lead to

$$P^{(0)} W P^{(0)} \chi_0^i = \varepsilon_{1,i} \chi_0^i \quad \forall i = 1, \ldots, d, \tag{7.4.20}$$

which is an eigenvalue equation for the restriction to $\mathcal{H}^{(0)}$ of the self-adjoint operator $P^{(0)} W P^{(0)}$. One then finds the secular equation already encountered in (7.4.4), and we first assume that $P^{(0)} W P^{(0)}$ has a non-degenerate spectrum when restricted to $\mathcal{H}^{(0)}$. The first-order correction to the eigenfunctions is then obtained from Eq. (7.4.19) in the form

$$\left(\Phi_b^\beta, \chi_1^i\right) = -\frac{\left(\Phi_b^\beta, W \chi_0^i\right)}{\left(E_b^{(0)} - \varepsilon_0\right)}. \tag{7.4.21}$$

This equation determines the components of χ_1^i (see Eq. (7.4.8)) in the orthogonal complement of $\mathcal{H}^{(0)}$. A basis-independent form of Eq. (7.4.21) is obtained upon multiplying both sides by the unperturbed eigenfunction Φ_b^β. Summation over all values of b and β and use of (7.4.13) therefore yields

$$\begin{aligned} Q^{(0)} \chi_1^i &= -(H_0 - \varepsilon_0 \mathbb{1})^{-1} Q^{(0)} W \chi_0^i \\ &= -Q^{(0)} (H_0 - \varepsilon_0 \mathbb{1})^{-1} Q^{(0)} W \chi_0^i, \end{aligned} \tag{7.4.22}$$

where the right-hand side is entirely known, and insertion of $Q^{(0)}$ on the last line occurs for later convenience (to avoid apparent singularities in subsequent calculations).

Second-order corrections to the eigenvalues are obtained from (7.4.18), because this leads to

$$P^{(0)} W \chi_1^i - \varepsilon_{1,i} P^{(0)} \chi_1^i - \varepsilon_{2,i} \chi_0^i = 0, \tag{7.4.23}$$

by virtue of (7.4.12). On taking the scalar product of Eq. (7.4.23) with the unperturbed eigenfunctions χ_0^j one finds

$$\varepsilon_{2,i} \delta_{ij} = \left(\chi_0^j, P^{(0)} W \chi_1^i\right) - \varepsilon_{1,i} \left(\chi_0^j, \chi_1^i\right), \tag{7.4.24}$$

by virtue of (7.4.12). Now we also use the identity (7.4.14) to express

$$P^{(0)} W = P^{(0)} W \left(P^{(0)} + Q^{(0)}\right)$$

on the right-hand side of Eq. (7.4.24). This eventually leads to

$$\varepsilon_{2,i} \delta_{ij} = (\varepsilon_{1,j} - \varepsilon_{1,i}) \left(\chi_0^j, \chi_1^i\right) + \left(\chi_0^j, W Q^{(0)} \chi_1^i\right), \tag{7.4.25}$$

by virtue of the secular equation (7.4.20) with i replaced by j, and after again using the identity (7.4.12). Thus, when $i = j$, Eqs. (7.4.22) and

(7.4.25) lead to

$$\varepsilon_{2,i} = -\left(\chi_0^i, P^{(0)} W Q^{(0)} (H_0 - \varepsilon_0 \mathbb{I})^{-1} Q^{(0)} W \chi_0^i\right), \tag{7.4.26}$$

for all $i = 1, \ldots, d$. Moreover, when $i \neq j$, Eqs. (7.4.22) and (7.4.25) yield

$$\left(\chi_0^j, \chi_1^i\right) = \frac{\left(\chi_0^j, W Q^{(0)} (H_0 - \varepsilon_0 \mathbb{I})^{-1} Q^{(0)} W \chi_0^i\right)}{(\varepsilon_{1,j} - \varepsilon_{1,i})}. \tag{7.4.27}$$

This formula, jointly with the orthogonality $\left(\chi_0^i, \chi_1^i\right) = 0$, resulting from (7.4.16), determines completely all components of the first-order correction χ_1^i in the eigenspace $\mathcal{H}^{(0)}$. For generic values of $p \geq 2$, the method determines therefore the correction $\varepsilon_{p,i}$ of order p to the eigenvalues, the components of the vector χ_{p-1}^i in the eigenspace $\mathcal{H}^{(0)}$, and the components of the vector χ_p^i in the orthogonal complement $\mathcal{H}_\perp^{(0)}$, for all $i = 1, \ldots, d$.

Last, if the secular equation (7.4.20) has multiple roots, one faces the problem of finding eigenvalues and eigenfunctions of the Hamiltonian

$$H = H_{(0,0)} + \lambda W^{(1)}, \tag{7.4.28}$$

where

$$H_{(0,0)} \equiv H_0 + P^{(0)} W P^{(0)}, \tag{7.4.29}$$

$$W^{(1)} \equiv W - P^{(0)} W P^{(0)}. \tag{7.4.30}$$

One can then apply all previous formulae for perturbations of degenerate eigenvalues, with a degeneracy $\widetilde{d} < d$.

7.5 Stark effect

The Stark effect consists in the splitting of the energy levels of the hydrogen atom (or other atoms) resulting from the application of an electric field (Stark 1914). If an electric field is applied, its effect is viewed as a perturbation, and is evaluated within the framework of time-independent perturbation theory for degenerate levels.

Here we are interested in the first-order Stark effect on the $n = 2$ states of the hydrogen atom. Indeed, we know from section 5.4 that, when the quantum number $n = 2$ for the hydrogen atom, the quantum number l takes the values 0 and 1, and the quantum number m takes the values $-1, 0, 1$. Thus, there exist four bound states with $n = 2$: $u_{2,0,0}, u_{2,1,-1}, u_{2,1,0}$ and $u_{2,1,1}$, and they all have the same energy E^0. The perturbation Hamiltonian $H^{(1)}$ for an electron of charge $-e_0$ in an electric field $\vec{E}$ directed along the z-axis is $H^{(1)} = e_0 E z$. Note that the first-order correction of the energy for any bound state $u_{n,l,m}$ vanishes (hereafter, $\mathrm{d}V$ is the integration measure in spherical coordinates):

$$E^{(1)} = e_0 E \int u_{n,l,m}^* \, z \, u_{n,l,m} \, \mathrm{d}V = 0. \tag{7.5.1}$$

This happens because the state $u_{n,l,m}$ has either positive parity (if l is even) or negative parity (if l is odd), while $z = r\cos\theta$ is an odd function, so that the integral (7.5.1), evaluated over the whole of $\mathbf{R}^3$, vanishes by construction (the integrand being an odd function, just as z is).

Non-vanishing values for first-order corrections of the energy levels can only be obtained from states that are neither even nor odd, and hence are linear combinations of states of opposite parity. The selection rule for matrix elements of the perturbation $H^{(1)}$ is obtained by virtue of the general formula (here C is a parameter depending on l, m, l', m' and P_l^m denotes Legendre polynomials)

$$\begin{aligned}\left(u_{n',l',m'}, \hat{z} u_{n,l,m}\right) &= C \int_0^\infty R_{n',l'}^* R_{nl} r^3 \, \mathrm{d}r \\ &\quad \times \int_{-1}^{1} \xi P_{l'}^{m'}(\xi) P_l^m(\xi) \, \mathrm{d}\xi \int_0^{2\pi} \mathrm{e}^{\mathrm{i}(m-m')\varphi} \, \mathrm{d}\varphi \\ &= \left(R_{n',l'}, \hat{r} R_{n,l}\right) \delta_{m,m'} \left(C^A \delta_{l',l+1} + C^B \delta_{l',l-1}\right),\end{aligned} \tag{7.5.2}$$

where

$$C^A \equiv \sqrt{\frac{(l+m+1)(l-m+1)}{(2l+1)(2l+3)}}, \tag{7.5.3}$$

$$C^B \equiv \sqrt{\frac{(l+m)(l-m)}{(2l-1)(2l+1)}}. \tag{7.5.4}$$

This implies that $H^{(1)}$ has non-vanishing matrix elements if and only if $m' - m = 0$ and $l' - l = \pm 1$.

As a first step, we compute the matrix elements of $H^{(1)}$ on the basis given by the states $u_{2,0,0}, u_{2,1,-1}, u_{2,1,0}$ and $u_{2,1,1}$. By virtue of (7.5.1), all diagonal matrix elements vanish in this case: $H_{ii}^{(1)} = 0$. The orthogonality properties of the eigenfunctions with different values of m lead to a further simplification of the calculation, and the only non-vanishing matrix elements of $H^{(1)}$ turn out to be those particular off-diagonal elements corresponding to bound states with the same value of m: $u_{2,1,0}$ and $u_{2,0,0}$. Recall now that

$$u_{2,0,0}(r,\theta,\varphi) = R_{2,0}(r) Y_{0,0}(\theta,\varphi), \tag{7.5.5}$$

$$u_{2,1,0}(r,\theta,\varphi) = R_{2,1}(r) Y_{1,0}(\theta,\varphi), \tag{7.5.6}$$

with right-hand sides evaluated from Eqs. (5.5.29), (5.5.30) and (5.6.49), (5.6.50). This leads to (Squires 1995)

$$
\begin{aligned}
H_{34}^{(1)} = H_{43}^{(1)} &= e_0 E \int u_{2,1,0}^{*} \, z \, u_{2,0,0} \, \mathrm{d}V \\
&= \frac{e_0 E}{16} \left(\frac{Z}{a_0} \right)^4 \\
&\times \int_0^{\infty} r^4 \left(2 - \frac{Zr}{a_0} \right) \mathrm{e}^{-Zr/a_0} \, \mathrm{d}r \int_0^{\pi} \cos^2 \theta \sin \theta \, \mathrm{d}\theta. \qquad (7.5.7)
\end{aligned}
$$

It is now convenient to introduce a new integration variable y defined by

$$
rZ \equiv y. \qquad (7.5.8)
$$

Moreover, we note that $\int_0^{\pi} \cos^2 \theta \sin \theta \, \mathrm{d}\theta = \frac{2}{3}$, by exploiting integration by parts. Thus, the formula for $H_{34}^{(1)}$ reduces to

$$
H_{34}^{(1)} = \frac{2}{3} e_0 E \left(\frac{1}{2a_0} \right)^4 \frac{1}{Z} \int_0^{\infty} y^4 \left(2 - \frac{y}{a_0} \right) \mathrm{e}^{-y/a_0} \, \mathrm{d}y. \qquad (7.5.9)
$$

At this stage, we can use the formula

$$
\int_0^{\infty} y^n \mathrm{e}^{-y/a_0} \, \mathrm{d}y = n! a_0^{n+1}. \qquad (7.5.10)
$$

Therefore, defining

$$
\varepsilon \equiv 3 e_0 E \frac{a_0}{Z}, \qquad (7.5.11)
$$

the matrix of $H^{(1)}$ on the basis of the four states with $n = 2$ is found to be

$$
H_{ij}^{(1)} = \begin{pmatrix} 0 & 0 & 0 & 0 \\ 0 & 0 & 0 & 0 \\ 0 & 0 & 0 & -\varepsilon \\ 0 & 0 & -\varepsilon & 0 \end{pmatrix}. \qquad (7.5.12)
$$

As we mentioned before, since the four unperturbed states have the same energy, we are dealing with perturbation theory in the degenerate case. It is clear from (7.5.12) that two eigenvalues of $H_{ij}^{(1)}$ vanish, and their eigenvectors are $u_{2,1,1}$ and $u_{2,1,-1}$, respectively. The non-trivial contribution to the energy spectrum results from the 2×2 sub-matrix $\begin{pmatrix} 0 & -\varepsilon \\ -\varepsilon & 0 \end{pmatrix}$. Its eigenvalues are $\lambda_+ = +\varepsilon$ and $\lambda_- = -\varepsilon$. In the former case, the eigenvector solves the equation

$$
\begin{pmatrix} 0 & -\varepsilon \\ -\varepsilon & 0 \end{pmatrix} \begin{pmatrix} u_1 \\ u_2 \end{pmatrix} = \begin{pmatrix} \varepsilon u_1 \\ \varepsilon u_2 \end{pmatrix},
$$

and hence is proportional to $\begin{pmatrix} 1 \\ -1 \end{pmatrix}$, while in the latter case it solves the equation

$$\begin{pmatrix} 0 & -\varepsilon \\ -\varepsilon & 0 \end{pmatrix} \begin{pmatrix} v_1 \\ v_2 \end{pmatrix} = \begin{pmatrix} -\varepsilon v_1 \\ -\varepsilon v_2 \end{pmatrix},$$

and is therefore proportional to $\begin{pmatrix} 1 \\ 1 \end{pmatrix}$. The corresponding normalized eigenvectors to first order are (see Eq. (7.4.5))

$$u_{\lambda_+} = \frac{1}{\sqrt{2}} \begin{pmatrix} 1 \\ -1 \end{pmatrix} \Big(u_{2,1,0}, u_{2,0,0} \Big) = \frac{1}{\sqrt{2}} \Big(u_{2,1,0} - u_{2,0,0} \Big), \tag{7.5.13}$$

$$u_{\lambda_-} = \frac{1}{\sqrt{2}} \begin{pmatrix} 1 \\ 1 \end{pmatrix} \Big(u_{2,1,0}, u_{2,0,0} \Big) = \frac{1}{\sqrt{2}} \Big(u_{2,1,0} + u_{2,0,0} \Big). \tag{7.5.14}$$

Note that $u_{2,1,0}$ has negative parity, while $u_{2,0,0}$ has positive parity. Thus, in agreement with our initial remarks, the eigenfunctions of the problem turn out to have mixed parity. To first order in perturbation theory, the energies are then found to be

$$E^0, \; E^0, \; E^0 + 3e_0 E \frac{a_0}{Z}, \; E^0 - 3e_0 E \frac{a_0}{Z}.$$

Interestingly, this simple calculation shows that there can only be a first-order Stark effect when degenerate states exist with different values of the quantum number l. By virtue of the first-order Stark effect, the hydrogen atom behaves as if it had a permanent electric dipole moment of magnitude $3e_0 a_0$. This dipole moment can be parallel, or anti-parallel, or orthogonal to the external electric field. In general, however, the ground states of atoms and nuclei are non-degenerate, and hence do not possess such a permanent electric dipole moment (Squires 1995). To second order in perturbation theory, the Stark effect provides a correction to the energy levels proportional to the square of the magnitude of the electric field. This term corresponds to an induced electric dipole moment.

We cannot conclude this section, however, without emphasizing a crucial point. The perturbation occurring in the Stark effect is unbounded from below and unbounded with respect to the comparison Hamiltonian, and this is a source of non-trivial features. Indeed, in suitable units, the Stark effect on a hydrogen-like atom is described by the Hamiltonian operator

$$H(E) = -\triangle - \frac{Z}{r} + 2Ex_3 \tag{7.5.15}$$

acting on $\mathcal{L}^2(\mathbf{R}^3)$, where $2E > 0$ is the uniform electric field directed along the x_3-axis (either x or y or z), Z is the atomic number and $r \equiv \sqrt{x^2 + y^2 + z^2}$. The spectrum of $H(E)$ is *absolutely continuous* in

$(-\infty, \infty)$, while the spectrum of the unperturbed operator $H(0)$ for the hydrogen atom is *discrete* along $(-\infty, 0)$. The spectral theory of this problem is then treated within the framework of asymptotic perturbation theory based upon strong convergence of resolvents as $E \rightarrow 0$ (see section 7.10 and appendix 7.A).

It should be stressed that usual perturbation theory fails when there is an energy degeneracy. We have just seen that the Stark effect requires a suitable choice of the unperturbed states to get meaningful results. Quite a different problem arises when the degeneracy cannot be removed by such a choice. This occurs in the case of scattering of the continuum states by a potential (chapter 8). The choice of the boundary conditions for the actual physical state determines whether there are spherically diverging or converging waves. These boundary conditions define the 'in' and 'out' states and can be incorporated into the perturbation theory by making the energy denominators have an infinitesimal negative or positive imaginary part. This choice is called the 'Sommerfeld radiation condition' (Sommerfeld 1964) and resolves the Poincaré catastrophe which is encountered in the study of perturbations of periodic orbits (Poincaré 1908). We will see related problems in the study of harmonic perturbations in subsection 7.8.1. More generally, whenever there is a coupling of discrete states with continuum states.

7.6 Zeeman effect

Another good example of how an external field may remove the degeneracy of the original problem is provided by the Zeeman effect (Zeeman 1897a,b). In this case, the lines of the emission spectrum of an atomic system are split into several nearby components, when a magnetic field of sufficiently high intensity (e.g. of the order of 10^3 G) is switched on. This phenomenon may occur in hydrogen-like atoms and alkali metals. The quantum-mechanical interpretation requires that the electron for which the Schrödinger equation in an external electromagnetic field is written is affected by the potential $U(r)$ describing the effects of the nucleus and of the 'interior electrons'. Moreover, an external magnetic field is applied. The resulting canonical momenta contain the kinematic momenta and a term proportional to the vector potential as we know from section 6.3, so that the Hamiltonian operator for an electron of charge $e = -e_0$ reads

$$\hat{H} = \sum_{k=1}^{3} \frac{1}{2m_{\rm e}} \left(\hat{p}_k - \frac{e}{c} \hat{A}_k \right)^2 + e\hat{V} + \hat{U}, \qquad (7.6.1)$$

where $\hat{V}$ is the operator corresponding to the scalar potential of the classical theory. We now use the operator identity (no summation over k)

$$\left(\hat{p}_k - \frac{e}{c}\hat{A}_k\right)^2 = \hat{p}_k^2 - \frac{e}{c}\left(\hat{p}_k\hat{A}_k + \hat{A}_k\hat{p}_k\right) + \frac{e^2}{c^2}\hat{A}_k^2, \tag{7.6.2}$$

and another basic result:

$$\hat{p}\cdot\hat{A} = \frac{\hbar}{\mathrm{i}}\sum_{k=1}^{3}\frac{\partial}{\partial x^k}\hat{A}_k\cdot = \hat{A}\cdot\hat{p} - \mathrm{i}\hbar\,\mathrm{div}\hat{A}. \tag{7.6.3}$$

The Hamiltonian operator (7.6.1) is thus found to read as

$$\hat{H} = \frac{\hat{p}^2}{2m_\mathrm{e}} + \frac{e_0}{m_\mathrm{e}c}\hat{A}\cdot\hat{p} - \frac{\mathrm{i}e_0\hbar}{2m_\mathrm{e}c}\mathrm{div}\hat{A} + \frac{e_0^2}{2m_\mathrm{e}c^2}\hat{A}^2 - e_0\hat{V} + \hat{U}. \tag{7.6.4}$$

Now we assume, for simplicity, that the scalar potential vanishes:

$$\hat{V} = 0. \tag{7.6.5}$$

Moreover, since the external magnetic field is generated by bodies of macroscopic dimensions, it can be taken to be uniform, in a first approximation, over distances of the order of atomic dimensions: $\vec{B}(\vec{x},t) = \vec{B}_0$. The form (7.6.4) suggests choosing the Coulomb gauge to further simplify the calculations:

$$\mathrm{div}\hat{A} = 0. \tag{7.6.6}$$

The form of the vector potential compatible with the above assumptions is thus found to be

$$\vec{A} = \frac{1}{2}\vec{B}_0 \wedge \vec{x}. \tag{7.6.7}$$

Its insertion into (7.6.4) makes it necessary to derive the identity

$$\left(\vec{B}_0 \wedge \vec{x}\right)\cdot\vec{p} = \varepsilon_{klm}B_l x_m p_k = -B_l\varepsilon_{lkm}x_m p_k = B_l\varepsilon_{lmk}x_m p_k = B_l\left(\vec{x}\wedge\vec{p}\right)_l = \vec{B}_0\cdot\vec{L}. \tag{7.6.8}$$

The desired Hamiltonian operator therefore reads as

$$\hat{H} = \frac{\hat{p}^2}{2m_\mathrm{e}} + \frac{e_0}{2m_\mathrm{e}c}\vec{B}_0\cdot\vec{L} + \frac{e_0^2}{8m_\mathrm{e}c^2}\left(\vec{B}_0\wedge\vec{x}\right)^2 + \hat{U}(r). \tag{7.6.9}$$

In a first approximation we now neglect the term quadratic in $\vec{B}_0$ and we choose

$$\vec{B}_0 = (0, 0, B_0), \tag{7.6.10}$$

i.e. an external magnetic field directed along the z-axis. By virtue of our assumptions, the Hamiltonian operator reduces to

$$\hat{H} = \frac{\hat{p}^2}{2m_e} + \hat{U}(r) + \frac{e_0}{2m_e c} B_0 \hat{L}_z = \hat{H}_0 + \frac{e_0}{2m_e c} B_0 \hat{L}_z. \tag{7.6.11}$$

The operators $\hat{H}_0, \hat{L}^2$ and $\hat{L}_z$ have common eigenvectors, here denoted by u_{nlm}, with eigenvalue equations

$$\hat{H}_0 u_{nlm} = W_{nl}^{(0)} u_{nlm}, \tag{7.6.12}$$

$$\hat{L}^2 u_{nlm} = l(l+1)\hbar^2 u_{nlm}, \tag{7.6.13}$$

$$\hat{L}_z u_{nlm} = m\hbar u_{nlm}. \tag{7.6.14}$$

The resulting eigenvalue equation for the full Hamiltonian is

$$\hat{H} u_{nlm} = \left(W_{nl}^{(0)} + m\mu_B B_0 \right) u_{nlm}. \tag{7.6.15a}$$

Thus, on neglecting contributions quadratic in B_0, the external magnetic field does not affect the eigenfunctions but modifies the eigenvalues, which are given by

$$W_{nlm} = W_{nl}^{(0)} + m\mu_B B_0. \tag{7.6.15b}$$

This means that the original invariance under rotations has been spoiled by the magnetic field, which has introduced a privileged direction. Each 'unperturbed' energy level $W_{nl}^{(0)}$ is therefore split into $2l+1$ distinct components. For example, in the transition $(n, l, m) \rightarrow (n', l', m')$ one now has the frequency

$$\begin{aligned} \nu_{nl \rightarrow n'l'} &= \frac{W_{nlm} - W_{n'l'm'}}{h} \\ &= \nu_{nl \rightarrow n'l'}^{(0)} - (\delta m) \frac{\mu_B B_0}{h}. \end{aligned} \tag{7.6.16}$$

For $(\delta m) = 0, \pm 1$ this leads to the three spectral lines

$$\nu_{nl \rightarrow n'l'}^{(0)} - \frac{\mu_B B_0}{h}, \ \nu_{nl \rightarrow n'l'}^{(0)}, \ \nu_{nl \rightarrow n'l'}^{(0)} + \frac{\mu_B B_0}{h}.$$

Such a theoretical model provides results in good agreement with observation only if the intensity of the magnetic field is so high that fine-structure corrections are negligible. The latter result from a relativistic evaluation of the Hamiltonian operator in a central potential, and lead to energy levels depending both on n and l. For example, fine-structure corrections cannot be neglected in the analysis of the D_1 and D_2 spectral lines of sodium, which are split into six and four components, respectively.

The so-called *normal Zeeman effect* inherits its name from the possibility of obtaining a classical description of the effect, resulting from the precession of the orbital angular momentum associated with the electron while it moves around the nucleus (cf. section 6.1). The *anomalous Zeeman effect* is observed in alkali metals under the influence of weak magnetic fields, and does not admit a classical interpretation. The transition from the anomalous to the normal Zeeman effect as the intensity of the external magnetic field is increased is called the *Paschen–Back effect* (Paschen and Back 1912, 1913).

7.7 Variational method

The variational method can be used to evaluate to high accuracy the lowest eigenvalues of the Hamiltonian operator, and hence deserves a brief description in our chapter. At a deeper level, it is possible to consider the perturbation theory of self-adjoint families of operators as a special case of the variational method (Harrell 1977), but we shall not be concerned with such advanced aspects.

The method relies on the following theorem: given a normalized vector ψ of a Hilbert space $\mathcal{H}$, the mean value of the Hamiltonian operator (taken to be self-adjoint on a suitable domain, or essentially self-adjoint) in the state ψ satisfies the inequality

$$(\psi, H\psi) \geq E_0, \tag{7.7.1}$$

where E_0 is the lowest eigenvalue of H. The proof (we assume, for simplicity, that H has a purely discrete spectrum) is obtained by considering a complete orthonormal set of eigenvectors of H, here denoted by $\{\varphi_n\}$:

$$H\varphi_n = E_n\varphi_n, \tag{7.7.2}$$

and then expanding ψ in terms of φ_n: $\psi = \sum_{n=0}^{\infty} a_n\varphi_n$. Bearing in mind that ψ is normalized, one then finds that

$$(\psi, H\psi) = \sum_{n=0}^{\infty} |a_n|^2 E_n \geq \sum_{n=0}^{\infty} |a_n|^2 E_0 = E_0,$$

i.e. the inequality (7.7.1). The variational method aims to find the minimum of $(\psi, H\psi)$ with $\psi \in \mathcal{H}$, and this is achieved by introducing a number of unknown parameters. If the minimum found in this way is the absolute minimum, the problem is solved and one has

$$E_0 = \min(\psi, H\psi). \tag{7.7.3}$$

The search for the minimum of the functional $(\psi, H\psi)$, subject to the normalization condition

$$(\psi, \psi) = 1, \tag{7.7.4}$$

is performed by requiring that the following variation should vanish:

$$\delta\Big[(\psi, H\psi) - \lambda(\psi, \psi)\Big] = 0, \tag{7.7.5}$$

where λ is a Lagrange multiplier related to the constraint (7.7.4). This leads to the equation

$$(\delta\psi, H\psi) + (\psi, H\delta\psi) - \lambda(\delta\psi, \psi) - \lambda(\psi, \delta\psi) = 0. \tag{7.7.6}$$

Thus, if one assumes that the variation of the parameters occurring in ψ makes it possible to regard $\delta\psi$ and its dual as independent of each other, Eq. (7.7.6) leads to the two equations

$$(\delta\psi, H\psi) - \lambda(\delta\psi, \psi) = 0, \tag{7.7.7}$$

$$(\psi, H\delta\psi) - \lambda(\psi, \delta\psi) = 0, \tag{7.7.8}$$

and hence

$$H\psi = \lambda\psi. \tag{7.7.9}$$

Once ψ is expanded in terms of φ_n, the diagonalization of H in the basis $\{\varphi_n\}$ makes it possible to determine the coefficients a_n, which can be taken as the variational parameters. For practical purposes, only a finite number of terms can be taken into account in such an expansion, and hence, at least in principle, the test vector ψ is no longer able to span the whole Hilbert space $\mathcal{H}$. Nevertheless, for a careful choice of the resulting sub-space (in particular, the exact ground state should belong to the sub-space), the variational method turns out to be of practical utility.

It should be stressed that, in the search for the minimum of $(\psi, H\psi)$, the variational method provides an estimate of the eigenvalue which is more accurate than the estimate of the corresponding eigenfunction. In other words, if the error in the evaluation of $\delta\psi$ is taken to be of first order, this leads to a second-order error in the estimate of E_0. This is clearly seen by writing

$$E_0 = (\psi_0, H\psi_0), \tag{7.7.10}$$

with the corresponding error

$$E_\delta \equiv (\psi, H\psi) - E_0, \tag{7.7.11}$$

because the method has provided the approximate solution

$$\psi = \psi_0 + \delta\psi, \tag{7.7.12}$$

which should be inserted into the formula (7.7.11) for the error E_δ.

Once the exact eigenvalue E_0 for the ground state has been found, the variational method can be applied again to determine the energy of the first excited level. One has then to make sure, however, that only the part of the Hilbert space consisting of all vectors orthogonal to ψ_0 is being 'explored', so that in the resulting sub-space the lowest eigenvalue is indeed the first excited level of the Hamiltonian. To rule out ψ_0 from the test vector, it is enough to choose ψ and χ in such a way that

$$\psi = \chi - \psi_0(\psi_0, \chi). \tag{7.7.13}$$

This test state vector is, by construction, orthogonal to ψ_0, because

$$(\psi_0, \psi) = (\psi_0, \chi) - (\psi_0, \psi_0)(\psi_0, \chi) = (\psi_0, \chi) - (\psi_0, \chi) = 0. \tag{7.7.14}$$

As an example, let us consider the Hamiltonian of the one-dimensional harmonic oscillator:

$$H = -\frac{\hbar^2}{2m}\frac{\mathrm{d}^2}{\mathrm{d}x^2} + \frac{1}{2}m\omega^2 x^2. \tag{7.7.15}$$

With our choice of test function, we try to work with just one parameter, and we bear in mind that, as $x \rightarrow \pm\infty$, the wave function should approach zero. We are thus led to consider an even function of x of the kind

$$\psi_0(x, \alpha) = A \exp\left(-\frac{1}{2}\alpha x^2\right). \tag{7.7.16}$$

The normalization of ψ_0 determines $A = (\alpha/\pi)^{1/4}$, and the integral in (7.7.1) becomes in our case

$$J(\alpha) = \int_{-\infty}^{\infty} \psi_0^*(x, \alpha) H \psi_0(x, \alpha)\,\mathrm{d}x = \frac{1}{4}\left(\frac{\hbar^2\alpha}{m} + \frac{m\omega^2}{\alpha}\right), \tag{7.7.17}$$

where we have used the well-known result

$$\int_{-\infty}^{\infty} \mathrm{e}^{-\alpha x^2}\,\mathrm{d}x = \sqrt{\frac{\pi}{\alpha}}. \tag{7.7.18}$$

The minimum of J is obtained for $\alpha = \alpha_0 = m\omega/\hbar$, which implies

$$E_0 = J(\alpha_0) = \frac{\hbar\omega}{2}, \tag{7.7.19}$$

$$\psi_0(x, \alpha_0) = (m\omega/\pi\hbar)^{1/4} \exp\left(-\frac{m\omega x^2}{2\hbar}\right). \tag{7.7.20}$$

Note that the results (7.7.19) and (7.7.20) coincide with the exact values of the ground-state energy and ground-state wave function, respectively.

Remarkably, thanks to a result of Barnes *et al.* (1976), it is possible to also find a *lower limit* for the energy E_0 of the ground state (unlike the upper limit provided by the Rayleigh–Ritz method). To outline the result, it is necessary to recall the classical partition function (V being the potential for a neutral scalar particle in a static field)

$$I(\beta) \equiv \int \mathrm{e}^{-\beta V(x)} \, \mathrm{d}^n x \tag{7.7.21}$$

and the quantum partition function (cf. chapter 14)

$$Z(\beta) \equiv \mathrm{Tr} \mathrm{e}^{-\beta H} = \sum_{k=0}^{\infty} \left(\varphi_k, \mathrm{e}^{-\beta H} \varphi_k \right). \tag{7.7.22}$$

At this stage, starting from the relation between E_0 and $Z(\beta)$:

$$E_0 = - \lim_{\beta \to \infty} \frac{1}{\beta} \log Z(\beta), \tag{7.7.23}$$

one can prove the inequality

$$E_0 \geq \max_{\omega > 0} \left\{ \omega \left[n + \frac{n}{2} \log(\pi/\omega) - \log(I(\omega^{-1})) \right] \right\}, \tag{7.7.24}$$

after finding a suitable majorization for $Z(\beta)$.

7.8 Time-dependent formalism

The time-dependent formalism in perturbation theory is appropriate for the description of quantum systems that are weakly interacting with other physical systems, so that the full Hamiltonian reads

$$H = H_0 + \varepsilon V(t), \tag{7.8.1}$$

where H_0 is an essentially self-adjoint comparison Hamiltonian independent of time, V is a given function of time and ε is a small dimensionless parameter. For example, this may happen for an atom in an electromagnetic field. One is then mainly interested in the probability that the system, initially in an eigenstate of the unperturbed Hamiltonian H_0, performs a transition to any other eigenstate. Typically, these are the probabilities that an atom can receive or emit energy by virtue of the interaction with electromagnetic radiation. The mathematical description of such processes is as follows.

Let $\psi(t)$ be the state of the system at time t. In quantum mechanics, this solves the equation

$$\mathrm{i}\hbar \frac{\mathrm{d}}{\mathrm{d}t} \psi(t) = H(t) \psi(t), \tag{7.8.2}$$

for a given initial condition $\psi(t_0)$. The wave function $\psi(\vec{x}, t)$ studied so far is recovered from the state vector $\psi(t)$ in a way that becomes clear with the help of the Dirac notation defined in appendix 4.A. On denoting by $|x\rangle$ a generalized solution of the eigenvalue equation for the position operator $\hat{x}$, with associated bra written as $\langle x|$, and writing $|\ \rangle_t$ for the state vector $\psi(t)$ at time t, one has

$$\psi(\vec{x},t) = \langle x|\ \rangle_t = \int_{\mathbf{R}^3} \delta^3(\vec{x}-\vec{\xi})\psi(\vec{\xi},t)\, \mathrm{d}^3\xi. \tag{7.8.3}$$

The propagator $U(t,t_0)$ maps, by definition, $\psi(t_0)$ into $\psi(t)$:

$$\psi(t) = U(t,t_0)\psi(t_0), \tag{7.8.4}$$

and hence satisfies the equation

$$\mathrm{i}\hbar \frac{\mathrm{d}}{\mathrm{d}t} U(t,t_0) = H(t)U(t,t_0), \tag{7.8.5}$$

with initial condition $U(t_0,t_0) = 1$. Since the time-dependent part $\varepsilon V(t)$ in the Hamiltonian is supposed to be a 'small' perturbation of H_0 (see below), the idea is that U should differ by a small amount from the unperturbed propagator:

$$U_0(t,t_0) = U_0(t-t_0) = \exp\Big[-\mathrm{i}(t-t_0)H_0/\hbar\Big]. \tag{7.8.6}$$

One is thus led to set

$$U = U_0\, W, \tag{7.8.7}$$

where W is another unitary operator, which encodes all the effects resulting from the interaction. In other words, since the full Hamiltonian depends on time, the propagator $U(t,t_0)$ cannot depend on the difference $t-t_0$, but its evaluation is reduced to a series of 'corrections' of the unperturbed propagator $U_0(t,t_0) = U_0(t-t_0)$, thanks to the introduction of the unitary operator W. The insertion of Eq. (7.8.7) into Eq. (7.8.5) yields

$$\mathrm{i}\hbar \frac{\mathrm{d}}{\mathrm{d}t} W(t,t_0) = \varepsilon U_0^{\dagger}(t-t_0)V(t)U_0(t-t_0)W(t,t_0). \tag{7.8.8}$$

We now look for a solution in the form

$$W(t,t_0) \sim \sum_{n=0}^{\infty} \varepsilon^n W_n(t,t_0). \tag{7.8.9}$$

Once more, we should stress that we are dealing with formal series. The right-hand side may or may not converge for a given form of $V(t)$, and we are assuming that $W(t,t_0)$ has an asymptotic expansion. With this understanding, and defining

$$V_{\mathrm{int}}(t,t_0) \equiv U_0^{\dagger}(t,t_0)V(t)U_0(t,t_0), \tag{7.8.10}$$

one finds, $\forall n \geq 0$, a recursive set of differential equations

$$\mathrm{i}\hbar \frac{\mathrm{d}}{\mathrm{d}t} W_{n+1}(t,t_0) = V_{\mathrm{int}}(t,t_0)W_n(t,t_0). \tag{7.8.11}$$

The initial condition $W(t_0, t_0) = 1$ leads to the initial data

$$W_0(t_0, t_0) = 1, \tag{7.8.12}$$

$$W_n(t_0, t_0) = 0, \quad \forall n > 0, \tag{7.8.13}$$

and hence the various terms are given by

$$W_1(t, t_0) = -\frac{\mathrm{i}}{\hbar} \int_{t_0}^{t} V_{\text{int}}(\tau, t_0)\, \mathrm{d}\tau, \tag{7.8.14}$$

$$W_2(t, t_0) = -\frac{\mathrm{i}}{\hbar} \int_{t_0}^{t} V_{\text{int}}(\tau, t_0) W_1(\tau, t_0)\, \mathrm{d}\tau, \tag{7.8.15}$$

$$\vdots$$

$$W_{n+1}(t, t_0) = -\frac{\mathrm{i}}{\hbar} \int_{t_0}^{t} V_{\text{int}}(\tau, t_0) W_n(\tau, t_0)\, \mathrm{d}\tau. \tag{7.8.16}$$

We can now insert into each integral the form of W_n given by the previous equation. The resulting series is the Dyson series (Dyson 1949), with

$$W_2(t, t_0) = (-i/\hbar)^2 \int_{t_0}^{t} \int_{t_0}^{t_1} V_{\text{int}}(t_1, t_0) V_{\text{int}}(t_2, t_0)\, \mathrm{d}t_1\, \mathrm{d}t_2, \tag{7.8.17}$$

$$W_3(t, t_0) = (-\mathrm{i}/\hbar)^3 \times \int_{t_0}^{t} \int_{t_0}^{t_1} \int_{t_0}^{t_2} V_{\text{int}}(t_1, t_0) V_{\text{int}}(t_2, t_0) V_{\text{int}}(t_3, t_0)\, \mathrm{d}t_1\, \mathrm{d}t_2\, \mathrm{d}t_3, \tag{7.8.18}$$

and so on. Note that, in the multiple integral defining $W_n(t, t_0)$, the operators $V_{\text{int}}(t, t_0)$ are ordered chronologically in that, if $t > t'$, $V_{\text{int}}(t, t_0)$ is to the left of $V_{\text{int}}(t', t_0)$. Thus, on using the symbol of chronological ordering for which

$$TH(t_1)H(t_2)\cdots H(t_n) = \theta(t_1, t_2, \ldots, t_n) H(t_1)H(t_2)\cdots H(t_n), \tag{7.8.19}$$

with

$$\theta(t_1, t_2, \ldots, t_n) = 1 \quad \text{if } t_1 > t_2 > \cdots > t_n, \ 0 \text{ otherwise}, \tag{7.8.20}$$

one finds

$$W(t, t_0) = T \exp\left[-\frac{\mathrm{i}\varepsilon}{\hbar} \int_{t_0}^{t} V_{\text{int}}(\tau, t_0)\, \mathrm{d}\tau\right]. \tag{7.8.21}$$

Needless to say, the calculation of the right-hand side of Eq. (7.8.21) is, in general, quite cumbersome, despite the elegance of the formula.

7.8.1 Harmonic perturbations

We can now evaluate, approximately, the transition amplitude from an eigenstate φ_n of H_0 to another eigenstate φ_m under the action of the perturbation $\varepsilon V(t)$. To first order in ε, one has

$$A_{n\rightarrow m}(t,t_0) = \Big(\varphi_m, U(t,t_0)\varphi_n\Big) = \exp\Big[-\frac{\mathrm{i}}{\hbar}E_m(t-t_0)\Big]\Big(\varphi_m, W(t,t_0)\varphi_n\Big), \quad (7.8.22)$$

where

$$\Big(\varphi_m, W(t,t_0)\varphi_n\Big) \sim \delta_{nm} - \frac{\mathrm{i}\varepsilon}{\hbar}\int_{t_0}^{t}\Big(\varphi_m, V_{\rm int}(\tau,t_0)\varphi_n\Big)\,{\rm d}\tau + {\rm O}(\varepsilon^2)$$
$$= \delta_{nm} - \frac{\mathrm{i}\varepsilon}{\hbar}\int_{t_0}^{t} {\rm e}^{[(\mathrm{i}/\hbar)(E_m-E_n)(\tau-t_0)]}\Big(\varphi_m, V(\tau)\varphi_n\Big)\,{\rm d}\tau + {\rm O}(\varepsilon^2). \quad (7.8.23)$$

It is quite interesting to consider the case of *harmonic perturbations*, for which $V(t)$ takes the form

$$V(t) = B\,{\rm e}^{-\mathrm{i}\omega t} + B^\dagger\,{\rm e}^{\mathrm{i}\omega t}, \quad (7.8.24)$$

where B is a given operator in the Hilbert space of the problem. Note that, if one studies such a perturbation in a finite interval $(0,T)$ one finds, for $m \neq n$,

$$A_{n\rightarrow m}(T) = \Big(\varphi_m, \varepsilon B\varphi_n\Big)\frac{1-{\rm e}^{\mathrm{i}(\omega_{mn}-\omega)T}}{\hbar(\omega_{mn}-\omega)} + \Big(\varphi_m, \varepsilon B^\dagger\varphi_n\Big)\frac{1-{\rm e}^{\mathrm{i}(\omega_{mn}+\omega)T}}{\hbar(\omega_{mn}+\omega)}, \quad (7.8.25)$$

having defined $\omega_{mn} \equiv \frac{E_m-E_n}{\hbar}$. Interestingly, this shows that the perturbation may lead to transitions even between states with Bohr frequencies ω_{mn} that differ from the frequency ω in Eq. (7.8.24). However, the transition probability $|A_{n\rightarrow m}(T)|^2$ receives the dominant contribution either from the region $\omega_{mn} \cong \omega$, or from the region $\omega_{mn} \cong -\omega$. The former corresponds to *resonant absorption* (the quantum system receives energy from the external perturbation, and the final state has energy $E_m > E_n$), while the latter corresponds to *resonant emission*. It should be stressed that, when first-order effects provide a good approximation, one finds

$$\Big(\varphi_m, \varepsilon B\varphi_n\Big) \ll \hbar\omega \text{ for resonant absorption}, \quad (7.8.26)$$

$$\Big(\varphi_m, \varepsilon B^\dagger\varphi_n\Big) \ll \hbar\omega \text{ for resonant emission}. \quad (7.8.27)$$

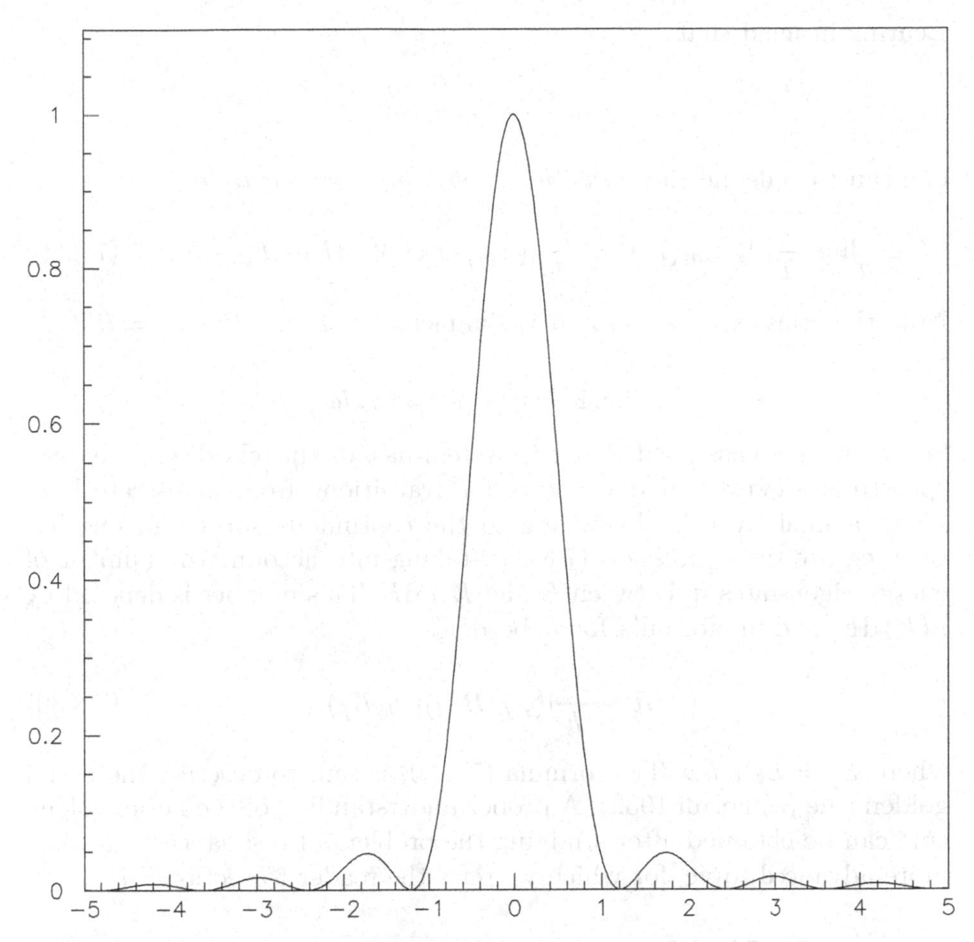

Fig. 7.1. The curve shows the behaviour of the transition probability P_{nm} as a function of ω at a fixed instant of time t in the presence of harmonic perturbations $V_0 e^{i\omega t}$. Such a transition probability is substantially different from zero when the energy E_m of the final state lies in a narrow neighbourhood of $(E_n \pm \hbar\omega)$. When a resonance occurs, P_{nm} attains a maximum equal to $\frac{|(\varphi_m, V_0 \varphi_n)|^2}{\hbar^2}$. When the difference $|\omega - \omega_{mn}|$ increases, P_{nm} oscillates in between 0 and the value $\frac{4|(\varphi_m, V_0 \varphi_n)|^2}{\hbar^2 (\omega - \omega_{mn})^2}$.

In the case of resonant absorption one can keep only the first term on the right-hand side of Eq. (7.8.25), and this leads to the following formula for the transition probability (see figure 7.1):

$$|A_{n \to m}(T)|^2 \sim \frac{|(\varphi_m, \varepsilon B \varphi_n)|^2}{\hbar^2} \left\{ \frac{\sin[(\omega - \omega_{mn})T/2]}{(\omega - \omega_{mn})/2} \right\}^2 . \tag{7.8.28}$$

Bearing in mind that

$$\lim_{\tau\to\infty} \frac{\sin^2 \alpha\tau}{\alpha^2\tau} = \pi\delta(\alpha),$$

one can then define the *transition probability per unit time*:

$$R \equiv \lim_{T\to\infty} \frac{1}{T}|A_{n\to m}(T)|^2 \sim \frac{2\pi}{\hbar}|(\varphi_m, \varepsilon B\varphi_n)|^2\delta(\hbar\omega - E_m + E_n). \quad (7.8.29)$$

Note that this expression for R is symmetric in T, i.e. $R(-T) = R(T)$.

7.8.2 Fermi golden rule

So far, we have assumed that the system has an entirely discrete energy spectrum. However, if one allows for transitions from a discrete level E_i to a final level E_f belonging to the continuous spectrum, one has to integrate the expression (7.8.29), taking into account the number of energy eigenstates in between E and $E + \mathrm{d}E$. This number is denoted by $n(E)\,\mathrm{d}E$, and the formula for R becomes

$$R \sim \frac{2\pi}{\hbar}|(\varphi_f, B\varphi_i)|^2 n(E_f), \quad (7.8.30)$$

where $E_f = E_i + \hbar\omega$. The formula (7.8.30) is said to describe the Fermi golden rule (cf. Fermi 1932). A proper understanding of the Fermi golden rule can be obtained after studying the problem of resonances. This is a more advanced topic, for which we refer the reader to section 8.7.

7.9 Limiting cases of time-dependent theory

If the eigenvalue problem for the unperturbed Hamiltonian H_0 is completely solved, with eigenvectors φ_n and eigenvalues $E_n^{(0)}$, the previous calculations are equivalent to expanding the state vector $\psi(t)$ for the full Hamiltonian H in the form (assuming, for simplicity, a purely discrete spectrum of H_0)

$$\psi(t) = \sum_{n=1}^{\infty} c_n(t) \mathrm{e}^{-\mathrm{i}E_n^{(0)}t/\hbar}\varphi_n, \quad (7.9.1)$$

with coefficients c_n representing the 'weight' at time t with which the stationary state φ_n contributes to the superposition that makes it possible to build $\psi(t)$. Such coefficients solve the system

$$\mathrm{i}\hbar\frac{\mathrm{d}c_n(t)}{\mathrm{d}t} = \sum_{m=1}^{\infty} c_m(t)\Big(\varphi_n, \varepsilon V_{\mathrm{int}}(t)\varphi_m\Big), \quad (7.9.2)$$

and hence are given by

$$c_n(t) = c_n(0) - \frac{\mathrm{i}}{\hbar} \sum_{m=1}^{\infty} \int_0^t c_m(t') \Big(\varphi_n, \varepsilon V_{\text{int}}(t') \varphi_m \Big) \, \mathrm{d}t'. \tag{7.9.3}$$

The Hamiltonians H_0 and H are here taken to be self-adjoint on the same domain, with the perturbation modifying the expansion of ψ valid for the stationary theory and breaking the invariance under time translations. The recurrence formulae for the various terms in the operator W lead to the following perturbative scheme for the evaluation of $c_n(t)$:

$$c_n(t) = c_n^{(0)} + c_n^{(1)}(t) + c_n^{(2)}(t) + \cdots, \tag{7.9.4}$$

where, for all $r = 0, 1, 2, \ldots$, one has

$$c_n^{(r+1)}(t) = -\frac{\mathrm{i}}{\hbar} \sum_{m=1}^{\infty} \int_0^t c_m^{(r)}(t') \Big(\varphi_n, \varepsilon V_{\text{int}}(t') \varphi_m \Big) \, \mathrm{d}t'. \tag{7.9.5}$$

According to the scheme outlined in section 7.8, if the system is initially in the state φ_i, the quantity

$$P_{if}(t) = |c_f(t)|^2 \tag{7.9.6}$$

can be interpreted as the probability that the system performs the transition from φ_i to the state φ_f as a result of the perturbation switched on in between the instants of time 0 and t. One can obtain a picture of the transition from the state φ_i to the state φ_f by considering the probability amplitude at the various orders of the perturbation expansion. The diagram in figure 7.2 represents the amplitude $c_f(t) \exp[-iE_f^{(0)} t/\hbar]$ relative to the state φ_f in the superposition (7.9.1), reached at time t and to first order according to the formula

$$c_f^{(1)}(t) = -\frac{\mathrm{i}}{\hbar} c_i^{(0)} \int_0^t \Big(\varphi_f, \varepsilon V_{\text{int}}(t') \varphi_i \Big) \, \mathrm{d}t'. \tag{7.9.7}$$

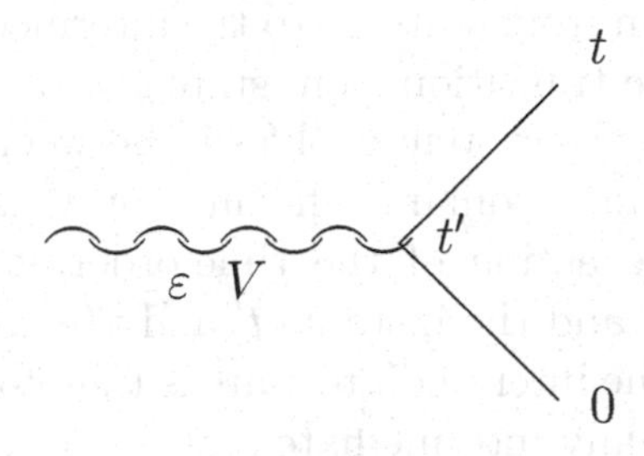

Fig. 7.2 Transition amplitude to first order in perturbation theory.

The segments in figure 7.2 describe the temporal evolution of the system according to the unperturbed Hamiltonian H_0: from time 0 to time t', with t' in between 0 and t, the system remains in state φ_i, and its state vector is simply multiplied by the phase factor $\mathrm{e}^{-\mathrm{i}E_i^{(0)}t/\hbar}$. At time t' the perturbation $\varepsilon V(t')$ leads to the transition from the state φ_i to the state φ_f, hence the matrix element $\left(\varphi_f, \varepsilon V_{\mathrm{int}} \varphi_i\right)$ and the factor $-\frac{\mathrm{i}}{\hbar}$. Lastly, the system evolves towards the final state φ_f from the instant t' to the instant t according to the unperturbed Hamiltonian H_0. This yields an 'evolution factor' $\mathrm{e}^{-\mathrm{i}E_f^{(0)}(t-t')/\hbar}$. Since the instant t' is a generic instant of time in between 0 and t, one has to 'sum' over all possible values of t'. We therefore consider

$$
\begin{aligned}
& a_f^{(1)}(t)\mathrm{e}^{-\mathrm{i}E_f^{(0)}t/\hbar} \\
& \quad = -\frac{\mathrm{i}}{\hbar}\int_0^t \mathrm{e}^{-\mathrm{i}E_f^{(0)}(t-t')/\hbar}\left(\varphi_f, \varepsilon V(t')\varphi_i\right)\mathrm{e}^{-\mathrm{i}E_i^{(0)}t'/\hbar}\,\mathrm{d}t',
\end{aligned}
$$

which is equal to

$$
-\frac{\mathrm{i}}{\hbar}\int_0^t \left(\varphi_f, \varepsilon V_{\mathrm{int}}(t')\varphi_i\right)\mathrm{e}^{-\mathrm{i}E_f^{(0)}t/\hbar}\,\mathrm{d}t'.
$$

In the same way, the diagram in figure 7.3 represents the same amplitude evaluated at second order by using the formula

$$
\begin{aligned}
a_f^{(2)}(t) = {} & \frac{1}{2!}(-\mathrm{i}/\hbar)^2 \sum_{m=1}^{\infty} \int_0^t \mathrm{d}t' \int_0^t \mathrm{d}t'' \\
& T\left(\varphi_f, \varepsilon V_{\mathrm{int}}(t')\varphi_m\right)\left(\varphi_m, \varepsilon V_{\mathrm{int}}(t'')\varphi_i\right).
\end{aligned}
\tag{7.9.8}
$$

In this case, two interactions with the perturbation occur, at the instant t'' and at the instant $t' \geq t''$. The first interaction is responsible for the transition of the system from state φ_i to the intermediate state φ_m, while the second leads to the transition from state φ_m to the final state φ_f. In the intermediate state the system evolves in between t'' and t' according to the unperturbed Hamiltonian. The instant t' is always subsequent to t'' by virtue of the action of the time-ordering operator. Both the intermediate state φ_m and the instants t' and t'' should be summed over in all possible ways. The intermediate state is therefore a *virtual state*, i.e. one of the infinitely many intermediate states through which the system can pass on its way towards the final state φ_f.

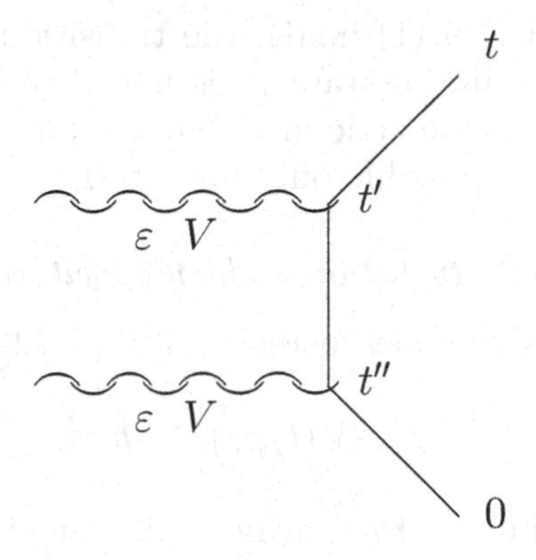

Fig. 7.3 Transition amplitude to second order in perturbation theory.

In the first-order formula for the transition probability (see Eq. (7.8.23))

$$P_{if}(t) = \frac{1}{\hbar^2}\left|\int_0^t \Big(\varphi_f, \varepsilon V_{\rm int}(t')\varphi_i\Big)\,{\rm d}t'\right|^2$$
$$= \frac{1}{\hbar^2}\left|\int_0^t \Big(\varphi_f, \varepsilon V(t')\varphi_i\Big){\rm e}^{{\rm i}\omega_{fi}t'}\,{\rm d}t'\right|^2, \qquad (7.9.9)$$

if the perturbation has a finite duration, i.e. $V(t) \neq 0$ only if $t \in]0, \tau[$, one can integrate by parts to find

$$P_{if}(\tau) = \frac{1}{\hbar^2\omega_{fi}^2}\left|\int_0^\tau {\rm e}^{{\rm i}\omega_{fi}t}\frac{\rm d}{{\rm d}t}\Big(\varphi_f, \varepsilon V(t)\varphi_i\Big)\,{\rm d}t\right|^2. \qquad (7.9.10)$$

The following limiting cases turn out to be of physical interest.

7.9.1 Adiabatic switch on and off of the perturbation

The variation of interaction energy during an oscillation period of the system is small with respect to the energy jump in between initial and final state, i.e.

$$\left|\frac{\rm d}{{\rm d}t}\Big(\varphi_f, \varepsilon V(t)\varphi_i\Big)\right| \ll \hbar\omega_{fi}^2. \qquad (7.9.11)$$

Since the time derivative of the matrix element of the perturbation remains basically constant during the interval $]0, \tau[$, the factor ${\rm e}^{{\rm i}\omega_{fi}t}$ in the integrand is the only important one, and hence one finds

$$P_{if}(\tau) = \frac{4}{\hbar^2\omega_{fi}^4}\left|\frac{\rm d}{{\rm d}t}\Big(\varphi_f, \varepsilon V(t)\varphi_i\Big)\right|^2 \sin^2\left(\frac{1}{2}\omega_{fi}\tau\right). \qquad (7.9.12)$$

Thus, if the condition (7.9.11) holds, the transition probability is much smaller than 1, i.e. the initial state φ_i is not abandoned after a time τ. This expression is also symmetric in τ, but the physical interpretation of it as a transition rate is possible only for $\tau > 0$.

7.9.2 Perturbation suddenly switched on

Under such conditions, one has instead (cf. Eq. (7.9.11))

$$\left| \frac{\mathrm{d}}{\mathrm{d}t} \Bigl(\varphi_f, \varepsilon V(t) \varphi_i \Bigr) \right| \gg \hbar \omega_{fi}^2. \tag{7.9.13}$$

The dominant contribution to the integral occurring in (7.9.10) is obtained from the integrand when the perturbation is switched on. If V_{fi} is the peak value of the matrix element of the perturbation, one obtains

$$P_{if}(\tau) = \frac{1}{\hbar^2 \omega_{fi}^2} |V_{fi}|^2. \tag{7.9.14}$$

It should be stressed that we are working under the assumption that first-order perturbation theory can be successfully applied. For example, for a harmonic perturbation

$$V(t) \equiv V_0 e^{i\omega t}, \tag{7.9.15}$$

this means that (cf. Eq. (7.8.26))

$$\left| \Bigl(\varphi_f, V_0 \varphi_i \Bigr) \right| \ll \hbar \omega_{fi}. \tag{7.9.16}$$

7.9.3 Two-level system

For a two-level system, however, the perturbation

$$V(t) \equiv V_0(x) \Bigl[\mathrm{e}^{\mathrm{i}\omega t} \varphi_1(\varphi_2, \cdot) + \mathrm{e}^{-\mathrm{i}\omega t} \varphi_2(\varphi_1, \cdot) \Bigr] \tag{7.9.17}$$

can be treated exactly. In such a case, only two eigenvectors of the unperturbed Hamiltonian exist, for which

$$H_0 \varphi_1 = E_1^{(0)} \varphi_1, \tag{7.9.18a}$$

$$H_0 \varphi_2 = E_2^{(0)} \varphi_2, \tag{7.9.18b}$$

and the expansion (7.9.1) reduces to

$$\psi(t) = c_1(t) \mathrm{e}^{-\mathrm{i}E_1^{(0)} t/\hbar} \varphi_1 + c_2(t) \mathrm{e}^{-\mathrm{i}E_2^{(0)} t/\hbar} \varphi_2. \tag{7.9.19}$$

Its insertion into the Schrödinger equation leads to the system

$$\mathrm{i}\hbar \frac{\mathrm{d}c_1(t)}{\mathrm{d}t} = \varepsilon \gamma_{11} c_2(t) \mathrm{e}^{\mathrm{i}(\delta\omega)t} + \varepsilon \gamma_{12} c_1(t) \mathrm{e}^{-\mathrm{i}\omega t}, \tag{7.9.20}$$

$$i\hbar \frac{dc_2(t)}{dt} = \varepsilon \gamma_{12}^* c_2(t) e^{i\omega t} + \varepsilon \gamma_{22} c_1(t) e^{-i(\delta\omega)t}, \tag{7.9.21}$$

where $\gamma_{\alpha\beta} \equiv \left(\varphi_\alpha, V_0 \varphi_\beta\right)$ (which reduces to $\gamma \delta_{\alpha\beta}$ if V_0 = constant γ), $(\delta\omega) \equiv \omega - \omega_0$ and ω_0 is defined by $\hbar\omega_0 \equiv E_2^{(0)} - E_1^{(0)}$. The solution of such a system of equations becomes easier if the frequency of the perturbation is close to the proper frequency of the system, i.e. if

$$|(\delta\omega)| \ll \omega_0. \tag{7.9.22}$$

In such a case the terms oscillating with frequency ω have vanishing average over time intervals comparable with the oscillation period $2\tau = \frac{2\pi}{\omega}$, and hence can be neglected with respect to the terms oscillating slowly with frequency $(\delta\omega)$. Thus, on defining the variables

$$a_{1,2}(t) \equiv \frac{1}{2\tau} \int_{t-\tau}^{t+\tau} c_{1,2}(t') \, dt', \tag{7.9.23}$$

one finds the differential equations

$$i\hbar \frac{da_1(t)}{dt} = \varepsilon \gamma_{11} a_2(t) e^{i(\delta\omega)t}, \tag{7.9.24}$$

$$i\hbar \frac{da_2(t)}{dt} = \varepsilon \gamma_{22} a_1(t) e^{-i(\delta\omega)t}, \tag{7.9.25}$$

which imply

$$\left[\frac{d^2}{dt^2} - i(\delta\omega)\frac{d}{dt} + \Omega^2\right] a_1(t) = 0, \tag{7.9.26}$$

$$\left[\frac{d^2}{dt^2} + i(\delta\omega)\frac{d}{dt} + \Omega^2\right] a_2(t) = 0, \tag{7.9.27}$$

having defined

$$\hbar^2 \Omega^2 \equiv \varepsilon^2 \gamma_{11} \gamma_{22}. \tag{7.9.28}$$

The solutions of Eqs. (7.9.26) and (7.9.27) are therefore of the type

$$a_1(t) = e^{i(\delta\omega)t/2} \left(A \cos\frac{\alpha t}{2} + B \sin\frac{\alpha t}{2}\right), \tag{7.9.29}$$

$$a_2(t) = e^{-i(\delta\omega)t/2} \left(C \cos\frac{\alpha t}{2} + D \sin\frac{\alpha t}{2}\right), \tag{7.9.30}$$

where

$$\alpha \equiv \sqrt{(\delta\omega)^2 + 4\Omega^2}. \tag{7.9.31}$$

If V_0 reduces to a constant γ as we mentioned earlier, so that $\gamma_{11} = \gamma_{22} = \gamma$, and if the system is initially in its ground state, for which

$$a_1(0) = 1, \quad a_2(0) = 0, \tag{7.9.32}$$

one has

$$A = 1, \quad C = 0. \tag{7.9.33}$$

On inserting these solutions into the first-order system one obtains

$$B = -\mathrm{i}\frac{(\delta\omega)}{\alpha}, \quad D = -2\mathrm{i}\frac{\gamma}{\hbar\alpha}. \tag{7.9.34}$$

The probability of finding the system in its excited state is therefore

$$|a_2(t)|^2 = \frac{4\Omega^2}{[(\delta\omega)^2 + 4\Omega^2]} \sin^2\left(\frac{\alpha t}{2}\right), \tag{7.9.35}$$

while the probability of again finding the system in the ground state is given by

$$|a_1(t)|^2 = \cos^2\left(\frac{\alpha t}{2}\right) + \frac{(\delta\omega)^2}{[(\delta\omega)^2 + 4\Omega^2]} \sin^2\left(\frac{\alpha t}{2}\right). \tag{7.9.36}$$

The evolution in time of $|a_2(t)|^2$ exhibits a clearly visible maximum in the limiting case for which $(\delta\omega) \rightarrow 0$. This is therefore a typical resonating behaviour. As time passes, the system oscillates in between the ground state and the excited state with frequency $\frac{\alpha}{2\pi}$.

7.10 The nature of perturbative series

The careful reader might have been worrying about the nature of the perturbative expansions used so far. Indeed, at least four cases may occur.

7.10.1 Regular perturbation theory

Let H_0 be the unperturbed Hamiltonian, which is assumed to admit a domain of self-adjointness $D(H_0)$. Moreover, let V be the perturbation potential, with domain $D(V)$ including $D(H_0)$, and let us define the Hamiltonian $H(\beta) \equiv H_0 + \beta V$ with domain $D(H_0) \cap D(V)$. The series for the eigenvalues and the eigenfunctions are convergent Taylor series (Kato 1995) provided that V (and hence βV) is bounded with respect to H_0. This means there exist some real parameters a and b such that

$$\|V\varphi\| \leq a\,\|H_0\varphi\| + b\,\|\varphi\| \quad \forall \varphi \in D(H_0). \tag{7.10.1}$$

More precisely, for H_0 a closed operator with non-empty resolvent set, the condition $D(H_0) \subset D(V)$, jointly with the H_0-boundedness property for V expressed by (7.10.1), is a necessary and sufficient condition to realize $H(\beta)$ as an *analytic family of type (A)*. This means that the domain of $H(\beta)$ is independent of β, and that, $\forall \psi \in D(H_0) \cap D(V)$, $H(\beta)\psi$ is a vector-valued analytic function of β. Remarkably, one can then prove that the isolated non-degenerate eigenvalue $E(\beta)$ of $H(\beta)$ is analytic near $\beta = 0$ (Reed and Simon 1978, Kato 1995). The Kato–Rellich theorem ensures that, if H_0 is self-adjoint, and if βV is symmetric and bounded with respect to H_0 with $a < 1$, then $H_0 + \beta V$ is self-adjoint on $D(H_0)$.

7.10.2 Asymptotic perturbation theory

It may happen that the series for the eigenvalues provides only an asymptotic expansion. For example, this is the case for a perturbative evaluation of the ground-state energy of the anharmonic oscillator, with the Hamiltonian (cf. Kunihiro 1998)

$$H \equiv p^2 + x^2 + \beta x^4, \quad \beta > 0, \tag{7.10.2}$$

where dimensionless units are used for simplicity. To be more precise, we need to define what is a strongly asymptotic series for an analytic function. Following Reed and Simon (1978), we say that the function E, analytic in the sector

$$S_\beta \equiv \left\{ \beta : |\beta| \in]0, B[, \ |\arg\beta| < \frac{\pi}{2} + \epsilon \right\}, \tag{7.10.3}$$

obeys a strong asymptotic condition and has

$$\sum_{n=0}^{\infty} a_n \beta^n$$

as a *strongly asymptotic series* if there exist some constants C and σ such that

$$\left| E(\beta) - \sum_{n=0}^{N} a_n \beta^n \right| \leq C \sigma^{N+1} (N+1)! |\beta|^{N+1}, \tag{7.10.4}$$

for all N, and for all β in S_β. Interestingly, unlike the case of ordinary asymptotic expansions (cf. appendix 4.C), if the series $\sum_{n=0}^{\infty} a_n \beta^n$ is a strongly asymptotic series for two analytic functions f and g, then f and g do actually coincide: $f = g$. The energy levels for the Hamiltonian (7.10.2) obey a strong asymptotic condition.

A theorem of Watson (1912) makes it possible to obtain a powerful summability method (see below). According to such a theorem, if $E(\beta)$ has $\sum_{n=0}^{\infty} a_n \beta^n$ as a strongly asymptotic series in the sector

$$\widetilde{S}_\beta \equiv \left\{ \beta : |\beta| \in]0, B[, \ |\arg\beta| \leq \frac{\pi}{2} + \epsilon \right\}, \tag{7.10.5}$$

then, on considering the function g defined by

$$g(z) \equiv \sum_{n=0}^{\infty} \frac{a_n}{n!} z^n, \tag{7.10.6}$$

which, by virtue of the strongly asymptotic condition, is analytic in a circle around $z = 0$, the following properties hold:

(1) g has an analytic continuation to the region

$$\{z : |\arg z| < \epsilon\}.$$

(2) If $|\beta| < B$ and $|\arg \beta| < \epsilon$, then

$$\int_0^{\infty} |g(\beta x)| \mathrm{e}^{-x} \, \mathrm{d}x < \infty. \tag{7.10.7}$$

(3) If $|\beta| < B$ and $|\arg \beta| < \epsilon$, then

$$E(\beta) = \int_0^{\infty} g(\beta x) \mathrm{e}^{-x} \, \mathrm{d}x. \tag{7.10.8}$$

This is the precise meaning of the statement according to which a finite result is obtained from a divergent series. The function g is called the Borel transform of the sequence $\{a_n\}_{n=0}^{\infty}$, and the right-hand side of (7.10.8) is the corresponding *inverse Borel transform.*

When the Hamiltonian (7.10.2) is considered, the inverse Borel transform converges to E for any β with a positive real part. However, if one studies the perturbed Hamiltonian

$$\widetilde{H} = p^2 + x^2 + \beta x^{2m}, \tag{7.10.9}$$

one has to consider a modified Borel transform, i.e.

$$g(z) \equiv \sum_{n=0}^{\infty} \frac{a_n}{[n(m-1)]!} z^n. \tag{7.10.10}$$

If β is positive and small, one then finds (cf. Eq. (7.10.8))

$$E(\beta) = \int_0^{\infty} g(\beta x^{m-1}) \mathrm{e}^{-x} \, \mathrm{d}x. \tag{7.10.11}$$

7.10.3 Spectral concentration

It may happen that, to an isolated unperturbed eigenvalue, there corresponds a perturbative series which is finite term by term, but *with no perturbed eigenvalue.* This happens if all the eigenvalues of H_0 are drowned into the continuous spectrum when the perturbation βV is turned on. A relevant example is provided by the Stark effect studied in section

7.5 (Graffi and Grecchi 1978). To spell out what we mean, we first define pseudo-eigenvalues and pseudo-eigenvectors. For this purpose, one considers a family $H(\beta)$ of self-adjoint operators defined for β real and small, so that $H(\beta) \rightarrow H_0$ as $\beta \rightarrow 0$ in the strong resolvent sense (see appendix 7.A). Let E_0 be an isolated non-degenerate eigenvalue of H_0, with normalized eigenvector ψ_0. One then says that a family of vectors $\psi(\beta), \beta \in \mathbf{R}$, and numbers $E_0 + \beta E_1$ is a *first-order pseudo-eigenvector* and a *first-order pseudo-eigenvalue*, respectively, if and only if

$$\lim_{\beta \to 0} \| \psi(\beta) - \psi_0 \| = 0, \tag{7.10.12}$$

$$\lim_{\beta \to 0} \frac{1}{\beta} \| (H(\beta) - E_0 - \beta E_1) \psi(\beta) \| = 0. \tag{7.10.13}$$

In our calculations of section 7.5, the first eigenvalue above the ground state is four-fold degenerate, and splits into one two-dimensional pseudo-eigenvalue and two one-dimensional pseudo-eigenvalues, to first order. Our calculations therein mean that we have found four linearly independent first-order pseudo-eigenvectors converging to vectors ψ_i satisfying $H_0 \psi_i = -\frac{1}{16} \psi_i$ (in suitable units).

In colloquial language, one can say that the spectrum of $H_0 + \beta V$ *bunches up* in the neighbourhood of the unperturbed eigenvalue, and that the centre of this spectral concentration is given by an asymptotic series (Reed and Simon 1978). To obtain a precise formulation of this concept, we need to state the following theorem (Reed and Simon 1978).

If the sequence of operators $H(\beta) \rightarrow H_0$ in a strong resolvent sense as $\beta \rightarrow 0$, with all the $H(\beta)$ self-adjoint, and if E_0 is an isolated non-degenerate eigenvalue of H_0, and I is an interval such that

$$\overline{I} \cap \sigma(H_0) = \{E_0\}, \tag{7.10.14}$$

then there exists a function f, obeying

$$\lim_{\beta \to 0} \frac{f(\beta)}{\beta} = 0, \tag{7.10.15}$$

so that part of the spectrum of $H(\beta)$ in I is asymptotically in the interval

$$I_\beta \equiv \Big(E_0 + \beta E_1 - f(\beta), E_0 + \beta E_1 + f(\beta) \Big) \tag{7.10.16}$$

if and only if $E_0 + \beta E_1$ is a first-order pseudo-eigenvalue for $H(\beta)$.

7.10.4 Singular perturbation theory

There exists a family of perturbations of the discrete spectrum that leaves the spectrum discrete but is more singular than all previous cases. A typical example is provided by the Hamiltonian $H_0 + \beta V$, where

$$H_0 \equiv -\frac{\mathrm{d}^2}{\mathrm{d}x^2} + x^2 \ \text{ on } \ \mathcal{L}^2(\mathbf{R}, \mathrm{d}x), \tag{7.10.17}$$

$$V(x) \equiv x^{-\alpha}. \tag{7.10.18}$$

If $\alpha > 1$, this family is discontinuous at $\beta = 0$. In other words, $H_0 + \beta V$ converges as $\beta \to 0$, in a strong resolvent sense, to an operator $\widetilde{H_0}$ different from H_0 (Esposito 2000, Klauder 2000).

If $H(\beta)$ is set equal to $\widetilde{H_0}$ when $\beta = 0$, the resulting family is analytic at $\beta = 0$ if $\alpha \in [1, 2[$. Moreover, if $\alpha \in [2, 3[$, the eigenvalues are given by asymptotic series to first order provided that $\beta > 0$. Furthermore one finds, for some $c \neq 0$ (Klauder 1973, Simon 1973a, De Facio and Hammer 1974, Klauder and Detwiler 1975, Harrell 1977),

$$E(\beta) - E(0) = c\beta \log(\beta) + \mathrm{O}(\beta) \quad \text{if} \ \ \alpha = 3, \tag{7.10.19}$$

$$E(\beta) - E(0) = c\beta^{1/(\alpha-2)} + \mathrm{o}(\beta^{1/(\alpha-2)}) \quad \text{if} \ \ \alpha > 3. \tag{7.10.20}$$

7.11 More about singular perturbations

In the literature on quantum-mechanical problems, an important role is played by the 'spiked' harmonic oscillator in three spatial dimensions. This is a system where the radial part $\psi(r)$ of the wave function is ruled by the Hamiltonian operator

$$\widetilde{H}(\alpha, \lambda) \equiv -\frac{\mathrm{d}^2}{\mathrm{d}r^2} + r^2 + \frac{l(l+1)}{r^2} + \frac{\lambda}{r^\alpha}, \tag{7.11.1}$$

in the sense that $\widetilde{H}(\alpha, \lambda)$ acts on $\varphi(r) \equiv r\psi(r)$. In the s-wave case (i.e. when the angular momentum quantum number l vanishes) $\widetilde{H}(\alpha, \lambda)$ reduces to

$$H(\alpha, \lambda) \equiv -\frac{\mathrm{d}^2}{\mathrm{d}r^2} + r^2 + \frac{\lambda}{r^\alpha} \equiv H(\alpha, 0) + \lambda V, \tag{7.11.2}$$

where $H(\alpha, 0)$ is formally equal to the simple harmonic oscillator Hamiltonian, r belongs to the interval $[0, \infty]$ and $V \equiv r^{-\alpha}$. For any fixed value of λ, the potential term in (7.11.2) diverges as $r \to 0$ in such a way that the operator $H(\alpha, \lambda)$ acts on wave functions that vanish at the origin:

$$\varphi(0) = 0. \tag{7.11.3}$$

More precisely, the imposition of Eq. (7.11.3) is necessary since not all functions in the domain of $H(\alpha,0)$ are in the domain of V. Thus, when $\lambda \rightarrow 0$ and α is fixed, the operator $H(\alpha,\lambda)$ converges to an operator formally equal to the unperturbed operator $H_0 \equiv -\frac{d^2}{dr^2}+r^2$, but supplemented by the Dirichlet boundary condition (7.11.3) for all functions in its domain. This means that the unperturbed operator H_0, for which the boundary condition (7.11.3) is not necessary, differs from the limiting operator $H(\alpha,0)$, for which Eq. (7.11.3) is instead necessary to characterize the domain.

The full potential in (7.11.1) or (7.11.2) inherits the name *spiked* from a pronounced peak near the origin for $\lambda > 0$, and its consideration is suggested by many concrete problems in chemical, nuclear and particle physics. Note now that for $\alpha > 2$ and integer, the Hamiltonian operators (7.11.1) or (7.11.2) lead to a non-Fuchsian singularity at $r = 0$ of the stationary Schrödinger equation, because their potential term has a pole of order > 2 therein. For α not integer the singularity is not a pole but a branch point. Subsection 7.11.1 outlines the method developed in Harrell (1977) for dealing with such singularities in a perturbative analysis. A non-trivial extension is studied in subsection 7.11.2, i.e. a model where the perturbation potential contains both an inverse power and a linear term. Concluding remarks are presented in subsection 7.11.3.

7.11.1 The Harrell method

The method developed by Harrell relies on the choice of suitable trial functions for self-adjoint operators (i.e. normalized vectors in their domain) and on the following lemma (Harrell 1977).

Lemma. If ψ_λ is a trial function for the self-adjoint operator $T + \lambda T'$, where both T and T' are self-adjoint and $E(0)$ is an isolated, non-degenerate stable eigenvalue of T, and $E(\lambda)$ is a continuous function such that the scalar product $\left(\psi_\lambda, [T+\lambda T' - E(\lambda)]\psi_\lambda\right)$ tends to 0 as λ tends to 0, and

$$\| [T+\lambda T' - E(\lambda)]\psi_\lambda \| = \mathrm{o}\left(\sqrt{(\psi_\lambda, [T+\lambda T' - E(\lambda)]\psi_\lambda)}\right), \qquad (7.11.4)$$

then the eigenvalue of $T+\lambda T'$ which converges to $E(0)$ is

$$E(\lambda) = \left(\psi_\lambda, [T+\lambda T']\psi_\lambda\right) + \mathrm{O}\left(\| [T+\lambda T' - E(\lambda)]\psi_\lambda \|^2\right). \qquad (7.11.5)$$

In a one-dimensional example, hereafter chosen for simplicity, let the full Hamiltonian be $H_0 + \lambda V$, where

$$H_0 \equiv -\frac{d^2}{dx^2} + x^2, \qquad (7.11.6)$$

and

$$V(x) \equiv x^{-\alpha}. \tag{7.11.7}$$

If $\alpha = 4$, on denoting by u_i the unperturbed eigenfunctions, a trial function can be chosen in the form

$$\psi(x) = W(x;\lambda)u_i(x), \tag{7.11.8}$$

where (inspired by the JWKB approximation)

$$W(x;\lambda) = N(\lambda)\exp\left(-\int_x^\infty \sqrt{\lambda V(\xi)}\,\mathrm{d}\xi\right) = N(\lambda)\exp\left(-\frac{\sqrt{\lambda}}{x}\right), \tag{7.11.9}$$

with $N(\lambda)$ a normalization factor that approaches 1 as λ tends to 0. One then finds the formula

$$[H_0 - E_i + \lambda V]Wu_i = 2W\frac{\sqrt{\lambda}}{x^2}\left(\frac{1}{x} - \frac{\mathrm{d}}{\mathrm{d}x}\right)u_i. \tag{7.11.10}$$

By virtue of Eq. (7.11.10) and of the Lemma at the beginning of this section one obtains the following formula for the energy eigenvalues of the spiked harmonic oscillator in one dimension with $\alpha = 4$:

$$E_i(\lambda) = E_i(0) + 2\left(u_i, x^{-2}\left(\frac{1}{x} - \frac{\mathrm{d}}{\mathrm{d}x}\right)u_i\right)\sqrt{\lambda} + \mathrm{O}(\lambda). \tag{7.11.11}$$

When $\alpha \neq 4$, one can use the identity (hereafter W_α replaces W)

$$\frac{\mathrm{d}^2}{\mathrm{d}x^2}(W_\alpha u_i) = W_\alpha u_i'' + 2W_\alpha' u_i' + W_\alpha'' u_i, \tag{7.11.12}$$

which implies

$$(H_0 - E_i + \lambda V)W_\alpha u_i = \left(-\frac{\mathrm{d}^2 W_\alpha}{\mathrm{d}x^2} - 2\frac{\mathrm{d}W_\alpha}{\mathrm{d}x}\frac{\mathrm{d}}{\mathrm{d}x} + \lambda V W_\alpha\right)u_i. \tag{7.11.13}$$

The idea is now to choose W_α in such a way that the action of $(H_0 - E_i + \lambda V)$ on $W_\alpha u_i$ again involves the action of the operator $\left(\frac{1}{x} - \frac{\mathrm{d}}{\mathrm{d}x}\right)$ on u_i (see Eq. (7.11.10)). For this purpose, Harrell imposed the differential equation

$$\left(\frac{\mathrm{d}^2}{\mathrm{d}x^2} + \frac{2}{x}\frac{\mathrm{d}}{\mathrm{d}x} - \frac{\lambda}{x^\alpha}\right)W_\alpha = 0, \tag{7.11.14}$$

so that Eq. (7.11.13) reduces indeed to (cf. Eq. (7.11.10))

$$(H_0 - E_i + \lambda V)W_\alpha u_i = 2\frac{\mathrm{d}W_\alpha}{\mathrm{d}x}\left(\frac{1}{x} - \frac{\mathrm{d}}{\mathrm{d}x}\right)u_i. \tag{7.11.15}$$

On defining

$$\nu \equiv \frac{1}{(\alpha - 2)}, \tag{7.11.16}$$

the solution of Eq. (7.11.14) can be expressed in the form

$$W_\alpha(x;\lambda) = \frac{2\nu^\nu \lambda^{\nu/2}}{\Gamma(\nu)} x^{-1/2} K_\nu(2\nu\sqrt{\lambda}x^{-1/2\nu}). \tag{7.11.17}$$

The energy eigenvalues of the spiked oscillator in one dimension with $\alpha \geq 4$ are then found to be

$$E_i(\lambda) = E_i(0) + 2\frac{\Gamma(1-\nu)}{\Gamma(1+\nu)}\nu^{2\nu}\left(u_i, x^{-2}\left(\frac{1}{x} - \frac{\mathrm{d}}{\mathrm{d}x}\right)u_i\right)\lambda^\nu + \mathrm{O}(\lambda^{2\nu}). \tag{7.11.18}$$

7.11.2 Extension to other singular potentials

When the perturbation potential is not an inverse power, Eq. (7.11.14) is replaced by an equation which cannot generally be solved. Harrell has however shown that, if V is bounded away from $x = 0$ and lies in between $x^{-\alpha}$ and $x^{-\beta}$, with $0 < \alpha < \beta$, then the effect of V on the eigenvalues is not essentially different.

It therefore appears to be of interest to study cases in which V is not (a pure) inverse power, but does not obey the restrictions considered in section 5 of Harrell (1977). For this purpose, we assume that (cf. Eq. (7.11.7))

$$V(x) \equiv x^{-\alpha} + \kappa x. \tag{7.11.19}$$

By doing so we study a model that reduces to the spiked oscillator in the neighbourhood of the origin, whereas at large x it approaches an oscillator perturbed by a 'Stark-like' term. The two terms in (7.11.19) are separately well studied in the literature, so that their joint effect provides a well-motivated departure from the scheme studied in Harrell (1977).

For this purpose we study an integral equation, the construction of which is as follows. On assuming the validity of Eq. (7.11.19) for the perturbation potential, Eq. (7.11.14) is replaced by the inhomogeneous equation

$$LW_\alpha(x;\lambda) = f_\alpha(x;\lambda), \tag{7.11.20}$$

where

$$L \equiv \frac{\mathrm{d}^2}{\mathrm{d}x^2} + \frac{2}{x}\frac{\mathrm{d}}{\mathrm{d}x} - \frac{\lambda}{x^\alpha}, \tag{7.11.21}$$

$$f_\alpha(x;\lambda) \equiv \lambda\kappa x W_\alpha(x;\lambda). \tag{7.11.22}$$

Since $V(x)$ approaches $x^{-\alpha}$ as $x \rightarrow 0$, we require again the boundary condition studied in Harrell (1977), i.e.

$$W_\alpha(0) = 0. \tag{7.11.23}$$

In our problem, we study one-dimensional Schrödinger operators with Dirichlet boundary conditions at 0 only on the space $L^2([0,\infty))$, to remove the degeneracy resulting from the decoupling of the two half-lines $(-\infty, 0)$ and $(0, \infty)$ (only odd eigenfunctions of the ordinary harmonic oscillator obey the Dirichlet condition at 0). Moreover, to be able to use all known results on one-dimensional boundary-value problems on closed intervals of the real line, we first work on the interval $(0, b)$ and then take the limit as $b \rightarrow \infty$. More precisely, we start from a problem for which the Green function $G(x, \xi; \lambda)$ satisfies the equation

$$LG = 0 \ \text{ for } x \in (0, \infty) \ \text{ and } \ \xi \in (0, \infty), \tag{7.11.24}$$

the boundary condition

$$G(0, \xi; \lambda) = 0, \tag{7.11.25}$$

the summability condition (in that the integral of G over a closed interval of values of λ yields a square-integrable function of x)

$$G(x, \xi; \lambda) \in L_2^{(c)}(0, \infty), \tag{7.11.26}$$

the continuity condition

$$\lim_{x \rightarrow \xi^+} G(x, \xi; \lambda) = \lim_{x \rightarrow \xi^-} G(x, \xi; \lambda) \tag{7.11.27}$$

and the jump condition

$$\lim_{x \rightarrow \xi^+} \frac{\partial G}{\partial x} - \lim_{x \rightarrow \xi^-} \frac{\partial G}{\partial x} = 1. \tag{7.11.28}$$

As one knows from the general theory, $G(x, \xi; \lambda)$ is recovered on first studying the Green function $G_b(x, \xi; \lambda)$ for a regular problem on the interval $(0, b)$, and then taking the limit as $b \rightarrow \infty$, i.e.

$$\lim_{b \rightarrow \infty} G_b(x, \xi; \lambda) = G(x, \xi; \lambda). \tag{7.11.29}$$

Such a relation holds independently of the boundary condition imposed on $G_b(x, \xi; \lambda)$ at $x = b$ (Stakgold 1979). With this understanding, the full solution of the inhomogeneous equation (7.11.20) with boundary condition (7.11.23) is given by (γ being a constant)

$$\begin{aligned} W_\alpha(x; \lambda) &= \gamma \frac{2\nu^\nu \lambda^{\nu/2}}{\Gamma(\nu)} x^{-1/2} K_\nu(2\nu\sqrt{\lambda} x^{-1/2\nu}) \\ &\quad + \lim_{b \rightarrow \infty} \lambda\kappa \int_0^b G_b(x, \xi; \lambda) \xi W_\alpha(\xi; \lambda) \, d\xi, \end{aligned} \tag{7.11.30}$$

where the first term on the right-hand side of (7.11.30) is the regular solution (7.11.17) of the homogeneous equation $LW_\alpha = 0$. The Green function $G_b(x,\xi;\lambda)$ has to obey the differential equation

$$LG_b = 0 \text{ for } x \in (0,\xi) \text{ and } x \in (\xi,b), \tag{7.11.31}$$

the homogeneous boundary conditions

$$G_b(0,\xi;\lambda) = 0, \quad G_b(b,\xi;\lambda) = 0, \tag{7.11.32}$$

the continuity condition

$$\lim_{x\to\xi+} G_b(x,\xi;\lambda) = \lim_{x\to\xi-} G_b(x,\xi;\lambda), \tag{7.11.33}$$

and the jump condition

$$\lim_{x\to\xi+} \frac{\partial G_b}{\partial x} - \lim_{x\to\xi-} \frac{\partial G_b}{\partial x} = 1. \tag{7.11.34}$$

To obtain the explicit form of $G_b(x,\xi;\lambda)$ one has to consider a non-trivial solution $u_0(x;\lambda)$ of the homogeneous equation $Lu = 0$ satisfying $u(0) = 0$, and a non-trivial solution $u_b(x;\lambda)$ of $Lu = 0$ satisfying $u(b) = 0$. By virtue of (7.11.31) and (7.11.32) one then finds

$$G_b(x,\xi;\lambda) = A(\xi;\lambda)u_0(x;\lambda) \quad \text{if } x \in (0,\xi), \tag{7.11.35}$$

$$G_b(x,\xi;\lambda) = B(\xi;\lambda)u_b(x;\lambda) \quad \text{if } x \in (\xi,b), \tag{7.11.36}$$

where u_0 and u_b are independent. The conditions (7.11.33) and (7.11.34) imply that $A(\xi;\lambda)$ and $B(\xi;\lambda)$ are obtained by solving the inhomogeneous system

$$A(\xi;\lambda)u_0(\xi;\lambda) - B(\xi;\lambda)u_b(\xi;\lambda) = 0, \tag{7.11.37}$$

$$B(\xi;\lambda)u_b'(\xi;\lambda) - A(\xi;\lambda)u_0'(\xi;\lambda) = 1, \tag{7.11.38}$$

which yields

$$A(\xi;\lambda) = \frac{u_b(\xi;\lambda)}{\Omega(u_0,u_b;\xi;\lambda)}, \tag{7.11.39}$$

$$B(\xi;\lambda) = \frac{u_0(\xi;\lambda)}{\Omega(u_0,u_b;\xi;\lambda)}, \tag{7.11.40}$$

where Ω is the Wronskian of u_0 and u_b. We now recall the Abel formula for Ω, according to which

$$\Omega(u_0,u_b;x;\lambda) = C(\lambda)\mathrm{e}^{-v(x)}, \tag{7.11.41}$$

where $C(\lambda)$ is a constant and, for the operator L in (7.11.21), v is a particular solution of the equation $\frac{\mathrm{d}v}{\mathrm{d}x} = \frac{2}{x}$, that is,

$$v(x) = \log x^2, \tag{7.11.42}$$

which implies

$$\Omega = \frac{C}{x^2}. \tag{7.11.43}$$

Thus, on defining as usual $x_< \equiv \min(x, \xi), x_> \equiv \max(x, \xi)$, the formulae (7.11.35), (7.11.36), (7.11.39), (7.11.40) and (7.11.43) imply that the Green function is expressed by

$$G_b(x, \xi; \lambda) = \frac{\xi^2}{C(\lambda)} u_0(x_<; \lambda) u_b(x_>; \lambda). \tag{7.11.44}$$

The integral equation (7.11.30) for W_α therefore becomes

$$\begin{aligned} W_\alpha(x; \lambda) &= \gamma \frac{2\nu^\nu \lambda^{\nu/2}}{\Gamma(\nu)} x^{-1/2} K_\nu(2\nu\sqrt{\lambda} x^{-1/2\nu}) \\ &+ \lim_{b\to\infty} \frac{\lambda\kappa}{C} \left[u_b(x; \lambda) \int_0^x u_0(\xi; \lambda) \xi^3 W_\alpha(\xi; \lambda)\, \mathrm{d}\xi \right. \\ &\left. + u_0(x; \lambda) \int_x^b u_b(\xi; \lambda) \xi^3 W_\alpha(\xi; \lambda)\, \mathrm{d}\xi \right]. \end{aligned} \tag{7.11.45}$$

In Eq. (7.11.45), $u_0(x; \lambda)$ and $u_b(x; \lambda)$ can be chosen to be of the form

$$u_0(x; \lambda) = C_0(\nu) x^{-1/2} K_\nu(2\nu\sqrt{\lambda} x^{-1/2\nu}), \tag{7.11.46}$$

$$\begin{aligned} u_b(x; \lambda) = x^{-1/2} \Big[& C_b^{(1)}(\nu) I_\nu(2\nu\sqrt{\lambda} x^{-1/2\nu}) \\ & + C_b^{(2)}(\nu) K_\nu(2\nu\sqrt{\lambda} x^{-1/2\nu}) \Big], \end{aligned} \tag{7.11.47}$$

in agreement with the homogeneous Dirichlet conditions at 0 and at b, respectively. Before taking the limit as $b \to \infty$ in Eq. (7.11.45) we can now regard the perturbation parameter λ as an eigenvalue. We are therefore studying, for finite b, a Fredholm integral equation of second kind, for which the general form is (the parameter a vanishes in our problem)

$$\varphi(s) = f(s) + \lambda \int_a^b K(s, t) \varphi(t)\, \mathrm{d}t. \tag{7.11.48}$$

If λ is an eigenvalue, a necessary and sufficient condition for the existence of solutions of Eq. (7.11.48) is that, for any solution χ of the equation

$$\chi(s) = \lambda \int_a^b K(t, s) \chi(t)\, \mathrm{d}t, \tag{7.11.49}$$

the known term $f(s)$ should satisfy the condition

$$\int_a^b f(s) \chi(s)\, \mathrm{d}s = 0. \tag{7.11.50}$$

In our problem, f is the first term on the right-hand side of Eq. (7.11.45), the kernel K is given by G_b in (7.11.44), and Eq. (7.11.50) provides a powerful operational criterion. Indeed, one might try to use the theory of integral equations directly on the interval $(0, \infty)$ instead of the limiting procedure in Eq. (7.11.45), but the necessary standard of rigour goes beyond our purposes.

7.11.3 Concluding remarks

We have exploited the fact that if a spiked harmonic oscillator is modified by the addition of a linear term, the full perturbation potential may be seen as consisting of an inverse power plus a term linear in the independent variable. It is then possible to evaluate the function $W(x; \lambda)$ occurring in the trial function (7.11.8) by solving the inhomogeneous equation (7.11.20), which leads to the integral equation (7.11.45). This is involved, but leads in principle to a complete calculational scheme. Note that, if one tries to combine the term x^2 in the unperturbed Hamiltonian with the linear term in the perturbation potential (7.11.19), one eventually moves the singular point away from the origin, whereas the spiked oscillator is (normally) studied by looking at the singular point at the origin.

It would be rather interesting, as a subject for further research, to consider suitable changes of the independent variable in the investigation of non-Fuchsian singularities. For example, given the Hamiltonian operator

$$H \equiv -\frac{\mathrm{d}^2}{\mathrm{d}r^2} + \frac{b}{r^2} + \frac{a}{r^p}, \tag{7.11.51}$$

if one defines the new independent variable

$$\rho \equiv r^\gamma, \tag{7.11.52}$$

for a suitable parameter γ, the stationary Schrödinger equation becomes

$$\left[\frac{\mathrm{d}^2}{\mathrm{d}\rho^2} + \left(1 - \frac{1}{\gamma}\right)\frac{1}{\rho}\frac{\mathrm{d}}{\mathrm{d}\rho} + \frac{1}{\gamma^2\rho^2}\left(E\rho^{2/\gamma} - a\rho^{(2-p)/\gamma} - b\right)\right]\varphi(\rho) = 0. \tag{7.11.53}$$

By construction, the larger γ is, the more Eq. (7.11.53) tends to its Fuchsian limit,

$$\left[\frac{\mathrm{d}^2}{\mathrm{d}\rho^2} + \left(1 - \frac{1}{\gamma}\right)\frac{1}{\rho}\frac{\mathrm{d}}{\mathrm{d}\rho} + \frac{(E - a - b)}{\gamma^2}\frac{1}{\rho^2}\right]\varphi(\rho) = 0, \tag{7.11.54}$$

for all values of ρ. This remark can be made precise by defining

$$\varepsilon \equiv \frac{1}{\gamma}, \tag{7.11.55}$$

$$F(\varepsilon) \equiv \varepsilon^2 \Big[E\rho^{2\varepsilon} - a\rho^{(2-p)\varepsilon} - b \Big], \tag{7.11.56}$$

and considering the asymptotic expansion at small ε (and hence large γ)

$$F(\varepsilon) \sim \varepsilon^2 \Big\{ E - a - b + \varepsilon[2E - (2-p)a] \log \rho + \mathrm{O}(\varepsilon^2) \Big\}. \tag{7.11.57}$$

The first 'non-Fuchsian correction' of Eq. (7.11.54) is therefore

$$\bigg\{ \frac{\mathrm{d}^2}{\mathrm{d}\rho^2} + (1-\varepsilon)\frac{1}{\rho}\frac{\mathrm{d}}{\mathrm{d}\rho} + \frac{1}{\rho^2}\Big[\varepsilon^2 \{ (E-a-b) + \varepsilon[2E - (2-p)a] \log \rho \} \Big] \bigg\} \varphi(\rho) = 0. \tag{7.11.58}$$

Interestingly, logarithmic terms in the potential can therefore be seen to result from a sequence of approximations relating Eq. (7.11.53) to its Fuchsian limit (7.11.54). Moreover, all equations with Fuchsian singularities such as Eq. (7.11.54) might be seen as non-trivial limits of stationary Schrödinger equations with non-Fuchsian singular points. It remains to be seen whether such properties can be useful in the investigation of the topics discussed in the previous sections.

Another topic for further research is the Schrödinger equation for perturbed stationary states of an isotropic oscillator in three dimensions written in the form

$$\left[\frac{\mathrm{d}^2}{\mathrm{d}r^2} + k^2 - \mu^2 r^2 - \frac{l(l+1)}{r^2} - S(r) \right] \varphi(r) = 0, \tag{7.11.59}$$

where, having set (here $\mu \equiv \frac{m\omega}{\hbar}$)

$$V(r) \equiv \frac{2m}{\hbar^2} U(r) = \mu^2 r^2 + S(r), \tag{7.11.60}$$

the function S represents the 'singular' part of the potential according to our terminology. We look for exact solutions of Eq. (7.11.59), which can be written as

$$\varphi(r) = A(r) \mathrm{e}^{B(r)} \mathrm{e}^{-\mu r^2/2}. \tag{7.11.61}$$

The second exponential in (7.11.61) takes into account that, at large r, the term $\mu^2 r^2$ dominates over all other terms in the potential (including, of course, $\frac{l(l+1)}{r^2}$), and has not been absorbed into $B(r)$ for later convenience. It is worth stressing that Eq. (7.11.61) is not a JWKB ansatz but rather a convenient factorization of the exact solution of Eq. (7.11.59). We determine $B(r)$ from a non-linear equation by straightforward integration (see below), while the corresponding second-order equation for A is rather involved.

Indeed, insertion of (7.11.61) into Eq. (7.11.59) leads to

$$\left\{ \frac{\mathrm{d}^2}{\mathrm{d}r^2} + 2(B' - \mu r)\frac{\mathrm{d}}{\mathrm{d}r} + \left[k^2 - \mu - \frac{l(l+1)}{r^2} \right.\right.$$
$$\left.\left. - 2\mu r B' + B'' + B'^2 - S(r) \right] \right\} A(r) = 0. \qquad (7.11.62)$$

To avoid having coefficients of this equation which depend in a non-linear way on B we *choose* the function B so that

$$B'^2 - S(r) = 0, \qquad (7.11.63)$$

which implies (up to a sign, here implicitly absorbed into the square root)

$$B(r) = \int \sqrt{S(r)}\, \mathrm{d}r. \qquad (7.11.64)$$

Hence one finds the following second-order equation for the function A:

$$\left\{ \frac{\mathrm{d}^2}{\mathrm{d}r^2} + 2(\sqrt{S} - \mu r)\frac{\mathrm{d}}{\mathrm{d}r} + \left[k^2 - \mu - \frac{l(l+1)}{r^2} \right.\right.$$
$$\left.\left. - 2\mu r\sqrt{S} + \frac{S'}{2\sqrt{S}} \right] \right\} A(r) = 0. \qquad (7.11.65)$$

It should be stressed that the step leading to Eq. (7.11.63) is legitimate but not mandatory. For each choice of $B(r)$ there will be a different equation for $A(r)$, but in such a way that $\varphi(r)$ remains the same (see Eq. (7.11.61)). Unfortunately, Eq. (7.11.65) remains too difficult, as far as we can see.

7.12 Problems

7.P1. Consider the matrix

$$\begin{pmatrix} E_1 & 0 & a \\ 0 & E_1 & b \\ a^* & b^* & E_2 \end{pmatrix},$$

which represents the Hamiltonian operator of a physical system. Compute the eigenvalues to second order in perturbation theory and compare with the exact result, assuming that

$$E_2 > E_1, \quad |a|, |b| \ll |E_2 - E_1|.$$

7.P2. A particle of mass m and charge e oscillates in a harmonic potential in one dimension with angular frequency ω.

(i) Compute, to first and second order in perturbation theory, the effect on the energy levels resulting from a constant electric field.

(ii) Compare with the predictions resulting from the classical analysis.

7.P3. A particle of mass m is subject to a one-dimensional harmonic potential $V^{(0)} = \frac{1}{2}Kx^2$, with angular frequency $\omega = \sqrt{\frac{k}{m}}$. A small perturbative term $V^{(1)} = \frac{1}{2}(\delta k)x^2$ is added to $V^{(0)}$. Compute the first- and second-order corrections to the ground-state energy.

7.P4. Consider the Hamiltonian operator $H_0 \equiv \vec{\sigma} \cdot \vec{p}$ in the Hilbert space $\mathbf{C}^2$ of spin $\frac{1}{2}$, and a perturbation $\varepsilon V = \varepsilon \sigma_z$. Compute eigenvectors and eigenvalues of H_0 and of $H_0 + \varepsilon V$. Then apply perturbation theory to the first non-trivial order and compare the results.

7.P5. Compute the first-order Stark effect on the $n = 2$ states of the hydrogen atom, when a constant electric field is applied along the z-axis. After this, consider the first-order Stark effect on the $n = 3$ states of the hydrogen atom. Bearing in mind the properties of the matrix elements of the $\hat{z}$ operator, pick out the non-vanishing contributions to the first-order calculation.

7.P6. Compute the first-order Stark effect on the $n = 2$ states of the hydrogen atom, in the three cases of uniform electric field along the x-, y- and z-axes, respectively.

7.P7. An isotropic harmonic oscillator in three dimensions is subject to a constant electric field along the z-axis. Compute the first- and second-order corrections to the first excited level, and interpret the result.

7.P8. A one-dimensional harmonic oscillator of mass m is given.

(i) Compute, to first order in perturbation theory, the correction to the ground-state energy resulting from the following perturbation term in the Hamiltonian operator: $-\frac{1}{8}\frac{\hat{p}^4}{m^3c^2}$, on assuming that $\hbar\omega$ is much smaller than mc^2.

(ii) Compute the matrix elements $\langle k|\hat{p}^2|r\rangle$.

7.P9. A one-dimensional harmonic oscillator is subject to the perturbation

$$\hat{V}(t) = a\hat{x}^3 + b\hat{x}^4 \quad \text{if } t \in [0, T], \tag{7.12.1a}$$

$$\hat{V}(t) = 0 \quad \text{if } t > T. \tag{7.12.1b}$$

(i) Compute, to the first non-vanishing order in perturbation theory, the transition probability from the first excited level to the ground state. What is the role played by the term $b\hat{x}^4$ in such an approximation?

(ii) On considering perturbations of stationary states, determine the relation between the dimensionful parameters a and b that is necessary to ensure that the sign of the energy eigenvalues is not affected by the perturbation, after including second-order effects.

(iii) If $a = 0$ and $b \neq 0$, discuss the properties of the perturbative series for the energy eigenvalues in the stationary theory. How should one interpret the calculation if, instead, $a \neq 0$ and $b = 0$?

7.P10. A particle of spin J is subject to a static magnetic field $\vec{B}$ parallel to the z-axis, and to another one of magnitude B_1, orthogonal to $\vec{B}$, and rotating with frequency ω. The spin part of the

Hamiltonian operator reads as

$$H(t) = -\gamma B J_z - \gamma B_1 \left[J_x \cos(\omega t) + J_y \sin(\omega t) \right], \tag{7.12.2}$$

where γ is a constant, and $\hbar$ is set to 1 for convenience.

(i) Prove that the corresponding propagator can be written as

$$U(t) = \mathrm{e}^{-\mathrm{i}\omega J_z t} \, \mathrm{e}^{-\mathrm{i}At}, \tag{7.12.3}$$

where

$$A \equiv \omega_1 J_x + (\omega_0 - \omega) J_z, \tag{7.12.4}$$

having defined $\omega_0 \equiv -\gamma B$ and $\omega_1 \equiv -\gamma B_1$.

(ii) Compute the transition probability from the initial state $\varphi_{J,J}$ to the final state $\varphi_{J,-J}$ after a time t (see, if necessary, the work in Balasubramanian (1972)).

7.P11. Consider a spin-$\frac{1}{2}$ particle in a magnetic field having the form

$$\vec{B}(t) = \Big(B_1 \cos(\omega t), B_1 \sin(\omega t), 0 \Big). \tag{7.12.5}$$

The Hamiltonian operator of this system is

$$H(t) = -\gamma \vec{s} \cdot \vec{B}(t), \tag{7.12.6}$$

where γ is a constant. Compute, in the adiabatic approximation, the probability that at time $T = \frac{\pi}{\omega}$ the spin lies along the x-axis, assuming that this was also its initial location. Compare the result with the exact calculation.

7.P12. Consider the following Hamiltonian operator in one spatial dimension (with $x \in \mathbf{R}$):

$$\hat{H} \equiv -\frac{\hbar^2}{2m} \frac{\mathrm{d}^2}{\mathrm{d}x^2} + \gamma x^4, \tag{7.12.7}$$

where the parameter γ is taken to be positive. On using the variational method, give an estimate of the ground-state energy.

Appendix 7.A

Convergence in the strong resolvent sense

One of the main technical difficulties with unbounded operators is that they are only densely defined. This is especially serious when one tries to define the convergence of a sequence $\{A_n\}$ of unbounded operators, since the domains of the operators A_n may have no vector in common. For example, if (Reed and Simon 1980)

$$A_n \equiv \left(1 - \frac{1}{n} \right) x \tag{7.A.1}$$

on $\mathcal{L}^2(\mathbf{R})$, it is clear that, in a suitable sense, $\{A_n\}$ tends to the operator A of multiplication by x. Still, it is possible to give domains $D(A_n)$ and $D(A)$ of essential self-adjointness for these operators which have no non-zero eigenvector in common (Reed and Simon 1980). Indeed, in this simple case the closures of A_n and A all have the same domain, but in general this is not the case, and one is often forced to deal with domains of essential self-adjointness, since closures of operators may be difficult to compute. The basic idea is that self-adjoint operators are 'close' if some bounded functions of such operators are close. For our purposes, we only need to recall the following definitions.

D.1 Let T be a closed operator on a Hilbert space $\mathcal{H}$. A complex number λ is in the *resolvent set*, $\rho(T)$, if $\lambda\mathbb{1} - T$ is a bijection of $D(T)$ onto $\mathcal{H}$, with a bounded inverse. If $\lambda \in \rho(T)$, then

$$R_\lambda(T) \equiv (\lambda\mathbb{1} - T)^{-1} \tag{7.A.2}$$

is called the *resolvent* of T at λ.

D.2 Let $A_n, n = 1, 2, \ldots,$ and A be self-adjoint operators. Then A_n is said to converge to A in the *norm resolvent sense* if

$$R_\lambda(A_n) \longrightarrow R_\lambda(A) \text{ in norm, } \forall\lambda : \mathrm{Im}(\lambda) \neq 0. \tag{7.A.3}$$

D.3 The sequence $\{A_n\}$ is said to converge to A in the *strong resolvent sense* if

$$R_\lambda(A_n) \longrightarrow R_\lambda(A) \text{ strongly, } \forall\lambda : \mathrm{Im}(\lambda) \neq 0. \tag{7.A.4}$$

D.4 If a family H_n of self-adjoint operators tends to H in a strong resolvent sense, with $\{S_n\}$ and T subsets of $\mathbb{R}$, one says that the part of the spectrum of H_n in T is asymptotically in S_n if and only if

$$\text{strong} - \lim_{n\to\infty} P_n(T/S_n) = 0, \tag{7.A.5}$$

where $\{P_n\}$ is the family of spectral projections of H_n.

8
Scattering theory

Scattering theory studies the behaviour of (quantum) systems over time and length scales which are very large, compared with the time or length that are characteristic of the interaction which affects them. The chapter begins with an outline of the basic problems of scattering theory, i.e. the existence of scattering states, their uniqueness, weak asymptotic completeness the existence of the scattering transformation, S, the reduction of S due to symmetries, the analyticity of S, asymptotic completeness and the existence of wave operators.

After considering the Schrödinger equation for stationary states, the integral equation for scattering problems is derived and studied. The Born approximation and the Born series are presented, and the conditions which ensure convergence of the Born series are also discussed. Further topics are partial wave expansion, its convergence and the uniqueness of the solution satisfying the asymptotic condition, the Levinson theorem. Lastly, in the first part of the chapter, scattering theory from singular potentials is introduced, with emphasis on the polydromy of the wave function, following early work in the literature, and the general problems in the theory of resonances are studied.

In the second part, we examine in detail a separable potential model, the occurrence of bound states in the completeness relationship, an excitable potential model and the unitarity of the Möller wave operator. Lastly, we study the survival amplitude associated with quantum decay transitions.

8.1 Aims and problems of scattering theory

Scattering theory is the branch of physics that is concerned with interacting systems on a scale of time and/or distance that is large compared with the scale of the interaction itself, and it provides the most powerful tool for studying the microscopic nature of the world. In quantum mechanics, a typical scattering process involves a beam of particles prepared with a given energy and with a more or less well-defined linear momentum. One then studies either the collision with a similar wave packet, or the collision of the given wave packet against a fixed target. In particular, for a two-body elastic scattering problem with conservative forces, one works in the centre of mass frame, hence reducing the original problem to the analysis of a particle in an external field of forces. Indeed, in section 5.5 we have already used such a technique in the investigation of the hydrogen atom. The crucial difference with respect to section 5.5 is that, in scattering problems, one studies the continuous spectrum and its perturbations, whereas the Balmer formula has to do with bound states.

The dynamics we are interested in is given by a set of transformations acting on physical states, and we consider some maps T_t describing the *interacting* dynamical transformations, and other maps $T_t^{(0)}$ corresponding instead to a comparison (free) dynamical transformation. All of these maps act on the set Σ of states, where Σ may represent points in phase space, or vectors in a Hilbert space, or Cauchy data for acoustic and optical problems. The basic problems of scattering theory can therefore be described as follows (Reed and Simon 1979, Berezin and Shubin 1991).

(i) **Existence of scattering states**. The interacting system is prepared in such a way that some of its constituents are so far from each other that their interaction is negligible. At this stage the interacting dynamics is allowed to act upon the system for a sufficiently long time, and after this process one tries to understand what happened. One expects that *any free state can be prepared*, i.e. $\forall \rho_- \in \Sigma$, there should exist a $\rho \in \Sigma$ such that

$$\lim_{t \rightarrow -\infty} \left(T_t \rho - T_t^{(0)} \rho_- \right) = 0. \tag{8.1.1}$$

The problem is to prove that this is the case. Note that such a weak limit does not preserve normalization in the presence of bound states (Chiu *et al.* 1992, Varma and Sudarshan 1996).

(ii) **Uniqueness of scattering states**. To describe the state prepared in terms of free states, one has to make sure that every free state corresponds to a unique interacting state, i.e. $\forall \rho_- \in \Sigma$, there exists, at the very best, one element $\rho \in \Sigma$ such that

$$\lim_{t \rightarrow -\infty} \left(T_t^{(0)} \rho_- - T_t \rho \right) = 0. \tag{8.1.2}$$

Thus, the free states should be isospectral with respect to the interacting states.

(iii) **Weak asymptotic completeness.** One deals with an interacting state, ρ, and one requires that this should approximate a free state both for $t \rightarrow -\infty$ and $t \rightarrow \infty$. One then has to prove that the following subsets of Σ:

$$\Sigma_{\rm in} \equiv \left\{ \rho \in \Sigma : \exists \rho_- \in \Sigma : \lim_{t \rightarrow -\infty} \left(T_t^{(0)} \rho_- - T_t \rho \right) = 0 \right\}, \tag{8.1.3}$$

$$\Sigma_{\rm out} \equiv \left\{ \rho \in \Sigma : \exists \rho_+ \in \Sigma : \lim_{t \rightarrow +\infty} \left(T_t^{(0)} \rho_+ - T_t \rho \right) = 0 \right\}, \tag{8.1.4}$$

do actually coincide. If this is the case, so that $\Sigma_{\rm in} = \Sigma_{\rm out}$, one says that the physical system under consideration is *weakly asymptotically complete*. This presupposes that there are *no* unstable particles in the problem.

(iv) **The S transformation.** Suppose one has a pair of dynamical systems for which existence and uniqueness of scattering states can be proved both for $t \rightarrow -\infty$ and $t \rightarrow +\infty$, and for which weak asymptotic completeness holds. The existence and uniqueness of scattering states implies that

$$\exists \Omega^- \rho \in \Sigma_{\rm in} : \lim_{t \rightarrow -\infty} \left(T_t(\Omega^- \rho) - T_t^{(0)} \rho \right) = 0, \tag{8.1.5}$$

$$\exists \Omega^+ \rho \in \Sigma_{\rm out} : \lim_{t \rightarrow +\infty} \left(T_t(\Omega^+ \rho) - T_t^{(0)} \rho \right) = 0. \tag{8.1.6}$$

Note that the generic $\Omega^\pm$ are *not* norm preserving. Hereafter, the maps Ω^- and Ω^+ work as follows:

$$\Omega^- : \Sigma \rightarrow \Sigma_{\rm in}, \tag{8.1.7}$$

$$\Omega^+ : \Sigma \rightarrow \Sigma_{\rm out}. \tag{8.1.8}$$

In other words, let H, H_0 be self-adjoint operators in the Hilbert space $\mathcal{H}$, and let $\mathcal{D}_+$ and $\mathcal{D}_-$ be linear sub-manifolds of $\mathcal{H}$ such that $\mathcal{D}_+$ (respectively, $\mathcal{D}_-$) consists of those vectors ψ_+ (respectively, ψ_-) in $\mathcal{H}$ for which there exists a vector $\psi \in \mathcal{H}$ fulfilling the condition

$$e^{-itH} \psi = e^{-itH_0} \psi_\pm + \xi_\pm(t), \tag{8.1.9}$$

where the norm of $\xi_\pm$ tends to 0 as $t \rightarrow \pm\infty$. The wave operators $\Omega^+(H, H_0)$ and $\Omega^-(H, H_0)$ are operators with domains $\mathcal{D}_+$ and $\mathcal{D}_-$, respectively, defined by the conditions (cf. Möller 1945)

$$\Omega^\pm \psi_\pm = \psi, \tag{8.1.10}$$

where $\psi_\pm$ and ψ are related by Eq. (8.1.9). This means that

$$\Omega^\pm \psi_\pm = \lim_{t \rightarrow \pm\infty} e^{itH} e^{-itH_0} \psi_\pm, \tag{8.1.11}$$

which is also expressed, using s-lim for the limit in strong operator topology,

$$\Omega^{\pm} = s - \lim_{t \to \pm\infty} e^{itH} e^{-itH_0}. \tag{8.1.12}$$

Note that H_0 is no longer *the* free Hamiltonian but *a* free Hamiltonian isospectral with H. By virtue of Eq. (8.1.9), the limits in (8.1.11) exist exactly on $\mathcal{D}_{\pm}$. Moreover, the wave operators $\Omega^{\pm}$ are isometric (i.e. they preserve distances) since they are strong limits of unitary operators. The *scattering operator* defined by

$$S \equiv (\Omega^{+})^{-1} \Omega^{-} \tag{8.1.13}$$

converts the vector ψ_{-} into the vector ψ_{+}, and it is an isometric operator defined on vectors $\psi_{-} \in \mathcal{D}_{-}$. Moreover, the operator S is unitary if and only if

$$\mathcal{D}_{+} = \mathcal{D}_{-} = \mathcal{H}, \quad \Omega^{+}\mathcal{D}_{+} = \Omega^{-}\mathcal{D}_{-}. \tag{8.1.14}$$

The idea behind the introduction of the scattering operator is as follows. The scattering operator, which converts ψ_{-} into ψ_{+}, determines how the interaction V affects the evolution of a particle. If we are only interested in the time evolution from $-\infty$ to $+\infty$, we can first consider the motion of a free particle in the time interval $(-\infty, 0)$, then apply the scattering operator to the state obtained, and eventually, again regard the particle as free for $t \in (0, +\infty)$. The whole interaction with an external field of forces can therefore be replaced by the action of the scattering operator at $t = 0$ (or, more generally, at any $t = t_0$), which relates the past and future asymptotics of 'interacting histories'.

(v) **Reduction of S due to symmetries**. In several problems, both the free and the interacting dynamics possess an underlying symmetry. This makes it possible to conclude, *a priori*, that S has a special form, without any detailed calculation.

(vi) **Analyticity of S**. The operator S, or the kernel of some integral operator related to it, can be realized as the boundary value of an analytic function. Schematically, S describes the reaction R of a system to some input in the following form:

$$R(t) = \int_{-\infty}^{t} f(t-y) I(y) \, dy. \tag{8.1.15}$$

Thus, $R(t)$ is *invariant under temporal translations*, because f depends only on the difference $t-y$. Moreover, *causality* holds, since $R(t)$ depends on $I(y)$ only for $y \leq t$, which means that f is defined on $[0, \infty[$, and hence its Fourier transform is the boundary value of an analytic function.

This causality argument is always present in our mind, when we discuss analytic properties.

(vii) **Asymptotic completeness**. Consider a system with forces among its components which decay when such constituents are moved away from each other. On physical grounds, one expects that a state of such a system can decay in freely moving clusters or, otherwise, is going to remain bound. In several cases, there exists a natural set of bound states, in that $\Sigma_{\rm bound} \subset \Sigma$. Usually, one manages to prove that $\Sigma_{\rm bound}$ has an empty intersection with $\Sigma_{\rm in}$, and one has

$$\Sigma_{\rm bound} \oplus \Sigma_{\rm in} = \Sigma = \Sigma_{\rm bound} \oplus \Sigma_{\rm out}, \tag{8.1.16}$$

where the symbol $\oplus$ denotes a direct sum of Hilbert spaces. The equality (8.1.16) is said to express the condition of *asymptotic completeness.* The states just mentioned are *not* states of the free Hamiltonian, but of a modified Hamiltonian H_0' isospectral with H. This H_0' should be used in computing the limit (8.1.12). A number of remarks are now in order.

(R1) The condition (8.1.16) does not always hold. For example, in deuteron–proton scattering, the continuum state contains $d-p$ states that violate (8.1.16). More precisely, one deals with a state which is in $\Sigma_{\rm in,out}$ but the deuteron itself is not! The $d-p$ state suffers a process which is omitted in the 'axiomatic' discussion.

(R2) Of course, if asymptotic completeness holds, the condition of weak asymptotic completeness is also respected.

(R3) The S-matrix can also be defined when the condition of weak asymptotic completeness fails to hold.

(R4) There exist physical theories which have no unperturbed dynamics that can be compared with the interacting dynamics. In such cases, one first picks out some especially simple solutions of the interacting system. The interactions among such simple solutions are then used to describe the asymptotic behaviour of the full interacting system.

(viii) **Existence of wave operators**. Let $\mathrm{e}^{-\mathrm{i}At}$ and $\mathrm{e}^{-\mathrm{i}Bt}$ be unitary groups such that $\mathrm{e}^{-\mathrm{i}At}$ describes an interacting dynamics, and $\mathrm{e}^{-\mathrm{i}Bt}$ corresponds instead to a free dynamics. The state $\mathrm{e}^{-\mathrm{i}At}\varphi$ is said to be *asymptotically free* as $t \rightarrow -\infty$ if there exists a φ_- such that

$$\lim_{t \rightarrow -\infty} \|\mathrm{e}^{-\mathrm{i}Bt}\varphi_- - \mathrm{e}^{-\mathrm{i}At}\varphi\| = 0, \tag{8.1.17a}$$

where the norms of φ and φ_- need not be equal. Such a condition may be re-expressed in the form

$$\lim_{t\rightarrow -\infty} \| e^{-iAt}\Big(\varphi - e^{iAt}e^{-iBt}\varphi_-\Big)\| = 0, \tag{8.1.17b}$$

which implies

$$\lim_{t\rightarrow -\infty} \| e^{iAt}e^{-iBt}\varphi_- - \varphi\| = 0. \tag{8.1.18}$$

Thus, one has to prove the existence of such strong limits. One should stress that all of this is only valid for two particles interacting through a potential. In several applications, B has only an absolutely continuous spectrum; if this is not the case, however, one has to choose φ_- in such a way that it lies in the absolutely continuous subspace of B. For example, if φ_+ is an eigenvector of B, then the strong limit exists only if φ_+ is also an eigenvector of A *with the same eigenvalue*. Thus, the wave operators are defined by first projecting on the absolutely continuous subspace of B. More precisely, if A and B are self-adjoint operators on a Hilbert space $\mathcal{H}$, and if $P_{ac}(B)$ denotes the projection on the absolutely continuous subspace of B, the *generalized wave operators* $\Omega^{\pm}(A, B)$ are said to exist if the strong limits (see Eq. (8.1.12))

$$\Omega^{\pm}(A, B) \equiv s - \lim_{t\rightarrow \pm\infty} e^{iAt}e^{-iBt}P_{ac}(B) \tag{8.1.19}$$

turn out to exist. When $\Omega^{\pm}(A, B)$ exist, one defines

$$\mathcal{H}_{\rm in} \equiv {\rm Ran}(\Omega^-) \equiv \mathcal{H}_-, \tag{8.1.20}$$

$$\mathcal{H}_{\rm out} \equiv {\rm Ran}(\Omega^+) \equiv \mathcal{H}_+. \tag{8.1.21}$$

Nature is in general more complicated. For example, condition (8.1.19) is not satisfied for deuteron–proton states. However, wave operators can be defined unitarily if all states are involved (Sudarshan 1962, Jordan *et al.* 1964).

8.2 Integral equation for scattering problems

We now specialize the generic considerations of the previous section to an important class of special problems, i.e. the scattering of a particle by a (time-independent) potential. For this purpose, we focus on a wave packet $\psi(\vec{x}, t)$ that is strongly peaked about the value $\vec{p} = \hbar\vec{k}$ in the space of momenta. We assume that it interacts with a short-range potential with compact support, and we analyse the motion of the packet in terms of stationary solutions. Indeed, after setting (recall that, in scattering problems, E is always positive)

$$k \equiv |\vec{k}| \equiv \frac{\sqrt{2mE}}{\hbar}, \tag{8.2.1}$$

$$\mathcal{U}(\vec{x}) \equiv \frac{2m}{\hbar^2} V(\vec{x}), \tag{8.2.2}$$

$$\phi_k(\vec{x}) \equiv \mathcal{U}(\vec{x}) \psi_k(\vec{x}), \tag{8.2.3}$$

the Schrödinger equation for stationary states reads

$$\left(\triangle + k^2\right) \psi_k(\vec{x}) = \phi_k(\vec{x}). \tag{8.2.4}$$

The full integral of Eq. (8.2.4) consists of the general integral of the homogeneous equation

$$\left(\triangle + k^2\right) \psi_k(\vec{x}) = 0,$$

i.e. $\psi_k(\vec{x}) = e^{\pm i \vec{k} \cdot \vec{x}}$, plus a particular solution of Eq. (8.2.4). For this purpose, we need to invert the operator $\left(\triangle + k^2\right)$, so that $\psi_k(\vec{x})$ may be expressed by an integral operator acting on $\phi_k(\vec{x})$. This is achieved by first finding the Green function G_k of Eq. (8.2.4). For this purpose, we may introduce polar coordinates, assuming that G_k depends only on the modulus $r = |\vec{x}|$. In other words, we first consider, *for simplicity of notation*, the problem of finding a $G_k(r)$ such that

$$\left(\frac{d^2}{dr^2} + \frac{2}{r} \frac{d}{dr} + k^2 \right) G_k(r) = 0 \quad \forall r > 0, \tag{8.2.5}$$

with the understanding that, in the integral equation we are going to derive, we shall replace r by the modulus

$$|\vec{x} - \vec{x}'| \equiv \sqrt{x^2 + x'^2 - 2xx' \cos \alpha},$$

where α is the angle between the vectors $\vec{x}$ and $\vec{x}'$. The solution of Eq. (8.2.5) is found to be

$$G_k(r) = \frac{e^{i\varepsilon kr}}{\beta r}, \tag{8.2.6}$$

where $\varepsilon = \pm 1$, and β takes the value -4π, which is fixed by the well-known property of the Laplace operator

$$\triangle(r^{-1}) = 0 \quad \forall r > 0, \tag{8.2.7}$$

jointly with the condition

$$\int_{\mathcal{A}} \triangle(r^{-1}) \, d^3x = -4\pi, \tag{8.2.8}$$

if the origin lies in $\mathcal{A}$. The formula (8.2.6) is obtained by re-expressing the Helmholtz equation (8.2.5) for the Green function in the form

$$\left(\frac{d^2}{dr^2} + k^2 \right) (rG) = 0 \quad \forall r > 0, \tag{8.2.9}$$

which is solved by

$$G = A\frac{e^{ikr}}{r} + B\frac{e^{-ikr}}{r},$$

for some constants A and B.

Hence we obtain the integral equation for the continuous spectrum

$$\psi_k^{\pm}(\vec{x}) = e^{\pm i\vec{k}\cdot\vec{x}} - \frac{1}{4\pi}\int \frac{e^{\pm ik|\vec{x}-\vec{x}'|}}{|\vec{x}-\vec{x}'|}\mathcal{U}(\vec{x}')\psi_k^{\pm}(\vec{x}')\, d^3x'. \tag{8.2.10}$$

The structure of this integral equation is quite general, and can be used in all cases where, to the free Hamiltonian H_0, an interaction term V is added. The advantages of the integral formulation are twofold:

(i) the boundary conditions on the scattering solution are automatically encoded;

(ii) a series solution can be found (see section 8.3).

An equivalent way to express Eq. (8.2.10) is

$$\psi_E^{\pm} = \psi_E + G_E^{\pm}V\psi_E^{\pm}, \tag{8.2.11}$$

which is called the Lippmann–Schwinger equation (Lippmann and Schwinger 1950).

If one deals with a short-range potential, one can approximate

$$|\vec{x}-\vec{x}'| \sim r - \vec{e}_x\cdot\vec{x}', \tag{8.2.12}$$

where $\vec{e}_x$ is the unit vector $\frac{\vec{x}}{|\vec{x}|}$, and hence the integral on the right-hand side of Eq. (8.2.10) can be approximated, for $\psi_k^+(\vec{x})$, by

$$\frac{e^{ikr}}{r}\int e^{-i\vec{\kappa}\cdot\vec{x}'}\mathcal{U}(\vec{x}')\psi_k^+(\vec{x}')\, d^3x',$$

where we have defined $\vec{\kappa} \equiv k\vec{e}_x$. One then finds for the first set of solutions, at large distances from the scattering centre,

$$\psi_k^+(\vec{x}) \sim e^{i\vec{k}\cdot\vec{x}} + f_{k\kappa}\frac{e^{ikr}}{r}, \tag{8.2.13}$$

where

$$f_{k\kappa} \equiv -\frac{1}{4\pi}\int e^{-i\vec{\kappa}\cdot\vec{x}'}\mathcal{U}(\vec{x}')\psi_k^+(\vec{x}')\, d^3x'. \tag{8.2.14}$$

This describes the joint effect of a plane wave, and of a spherical wave modulated by the factor $f_{k\kappa}$. To recover the time-dependent formulation, one has to build the wave packet

$$\psi(\vec{x},t) = \int C(k)\psi_k^+(\vec{x})e^{-i\frac{\hbar k^2}{2m}t}\, d^3k, \tag{8.2.15}$$

where C is a function with compact support. Thus, on applying the stationary phase method (see appendix 5.A) one finds that, when $t < 0$, the spherical wave term does not contribute, whereas, for $t > 0$, it plays a crucial role, and represents a spherical wave propagating from the centre with amplitude depending on $f_{k\kappa}$. The term $e^{i\vec{k}\cdot\vec{x}}$ represents, instead, the motion of a free wave packet.

It should be stressed that the solutions ψ_k^- can be legitimately considered, from a purely mathematical point of view. However, on applying the stationary-phase method, they give rise to a wave packet which, for $t \rightarrow -\infty$, consists of a plane wave and a wave directed towards the centre, whereas, for $t \rightarrow \infty$, they give rise to a plane wave propagating in the direction $-\vec{k}$. It remains unclear how to realize such a situation in the actual experiments, since the coherence properties of the wave function are required to hold on macroscopic length scales. Thus, on physical grounds, the stationary states ψ_k^- are discarded, and one works with the states ψ_k^+ only (they are in an extended space which spans the Hilbert space) to build a scattered wave packet. Of course, the full Hilbert space of the problem might include bound states as well.

8.3 The Born series and potentials of the Rollnik class

After having decided that we only consider scattering with mutual interaction in a two-particle system, we focus on the stationary states ψ_k^+, for which the Lippmann–Schwinger equation (8.2.11) can be written in the form

$$\psi_E^+ = \psi_E + G_E^+ V \psi_E^+, \tag{8.3.1}$$

which implies

$$\left(\mathbb{I} - G_E^+ V\right)\psi_E^+ = \psi_E, \tag{8.3.2}$$

and hence, upon using the inverse operator $\left(\mathbb{I} - G_E^+ V\right)^{-1}$,

$$\psi_E^+ = \left(\mathbb{I} - G_E^+ V\right)^{-1} \psi_E. \tag{8.3.3}$$

This applies only to the continuous spectrum (more generally for states labelled by the same labels as for the 'free' system). To lowest order, one has $\psi_E^+ \sim \psi_E$, and the resulting scattering amplitude is the one of the Born approximation, i.e. (cf. Eq. (8.2.14))

$$f_{k\kappa}^{\rm Born} = -\frac{m}{2\pi\hbar^2} \int e^{i(\vec{k}-\vec{\kappa})\cdot\vec{x}} V(\vec{x})\, d^3x, \tag{8.3.4}$$

which is proportional to the Fourier transform of the potential, evaluated for $\vec{q} = \vec{\kappa} - \vec{k}$.

In general, one can try to express the inverse operator $\left(\mathbb{I}-G_E^+V\right)^{-1}$ as a (formal) series

$$\left(\mathbb{I}-G_E^+V\right)^{-1} \sim \sum_{n=0}^{\infty}(G_E^+V)^n, \tag{8.3.5}$$

where the right-hand side is called the Born series. There exist, actually, some rigorous results which provide sufficient conditions for the convergence of the Born series. They are as follows (Reed and Simon 1979).

Theorem 8.1 Let V be Lebesgue-summable on $\mathbf{R}^3$:

$$\int_{\mathbf{R}^3} |V(\vec{x})| \, \mathrm{d}^3x < \infty \tag{8.3.6}$$

and in the Rollnik class$_R$, i.e. such that (Rollnik 1956)

$$\|V\|_R^2 \equiv \int\int_{\mathbf{R}^6} \frac{|V(\vec{x})||V(\vec{y})|}{|\vec{x}-\vec{y}|^2} \, \mathrm{d}^3x \, \mathrm{d}^3y < \infty. \tag{8.3.7}$$

Under such conditions, there exists an energy E_0 such that the Born series is convergent $\forall E > E_0$.

Theorem 8.2 If the assumptions of theorem 8.1 hold, and the Rollnik norm of V satisfies the inequality

$$\|V\|_R < 4\pi, \tag{8.3.8}$$

the Born series converges $\forall E > 0$.

For the detailed proof of such theorems we refer the reader to Reed and Simon (1979). However, we remark that, if both V and V^{-1} are bounded functions, the integral operator G_E^+V is related by the following transformation to

$$\widetilde{G} \equiv \sqrt{V} \, G_E^+ \, \sqrt{V}. \tag{8.3.9}$$

One then uses the property

$$\|\widetilde{G}\|^2 = (4\pi)^{-2}\|V\|_R^2 \tag{8.3.10}$$

to prove theorem 8.2, because the maximal eigenvalue of G_E^+V is smaller than 1 if (8.3.8) is satisfied, by virtue of (8.3.9) and (8.3.10).

Potentials V that belong to $\mathcal{L}^1(\mathbf{R}^3)$ and are in the Rollnik class play an important role in proving uniqueness of the solution of the Lippmann–Schwinger equation. More precisely, if $H_0 = -\triangle$ on $\mathcal{L}^2(\mathbf{R}^3)$, and if $H = H_0 + V$, with V satisfying (8.3.6) and (8.3.7), one can prove that there exists a subset $\mathcal{E}$ of $\mathbf{R}_+$, closed and with zero Lebesgue measure such that, if $k^2 \notin \mathcal{E}$, there exists a unique solution of Eq. (8.2.11) (Reed and Simon 1979).

8.4 Partial wave expansion

As we know from section 5.6, in a central potential the self-adjoint operators $\hat{H}, \hat{L}^2, \hat{L}_z$ have common eigenfunctions $u_{n,l,m}(r,\theta,\varphi)$ (the effects of spin are neglected here). Moreover, one can arrange the z-axis along the direction of the incoming beam. Such a beam has, in general, cylindrical symmetry about the propagation direction, which implies we are eventually dealing with eigenfunctions of $\hat{L}_z$ belonging to the eigenvalue $m = 0$. Thus, bearing in mind that the spherical harmonic Y_l^0 is proportional to the Legendre polynomial $P_l(\cos\theta)$, one can write the stationary states in the form

$$\psi_k^+(\vec{x}) = \sum_{l=0}^{\infty} C_l R_{kl}(r) P_l(\cos\theta). \tag{8.4.1}$$

If the potential has compact support:

$$V(r) = 0 \quad \forall r > a, \tag{8.4.2}$$

one finds

$$R_{kl}(r) = A_l j_l(kr) + B_l n_l(kr) \quad \forall r > a, \tag{8.4.3}$$

in terms of the spherical Bessel functions (see appendix 5.B). Thus, by virtue of the well-known asymptotic formulae valid as $r \to \infty$,

$$j_l(kr) \sim \frac{1}{kr} \sin\left(kr - \frac{l\pi}{2}\right), \tag{8.4.4}$$

$$n_l(kr) \sim -\frac{1}{kr} \cos\left(kr - \frac{l\pi}{2}\right), \tag{8.4.5}$$

and defining

$$\frac{B_l}{A_l} \equiv -\tan(\delta_l), \tag{8.4.6}$$

$$\frac{A_l}{\cos(\delta_l)} \equiv a_l, \tag{8.4.7}$$

one finds

$$\begin{aligned} R_{kl}(r) &= a_l \Big[\cos(\delta_l) j_l(kr) - \sin(\delta_l) n_l(kr)\Big] \\ &\sim \frac{a_l}{kr} \sin\left(kr - \frac{l\pi}{2} + \delta_l\right). \end{aligned} \tag{8.4.8}$$

The insertion of (8.4.8) into the expansion (8.4.1) therefore yields the asymptotic formula

$$\psi_k^+(r) \sim \sum_{l=0}^{\infty} \frac{a_l(k)}{kr} P_l(\cos\theta) \sin\left(kr - \frac{l\pi}{2} + \delta_l\right). \tag{8.4.9}$$

On the other hand, from Eq. (8.2.13) one finds another useful formula for $\psi_k^+(r)$, i.e. (with $z \equiv r\cos\theta$)

$$\psi_k^+(r) \sim e^{ikz} + f_k(\theta)\frac{e^{ikr}}{r}. \tag{8.4.10}$$

Further progress is made using the expansion of a plane wave, in spherical coordinates, in terms of Legendre polynomials, i.e.

$$e^{ikr\cos\theta} = \sum_{l=0}^{\infty}(2l+1)i^l \frac{\sin(kr - l\pi/2)}{kr} P_l(\cos\theta). \tag{8.4.11}$$

The next step is now to compare the asymptotic formulae (8.4.9) and (8.4.10), and re-express the sin functions with the help of the identity $e^{ix} = \cos(x) + i\sin(x)$. This leads to the equation

$$\begin{aligned} f_k(\theta)\frac{e^{ikr}}{r} &+ \sum_{l=0}^{\infty}(2l+1)\frac{i^l}{2ikr}e^{-il\pi/2}P_l(\cos\theta)e^{ikr} \\ &+ \sum_{l=0}^{\infty}(2l+1)\frac{i^l}{2kr}ie^{il\pi/2}P_l(\cos\theta)e^{-ikr} \\ &= \sum_{l=0}^{\infty}a_l(k)\frac{1}{2ikr}e^{-il\pi/2}e^{i\delta_l}P_l(\cos\theta)e^{ikr} \\ &+ \sum_{l=0}^{\infty}a_l(k)\frac{i}{2kr}e^{il\pi/2}e^{-i\delta_l}P_l(\cos\theta)e^{-ikr}. \end{aligned} \tag{8.4.12}$$

This can be re-expressed in the form

$$T_{1,k}e^{-ikr} + T_{2,k}e^{ikr} = 0, \tag{8.4.13}$$

which is satisfied $\forall r$ if and only if

$$T_{1,k} = 0, \quad T_{2,k} = 0. \tag{8.4.14}$$

Now from (8.4.12) and (8.4.14) one finds

$$a_l(k) = (2l+1)i^l e^{i\delta_l}, \tag{8.4.15}$$

$$f_k(\theta) = \sum_{l=0}^{\infty}\frac{(2l+1)}{k}e^{i\delta_l}\sin(\delta_l)P_l(\cos\theta). \tag{8.4.16}$$

Thus, the *phase shifts* $\delta_l(k)$ determine the scattering amplitude completely, and one finds, for the total cross section,

$$\sigma = \int |f_k(\theta)|^2 \, d\Omega = \frac{4\pi}{k^2}\sum_{l=0}^{\infty}(2l+1)\sin^2\delta_l(k). \tag{8.4.17}$$

A number of rigorous results exist on the partial wave expansion. In particular, we would like to emphasize what follows (Reed and Simon 1979).

Theorem 8.3 Let V be a central potential with $e^{\alpha|\vec{x}|}V \in R$, for some $\alpha > 0$. Given $E \in \mathbf{R}_+/\mathcal{E}$, the partial wave expansion

$$F(E,\cos\theta) = \sum_{l=0}^{\infty}(2l+1)f_l(E)P_l(\cos\theta) \tag{8.4.18}$$

converges uniformly for $\theta \in [0, 2\pi]$.

Theorem 8.4 Let $V(r)$ be a central potential, piecewise continuous on $[0, \infty[$. Suppose that

$$\int_0^1 r|V(r)|\,\mathrm{d}r < \infty, \quad \int_1^{\infty} |V(r)|\,\mathrm{d}r < \infty. \tag{8.4.19}$$

Let E be positive and $l \geq 0$. If these conditions hold, there exists a unique function $\varphi_{l,E}(r)$ on $(0, \infty)$, which is of class C^1, piecewise C^2, and which satisfies the equation

$$\left[-\frac{\mathrm{d}^2}{\mathrm{d}r^2} + V_l(r)\right]\varphi_{l,E}(r) = E\varphi_{l,E}(r), \tag{8.4.20}$$

where

$$V_l(r) \equiv V(r) + \frac{l(l+1)}{r^2}, \tag{8.4.21}$$

jointly with the boundary conditions

$$\lim_{r\to 0}\varphi_{l,E}(r) = 0, \tag{8.4.22a}$$

$$\lim_{r\to 0} r^{-l-1}\varphi_{l,E}(r) = 1. \tag{8.4.22b}$$

Moreover, there exists a constant b such that

$$\lim_{r\to\infty}\left[b\varphi_{l,E}(r) - \sin\left(kr - \frac{l\pi}{2} + \delta_l(E)\right)\right] = 0. \tag{8.4.23}$$

Note that we are concerned with $\varphi_{l,E}(r)$, which is obtained from the radial wave function by the relation (cf. Eq. (5.4.37) and the discussion thereafter) $R_{l,E}(r) = \frac{\varphi_{l,E}(r)}{r}$. We should also acknowledge that the condition (8.4.22a) is implied by the limit (8.4.22b), whereas the converse does not hold. They are both written explicitly to make it easier to perform a comparison with other kinds of boundary-value problems occurring in theoretical physics.

8.5 The Levinson theorem

Here we depart from the previous restriction to the continuous spectrum and deal with possible discrete states. For convenience, only bound states with $E < 0$ are considered, although discrete normalizable states with $E > 0$ are possible (Simon 1969). All of these satisfy the homogeneous Lippmann–Schwinger equation.

Consider the stationary Schrödinger equation for a central potential in $\mathbf{R}^3$, here written in the form (with $y(r)$ corresponding to $\varphi_{l,E}(r)$ in (8.4.20))

$$\left[\frac{\mathrm{d}^2}{\mathrm{d}r^2} + k^2 - \frac{\left(\lambda^2 - \frac{1}{4}\right)}{r^2} - V(r)\right] y(r) = 0, \tag{8.5.1}$$

where $k^2 \equiv \frac{2mE}{\hbar^2}, \lambda \equiv l + \frac{1}{2}, V(r) \equiv \frac{2m}{\hbar^2}U(r)$, and the potential is taken to satisfy the conditions

$$\int_0^{r_0} r|V(r)|\, \mathrm{d}r < \infty, \tag{8.5.2}$$

$$\int_{r_0}^{\infty} r^2|V(r)|\, \mathrm{d}r < \infty, \tag{8.5.3}$$

which are automatically satisfied by potentials with compact support and everywhere finite. If the logarithmic derivative of $y(r)$ at r_0 is continuous at a definite energy $E < 0$, there is a bound state with this energy. Moreover, the logarithmic derivative $\frac{y'}{y}$ is monotonic with respect to the energy (see below), and hence the continuity of the logarithmic derivative at zero energy determines whether there are bound states or not. On the other hand, the logarithmic derivative for $E \geq 0$ determines the phase shift at zero energy $\delta_l(0)$. The Levinson theorem shows the link between $\delta_l(0)$ and the number n_l of bound states (Levinson 1953).

In our proof we assume, for simplicity, that condition (8.5.3) is fulfilled by a potential with compact support:

$$V(r) = 0 \quad \forall r \geq r_0. \tag{8.5.4}$$

First, we introduce a real parameter μ for which one can write

$$V(r, \mu) = \mu V(r). \tag{8.5.5}$$

The idea is that, as μ ranges from 0 to 1, the rescaled potential $V(r, \mu)$ ranges from 0 to the original value $V(r)$. The radial equation (8.5.1) is hence replaced by

$$\left[\frac{\partial^2}{\partial r^2} + k^2 - \frac{(\lambda^2 - \frac{1}{4})}{r^2} - V(r, \mu)\right] y_{k,\lambda}(r, \mu) = 0. \tag{8.5.6}$$

We now consider Eq. (8.5.6) for two different values, e.g. k (with solution $y_{k,\lambda}$) and $\overline{k}$ (with solution $\overline{y}_{k,\lambda}$), and multiply the equations for $y_{k,\lambda}(r,\mu)$ and $\overline{y}_{k,\lambda}(r,\mu)$ by $\overline{y}_{k,\lambda}(r,\mu)$ and $y_{k,\lambda}(r,\mu)$, respectively. On taking the difference of the resulting equations, one finds

$$\frac{\partial}{\partial r}\left(y_{k,\lambda}\frac{\partial}{\partial r}\overline{y}_{k,\lambda} - \overline{y}_{k,\lambda}\frac{\partial}{\partial r}y_{k,\lambda}\right) + \left(\overline{k}^2 - k^2\right)y_{k,\lambda}\,\overline{y}_{k,\lambda} = 0. \tag{8.5.7}$$

Since, by regularity, both $y_{k,\lambda}$ and $\overline{y}_{k,\lambda}$ have a vanishing limit as $r \to 0$, the integration of Eq. (8.5.7) over the interval $[0, r_0]$ yields

$$\left[y_{k,\lambda}(r,\mu)\frac{\partial}{\partial r}\overline{y}_{k,\lambda}(r,\mu) - \overline{y}_{k,\lambda}(r,\mu)\frac{\partial}{\partial r}y_{k,\lambda}(r,\mu)\right]_{r=r_0^-} + \left(\overline{k}^2 - k^2\right)\int_0^{r_0} y_{k,\lambda}(r',\mu)\overline{y}_{k,\lambda}(r',\mu)\,\mathrm{d}r' = 0, \tag{8.5.8}$$

where r_0^- denotes $\lim_{\varepsilon\to 0}(r_0 - \varepsilon)$. Since $\overline{k} \neq k$ by hypothesis, we can multiply both sides of (8.5.8) by $\frac{1}{y_{k,\lambda}(r_0,\mu)\overline{y}_{k,\lambda}(r_0,\mu)}\frac{1}{(\overline{k}^2-k^2)}$, which yields

$$\frac{1}{(\overline{k}^2 - k^2)}\left[\frac{1}{\overline{y}_{k,\lambda}(r_0,\mu)}\frac{\partial}{\partial r}\,\overline{y}_{k,\lambda}(r,\mu)\Big|_{r_0} - \frac{1}{y_{k,\lambda}(r_0,\mu)}\frac{\partial}{\partial r}\,y_{k,\lambda}(r,\mu)\Big|_{r_0}\right] = -\int_0^{r_0}\frac{y_{k,\lambda}(r',\mu)\,\overline{y}_{k,\lambda}(r',\mu)}{y_{k,\lambda}(r_0,\mu)\,\overline{y}_{k,\lambda}(r_0,\mu)}\,\mathrm{d}r'. \tag{8.5.9}$$

At this stage, on taking the limit of both sides of (8.5.9) as $\overline{k} \to k$, one finds

$$\frac{\partial}{\partial E}\left[\frac{1}{y_{E,\lambda}(r,\mu)}\frac{\partial}{\partial r}y_{E,\lambda}(r,\mu)\right]_{r=r_0^-} = -y_{E,\lambda}^{-2}(r_0,\mu)\int_0^{r_0} y_{E,\lambda}^2(r',\mu)\,\mathrm{d}r' < 0, \tag{8.5.10}$$

where the subscript k for $y(r,\mu)$ has been replaced by $E = \frac{\hbar^2}{2m}k^2$, which is more convenient from now on. Similarly, one finds

$$\frac{\partial}{\partial E}\left[\frac{1}{y_{E,\lambda}(r,\mu)}\frac{\partial}{\partial r}y_{E,\lambda}(r,\mu)\right]_{r=r_0^+} = y_{E,\lambda}^{-2}(r_0,\mu)\int_{r_0}^{\infty} y_{E,\lambda}^2(r',\mu)\,\mathrm{d}r' > 0, \tag{8.5.11}$$

where r_0^+ means $\lim_{\varepsilon\to 0}(r_0 + \varepsilon)$. Equations (8.5.10) and (8.5.11) prove that, as the energy increases, the logarithmic derivative of the radial function at r_0^- *decreases monotonically,* whereas that at r_0^+ *increases monotonically.* This expresses the Sturm–Liouville theorem (cf. Sturm 1836).

The matching condition at r_0 for the logarithmic derivative of the radial function is, of course,

$$A_\lambda(E,\mu) \equiv \left[\frac{1}{y_{E,\lambda}(r,\mu)}\frac{\partial}{\partial r}y_{E,\lambda}(r,\mu)\right]_{r=r_0^-}$$
$$= \left[\frac{1}{y_{E,\lambda}(r,\mu)}\frac{\partial}{\partial r}y_{E,\lambda}(r,\mu)\right]_{r=r_0^+}. \quad (8.5.12)$$

Some limiting cases of Eq. (8.5.6) are easily dealt with. For example, for a free particle, which corresponds to a vanishing value of μ, one finds (J_λ being the Bessel function of first kind and order λ)

$$y_{E,\lambda}(r,0) = \sqrt{\frac{1}{2}\pi kr}\; J_\lambda(kr), \quad (8.5.13a)$$

when $E > 0$, and

$$y_{E,\lambda}(r,0) = \mathrm{e}^{-\mathrm{i}\lambda\pi/2}\sqrt{\frac{1}{2}\pi\kappa r}\; J_\lambda(\mathrm{i}\kappa r), \quad (8.5.13b)$$

with $\kappa \equiv \frac{\sqrt{-2mE}}{\hbar}$, if $E \leq 0$.

In the interval $[r_0, \infty[$ the potential V vanishes, and for positive values of E two oscillating solutions of Eq. (8.5.6) exist, so that the general solution reads

$$y_{E,\lambda}(r,\mu) = \sqrt{\frac{1}{2}\pi kr}\Big[J_\lambda(kr)\cos\delta_\lambda(k,\mu) - N_\lambda(kr)\sin\delta_\lambda(k,\mu)\Big], \quad (8.5.14)$$

where $\delta_\lambda(k,\mu)$ is the phase shift and N_λ is the Neumann function of order λ. The matching condition (8.5.12) leads to a very useful formula for the phase shift, upon using the result (8.5.14), i.e.

$$\tan\delta_\lambda(k,\mu) = \frac{J_\lambda(kr_0)}{N_\lambda(kr_0)}\frac{\left[A_\lambda(E,\mu) - k\frac{J'_\lambda(kr_0)}{J_\lambda(kr_0)} - \frac{1}{2r_0}\right]}{\left[A_\lambda(E,\mu) - k\frac{N'_\lambda(kr_0)}{N_\lambda(kr_0)} - \frac{1}{2r_0}\right]}. \quad (8.5.15)$$

Equation (8.5.15) provides the key tool for proving the Levinson theorem, jointly with a careful analysis of matching conditions. Here we shall require that the phase shift is determined with respect to the phase shift $\delta_\lambda(k,0)$ for a free particle, where, by definition, one chooses $\delta_\lambda(k,0) = 0$. With such a convention, the phase shift is determined completely as μ increases from 0 to 1.

If $E \leq 0$, the only square-integrable solution of Eq. (8.5.6) is

$$y_{E,\lambda} = \mathrm{e}^{\mathrm{i}(\lambda+1)\pi/2}\sqrt{\frac{1}{2}\pi\kappa r}\; H_\lambda^{(1)}(\mathrm{i}\kappa r), \quad (8.5.16)$$

where $H_\lambda^{(1)}$ is the standard notation for Hankel functions of first kind and order λ. Thus, the right-hand side of the matching condition (8.5.12) reads as

$$\left[\frac{1}{y_{E,\lambda}(r,0)}\frac{\partial}{\partial r}y_{E,\lambda}(r,0)\right]_{r=r_0^+} = \frac{i\kappa H_\lambda^{(1)'}(i\kappa r_0)}{H_\lambda^{(1)}(i\kappa r_0)} + \frac{1}{2r_0}. \tag{8.5.17}$$

As $E \rightarrow 0^-$, the right-hand side of (8.5.17) reduces to

$$\left(-\lambda + \frac{1}{2}\right)\frac{1}{r_0} \equiv \rho_\lambda, \tag{8.5.18}$$

whereas it tends to $-\kappa$ as $E \rightarrow -\infty$. Moreover, the solution (8.5.13*b*) satisfies the condition

$$\left[\frac{1}{y_{E,\lambda}(r,0)}\frac{\partial}{\partial r}y_{E,\lambda}(r,0)\right]_{r=r_0^-} = \frac{i\kappa J_\lambda'(i\kappa r_0)}{J_\lambda(i\kappa r_0)} + \frac{1}{2r_0}, \tag{8.5.19}$$

for which the right-hand side tends to

$$\left(\lambda + \frac{1}{2}\right)\frac{1}{r_0} \equiv \tilde{\rho}_\lambda \quad \text{as } E \rightarrow 0^-. \tag{8.5.20}$$

Thus, no bound state exists when $\mu = 0$.

If $A_\lambda(0,\mu)$ decreases across the value ρ_λ as μ increases, an overlap between two ranges of variation of the logarithmic derivative on the two sides of r_0 occurs. Bearing in mind the Sturm–Liouville theorem expressed by (8.5.10) and (8.5.11), such an overlap means that the matching condition (8.5.12) can only be satisfied by one particular value of the energy, and hence a scattering state is turned into a bound state. In general, as μ increases from 0 to 1, each time A_λ decreases across ρ_λ, a scattering state is turned into a bound state for the above reasons. In contrast, each time $A_\lambda(0,\mu)$ increases across ρ_λ, a bound state is turned into a scattering state. The number of bound states is then equal to the number of times that $A_\lambda(0)$ decreases across ρ_λ as μ ranges from 0 to 1, minus the number of times that $A_\lambda(0)$ increases across the value ρ_λ. The next task is now to prove that this difference equals $\delta_\lambda(0)$, the phase shift at zero momentum, divided by π.

For this purpose, we have to evaluate $\tan \delta_\lambda(k,\mu)$ when $k \ll \frac{1}{r_0}$, because

$$\delta_\lambda(0,\mu) = \lim_{k\rightarrow 0} \delta_\lambda(k,\mu).$$

By virtue of the exact result (8.5.15) and of the limiting behaviour of Bessel functions at small argument, one finds, to lowest order in kr_0,

$$\tan\delta_\lambda(k,\mu) \sim -\frac{\pi(kr_0)^{2\lambda}}{2^{2\lambda}\lambda\Gamma^2(\lambda)} \frac{\left[A_\lambda(0,\mu) - (\lambda+\frac{1}{2})\frac{1}{r_0}\right]}{\left[A_\lambda(0,\mu) + (\lambda-\frac{1}{2})\frac{1}{r_0}\right]}$$

$$= -\frac{\pi(kr_0)^{2\lambda}}{2^{2\lambda}\lambda\Gamma^2(\lambda)} \frac{\left[A_\lambda(0,\mu) + (\rho_\lambda - \frac{1}{r_0})\right]}{[A_\lambda(0,\mu) - \rho_\lambda]}, \qquad (8.5.21)$$

where we have neglected, for simplicity, higher-order terms in kr_0 in the denominator.

Note that $\tan\delta_\lambda(k,\mu)$ tends to 0 as $k \to 0$ (since $\lambda \geq 0$), and hence $\delta_\lambda(0,\mu)$ is always equal to an integer multiple of π. This means that the phase shift changes discontinuously. Besides, the exact formula (8.5.15) can be used to prove that the phase shift increases monotonically as the logarithmic derivative $A_\lambda(E,\mu)$ decreases. Thus, $\delta_\lambda(0,\mu)$ jumps by π if, for k sufficiently small, $\tan\delta_\lambda(k,\mu)$ changes sign as $A_\lambda(E,\mu)$ decreases. In summary, when μ ranges from 0 to 1, whenever $A_\lambda(0,\mu)$ decreases from near and larger than the value ρ_λ to smaller than ρ_λ, the denominator in (8.5.21) changes sign from positive to negative, leading to a jump of $\delta_\lambda(0,\mu)$ equal to π. In contrast, whenever $A_\lambda(0,\mu)$ increases across ρ_λ, $\delta_\lambda(0,\mu)$ jumps by $-\pi$. Bearing in mind what we said after Eq. (8.5.20), we conclude that $\delta_\lambda(0)$ divided by π is indeed equal to the number n_λ of bound states:

$$\delta_\lambda(0) = n_\lambda\,\pi, \qquad (8.5.22)$$

which is the form of the Levinson theorem for central potentials in $\mathbf{R}^3$ obeying (8.5.2) and (8.5.3) (Ma 1985).

8.6 Scattering from singular potentials

The consideration of singular potential scattering equations was motivated, in the 1960s, by the need to obtain new ideas and techniques that could be extended to quantum field theory (Bastai *et al.* 1963, Khuri and Pais 1964, De Alfaro and Regge 1965, Calogero 1967, Frank *et al.* 1971, Graeber and Dürr 1977, Enss 1979, Dolinszky 1980), especially for the case of field theories where conventional perturbative methods fail to provide a consistent picture. Our reader, however, is not assumed to know quantum field theory, and he/she will not need it to understand what we are going to say.

The remarkable feature of *singular potentials* (which can be used to define them) is that they lead to differential equations with non-Fuchsian singularities (cf. appendix 5.B) *both* as $r \to 0$ *and* as $r \to \infty$. We are concerned with a Schrödinger equation for stationary states in a central potential in two spatial dimensions, which therefore reads as

$$\left(\frac{\mathrm{d}^2}{\mathrm{d}r^2} + \frac{1}{r}\frac{\mathrm{d}}{\mathrm{d}r} - \frac{\lambda^2}{r^2} + k^2\right)\psi(r) = V(r)\psi(r). \tag{8.6.1}$$

Equation (8.6.1) can also be obtained from the stationary Schrödinger equation in three dimensions, after a suitable rescaling of the radial part of the wave function (Frank *et al.* 1971). What is crucial is the study of the polydromy (i.e. many-valuedness) of the wave function in the r variable. Interestingly, both singular and regular potentials have very simple and general features within this framework (Fubini and Stroffolini 1965, Stroffolini 1971, Esposito 1998a). Indeed, if the potential $V(r)$ is a single-valued function of r, one can find two independent solutions

$$\psi_1(r) = r^{\gamma}\chi_1(r), \tag{8.6.2}$$

$$\psi_2(r) = r^{-\gamma}\chi_2(r), \tag{8.6.3}$$

where χ_1 and χ_2 are single-valued functions of r, and γ is a (complex) parameter, the fractional part of which can be determined from an eigenvalue problem (see below). Note also that the operator (in $\hbar = 1$ units)

$$M \equiv \exp(-2\pi\vec{x}\cdot\vec{p}) = \exp\left(2\pi\mathrm{i}r\frac{\partial}{\partial r}\right) \tag{8.6.4}$$

performs a rotation of $2\pi\mathrm{i}$ in the r-space, and hence its eigenfunctions of the form (8.6.2) and (8.6.3) have well-defined many-valuedness properties. If $V(r)$ is single-valued, the operator M commutes with the Hamiltonian, and Eqs. (8.6.2) and (8.6.3) simply state that the solutions of Eq. (8.6.1) can be classified by means of the operator M. The general solution of Eq. (8.6.1) is therefore of the form

$$\psi(r) = \alpha_1\psi_1(r) + \alpha_2\psi_2(r). \tag{8.6.5}$$

Remarkably, one can compute directly $\chi_1(r)$ and $\chi_2(r)$ and study their behaviour as $r \to 0$ and as $r \to \infty$. For this purpose, the following Laurent expansions are used:

$$W(r) \equiv r^2\Big[V(r) - k^2\Big] = \sum_{n=-\infty}^{\infty} w_n r^n \quad r \in]0,\infty[, \tag{8.6.6}$$

$$\chi(r) = \sum_{n=-\infty}^{\infty} c_n r^n \quad r \in]0,\infty[. \tag{8.6.7}$$

These expansions hold because $V(r)$ is assumed to be an analytic function in the complex-r plane, with singularities only at infinity and at the origin

(Forsyth 1959). The Laurent series (8.6.6) and (8.6.7) are now inserted into Eq. (8.6.1), which is equivalent to the differential equation

$$\left[r^2\frac{\mathrm{d}^2}{\mathrm{d}r^2}+(2\gamma+1)r\frac{\mathrm{d}}{\mathrm{d}r}+(\gamma^2-\lambda^2)\right]\chi(r)=W(r)\chi(r). \tag{8.6.8}$$

One thus finds the following infinite system of equations for the coefficients (Stroffolini 1971):

$$\left[(\gamma+n)^2-\overline{\lambda}^2\right]c_n=\sum_{m=-\infty}^{\infty}\overline{u}_{n-m}c_m, \tag{8.6.9}$$

where

$$\overline{\lambda}^2\equiv\lambda^2+w_0, \tag{8.6.10}$$

$$\overline{u}_n\equiv w_n-w_0\delta_{n,0}. \tag{8.6.11}$$

To solve the system (8.6.9) one first writes an equivalent system for which the determinant of the matrix of coefficients is well defined. Such a new system is obtained from (8.6.9) by dividing the nth equation by $(\gamma+n)^2-\overline{\lambda}^2$. The resulting matrix of coefficients has elements

$$H_{n,m}=\delta_{n,m}-\frac{\overline{u}_{n-m}}{[(\gamma+n)^2-\overline{\lambda}^2]}, \tag{8.6.12}$$

where $\det(H)$ exists since the double series

$$\sum_{n,m}\frac{\overline{u}_{n-m}}{[(\gamma+n)^2-\overline{\lambda}^2]}$$

converges for all values of γ that do not correspond to zeros of the denominator. At this stage one can appreciate the substantial difference between regular and singular potentials. In the former case, u_n is non-vanishing only for positive n. In the singular case, however, the presence of negative powers in the Laurent series (8.6.6) gives γ as the solution of a transcendental equation, i.e. (the vanishing of $\det(H)$ being necessary to find non-trivial solutions of the system (8.6.9))

$$F(\gamma)\equiv\det(H)=0, \tag{8.6.13}$$

which can be cast in the form (Stroffolini 1971)

$$\cos(2\pi\gamma)=1-F(0)[1-\cos(2\pi\overline{\lambda})]. \tag{8.6.14}$$

The work in Stroffolini (1971) provides a detailed application of such a technique to the analysis of potentials of the form $\frac{g}{r^q}$ with q even, or consisting of finitely many terms proportional to r^{-q} with q even and odd.

(To deal with them, one can take the limiting case of a Dirac-δ potential (Gottfried 1966), using the continuity of the probability density.)

In fairly recent times, very important results in scattering theory are as follows.

(i) A proof of asymptotic completeness for two-body quantum scattering, and extension to three-body quantum scattering (Enss 1978, 1983).

(ii) The first general proof of asymptotic completeness for short-range potentials (Sigal and Soffer 1987).

(iii) Asymptotic completeness for long-range potentials falling off as $r^{-\mu}$, with $\mu > \sqrt{3} - 1$ (Derezinski 1993).

Yet other developments deal with scattering from potentials involving the Dirac delta functional and its first derivative (Boya and Sudarshan 1996).

8.7 Resonances

There exist physical systems where it is not only true that there is no perturbed eigenvalue (see the Stark effect in section 7.5), but where the unperturbed eigenvalue is not isolated. For example, for the helium atom (subsection 14.9.1), the majority of the eigenvalues of the unperturbed Hamiltonian H_0 (which is the sum of two hydrogen-like Hamiltonians) are in the continuous spectrum. When the perturbation V is turned on, the very concept of perturbed eigenvalues becomes meaningless (Reed and Simon 1978), and the eigenvalues in the continuous spectrum are found to 'dissolve'. However, a 'memory' of the eigenvalues of H_0 is observed in the physics of the helium atom. In the scattering of electrons off helium ions, one observes 'bumps' in the scattering cross section for total $\mathrm{He}^+ + \mathrm{e}^-$ energies near the unperturbed energies (see figure 8.1). Similar bumps are found in the absorption of light by helium, i.e. at frequencies of incident light for which the energy of a light quantum is near the difference of a given unperturbed eigenvalue and the energy of the ground state.

Not only are the bumps, called Auger or autoionizing states, observed, but their widths are fairly well described by the Fermi golden rule (7.8.30). We are now aiming to isolate the mathematical quantity corresponding to the widths of the bumps, following Simon (1973b) and Reed and Simon (1978). First, we say that the scattering amplitude can be shown to be a complex-valued function of the energy and scattering angle, and is typically an analytic function of energy in a cut plane $\mathbf{C}/\sigma(H)$, where H is the Hamiltonian of the interacting quantum system. Suppose now that the scattering amplitude $f(E)$ has an analytic continuation onto a second

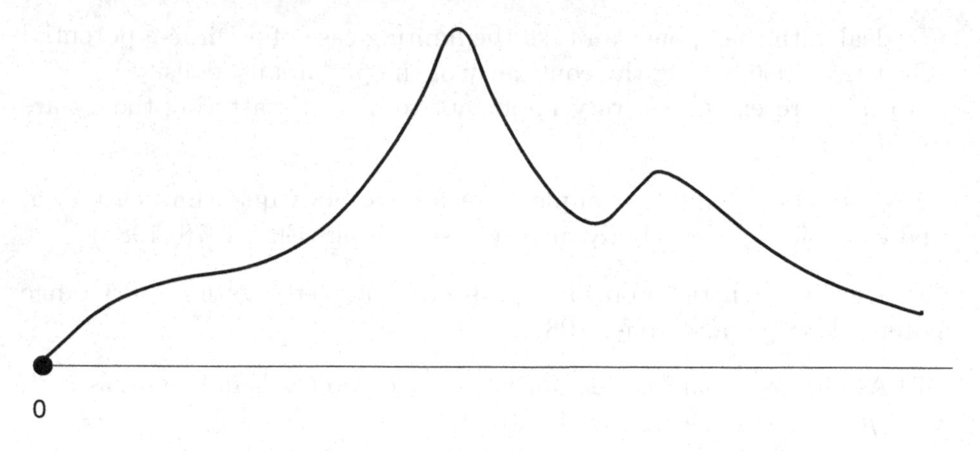

Fig. 8.1. Schematic elastic scattering cross section for the process $e^- + He^+ \rightarrow e^- + He^+$. The quantity on the abscissa is the total kinetic energy.

sheet and that there is a simple pole on the second sheet at a position $E_r - i\frac{\Gamma}{2}$ very near the real axis. We are thus assuming that

$$f(E) = \frac{C}{(E - E_r + i\frac{\Gamma}{2})} + f_b(E), \tag{8.7.1}$$

where the background f_b is analytic at $E = E_r - i\frac{\Gamma}{2}$. If the pole is very near the real axis and if $|f_b(E_r)|$ is not too large, then

$$|f(E)|^2 = \frac{|C|^2}{(E - E_r)^2 + \Gamma^2/4} + R, \tag{8.7.2}$$

where the remainder R is small near $E = E_r$. The so-called Breit–Wigner resonance shape (Breit and Wigner 1936) has the exact form (see figure 8.2)

$$|f_{BW}(E)|^2 = \frac{|C|^2}{(E - E_r)^2 + \Gamma^2/4}, \tag{8.7.3}$$

which holds only for $E > 0$. Its width at half-maximum is Γ. The pole in $f(E)$ is called a *resonance pole* and Γ is called the width of the resonance. If the pole term is much larger than the background term for $E = E_r$, Γ approximates the width of a bump in $|f(E)|^2$.

We should now use a technical result, which states that the scattering amplitude is related to the boundary values of the resolvent $(H - E\mathbb{I})^{-1}$ as E approaches the real axis from the upper half-plane. We therefore look for a method of continuing the expectation value

$$R_\psi(E) \equiv \Big(\psi, (H - E\mathbb{I})^{-1}\psi\Big) \tag{8.7.4}$$

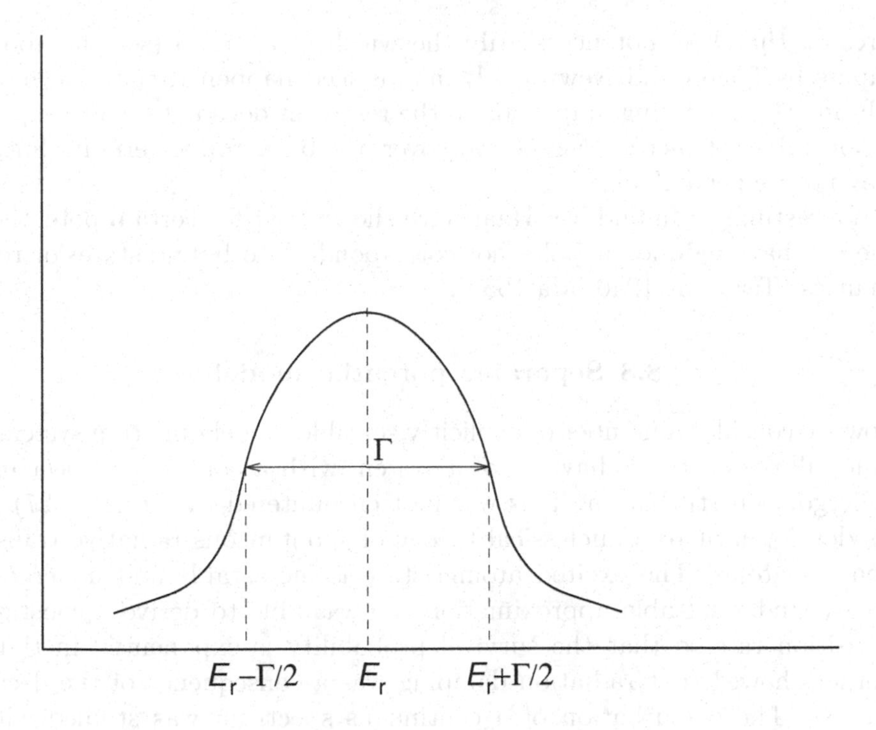

Fig. 8.2 A Breit–Wigner resonance shape.

to the second sheet. If we find that such a continuation is possible for a dense set of $\psi \in \mathcal{H}$ and that for this dense set, the function $R_\psi(E)$ has a pole at $E = E_r - i\frac{\Gamma}{2}$, we will associate E with a resonance pole and Γ with a resonance width. Of course, a reason is needed to think that the pole should be associated with H rather than the particular dense set of ψ, and hence only those ψ are considered for which

$$R_\psi^{(0)}(E) \equiv \Big(\psi, (H_0 - E\mathbb{1})^{-1}\psi\Big) \tag{8.7.5}$$

also has a continuation to the second sheet but without a pole at $E_r - i\frac{\Gamma}{2}$. We are now in a position to give a rigorous definition of resonance poles (Reed and Simon 1978).

Definition 8.1. Suppose that there is a dense set of vectors $D \subset \mathcal{H}$ such that for all $\psi \in D$, both $R_\psi(E)$ and $R_\psi^{(0)}(E)$ have analytic continuations onto the second sheet across the real axis from the upper half-plane of the first sheet. If $R_\psi^{(0)}(E)$ is analytic at $E_0 \equiv E_r - i\frac{\Gamma}{2}$ and $R_\psi(E)$ has a pole at E_0 for some ψ, the point E_0 is said to be a *resonance pole.* Γ is then called the width of the resonance.

Work in Fonda and Newton (1959, 1960) and Newton and Fonda (1960) has shown remarkable behaviour in multichannel processes. It should be stressed that Γ is not necessarily the width of a bump (see the above papers by Fonda and Newton). It may indeed happen that the formula relating the scattering amplitude to the resolvent does not involve expectation values of vectors $\psi \in D$. Moreover, the background term in (8.7.1) may not be negligible.

Interestingly, Ma and Ter Haar have shown that for certain potentials one can have redundant poles not corresponding to bound states or resonances (Ter Haar 1946, Ma 1952).

8.8 Separable potential model

Now we consider a number of explicitly solvable model quantum systems, which illustrate the behaviour of a system with a continuous spectrum undergoing perturbation. This was first encountered by Dirac (1927) in his development of semiclassical theory of spontaneous radiative transitions in atoms. The excited atomic state is metastable and undergoes decay. Under suitable approximations he was able to derive a constant transition rate so that the survival probability is exponential in time. Heitler showed that radiation damping was a consequence of the decay process. The perturbation of a continuous spectrum was studied with mathematical rigour by Friedrichs (1948). Formation of bound states in the interaction between two particles was studied in terms of separable potentials in Yamaguchi (1954), Yamaguchi and Yamaguchi (1954), Ernst *et al.* (1973), and was applied systematically to nuclear physics in Mitra (1961).

The use of perturbation theory carried out to all orders can obtain the exact results when there are no new bound states or metastable states. But even when this is the case, the perturbation calculations carried out in the complex energy plane can be used even where metastable states are included. These explicit model solutions are used as a test of standard scattering theories. The 'asymptotic condition' postulated in these formulations is found to be incorrect when composite particles occur. Suitable amendments are suggested.

We begin by studying the stationary Schrödinger equation for scattering problems, with the Hamiltonian operator

$$H_{\omega\omega'} = \omega\delta(\omega - \omega') + \eta G(\omega)G(\omega'), \tag{8.8.1}$$

with G real and $\eta = \pm 1$. To find the improper eigenfunctions ϕ_λ of H we must solve

$$\int H(\omega, \omega')\phi_\lambda(\omega')\,\mathrm{d}\omega' = \lambda\phi_\lambda(\omega), \tag{8.8.2}$$

which implies

$$(\lambda-\omega)\phi_\lambda(\omega)=\eta G(\omega)\int G(\omega')\phi_\lambda(\omega')\,\mathrm{d}\omega'=\eta G(\omega)g(\lambda), \qquad (8.8.3)$$

the last equality defining $g(\lambda)$. The eigenvalue equation in this last form can be solved immediately to give

$$\phi_\lambda(\omega)=\frac{\eta G(\omega)}{(\lambda-\omega)}g(\lambda), \qquad (8.8.4)$$

but the problem remains of making $\phi_\lambda(\omega)$ well defined for $\omega \to \lambda$. This can be done by adding the proper multiple of the solution of the homogeneous equation to obtain

$$\phi_\lambda(\omega)=\frac{\eta G(\omega)}{(\lambda-\omega-\mathrm{i}\varepsilon)}g(\lambda)+\delta(\lambda-\omega), \qquad (8.8.5)$$

which represents a plane wave plus incoming spherical waves. The function g can be found from

$$g(\lambda)=\int G(\omega)\phi_\lambda(\omega)\,\mathrm{d}\omega=\eta\int\frac{G^2(\omega)\,\mathrm{d}\omega}{(\lambda-\omega-\mathrm{i}\varepsilon)}g(\lambda)+G(\lambda), \qquad (8.8.6)$$

which yields

$$g(\lambda)=\frac{G(\lambda)}{\beta^-(\lambda)}, \qquad (8.8.7)$$

where

$$\beta^-(\lambda)\equiv 1-\eta\int\frac{G^2(x)}{(\lambda-x-\mathrm{i}\varepsilon)}\,\mathrm{d}x. \qquad (8.8.8)$$

With our notation, $\beta^-(\lambda)=\beta(\lambda-\mathrm{i}\varepsilon)$, where

$$\beta(z)\equiv 1-\eta\int_1^\infty\frac{G^2(x)}{(z-x)}\,\mathrm{d}x, \qquad (8.8.9)$$

and ϕ_λ reads eventually

$$\phi_\lambda(\omega)=\delta(\lambda-\omega)+\frac{\eta G(\lambda)G(\omega)}{\beta^-(\lambda)(\lambda-\omega-\mathrm{i}\varepsilon)}. \qquad (8.8.10)$$

The lower limit of integration in Eq. (8.8.9) is the energy of the lowest-energy state of the continuum allowed by the unperturbed Hamiltonian H_0. That this lower limit is unchanged by the interaction is a result of theorems proved by Riesz and Nagy. It is clear from (8.8.9), assuming G becomes zero sufficiently rapidly at ∞, that β is analytic in the whole complex plane except for one branch point at $z=1$ from which there is a cut running along the positive real axis. Thus,

$$\beta(\lambda+\mathrm{i}\varepsilon)-\beta(\lambda-\mathrm{i}\varepsilon)=0 \quad \text{if } \lambda<1,$$

and

$$\beta(\lambda + \mathrm{i}\varepsilon) - \beta(\lambda - \mathrm{i}\varepsilon) = \eta 2\pi \mathrm{i} G^2(\lambda) \tag{8.8.11}$$

if $\lambda > 1$, where we have used the identity

$$\frac{1}{(\lambda - \omega \pm \mathrm{i}\varepsilon)} = \frac{\mathrm{P}}{(\lambda - \omega)} \mp \mathrm{i}\pi\delta(\lambda - \omega). \tag{8.8.12}$$

The function β is therefore real analytic, i.e. $\beta(z^*) = \beta^*(z)$. For real $\lambda < 1$ one has

$$\frac{\mathrm{d}\beta}{\mathrm{d}\lambda} = \eta \int_1^\infty \frac{G^2(x)}{(\lambda - x)^2}\,\mathrm{d}x, \tag{8.8.13}$$

which is positive or negative according to the sign of η. Note also from the definition (8.8.9) that $\beta(-\infty) = 1$.

These two facts imply that β can have a real zero for $\lambda < 1$ only if $\eta < 0$, corresponding to an attractive interaction. There can be at most one such zero since by (8.8.13) β is monotonic and continuous for $\lambda < 1$. Furthermore, the existence of the zero depends on G being large enough so that

$$\eta \int_1^\infty \frac{G^2(x)}{(1 - x)}\,\mathrm{d}x > 1. \tag{8.8.14}$$

We now show that a zero of β at some real $\Lambda < 1$ corresponds to a bound-state solution of (8.8.2), i.e. a discrete eigenstate with eigenvalue Λ. If such a state exists the general solution (8.8.5) reduces to the particular solution (8.8.4) since $(\Lambda - \omega)$ is never zero for $\Lambda < 1$ and $1 \le \omega \le \infty$.

The definition of $g(\lambda)$ requires that

$$g(\Lambda) = \int G(\omega')\phi_\Lambda(\omega')\,\mathrm{d}\omega' = g(\Lambda)\eta \int_1^\infty \frac{G^2(\omega')}{(\Lambda - \omega')}\,\mathrm{d}\omega'$$

or

$$g(\Lambda)\left[1 - \eta \int_1^\infty \frac{G^2(\omega)}{(\Lambda - \omega)}\,\mathrm{d}\omega\right] = g(\Lambda)\beta(\Lambda) = 0, \tag{8.8.15}$$

which implies that either β has a zero or else g and hence ϕ are zero and there is no discrete solution. We now check the orthonormality of the solutions of (8.8.2) found above. First, we look at the normalization of the bound state:

$$\int_1^\infty |\phi_\Lambda(\omega)|^2\,\mathrm{d}\omega = 1 \tag{8.8.16a}$$

implies

$$\eta^2 g^2(\Lambda) \int_1^\infty \frac{G^2(\omega)}{(\Lambda - \omega)^2}\,\mathrm{d}\omega = 1, \tag{8.8.16b}$$

hence the condition

$$g(\Lambda) = \frac{1}{\sqrt{\beta'(\Lambda)}}. \tag{8.8.17}$$

Thus, the bound-state wave function is

$$\phi_\Lambda(\omega) = \frac{\eta}{\sqrt{\beta'(\Lambda)}} \frac{G(\omega)}{(\Lambda - \omega)}. \tag{8.8.18}$$

Secondly, the bound state is orthogonal to all continuum states:

$$\begin{aligned}(\phi_\lambda, \phi_\Lambda) &= \frac{G(\lambda)}{(\Lambda - \lambda)} + \eta \frac{G(\lambda)}{\beta^+(\lambda)} \int_1^\infty \frac{G^2(\omega)\,\mathrm{d}\omega}{(\lambda + \mathrm{i}\varepsilon - \omega)(\Lambda - \omega)} \\ &= \frac{G(\lambda)}{(\Lambda - \lambda)} + \eta \frac{G(\lambda)}{\beta^+(\lambda)} \frac{1}{\eta} \frac{\beta^+(\lambda)}{(\lambda - \Lambda)} = 0, \end{aligned} \tag{8.8.19}$$

where we have exploited Eq. (8.8.11), the residue theorem and suitable integration contours in the complex plane (see problem 8.P6). Similarly, one finds

$$(\phi_\lambda, \phi_{\lambda'}) = \delta(\lambda - \lambda'). \tag{8.8.20}$$

8.9 Bound states in the completeness relationship

Having demonstrated orthonormality it is now natural to prove the dual relationship of completeness. Without considering, for the moment, the possible existence of a bound state we look at the integral

$$\begin{aligned}\int \phi_\lambda^*(\omega')\phi_\lambda(\omega)\,\mathrm{d}\lambda &= \int \mathrm{d}\lambda \left[\delta(\lambda - \omega') + \frac{\eta G(\lambda) G(\omega')}{\beta^+(\lambda)(\lambda - \omega' + \mathrm{i}\varepsilon)}\right] \\ &\quad \times \left[\delta(\lambda - \omega) + \frac{\eta G(\lambda) G(\omega)}{\beta^-(\lambda)(\lambda - \omega - \mathrm{i}\varepsilon)}\right] \\ &= \delta(\omega - \omega') + \frac{\eta G(\omega') G(\omega)}{\beta^-(\omega')(\omega' - \omega - \mathrm{i}\varepsilon)} + \frac{\eta G(\omega) G(\omega')}{\beta^+(\omega)(\omega - \omega' + \mathrm{i}\varepsilon)} \\ &\quad + \eta^2 G(\omega') G(\omega) \int \mathrm{d}\lambda \frac{G^2(\lambda)}{\beta^+(\lambda)\beta^-(\lambda)(\lambda - \omega' + \mathrm{i}\varepsilon)(\lambda - \omega - \mathrm{i}\varepsilon)}. \end{aligned} \tag{8.9.1}$$

Using (8.8.11) to rewrite $G^2(\lambda)$ in terms of β^+ and β^-, the last term in (8.9.1) becomes

$$-\frac{\eta G(\omega') G(\omega)}{2\pi \mathrm{i}} \int_1^\infty \mathrm{d}\lambda \left[\frac{1}{\beta^+(\lambda)} - \frac{1}{\beta^-(\lambda)}\right] \frac{1}{(\lambda - \omega' + \mathrm{i}\varepsilon)(\lambda - \omega - \mathrm{i}\varepsilon)}.$$

Two cases may now occur depending on whether there is, or is not, a bound state; which is to say whether there is, or is not, a zero in β.

If there is no zero of β, the right-hand side of Eq. (8.9.1) reduces to $\delta(\omega - \omega')$ (Sudarshan 1962). On the other hand, if there is a zero of β, for $\lambda < 1$ the zero must be a first-order one since $\beta'(\lambda)$ is positive for $\lambda < 1$. One then finds that the right-hand side of (8.9.1) does not reduce to $\delta(\omega-\omega')$, and the scattering solutions are not complete. This, however, should be expected since in the sum over all states the bound state should be included, leading to the 'amended' completeness relation

$$\int \phi_\lambda^*(\omega')\phi_\lambda(\omega)\,\mathrm{d}\lambda + \phi_\Lambda^*(\omega')\phi_\Lambda(\omega) = \delta(\omega - \omega'), \tag{8.9.2}$$

where the second term on the left-hand side of (8.9.2) appears only if the solution $\phi_\Lambda(\omega)$ exists. In other words, just when the physics of the model changes in such a way that a bound-state contribution must be added, the mathematics of the model also changes, through the introduction of a pole, in such a way that the completeness relation holds in the form (8.9.2). If we were forgetting about the bound state, the completeness requirement would tell us that we cannot do so.

8.10 Excitable potential model

Consider now a Hamiltonian which, in matrix form, reads as

$$H = \begin{pmatrix} M & F(\omega') \\ F(\omega) & \omega\delta(\omega - \omega') \end{pmatrix}. \tag{8.10.1}$$

We are aiming to solve the eigenvalue equation

$$H\phi_\lambda = \lambda\phi_\lambda. \tag{8.10.2}$$

From the form of (8.10.1) we see that the solution of Eq. (8.10.2) looks like

$$\phi_\lambda = \begin{pmatrix} \phi_\lambda^M \\ \phi_\lambda(\omega) \end{pmatrix}. \tag{8.10.3}$$

Inserting (8.10.3) and (8.10.1) in (8.10.2) and taking components gives the equations

$$(\lambda - M)\phi_\lambda^M = \int \phi_\lambda(\omega')F(\omega')\,\mathrm{d}\omega', \tag{8.10.4}$$

$$(\lambda - \omega)\phi_\lambda(\omega) = F(\omega)\phi_\lambda^M. \tag{8.10.5}$$

There are two cases according to whether $(\lambda - \omega)$ can, or cannot, have a zero. If it cannot vanish one can write immediately

$$\phi_\lambda(\omega) = \frac{F(\omega)\phi_\lambda^M}{(\lambda - \omega)} \tag{8.10.6}$$

and

$$(\lambda - M)\phi_\lambda^M = \int \frac{F(\omega')\phi_\lambda^M}{(\lambda - \omega')}\,d\omega', \tag{8.10.7}$$

which can be rewritten as

$$\left[\lambda - M - \int \frac{F^2(\omega')}{(\lambda - \omega')}\,d\omega'\right]\phi_\lambda^M = 0, \tag{8.10.8}$$

i.e.

$$\alpha(\lambda)\phi_\lambda^M = 0, \tag{8.10.9}$$

where the function α is defined by

$$\alpha(z) \equiv z - M - \int_1^\infty \frac{F^2(\omega')}{(z - \omega')}\,d\omega'. \tag{8.10.10}$$

In analogy to (8.8.11) we note

$$\alpha(\lambda + i\varepsilon) - \alpha(\lambda - i\varepsilon) = 2\pi i F^2(\lambda), \tag{8.10.11}$$

if $1 < \lambda < \infty$. There will thus be a solution for $\lambda = \Lambda < 1$ if and only if $\alpha(z)$ has a zero on the real axis at an energy less than that of the start of the continuum. To see whether such a bound state exists we notice that $\lambda < 1$ implies

$$\frac{d\alpha}{d\lambda} = 1 + \int_1^\infty \frac{F^2(\omega')}{(\lambda - \omega')^2}\,d\omega' > 1, \tag{8.10.12}$$

so that α is strictly increasing for $\lambda < 1$. Furthermore, from the definition one finds $\alpha(-\infty) = -\infty$. There can therefore be at most one zero (Λ) and it may or may not occur, according to the sizes of F and M. The process of increasing (decreasing) F and decreasing (increasing) M makes the zero in α more (less) 'likely' to appear.

The physics is as follows (Sudarshan 1962): we are considering the scattering of a particle of mass μ from one of mass $1 - \mu$; before we 'turn on' the interaction there is a particle of mass M, which may be greater or less than 1, with the same quantum numbers as the system with mass 1. After the interaction is turned on there is a change in the energy of the particle M to an energy Λ. If $\Lambda < 1$ it appears as a bound state of the particles $1 - \mu$ and μ and we have a model analogous to the Lee model (Lee 1954). Otherwise we have the Lee model with unstable V particle.

Here we assume that there is a zero in α for some $\Lambda < 1$. Then, letting $\phi_\Lambda^M = C$ we have $\phi_\Lambda(\omega) = \frac{CF(\omega)}{(\Lambda - \omega)}$ and we normalize using

$$1 = (\phi_\Lambda, \phi_\Lambda) = |\phi_\Lambda^M|^2 + \int |\phi_\Lambda(\omega)|^2\,d\omega$$

$$= C^2 + |C|^2 \int \frac{F^2(\omega)}{(\Lambda - \omega)^2}\, \mathrm{d}\omega = |C|^2 \alpha'(\Lambda), \tag{8.10.13}$$

so that, if a real phase is chosen for C, one finds

$$C = \frac{1}{\sqrt{\alpha'(\Lambda)}}. \tag{8.10.14}$$

Thus, the bound-state solution is

$$\phi_\Lambda^M = \frac{1}{\sqrt{\alpha'(\Lambda)}}, \tag{8.10.15}$$

$$\phi_\Lambda(\omega) = \frac{F(\omega)}{\sqrt{\alpha'(\Lambda)}} \frac{1}{(\Lambda - \omega)}. \tag{8.10.16}$$

Moreover, we choose

$$\phi_\lambda(\omega) = \frac{F(\omega)\phi_\lambda^M}{(\lambda - \omega - \mathrm{i}\varepsilon)} + \delta(\lambda - \omega), \tag{8.10.17}$$

and substituting in (8.10.4)

$$(\lambda - M)\phi_\lambda^M = \int F(\omega') \left[\delta(\lambda - \omega') + \frac{F(\omega')\phi_\lambda^M}{(\lambda - \omega' - \mathrm{i}\varepsilon)} \right] \mathrm{d}\omega' \tag{8.10.18}$$

or

$$F(\lambda) = \left[\lambda - M - \int \frac{F^2(\omega')}{(\lambda - \omega' - \mathrm{i}\varepsilon)}\, \mathrm{d}\omega' \right] \phi_\lambda^M = \alpha^-(\lambda)\phi_\lambda^M, \tag{8.10.19}$$

where $\alpha^-(\lambda) = \alpha(\lambda - \mathrm{i}\varepsilon)$ and $\alpha(z)$ is defined by (8.10.10) from which it is apparent that $\alpha^-(\lambda) \neq 0$ for $1 < \lambda < \infty$ since, in particular, $\mathrm{Im}\, \alpha^-(\lambda) = -\pi F^2(\omega) \neq 0$. Therefore we may divide by $\alpha^-(\lambda)$ and our solution is

$$\phi_\lambda^M = \frac{F(\lambda)}{\alpha^-(\lambda)}, \tag{8.10.20}$$

$$\phi_\lambda(\omega) = \delta(\lambda - \omega) + \frac{F(\lambda)F(\omega)}{\alpha^-(\lambda)(\lambda - \omega - \mathrm{i}\varepsilon)}. \tag{8.10.21}$$

It is possible to verify, as in the separable potential model, that the solutions (8.10.15), (8.10.16) and (8.10.20), (8.10.21) form an orthonormal and complete set.

8.11 Unitarity of the Möller operator

The orthonormality relations

$$\int \phi_\lambda^*(\omega)\phi_{\lambda'}(\omega)\,\mathrm{d}\omega + \phi_\lambda^*(0)\phi_\lambda(0) = \delta(\lambda - \lambda'), \tag{8.11.1}$$

$$\int \phi_M^*(\omega)\phi_{\lambda'}(\omega)\,\mathrm{d}\omega + \phi_M^*(0)\phi_\lambda(0) = 0, \tag{8.11.2}$$

$$\int \phi_\lambda^*(\omega)\phi_M(\omega)\,\mathrm{d}\omega + \phi_\lambda^*(0)\phi_M(0) = 0, \tag{8.11.3}$$

$$\int \phi_M^*(\omega)\phi_M(\omega)\,\mathrm{d}\omega + \phi_M^*(0)\phi_M(0) = 1, \tag{8.11.4}$$

and completeness relations

$$\int_0^\infty \phi_\lambda(\omega)\phi_\lambda^*(\omega')\,\mathrm{d}\lambda + \phi_M(\omega)\phi_M^*(\omega') = \delta(\omega - \omega'), \tag{8.11.5}$$

$$\int_0^\infty \phi_\lambda(\omega)\phi_\lambda^*(0)\,\mathrm{d}\lambda + \phi_M(\omega)\phi_M^*(0) = 0, \tag{8.11.6}$$

$$\int_0^\infty \phi_\lambda(0)\phi_\lambda^*(\omega')\,\mathrm{d}\lambda + \phi_M(0)\phi_M^*(\omega') = 0, \tag{8.11.7}$$

$$\int_0^\infty \phi_\lambda(0)\phi_\lambda^*(0)\,\mathrm{d}\lambda + \phi_M(0)\phi_M^*(0) = 1, \tag{8.11.8}$$

are precisely the condition that the operator

$$\Omega \equiv \begin{pmatrix} \Omega(\lambda,\omega) & \Omega(\lambda,0) \\ \Omega(M,\omega) & \Omega(M,0) \end{pmatrix} \tag{8.11.9}$$

is unitary. This is thc extended Möller wave operator, for which

$$\Omega^\dagger\Omega = \Omega\Omega^\dagger = \mathbb{I}. \tag{8.11.10}$$

Note that, depending on whether $\alpha(M) = 0$ for $M < 0$, the 'matrix' has a discrete row. But in all cases this matrix is unitary.

The wave operator Ω has the remarkable property that it intertwines the fully interacting Hamiltonian H and a comparison Hamiltonian H_c. The comparison Hamiltonian is the same as the free Hamiltonian except for the replacement $m \to M$. Hence both H and H_c have the same spectrum (while H_0 does not). By no asymptotic limit does H go into H_0. One can explicitly verify the unitary equivalence of H and H_c by computing integrals such as $\Omega H \Omega^\dagger$ and $\Omega^\dagger H_\mathrm{c}\Omega$ to obtain H_c and H, respectively. The interacting system appears as a 'free' comparison Hamiltonian in a canonically transformed system.

8.12 Quantum decay and survival amplitude

The time development of a quantum system with an analytic Hamiltonian is now studied with special reference to decays and resonances. The use of complex contours to evaluate survival probability is exhibited (Sudarshan *et al.* 1978). For every allowed contour there is a vector space but the spaces for different contours are not isomorphic to that for another, but any one relative to the other is dense. Point eigenvalues in the complex plane can be made to correspond to an extended contour by deforming the Cauchy circle around the point.

We also study several solvable models in which not only the scattering amplitude but also the 'in' and 'out' state vectors can be evaluated. It is shown that these are analytically continuable and survival and scattering amplitudes may be computed for any contour. The Möller matrices can also be analytically continued and they are unitary in all cases whether bound states or unstable states exist. A generalized perturbation theory is developed and used to find exact solutions by summing the perturbation series.

The 'decay' is the decrease of the survival amplitude as the time difference increases. The vectors at each time are normalized. Finally, no irreversibility is implied since the same decrease of the 'survival' probability is also obtained for negative times. A detailed study of the survival amplitude shows Zeno and Khalfin modifications for all systems.

8.12.1 Law of radioactive decay: Poisson distribution

If we have a number N of radioactive atoms of a particular kind, the probability that one particle would decay in a small time δt is $\gamma N \delta t$ so that

$$\frac{\mathrm{d}N}{\mathrm{d}t} = -\gamma N(t). \tag{8.12.1}$$

The solution of this differential equation is

$$N(t) = N(0)\mathrm{e}^{-\gamma t}. \tag{8.12.2}$$

Since γ is independent of time we have a pure death process. Given this we can find the probability of n decays in a time interval T. These probabilities may be derived by considering the probabilities for T_1 and T_2 with $T = T_1 + T_2$:

$$\Pi_n(T_1 + T_2) = \sum_{m=0}^{n} \Pi_m(T_1)\Pi_{n-m}(T_2). \tag{8.12.3}$$

The solution of this functional equation is unique:

$$\Pi_n(T) = \mathrm{e}^{-\gamma T}\frac{(\gamma T)^n}{n!}, \tag{8.12.4}$$

except for the parameter γ of decay. This probability distribution is called the Poisson distribution. It may be verified by inspection that

$$\sum_{n=0}^{\infty} \Pi_n(T) = 1, \quad \Pi_n(T) \geq 0. \tag{8.12.5}$$

The average number of decays is

$$\langle n \rangle = \sum_{n=0}^{\infty} n\Pi_n(T) = \gamma T, \tag{8.12.6}$$

and the 'mean squared' decay is

$$\begin{aligned} \langle n^2 \rangle &= \sum_{n=0}^{\infty} n^2 \Pi_n(T) = \sum_{n=0}^{\infty} \mathrm{e}^{-\gamma T} \frac{n(n-1)+n}{n!} (\gamma T)^n \\ &= (\gamma T)^2 + \gamma T = \langle n \rangle^2 + \langle n \rangle. \end{aligned} \tag{8.12.7}$$

Hence the mean square fluctuation is given by the mean

$$\langle n^2 \rangle - \langle n \rangle^2 = \langle n \rangle. \tag{8.12.8}$$

This is a characteristic property of the Poisson distribution. In deriving these results we have assumed that the decay constant γ is a definite constant value. But there are processes in which one could consider a probabilistic version where γ itself is distributed according to some law. One particular distribution is the Rayleigh distribution that gives the probability of a value of γ between γ and $\gamma + \delta\gamma$ is

$$(\delta\gamma)\alpha_0^{-1}\mathrm{e}^{-\gamma T/\alpha_0},$$

which is non-negative and normalized. Then the average of $\Pi_n(T)$ over the Rayleigh distribution of parameters is

$$p_n(T) = \frac{1}{\alpha_0} \int_0^{\infty} \mathrm{e}^{-\gamma T/\alpha_0} \frac{\gamma^n T^n}{n!} \,\mathrm{d}\gamma = \alpha_0^n (1 - \alpha_0). \tag{8.12.9}$$

Again $p_n \geq 0, \sum_n p_n = 1$ assuming that $p_n(T)$ is a probability distribution. It is called the Bose distribution. For this case

$$\langle n \rangle = \langle n^2 \rangle = 2\langle n \rangle^2 + \langle n \rangle,$$

so that

$$\langle n^2 \rangle - \langle n \rangle^2 = \langle n \rangle (1 + \langle n \rangle). \tag{8.12.10}$$

This excess fluctuation is characteristic of waves. The Rayleigh distribution is appropriate to describing the ensemble of complex wave fields, which are Gaussian distributed. Then the intensity (which is the absolute value of the wave amplitude) for a wave ensemble obtained by a very large number of independent sources contributing incoherently is Rayleigh distributed. Thus, the photoelectric counts for an intensity which is Rayleigh distributed obey the Bose distribution (Mandel *et al.* 1964, Klauder and Sudarshan 1968).

8.12.2 Quantum decay transitions. The survival amplitude

The rate of decay of a metastable quantum system was first used by Gamow (1928) for α radioactivity; and it was derived for radiative transitions in atoms using time-dependent perturbation theory by Dirac; many years later Fermi renamed the transition rate formula the 'Golden Rule' in terms of the square of the transition matrix element. In quantum theory the quantity that obeys the linear equation of motion is the amplitude. The transition from the excited state to the ground state for the amplitude is proportional to t. Therefore the probability is proportional to t^2. By integrating over a set of extremely close final states, Dirac was able to approximate the decay rate to be proportional to t and the density of levels. However the approximation fails for very short times when the probability goes as t^2 and hence the rate itself tends to zero as t. Thus, repeated observations on a metastable state make the decay go to zero. This is the quantum Zeno effect (Misra and Sudarshan 1977).

To describe decay we need to compute the 'survival amplitude', i.e. the overlap between a state $|\psi(t_1)\rangle$ and its evolute $|\psi(t_2)\rangle$:

$$A(t_1, t_2) \equiv \langle\psi(t_1)|\psi(t_2)\rangle = \langle\psi(0)|e^{iHt_1}e^{-iHt_2}|\psi(0)\rangle. \tag{8.12.11}$$

By introducing a complete set $\{|\lambda\rangle\}$ of eigenstates of the total Hamiltonian H one can write

$$\begin{aligned} A(t_1, t_2) &= A(0, t_2 - t_1) \\ &= \int_0^\infty \langle\psi(0)|\lambda\rangle e^{-i\lambda(t_2-t_1)}\langle\lambda|\psi(0)\rangle \, d\lambda \\ &= \int_0^\infty |c(\lambda)|^2 e^{-i\lambda(t_2-t_1)} \, d\lambda. \end{aligned} \tag{8.12.12}$$

The expectation values of H and H^2 in the state $|\psi\rangle$ are

$$\langle\psi|H|\psi\rangle = \int_0^\infty |c(\lambda)|^2 \lambda \, d\lambda = \varepsilon, \tag{8.12.13}$$

$$\langle\psi|H^2|\psi\rangle = \int_0^\infty |c(\lambda)|^2 \lambda^2 \, d\lambda = \mathcal{D}^2 + \varepsilon^2, \tag{8.12.14}$$

which may or may not be finite. If $H|\psi\rangle$ is in the Hilbert space, such expectation values are finite. From the expression for the amplitude A

one can recognize that $A(t_2 - t_1)$ is analytic in $(t_2 - t_1)$ and can be expanded in a power series

$$A(t) = 1 - \mathrm{i}\varepsilon t - \frac{t^2}{2}\left(\varepsilon^2 + \mathcal{D}^2\right) + \cdots . \tag{8.12.15}$$

The resulting squared modulus is

$$|A(t)|^2 = 1 - t^2\left(\varepsilon^2 + \mathcal{D}^2\right) + \cdots . \tag{8.12.16}$$

Therefore

$$\begin{aligned} |A(t/N)|^{2N} &= \left[1 - \frac{t^2}{N^2}\left(\varepsilon^2 + \mathcal{D}^2\right) + \cdots\right]^N \\ &= 1 - \frac{t^2}{N}\left(\varepsilon^2 + \mathcal{D}^2\right) + \cdots , \end{aligned} \tag{8.12.17}$$

which tends to reduce to unity as $N \to \infty$. Hence one may say that 'a watched pot never boils'.

We already saw that the survival amplitude is the Fourier transform of the spectral density. Since the spectral density $|c(\lambda)|^2$ integrates to unity, the survival amplitude is the characteristic function. It is analytic in the whole complex plane for t. We saw that at small times its power-series expansion leads to a survival probability, which goes as $1 - \beta t^2$ for small values of t. The analyticity of $A(t)$ implies that (except for the zeros of $A(t)$) $\log A(t)$ is also analytic. The Paley–Wiener criterion then demands that

$$\int \frac{\log |A(t)|}{1 + t^2}\,\mathrm{d}t < \infty. \tag{8.12.18}$$

But for an exponential law

$$|A(t)| = \mathrm{e}^{-\gamma t/2} \tag{8.12.19}$$

this integral diverges. Thus, for long times one cannot have an exponential law (Khalfin 1958).

From the formula for $A(t)$ we can see that for large values of t the integral depends only on $|c(\lambda)|^2$ for small λ. With a three-dimensional space $|c(\lambda)|^2\,\mathrm{d}\lambda = |b(\lambda)|^2\sqrt{\lambda}\,\mathrm{d}\lambda$, and therefore

$$A(t) \approx \int_0^\infty \mathrm{e}^{-\mathrm{i}\lambda t}\sqrt{\lambda}\,\mathrm{d}\lambda = \frac{\sqrt{\pi}}{2}\mathrm{i}^{-3/2}t^{-3/2}. \tag{8.12.20}$$

So this dependence is purely kinematic.

The generic forms of the survival probability over the whole range of t may be studied using complex variable integration techniques. For intermediate times the behaviour of the survival amplitude is exponential. It is as if the state had a unique complex energy (see section 8.7)

$$E_{\rm r} = E - {\rm i}\frac{\gamma}{2}, \tag{8.12.21}$$

so that the state vector has the non-unitary evolution

$$\psi(t) \approx {\rm e}^{-{\rm i}E_{\rm r}t}{\rm e}^{-\gamma t/2}\psi(0). \tag{8.12.22}$$

The 'resonance line shape' reads, with the present notation,

$$|c(\lambda)|^2 = \frac{1}{(\lambda - E_{\rm r})^2 + \gamma^2/4}. \tag{8.12.23}$$

Such a state, however, cannot occur physically since the energy is bounded from below, and we must develop a formalism in which complex energy states arise naturally. Since the Hamiltonian is self-adjoint in Hilbert space its spectrum in this space must be real. Hence we must go beyond the Hilbert space formalism.

Digression. The process of quantum decay contains some curious results. The normal expectation is that the decay product of one atom cannot excite another atom a distance D away until a time $\frac{D}{c}$ has elapsed (c being the velocity of light). But Hegerfeldt has shown that the excitation amplitude for the second atom to be excited is non-zero for any interval $t > 0$ (Hegerfeldt 1994). This is a consequence of the analyticity of the amplitude in (complex) time stemming from the energy being bounded from below.

Some authors have treated decay phenomenologically using a state of complex energy (8.12.21) and then argued that there is such a state and its probability decreases exponentially. This is absurd in a theory with a self-adjoint time-independent Hamiltonian. If we must treat generalization to complex energies by analytic continuation, any discrete complex energy state must be supplemented by a complex background integral that restores the long- and short-time modifications. When this is properly carried out, the normalization of the state does *not* change in time but the survival amplitude decreases with time.

The error comes in when one takes the analytic continuation of the wave function $\psi(\omega)$ to complex values of ω. The dual to $\psi(\omega)$ for a complex argument is not $\psi^*(\omega)$, which is *not* an analytic function of ω but rather $\psi^*(\omega^*)$. Once this is properly taken care of, the norm of the state

$$\|\psi\|^2 = \int \psi^*(\omega^*)\psi(\omega)\, {\rm d}\omega$$

remains constant in time (for a time-independent Hamiltonian), but the survival amplitude (and hence the probability) has the expected 'decay' form:

$$A(t) = \int \psi^*(\omega^*) e^{-iHt} \psi(\omega)\, d\omega.$$

The quantity $\int \psi^*(\omega)\psi(\omega)\, d\omega$ is an improper choice which then leads to the absurd conclusion that $\|\psi(t)\|$ decreases with time. The unstable (resonant) state corresponds to a pole of $\psi(\omega)$ (and hence of $\psi^*(\omega^*)$) in the lower half-plane and the contour involves a small circle around it when the complex contour snags this pole.

8.12.3 Decay amplitude under a Lorentz transformation

Let $|M\rangle$ be a discrete normalized state (the eigenfunction of the unperturbed Hamiltonian, the 'unstable particle') and let H be the exact Hamiltonian. Let H have only a continuous spectrum (bounded from below): $0 < \lambda < \infty$. Then,

$$|M\rangle = \int \langle\lambda|M\rangle|\lambda\rangle\, d\lambda = \int a(\lambda)|\lambda\rangle\, d\lambda.$$

The time dependence of this state gives

$$e^{-iHt}|M\rangle = \int a(\lambda) e^{-i\lambda t}|\lambda\rangle\, d\lambda.$$

Hence the survival amplitude $A(t)$ is

$$A(t) = \langle M|e^{-iHt}|M\rangle = \int |a(\lambda)|^2 e^{-i\lambda t}\, d\lambda.$$

If we observe the 'decaying particle' from a moving Lorentz frame, we find $|M\rangle \to B(\gamma)|M\rangle$, where $B(\gamma)$ is the boost operator which is unitary. Hence the new survival amplitude $A_\gamma(t)$ is

$$A_\gamma(t) = \langle M|B^\dagger e^{-iHt} B|M\rangle = \langle M|e^{-iHt/\gamma}|M\rangle,$$

since $B^\dagger = B^{-1}$, $B^{-1}e^{-iHt}B = e^{-iHt/\gamma}$. Consequently $A_\gamma(t) \equiv A(t/\gamma)$. In the approximation of an exponential decay, i.e. $A(t) \cong e^{-iE_0 t - \frac{\Gamma}{2}t}$, we obtain $A_\gamma(t) \cong e^{-i\frac{E_0}{\gamma}t - \frac{\Gamma}{2\gamma}t}$. The lifetime is extended by a factor of γ, so that

$$P(t) = |A(t)|^2 = e^{-t/\tau},$$

$$P_\gamma(t) = |A_\gamma(t)|^2 = e^{-t/\gamma\tau}.$$

8.12.4 Quantum mechanics in dual spaces

Let a quantum system be labelled in terms of energy eigenstates (including the continuous spectrum). We represent the state $|\psi\rangle$ by the distribution $\psi(\omega)$ with ω real and

$$\int \psi^*(\omega)\psi(\omega)\,\mathrm{d}\omega = 1. \tag{8.12.24}$$

The integration over the energy is a real contour along the real line with a lower limit ω_0, and without loss of generality we may choose ω_0 to be zero. The Hamiltonian operator H is represented by a Hermitian kernel $H(\omega, \omega')$. Then the time development gives

$$\psi(\omega, t) = \int [\exp(-\mathrm{i}Ht)]_{\omega\omega'}\psi(\omega')\,\mathrm{d}\omega'. \tag{8.12.25}$$

We may approximate $\psi(\omega)$ arbitrarily closely by an analytic function in ω. That is to say analytic vectors $\psi(\omega)$ are dense in the Hilbert space of square-integrable functions. If in addition $H(\omega, \omega')$ is analytic in ω and in ω' then analytic vectors are taken into analytic vectors under time displacement. We may then define the analytic state vector $\psi(z)$ and analytic Hamiltonian $H(z, z')$ with the integration contour entirely in the domain of analyticity:

$$\psi(z, t) = \int_C (\exp -\mathrm{i}Ht)_{z,z'}\psi(z', 0)\,\mathrm{d}z'. \tag{8.12.26}$$

Then,

$$\int_R \psi^*(\omega)\phi(\omega)\,\mathrm{d}\omega = (\psi, \phi) = \int_C \psi^*(z^*)\phi(z)\,\mathrm{d}z, \tag{8.12.27}$$

where $\psi(\omega)$ is expressed in terms of the spectral resolution of the Hamiltonian. In particular, the survival amplitude reads as

$$A(t) = \int_C \psi^*(z^*)\mathrm{e}^{-\mathrm{i}zt}\psi(z)\,\mathrm{d}z. \tag{8.12.28}$$

The time-evolved state $|\psi(t)\rangle$ has the representative $\psi(z)\mathrm{e}^{-\mathrm{i}zt}$, while the survival amplitude is an open contour integral

$$\int_C \psi^*(z^*)\psi(z)\mathrm{e}^{-\mathrm{i}zt}\,\mathrm{d}z. \tag{8.12.29}$$

If it turns out that beyond the domain of analyticity the quantity

$$\psi^*(z^*)\psi(z)$$

has an isolated pole but is otherwise analytic, we can use a new contour C' provided we draw a mini-circle contour C'' around the pole, and hence

(by evaluating residua)

$$
\begin{aligned}
A(t) &= \int_{C' \cup C''} \psi^*(z^*)\psi(z) e^{-izt}\, dz \\
&= 2\pi i a e^{-iat} + \int_{C'} \psi^*(z^*)\psi(z) e^{-izt}\, dz. \qquad (8.12.30)
\end{aligned}
$$

Thus, we now have a discrete 'resonance' contribution to the survival amplitude. Note that since $a = E_r - i\frac{\gamma}{2}$ lies in the lower half-plane we have a decreasing amplitude as time goes forward. However, we must include the integral along the contour C' to recover the correct survival amplitude. The integrand along C' also has a decreasing dependence on time. If we had deformed into the upper half-plane and located a pole at $a^* = E_r + i\frac{\gamma}{2}$, its contribution would have increased for t increasing, but so would the contribution from the contour, and they would together give the same result as the contours in the lower half-plane.

Can we take the calculations for $t < 0$? This also decreases with increasing $|t|$. Hence the 'decay', unlike the classical 'death process' implies no irreversibility. Note also that the contour integral should be included along with the pole in all cases.

The vector space of functions analytic in a domain around the real axis form a vector space but it is not complete since the limit of a sequence of analytic functions need not be analytic. But analytic square-integrable functions are dense in the set of all square-integrable functions. The eigenvalue equations have right eigenvectors and distinct left eigenvectors, and they are pairwise orthogonal. The vector $\psi^*(z^*)$ is the dual to the vectors $\psi(z)$, and they obey products that are defined as between vectors in the space of analytic functions $f(z)$ and their duals $g^*(z^*)$.

8.13 Problems

8.P1. Prove that the wave operators satisfy the following equations (Berezin and Shubin 1991):

$$
\Omega_+ e^{itH_0} = e^{itH}\Omega_+, \quad \Omega_- e^{itH_0} = e^{itH}\Omega_-, \qquad (8.13.1)
$$

for all real values of t.

8.P2. In the analysis of scattering from singular potentials in an arbitrary number n of spatial dimensions, start from the form (4.10.3) of the Schrödinger equation for stationary states.

(i) Looking for solutions in the form (8.6.2), prove that one obtains an equation formally analogous to Eq. (8.6.8), but with γ replaced by (Esposito 1998a)

$$
\tilde{\gamma} \equiv \gamma + \frac{1}{2}(n-2), \qquad (8.13.2)
$$

and λ equal to

$$
\lambda \equiv l + \frac{1}{2}(n-2). \qquad (8.13.3)
$$

(ii) Prove that the matrix elements (8.6.12) now read

$$H_{s,m} = \delta_{s,m} - \frac{\overline{u}_{s-m}}{\left[(s+\widetilde{\gamma})^2 - \widetilde{\lambda}^2\right]}, \tag{8.13.4}$$

where

$$\overline{u}_s \equiv w_s - w_0 \delta_{s,0}, \tag{8.13.5}$$

$$\widetilde{\lambda}^2 \equiv \lambda^2 + w_0. \tag{8.13.6}$$

(iii) Prove that the resulting Hill-type equation which leads, in principle, to the evaluation of the fractional part of the polydromy parameter γ, involves an even periodic function of the parameter $\widetilde{\gamma}$ defined in Eq. (8.13.2).

8.P3. On studying again Eq. (4.10.3), look for solutions in the form

$$\psi(r) = r^\beta y(r). \tag{8.13.7}$$

(i) Find how β should depend on n to get a second-order differential equation for $y(r)$ where the coefficient of the first derivative vanishes (Esposito 1998b).

(ii) Prove that on defining λ as in (8.13.3) the resulting equation for stationary states reads (Esposito 1998b)

$$\left[\frac{\mathrm{d}^2}{\mathrm{d}r^2} + k^2 - \frac{(\lambda^2 - \frac{1}{4})}{r^2} - V(r)\right] y(r) = 0. \tag{8.13.8}$$

(iii) Write the form of the two fundamental solutions $\varphi_1(r)$ and $\varphi_2(r)$ of Eq. (8.13.8) in the neighbourhood of the Fuchsian singularity at $r = 0$.

(iv) Taking into account that the operator in Eq. (8.13.8) is even in λ, write $\varphi(\lambda, k, r)$ and $\varphi(-\lambda, k, r)$ in place of $\varphi_1(r)$ and $\varphi_2(r)$, where λ is allowed to be freely specifiable in the complex plane. Prove that $\varphi(\lambda, k, r)$ is entire (i.e. analytic in the whole complex plane) in k^2 and analytic in the domain $\mathrm{Re}(\lambda) > 0$ (De Alfaro and Regge 1965).

8.P4. Consider a singular potential with a δ or δ' singularity. Using the continuity of the probability density and the probability current compute the reflection and transmission coefficients and also the phase shifts (Boya and Sudarshan 1996).

8.P5. Evaluate the partial wave series for the Coulomb scattering amplitude in three dimensions in spherical coordinates, following the method in Lin (2000).

8.P6. Prove Eq. (8.8.19) by choosing a contour C and adding to it the negative of suitable contours C', Γ and B in such a way that the result is equivalent to integrating over the boundary of a simply connected region inside which the integrand is analytic (Sudarshan 1962).

Part II

Weyl quantization and algebraic methods

9
Weyl quantization

The commutator of position and momentum operators is first considered in the coordinate and momentum representations, and then in an abstract Hilbert space. The reader is then introduced to canonical operators with the associated canonical quantization of commutation rules, which consists of maps defined on a symplectic space and taking values in the set of unitary operators on a Hilbert space. If the symplectic space is finite-dimensional, all of its irreducible representations are unitarily equivalent. The Weyl exponentiated form of the commutation relations is described and investigated. One then arrives at the definition of Weyl systems, and it is shown how to recover from them the formulation in terms of self-adjoint operators which satisfy a generalized form of commutation relations.

Further topics are the Heisenberg representation for temporal evolution and the generalized uncertainty relations. Unitary operators associated with symplectic linear transformations are then considered, and, within this framework, translations, rotations and the harmonic oscillator are considered. The Weyl programme is then reassessed.

Lastly, the basic postulates of modern quantum theory are eventually studied: the probabilistic nature of the predictions of results of measurements; the correspondence between observables and self-adjoint operators on Hilbert spaces; the expected result of measuring an observable quantity; and the evolution of the state vector in systems which are not affected by any external influence. The chapter ends with an outline of rigged Hilbert space formalism and its applications.

9.1 The commutator in wave mechanics

In the first part of this book we studied wave mechanics within the framework of the Schrödinger equation. The reader is therefore familiar, from chapter 4, with the coordinate representation in which the position operator acts as a multiplication on square-integrable functions:

$$\hat{x}\varphi(x) = x\varphi(x),$$

and the momentum operator is a first-order differential operator on such functions:

$$\hat{p}\varphi(x) = \frac{\hbar}{\mathrm{i}}\mathrm{grad}\ \varphi(x).$$

Moreover, one may equally well consider the momentum representation, where the roles of $\hat{x}$ and $\hat{p}$ are interchanged: $\hat{x}$ acts as a first-order operator on square-integrable functions with respect to the p variable, i.e.

$$\hat{x}\phi(p) = \mathrm{i}\hbar\frac{\partial}{\partial p}\phi(p),$$

while $\hat{p}$ acts as a multiplication operator on this class of functions:

$$\hat{p}\phi(p) = p\phi(p).$$

In both representations, one finds

$$\hat{x}\,\hat{p} - \hat{p}\,\hat{x} = \mathrm{i}\hbar\mathbb{I}, \tag{9.1.1}$$

which is an operator equation where both sides are meant to act on a suitable class of functions.

9.2 Abstract version of the commutator

We may first have in mind a concrete realization, where $\hat{x}$ and $\hat{p}$ act on functions $\varphi : x \rightarrow \varphi(x) \in \mathcal{L}^2(\mathbf{R}^3)$, or on $\phi : p \rightarrow \phi(p) \in \mathcal{L}^2(\mathbf{R}^3)$. However, we may take a more abstract point of view, where we refrain ourselves from considering a particular realization of either $\hat{x}$ or $\hat{p}$ as differential operators, and in which they act on elements of an abstract Hilbert space $\mathcal{H}$. We still require, however, that the commutator defined by

$$[\hat{x}, \hat{p}] \equiv \hat{x}\,\hat{p} - \hat{p}\,\hat{x} \tag{9.2.1}$$

should satisfy the fundamental condition (9.1.1), where $\mathbb{I}$ should be viewed as the identity in $\mathcal{H}$. Equation (9.1.1) is then the defining relation of the Heisenberg algebra A_{H} of the position operator $\hat{x}$, momentum operator $\hat{p}$ and unity $\mathbb{I}$. The work in Heisenberg (1925), Dirac (1926a) led eventually to the postulate Eq. (9.1.1), and its extension to systems with n position and momentum coordinates:

$$\hat{x}^r\,\hat{p}_s - \hat{p}_s\,\hat{x}^r = \mathrm{i}\hbar\delta^r{}_s\,\mathbb{I}, \tag{9.2.2}$$

for $r, s \in 1, \ldots, n$.

Here we do not follow the historical path, but we stress that it would be desirable to have the abstract operators $\hat{x}$ and $\hat{p}$ acting on a normed vector space. It is not necessary to choose coordinate or momentum representations. If the phase space is a vector space endowed with a symplectic structure ω, the commutation relations can be written for any u, v with the symplectic form $\omega(u, v)$:

$$\hat{u}\,\hat{v} - \hat{v}\,\hat{u} = \mathrm{i}\hbar\omega(u, v), \tag{9.2.3}$$

where $\hat{u}$ and $\hat{v}$ are the operators corresponding to elements in the linear phase space. If $\omega(u_1, u_2)$ vanishes, then u_1 and u_2 are in a Lagrangian sub-space and the operators associated with them commute. We can take any even-dimensional space for the phase space. The quantization for finitely many degrees of freedom is structurally similar and leads again to Eq. (9.2.3). It has been demonstrated in von Neumann (1931) that the operator algebra so defined is unique within isomorphism, and its implementations on any phase space are unitarily equivalent. When the number of degrees of freedom is (countably) infinite the quantization can again be carried out with $\omega(u, v)$ being the symplectic form on this infinite-dimensional space. This implementation is called second quantization (the first quantization obtains by equipping the space of u, v with a positive-definite sesquilinear form, the symplectic form being the anti-symmetric imaginary part of the scalar product). However, in this case there are many distinct representations that are not unitarily equivalent.

9.3 Canonical operators and the Wintner theorem

Now we begin a more accurate analysis of the mathematical foundations of the quantization procedure from the point of view of canonical quantization. The first step begins with the commutation relations for position and momentum operators.

If P and Q are operators satisfying the *canonical commutation relations* (unlike Eq. (9.1.1), here we work in $\hbar = 1$ units)

$$Q\,P\ - P\,Q = \mathrm{i}\mathbb{1}, \tag{9.3.1}$$

they cannot both be bounded. Indeed, what happens is that repeated application of (9.3.1) yields, for any positive integer n,

$$Q^n\,P - P\,Q^n = \mathrm{i}nQ^{n-1}. \tag{9.3.2}$$

Thus, if both P and Q were bounded, one could write (Reed and Simon 1980)

$$n\|Q\|^{n-1} = n\|Q^{n-1}\| \leq 2\|P\|\,\|Q\|^n, \tag{9.3.3}$$

which implies that

$$\|P\|\,\|Q\| \geq \frac{n}{2}, \tag{9.3.4}$$

for all n. Condition (9.3.4) therefore proves that P or Q are not bounded, contrary to the hypothesis. This result is also known as the Wintner theorem (Wintner 1947).

We can now use this property to gain a proper understanding of Eq. (9.3.1). The right-hand side is a multiple of the identity and therefore is defined over the whole Hilbert space; on the other hand, the commutator on the left-hand side is not defined over the whole Hilbert space, hence it cannot be meaningful without further qualifications. To avoid these problems resulting from domains of unbounded operators, Weyl suggested replacing Eq. (9.3.1) with its 'exponentiated version', so that one deals with unitary operators defined over all of $\mathcal{H}$.

In other words, the identification of momentum and position with differential operators acting on complex-valued functions (e.g. with domain the C^∞ functions on $(-\infty, 0)$ or $(0, \infty)$ with compact support away from the origin, or the absolutely continuous functions on the closed interval $[0, 1]$, with suitable boundary conditions) leads to the commutation relations (9.3.1), where $\hbar = 1$ units are used as we just mentioned (such a choice is more convenient if one is interested in the general mathematical structures, and provided that one has a good understanding of physical dimensions). Even if P and Q are realized as self-adjoint operators on a Hilbert space, the outstanding technical problem remains to make sense of the commutator (9.3.1), since one is dealing with unbounded operators. For this purpose, one considers the one-parameter groups (see section 9.4)

$$U(s) \equiv \mathrm{e}^{\mathrm{i}sP}, \quad V(t) \equiv \mathrm{e}^{\mathrm{i}tQ}, \tag{9.3.5}$$

subject to the relation

$$U(s)V(t) = \mathrm{e}^{\mathrm{i}st} \, V(t)U(s). \tag{9.3.6}$$

Equation (9.3.6) does not express a generic property of continuous one-parameter unitary groups, but only holds for groups generated by *canonical operators* (i.e. operators satisfying Eq. (9.3.1)). If both P and Q were bounded operators, one could express the one-parameter groups $U(s)$ and $V(t)$ in terms of their Taylor series, and the insertion into (9.3.6) enables one to recover the 'ill-defined' commutation relations (9.3.1). However, if one takes Eq. (9.3.6) as the starting point of our formulation of commutation relations, the resulting analysis is not affected by the operators P or Q being unbounded, and the reduction to the form (9.3.1) has only a heuristic value. To understand this last point, we need the material discussed in the following section.

9.4 Canonical quantization of commutation relations

The exponentiated version of the canonical commutation relations proposed in Weyl (1927) relies on the following theorem in Stone (1932), which relates a one-parameter group of unitary transformations (see below) with its infinitesimal generator, to give back, under the appropriate conditions, the canonical commutation relations proposed by Dirac. To obtain a precise formulation of these concepts we begin by giving the following definition of strong continuity: if $\mathcal{H}$ is a Hilbert space, the operator-valued function $U : \mathbf{R} \rightarrow \mathcal{U}(\mathcal{H})$ satisfying the following conditions:

(i) $\forall t \in \mathbf{R}$, $U(t)$ is a unitary operator, and $U(t+s) = U(t)U(s)$, $\forall s, t \in \mathbf{R}$;

(ii) $\forall \varphi \in \mathcal{H}$, if $t \rightarrow t_0$, then $U(t)\varphi \rightarrow U(t_0)\varphi$;

is called a strongly continuous one-parameter unitary group. For those vectors ψ such that the following limit exists:

$$\lim_{t \to 0} \frac{U(t)\psi - \psi}{t} \equiv \mathrm{i}A\psi,$$

one defines the 'infinitesimal generator' A of the one-parameter group of unitary transformations under consideration. The set of all ψ for which the limit exists is the domain of A. Within this domain, A turns out to be essentially self-adjoint (Stone theorem).

We now assume that a symplectic space (V, ω) is given. For the time being, V is not necessarily finite-dimensional. By definition, the *canonical quantization* of commutation relations consists of the maps

$$U : V \longrightarrow \mathcal{U}(\mathcal{H}),$$

where $\mathcal{U}(\mathcal{H})$ is the set of unitary operators on a Hilbert space $\mathcal{H}$, and the maps are *strongly continuous on all finite-dimensional sub-spaces* of V and are such that

$$\begin{aligned} U(X+Y) &= \mathrm{e}^{\mathrm{i}\omega(X,Y)}\, U(X)U(Y) = U(Y+X) \\ &= \mathrm{e}^{\mathrm{i}\omega(Y,X)}\, U(Y)U(X), \end{aligned} \tag{9.4.1}$$

for all $X, Y \in V$. With our notation, $\omega(X, Y)$ is the quadratic form on V associated with the symplectic structure ω.

An alternative formulation relies on the possibility of decomposing V as the direct sum of closed Lagrangian sub-spaces

$$V = V_1 \oplus V_2, \tag{9.4.2}$$

i.e. sub-spaces V_1 and V_2 such that the symplectic form vanishes on them (section 2.3). This operation is well defined in the case of a finite number of degrees of freedom, which is the case studied in ordinary quantum mechanics, and may be extended, with some care, to an infinite number. As we just mentioned, the sub-spaces V_1 and V_2 are such that the Lagrangian requirement

$$\omega|V_1 = 0, \quad \omega|V_2 = 0, \tag{9.4.3}$$

is satisfied, and these properties correspond to the vanishing Poisson brackets of positions among themselves and momenta among themselves, respectively. One then looks for the unitary representations $U_1 : V_1 \longrightarrow \mathcal{U}(\mathcal{H})$, $U_2 : V_2 \longrightarrow \mathcal{U}(\mathcal{H})$, strongly continuous on all finite-dimensional sub-spaces, and such that (cf. Eq. (9.3.7))

$$U_1(x_1)U_2(x_2) = \mathrm{e}^{\mathrm{i}B(x_1,x_2)}\, U_2(x_2)U_1(x_1), \tag{9.4.4}$$

where we have defined

$$B(x_1, x_2) \equiv -\omega[(x_1, 0), (0, x_2)]. \tag{9.4.5}$$

This setting can be modified by requiring that an isomorphism $F : V_1 \rightarrow V_2$ should exist such that

$$\omega[(x, 0), (0, F(x))] \neq 0, \quad \forall x \in V_1, \quad x \neq 0. \tag{9.4.6}$$

For all $x \in V_1$, let $P(x)$ and $Q(x)$ be the self-adjoint infinitesimal generators of the one-parameter groups $t \rightarrow U(tx)$, and $t \rightarrow U(tF(x))$. The operators

$$a(x) \equiv \frac{1}{\sqrt{2}}\Big[Q(x) + \mathrm{i}P(x)\Big], \tag{9.4.7}$$

$$a^{\dagger}(x) \equiv \frac{1}{\sqrt{2}}\Big[Q(x) - \mathrm{i}P(x)\Big], \tag{9.4.8}$$

are then said to be the *annihilation* and *creation* operators, respectively, associated with the specific decomposition of V into $V_1 \oplus V_2$.

If the symplectic vector space (V, ω) is endowed with a symplectic basis $e_1, e_2, \ldots,\ e'_1, e'_2, \ldots$, which implies that (see Eq. (9.4.3))

$$\omega(e_n, e_p) = \omega(e'_n, e'_p) = 0, \tag{9.4.9}$$

$$\omega(e_n, e'_p) = \delta_{np}, \tag{9.4.10}$$

the problem of canonical quantization may be formulated in a way that resembles Eq. (9.3.1), i.e. it becomes the problem of finding a family of unitary operators $U_{1,j}$ and $U_{2,j}$ such that

$$U_{1,j}\, U_{1,k} = U_{1,k}\, U_{1,j}, \tag{9.4.11}$$

$$U_{2,j}\, U_{2,k} = U_{2,k}\, U_{2,j}, \tag{9.4.12}$$

$$U_{1,j}\, U_{2,k} = \mathrm{e}^{-\mathrm{i}\delta_{jk}}\, U_{2,k}\, U_{1,j}. \tag{9.4.13}$$

The corresponding annihilation (a_j) and creation ($a_j^\dagger$) operators satisfy the relations

$$\left[a_j, a_k\right] = \left[a_j^\dagger, a_k^\dagger\right] = 0, \tag{9.4.14}$$

$$\left[a_j, a_k^\dagger\right] = \delta_{jk}. \tag{9.4.15}$$

A fundamental theorem due to von Neumann states that the irreducible representations of (V, ω) are all unitarily equivalent, *if* the symplectic space under consideration is *finite-dimensional.* Within this framework, one deals with Weyl quantization, which is studied in detail in the following sections.

9.5 Weyl quantization and Weyl systems

This section defines the concept of Weyl quantization and Weyl systems. The key logical steps are as follows.

9.5.1 Representations

Let G be a topological group (appendix 2.B), and let $\mathcal{V}$ be a topological vector space. A *representation* T of G on $\mathcal{V}$ is a continuous map

$$T : G \rightarrow \text{Aut}(\mathcal{V}) : a \rightarrow T(a),$$

which associates to every element of G a continuous linear operator $T(a)$ on $\mathcal{V}$, and such that

$$T(ab) = T(a)T(b), \tag{9.5.1}$$

$$T(e) = 1\!\!1, \tag{9.5.2}$$

for all $a, b \in G$, where e is the identity element of G. If $\mathcal{V}$ is a Hilbert space $\mathcal{H}$, and if $T(a)$ is unitary $\forall a \in G$, our continuous representations will be taken to be strongly continuous (cf. section 9.4). We further assume that T is unitary, and we then say that T is a unitary representation.

9.5.2 Unitary equivalence

If T and T' are representations of G on the Hilbert spaces $\mathcal{H}$ and $\mathcal{H}'$, T and T' are said to be *unitarily equivalent* if a unitary map U from $\mathcal{H}$ into $\mathcal{H}'$ exists such that, for all $a \in G$,

$$UT(a) = T'(a)U. \tag{9.5.3}$$

The statement of formula (9.5.3) is often rephrased by saying that the map U *interwines* the action of G on the two Hilbert spaces. A continuous unitary representation T is *irreducible* if no non-empty sub-spaces of $\mathcal{H}$ exist that are invariant under the action of the whole set of $T(a)$ transformations.

9.5.3 Weyl quantization

We are now going to define the *Weyl* (exponentiated) *form* of the commutation relations (Weyl 1931). For this purpose, let us try to be more precise about the geometrical objects occurring in our analysis. We consider a *finite-dimensional* vector space L, and a representation of L in terms of unitary operators on a Hilbert space $\mathcal{H}$:

$$U : L \longrightarrow \mathcal{U}(\mathcal{H}).$$

Moreover, we also consider the representation

$$V : L^* \longrightarrow \mathcal{U}(\mathcal{H}),$$

L^* being the dual of L. Here, both U and V are continuous unitary representations of L and L^*, respectively, on the set $\mathcal{U}(\mathcal{H})$. We say that the pair (U, V) represents a *Weyl system* (cf. Dubin *et al.* 2000) if

$$V(f)U(x) = \mathrm{e}^{\mathrm{i}f(x)}\, U(x)V(f), \quad x \in L, \quad f \in L^*, \tag{9.5.4}$$

and

$$U(x)U(x') = U(x')U(x), \tag{9.5.5}$$

$$V(f)V(f') = V(f')V(f). \tag{9.5.6}$$

Note that the direct sum of L and its dual carries a natural symplectic structure, for if (x, f) and (x', f') are any two elements of $L \oplus L^*$, one finds

$$\omega\Big((x, f), (x', f')\Big) = f(x') - f'(x). \tag{9.5.7}$$

It is now convenient to set $z \equiv (x, f)$ and $z' \equiv (x', f')$. We can then consider a unitary representation of the symplectic vector space in the form

$$W(z) \equiv \mathrm{e}^{\frac{\mathrm{i}}{2}f(x)}\, U(x)V(f), \tag{9.5.8}$$

which implies that

$$W(z)W(z') = \mathrm{e}^{\frac{\mathrm{i}}{2}f(x)}\, U(x)V(f)\, \mathrm{e}^{\frac{\mathrm{i}}{2}f'(x')}\, U(x')V(f'). \tag{9.5.9}$$

On the other hand, the evaluation of $W(z+z')$ yields (by virtue of (9.5.4)–(9.5.8))

$$\begin{aligned} W(z + z') &= \mathrm{e}^{\frac{\mathrm{i}}{2}(f+f')(x+x')}\, U(x + x')V(f + f') \\ &= \mathrm{e}^{\frac{\mathrm{i}}{2}(f+f')(x+x')}\, U(x)U(x')V(f)V(f') \\ &= \mathrm{e}^{\frac{\mathrm{i}}{2}(f+f')(x+x')}\, U(x)\, \mathrm{e}^{-\mathrm{i}f(x')}\, V(f)U(x')V(f') \\ &= \mathrm{e}^{\frac{\mathrm{i}}{2}(f'(x)-f(x'))} \\ &\quad \times \Big[\mathrm{e}^{\frac{\mathrm{i}}{2}f(x)}\, U(x)V(f)\Big]\Big[\mathrm{e}^{\frac{\mathrm{i}}{2}f'(x')}\, U(x')V(f')\Big]. \end{aligned} \tag{9.5.10}$$

A comparison of (9.5.9) and (9.5.10) yields (see Eq. (9.5.7))

$$W(z+z') = \mathrm{e}^{-\frac{\mathrm{i}}{2}\omega(z,z')}\, W(z)W(z'). \tag{9.5.11}$$

Note that we might have started from a symplectic space S to define a Weyl system, without using the decomposition of S into $L \oplus L^*$ (Segal 1959). It would have then been possible to re-obtain the representations of L and L^*, respectively, by setting

$$U \equiv W|L, \quad V \equiv W|L^*, \tag{9.5.12}$$

where L is a Lagrangian sub-space and L^* is defined using ω evaluated on vectors of L, i.e. $\forall u \in L, \omega(u) \in L^*$. In general, a *Weyl system* is then a continuous, unitary representation of a symplectic space:

$$W : S \rightarrow \mathcal{U}(\mathcal{H}),$$

such that

(i) W is strongly continuous as a function of z;

(ii) $W(z+z') = \mathrm{e}^{-\frac{\mathrm{i}}{2}\omega(z,z')}\, W(z)W(z')$.

The scheme described is due to Irving Segal (1959).

9.6 The Schrödinger picture

If one considers a split of the symplectic space S into $L \oplus L^*$, it is possible to give a realization of $\mathcal{H}$ and of the unitary operators acting on it, associated with L and L^*. Let $\mathcal{H}$ be the Hilbert space $\mathcal{L}^2(L)$, and let us consider the unitary operators associated with L and L^*, respectively:

$$U : L \longrightarrow \mathcal{U}(\mathcal{L}^2(L)) : x \rightarrow \mathrm{e}^{\mathrm{i}xP},$$

$$\left(\mathrm{e}^{\mathrm{i}xP}\psi\right)(y) \equiv \psi(y+x), \tag{9.6.1}$$

and

$$V : L^* \longrightarrow \mathcal{U}(\mathcal{L}^2(L)) : f \rightarrow \mathrm{e}^{\mathrm{i}fQ},$$

$$\left(\mathrm{e}^{\mathrm{i}fQ}\psi\right)(y) \equiv \mathrm{e}^{\mathrm{i}f(y)}\, \psi(y). \tag{9.6.2}$$

To check the understanding of the rules given above, it is instructive to perform the following calculation:

$$\left(\mathrm{e}^{\mathrm{i}fQ}\mathrm{e}^{\mathrm{i}xP}\psi\right)(y) = \mathrm{e}^{\mathrm{i}f(y)}\left(\mathrm{e}^{\mathrm{i}xP}\psi\right)(y) = \mathrm{e}^{\mathrm{i}f(y)}\psi(y+x). \tag{9.6.3}$$

On the other hand, on inverting the order of the operations, one finds

$$\left(\mathrm{e}^{\mathrm{i}xP}\mathrm{e}^{\mathrm{i}fQ}\psi\right)(y) = \left(\mathrm{e}^{\mathrm{i}fQ}\psi\right)(y+x) = \mathrm{e}^{\mathrm{i}f(y+x)}\psi(y+x). \tag{9.6.4}$$

In other words, one has

$$\begin{aligned}\left[\left(e^{ifQ}e^{ixP}e^{-ifQ}e^{-ixP}\right)\psi\right](y) &= e^{if(y)}\left[\left(e^{ixP}e^{-ifQ}e^{-ixP}\right)\psi\right](y) \\ &= e^{if(y)}\left(e^{-ifQ}e^{-ixP}\psi\right)(y+x) \\ &= e^{if(y)}e^{-if(y+x)}\left(e^{-ixP}\psi\right)(y+x) \\ &= e^{-if(x)}\psi(y). \qquad (9.6.5)\end{aligned}$$

In the following section we will state and discuss a fundamental theorem of von Neumann, which clarifies why the Schrödinger picture defined here can be regarded as the 'prototype' realization in quantum mechanics. (Dirac points out that in the explicit operator representation, one could either use the Heisenberg 'picture' or the Schrödinger 'picture'. Picture therefore corresponds to protocol.)

9.7 From Weyl systems to commutation relations

Given a Weyl system, our problem is now to recover the formulation in terms of self-adjoint operators which satisfy a generalized form of commutation relations (cf. Cavallaro *et al.* (1999)). Indeed, the Stone theorem described in section 9.4 makes it possible to consider the self-adjoint operator $R(z)$ in $\mathcal{H}$ such that

$$W(z) = e^{iR(z)}. \qquad (9.7.1)$$

By virtue of the requirements in the definition of a Weyl system, $R(z)$ is found to depend *linearly* on $z \in S$ in that

$$R(tz) = tR(z), \quad \forall t \in \mathbf{R}, \forall z \in S. \qquad (9.7.2)$$

For the linear combination of z and z' one finds from (9.5.11)

$$e^{iR(\alpha z+\beta z')} = e^{iR(\alpha z)}\, e^{iR(\beta z')}\, e^{-\frac{i}{2}\alpha\beta\omega(z,z')}. \qquad (9.7.3)$$

Since $tz + t'z'$ is the same as $t'z' + tz$, one has

$$W(tz + t'z') = W(t'z' + tz). \qquad (9.7.4)$$

By virtue of (9.5.11) and (9.7.1), this implies

$$e^{itR(z)}\, e^{it'R(z')} = e^{-itt'\omega(z',z)}\, e^{it'R(z')}\, e^{itR(z)}. \qquad (9.7.5)$$

The series expansion of both sides of (9.7.5) up to second order yields

$$\left[R(z), R(z')\right] = -i\omega(z,z'), \qquad (9.7.6)$$

bearing in mind that $\Big[R(z), R(z')\Big] \equiv R(z)R(z') - R(z')R(z)$, and the antisymmetry of $\omega(z, z')$.

A careful statement of the von Neumann theorem (von Neumann 1931, 1955) is now in order. In the language of functional analysis, such a theorem states that, if $U(\alpha)$ and $V(\beta)$ are continuous, one-parameter unitary groups on a separable Hilbert space $\mathcal{H}$, and satisfy the Weyl relations (9.3.6), there exist closed sub-spaces such that:

(1) $\mathcal{H} = \oplus_{l=1}^{N} \mathcal{H}_l$, $N > 0$, $N \leq \infty$;

(2) each $\mathcal{H}_l$ is invariant under $U(\alpha)$ and $V(\beta)$, $\forall \alpha, \beta \in \mathbf{R}$, i.e. $U(\alpha) : \mathcal{H}_l \rightarrow \mathcal{H}_l$, $V(\beta) : \mathcal{H}_l \rightarrow \mathcal{H}_l$;

(3) for all l, there exists a unitary operator $T_l : \mathcal{H}_l \rightarrow \mathcal{L}^2(\mathbf{R})$ such that $T_l U(\alpha) T_l^{-1}$ is translation to the left by α, and $T_l V(\beta) T_l^{-1}$ is multiplication by $\mathrm{e}^{\mathrm{i}\beta x}$.

Then, if P and Q are the infinitesimal generators of $U(\alpha)$ and $V(\beta)$, respectively (see Eq. (9.3.5)), one can prove, as a corollary, that there exists a dense domain $\mathcal{D} \subset \mathcal{H}$ such that:

(i) both P and Q act as maps $\mathcal{D} \rightarrow \mathcal{D}$;

(ii) $QP\varphi - PQ\varphi = \mathrm{i}\varphi$, $\forall \varphi \in \mathcal{D}$;

(iii) P and Q are essentially self-adjoint on $\mathcal{D}$.

This means that each solution of the Weyl relations has infinitesimal generators satisfying the canonical commutation relations in the form specified by conditions (i)–(iii). *The converse, in general, does not hold.* In the applications, whenever one writes commutators, it should be clear that, strictly, *they only hold on a dense domain of a separable Hilbert space*, and involve the infinitesimal generators of continuous, one-parameter unitary groups which satisfy the Weyl relations (9.3.6).

In the language of the previous sections, the von Neumann theorem states that any irreducible representation of (S, ω) is unitarily equivalent to the Schrödinger representation.

Weyl quantization is extended to all functions on V by using the Fourier transform $\widetilde{F}$ of F, for which

$$F(q,p) = \int_{\mathbf{R}^{2n}} \widetilde{F}(x,f) \mathrm{e}^{\mathrm{i}\frac{(xp+fq)}{\hbar}} \, \mathrm{d}x \, \mathrm{d}f, \tag{9.7.7}$$

and associating with it the following operator:

$$F(\hat{Q}, \hat{P}) = \int_{\mathbf{R}^{2n}} \widetilde{F}(x,f) \mathrm{e}^{\mathrm{i}(x\hat{P}+f\hat{Q})} \, \mathrm{d}x \, \mathrm{d}f. \tag{9.7.8}$$

The resulting formulation of quantum mechanics will be outlined in subsection 15.5.1.

9.8 Heisenberg representation for temporal evolution

If one is given the Schrödinger equation for the state vector, which is a first-order differential equation on an infinite-dimensional Hilbert space (see Eq. (7.8.2)), we know from (7.8.4) that its solution may be expressed in the form $\psi(t) = U(t, t_0)\psi(t_0)$, where the unitary operator U satisfies the first-order equation

$$\mathrm{i}\hbar \frac{\mathrm{d}}{\mathrm{d}t} U(t, t_0) = H U(t, t_0), \tag{9.8.1}$$

with the initial condition

$$U(t_0, t_0) = 1. \tag{9.8.2}$$

This is sharply different from the 'space–time picture' associated with the Schrödinger equation for the wave function, which is instead a partial differential equation, the solution of which is defined over the whole Minkowski space–time.

The mean value of the observable A in the state ψ is

$$\langle A \rangle_\psi \equiv \frac{(\psi, A\psi)}{(\psi, \psi)}, \tag{9.8.3}$$

but if one considers a new state vector (where $V(t)$ is a suitably chosen operator)

$$\Phi(t) \equiv V(t)\psi(t), \tag{9.8.4}$$

its mean value becomes

$$\langle A \rangle_\psi = \frac{(\Phi, VAV^{-1}\Phi)}{(\Phi, \Phi)}. \tag{9.8.5}$$

One is thus led to define

$$A(t, t_0) \equiv V(t) A V^{-1}(t), \tag{9.8.6}$$

and in the *Heisenberg picture* one chooses

$$V(t) \equiv U^{-1}(t, t_0), \tag{9.8.7}$$

which implies

$$A(t, t_0) = U^{-1}(t, t_0) A(t_0) U(t, t_0), \tag{9.8.8}$$

$$\Phi(t) = U^{-1}(t, t_0) U(t, t_0) \psi(t_0) = \psi(t_0). \tag{9.8.9}$$

Remarkably, this means that *in the Heisenberg picture the operators evolve in time, whereas the state vectors remain fixed.* Moreover, by virtue of Eq. (9.8.1) and of its conjugate equation,

$$\mathrm{i}\hbar \frac{\mathrm{d}}{\mathrm{d}t} U^{\dagger}(t, t_0) = -U^{-1}(t, t_0) H, \tag{9.8.10}$$

one finds

$$\mathrm{i}\hbar \frac{\mathrm{d}}{\mathrm{d}t} A(t, t_0) = U^{-1}(t, t_0)(AH - HA)U(t, t_0) = \Big[A(t, t_0), H\Big]. \tag{9.8.11}$$

This makes it clear that the operator $A(t, t_0)$ is a constant of motion if and only if it commutes with the Hamiltonian. At a deeper mathematical level, we should stress that Eq. (9.8.1) is written on the *automorphism group* of an Hilbert space (see appendix 4.A), whereas the Heisenberg equation (9.8.11) is written on the *algebra of the automorphism group*.

9.9 Generalized uncertainty relations

In the introductory presentations of quantum mechanics, emphasis is put on the uncertainty relations for position and momentum operators. We are now going to derive a general result which includes, as a particular case, the formula involving position and momentum operators.

Let A and B be two self-adjoint operators on a Hilbert space $\mathcal{H}$, with domains $D(A)$ and $D(B)$, respectively, and satisfying the commutation relation

$$AB - BA = \mathrm{i}C. \tag{9.9.1}$$

Note that

$$\begin{aligned}(\mathrm{i}C)^{\dagger} = -\mathrm{i}C^{\dagger} &= (AB)^{\dagger} - (BA)^{\dagger} = B^{\dagger}A^{\dagger} - A^{\dagger}B^{\dagger} \\ &= BA - AB = -(AB - BA) = -\mathrm{i}C,\end{aligned} \tag{9.9.2}$$

and hence $C^{\dagger} = C$. However, such operations do not consider domains and therefore do not prove self-adjointness of C. Now bearing in mind the Schwarz inequality (see Eq. (4.A.6))

$$|(f, g)|^2 \leq \|f\|^2 \|g\|^2, \tag{9.9.3}$$

one can apply it to the vectors defined by

$$f \equiv \Big(A - \langle A \rangle\Big)\psi, \tag{9.9.4}$$

$$g \equiv \Big(B - \langle B \rangle\Big)\psi, \tag{9.9.5}$$

for $\psi \in \mathcal{H}$. One thus finds that

$$\alpha \equiv |((A - \langle A \rangle)\psi, (B - \langle B \rangle)\psi)|^2$$
$$= |(\psi, (A - \langle A \rangle)(B - \langle B \rangle)\psi)|^2 \leq (\triangle A)_\psi^2 (\triangle B)_\psi^2. \quad (9.9.6)$$

On the other hand, by adding and subtracting the same operator, one finds a useful identity for the operator occurring on the second line of (9.9.6):

$$(A - \langle A \rangle)(B - \langle B \rangle) = \frac{1}{2}\Big[(A - \langle A \rangle)(B - \langle B \rangle) + (B - \langle B \rangle)(A - \langle A \rangle)\Big]$$
$$+ \frac{1}{2}\Big[(A - \langle A \rangle)(B - \langle B \rangle) - (B - \langle B \rangle)(A - \langle A \rangle)\Big]$$
$$= F + \frac{\mathrm{i}}{2}C, \quad (9.9.7)$$

where we have defined

$$F \equiv \frac{1}{2}\Big[(A - \langle A \rangle)(B - \langle B \rangle) + (B - \langle B \rangle)(A - \langle A \rangle)\Big]. \quad (9.9.8)$$

In the light of all of these properties, one finds the inequality

$$(\triangle A)_\psi^2 (\triangle B)_\psi^2 \geq \left|(\psi, F\psi) + \frac{\mathrm{i}}{2}(\psi, C\psi)\right|^2, \quad (9.9.9)$$

and hence

$$(\triangle A)_\psi^2 (\triangle B)_\psi^2 \geq \langle F \rangle_\psi^2 + \frac{1}{4}\langle C \rangle_\psi^2 \geq \frac{1}{4}\langle C \rangle_\psi^2, \quad (9.9.10)$$

which leads to the weaker condition

$$(\triangle A)_\psi (\triangle B)_\psi \geq \frac{1}{2}\langle C \rangle_\psi. \quad (9.9.11)$$

This represents the *generalized uncertainty principle*. For the position and momentum operators satisfying the commutation relation

$$x_k p_l - p_l x_k = \mathrm{i}\hbar\delta_{kl}, \quad (9.9.12)$$

it reduces to

$$(\triangle x_k)_\psi (\triangle p_l)_\psi \geq \frac{\hbar}{2}\delta_{kl}. \quad (9.9.13)$$

This minorization is not canonically invariant. Lack of canonical invariance of (9.9.13) prompts us to obtain the stronger Robertson–Schrödinger inequality (Robertson 1930, Sudarshan *et al.* 1995)

$$(\triangle x)_\psi^2 (\triangle p)_\psi^2 - \triangle_\psi^2(xp) \geq \frac{\hbar^2}{4}, \quad (9.9.14)$$

where

$$\triangle_{\psi}(xp) \equiv \left\langle \frac{xp+px}{2} \right\rangle_{\psi} - \langle x \rangle_{\psi} \langle p \rangle_{\psi}. \tag{9.9.15}$$

The relation (9.9.14) is canonically invariant and is the only quadratic symplectic invariant. It should be stressed that the generalized uncertainty relations (9.9.11) *only hold* if $A\psi \in D(B)$ *and* $B\psi \in D(A)$. More precisely, the domain of AB is given by

$$D(AB) = \{\psi \in D(B) : B\psi \in D(A)\}, \tag{9.9.16}$$

and the domain of BA consists of

$$D(BA) = \{\psi \in D(A) : A\psi \in D(B)\}. \tag{9.9.17}$$

Thus, the commutator $[A, B]$ is defined on

$$D(AB) \cap D(BA) \subset D(A) \cap D(B).$$

The commutator $[A, B]$ is usually not closed, and its domain need not be dense in $\mathcal{H}$ (it may even be the empty set!).

If the above conditions are not satisfied, one can find counterexamples. For this purpose, let us consider the position and momentum operators on the space $\mathcal{L}^2(S^1)$ (i.e. on the space of square-integrable functions on the circle):

$$p : f \rightarrow \frac{\hbar}{\mathrm{i}} \frac{\mathrm{d}f}{\mathrm{d}x},$$

$$q : f \rightarrow xf,$$

for $x \in [0, 2\pi[$. But then $xf(x)$ does not belong to the domain of p, because p is self-adjoint only upon requiring proportionality of the boundary values via a phase factor: $f(2\pi) = \mathrm{e}^{\mathrm{i}\alpha} f(0), \alpha \in \mathbf{R}$. Thus, the commutator $[q, p]$ is not defined on the eigenfunctions of p, and no contradiction is obtained upon remarking that, for such eigenfunctions, $\triangle p$ vanishes, whereas $\triangle q$ is finite by construction (see problem 9.P1). One should also bear in mind that, in (9.9.11), $(\triangle A)_{\psi}$ and $(\triangle B)_{\psi}$ refer to the dispersion of the measured values of A and B in sequences of *identical* and *independent* experiments. During such experiments, the system is prepared in the same state, described by the vector ψ, and hence a measurement is performed of A, or B, or $\mathrm{i}[A, B]$ (but not of A and B simultaneously!).

The inequality (9.9.11) reduces to an equality (so that A and B have the minimal dispersion) if the following two conditions hold.

(i) There exists $\lambda \in \mathbf{C}$ such that

$$\lambda(A - \langle A \rangle)\psi = (B - \langle B \rangle)\psi. \tag{9.9.18}$$

(ii) The operator F annihilates ψ:

$$F\psi = 0. \tag{9.9.19}$$

By imposing Eqs. (9.9.18) and (9.9.19), and using the definition (9.9.8), one finds

$$\lambda(\triangle A)^2_\psi + \frac{1}{\lambda}(\triangle B)^2_\psi = 0. \tag{9.9.20}$$

On the other hand, by evaluating the operator identity (9.9.1) on the vector ψ, one finds

$$\lambda(\triangle A)^2_\psi - \frac{1}{\lambda}(\triangle B)^2_\psi = \mathrm{i}\langle C\rangle_\psi. \tag{9.9.21}$$

On adding to each other Eqs. (9.9.20) and (9.9.21) one finds

$$\lambda = \frac{\mathrm{i}}{2}\frac{\langle C\rangle_\psi}{(\triangle A)^2_\psi}. \tag{9.9.22}$$

In the particular case when A is the position operator x, and B is the momentum operator $\frac{\hbar}{\mathrm{i}}\frac{\partial}{\partial x}$ in one-dimensional problems, the normalized solution of Eq. (9.9.18) turns out to be (by separation of variables and subsequent integration)

$$f(x) = \frac{1}{\sqrt{2\pi(\triangle x)^2}} \exp\left[-\frac{(x-\langle x\rangle)^2}{4(\triangle x)^2} + \frac{\mathrm{i}}{\hbar}\langle p_x\rangle x\right], \tag{9.9.23}$$

i.e. a Gaussian. This was already found in Eq. (4.2.7).

9.9.1 Time–energy uncertainty relation

In the literature, the so-called time–energy uncertainty relation is also discussed:

$$(\triangle E)(\triangle t) \geq \frac{\hbar}{2}. \tag{9.9.24}$$

This inequality, however, should be handled with great care. The reason is, that the time variable in (relativistic or non-relativistic) quantum mechanics should be viewed as a parameter and not as an operator, so that it does not correspond to any observable. One can however define a 'time operator' from the dynamical variables in the Heisenberg picture of section 9.8. This operator is 'dynamic' rather than 'kinematic' in that it depends on the interaction (and hence the Hamiltonian). This time operator $\tau(p,q)$ is defined almost everywhere and is canonically conjugate to the Hamiltonian. For this variable one has the uncertainty relation (9.9.24) with t replaced by τ. For a non-relativistic free particle $\tau = \frac{m\vec{q}\cdot\vec{p}}{p^2}$ (cf. subsection 2.3.2).

The interpretation of (9.9.24) is suggested by the analysis of unstable systems that decay with a certain mean lifetime, here denoted by τ; one can then set $\Delta t = \tau$ for an unstable system. In such a state, E is not well defined, but it only makes sense as a parameter belonging to an energy band

$$E \in \Big[E_0 - \Delta E, E_0 + \Delta E\Big]. \tag{9.9.25}$$

One can think, for example, of an excited atom which decays by emitting electromagnetic radiation: the corresponding spectral line is never absolutely monochromatic, but it has a certain 'line width' determined by the dispersion of the energy of the atomic levels. On denoting by Γ the width of the energy band of the unstable state, $\Gamma \equiv 2\,\Delta E$, one has

$$\tau\Gamma \cong \hbar. \tag{9.9.26}$$

A method for the derivation of the inequality (9.9.24) is as follows. Let ψ be a normalized physical state which is not an energy eigenstate. One then has the expectation value $E_0 = (\psi, H\psi)$ with indeterminacy

$$\Delta E = \Big(\langle H^2\rangle_\psi - \langle H\rangle_\psi^2\Big)^{1/2}. \tag{9.9.27}$$

Let now F be a generic physical observable that does not depend explicitly on time. The mean value of F in the state ψ evolves in time according to the equation

$$\frac{\mathrm{d}}{\mathrm{d}t}\langle F\rangle_\psi = \frac{1}{\mathrm{i}\hbar}(\psi, [F, H]\psi). \tag{9.9.28}$$

The mean quadratic deviation for the observation of F in the state ψ is

$$\Delta F = \Big(\langle F^2\rangle_\psi - \langle F\rangle_\psi^2\Big)^{1/2}, \tag{9.9.29}$$

and one has

$$(\Delta E)(\Delta F) \geq \frac{1}{2}\left|(\psi, [F, H]\psi)\right|. \tag{9.9.30}$$

By virtue of Eq. (9.9.28), the inequality (9.9.30) can be re-expressed in the form

$$(\Delta E)(\Delta F) \geq \frac{\hbar}{2}\left|\frac{\mathrm{d}}{\mathrm{d}t}\langle F\rangle_\psi\right|. \tag{9.9.31}$$

If one denotes by Δt the time interval during which the mean value of F changes by the amount ΔF, one can write the formula

$$\Delta t = \frac{\Delta F}{\left|\frac{\mathrm{d}}{\mathrm{d}t}\langle F\rangle_\psi\right|} = \tau_\psi(F), \tag{9.9.32}$$

which, upon insertion into (9.9.31), yields the desired proof of the inequality (9.9.24). Such a $\tau_\psi(F)$ provides the time that should elapse, starting from t, for the average of distribution of values of F to change by the amount $\triangle_\psi(F)$. With this understanding, one can view $\tau_\psi(F)$ as a time interval that is characteristic of the evolution of the system. Of course, if F is a constant of motion, one has

$$\left|\frac{\mathrm{d}}{\mathrm{d}t}\langle F\rangle_\psi\right| = 0 \Longrightarrow \tau_\psi(F) = \infty.$$

One has to assume that $\tau_\psi(F)$ becomes infinite to interpret the above operations correctly. By considering the family $\{F\}$ of all observables which do not depend explicitly on time, one can define

$$\tau_\psi \equiv \inf_{F\in\{F\}} \{\tau_\psi(F)\}. \tag{9.9.33}$$

In such a way τ_ψ becomes a characteristic time interval for the evolution of the system, independently of the observable that is under consideration. This implies that (9.9.31) reads as

$$\tau_\psi \; \triangle_\psi H \geq \frac{\hbar}{2}. \tag{9.9.34}$$

For example, for a two-level system with state vector initially equal to

$$\psi(t_0) = c_1\varphi_1 + c_2\varphi_2, \tag{9.9.35}$$

with φ_1 and φ_2 eigenstates of the Hamiltonian with eigenvalues E_1 and E_2, respectively, one has the time evolution

$$\psi(t) = c_1 \mathrm{e}^{-\mathrm{i}E_1(t-t_0)/\hbar}\varphi_1 + c_2 \mathrm{e}^{-\mathrm{i}E_2(t-t_0)/\hbar}\varphi_2. \tag{9.9.36}$$

The corresponding quadratic deviation of the Hamiltonian in the state ψ reduces to $|E_1 - E_2|$ if $|c_1|$ and $|c_2|$ are nearly equal, otherwise

$$\triangle_\psi H = |c_1 c_2||E_1 - E_2|.$$

The probability of finding the value $f_n \in \sigma_p(F)$ in the measurement of F on the state ψ reads (in the absence of degeneracy)

$$\begin{aligned} P_{F,\psi}(f_n;t) &= |(f_n,\psi(t))|^2 \\ &= |c_1|^2|(f_n,\varphi_1)|^2 + |c_2|^2|(f_n,\varphi_2)|^2 \\ &\quad + 2\mathrm{Re}\left[c_1 c_2^* \mathrm{e}^{-\mathrm{i}(E_1-E_2)(t-t_0)/\hbar}(f_n,\varphi_1)(f_n,\varphi_2)^*\right]. \end{aligned} \tag{9.9.37}$$

Such a probability oscillates between two extremal values with frequency $\frac{|E_1-E_2|}{\hbar}$. The characteristic interval for the evolution of the observable is

$$\tau_\psi \equiv \int_0^\infty |(\psi(0),\psi(t))|^2 \, \mathrm{d}t, \tag{9.9.38}$$

and it satisfies the minorization

$$\tau_\psi \geq \frac{\hbar}{|E_1 - E_2|}. \tag{9.9.39}$$

9.10 Unitary operators and symplectic linear maps

In this section we are going to recover the Dirac prescription (9.2.2) for linear and quadratic functions. To begin, if the map $W : S \to \mathcal{U}(\mathcal{H})$ is a Weyl system, then for any linear map $T : S \to S$ one can write that

$$W(T(z+z')) = \mathrm{e}^{-\frac{\mathrm{i}}{2}\omega(Tz,Tz')} W(Tz)W(Tz'), \tag{9.10.1}$$

by virtue of (9.5.11). Moreover, if the map T is also symplectic, one has

$$W(T(z+z')) = \mathrm{e}^{-\frac{\mathrm{i}}{2}\omega(z,z')} W(Tz)W(Tz'), \tag{9.10.2}$$

since $\omega(Tz, Tz') = \omega(z, z')$ for symplectic maps. The above equation suggests associating to the Weyl system W a new map

$$W_T : S \to \mathcal{U}(\mathcal{H}) : z \to W(Tz). \tag{9.10.3}$$

Then with T we associate an automorphism

$$\nu_T : \mathcal{U}(\mathcal{H}) \to \mathcal{U}(\mathcal{H})$$

by setting

$$W(Tz) \equiv \nu_T(W(z)). \tag{9.10.4}$$

This automorphism can be written in the form

$$\nu_T(W(z)) = U_T^{-1} W(z) U_T, \tag{9.10.5}$$

with U_T being a unitary transformation acting on $\mathcal{H}$.

For any symplectic linear transformation $T : S \to S$ we therefore define

$$W_T(z) \equiv W(Tz), \tag{9.10.6}$$

with the associated definition for generators, i.e.

$$R_T(z) \equiv R(Tz). \tag{9.10.7}$$

We are going to consider symplectic linear transformations on S and derive the corresponding transformations on $\mathcal{U}(\mathcal{H})$ or on their infinitesimal generators (observables). Using the von Neumann theorem on the uniqueness of the commutation relations up to unitary equivalence, whenever useful we shall perform computations in the Schrödinger picture. In particular, we require that T maps L into itself and is a canonical transformation such that $T^*(\theta) - \theta = \mathrm{d}F_T$, and hence

$$\left(U_T\psi\right)(y) = \mathrm{e}^{\mathrm{i}F_T(y)}\, \psi(Ty),$$

which defines a unitary operator if T preserves the Lebesgue measure on L. Moreover, we shall look for implementations of symplectic transformations in terms of unitary operators, by solving the equation (see Eq. (9.10.5))

$$U_T^{-1} W(z) U_T = W(Tz), \tag{9.10.8}$$

or also

$$U_T^{-1} R(z) U_T = R(Tz) = R_T(z) = xP_T + fQ_T. \tag{9.10.9}$$

Note that $R_T(z)$ defines $xP_T + fQ_T$ as a 'whole'; whether or not we may consider it as a sum of operators defined autonomously is a different matter. Very often, it is convenient to consider one-parameter groups of symplectic transformations and to solve for U_T using the corresponding linear differentiated version. We shall consider three relevant cases: (i) translations; (ii) rotations; (iii) linear dynamical evolution.

9.10.1 Translations

Here we use the Schrödinger picture, with Hilbert space $\mathcal{H} = \mathcal{L}^2(L)$, as we know from section 9.6. The unitary operator associated with translations is defined by

$$\Big(U_a \psi\Big)(y) \equiv \psi(y+a). \tag{9.10.10}$$

From the Weyl quantization one finds (hereafter, position and momentum operators are written without the 'hat' symbol for simplicity of notation)

$$U_a = e^{iaP}. \tag{9.10.11}$$

Let us look at the induced transformations on observables (see Eq. (9.10.8))

$$U_a^{-1} W(z) U_a = W(z+a) = W_a(z), \tag{9.10.12}$$

along with (see Eq. (9.10.9))

$$iU_a^{-1}(xP + fQ)U_a = i\Big(xP_a + fQ_a\Big). \tag{9.10.13}$$

One therefore finds

$$P = P_a, \quad Q = Q_a + a. \tag{9.10.14}$$

9.10.2 Rotations

We are already familiar with the concept of rotations from appendix 2.B and chapter 5. In the present framework, rotations are maps of S into S: $\mathcal{R} : S \to S$, such that

$$\mathcal{R}\,\mathcal{R}^{\mathrm{T}} = \mathbb{I}. \tag{9.10.15}$$

We look for

$$U_{\mathcal{R}}^{-1} W(z) U_{\mathcal{R}} = W(\mathcal{R}z) = W_{\mathcal{R}}(z), \tag{9.10.16}$$

and we consider the transformation

$$x_\alpha = \mathcal{R}_\alpha x, \quad f_\alpha \equiv \left(\mathcal{R}_\alpha^{-1}\right)^{\mathrm{T}} f = \mathcal{R}_\alpha f. \tag{9.10.17}$$

Of course, such a transformation is symplectic, because

$$\omega\Big((x_\alpha, f_\alpha), (y_\alpha, g_\alpha)\Big) = \omega((x, f), (y, g)), \tag{9.10.18a}$$

i.e.

$$f_\alpha(y_\alpha) - g_\alpha(x_\alpha) = f(y) - g(x), \tag{9.10.18b}$$

as follows from (9.10.17). To find the associated unitary transformation, one has to solve the equation

$$U_\alpha^{-1} \widehat{\mathcal{R}}(z) U_\alpha = \widehat{\mathcal{R}}_\alpha(z) = x P_\alpha + f Q_\alpha = x_\alpha P + f_\alpha Q. \tag{9.10.19}$$

On taking the derivative with respect to the parameter α one finds

$$x \frac{\mathrm{d}P_\alpha}{\mathrm{d}\alpha} + f \frac{\mathrm{d}Q_\alpha}{\mathrm{d}\alpha} = \frac{\mathrm{d}x_\alpha}{\mathrm{d}\alpha} P + \frac{\mathrm{d}f_\alpha}{\mathrm{d}\alpha} Q, \tag{9.10.20}$$

and also

$$\left[\widehat{\mathcal{R}}(z), \frac{\mathrm{d}U_\alpha}{\mathrm{d}\alpha}\right]_{\alpha=0} = \left.\frac{\mathrm{d}x_\alpha}{\mathrm{d}\alpha}\right|_{\alpha=0} P + \left.\frac{\mathrm{d}f_\alpha}{\mathrm{d}\alpha}\right|_{\alpha=0} Q, \tag{9.10.21}$$

which should be considered as an equation for $\left.\frac{\mathrm{d}U_\alpha}{\mathrm{d}\alpha}\right|_{\alpha=0}$. Now restricting to rotations in a plane:

$$\begin{pmatrix} x_{1\alpha} \\ x_{2\alpha} \end{pmatrix} = \begin{pmatrix} \cos\alpha & -\sin\alpha \\ \sin\alpha & \cos\alpha \end{pmatrix} \begin{pmatrix} x_1 \\ x_2 \end{pmatrix}, \tag{9.10.22}$$

$$\begin{pmatrix} f_{1\alpha} \\ f_{2\alpha} \end{pmatrix} = \begin{pmatrix} \cos\alpha & -\sin\alpha \\ \sin\alpha & \cos\alpha \end{pmatrix} \begin{pmatrix} f_1 \\ f_2 \end{pmatrix}, \tag{9.10.23}$$

for which

$$\left.\frac{\mathrm{d}x_\alpha}{\mathrm{d}\alpha}\right|_{\alpha=0} = \begin{pmatrix} 0 & -1 \\ 1 & 0 \end{pmatrix} x_0, \tag{9.10.24a}$$

$$\left.\frac{\mathrm{d}f_\alpha}{\mathrm{d}\alpha}\right|_{\alpha=0} = \begin{pmatrix} 0 & -1 \\ 1 & 0 \end{pmatrix} f_0, \tag{9.10.24b}$$

one finds from (9.10.21)

$$\begin{aligned} &\mathrm{i}\left[x_1 P_1 + x_2 P_2 + f_1 Q_1 + f_2 Q_2, \widehat{\mathcal{R}}_\alpha\right]_{\alpha=0} \\ &= (P_1, P_2) \begin{pmatrix} 0 & -1 \\ 1 & 0 \end{pmatrix} \begin{pmatrix} x_1 \\ x_2 \end{pmatrix} \end{aligned}$$

$$+ (Q_1, Q_2) \begin{pmatrix} 0 & -1 \\ 1 & 0 \end{pmatrix} \begin{pmatrix} f_1 \\ f_2 \end{pmatrix}. \tag{9.10.25}$$

We look for a solution $\widehat{\mathcal{R}}_\alpha$ by writing the most general quadratic operator in P_j and Q_k. It has to be quadratic because the bracket is homogeneous of degree -2, i.e. with P, Q it associates a multiple of the identity. Indeed, Eq. (9.10.25) leads to

$$P_2 = \mathrm{i}\Big[P_1, \widehat{\mathcal{R}}_\alpha\Big]_{\alpha=0}, \tag{9.10.26}$$

$$-P_1 = \mathrm{i}\Big[P_2, \widehat{\mathcal{R}}_\alpha\Big]_{\alpha=0}, \tag{9.10.27}$$

$$Q_2 = \mathrm{i}\Big[Q_1, \widehat{\mathcal{R}}_\alpha\Big]_{\alpha=0}, \tag{9.10.28}$$

$$-Q_1 = \mathrm{i}\Big[Q_2, \widehat{\mathcal{R}}_\alpha\Big]_{\alpha=0}. \tag{9.10.29}$$

By virtue of the irreducibility of the canonical commutation relations, one finds

$$\widehat{\mathcal{R}}_{\alpha=0} = Q_1 P_2 - Q_2 P_1 + \lambda 1\!\!\mathrm{I}. \tag{9.10.30}$$

In the coordinate representation on $\mathbf{R}^3$, one usually denotes by $\vec{L}$ the *orbital angular momentum*, and one has

$$\vec{L} \equiv \vec{Q} \wedge \vec{P}. \tag{9.10.31}$$

Here, in analogy, we have found the quantum-mechanical angular momentum as the generator of rotations in the space of operators. The following commutation relations are easily checked (with $T = L, P, Q$, without 'hat' symbols as before):

$$\Big[L_m, T_s\Big] = \mathrm{i}\varepsilon_{msr} T_r, \tag{9.10.32}$$

$$\Big[L_m, L^2\Big] = 0, \tag{9.10.33}$$

$$\Big[L_m, Q^2\Big] = 0, \tag{9.10.34}$$

$$\Big[L_m, P^2\Big] = 0. \tag{9.10.35}$$

Within the framework of Weyl quantization, one can also study the *parity operator* P, which relates the two connected components of $O(3)$, with an action defined by the equation (Thirring 1981)

$$P^{-1} W(z) P \equiv W(-z). \tag{9.10.36}$$

The operator P is sometimes fixed in the literature by imposing the condition

$$(P\psi)(x) = \psi(-x), \quad \psi \in \mathcal{L}^2(L). \tag{9.10.37}$$

Hence one finds

$$P^2 = \mathbb{1}, \tag{9.10.38}$$

$$P^{-1} = P^* = P. \tag{9.10.39}$$

Remark. The choice (9.10.38) is not good on real (Majorana) spinor wave functions (nor on pseudoscalar particles where there is an additional odd intrinsic parity), nor is (9.10.39), since

$$(P\psi)(x) = \mathrm{i}\beta\psi(-x), \quad \beta^2 = \mathbb{1}, \quad \beta^* = -\beta.$$

The phase factor is the intrinsic parity η_P and hence P is fixed by (Marshak and Sudarshan 1962)

$$(P\psi)(x) = \eta_P \psi(-x), \quad \psi \in \mathcal{L}^2(L).$$

The parity operator commutes with $\vec{L}$, i.e.

$$P\,\vec{L}\,P^{-1} = \vec{L}, \tag{9.10.40}$$

and can be expressed in the form (Thirring 1981)

$$P = \eta \mathrm{e}^{\mathrm{i}\pi\left(\sqrt{L^2+\frac{1}{4}}-\frac{1}{2}\right)}. \tag{9.10.41}$$

9.10.3 Harmonic oscillator

To study the harmonic oscillator, which, unlike translations and rotations, does not preserve the split $L \oplus L^*$, we take $S = \mathbf{R}^2$ (which coincides with $\mathbf{C} = T^*\mathbf{R}$), with

$$\omega\Big((\alpha,\beta);(\alpha',\beta')\Big) \equiv -\alpha\beta' + \alpha'\beta, \tag{9.10.42}$$

$$R(\alpha,\beta) \equiv \alpha P + \beta Q. \tag{9.10.43}$$

The classical Hamiltonian

$$H = \frac{\beta^2}{2m} + \frac{m}{2}\omega^2 x^2, \tag{9.10.44}$$

by evolution on α and β, induces a one-parameter group of automorphisms on the observables P_t, Q_t which is found to be

$$P_t \equiv P\cos(\omega t) - Qm\omega\sin(\omega t), \tag{9.10.45}$$

$$Q_t \equiv \frac{P}{m\omega}\sin(\omega t) + Q\cos(\omega t), \tag{9.10.46}$$

resulting from the action of $\nu_t(W(z)) \equiv W(T_t z)$ on $R(\alpha, \beta)$. Specifically,

$$\begin{aligned}\nu_t(R(\alpha,\beta)) &= \alpha P_t + \beta Q_t \\ &= \left[\alpha \cos(\omega t) + \frac{\beta}{m\omega}\sin(\omega t)\right] P + \left[-m\omega\alpha \sin(\omega t) + \beta\cos(\omega t)\right] Q \\ &= R(\alpha_t, \beta_t),\end{aligned} \tag{9.10.47}$$

where

$$\begin{pmatrix} \alpha_t \\ \beta_t \end{pmatrix} \equiv \begin{pmatrix} \cos(\omega t) & \frac{1}{m\omega}\sin(\omega t) \\ -m\omega \sin(\omega t) & \cos(\omega t) \end{pmatrix} \begin{pmatrix} \alpha \\ \beta \end{pmatrix}. \tag{9.10.48}$$

In other words, one has

$$\nu_t(R(\alpha,\beta)) = \alpha P_t + \beta Q_t = \alpha_t P + \beta_t Q = U_t^{-1} R(\alpha,\beta) U_t. \tag{9.10.49}$$

Formulae (9.10.45) and (9.10.46) for the evolution of P and Q are solutions of the Heisenberg equations of motion:

$$\mathrm{i}\hbar \frac{\mathrm{d}}{\mathrm{d}t} A_t = \left[A_t, H\right], \tag{9.10.50}$$

with the Hamiltonian H taking the form $\frac{P^2}{2m} + \frac{m}{2}\omega^2 Q^2 + \lambda \mathbb{1}$.

The induced automorphism reads

$$\nu_t(A) = A_t = \mathrm{e}^{\mathrm{i}Ht}\, A\, \mathrm{e}^{-\mathrm{i}Ht}, \tag{9.10.51}$$

and T_t preserves the complex structure (i.e. a map for which the square is minus the identity)

$$J = \begin{pmatrix} 0 & \frac{1}{m\omega} \\ -m\omega & 0 \end{pmatrix}. \tag{9.10.52}$$

It is also possible to give a deep formula for the annihilation and creation operators, introduced in (9.4.7) and (9.4.8) (here $J|_{V_1} = F$), as

$$a(z) \equiv \frac{1}{\sqrt{2}}\left[R(z) + \mathrm{i}R(Jz)\right], \tag{9.10.53}$$

$$a^\dagger(z) \equiv \frac{1}{\sqrt{2}}\left[R(z) - \mathrm{i}R(Jz)\right]. \tag{9.10.54}$$

It should be stressed that the Schrödinger picture depends on the direct-sum decomposition $S = L \oplus L^*$, with Hilbert space $\mathcal{H} = \mathcal{L}^2(L)$, whereas, if the definitions (9.10.53) and (9.10.54) are used, we only rely on the complex structure J, which is a map $J : S \longrightarrow S$ such that $J^2 = -\mathbb{1}$. Once that (9.10.53) and (9.10.54) are given, one can obtain the Weyl system

$$W(z) \equiv \mathrm{e}^{z\hat{a}^\dagger - z^*\hat{a}}, \tag{9.10.55}$$

which will be studied in detail in section 10.4.

9.11 On the meaning of Weyl quantization

In his development of quantum mechanics, Heisenberg insisted on having quantum mechanical equations of motion in complete formal analogy with the equations of classical mechanics, and Dirac obtained his 'quantum Poisson brackets' after a careful examination of the properties of Poisson brackets in classical mechanics. Both Heisenberg and Dirac were therefore led by taking account of classical mechanics.

The Weyl programme studied in this chapter leads to a deep conceptual revolution, because the emphasis on group-theoretical methods provides a scheme where Weyl systems are considered in the first place, and classical mechanics is eventually recovered. Unlike what we say in the title of the present monograph, the modern point of view might therefore take quantum mechanics as the starting point, and investigate the conditions under which classical physics on the macroscopic scale is recovered. It is still too early, for the general reader, to elaborate on this point, but a summary of the whole Weyl programme can be quite appropriate. Following Mackey (1998) and our previous presentation, the key elements can be summarized as follows (at the risk of slight repetition).

It is well known in mathematics that a one-to-one correspondence exists between self-adjoint operators in a Hilbert space $\mathcal{H}$ and the unitary representations of the additive group of the real line in the same Hilbert space. If $\mathcal{H}$ is finite-dimensional, the representation $t \rightarrow U_t^A$ corresponding to the self-adjoint operator A is given by the formula

$$U_t^A = \mathrm{e}^{\mathrm{i}At} = \sum_{n=0}^{\infty} \frac{(\mathrm{i}At)^n}{n!}, \tag{9.11.1}$$

where the series on the right-hand side converges uniformly. The one-to-one correspondence is readily established and one can show that $\frac{1}{\mathrm{i}}\frac{\mathrm{d}}{\mathrm{d}t}U_t^A$ is equal to A when t is set to zero. Weyl conjectured that Hilbert's spectral theorem would soon be used to extend this correspondence to the general case, including unbounded self-adjoint operators, and proceeded as though this was done (such a task was accomplished in Stone (1930)).

It is also well known that two self-adjoint operators Q and P commute with one another if and only if the associated strongly continuous one-parameter unitary groups satisfy Eq. (9.3.6). Since the operators $\mathrm{e}^{\mathrm{i}sP}$ and $\mathrm{e}^{\mathrm{i}tQ}$ are everywhere defined bounded operators for all s and t, they are easier to work with in a rigorous manner than P and Q, which are always unbounded and only densely defined. Weyl accordingly proposed that one should take the unitary representations of the real line $s \rightarrow \mathrm{e}^{\mathrm{i}sP}$ and $t \rightarrow \mathrm{e}^{\mathrm{i}tQ}$ as the fundamental objects, instead of the operators P and Q. Within this framework the Heisenberg commutation relations for a system with N coordinates and momenta can be written in the form

$$e^{itQ_k}e^{isQ_j} - e^{isQ_j}e^{itQ_k} = 0 = e^{isP_k}e^{itP_j} - e^{itP_j}e^{isP_k}, \tag{9.11.2}$$

$$e^{isP_k}e^{itQ_j} = e^{i\frac{st}{\hbar}\delta_j^i}\, e^{itQ_j}e^{isP_k}, \tag{9.11.3}$$

for all s, t, i, j, k. At this stage, Weyl pointed out that the various e^{isP_k} can be combined together into a single unitary representation U of the commutation group $\mathbf{R}^N$ of all N-tuples of real numbers under addition, i.e.

$$U_{s_1,s_2,\ldots,s_N} \equiv e^{is_1P_1}e^{is_2P_2}\ldots e^{is_NP_N}, \tag{9.11.4}$$

and similarly with e^{itQ_j}, i.e.

$$V_{t_1,t_2,\ldots,t_N} \equiv e^{it_1Q_1}e^{it_2Q_2}\ldots e^{it_NQ_N}. \tag{9.11.5}$$

More significantly, Weyl defined a unitary operator-valued function of the additive group of $\mathbf{R}^{2N}$ by the formula

$$W_{s_1,s_2,\ldots,s_N,t_1,t_2,\ldots,t_N} \equiv U_{s_1,s_2,\ldots,s_N}\, V_{t_1,t_2,\ldots,t_N}. \tag{9.11.6}$$

He realized that, although W is not a representation of $\mathbf{R}^{2N}$, it is 'almost' a representation in that, for any two elements x and y of $\mathbf{R}^{2N}$, W_{x+y} differs from W_xW_y only by a multiplicative factor depending on x and y. More precisely, one has

$$W_{x+y} = \sigma(x,y)W_xW_y, \tag{9.11.7}$$

where the map $\sigma : \mathbf{R}^{2N} \rightarrow \mathbf{C}$ has unit modulus. This means that W is a *projective representation* with multiplier σ. In the construction (9.11.6), the associated multiplier σ is defined by the equation

$$\sigma\Big((s_1,s_2,\ldots,s_N,t_1,t_2,\ldots,t_N)(s_1',s_2',\ldots,s_N',t_1',t_2',\ldots,t_N')\Big) \equiv e^{\frac{i}{\hbar}(t_1s_1'+t_2s_2'+\cdots+t_Ns_N')}. \tag{9.11.8}$$

This picture was extended by I. E. Segal, by replacing $\mathbf{R}^{2n}$ with a complex Hilbert space, while $\sigma(x,y)$ is replaced by the exponential of the antisymmetric part of $\langle x|y\rangle$.

The global form of the Heisenberg commutation relations is therefore equivalent to the statement that W is a projective unitary representation with multiplier σ defined in Eq. (9.11.8). Weyl then pointed out that σ is the only possible non-degenerate multiplier for $\mathbf{R}^{2N}$, and presented a heuristic argument in support of the conjecture that, for each non-degenerate σ, there exists within equivalence a unique irreducible projective unitary representation of $\mathbf{R}^{2N}$ with the multiplier σ. These remarks by Weyl were possibly the first step in the programme of deriving fundamental principles in quantum mechanics from group-theoretical symmetry principles. For an account of modern developments, the reader is referred again to Mackey (1998). We note that any symplectic structure on $\mathbf{R}^{2n}$ invariant under given translations of $\mathbf{R}^{2n}$ on itself reduces to the standard one. However, there might be alternative realizations of the translation group.

9.12 The basic postulates of quantum theory

We now present the basic postulates of quantum mechanics. For this purpose, we first state the four rules that deal with the mathematical framework within which all quantum-mechanical systems can be described (but the reader should be aware that the scheme is, by no means, unique. There exist yet other mathematical frameworks for the development of quantum theory, while some early attempts of applying it to the whole universe are described in DeWitt and Graham (1973)). The four basic postulates are as follows (Bohm 1958, Mackey 1963, Beltrametti and Cassinelli 1976, 1981, Prugovecki 1981, Varadarajan 1985, Weinberg 1989, Peres 1993, Isham 1995, Klauder 1997, Auletta 2000).

(i) To each physical system S there corresponds a suitable Hilbert space $\mathcal{H}_S$. Every state of S is represented by a normalized vector $\psi \in \mathcal{H}_S$, called the *state vector*, which contains all possible informations on the system (see, however, remark 1 following Eq. (9.12.6)). The time evolution of ψ is ruled by the time-dependent Schrödinger equation (see Eqs. (7.8.2) and (7.8.3))

$$i\hbar \frac{\mathrm{d}\psi}{\mathrm{d}t} = \hat{H}\psi, \tag{9.12.1}$$

where $\hat{H}$ is an essentially self-adjoint operator in $\mathcal{H}_S$.

(ii) To each observable quantity A there corresponds a self-adjoint operator $\hat{A}$ in $\mathcal{H}_S$. The discrete spectrum $\sigma_{\mathrm{d}}(\hat{A})$ and the continuous spectrum $\sigma_{\mathrm{c}}(\hat{A})$ of $\hat{A}$ represent the set of all possible values taken by A (see comments concerning the singular spectrum made after Eq. (4.4.23)). Given the eigenvalue equations

$$\hat{A}\varphi_{ru} = \alpha_r \varphi_{ru}, \tag{9.12.2}$$

$$\hat{A}\varphi_{\alpha u} = \alpha \varphi_{\alpha u}, \tag{9.12.3}$$

where φ_{ru} and $\varphi_{\alpha u}$ are the proper and improper eigenvectors, respectively, so that the state vector can be expanded in the form

$$\psi(t) = \sum_{r,u} c_{ru}(t)\varphi_{ru} + \sum_u \int_{\sigma_{\mathrm{c}}} c_u(\alpha, t)\varphi_{\alpha u}\, \mathrm{d}\alpha, \tag{9.12.4}$$

the *probability* that an observation of A at time t yields the value $\alpha_r \in \sigma_{\rm d}$ or a value lying in the interval $(\alpha, \alpha + {\rm d}\alpha) \subset \sigma_{\rm c}$, is given by

$$P(A = \alpha_r | t) = \sum_u |c_{ru}(t)|^2 = \sum_u |(\varphi_{ru}, \psi)|^2, \tag{9.12.5}$$

$$P(\alpha \leq A \leq \alpha + {\rm d}\alpha | t) = \sum_u |c_u(\alpha, t)|^2 \, {\rm d}\alpha = \sum_u |(\varphi_{\alpha u}, \psi)|^2 \, {\rm d}\alpha. \tag{9.12.6}$$

With our notation, the summation over u takes into account that degeneracies may occur, i.e. several eigenvectors belonging to the same eigenvalue.

Remark 1. This implies that the *physical predictions* of the theory remain unaffected if the state vector ψ is multiplied by an arbitrary complex number μ such that $|\mu| = 1$, i.e. $\mu = {\rm e}^{{\rm i}\rho}$, with $\rho \in {\bf R}$, and $\psi {\rm e}^{{\rm i}\rho}$ leads to the same physics. By virtue of this invariance property, one might choose to deal with a space of *rays*. Note, however, that the sum of state vectors $\psi_1 = \rho_1 {\rm e}^{{\rm i}\varphi_1}$ and $\psi_2 = \rho_2 {\rm e}^{{\rm i}\varphi_2}$ obeys the law

$$\psi_1 + \psi_2 = {\rm e}^{{\rm i}\varphi_1} \left[\rho_1 + {\rm e}^{{\rm i}(\varphi_2 - \varphi_1)} \rho_2 \right],$$

and hence is more convenient if one wants to describe interference experiments, where the relative phase $(\varphi_2 - \varphi_1)$ plays a role (Man'ko *et al.* 2000). In contrast, all phases are factored out in the space of rays and the sum of two rays is not defined, which makes it more difficult to account for interference.

Remark 2. By virtue of the postulate (ii), if, for $t = t_0$, the state vector is a proper eigenvector of $\hat{A}$ corresponding to the eigenvalue α_r, i.e. if

$$\psi(t_0) = \sum_u c_u \varphi_{ru}, \tag{9.12.7}$$

one finds

$$P(A = \alpha_r | t_0) = 1. \tag{9.12.8}$$

This is why the proper eigenvectors of $\hat{A}$ are also called the *eigenstates* of A. The postulate (ii) is a non-trivial generalization to any observable of what is known from wave mechanics about energy measurements on a given quantum state.

Remark 3. It is physically impossible to measure with *absolute accuracy* an observable C having a continuous spectrum. However, one can measure such an observable with an *arbitrarily good accuracy*. For this purpose, it is sufficient to choose a step function θ, which approximates the function f with the desired accuracy, and measure $\theta(C)$, which is an observable with a discrete spectrum. It should be stressed that the introduction of $\theta(C)$ *is not a mathematical trick*, but is essential for reasons of principle (Daneri *et al.* 1962).

(iii) Any two *compatible observables* A, B are represented by the self-adjoint operators $\hat{A}, \hat{B}$ with a complete orthonormal set of common eigenvectors, and the probability of finding in the simultaneous measurement of A and B the result $A = \alpha_r, B = \beta_s$ is expressed by

$$P(A = \alpha_r, B = \beta_s | t) = \sum_u |c_{rsu}(t)|^2 = \sum_u |(\varphi_{rsu}, \psi)|^2. \tag{9.12.9}$$

Remark 4. The postulate (iii) can be generalized to any number of compatible observables.

Remark 5. One can prove that a necessary and sufficient condition for two self-adjoint operators $\hat{A}$ and $\hat{B}$ to have a complete set of common eigenvectors is that $\hat{A}$ and $\hat{B}$ commute.

Remark 6. If $\hat{A}$ and $\hat{B}$ are unbounded operators, which is quite often the case, the expression $[\hat{A}, \hat{B}]$ for their commutator should be replaced by the group 'commutator'

$$C_{\rm g}(\hat{A}, \hat{B}) \equiv {\rm e}^{{\rm i}t\hat{A}} \, {\rm e}^{{\rm i}s\hat{B}} \, {\rm e}^{-{\rm i}t\hat{A}} \, {\rm e}^{-{\rm i}s\hat{B}}. \tag{9.12.10}$$

Here we are using the Stone theorem (Stone 1932) already introduced, according to which, given a strongly continuous one-parameter unitary group on a Hilbert space $\mathcal{H}$, there exists a self-adjoint operator $\hat{A}$ on $\mathcal{H}$ so that $U(t) = {\rm e}^{{\rm i}t\hat{A}}$.

To state the following postulate, we have to define what we mean by observations of the first or second kind (a classification due to Pauli).

Definition 1. An *observation of the first kind* perturbs an observed quantity by an amount which is either *negligible* (from the experimental point of view) or *well known*, in that, upon repeating the measurement immediately afterwards, one finds the same result, or the outcome of the second measurement is exactly predictable. This is indeed the case for energy measurements in a bubble chamber, performed by measuring the curvature of part of the path of the particle.

Definition 2. An *observation of the second kind* perturbs the observed quantity in a way which is both *substantial* and *unpredictable*. This is where a stochastic element enters. The phenomenon occurs, for example, when a photon is detected in a photomultiplier, because in such a device the photon is completely absorbed, so that no further measurements can be performed.

We are now in a position to state the *projection postulate*, which plays a crucial role in quantum mechanics.

(iv) Suppose that, at $t = t_0$, an observation of the first kind is performed on the system, consisting of the measurement of a certain set of compatible observables A and B, and leading to the result $A = \alpha_r, B = \beta_s$. If $\psi(t_0)$ is the state vector of the system immediately prior to the observation, then the state vector immediately after the observation, denoted by $\psi(t_0 + \tau)$, is given by the projection of $\psi(t_0)$ on the eigenspace corresponding to the pair of eigenvalues α_r, β_s. Thus, on expressing $\psi(t_0)$ as the linear combination

$$\psi(t_0) = \sum_{r,s,u} c_{rsu}\varphi_{rsu}, \tag{9.12.11}$$

one should find

$$\psi(t_0 + \tau) = \frac{\sum_u c_{rsu}\varphi_{rsu}}{\sqrt{\sum_u |c_{rsu}|^2}}. \tag{9.12.12}$$

Remark 7. This means that, at the moment when the observation of the first kind is performed, the state vector undergoes a change which is both *discontinuous* and *irreversible* (often called the 'collapse' of the state vector), unlike the undisturbed evolution described by the Schrödinger equation. This property becomes acceptable if one bears in mind that the state vector is *just a mathematical tool to formulate predictions of a statistical nature.* However, one should acknowledge that a very rich literature exists on the applications and on the ultimate meaning of the projection postulate (D'Espagnat 1976, Wheeler and Zurek 1983, Isham 1995).

With hindsight, we should also say that two interpretations of the state vector have emerged in the literature. They are as follows (Isham 1995).

(i) *Minimal, pragmatic approach*: the state vector refers only to a *large collection* of suitably prepared systems, on which repeated measurements are performed. Any actual collection of copies of a system on which measurements are made is always finite in number, and hence the relative frequencies of the outcomes can only approximate the theoretical probabilities. Repeated measurements may be made on many copies of the same system, which are all measured at (essentially) the same time; or the copies may be measured sequentially; or the experiment may be repeated several times, using the same system suitably prepared on each occasion. It is then crucial to understand how many times an experiment should be repeated, for the measured relative frequencies of different outcomes to be an acceptable approximation of the theoretically predicted probabilities. Moreover, one has to make sure that the appropriate copy of the system is in the 'correct state' when a repeated measurement is performed. In the pragmatic approach, quantum theory is viewed as a scheme aimed at predicting the probabilistic distribution of results of measurements performed on copies of a suitably prepared physical system. To emphasize the measurement aspect, the term 'physical quantity' is replaced by 'observable quantity' or 'observable'. Probabilities are interpreted in a statistical way, by referring to relative frequencies with which the various results are obtained, if measurements are repeated a sufficiently large number of times.

(ii) *State vectors of individual systems*: the state vector refers to a *single system*, and it leads to probabilistic predictions of the results of repeated measurements. Indeed, the actual state preparations are almost always performed on individual systems. To say that a system is in a certain state means that it has been subjected to a preparation procedure, and hence it is quite natural to associate a state with an individual system. However, one should then bear in mind that physical quantities do not necessarily have values for each such system. In other words, even if a quantum state is associated to an individual system, it is impossible to assign definite values to all physical quantities.

We have adopted the latter point of view, interpretation (ii) above, because it makes it possible to deal properly with many experimental situations where individual systems are actually studied and their quantum states are prepared and manipulated.

9.12.1 *Rigged Hilbert spaces*

A Hilbert space has all vectors in it of finite norm. For a finite-dimensional space all vectors with finite components have finite length. But when the number of dimensions is (countably) infinite we may have vectors of infinite length, which can have all components in a suitable basis being finite. For example a vector

$$V = v_1 e_1 + \cdots + v_n e_n + \cdots \tag{9.12.13}$$

has length squared

$$|v_1|^2 + |v_2|^2 + \cdots + |v_n|^2 + \cdots .$$

Even such vectors with infinite length are useful; and with the components $\{v_1, v_2, \ldots\}$ in the basis $\{e_1, e_2, \ldots\}$ all finite the vector is well defined.

But even though the vector is outside the Hilbert space there are a subset of vectors

$$U = u_1 e_1 + \cdots + u_n e_n + \cdots \tag{9.12.14}$$

that have finite scalar products with the vector V provided

$$u_1^* v_1 + u_2^* v_2 + \cdots + u_n^* v_n + \cdots < \infty. \tag{9.12.15}$$

The vectors V form a vector space $\mathcal{H}'$ that is larger than $\mathcal{H}$ and the vectors U form a vector space $\mathcal{H}''$ such that

$$\mathcal{H}'' \subset \mathcal{H} \subset \mathcal{H}'. \tag{9.12.16}$$

The spaces $\mathcal{H}'$ and $\mathcal{H}''$ are dual to each other; the space $\mathcal{H}$ is self-dual.

For function spaces the $\mathcal{L}^2$ functions over any suitable interval constitute (on completion) a Hilbert space. The functions which increase no faster than $e^{x^2/2}$ constitute a bigger space, but most of them are not $\mathcal{L}^2$ functions. In contrast if one takes functions that fall off faster than $e^{-x^2/2}$ they form a sub-space of the Hilbert space. The functions increasing no faster than $e^{x^2/2}$ constitute $\mathcal{H}'$, while the functions falling off faster than $e^{-x^2/2}$ constitute $\mathcal{H}''$. This triple of vector spaces $\mathcal{H}'', \mathcal{H}, \mathcal{H}'$ was introduced by Gel'fand and are called *Gel'fand triples* or *rigged Hilbert spaces.* In the rigged Hilbert space formalism there are eigenvectors for a continuous spectrum, but these are in the space $\mathcal{H}'$ though not in $\mathcal{H}$. Note that a vector in $\mathcal{H}'$ can be completely characterized by its scalar products with all members of a basis, for which all of the unit vectors lie in $\mathcal{H}$.

Position and momentum operators

Another important example of an eigenvalue problem that leads to the analysis of Gel'fand triples is the eigenvalue problem for position and momentum operators, because the eigenvalue equations for Q and P admit weak (distributional) solutions. For example, the Dirac generalized function (distribution) with support in x_0, i.e. $\delta_{x_0}(x) \equiv \delta(x - x_0)$, is a weak solution of the eigenvalue equation

$$(Q\psi_{x_0})(x) = x_0 \psi_{x_0}(x), \tag{9.12.17}$$

as is shown by the following smeared relation with a test function $\varphi \in \mathcal{S}(\mathbf{R})$:

$$\int_{\mathbf{R}} dx \, x \delta_{x_0}(x) \varphi(x) = x_0 \varphi(x_0) = \int_{\mathbf{R}} dx \, x_0 \delta_{x_0}(x) \varphi(x). \tag{9.12.18}$$

The Dirac generalized function and the generalized function $x\delta_{x_0}$ do not belong to the domain of definition $\mathcal{S}(\mathbf{R})$ of Q; rather they belong to its dual space, i.e.

$$\mathcal{S}'(\mathbf{R}) \equiv \{\omega : \mathcal{S}(\mathbf{R}) \to \mathbf{C} \text{ linear and continuous}\}, \tag{9.12.19}$$

which is the space of tempered distributions on $\mathbf{R}$. Their rigorous definition is expressed by

$$\delta_{x_0} : \mathcal{S}(\mathbf{R}) \rightarrow \mathbf{C}, \quad \varphi \rightarrow \delta_{x_0}(\varphi) = \varphi(x_0), \tag{9.12.20}$$

jointly with

$$x\delta_{x_0} : \mathcal{S}(\mathbf{R}) \rightarrow \mathbf{C}, \quad \varphi \rightarrow (x\delta_{x_0})(\varphi) = \delta_{x_0}(x\varphi) = x_0\varphi(x_0). \tag{9.12.21}$$

The accurate form of Eq. (9.12.17) is therefore

$$(x\delta_{x_0})(\varphi) = (x_0\delta_{x_0})(\varphi), \quad \forall \varphi \in \mathcal{S}(\mathbf{R}). \tag{9.12.22}$$

Thus, the eigenvalue equation for the position operator Q admits a distributional solution ψ_{x_0} for every value $x_0 \in \mathbf{R}$ and the spectrum of Q is purely continuous.

In analogous way, the function $\psi_p(x) = \frac{1}{\sqrt{2\pi\hbar}} e^{i\frac{px}{\hbar}}$ defines a distribution l_p according to

$$l_p : \mathcal{S}(\mathbf{R}) \rightarrow \mathbf{C}, \tag{9.12.23}$$

$$\varphi \rightarrow l_p(\varphi) = \int_{\mathbf{R}} dx \; \psi_p^*(x)\varphi(x) \equiv (\mathcal{F}\varphi)(p), \tag{9.12.24}$$

where $\mathcal{F}\varphi$ denotes the Fourier transform of φ. The distribution l_p represents a solution of the eigenvalue equation $Pl_p = pl_p$.

In summary, the eigenvalue problem for the operators Q and P which admit a continuous spectrum suggests considering the following Gel'fand triple:

$$\mathcal{S}(\mathbf{R}) \subset \mathcal{L}^2(\mathbf{R}, dx) \subset \mathcal{S}'(\mathbf{R}). \tag{9.12.25}$$

Here, $\mathcal{S}(\mathbf{R})$ is a dense sub-space of $\mathcal{L}^2(\mathbf{R}, dx)$, and every function $\psi \in \mathcal{L}^2(\mathbf{R}, dx)$ defines a distribution $\omega_\psi \in \mathcal{S}'(\mathbf{R})$ according to

$$\omega_\psi : \mathcal{S}(\mathbf{R}) \rightarrow C, \quad \varphi \rightarrow \omega_\psi(\varphi) = \int_{\mathbf{R}} dx \; \psi^*(x)\varphi(x). \tag{9.12.26}$$

Note that $\mathcal{S}'(\mathbf{R})$ also contains distributions such as the Dirac distribution δ_{x_0} or the distribution l_p, which cannot be represented by means of a function $\psi \in \mathcal{L}^2(\mathbf{R}, dx)$ according to (9.12.26). The procedure of smearing out with a test function $\varphi \in \mathcal{S}(\mathbf{R})$ corresponds to the formation of wave packets studied in chapters 4 and 5.

In general, one can study a self-adjoint operator A on the Hilbert space $\mathcal{H}$. The improper eigenfunctions associated with elements of the continuous spectrum of A do not belong to the Hilbert space $\mathcal{H}$: one has then to equip $\mathcal{H}$ with an appropriate dense sub-space Ω and its dual Ω', which contains the generalized eigenfunctions of A so that one can write $\Omega \subset \mathcal{H} \subset \Omega'$. One has to choose the space Ω in a maximal way so as to ensure that Ω' is as 'close' as possible to $\mathcal{H}$; this provides the closest possible analogy with the finite-dimensional case, where $\Omega, \mathcal{H}$ and Ω' coincide. For further details, we refer the reader to the work in Gieres (2000) and de la Madrid *et al.* (2002).

Remark 8. Since the spectrum of the self-adjoint operator A continues to be the set of singular points of the resolvent of A, there are *no* complex eigenvalues or corresponding generalized eigenfunctions in Ω'. This formalism, like the heuristic extension by Dirac, cannot therefore deal with decaying states or complex resonance energies. One can introduce a family of vector spaces associated with an analytic self-adjoint operator A (with real spectrum). A family of analytic continuations exists that introduce vector spaces in which the analytically continued operator A has a complex spectrum, which is possibly continuous.

It should be emphasized that this process is defined only for analytic Hamiltonians that can be analytically continued from the Hilbert space. Any vector in the Hilbert space can be approximated by an analytically continuable function arbitrarily closely. But the correspondence is only between dense subsets of vectors. In particular, a discrete complex eigenvalue has an eigenfunction in the extended space which cannot be made to correspond to any vector in Hilbert space.

It would therefore be inconsistent to treat 'decaying states' using rigged Hilbert spaces. The 'decaying states' do not have a decreasing norm: that would be inconsistent with a time-independent Hamiltonian. What is involved is the survival amplitude (section 8.12) and 'survival probability', which decrease with time. Hence we must go beyond the rigged Hilbert space to deal with these states. The use of dual spaces for these problems is available in the literature, where both right and left eigenvectors are introduced and their scalar products used (Sudarshan 1994).

9.13 Problems

9.P1. Let $\mathcal{H}$ be the Hilbert space of square-integrable functions on $[0,1]$, and consider the operator A such that

$$Af \equiv \mathrm{i}\frac{\partial f}{\partial x} \tag{9.13.1}$$

on the domain of all absolutely continuous functions f on $[0,1]$ such that $f' \in \mathcal{L}^2[0,1]$ and $f(0) = f(1)$. Moreover, let B be the operator defined by

$$Bf(x) \equiv xf(x), \tag{9.13.2}$$

with the domain being the whole Hilbert space $\mathcal{H}$.

(i) Prove that the commutator $[A,B]$ is a multiple of the identity on the domain of all absolutely continuous functions such that $f' \in \mathcal{L}^2[0,1]$ and $f(0) = f(1) = 0$.

(ii) Bearing in mind that the above domain is dense in $\mathcal{H}$ and $[A,B]$ is bounded, prove that $C \equiv \overline{[A,B]}$ equals $\mathrm{i}\mathbb{1}$ on $\mathcal{H}$.

(iii) Now take a constant function u, for which $Au = 0$. Evaluate the modulus of the expectation value (u, Cu), compare with the generalized uncertainty relations of section 9.9 and interpret the result.

9.P2. Consider the position and momentum operators with the associated maximal domains of definition on the real line. Prove that symmetry of Q and P implies that the operator

$$A \equiv Q^3 P + P Q^3 \tag{9.13.3}$$

is symmetric as well. Next, consider the square-integrable function f defined by

$$f(x) = \frac{1}{\sqrt{2}} |x|^{-3/2} \exp\left(-\frac{1}{4x^2}\right) \quad \text{for } x \neq 0, \; f(0) = 0. \tag{9.13.4}$$

(i) Does f belong to the domain of A?

(ii) Compute Af and try to interpret the result (Gieres 2000).

9.P3. For a quantum-mechanical problem studied with spherical coordinates in $\mathbf{R}^3$, let $\widehat{\varphi}$ be the operator of multiplication of the wave function by $\varphi \in [0, 2\pi]$, and let $\widehat{L}_z$ be the z-component of the orbital angular momentum, realized as the first-order differential operator $\widehat{L}_z \equiv \frac{\hbar}{\mathrm{i}} \frac{\partial}{\partial \varphi}$.

(i) Do the eigenfunctions of $\widehat{L}_z$ belong to the domain of the commutator $[\widehat{\varphi}, \widehat{L}_z]$?

(ii) Prove the inequality

$$(\Delta_\psi \widehat{L}_z)(\Delta_\psi \widehat{\varphi}) \geq \frac{\hbar}{2} |1 - 2\pi \psi(2\pi)|^2 \quad \forall \psi \in D(\widehat{L}_z) \cap D(\widehat{\varphi}) \tag{9.13.5}$$

(Gieres 2000).

9.P4. Consider a particle of mass m in the one-dimensional infinite potential well of Eq. (4.6.27), with Hamiltonian H. Let

$$\psi(x, 0) = \frac{\sqrt{15}}{4a^{5/2}} (a^2 - x^2) \quad \text{if } x \in [-a, a] \tag{9.13.6}$$

be the normalized wave function of the particle at $t = 0$, which vanishes outside the interval $[-a, a]$.

(i) Compute the average value of the squared Hamiltonian H^2 in the state ψ.

(ii) Find under which boundary conditions the operators H and H^2 are self-adjoint (Gieres 2000).

9.P5. Let M be the Riemann surface of the square root function, and let $\mathcal{H}$ be the Hilbert space of square-integrable functions on M. Consider the operators (Reed and Simon 1980)

$$P \equiv -\mathrm{i} \frac{\partial}{\partial x}, \tag{9.13.7}$$

$$Q \equiv x - \mathrm{i} \frac{\partial}{\partial y}, \tag{9.13.8}$$

on the domain $\mathcal{D}$ of all C^∞ functions with compact support not containing the origin.

(i) Prove that P and Q satisfy all properties (i)–(iii) listed in section 9.7.

(ii) Do the groups generated by P and Q satisfy the Weyl relations?

9.P6. Solve, with given initial conditions, the Heisenberg equations of motion for a one-dimensional harmonic oscillator with position operator $q_{\rm H}(t)$ and momentum operator $p_{\rm H}(t)$. Evaluate the commutators at different times

$$\Big[q_{\rm H}(t_1), q_{\rm H}(t_2)\Big], \Big[q_{\rm H}(t_1), p_{\rm H}(t_2)\Big], \Big[p_{\rm H}(t_1), p_{\rm H}(t_2)\Big],$$

and interpret the result.

9.P7. By independently varying the wave function $\psi(q', t) = \langle q', t|\psi\rangle$ and its complex conjugate $\psi^\dagger$, prove that the action

$$S \equiv \int {\rm d}t \int {\rm d}q' \; L, \tag{9.13.9}$$

with ${\rm d}q' \equiv \prod_{i=1}^{n} {\rm d}q'^i$, and L the Lagrangian

$$L \equiv \frac{\rm i}{2}\left(\psi^\dagger \dot{\psi} - \dot{\psi}^\dagger \psi\right) - \psi^\dagger H_{q'}(t)\psi, \tag{9.13.10}$$

leads to the Schrödinger equation for a system having n degrees of freedom (DeWitt 1965). Recall that $H_{q'}(t)$, the Hamiltonian operator in the coordinate representation, is obtained from the Hamiltonian operator $H(t)$ in the Heisenberg representation by means of

$$H_{q'}(t)\langle q', t|\psi\rangle = \langle q', t|H(t)|\psi\rangle. \tag{9.13.11}$$

9.P8. Canonical symmetries are maps that preserve the canonical commutation relations. Prove that there exist canonical symmetries which are not expressed by unitary operators. Try to interpret this class of symmetries.

10
Harmonic oscillators and quantum optics

Starting from the definition of annihilation and creation operators, the basic properties of harmonic oscillators in quantum mechanics are first derived with algebraic methods: the existence of the ground state, the discrete nature of the spectrum of the Hamiltonian and extension to higher dimensions. A thorough investigation of the infinite degeneracy of Landau levels is then performed.

In the second part, emphasis is put on the applications to the analysis of coherent states. These are an overcomplete and non-orthogonal system of Hilbert-space vectors, which are very useful for describing coherent laser beams within the framework of quantum theory. They are also studied, here, from the point of view of the general theory of Weyl systems, defining eventually the Bargmann–Fock representation, which leads to a realization of the Hilbert space of states as a space of entire functions. Two-photon coherent states, which are a generalized form of coherent states relevant for the analysis of two-photon lasers, are also studied.

10.1 Algebraic formalism for harmonic oscillators

A one-dimensional harmonic oscillator of mass m and frequency ω has a classical Hamiltonian

$$H = \frac{p^2}{2m} + \frac{m}{2}\omega^2 x^2. \tag{10.1.1}$$

In section 4.7 we have performed the quantum analysis within the framework of wave mechanics. Now we are going to follow a more abstract path.

In quantum mechanics, according to the rules of section 9.11 for the observables, one wants to realize x and its conjugate momentum as self-adjoint operators, denoted by $\hat{x}$ and $\hat{p}$, respectively, and obeying the commutation relations (9.1.1), and then investigate the properties of the operators defined by the linear combinations

$$\hat{a} \equiv \frac{1}{\sqrt{2m\hbar\omega}}\Big(m\omega\hat{x} + \mathrm{i}\hat{p}\Big), \tag{10.1.2}$$

$$\hat{a}^\dagger \equiv \frac{1}{\sqrt{2m\hbar\omega}}\Big(m\omega\hat{x} - \mathrm{i}\hat{p}\Big). \tag{10.1.3}$$

Note that, so far, $\hat{a}^\dagger$ is only the *formal adjoint* of $\hat{a}$. We need to determine whether it really is the adjoint. For this purpose, we have to introduce a Hilbert space and a realization of our operators on it. We can use the approach of the previous chapter and look for a realization in terms of differential operators on square-integrable functions on a Lagrangian subspace. It is now convenient to introduce dimensionless units, which lead, eventually, to the operator

$$T \equiv \xi + \frac{\mathrm{d}}{\mathrm{d}\xi}, \tag{10.1.4}$$

for $\hat{a}$. On defining the scalar product (thus, the definition depends ultimately on the metric)

$$(u, v) \equiv \int_{-\infty}^{\infty} u^*(\xi)v(\xi)\,\mathrm{d}\xi, \tag{10.1.5}$$

one finds, after integration by parts,

$$(Tu, v) = \int_{-\infty}^{\infty} u^*(\xi)\left(\xi - \frac{\mathrm{d}}{\mathrm{d}\xi}\right)v(\xi)\,\mathrm{d}\xi + [u^*(\xi)v(\xi)]_{-\infty}^{\infty}. \tag{10.1.6}$$

Thus, the operator $\xi - \frac{\mathrm{d}}{\mathrm{d}\xi}$ is the adjoint $T^\dagger$ of T provided that, for $u \in D(T)$ and $v \in D(T^\dagger)$, the following boundary term vanishes:

$$u^*(\infty)v(\infty) - u^*(-\infty)v(-\infty) = 0. \tag{10.1.7}$$

This means that one has to look for functions that are absolutely continuous on the whole real line and vanish at $\pm\infty$. The former condition ensures that such functions are at least of class C^1, and their weak derivatives are Lebesgue summable on $\mathbf{R}$. More precisely, one has to look for u and v in the Schwarz space of C^∞ functions with rapid decrease (Reed and Simon 1980).

Starting from the definitions (10.1.2) and (10.1.3), we re-express the operators $\hat{x}$ and $\hat{p}$:

$$\hat{x} = \sqrt{\frac{\hbar}{2m\omega}}\Big(\hat{a} + \hat{a}^\dagger\Big), \tag{10.1.8}$$

$$\hat{p} = \mathrm{i}\sqrt{\frac{m\hbar\omega}{2}}\Big(\hat{a}^\dagger - \hat{a}\Big). \tag{10.1.9}$$

The Hamiltonian operator

$$\hat{H} \equiv \frac{\hat{p}^2}{2m} + \frac{m}{2}\omega^2 \hat{x}^2 \tag{10.1.10}$$

is thus found to be

$$\hat{H} = \frac{1}{2}\hbar\omega\Big(\hat{a}\hat{a}^\dagger + \hat{a}^\dagger\hat{a}\Big). \tag{10.1.11}$$

Moreover, by virtue of (9.1.1), one has

$$\Big[\hat{a}, \hat{a}^\dagger\Big] = \mathbb{I}, \tag{10.1.12}$$

and hence $\hat{H}$ takes the form

$$\hat{H} = \hbar\omega\Big(\hat{a}^\dagger\hat{a} + \frac{1}{2}\mathbb{I}\Big). \tag{10.1.13}$$

The form (10.1.13) of the Hamiltonian suggests defining the operator

$$\hat{N} \equiv \hat{a}^\dagger\hat{a}. \tag{10.1.14}$$

By construction, the operator $\hat{N}$ commutes with $\hat{H}$, and hence its eigenstates are also eigenstates of the Hamiltonian. Moreover, the property (10.1.12) implies that

$$\Big[\hat{N}, \hat{a}\Big] = -\hat{a}, \tag{10.1.15}$$

$$\Big[\hat{N}, \hat{a}^\dagger\Big] = \hat{a}^\dagger. \tag{10.1.16}$$

Another key property of the operator $\hat{N}$ is that it is positive-definite. Indeed, given a generic ket vector $|f\rangle$, one has

$$\langle f|\hat{N}|f\rangle = \langle f|\hat{a}^\dagger\hat{a}|f\rangle = \langle \hat{a}f|\hat{a}f\rangle = \| \hat{a}|f\rangle \|^2, \tag{10.1.17}$$

which proves that the spectrum of $\hat{N}$ consists of non-negative eigenvalues λ:

$$\hat{N}|v\rangle = \lambda|v\rangle. \tag{10.1.18}$$

Moreover, by virtue of (10.1.15) and (10.1.16), one finds

$$\hat{N}\hat{a}|v\rangle = (\lambda - 1)\hat{a}|v\rangle, \tag{10.1.19}$$

$$\hat{N}\hat{a}^\dagger|v\rangle = (\lambda + 1)\hat{a}^\dagger|v\rangle, \tag{10.1.20}$$

and, by repeated application of this property,

$$\hat{N}\hat{a}^n|v\rangle = (\lambda - n)\hat{a}^n|v\rangle, \tag{10.1.21}$$

$$\hat{N}(\hat{a}^\dagger)^s|v\rangle = (\lambda + s)(\hat{a}^\dagger)^s|v\rangle, \tag{10.1.22}$$

where n and s are non-negative integers. This means that the sequence of eigenvectors

$$|v\rangle,\ \hat{a}|v\rangle, \ldots, \hat{a}^n|v\rangle$$

cannot continue for ever, but an integer n_0 exists such that

$$\hat{a}^{n_0}|v\rangle \neq 0, \tag{10.1.23}$$

$$\hat{a}^{n_0+1}|v\rangle = 0, \tag{10.1.24}$$

since otherwise (10.1.21) would lead, after a finite number of steps, to negative eigenvalues, which is impossible by virtue of (10.1.17). It follows from (10.1.23) and (10.1.24) that

$$\hat{N}\hat{a}^{n_0}|v\rangle = 0, \tag{10.1.25}$$

whereas Eq. (10.1.21) implies that

$$\hat{N}\hat{a}^{n_0}|v\rangle = (\lambda - n_0)\hat{a}^{n_0}|v\rangle. \tag{10.1.26}$$

Comparison of (10.1.25) and (10.1.26) therefore yields

$$\lambda = n_0 \geq 0, \tag{10.1.27}$$

i.e. the spectrum of $\hat{N}$ consists of non-negative integers (hence $\hat{N}$ is called the *number operator*), and the spectrum of $\hat{H}$ has the form

$$E_n = \hbar\omega\left(n + \frac{1}{2}\right), \quad \forall n \geq 0. \tag{10.1.28}$$

Hereafter, the eigenstates of $\hat{N}$ and $\hat{H}$ will therefore be denoted by $|n\rangle$. If κ is a constant, one can define the ket

$$|0\rangle \equiv \kappa\, \hat{a}^{n_0}|v\rangle, \tag{10.1.29}$$

which, by virtue of (10.1.24), has the property of being annihilated by the operator $\hat{a}$:

$$\hat{a}|0\rangle = 0. \tag{10.1.30}$$

The properties (10.1.19) and (10.1.20) may now be expressed in the form

$$\hat{a}|n\rangle = c_n|n-1\rangle, \tag{10.1.31}$$

$$\hat{a}^\dagger|n\rangle = b_n|n+1\rangle, \tag{10.1.32}$$

where c_n and b_n are some numbers which can be determined by the following argument. The evaluation of $\langle n|\hat{a}^\dagger\hat{a}|n\rangle$ yields

$$\langle n|\hat{a}^\dagger\hat{a}|n\rangle = n\langle n|n\rangle. \tag{10.1.33}$$

On the other hand, by virtue of (10.1.31), one also has

$$\langle n|\hat{a}^\dagger \hat{a}|n\rangle = |c_n|^2 \langle n-1|n-1\rangle. \tag{10.1.34}$$

By comparison of (10.1.33) and (10.1.34), and imposing the normalization condition

$$\langle r|r\rangle = 1 \quad \forall r \geq 0, \tag{10.1.35}$$

one finds

$$c_n = \sqrt{n}\, e^{i2k\pi}, \tag{10.1.36}$$

for all $k = 0, 1, 2, \ldots$. We have exploited the fact that, since the eigenstates are defined only up to a phase factor, they can all be chosen so that $e^{i\varphi} = 1$. Similarly, one finds

$$\langle n|\hat{a}\hat{a}^\dagger|n\rangle = (n+1)\langle n|n\rangle = |b_n|^2 \langle n+1|n+1\rangle, \tag{10.1.37}$$

which implies

$$b_n = \sqrt{n+1}\, e^{i2k\pi}, \tag{10.1.38}$$

for all $k = 0, 1, 2, \ldots$. It is therefore clear from the rules

$$\hat{a}|n\rangle = \sqrt{n}|n-1\rangle, \tag{10.1.39}$$

$$\hat{a}^\dagger|n\rangle = \sqrt{n+1}|n+1\rangle, \tag{10.1.40}$$

that $\hat{a}$ maps an eigenstate of the number operator belonging to the eigenvalue n into an eigenstate of the number operator belonging to the eigenvalue $n-1$. The operator $\hat{a}^\dagger$ maps instead $|n\rangle$ into an eigenstate of $\hat{N}$ belonging to the eigenvalue $n+1$. This is why they are called the *annihilation* and the *creation* operator, respectively (the above equations were already obtained, in the Schrödinger picture, in section 4.7). The eigenstate $(\hat{a}^\dagger)^n|0\rangle$ therefore belongs to the eigenvalue $\lambda = n$ of $\hat{N}$. To obtain the appropriate normalization coefficient, we have to evaluate

$$\left\|(\hat{a}^\dagger)^n|0\rangle\right\|^2 = \langle(\hat{a}^\dagger)^n 0|(\hat{a}^\dagger)^n 0\rangle = \langle 0|\hat{a}^n(\hat{a}^\dagger)^n|0\rangle. \tag{10.1.41}$$

For this purpose, we need the identity

$$\hat{a}(\hat{a}^\dagger)^n - (\hat{a}^\dagger)^n\hat{a} = n(\hat{a}^\dagger)^{n-1}. \tag{10.1.42}$$

This is proved by repeated application of (10.1.12), i.e.

$$\begin{aligned}
\hat{a}(\hat{a}^\dagger)^n &= a\hat{a}^\dagger(\hat{a}^\dagger)^{n-1} = \hat{a}^\dagger\hat{a}(\hat{a}^\dagger)^{n-1} + (\hat{a}^\dagger)^{n-1} \\
&= \hat{a}^\dagger(\hat{a}^\dagger\hat{a}+1)(\hat{a}^\dagger)^{n-2} + (\hat{a}^\dagger)^{n-1} \\
&= (\hat{a}^\dagger)^2\hat{a}\hat{a}^\dagger(\hat{a}^\dagger)^{n-3} + 2(\hat{a}^\dagger)^{n-1} \\
&= (\hat{a}^\dagger)^3\hat{a}\hat{a}^\dagger(\hat{a}^\dagger)^{n-4} + 3(\hat{a}^\dagger)^{n-1} \\
&= \cdots = (\hat{a}^\dagger)^n\hat{a} + n(\hat{a}^\dagger)^{n-1}.
\end{aligned} \tag{10.1.43}$$

The action of the operator $\hat{a}^n (\hat{a}^\dagger)^n$ on the ket $|0\rangle$ in (10.1.41) is then found to be

$$\begin{aligned} \hat{a}^n (\hat{a}^\dagger)^n |0\rangle &= \hat{a}^{n-1} \hat{a} (\hat{a}^\dagger)^n |0\rangle \\ &= \hat{a}^{n-1} (\hat{a}^\dagger)^n \hat{a} |0\rangle + n \hat{a}^{n-1} (\hat{a}^\dagger)^{n-1} |0\rangle \\ &= n \hat{a}^{n-1} (\hat{a}^\dagger)^{n-1} |0\rangle. \end{aligned} \tag{10.1.44}$$

Repeated application of this algorithm therefore yields

$$\langle 0 | \hat{a}^n (\hat{a}^\dagger)^n | 0 \rangle = n! \langle 0 | 0 \rangle = n!, \tag{10.1.45}$$

which implies that the normalized eigenstates we are looking for are $\frac{(\hat{a}^\dagger)^n}{\sqrt{n!}} |0\rangle$. By construction, we only need to solve the differential equation for $|0\rangle$. Working in the Schrödinger representation, where

$$\hat{x} |\psi\rangle = x |\psi\rangle, \tag{10.1.46}$$

$$\hat{p} |\psi\rangle = \frac{\hbar}{\mathrm{i}} \frac{\mathrm{d}}{\mathrm{d}x} |\psi\rangle, \tag{10.1.47}$$

one finds

$$\hat{a} = \frac{1}{2\sigma} \left(x + 2\sigma^2 \frac{\mathrm{d}}{\mathrm{d}x} \right), \tag{10.1.48}$$

where $\sigma \equiv \sqrt{\frac{\hbar}{2m\omega}}$. Since the annihilation operator is realized here as a first-order differential operator, we should consider its action on the stationary state $\langle x | 0 \rangle$, i.e.

$$\left(x + 2\sigma^2 \frac{\mathrm{d}}{\mathrm{d}x} \right) \langle x | 0 \rangle = 0. \tag{10.1.49}$$

Equation (10.1.49) is solved by

$$\langle x | 0 \rangle = C \, \mathrm{e}^{-m\omega x^2 / 2\hbar}, \tag{10.1.50}$$

where C is determined by the normalization condition:

$$\langle x | 0 \rangle = \left(\frac{m\omega}{\pi\hbar} \right)^{1/4} \mathrm{e}^{-m\omega x^2 / 2\hbar}. \tag{10.1.51}$$

Combining (10.1.51) with the previous analysis, one finds

$$|n\rangle = \frac{1}{\sqrt{n!}} \left[\frac{1}{\sqrt{2}} \left(\xi - \frac{\mathrm{d}}{\mathrm{d}\xi} \right) \right]^n |0\rangle \quad \forall n \geq 0. \tag{10.1.52}$$

After this explicit construction is performed, one can take a different point of view, and start from an abstract Hilbert space $\mathcal{H}$ with a complete orthonormal set $|\psi_n\rangle$. The annihilation and creation operators are then defined axiomatically by the conditions

$$\hat{a}|\psi_n\rangle = \sqrt{n}|\psi_{n-1}\rangle, \tag{10.1.53}$$

$$\hat{a}^\dagger|\psi_n\rangle = \sqrt{n+1}|\psi_{n+1}\rangle, \tag{10.1.54}$$

with a *ground state* such that

$$\hat{a}|\psi_0\rangle = 0, \tag{10.1.55}$$

and Hamiltonian operator (10.1.13). These newly defined operators will satisfy the commutation relations of the oscillator algebra. The term $\frac{\hbar\omega}{2}$ plays a key role in the quantum theory of the electromagnetic field. If this is viewed as an infinite set of harmonic oscillators, one faces the problem of an infinite value of the zero-point energy, i.e. the energy obtained when all number operators for the various oscillators have a vanishing eigenvalue. What really matters, however, are *differences* in zero-point energies. These are finite (upon making suitable definitions) and produce measurable effects. The investigation of these problems goes well beyond the aims of our monograph, but the reader should be aware that many fascinating problems lie behind the zero-point energy $\frac{\hbar\omega}{2}$ (Casimir 1948, Grib *et al.* 1994).

In more than one degree of freedom, one deals with the commutation rules

$$\left[\hat{a}_r, \hat{a}_s^\dagger\right] = \delta_{rs}, \quad \left[\hat{a}_r, \hat{a}_s\right] = 0, \quad \left[\hat{a}_r^\dagger, \hat{a}_s^\dagger\right] = 0. \tag{10.1.56}$$

For each r, one can define a generalized 'number' operator

$$\hat{N}_r \equiv \hat{a}_r^\dagger \, \hat{a}_r, \tag{10.1.57}$$

and the rule for a matrix representation of the annihilation and creation operators is expressed by

$$\left(\hat{a}_r^\dagger\right)_{N_r, N_r'} = \sqrt{N_r} \ \text{ if } N_r = N_r' + 1, \ 0 \text{ otherwise}, \tag{10.1.58}$$

$$\left(\hat{a}_r\right)_{N_r, N_r'} = \sqrt{N_r + 1} \ \text{ if } N_r = N_r' - 1, \ 0 \text{ otherwise}. \tag{10.1.59}$$

These rules give rise to the matrix form of $\hat{a}, \hat{a}^\dagger$ written in the eigenstates of the number operator:

$$\hat{a} = \begin{pmatrix} 0 & \sqrt{1} & \cdot & \cdot & \cdot \\ \cdot & 0 & \sqrt{2} & \cdot & \cdot \\ \cdot & \cdot & 0 & \sqrt{3} & \cdot \\ \cdot & \cdot & \cdot & 0 & \cdot \\ \cdot & \cdot & \cdot & \cdot & \cdot \end{pmatrix}, \tag{10.1.60}$$

$$\hat{a}^\dagger = \begin{pmatrix} 0 & \cdot & \cdot & \cdot & \cdot \\ \sqrt{1} & 0 & \cdot & \cdot & \cdot \\ \cdot & \sqrt{2} & 0 & \cdot & \cdot \\ \cdot & \cdot & \sqrt{3} & \cdot & \cdot \\ \cdot & \cdot & \cdot & \cdot & \cdot \end{pmatrix}. \tag{10.1.61}$$

In the case of a system with many degrees of freedom, one has a Hilbert space which consists of the tensor product (see appendix 4.A) of the various Hilbert spaces:

$$\mathcal{H} = \mathcal{H}_1 \otimes \mathcal{H}_2 \otimes \mathcal{H}_3 \otimes \cdots . \tag{10.1.62}$$

A multi-particle state is then expressed as

$$\begin{aligned} |n_1, n_2, \ldots, n_k\rangle &= |n_1\rangle \otimes |n_2\rangle \otimes \cdots \otimes |n_k\rangle \\ &= |n_1\rangle\, |n_2\rangle\, \cdots |n_k\rangle. \end{aligned} \tag{10.1.63}$$

Moreover, the associated bras are of the form

$$\langle x| = \langle x_1| \otimes \cdots \otimes \langle x_k|, \tag{10.1.64}$$

so that

$$\langle x|n\rangle = \psi_{n_1}(x_1)\psi_{n_2}(x_2)\cdots\psi_{n_k}(x_k). \tag{10.1.65}$$

The Hamiltonian of the system is then expressed by the sum

$$\hat{H} = \sum_\alpha \hbar\omega_\alpha \left(\hat{a}^\dagger_\alpha \hat{a}_\alpha + \frac{1}{2}\mathbb{I}\right) = \sum_\alpha \frac{1}{2m_\alpha}\left(\hat{p}^2_\alpha + m^2_\alpha \omega^2_\alpha \hat{x}^2_\alpha\right), \tag{10.1.66}$$

and the *ground state* is defined by the condition

$$\hat{a}_\alpha |0\rangle = 0 \quad \forall \alpha. \tag{10.1.67}$$

A normalized state is obtained according to the algorithm

$$|n_1, n_2, \ldots, n_k\rangle = \frac{(\hat{a}^\dagger_1)^{n_1}\,(\hat{a}^\dagger_2)^{n_2} \ldots (\hat{a}^\dagger_k)^{n_k}}{\sqrt{n_1! n_2! \cdots n_k!}}\,|0\rangle, \tag{10.1.68}$$

which results from (10.1.63) and from the properties

$$|n_r\rangle = \frac{(\hat{a}^\dagger_r)^{n_r}}{\sqrt{n_r!}}|0_r\rangle \quad \forall r = 1, \ldots, k,$$

$$|0\rangle = |0_1\rangle\, |0_2\rangle \cdots\, |0_k\rangle.$$

By construction, this state contains n_1 excitations of the first oscillator, n_2 excitations of the second oscillator, up to n_k excitations of the kth oscillator.

10.2 A thorough understanding of Landau levels

We are now in a position to obtain a proper understanding of the result concerning Landau levels derived in section 6.5. For this purpose, we consider the following formulation of the problem. The electron moves in the whole xy plane, with Hamiltonian

$$\hat{H} = \frac{1}{2m}\left[\left(\hat{p}_x - \frac{e_0 B}{2c}\hat{y}\right)^2 + \left(\hat{p}_y + \frac{e_0 B}{2c}\hat{x}\right)^2\right] + \frac{e_0 B}{2mc}\hat{\sigma}_z, \tag{10.2.1}$$

because $\hat{A}_x = -\frac{B\hat{y}}{2}, \hat{A}_y = \frac{B\hat{x}}{2}$. This form of the Hamiltonian makes it possible to define two sets of operators (hereafter $\Omega \equiv \frac{e_0 B}{c}$):

$$\hat{b}_1 \equiv \hat{p}_x + \frac{\Omega}{2}\hat{y}, \quad \hat{b}_2 \equiv \hat{p}_y - \frac{\Omega}{2}\hat{x}, \tag{10.2.2}$$

$$\hat{\pi}_1 \equiv \hat{p}_x - \frac{\Omega}{2}\hat{y}, \quad \hat{\pi}_2 \equiv \hat{p}_y + \frac{\Omega}{2}\hat{x}, \tag{10.2.3}$$

which satisfy the commutation relations

$$\left[\hat{b}_1, \hat{b}_2\right] = \mathrm{i}\hbar\Omega \, \mathbb{I}, \tag{10.2.4}$$

$$\left[\hat{\pi}_1, \hat{\pi}_2\right] = -\mathrm{i}\hbar\Omega \, \mathbb{I}. \tag{10.2.5}$$

One can now take linear combinations of the operators defined in (10.2.2) and (10.2.3) to obtain yet new operators which satisfy the commutation relations for annihilation and creation operators. For this purpose, denoting by $\mathcal{N}$ a constant, we define

$$\hat{A} \equiv \mathcal{N}\left(\hat{\pi}_1 - \mathrm{i}\hat{\pi}_2\right), \quad \hat{A}^\dagger \equiv \mathcal{N}\left(\hat{\pi}_1 + \mathrm{i}\hat{\pi}_2\right), \tag{10.2.6}$$

$$\hat{F} \equiv \mathcal{N}\left(\hat{b}_1 + \mathrm{i}\hat{b}_2\right), \quad \hat{F}^\dagger \equiv \mathcal{N}\left(\hat{b}_1 - \mathrm{i}\hat{b}_2\right). \tag{10.2.7}$$

By virtue of (10.2.2)–(10.2.7), one finds

$$\left[\hat{A}, \hat{A}^\dagger\right] = \left[\hat{F}, \hat{F}^\dagger\right] = 2\mathcal{N}^2\Omega\hbar \, \mathbb{I}. \tag{10.2.8}$$

Thus, the value of $\mathcal{N}$ for which $\left[\hat{A}, \hat{A}^\dagger\right] = \left[\hat{F}, \hat{F}^\dagger\right] = \mathbb{I}$ is

$$\mathcal{N} = \frac{1}{\sqrt{2\Omega\hbar}}, \tag{10.2.9}$$

which implies that the Hamiltonian operator takes the form

$$\hat{H} = \hbar\Omega\left[\frac{1}{2m}\left(\hat{A}\hat{A}^\dagger + \hat{A}^\dagger\hat{A}\right) + \frac{1}{2m}\sigma_3\right]. \tag{10.2.10}$$

The infinite degeneracy of the spectrum of $\hat{H}$ results from the commutation property of both $\hat{F}$ and $\hat{F}^\dagger$ with $\hat{H}$:

$$\left[\hat{F}, \hat{H}\right] = \left[\hat{F}^\dagger, \hat{H}\right] = 0. \tag{10.2.11}$$

What happens is that the sub-space corresponding to a given energy level is invariant under the action of both $\hat{F}$ and $\hat{F}^\dagger$. On the other hand, these operators satisfy the commutation relations

$$\left[\hat{F}, \hat{F}^\dagger\right] = \mathbb{I}. \tag{10.2.12}$$

As we know from section 9.3, this implies that the operators $\hat{F}$ and $\hat{F}^\dagger$ cannot be both bounded, and hence the sub-space invariant under their action cannot be finite-dimensional.

To prove the result (10.2.11), note that

$$\left[\hat{F}, \hat{H}\right] = \frac{\hbar}{m}\Omega\left[\hat{F}, \hat{A}^\dagger \hat{A}\right], \tag{10.2.13}$$

where

$$\left[\hat{F}, \hat{A}^\dagger \hat{A}\right] = \mathcal{N}^3\left[\hat{b}_1 + \mathrm{i}\hat{b}_2, \hat{\pi}_1^2 + \hat{\pi}_2^2\right]. \tag{10.2.14}$$

By virtue of (10.2.2) and (10.2.3), one can re-express $\hat{\pi}_1$ and $\hat{\pi}_2$ in terms of $\hat{b}_1$ and $\hat{b}_2$:

$$\hat{\pi}_1 = \hat{b}_1 - \Omega\hat{y}, \tag{10.2.15}$$

$$\hat{\pi}_2 = \hat{b}_2 + \Omega\hat{x}, \tag{10.2.16}$$

which implies

$$\left[\hat{F}, \hat{A}^\dagger \hat{A}\right] = \mathcal{N}^3\left[\hat{b}_1 + \mathrm{i}\hat{b}_2, \hat{b}_1^2 + \hat{b}_2^2 + \Omega^2\left(\hat{x}^2 + \hat{y}^2\right) - 2\Omega\hat{b}_1\hat{y} + 2\Omega\hat{b}_2\hat{x}\right]. \tag{10.2.17}$$

The non-vanishing commutators occurring in (10.2.17) are found to be

$$\left[\hat{b}_1, \hat{b}_2^2\right] = 2\mathrm{i}\hbar\Omega\hat{b}_2, \quad \left[\hat{b}_1, \hat{x}^2 + \hat{y}^2\right] = -2\mathrm{i}\hbar\hat{x}, \tag{10.2.18}$$

$$\left[\hat{b}_1, \hat{b}_2\hat{x}\right] = -\mathrm{i}\hbar\hat{p}_y + \frac{3}{2}\mathrm{i}\hbar\Omega\hat{x}, \quad \left[\hat{b}_2, \hat{b}_1^2\right] = -2\mathrm{i}\hbar\Omega\hat{b}_1, \tag{10.2.19}$$

$$\left[\hat{b}_2, \hat{x}^2 + \hat{y}^2\right] = -2\mathrm{i}\hbar\hat{y}, \quad \left[\hat{b}_2, \hat{b}_1\hat{y}\right] = -\mathrm{i}\hbar\hat{p}_x - \frac{3}{2}\mathrm{i}\hbar\Omega\hat{y}. \tag{10.2.20}$$

Collecting together all terms, one thus finds

$$\left[\hat{F}, \hat{A}^\dagger \hat{A}\right] = \mathcal{N}^3\Omega\left[2\hbar(\hat{b}_1 + \mathrm{i}\hat{b}_2) + \mathrm{i}\hbar\Omega(\hat{x} + \mathrm{i}\hat{y}) - 2\hbar(\hat{p}_x + \mathrm{i}\hat{p}_y)\right] = 0. \tag{10.2.21}$$

The validity of (10.2.11) is therefore proven.

We would like to conclude this section by emphasizing that a deeper and more systematic way of deriving the various Landau levels actually exists, and this relies on the operator $\hat{A}$ just introduced (see Eq. (10.2.6)). For example, the first Landau level is, by definition, a solution of the equation

$$\hat{A}\,\psi = 0. \tag{10.2.22}$$

This leads to the first-order partial differential equation

$$\left(\frac{\hbar}{\mathrm{i}}\frac{\partial}{\partial x} - \frac{\Omega}{2}y - \hbar\frac{\partial}{\partial y} - \mathrm{i}\frac{\Omega}{2}x\right)\psi(x,y) = 0, \tag{10.2.23}$$

which implies

$$\left(\frac{\partial}{\partial x} - \mathrm{i}\frac{\partial}{\partial y}\right)\psi(x,y) = -\frac{\Omega}{2\hbar}(x - \mathrm{i}y)\psi(x,y), \tag{10.2.24}$$

and hence

$$\left(\frac{\partial}{\partial x} - \mathrm{i}\frac{\partial}{\partial y}\right)\log\psi(x,y) = -\frac{\Omega}{2\hbar}(x - \mathrm{i}y). \tag{10.2.25}$$

This form of the equation suggests considering the complex variables $\zeta \equiv x - \mathrm{i}y$ and $\overline{\zeta} \equiv x + \mathrm{i}y$. On taking into account the transformation rules for derivatives,

$$\frac{\partial}{\partial x} = \frac{\partial}{\partial \zeta} + \frac{\partial}{\partial \overline{\zeta}}, \tag{10.2.26}$$

$$\frac{\partial}{\partial y} = -\mathrm{i}\frac{\partial}{\partial \zeta} + \mathrm{i}\frac{\partial}{\partial \overline{\zeta}}, \tag{10.2.27}$$

Eq. (10.2.25) is found to take the form

$$\frac{\partial}{\partial \overline{\zeta}}\log\psi = -\frac{\Omega}{4\hbar}\zeta, \tag{10.2.28}$$

which is solved by

$$\log\psi = -\frac{\Omega}{4\hbar}|\zeta|^2 + \kappa, \tag{10.2.29}$$

where κ is independent of $\overline{\zeta}$: $\frac{\partial\kappa}{\partial\overline{\zeta}} = 0$. This makes it possible to express the first Landau level in the form

$$\psi(x,y) = f(x - \mathrm{i}y)\exp\left[-\frac{e_0 B}{4\hbar c}(x^2 + y^2)\right], \tag{10.2.30}$$

where f is an analytic function of the complex variable ζ.

10.3 Coherent states

Many important applications of the formalism of harmonic oscillators deal with the concept of coherent states. The key points are as follows.

(i) *Coherent states* are an overcomplete and non-orthogonal system of Hilbert-space vectors (Schrödinger 1926, Glauber 1963, Sudarshan 1963, Perelomov 1986). This means that at least one vector exists in the system which can be removed, while the system remains complete. A system of non-orthogonal wave functions to describe non-spreading wave packets for quantum oscillators was indeed first introduced in Schrödinger (1926).

(ii) In von Neumann (1955), an important subset of such wave functions was considered, related to the regular cell partition of the phase plane of a one-dimensional dynamical system. The aim of von Neumann was to study the position and momentum measurement processes in quantum mechanics.

(iii) In Glauber (1963), the author tried to apply the formalism of quantum theory to problems relevant for optics. For this purpose, he needed states which reduce the correlation functions of the electric field to factorized form, i.e. a product of eigenvalues for negative- and positive-frequency parts of the electric field. Such states were the coherent states, and turned out to be very useful for describing coherent laser beams within the framework of quantum theory (Klauder and Sudarshan 1968, Squires 1995, Scully and Zubairy 1997). However, one does not need a large number of photons to be able to apply the coherent-state formalism (Sudarshan 1963, Klauder and Sudarshan 1968).

(iv) Coherent states provide relevant applications of the formalism of Weyl systems (see section 10.4) and make it possible to realize the Hilbert space of states as a space of entire functions (i.e. functions analytic in the whole complex plane). This property leads, in turn, to simpler solutions of a number of problems, by exploiting the theory of entire functions (Perelomov 1986).

(v) Further developments deal with problems where it is necessary to minimize the quantum noise, while keeping fixed the product $(\triangle q)_\psi (\triangle p)_\psi$. This is achieved with the help of the so-called two-photon coherent states, also called squeezed states (section 10.5).

Since the general reader is not necessarily familiar with the items mentioned above, we begin our analysis in a simplified (but correct) way, i.e. by defining coherent states as a subset of the general set of quantum states of a one-dimensional harmonic oscillator. This means that our definition should give rise to square-integrable functions on the real line, with some peculiar properties. The latter result from the fact that, although it is important in quantum mechanics to have a theory of self-adjoint operators, the spectral theory of non-self-adjoint operators (for which eigenvalues are, in general, complex) plays an important role. Within this framework, we are interested in the annihilation operator for harmonic oscillators. Let us study a one-dimensional harmonic oscillator with mass m and angular frequency $\omega = 2\pi\nu$. Suppose that at time $t = 0$ the wave function $\psi(x,t)$ of the oscillator is an eigenfunction of the annihilation operator $\hat{a}$ with complex eigenvalue μ:

$$\hat{a}\psi(x,0) = \mu\, \psi(x,0). \tag{10.3.1}$$

We want to express $\psi(x,0)$ in terms of eigenfunctions of the Hamiltonian, and show that $\psi(x,t)$ is an eigenfunction of $\hat{a}$ with eigenvalue $\mu e^{-i\omega t}$. Moreover, we are going to prove that the square of the modulus of ψ is a Gaussian that undergoes harmonic motion, and its shape remains the same.

To begin our analysis, since the Hamiltonian operator is self-adjoint, we expand the initial condition $\psi(x,0)$ as an infinite sum of normalized eigenfunctions of the Hamiltonian with quantum number n (the convergence of the sum being in the $\mathcal{L}^2$ norm):

$$\psi(x,0) = \sum_{n=0}^{\infty} c_n u_n. \tag{10.3.2}$$

Since here we consider the realization of $\hat{a}$ and $\hat{a}^\dagger$ as first-order differential operators, it is convenient to use a notation which refers to wave functions acted upon by operators on a Hilbert space, rather than the Dirac notation of section 10.1. The well-known property (10.1.31) of annihilation operators, jointly with Eq. (10.3.1), leads to the equation (Squires 1995)

$$\sum_{n=1}^{\infty} \sqrt{n} c_n u_{n-1} = \sum_{n=0}^{\infty} \mu c_n u_n. \tag{10.3.3}$$

Comparison of coefficients of the u_n shows that

$$\sqrt{n} c_n = \mu c_{n-1}, \tag{10.3.4}$$

from which it follows that

$$c_n = \frac{\mu^n}{\sqrt{n!}} c_0, \tag{10.3.5}$$

which implies (see Eq. (10.3.2))

$$\psi(x,0) = \sum_{n=0}^{\infty} \frac{\mu^n}{\sqrt{n!}} c_0 u_n. \tag{10.3.6}$$

As we know from section 4.4, the wave function at time t is obtained by applying the time-evolution operator to the initial condition:

$$\psi(x,t) = \mathrm{e}^{-\mathrm{i}Ht/\hbar}\, \psi(x,0), \tag{10.3.7}$$

which implies (see Eq. (10.1.28))

$$\begin{aligned} \psi(x,t) &= \sum_{n=0}^{\infty} \frac{\mu^n}{\sqrt{n!}} c_0 u_n \mathrm{e}^{-\mathrm{i}(n+1/2)\omega t} \\ &= \mathrm{e}^{-\mathrm{i}\omega t/2} \sum_{n=0}^{\infty} \frac{(\mu \mathrm{e}^{-\mathrm{i}\omega t})^n}{\sqrt{n!}} c_0 u_n. \end{aligned} \tag{10.3.8}$$

The interpretation of this simple calculation is that, up to the inessential phase factor $\mathrm{e}^{-\mathrm{i}\omega t/2}$, $\psi(x,t)$ and $\psi(x,0)$ have the same functional form. This property leads in turn to the conclusion that $\psi(x,t)$ is an eigenfunction of the annihilation operator, with eigenvalue $\mu \mathrm{e}^{-\mathrm{i}\omega t}$.

Let us now use the expression (10.1.2) of the annihilation operator in terms of the position and momentum operators. In terms of the parameter $\sigma \equiv \sqrt{\frac{\hbar}{2m\omega}}$, this reads (see Eq. (10.1.48))

$$\hat{a} = \frac{1}{2\sigma}\hat{x} + \sigma \frac{\mathrm{d}}{\mathrm{d}x}, \tag{10.3.9}$$

and from the previous analysis we know that such an operator possesses the eigenvalue equation

$$\hat{a}\psi(x,t) = \gamma \psi(x,t). \tag{10.3.10a}$$

Moreover, we have shown above that the complex eigenvalue γ is $\mu \mathrm{e}^{-\mathrm{i}\omega t}$. Thus, writing $\mu = \lambda \mathrm{e}^{\mathrm{i}\rho}$, where λ is real, γ reads $\lambda \mathrm{e}^{\mathrm{i}\theta}$, with $\theta = \rho - \omega t$. In the light of (10.3.9), Eq. (10.3.10a) can be written in the form

$$\frac{\mathrm{d}\psi}{\psi} = \left(\frac{\gamma}{\sigma} - \frac{x}{2\sigma^2} \right) \mathrm{d}x. \tag{10.3.10b}$$

The left-hand side of (10.3.10b) is the logarithmic derivative of ψ, and hence one finds, by integration and subsequent exponentiation,

$$\psi = C \exp\!\left(-\frac{x^2}{4\sigma^2} + \frac{\gamma}{\sigma} x \right), \tag{10.3.11}$$

where C is a constant. It is therefore possible to express the square of the modulus of $\psi(x,t)$ as

$$|\psi(x,t)|^2 = |C|^2 \mathrm{e}^{-G}, \tag{10.3.12}$$

where, completing the square, one finds

$$G = \frac{x^2}{2\sigma^2} - \frac{x}{\sigma}(\gamma + \gamma^*) = \frac{1}{2\sigma^2}\Big(x - 2\sigma\lambda\cos\theta\Big)^2 - 2\lambda^2\cos^2\theta. \quad (10.3.13)$$

This leads in turn to the result

$$|\psi(x,t)|^2 = |C|^2 \, \mathrm{e}^{2\lambda^2\cos^2\theta} \, \mathrm{e}^{-(x-x_0)^2/2\sigma^2}, \quad (10.3.14)$$

where we have defined

$$x_0 \equiv 2\sigma\lambda\cos\theta = 2\sigma\lambda\cos(\rho - \omega t). \quad (10.3.15)$$

Of course, the constant C can be evaluated from the normalization condition

$$\int_{-\infty}^{\infty} |\psi|^2 \, \mathrm{d}x = 1. \quad (10.3.16)$$

This method yields the simple but fundamental result (Squires 1995)

$$|\psi|^2 = \frac{1}{\sqrt{2\pi\sigma^2}} \mathrm{e}^{-(x-x_0)^2/2\sigma^2}. \quad (10.3.17)$$

Such a formula shows that $|\psi(x,t)|^2$ is a Gaussian that performs harmonic motion. The corresponding plot is a sequence of Gaussian curves with the same standard deviation σ and the same amplitude $2\sigma\lambda$ (see Eq. (10.3.15) and figure 10.1).

The eigenfunction $\psi(x,t)$ described so far is known as a *coherent state.* The annihilation operator $\hat{a}$ *does not* correspond to any physical observable and is not self-adjoint. Hence its eigenvalues are in general complex. Another important property is that eigenfunctions corresponding to different eigenvalues μ and μ' are not orthogonal (see Eq. (10.4.11)). It should also be stressed that, unlike $\hat{a}$, the creation operator $\hat{a}^\dagger$ does not have eigenfunctions belonging to $\mathcal{L}^2(\mathbf{R}, \mathrm{d}x)$, because the solutions of the equation

$$\hat{a}^\dagger \psi = \rho\psi,$$

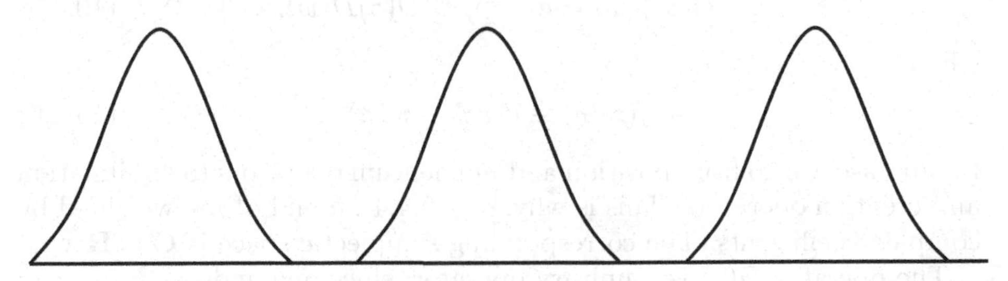

Fig. 10.1. Motion of a coherent state. The Gaussian wave packet undergoes harmonic motion while its shape remains unchanged.

which reads, upon realizing $\hat{a}^\dagger$ as a differential operator,

$$\frac{d\psi}{dx} = \left(\frac{\rho}{\sigma} + \frac{x}{2\sigma^2}\right)\psi,$$

increase exponentially at large x, unlike the case of Eq. (10.3.11).

Coherent states are well known to be minimum-uncertainty states, and this is clear in the light of Eq. (10.3.11) and of the result (9.9.21) on the general form of minimal uncertainty states.

10.4 Weyl systems for coherent states

On using the Dirac notation for 'ket' and 'bra' vectors, the eigenvalue equation for coherent states reads

$$\hat{a}|z\rangle = z|z\rangle \quad z \in \mathbf{C}, \tag{10.4.1}$$

and the result (10.3.8) takes the form

$$|z\rangle = e^{-\frac{1}{2}|z|^2} \sum_{n=0}^{\infty} \frac{z^n}{\sqrt{n!}}|n\rangle. \tag{10.4.2}$$

Bearing in mind the special case of the Baker–Campbell–Hausdorff formula when the commutator $[A, B]$ commutes with both A and B (Baker Jr 1958, Moyal 1949, Glauber 1963):

$$e^{A+B} = e^A \, e^B \, e^{-\frac{1}{2}[A,B]}, \tag{10.4.3}$$

we are now led to consider the operator

$$D(z) \equiv e^{z\hat{a}^\dagger - z^*\hat{a}}. \tag{10.4.4}$$

This operator provides a relevant application of the formalism of section 9.5, where we have studied the Weyl quantization and Weyl systems, and hence we have considered unitary representations of symplectic vector spaces. The explicit calculation shows, indeed, that

$$D(z+w) = e^{-\frac{i}{2}\omega(z,w)} D(z)D(w), \tag{10.4.5}$$

where

$$\omega(z,w) \equiv i(wz^* - w^*z). \tag{10.4.6}$$

In our case, we go from position and momentum operators to annihilation and creation operators. This is why, in (10.4.4), $\hat{a}$ and $\hat{a}^\dagger$ are weighted by complex coefficients. The corresponding symplectic space is $\mathbf{C} \cong \mathbf{R}^2$.

The operator $D(z)$ is a unitary operator, since one finds

$$D^\dagger(z) = e^{z^*\hat{a} - z\hat{a}^\dagger}, \tag{10.4.7}$$

and the formula (10.4.3) yields

$$D(z)D^{\dagger}(z) = D^{\dagger}(z)D(z) = \mathbb{1}, \tag{10.4.8}$$

as may be verified by direct computation. A further consequence of (10.4.3) is that the operator $D(z)$ takes the form

$$D(z) = \mathrm{e}^{-\frac{1}{2}|z|^2}\, \mathrm{e}^{z\hat{a}^{\dagger}} \mathrm{e}^{-z^* \hat{a}}. \tag{10.4.9}$$

This formula can be used to obtain the action of $D(z)$ on the ground state as follows:

$$\begin{aligned} D(z)|0\rangle &= \mathrm{e}^{-\frac{1}{2}|z|^2}\, \mathrm{e}^{z\hat{a}^{\dagger}}|0\rangle = \mathrm{e}^{-\frac{1}{2}|z|^2} \sum_{n=0}^{\infty} \frac{(z\hat{a}^{\dagger})^n}{n!}|0\rangle \\ &= \mathrm{e}^{-\frac{1}{2}|z|^2} \sum_{n=0}^{\infty} \frac{z^n}{\sqrt{n!}}|n\rangle = |z\rangle, \end{aligned} \tag{10.4.10}$$

where we have used the property

$$\mathrm{e}^{-z^* \hat{a}}\, |0\rangle = |0\rangle,$$

which results from the Taylor expansion of $\mathrm{e}^{-z^* \hat{a}}$, jointly with the definition of the annihilation operator. Interestingly, the result (10.4.10) shows that $D(z)$ is a unitary operator which turns the ground state into an eigenfunction of the annihilation operator. More generally,

$$D(z)|z_1\rangle = |z_1 + z\rangle, \quad D^{\dagger}(z)|z_1\rangle = |z_1 - z\rangle$$

apart from phase factors.

The lack of orthogonality of coherent states is easily proved, and one finds

$$\begin{aligned} \langle z|z'\rangle &= \mathrm{e}^{-\frac{1}{2}|z|^2 - \frac{1}{2}|z'|^2} \sum_{n,r=0}^{\infty} \frac{(z^*)^n (z')^r}{\sqrt{n!r!}} \langle n|r\rangle \\ &= \mathrm{e}^{-\frac{1}{2}|z|^2 - \frac{1}{2}|z'|^2 + z^* z'} \\ &= \mathrm{e}^{-\frac{1}{2}|z-z'|^2}. \end{aligned} \tag{10.4.11}$$

This scalar product never vanishes exactly (but tends to 0 if $|z-z'| \to \infty$).

Despite the lack of orthogonality, coherent states obey a remarkable 'completeness relation', in that (Glauber 1963)

$$\frac{1}{\pi}\int \mathrm{d}^2 z\, |z\rangle\langle z| = \sum_{n=0}^{\infty} |n\rangle\langle n| = 1. \tag{10.4.12}$$

For advanced applications of this overcompleteness relation, we refer the reader to the work in Mehta and Sudarshan (1965). Now let $|\psi\rangle$ be an arbitrary normalized vector in the Hilbert space of the theory. By virtue of (10.3.2) and (10.4.10), one finds

$$\langle z|\psi\rangle = \mathrm{e}^{-\frac{1}{2}|z|^2}\ \psi(z^*), \tag{10.4.13}$$

where

$$\psi(z) \equiv \sum_{n=0}^{\infty} c_n u_n(z), \tag{10.4.14}$$

$$u_n(z) \equiv \frac{z^n}{\sqrt{n!}}. \tag{10.4.15}$$

The series in (10.4.14) converges uniformly in any compact domain of the z-plane because of the normalization condition for $|\psi\rangle$:

$$\sum_{n=0}^{\infty} |c_n|^2 = 1. \tag{10.4.16}$$

Thus, $\psi(z)$ is an entire function in the complex-z plane (i.e. analytic over the whole complex plane), and

$$\|\psi\|^2 = \langle\psi|\psi\rangle = \int \mathrm{e}^{-|z|^2}\ |\psi(z)|^2\ \mathrm{d}^2 z < \infty, \tag{10.4.17}$$

with the measure $\mathrm{e}^{-|z|^2}$. The scalar product of two entire functions of z, with finite norm so that it is of growth order less than $\mathrm{e}^{-|z|^2/2}$, is defined by

$$\langle\psi_1|\psi_2\rangle \equiv \int \mathrm{e}^{-|z|^2}\ \psi_1^*(z)\psi_2(z)\ \mathrm{d}^2 z. \tag{10.4.18}$$

It was Bargmann who proved that the resulting functional space is a Hilbert space (Bargmann 1961).

One is thus led to a concrete *realization of the Hilbert space as a space of entire functions* of z, with finite norm. A physical interpretation as suggested, for example, in Perelomov (1986), regards Eq. (10.4.17) as a statistical average of the function $\psi(z)$ over the classical phase space, with $z \equiv q + \mathrm{i}p$, for a classical oscillator Hamiltonian

$$H = \frac{1}{2}(q^2 + p^2) = \frac{1}{2}|z|^2$$

at $\beta = \frac{1}{KT} = 2$, with the distribution being given by $\mathrm{e}^{-\beta H}$ (cf. chapter 14).

Fock was actually the first to propose the above realization of the Hilbert space, motivated by his interest in finding an operator solution of the Heisenberg commutation relations. In the modern literature, equations (10.4.13)–(10.4.18) describe the Bargmann–Fock representation, frequently used because it makes it possible to find simpler solutions of a number of problems, by exploiting the theory of entire functions, as we anticipated at the beginning of section 10.3.

10.5 Two-photon coherent states

The present section is the counterpart of section 9.10 when Weyl systems are considered on $\mathbf{C}^n \equiv \mathbf{R}^{2n}$.

In the applications to quantum optics, it is of considerable interest to construct the so-called *two-photon coherent states*, first introduced by Yuen (1976). For this purpose, one begins by studying the following linear combination of annihilation and creation operators (this is called the Bogoliubov–Valatin transformation of $\hat{a}, \hat{a}^\dagger$):

$$\hat{b} \equiv \mu \hat{a} + \nu \hat{a}^\dagger, \tag{10.5.1}$$

where μ and ν are complex coefficients. Equation (10.5.1) corresponds to (9.10.9) when the realization of the Heisenberg algebra in terms of Q, P is replaced with that in terms of $\hat{a}, \hat{a}^\dagger$. On requiring that $\hat{b}$ and its formal adjoint, $\hat{b}^\dagger$, should obey the standard commutation property for annihilation and creation operators, i.e.

$$[\hat{b}, \hat{b}^\dagger] = \mathbb{I}, \tag{10.5.2}$$

one finds that μ and ν are restricted by the condition

$$|\mu|^2 - |\nu|^2 = 1. \tag{10.5.3}$$

Equations (10.5.1) and (10.5.3) therefore describe an automorphism for the Heisenberg algebra. By definition, a *two-photon coherent state* is a solution of the eigenvalue equation for the operator $\hat{b}$ once that this is realized as a differential operator:

$$\hat{b}\psi = \beta\, \psi. \tag{10.5.4}$$

Yuen was led to the analysis of this equation in his investigation of the possible ways to generate minimum-uncertainty states. He then discovered a broad class of radiation states which include, in particular, the coherent states, which are recovered on setting $\mu = 1$ and $\nu = 0$ in Eq. (10.5.1). These states are the vacuum states for the squeezed operators $q \rightarrow \lambda q, p \rightarrow \lambda^{-1} p$.

For example, if $\hat{a}$ and $\hat{a}^\dagger$ are the annihilation and creation operators of a one-dimensional harmonic oscillator, Eq. (10.5.4) takes the form, at $t = 0$,

$$(\mu - \nu)\frac{\mathrm{d}\psi}{\mathrm{d}x} = \frac{\beta}{\sigma}\psi - \frac{(\mu + \nu)}{2\sigma^2} x\psi, \tag{10.5.5}$$

where $\sigma \equiv \sqrt{\frac{\hbar}{2m\omega}}$ as in Eq. (10.3.10). Thus, denoting by γ a constant that can be fixed by a normalization condition (see below), one finds, by separation of variables and integration, the solution

$$\psi(x) = \gamma \exp\left[\frac{\beta}{(\mu - \nu)\sigma}x - \frac{(\mu + \nu)}{(\mu - \nu)}\frac{x^2}{4\sigma^2}\right]. \tag{10.5.6}$$

Interestingly, this solution is normalizable provided that the condition (10.5.3) holds, because one then finds

$$\frac{(\mu + \nu)}{(\mu - \nu)} = \frac{(\mu + \nu)(\mu^* - \nu^*)}{(\mu - \nu)(\mu^* - \nu^*)} = \frac{1 + (\nu\mu^* - \mu\nu^*)}{|\mu - \nu|^2}, \tag{10.5.7}$$

where $\nu\mu^* - \mu\nu^*$ is purely imaginary:

$$\nu\mu^* - \mu\nu^* = 2\mathrm{i}\Big[\mathrm{Re}(\mu)\mathrm{Im}(\nu) - \mathrm{Im}(\mu)\mathrm{Re}(\nu)\Big]. \tag{10.5.8}$$

Hence one finds that the real part of the coefficient of x^2 in the exponential of Eq. (10.5.6) is always negative, which ensures that the condition (10.3.16) is fulfilled. Two-photon coherent states, often denoted by $|\beta\rangle_{\mathrm{g}}$, obey a completeness relation analogous to Eq. (10.4.12) (being canonical operators by construction):

$$\frac{1}{\pi}\int \mathrm{d}^2\beta \; |\beta\rangle_{\mathrm{g}} \; \langle\beta|_{\mathrm{g}} = 1, \tag{10.5.9}$$

and, similarly to Eq. (10.4.10), one finds

$$|\beta\rangle_{\mathrm{g}} = D_{\mathrm{g}}(\beta)|0\rangle_{\mathrm{g}}, \tag{10.5.10}$$

where the unitary operator $D_{\mathrm{g}}(\beta)$ takes the form

$$D_{\mathrm{g}}(\beta) = \mathrm{e}^{\beta\hat{b}^\dagger - \beta^*\hat{b}}, \tag{10.5.11}$$

and the ground state, $|0\rangle_{\mathrm{g}}$, can be defined by the condition

$$N_{\mathrm{g}}|0\rangle_{\mathrm{g}} = 0, \tag{10.5.12}$$

where the number operator reads

$$N_{\mathrm{g}} \equiv \hat{b}^\dagger\hat{b}. \tag{10.5.13}$$

This is obvious since $\hat{b}, \hat{b}^\dagger$ satisfy the same commutation rules of $\hat{a}, \hat{a}^\dagger$. We stress that any realization of the Heisenberg algebra will give rise to a displacement operator $D_{\mathrm{g}}(\beta)$ and hence to the associated coherent states. Such a situation is analogous to classical mechanics, where each symplectic chart makes it possible to define a harmonic oscillator.

If one looks for a unitary operator U that maps $\hat{a}, \hat{a}^\dagger$ into $\hat{b}, \hat{b}^\dagger$, i.e.

$$U\hat{a}U^\dagger = \hat{b} = \mu\hat{a} + \nu\hat{a}^\dagger, \tag{10.5.14}$$

one finds that, for a quadratic Hamiltonian

$$\hat{H} = \hbar\Big(f_1 \hat{a}^\dagger \hat{a} + f_2^* \hat{a}^2 + f_2 \hat{a}^{\dagger 2} + f_3^* \hat{a} + f_3 \hat{a}^\dagger\Big), \tag{10.5.15}$$

such an operator reads (Yuen 1976)

$$U = \exp\Big(\gamma_1 \hat{a}^{\dagger 2} + \gamma_2 \hat{a}^\dagger \hat{a} + \gamma_3 \hat{a}^2 + \gamma_4 \hat{a}^\dagger + \gamma_5 \hat{a}\Big), \tag{10.5.16}$$

where the γ-coefficients are c-numbers, i.e. they belong to an algebra of commuting variables. Now for radiation–matter interactions responsible for two-photon transitions, where two photons of the same frequency from the same radiation mode can be absorbed in a single atomic transition between two levels, the Hamiltonian reads

$$\hat{H}_I = \hbar\Big(p\hat{M}\hat{a}^{\dagger 2} + p^* \hat{M}^\dagger \hat{a}^2\Big). \tag{10.5.17}$$

In Eq. (10.5.17) p is a coupling coefficient and $\hat{M}$ is an operator which flips the state of the atom. Interestingly, when a two-photon laser operates far above threshold, $\hat{M}$ can be approximated by a c-number, and in this limit the Hamiltonian (10.5.17) reduces to a particular case of the Hamiltonian (10.5.15), for which the operator U mapping $\hat{a}, \hat{a}^\dagger$ into $\hat{b}, \hat{b}^\dagger$ can be constructed. This is why the eigenfunctions of $\hat{b}$ are called two-photon coherent states. In other words, physical arguments and some mathematical properties *suggest* that the output radiation of an ideal, monochromatic two-photon laser is in the state we therefore called a two-photon coherent state, following Yuen (1976).

10.6 Problems

10.P1. For an isotropic harmonic oscillator in three dimensions, the radial part of the stationary Schrödinger equation reads, in spherical coordinates (here n is an integer ≥ 0),

$$\left[-\frac{\hbar^2}{2m}\frac{d^2}{dr^2} + \frac{l(l+1)\hbar^2}{2mr^2} + \frac{m}{2}\omega^2 r^2\right] y_{nl} = \left(n + \frac{3}{2}\right)\hbar\omega y_{nl}, \tag{10.6.1}$$

after rescaling the radial wave function as in Eq. (5.4.38).

(i) On defining

$$\alpha \equiv \sqrt{\frac{m\omega}{\hbar}}, \tag{10.6.2}$$

$$\varepsilon_n \equiv 2n + 3, \tag{10.6.3}$$

$$\xi \equiv \alpha r, \tag{10.6.4}$$

check that the eigenvalue equation (10.6.1) may be cast in the form

$$\left[\frac{d^2}{d\xi^2} - \frac{l(l+1)}{\xi^2} + \varepsilon_n - \xi^2\right] y_{nl} = 0. \tag{10.6.5}$$

(ii) Bearing in mind that $\xi = 0$ is a Fuchsian singularity of the eigenvalue equation, look for a solution in the form

$$y_{nl} = e^{-\xi^2/2} \xi^{\rho} \sum_{p=0}^{\infty} a_p \xi^p, \tag{10.6.6}$$

and prove that (10.6.5) and (10.6.6) lead to the equation

$$\sum_{p=0}^{\infty} [(p+\rho)(p+\rho-1) - l(l+1)] a_p \xi^{p+\rho-2}$$
$$+ \sum_{p=0}^{\infty} [\varepsilon_n - 1 - 2(p+\rho)] a_p \xi^{p+\rho} = 0. \tag{10.6.7}$$

(iii) Use Eq. (10.6.7) to prove that the regular solution of Eq. (10.6.5) is obtained when $\rho = l+1$ if $a_0 \neq 0$ and $a_1 = 0$, and when $\rho = l$ if $a_0 = 0$ and $a_1 \neq 0$. Therefore,

$$l = n, n-2, \ldots, 0, \quad \text{if } n \text{ is even}, \tag{10.6.8}$$

$$l = n, n-2, \ldots, 1, \quad \text{if } n \text{ is odd}. \tag{10.6.9}$$

Can the series in (10.6.6) contain infinitely many non-vanishing terms?

10.P2. Consider a perturbation of the three-dimensional harmonic oscillator, in the isotropic case, with a term $\frac{\beta}{\xi^4}$ having $\beta > 0$, in the dimensionless units of Eq. (10.6.5).

(i) Find the appropriate boundary condition at $\xi = 0$.

(ii) Find the limit of the perturbed Hamiltonian as $\beta \to 0$.

10.P3. An electron is subject to a constant magnetic field, so that the effective Hamiltonian takes the form (10.2.1).

(i) Find the operators $\mathcal{A}$ and $\mathcal{A}^\dagger$ such that (use, for simplicity, $\hbar = 1$ units)

$$H = \frac{eB}{2mc} \left(\mathcal{A}^\dagger \mathcal{A} + \mathcal{A}\mathcal{A}^\dagger + \sigma_z \right). \tag{10.6.10}$$

(ii) Find the pair of operators $\mathcal{F}$ and $\mathcal{F}^\dagger$ which commute with H, and investigate their effect on the degeneracy of the energy levels of the Hamiltonian.

(iii) By definition, the first Landau level is the solution of the equation

$$\mathcal{A}\, \psi = 0. \tag{10.6.11}$$

Find the form taken by ψ, and interpret the result.

10.P4. Consider a one-dimensional harmonic oscillator. When $t = 0$, consider the eigenvalue equation for the annihilation operator, and find the explicit form of its normalized eigenfunctions. For positive values of the time variable, find the time dependence of the eigenvalues of $\hat{a}$, and the formula for the square of the modulus of the eigenfunctions of $\hat{a}$. Interpret the result. As a next step, one wants to build the unitary operator $D(z)$ that relates the ground state to the eigenfunctions of $\hat{a}$. What is the relation between $D(z+w)$ and $D(z)D(w)$? Try to interpret this formula. Lastly, consider the eigenvalue problem for the creation operator. Can one find normalizable eigenfunctions? Why? What is the role played by the measure (if any) in the course of performing this investigation?

10.P5. Consider the following linear combination of annihilation and creation operators for a one-dimensional harmonic oscillator: $\hat{b} \equiv \mu \hat{a} + \nu \hat{a}^\dagger$, where the parameters μ and ν are, in general, complex.

(i) Find for which values of μ and ν the operator $\hat{b}$ satisfies, jointly with the operator $\hat{b}^\dagger \equiv \mu^* \hat{a}^\dagger + \nu^* \hat{a}$, the commutation relations of annihilation and creation operators:

$$\left[\hat{b}, \hat{b}^\dagger\right] = \mathbb{I}. \tag{10.6.12}$$

(ii) Consider now the realization of the momentum operator as a first-order differential operator:

$$p : f \rightarrow \frac{\hbar}{\mathrm{i}} \frac{\mathrm{d}f}{\mathrm{d}x}, \tag{10.6.13}$$

and of the position operator as the operator which multiplies by x. At this stage the operators $\hat{a}, \hat{a}^\dagger$, and hence $\hat{b}, \hat{b}^\dagger$, become first-order differential operators. Study the eigenvalue equation for the operator $\hat{b}$. Does it have solutions in $\mathcal{L}^2(\mathbf{R}, dx)$? In the affirmative case, find their explicit form.

10.P6. Prove that the ground state of the harmonic oscillator is a cyclic vector (see the definition of cyclic vector in appendix 4.A).

10.P7. Let

$$H(t) \equiv -\frac{1}{2} \frac{\mathrm{d}^2}{\mathrm{d}x^2} + \frac{x^2}{2} - f(t)x \tag{10.6.14}$$

be the Hamiltonian operator of a forced harmonic oscillator. On denoting by $U(t, t_0)$ the associated propagator, prove that

$$U(t, t_0)|z(0)\rangle = \mathrm{e}^{i\varphi(t)}|z(t)\rangle, \tag{10.6.15}$$

where $|z(t)\rangle$ is a coherent state, and find how $\varphi(t)$ and $|z(t)\rangle$ depend on time. (The case when $f(t)$ is periodic in time with a pure harmonic dependence is an important physical Hamiltonian.)

11
Angular momentum operators

The general formalism of angular momentum operators is developed. The basic elements are an abstract Hilbert space on which a triple of self-adjoint operators is given, with assigned commutation relations. The spectra of the operators J^2 and J_3 are derived in detail, after the introduction of the raising and lowering operators. The latter turn out to be the building blocks of the analysis, and their matrix elements are also obtained. The spectrum of J^2 is of the form $j(j+1)\hbar^2$, where j is either an integer or a half-odd. Matrix representations are described when $j = \frac{1}{2}, 1, \frac{3}{2}$. Applications presented here deal with a two-dimensional harmonic oscillator, and with the analysis of lossless devices with two input ports and two output ports. One then finds that such devices may be viewed as measuring rotations of the angular momentum operators.

In the second part, Clebsch–Gordan coefficients, general properties of quantum mechanics including spin, the formulation of spin within the framework of Weyl systems and the theory of monopole harmonics are presented.

11.1 Angular momentum: general formalism

Section 5.6 has developed an introduction to angular momentum in quantum mechanics. Now we are ready for a more advanced treatment. In the general formulation of angular momentum operators, one deals with an abstract Hilbert space $\mathcal{H}$ on which three self-adjoint operators J_1, J_2, J_3 are given, which satisfy the commutation relations

$$\left[J_k, J_l\right] = \mathrm{i}\hbar\varepsilon_{klm} J_m. \tag{11.1.1}$$

Note that *hat* symbols for operators are omitted in this chapter for simplicity. These rules are fulfilled, for example, by the operator version of the orbital angular momentum $\vec{L} \equiv \vec{r} \times \vec{p}$. However, the framework that we are studying is more general, and we will see that (11.1.1) leads to many non-trivial properties.

11.1.1 Algebraic method for the spectrum

First, note that, defining

$$J^2 \equiv \sum_{i=1}^{3} J_i^2, \tag{11.1.2}$$

one finds

$$\left[J^2, J_i\right] = 0 \quad \forall i = 1, 2, 3, \tag{11.1.3}$$

and hence one may consider ket vectors which are simultaneous eigenstates of J^2 and J_3. Furthermore, one may define the operators

$$J_+ \equiv J_1 + \mathrm{i}J_2, \tag{11.1.4}$$

$$J_- \equiv J_1 - \mathrm{i}J_2. \tag{11.1.5}$$

By virtue of the Hermiticity of J_1, J_2 and J_3, one finds

$$\left(J_+\right)^\dagger = J_-, \quad \left(J_-\right)^\dagger = J_+. \tag{11.1.6}$$

Moreover, like J_1, J_2, J_3, the operators J_+ and J_- do not commute, and one has

$$J_+J_- = J^2 - J_3^2 + \hbar J_3, \tag{11.1.7}$$

$$J_-J_+ = J^2 - J_3^2 - \hbar J_3, \tag{11.1.8}$$

which implies

$$J_+J_- - J_-J_+ = 2\hbar J_3. \tag{11.1.9}$$

Further non-trivial properties are the commutators of J_3 with J_+ and J_-, respectively:

$$\left[J_3, J_+\right] = \hbar J_+, \quad \left[J_3, J_-\right] = -\hbar J_-. \tag{11.1.10}$$

Following previous remarks, let $|a, b\rangle$ denote the eigenstates of both J^2 and J_3:

$$J^2|a, b\rangle = a|a, b\rangle, \tag{11.1.11}$$

$$J_3|a, b\rangle = b|a, b\rangle, \tag{11.1.12}$$

where a and b are the eigenvalues that we are going to determine. The operators J_+ and J_- have a peculiar action on such eigenstates, because, by virtue of (11.1.10), one finds

$$J_3J_\pm|a, b\rangle = (b \pm \hbar)J_\pm|a, b\rangle. \tag{11.1.13}$$

This means that J_+ raises the eigenvalue b of J_3 by 1 in $\hbar$ units, while J_- lowers the eigenvalue of J_3 by 1 in $\hbar$ units. Thus, hereafter, we will call them the *raising operator* and the *lowering operator*, respectively. In contrast, they do not affect the eigenvalue a in (11.1.11), since (11.1.3) implies that

$$\left[J_+, J^2\right] = \left[J_-, J^2\right] = 0. \tag{11.1.14}$$

Another useful identity holds, i.e.

$$J^2 - J_3^2 = J_1^2 + J_2^2 = \frac{1}{2}\Big(J_+ J_- + J_- J_+\Big), \tag{11.1.15}$$

which completes the construction of all the original operators in terms of J_+ and J_- only. Note that, by virtue of (11.1.15), the matrix element

$$\langle a, b | J^2 - J_3^2 | a, b \rangle$$

is positive. On the other hand, one also has

$$\langle a, b | J^2 - J_3^2 | a, b \rangle = a - b^2, \tag{11.1.16}$$

and hence a cannot be negative, since b is real, J_3 being self-adjoint by hypothesis: $a \geq b^2 \geq 0$.

The property (11.1.13) may be expressed in the form

$$J_+ |a, b\rangle = C_+ |a, b + \hbar\rangle, \tag{11.1.17}$$

$$J_- |a, b\rangle = C_- |a, b - \hbar\rangle, \tag{11.1.18}$$

where the numbers C_+ and C_- will be determined up to a phase factor. First, we prove that the raising and lowering actions performed by J_+ and J_- do not go on for ever, but finite values $b_{\rm max}$ and $b_{\rm min}$ exist such that

$$J_+ |a, b_{\rm max}\rangle = 0, \tag{11.1.19}$$

$$J_- |a, b_{\rm min}\rangle = 0. \tag{11.1.20}$$

Indeed, by virtue of (11.1.6), one has

$$\langle a, b | J_+ J_- | a, b \rangle \geq 0. \tag{11.1.21}$$

On the other hand, Eq. (11.1.7) leads to

$$\langle a, b | J_+ J_- | a, b \rangle = \Big(a - b^2 + b\hbar\Big) \langle a, b | a, b \rangle. \tag{11.1.22}$$

Hence one finds the inequality $a - b^2 + b\hbar \geq 0$, which, adding and subtracting $\frac{1}{4}\hbar^2$, can be written in the form

$$a + \frac{1}{4}\hbar^2 \geq \left(b - \frac{1}{2}\hbar\right)^2 \geq 0. \tag{11.1.23}$$

Now let k be the parameter, depending on a, such that

$$k + \frac{1}{2}\hbar \equiv \left(a + \frac{1}{4}\hbar^2\right)^{1/2}. \tag{11.1.24}$$

This definition makes it possible to re-express (11.1.23) in the form $k + \frac{1}{2}\hbar \geq \left|b - \frac{1}{2}\hbar\right|$. Thus, when $b - \frac{1}{2}\hbar > 0$, one finds $k + \hbar \geq b$, whereas, if $b - \frac{1}{2}\hbar < 0$, one has

$$b \geq -k. \tag{11.1.25}$$

Combining the above inequalities one finds

$$k + \hbar \geq b \geq -k. \tag{11.1.26}$$

Similarly, from the property

$$\langle a, b|J_- J_+|a, b\rangle = \left(a - b^2 - b\hbar\right)\langle a, b|a, b\rangle, \tag{11.1.27}$$

one finds

$$k \geq b \geq -k - \hbar. \tag{11.1.28}$$

The joint effect of (11.1.26) and (11.1.28) is that

$$k(a) \geq b \geq -k(a), \tag{11.1.29}$$

where the equalities hold for b_{max} and b_{min}, respectively. Hence the sequence of eigenvalues

$$b, \; b + \hbar, \; b + 2\hbar, \ldots$$

has an upper bound, and b takes all values in between b_{max} and $-b_{\text{max}}$. The result according to which $b_{\text{max}} = -b_{\text{min}}$ is proved, bearing in mind (11.1.19) and (11.1.20), and imposing the conditions

$$J_- J_+|a, b_{\text{max}}\rangle = 0, \tag{11.1.30}$$

$$J_+ J_-|a, b_{\text{min}}\rangle = 0, \tag{11.1.31}$$

which hold as a corollary. One then finds by comparison that

$$a = b_{\text{max}}\left(b_{\text{max}} + \hbar\right) = b_{\text{min}}\left(b_{\text{min}} - \hbar\right), \tag{11.1.32}$$

which implies

$$b_{\text{max}} = -b_{\text{min}}. \tag{11.1.33}$$

On the other hand, Eq. (11.1.17) implies that, for some $n \in \mathbf{Z}^+$, one should have

$$b_{\text{max}} = b_{\text{min}} + n\hbar. \tag{11.1.34}$$

This result, combined with (11.1.33), leads to

$$b_{\rm max} = \frac{n}{2}\hbar, \tag{11.1.35}$$

and hence

$$j \equiv \frac{b_{\rm max}}{\hbar} = \frac{n}{2}. \tag{11.1.36}$$

Note that (11.1.36) is not necessarily an integer value. It is either an integer, if $n = 2p$ for some integer p, or a half-odd, if $n = 2q+1$ for some integer q. One can thus label the eigenstates of J^2 and J_3 in terms of j and m, where j is given in (11.1.36) and $m = -j, -j+1, \ldots, j-1, j$. The eigenvalue equations for J^2 and J_3 now read

$$J^2|j,m\rangle = j(j+1)\hbar^2|j,m\rangle, \tag{11.1.37}$$

$$J_3|j,m\rangle = m\hbar|j,m\rangle, \tag{11.1.38}$$

and the matrix elements of these operators are

$$\langle j',m'|J^2|j,m\rangle = j(j+1)\hbar^2\ \delta_{j,j'}\ \delta_{m,m'}, \tag{11.1.39}$$

$$\langle j',m'|J_3|j,m\rangle = m\hbar\ \delta_{j,j'}\ \delta_{m,m'}. \tag{11.1.40}$$

From these matrix elements, with the help of (11.1.8), one finds

$$\begin{aligned}\langle j,m|(J_+)^\dagger J_+|j,m\rangle &= \langle j,m|J_-J_+|j,m\rangle \\ &= \hbar^2\Big(j(j+1) - m^2 - m\Big) \\ &= \left|C^+_{j,m}\right|^2 \langle j,m+1|j,m+1\rangle,\end{aligned} \tag{11.1.41}$$

which implies (the eigenstates being normalized)

$$C^+_{j,m} = {\rm e}^{{\rm i}\varphi}\hbar\sqrt{j(j+1) - m^2 - m}. \tag{11.1.42}$$

Setting $\varphi = 0 \pmod{2\pi}$, adding and subtracting mj in the square root, one obtains

$$\begin{aligned}J_+|j,m\rangle &= \hbar\sqrt{(j-m)(j+m+1)}|j,m+1\rangle \\ &= \hbar\sqrt{j(j+1) - m(m+1)}|j,m+1\rangle.\end{aligned} \tag{11.1.43}$$

A similar analysis of $\langle j,m|J_+J_-|j,m\rangle$ yields

$$\begin{aligned}J_-|j,m\rangle &= \hbar\sqrt{(j+m)(j-m+1)}|j,m-1\rangle \\ &= \hbar\sqrt{j(j+1) - m(m-1)}|j,m-1\rangle.\end{aligned} \tag{11.1.44}$$

The matrix elements of J_1 and J_2 are thus, from (11.1.4) and (11.1.5):

$$\langle j', m'|J_1|j, m\rangle = \frac{\hbar}{2}\sqrt{(j-m)(j+m+1)}\,\delta_{j,j'}\delta_{m',m+1} + \frac{\hbar}{2}\sqrt{(j+m)(j-m+1)}\,\delta_{j,j'}\delta_{m',m-1}, \quad (11.1.45)$$

$$\langle j', m'|J_2|j, m\rangle = \frac{\hbar}{2\mathrm{i}}\sqrt{(j-m)(j+m+1)}\,\delta_{j,j'}\delta_{m',m+1} - \frac{\hbar}{2\mathrm{i}}\sqrt{(j+m)(j-m+1)}\,\delta_{j,j'}\delta_{m',m-1}. \quad (11.1.46)$$

The matrix elements

$$\langle j, m'|J_\pm|j, m\rangle = \hbar\sqrt{(j\mp m)(j\pm m+1)}\,\delta_{m',m\pm 1} \equiv D^{(j)}_{mm'}(J_\pm), \quad (11.1.47)$$

$$\langle j, m'|J_3|j, m\rangle = m\hbar\delta_{m'm} \equiv D^{(j)}_{mm'}(J_3), \quad (11.1.48)$$

define the $(2j+1)$-dimensional representation of J_1, J_2, J_3 in terms of Hermitian matrices. This is called the *weight-j* or *spin-j* representation, and is denoted by $D^{(j)}$. We are clearly dealing with an *irreducible* representation, since the $(2j+1)$-dimensional vector space formed by all linear combinations of the vectors $|j, m\rangle$ does not contain any sub-space invariant under the action of $J_\pm$ and J_3. To sum up, by restricting to each invariant sub-space we are dealing with linear operators acting on a finite-dimensional Hilbert space. Thus, on these sub-spaces they are represented as matrices.

11.1.2 Representations

On passing from the Hermitian matrices $D^{(j)}(J_k)$ which represent J_k, for $k = 1, 2, 3$, to their exponentiation, one obtains the weight-j unitary irreducible representation of the universal cover of $SO(3)$, i.e. the group $SU(2)$ (recall section 5.7). For any element u of $SU(2)$, such a representation is denoted again by $D^{(j)}$:

$$u \rightarrow D^{(j)}(u) \equiv \exp\left[-\mathrm{i}\alpha\vec{n}\cdot D^{(j)}(\vec{J})\right]. \quad (11.1.49)$$

If j is an integer, $D^{(j)}$ is a representation (in the usual sense) of $SO(3)$, whereas if $j = \frac{(2k+1)}{2}$, $D^{(j)}$ is a representation of $SU(2)$, but a *two-valued* representation of $SO(3)$. With this understanding, in section 11.4 we shall also use the notation

$$R \rightarrow D^{(j)}(R)$$

when $j = \frac{(2k+1)}{2}$.

It is useful to give a matrix representation of the angular momentum operators in some simple cases. For example, when $j = \frac{1}{2}$, one has the raising and lowering operators

$$J_+ = \hbar \begin{pmatrix} 0 & 1 \\ 0 & 0 \end{pmatrix}, \tag{11.1.50}$$

$$J_- = \hbar \begin{pmatrix} 0 & 0 \\ 1 & 0 \end{pmatrix}. \tag{11.1.51}$$

Equations (11.1.4), (11.1.5), (11.1.47) and (11.1.48) enable one to reconstruct the angular momentum operators, which are found to coincide with the spin operators for spin $\frac{1}{2}$ in section 6.2, i.e.

$$J_k = \frac{\hbar}{2}\sigma_k, \quad \forall k = 1, 2, 3,$$

with σ_k given by the Pauli matrices (5.7.10)–(5.7.12).

When $j = 1$, one has from (11.1.43) and (11.1.44) (where the kets on the first row are $|1,1\rangle, |1,0\rangle$ and $|1,-1\rangle$, respectively)

$$J_+ = \hbar\sqrt{2} \begin{pmatrix} 0 & 1 & 0 \\ 0 & 0 & 1 \\ 0 & 0 & 0 \end{pmatrix}, \quad J_- = \hbar\sqrt{2} \begin{pmatrix} 0 & 0 & 0 \\ 1 & 0 & 0 \\ 0 & 1 & 0 \end{pmatrix}, \tag{11.1.52}$$

and the corresponding angular momentum operators are found to be

$$J_1 = \frac{\hbar}{\sqrt{2}} \begin{pmatrix} 0 & 1 & 0 \\ 1 & 0 & 1 \\ 0 & 1 & 0 \end{pmatrix}, \quad J_2 = \frac{\hbar}{\sqrt{2}} \begin{pmatrix} 0 & -\mathrm{i} & 0 \\ \mathrm{i} & 0 & -\mathrm{i} \\ 0 & \mathrm{i} & 0 \end{pmatrix}, \tag{11.1.53}$$

$$J_3 = \hbar \begin{pmatrix} 1 & 0 & 0 \\ 0 & 0 & 0 \\ 0 & 0 & -1 \end{pmatrix}, \quad J^2 = 2\hbar^2 \begin{pmatrix} 1 & 0 & 0 \\ 0 & 1 & 0 \\ 0 & 0 & 1 \end{pmatrix}. \tag{11.1.54}$$

Similarly, if $j = \frac{3}{2}$, one has

$$J_+ = \hbar \begin{pmatrix} 0 & \sqrt{3} & 0 & 0 \\ 0 & 0 & 2 & 0 \\ 0 & 0 & 0 & \sqrt{3} \\ 0 & 0 & 0 & 0 \end{pmatrix}, \tag{11.1.55}$$

$$J_- = \hbar \begin{pmatrix} 0 & 0 & 0 & 0 \\ \sqrt{3} & 0 & 0 & 0 \\ 0 & 2 & 0 & 0 \\ 0 & 0 & \sqrt{3} & 0 \end{pmatrix}, \tag{11.1.56}$$

which imply that

$$J_1 = \frac{\hbar}{2} \begin{pmatrix} 0 & \sqrt{3} & 0 & 0 \\ \sqrt{3} & 0 & 2 & 0 \\ 0 & 2 & 0 & \sqrt{3} \\ 0 & 0 & \sqrt{3} & 0 \end{pmatrix}, \qquad (11.1.57)$$

$$J_2 = \frac{\mathrm{i}}{2}\hbar \begin{pmatrix} 0 & -\sqrt{3} & 0 & 0 \\ \sqrt{3} & 0 & -2 & 0 \\ 0 & 2 & 0 & -\sqrt{3} \\ 0 & 0 & \sqrt{3} & 0 \end{pmatrix}, \qquad (11.1.58)$$

$$J_3 = \frac{\hbar}{2} \begin{pmatrix} 3 & 0 & 0 & 0 \\ 0 & 1 & 0 & 0 \\ 0 & 0 & -1 & 0 \\ 0 & 0 & 0 & -3 \end{pmatrix}. \qquad (11.1.59)$$

While $j = 1$ yields a representation of $SO(3)$, the values $j = \frac{1}{2}, \frac{3}{2}$ apply only for $SU(2)$.

11.1.3 Hilbert space

A crucial question is how to actually realize the angular momentum operators as differential operators with the general properties we have just derived. An enlightening example is indeed provided by the operators of orbital angular momentum. More precisely, let us consider a spherically symmetric potential on $\mathbf{R}^n$ (see also section 5.4 for the case $n = 3$). One may regard each $\varphi \in \mathcal{L}^2(\mathbf{R}^n)$ as a function of r and $n-1$ variables ξ on the sphere S^{n-1}, and one can exploit the decomposition

$$\mathcal{L}^2(\mathbf{R}^n) = \mathcal{L}^2(\mathbf{R}_+, r^{n-1}\,\mathrm{d}r) \otimes \mathcal{L}^2(S^{n-1}, \mathrm{d}\Omega), \qquad (11.1.60)$$

where $\mathrm{d}\Omega$ is the usual area measure on the sphere. On functions of the form $f(r)g(\xi)$, the operator $-\triangle + V(r)$ acts by (Reed and Simon 1975)

$$\Big[-\triangle + V(r)\Big] f(r)g(\xi) = \left[-\frac{\mathrm{d}^2}{\mathrm{d}r^2} + V(r) - \frac{(n-1)}{r}\frac{\mathrm{d}}{\mathrm{d}r}\right] f(r)g(\xi) - \frac{1}{r^2} f(r) B g(\xi), \qquad (11.1.61)$$

where

$$B \equiv \frac{1}{\sqrt{\det g}} \partial_i \Big(\sqrt{\det g}\; g^{ij} \partial_j \Big) \qquad (11.1.62)$$

is the Laplace–Beltrami operator on $\mathcal{L}^2(S^{n-1})$, written therefore in the spherical coordinates appropriate for this geometry with metric g. The operator B turns out to be essentially self-adjoint and negative-definite on $C^{\infty}(S^{n-1})$, it only has a point spectrum of finite multiplicity, and the corresponding eigenfunctions are of class C^{∞}. On denoting by K_l the eigenspace corresponding to the lth eigenvalue κ_l, one finds the direct-sum decomposition (Reed and Simon 1975)

$$\mathcal{L}^2(\mathbf{R}_+, r^{n-1}\,dr) \otimes \mathcal{L}^2(S^{n-1}, d\Omega) = \oplus_{l=0}^{\infty} \mathcal{L}_l, \tag{11.1.63}$$

where

$$\mathcal{L}_l \equiv \mathcal{L}^2(\mathbf{R}_+, r^{n-1}\,dr) \otimes K_l. \tag{11.1.64}$$

The result (11.1.63) has an analogue for functions over more general Lie groups (Peter and Weyl 1927). In particular, when $n = 3$, the group $SO(3)$ of rotations, acting on $\mathcal{L}^2(S^2, d\Omega)$, induces a decomposition of $\mathcal{L}^2(S^2, d\Omega)$ into a direct sum $\oplus_{l=0}^{\infty} \mathcal{H}_l$, where $\mathcal{H}_l$ is the $(2l+1)$-dimensional sub-space spanned by the spherical harmonics of degree l. Each sub-space remains invariant under $SO(3)$, and the restriction of the action of $SO(3)$ to $\mathcal{H}_l$ is an irreducible representation. The representations are inequivalent for distinct l.

11.2 Two-dimensional harmonic oscillator

The angular momentum operators can be described in terms of a two-dimensional harmonic oscillator (for simplicity, here we are concerned with the isotropic case only). Indeed, in two dimensions, one has two sets of annihilation and creation operators, which obey the commutation relations

$$\left[a_1, a_1\right] = \left[a_1^{\dagger}, a_1^{\dagger}\right] = 0, \quad \left[a_1, a_1^{\dagger}\right] = \mathbb{1}, \tag{11.2.1}$$

$$\left[a_2, a_2\right] = \left[a_2^{\dagger}, a_2^{\dagger}\right] = 0, \quad \left[a_2, a_2^{\dagger}\right] = \mathbb{1}. \tag{11.2.2}$$

One can thus define the operators (hereafter we set $\hbar = 1$)

$$J_+ \equiv a_1^{\dagger} a_2, \tag{11.2.3}$$

$$J_- \equiv a_2^{\dagger} a_1, \tag{11.2.4}$$

$$J_3 \equiv \frac{1}{2}\Big(J_+ J_- - J_- J_+\Big) = \frac{1}{2}\Big(a_1^{\dagger} a_1 - a_2^{\dagger} a_2\Big), \tag{11.2.5}$$

$$N \equiv a_1^{\dagger} a_1 + a_2^{\dagger} a_2. \tag{11.2.6}$$

Their action on a multiparticle state (cf. Eq. (10.1.63)) is given by

$$N|n_1, n_2\rangle = (n_1 + n_2)|n_1, n_2\rangle, \tag{11.2.7}$$

$$J_+|n_1, n_2\rangle = \sqrt{(n_1 + 1)n_2}|n_1 + 1, n_2 - 1\rangle, \tag{11.2.8}$$

$$J_-|n_1,n_2\rangle = \sqrt{n_1(n_2+1)}|n_1-1,n_2+1\rangle, \quad (11.2.9)$$

$$J_+J_-|n_1,n_2\rangle = \sqrt{n_1(n_2+1)}J_+|n_1-1,n_2+1\rangle \\ = n_1(n_2+1)|n_1,n_2\rangle, \quad (11.2.10)$$

$$J_-J_+|n_1,n_2\rangle = \sqrt{(n_1+1)n_2}J_-|n_1+1,n_2-1\rangle \\ = (n_1+1)n_2|n_1,n_2\rangle, \quad (11.2.11)$$

$$J_3|n_1,n_2\rangle = \frac{1}{2}\Big[n_1(n_2+1) - n_2(n_1+1)\Big]|n_1,n_2\rangle \\ = \frac{1}{2}\Big(n_1-n_2\Big)|n_1,n_2\rangle. \quad (11.2.12)$$

Thus, after defining

$$j \equiv \frac{1}{2}\Big(n_1+n_2\Big), \quad m \equiv \frac{1}{2}\Big(n_1-n_2\Big), \quad (11.2.13)$$

where $j(j+1)$ are the eigenvalues of J^2, the ket vectors for this two-dimensional system can be written as $|j,m\rangle$, and represent eigenstates of the J^2 and J_3 operators, according to the rules of section 11.1. They can be obtained from the ground state as follows:

$$|j,m\rangle = \frac{(a_1^\dagger)^{j+m}(a_2^\dagger)^{j-m}}{\sqrt{(j+m)!(j-m)!}}|0\rangle. \quad (11.2.14)$$

11.2.1 Introduction of different bases

One can also consider the operators

$$A_\pm \equiv \frac{1}{\sqrt{2}}\Big(a_1 \mp ia_2\Big), \quad (11.2.15)$$

$$A_\pm^\dagger \equiv \frac{1}{\sqrt{2}}\Big(a_1^\dagger \pm ia_2^\dagger\Big), \quad (11.2.16)$$

which satisfy the commutation relations

$$\Big[A_m, A_n\Big] = \Big[A_m^\dagger, A_n^\dagger\Big] = 0, \quad (11.2.17)$$

$$\Big[A_m, A_n^\dagger\Big] = \delta_{m,n}. \quad (11.2.18)$$

If we introduce 'quanta of type + or −', the operators A_+ and $A_+^\dagger$ turn out to be annihilation and creation operators of quanta of type +, while A_- and $A_-^\dagger$ are annihilation and creation operators of quanta of type −.

The introduction of the operators

$$N \equiv N_+ + N_- \tag{11.2.19}$$

and

$$L \equiv N_+ - N_-, \tag{11.2.20}$$

which form a complete set of commuting observables, makes it possible to consider a different basis of eigenstates. Let us point out that the following commutation relations hold:

$$\left[L, A_\pm^\dagger\right] = \pm A_\pm^\dagger, \tag{11.2.21}$$

$$\left[L, A_\pm\right] = \mp A_\pm. \tag{11.2.22}$$

Thus, on the eigenstates of L, the operators $A_+^\dagger$ and A_- raise by one the eigenvalues, while $A_-^\dagger$ and A_+ lower by one unit the eigenvalues. Various interpretations of this property are possible. In the theory of charged fields, where the field is described by a set of isotropic oscillators in two dimensions, N_+ is the number of particles with positive charge, and N_- is the number of particles with negative charge. The operator L represents (up to a constant) the total charge. According to this interpretation, the operator $A_+^\dagger$ creates a positive charge while A_- absorbs a negative charge, and they both increase the charge by one unit. By analogy, the operators $A_-^\dagger$ and A_+ are viewed as reducing the charge by one unit.

In the theory of lattice vibrations the displacements of the lattice are equally represented by a set of isotropic two-dimensional oscillators, and the oscillation quanta are called phonons. The representation in terms of phonons of type 1 and 2 provides a classification in stationary waves. The use of phonons of type + and − corresponds instead to progressive waves that propagate in a given direction or in the opposite direction.

Similarly, in three dimensions one can define the operators

$$A_1 \equiv \frac{1}{\sqrt{2}}\Big(a_x - {\rm i}a_y\Big), \tag{11.2.23}$$

$$A_0 \equiv a_z, \tag{11.2.24}$$

$$A_{-1} \equiv \frac{1}{\sqrt{2}}\Big(a_x + {\rm i}a_y\Big), \tag{11.2.25}$$

which, jointly with their Hermitian conjugates, obey canonical commutation relations analogous to (11.2.17) and (11.2.18), but bearing in mind that m and n range now from 1 to 3. These operators can again be interpreted as annihilation and creation operators of quanta of type $-1, 0, 1$. The corresponding number operators are

$$N_{-1} \equiv A_{-1}^{\dagger} A_{-1}, \tag{11.2.26}$$

$$N_0 \equiv A_0^{\dagger} A_0, \tag{11.2.27}$$

$$N_1 \equiv A_1^{\dagger} A_1, \tag{11.2.28}$$

which form a maximal set of commuting observables. The Hamiltonian therefore reads as

$$H = \left(N_1 + N_0 + N_{-1} + \frac{3}{2} \right) \hbar\omega, \tag{11.2.29}$$

and the full number operator is

$$N = N_1 + N_0 + N_{-1}. \tag{11.2.30}$$

To each triple of eigenvalues (n_{-1}, n_0, n_1) there corresponds an eigenvector for the three observables given by

$$|n_1, n_0, n_{-1}\rangle = (n_1! n_0! n_{-1}!)^{-1/2} \left(A_1^{\dagger} \right)^{n_1} \left(A_0^{\dagger} \right)^{n_0} \left(A_{-1}^{\dagger} \right)^{n_{-1}} |0\rangle, \tag{11.2.31}$$

and one has

$$H|n_1, n_0, n_{-1}\rangle = \left(n_1 + n_0 + n_{-1} + \frac{3}{2} \right) \hbar\omega |n_1, n_0, n_{-1}\rangle. \tag{11.2.32}$$

The eigenvectors obtained in this way are not eigenvectors of L^2, but are eigenvectors of L_z, which takes the form

$$L_z = (N_1 - N_{-1})\hbar, \tag{11.2.33}$$

and hence the eigenvalues of L_z are $m \equiv n_1 - n_{-1}$. The whole triple $(L_x, L_y, L_z) = \vec{L}$ of angular momentum operators can be obtained from the equation $\vec{L} = A^{\dagger} \vec{S} A$, where $\vec{S}$ denotes the 3×3 spin-1 matrices.

11.3 Rotations of angular momentum operators

The theoretical investigation of interferometers shows the existence of a further deep link between the formalism of annihilation and creation operators and that for angular momentum (Yurke *et al.* 1986). Here we are interested in lossless devices with two input ports and two output ports. Let a_1 and a_2 be the annihilation operators for two light beams, e.g. the two light beams entering a beamsplitter, or the two light beams leaving such a device. The corresponding creation operators are denoted by $a_1^{\dagger}$ and $a_2^{\dagger}$. All of these operators obey the commutation relations

$$\left[a_i, a_j \right] = 0 \quad \forall i, j = 1, 2, \tag{11.3.1}$$

$$\left[a_i^\dagger, a_j^\dagger\right] = 0 \quad \forall i, j = 1, 2, \tag{11.3.2}$$

$$\left[a_i, a_j^\dagger\right] = \delta_{ij} \mathbb{I}. \tag{11.3.3}$$

One can now define the operators (cf. Eqs. (11.2.5)–(11.2.7))

$$J_x \equiv \frac{1}{2}\left(a_1^\dagger a_2 + a_2^\dagger a_1\right), \tag{11.3.4}$$

$$J_y \equiv -\frac{\mathrm{i}}{2}\left(a_1^\dagger a_2 - a_2^\dagger a_1\right), \tag{11.3.5}$$

$$J_z \equiv \frac{1}{2}\left(a_1^\dagger a_1 - a_2^\dagger a_2\right). \tag{11.3.6}$$

Moreover, a 'number' operator can also be defined as in (11.2.6). One can check, by repeated application of (11.3.1)–(11.3.3), that the following formulae hold:

$$\left[J_x, J_y\right] = \mathrm{i} J_z, \quad \left[J_y, J_z\right] = \mathrm{i} J_x, \quad \left[J_z, J_x\right] = \mathrm{i} J_y, \tag{11.3.7}$$

$$J^2 \equiv J_x^2 + J_y^2 + J_z^2 = \frac{N}{2}\left(\frac{N}{2} + 1\right). \tag{11.3.8}$$

In the course of deriving these identities, besides the many cancellations that occur, one has to re-express any term such as $a_i a_i^\dagger$ as $a_i^\dagger a_i + \mathbb{I}$, using (11.3.3). One thus finds that (11.3.4)–(11.3.6) are indeed angular momentum operators.

Let a_1 and a_2 be the annihilation operators for incoming light, while b_1 and b_2 denote the annihilation operators for the outgoing light. The *scattering matrix* for our interferometric device is defined by the equation (Yurke *et al.* 1986)

$$\begin{pmatrix} b_1 \\ b_2 \end{pmatrix} = \begin{pmatrix} U_{11} & U_{12} \\ U_{21} & U_{22} \end{pmatrix} \begin{pmatrix} a_1 \\ a_2 \end{pmatrix}. \tag{11.3.9}$$

If the commutation relations (11.3.1)–(11.3.3) are required to hold for both the ingoing and outgoing annihilation and creation operators, one finds that

$$U\, U^\dagger = U^\dagger\, U = \mathbb{I}, \tag{11.3.10}$$

i.e. the scattering matrix should be a *unitary* matrix. At a deeper level, the matrix U preserves both the symplectic and the complex structure, and hence it can only be unitary.

For example, a beamsplitter for which the scattering matrix takes the form

$$U = \begin{pmatrix} \cos\frac{\alpha}{2} & -\mathrm{i}\sin\frac{\alpha}{2} \\ -\mathrm{i}\sin\frac{\alpha}{2} & \cos\frac{\alpha}{2} \end{pmatrix}, \tag{11.3.11}$$

leads to

$$\begin{pmatrix} J_x \\ J_y \\ J_z \end{pmatrix}_{\text{out}} = \begin{pmatrix} 1 & 0 & 0 \\ 0 & \cos\alpha & -\sin\alpha \\ 0 & \sin\alpha & \cos\alpha \end{pmatrix} \begin{pmatrix} J_x \\ J_y \\ J_z \end{pmatrix}_{\text{in}}, \tag{11.3.12}$$

by virtue of Eqs. (11.3.4)–(11.3.6), (11.3.9) and (11.3.11). Equation (11.3.12) shows that the abstract angular momentum operators are rotated by an angle α about the x-axis. This result can be re-expressed in the form

$$\begin{pmatrix} J_x \\ J_y \\ J_z \end{pmatrix}_{\text{out}} = \mathrm{e}^{\mathrm{i}\alpha J_x} \begin{pmatrix} J_x \\ J_y \\ J_z \end{pmatrix}_{\text{in}} \mathrm{e}^{-\mathrm{i}\alpha J_x}. \tag{11.3.13}$$

This corresponds to a Heisenberg-like picture. However, one can also work in a Schrödinger-like picture, where J_x, J_y and J_z remain fixed, while the state vector, after interacting with the beamsplitter, is turned into

$$|\text{out}\rangle = \mathrm{e}^{-\mathrm{i}\alpha J_x} |\text{in}\rangle. \tag{11.3.14}$$

With our notation, $|\text{in}\rangle$ is the state vector for light before it interacts with the beamsplitter.

Another relevant example of scattering matrix is given by

$$U = \begin{pmatrix} \cos\frac{\beta}{2} & -\sin\frac{\beta}{2} \\ \sin\frac{\beta}{2} & \cos\frac{\beta}{2} \end{pmatrix}, \tag{11.3.15}$$

which leads to a change of the angular momentum operators according to

$$\begin{pmatrix} J_x \\ J_y \\ J_z \end{pmatrix}_{\text{out}} = \begin{pmatrix} \cos\beta & 0 & \sin\beta \\ 0 & 1 & 0 \\ -\sin\beta & 0 & \cos\beta \end{pmatrix} \begin{pmatrix} J_x \\ J_y \\ J_z \end{pmatrix}_{\text{in}}. \tag{11.3.16}$$

This means that a rotation by an angle β about the y-axis is performed:

$$\begin{pmatrix} J_x \\ J_y \\ J_z \end{pmatrix}_{\text{out}} = \mathrm{e}^{\mathrm{i}\beta J_y} \begin{pmatrix} J_x \\ J_y \\ J_z \end{pmatrix}_{\text{in}} \mathrm{e}^{-\mathrm{i}\beta J_y}. \tag{11.3.17}$$

In the corresponding Schrödinger picture, the state vector for light is turned into

$$|\text{out}\rangle = \mathrm{e}^{-\mathrm{i}\beta J_y} |\text{in}\rangle. \tag{11.3.18}$$

Lastly, if the light beams incur the phase shifts γ_1 and γ_2 respectively, the scattering matrix reads

$$U = \mathrm{e}^{\mathrm{i}\frac{\gamma_1+\gamma_2}{2}} \begin{pmatrix} \mathrm{e}^{\mathrm{i}\frac{\gamma_1-\gamma_2}{2}} & 0 \\ 0 & \mathrm{e}^{-\mathrm{i}\frac{\gamma_1-\gamma_2}{2}} \end{pmatrix}, \tag{11.3.19}$$

so that a rotation by an angle $\gamma_2 - \gamma_1$ about the z-axis is performed:

$$\begin{pmatrix} J_x \\ J_y \\ J_z \end{pmatrix}_{\text{out}} = e^{i(\gamma_2-\gamma_1)J_z} \begin{pmatrix} J_x \\ J_y \\ J_z \end{pmatrix}_{\text{in}} e^{-i(\gamma_2-\gamma_1)J_z}. \tag{11.3.20}$$

In the Schrödinger picture, one writes

$$|\text{out}\rangle = e^{i\frac{(\gamma_1+\gamma_2)N}{2}} e^{-i(\gamma_2-\gamma_1)J_z}|\text{in}\rangle. \tag{11.3.21}$$

These fairly simple equations express a deep property: a lossless device with two input ports and two output ports may be viewed as the process of measuring rotations of angular momentum operators. These (abstract) operators are defined in terms of annihilation and creation operators for incoming and outgoing light beams.

11.4 Clebsch–Gordan coefficients and the Regge map

Recall, from section 6.2, that the mathematical problem of adding angular momentum operators finds its motivation in the physical problem of studying particles possessing both orbital and spin angular momentum. Moreover, multiparticle states involving two spins and one relative angular momentum may also be relevant. Here we are interested in a more advanced treatment. Of course, we start with two angular momentum operators, J_A and J_B,

$$\left[J_{A,k}, J_{A,l}\right] = i\hbar\varepsilon_{klp}J_{A,p}, \tag{11.4.1}$$

$$\left[J_{B,k}, J_{B,l}\right] = i\hbar\varepsilon_{klp}J_{B,p}, \tag{11.4.2}$$

which commute:

$$\left[J_{A,k}, J_{B,l}\right] = 0, \quad \forall k, l = 1, 2, 3, \tag{11.4.3}$$

so that the total angular momentum

$$J \equiv J_A + J_B \tag{11.4.4}$$

generates a representation of rotations which is the direct product of the representations generated by J_A and J_B:

$$U(R) = U_A(R)U_B(R). \tag{11.4.5}$$

By hypothesis, $U_A(R)$ and $U_B(R)$ define two irreducible representations $D^{(j_1)}$ and $D^{(j_2)}$:

$$U_A(R) = D^{(j_1)}(R) \otimes \mathbb{I}, \tag{11.4.6}$$

$$U_B(R) = \mathbb{I} \otimes D^{(j_2)}(R). \tag{11.4.7}$$

The representation

$$R \to U(R) = D^{(j_1)}(R) \otimes D^{(j_2)}(R)$$

is *reducible*, and the problem arises *to decompose it into irreducible representations.* The corresponding complete set of commuting observables is $\{J_A^2, J_B^2, J^2, J_z\}$. The operators $U(R)$ act on the $(2j_1+1)(2j_2+1)$-dimensional space, linearly generated by the simultaneous eigenvectors of $J_A^2, J_B^2, J_{A,z}, J_{B,z}$, i.e.

$$|j_1, m_1\rangle \otimes |j_2, m_2\rangle.$$

The operator $J_z \equiv J_{A,z} + J_{B,z}$ is diagonal in this basis, with eigenvalues $m = m_1 + m_2$. Each eigenvalue m has its own degeneracy. Only the eigenvector $|j_1, j_1; j_2, j_2\rangle$ has the eigenvalue $m = j_1 + j_2$, whereas there exist two eigenvectors with eigenvalue $m = j_1 + j_2 - 1$, three eigenvectors with eigenvalue $m = j_1 + j_2 - 2, \ldots,$ up to $m = |j_1 - j_2|$, for which there are two $\min(j_1, j_2)$ eigenvectors. This is the maximal degeneracy, which remains constant until m reaches the negative value $-|j_1 - j_2|$. For lower values of m, the degeneracy starts decreasing until m reaches the lowest value, $-j_1 - j_2$, for which only one eigenvector exists: $|j_1, -j_1; j_2, -j_2\rangle$.

Our task is now to develop a technique to express the common eigenvectors of J_A^2, J_B^2, J^2, J_z in terms of the common eigenvectors of $J_A^2, J_{A,z}, J_B^2$ and $J_{B,z}$. For this purpose, we denote the former by

$$|j_1, j_2; j, m\rangle,$$

and the latter by

$$|j_1, m_1; j_2, m_2\rangle = |j_1, m_1\rangle \otimes |j_2, m_2\rangle = |j_1, m_1\rangle \; |j_2, m_2\rangle,$$

and we point out that these two sets of eigenvectors are related by an orthogonal transformation (Srinivasa Rao and Rajeswari 1993),

$$|j_1, j_2; j, m\rangle = \sum_{m_1, m_2} \langle j_1, m_1; j_2, m_2 | j, m\rangle \; |j_1, m_1; j_2, m_2\rangle, \qquad (11.4.8)$$

where the coefficients

$$\langle j_1, m_1; j_2, m_2 | j, m\rangle \equiv C_{j,m}^{j_1, m_1, j_2, m_2}$$

are the Clebsch–Gordan coefficients (after the work of Clebsch (1872) and Gordan (1875) on the invariant theory of algebraic forms). By construction, they are non-vanishing if and only if

$$m_1 + m_2 = m,$$

$$j \in \{j_1 + j_2, \ldots, |j_1 - j_2|\}.$$

A set of recurrence relations for the evaluation of all Clebsch–Gordan coefficients is obtained on applying the raising and lowering operators to Eq. (11.4.8). This yields

$$\begin{aligned} J_\pm |j_1, j_2; j, m\rangle &= \sqrt{(j \mp m)(j \pm m + 1)} |j_1, j_2; j, m \pm 1\rangle \\ &= \sum_{m_1, m_2} \Big(\sqrt{(j_1 \mp m_1)(j_1 \pm m_1 + 1)}\, |j_1, m_1 \pm 1; j_2, m_2\rangle \\ &\quad + \sqrt{(j_2 \mp m_2)(j_2 \pm m_2 + 1)}\, |j_1, m_1; j_2, m_2 \pm 1\rangle \Big) C_{j,m}^{j_1, m_1, j_2, m_2}. \end{aligned} \tag{11.4.9}$$

At this stage, it is convenient to set

$$m_1' \equiv m_1 \pm 1, \quad m_2' \equiv m_2 \pm 1. \tag{11.4.10}$$

On the right-hand side of Eq. (11.4.9) one can then express, for $k = 1, 2$,

$$j_k \mp m_k = j_k \mp (m_k' \mp 1) = j_k \mp m_k' + 1, \tag{11.4.11}$$

$$j_k \pm m_k = j_k \pm (m_k' \mp 1) = j_k \pm m_k' - 1. \tag{11.4.12}$$

Thus, bearing in mind that summation over all values of m_1' and m_2' makes it possible to re-label them, one finds from (11.4.8) and (11.4.9) the very useful formula

$$\begin{aligned} &\sum_{m_1, m_2} \sqrt{(j \mp m)(j \pm m + 1)} C_{j, m \pm 1}^{j_1, m_1, j_2, m_2} |j_1, m_1; j_2, m_2\rangle \\ &= \sum_{m_1, m_2} \sqrt{(j_1 \mp m_1 + 1)(j_1 \pm m_1)} C_{j,m}^{j_1, m_1 \mp 1, j_2, m_2} |j_1, m_1; j_2, m_2\rangle \\ &\quad + \sum_{m_1, m_2} \sqrt{(j_2 \mp m_2 + 1)(j_2 \pm m_2)} C_{j,m}^{j_1, m_1, j_2, m_2 \mp 1} |j_1, m_1; j_2, m_2\rangle. \end{aligned} \tag{11.4.13}$$

This leads in turn to the desired recursive algorithm:

$$\begin{aligned} &\sqrt{(j \mp m)(j \pm m + 1)} C_{j, m \pm 1}^{j_1, m_1, j_2, m_2} \\ &= \sqrt{(j_1 \mp m_1 + 1)(j_1 \pm m_1)} C_{j,m}^{j_1, m_1 \mp 1, j_2, m_2} \\ &\quad + \sqrt{(j_2 \mp m_2 + 1)(j_2 \pm m_2)} C_{j,m}^{j_1, m_1, j_2, m_2 \mp 1}. \end{aligned} \tag{11.4.14}$$

Interestingly, the orthogonal transformation (11.4.8) can be inverted, in the form

$$|j_1, m_1; j_2, m_2\rangle = \sum_{j,m} C_{j,m}^{j_1, m_1, j_2, m_2} |j_1, j_2; j, m\rangle, \tag{11.4.15}$$

by virtue of the orthogonality properties satisfied by the Clebsch–Gordan coefficients (Srinivasa Rao and Rajeswari 1993):

$$\sum_{m_1,m_2} C^{j_1,m_1,j_2,m_2}_{j,m} \; C^{j_1,m_1,j_2,m_2}_{j',m'} = \delta_{j,j'} \; \delta_{m,m'} \tag{11.4.16}$$

and

$$\sum_{j,m} C^{j_1,m_1,j_2,m_2}_{j,m} \; C^{j_1,m'_1,j_2,m'_2}_{j,m} = \delta_{m_1,m'_1} \; \delta_{m_2,m'_2}. \tag{11.4.17}$$

In 1940, Wigner defined the 3-j symbol as

$$\begin{pmatrix} j_1 & j_2 & j_3 \\ m_1 & m_2 & m_3 \end{pmatrix} \equiv \frac{(-1)^{j_1-j_2-m_3}}{[j_3]} C^{j_1,m_1,j_2,m_2}_{j_3,-m_3}, \tag{11.4.18}$$

where

$$[j_3] \equiv \sqrt{2j_3+1}, \tag{11.4.19}$$

and the m_k quantum numbers in the 3-j coefficients satisfy the condition

$$m_1 + m_2 + m_3 = 0. \tag{11.4.20}$$

In the current literature, the 3-j coefficient is defined as

$$\begin{aligned} \begin{pmatrix} j_1 & j_2 & j_3 \\ m_1 & m_2 & m_3 \end{pmatrix} &\equiv \delta_{m_1+m_2+m_3,0}(-1)^{j_1-j_2-m_3} f(j_1, j_2, j_3) \\ &\times \sum_t (-1)^t \left[t! \prod_{k=1}^{2} (t-\alpha_k)! \prod_{l=1}^{3} (\beta_l - t)! \right]^{-1} \\ &\times \prod_{i=1}^{3} \sqrt{(j_i+m_i)!(j_i-m_i)!}, \end{aligned} \tag{11.4.21}$$

where $t \in \left[t_{\min}, t_{\max}\right]$, and

$$t_{\min} \equiv \max(0, \alpha_1, \alpha_2), \; t_{\max} \equiv \min(\beta_1, \beta_2, \beta_3), \tag{11.4.22}$$

$$\alpha_1 \equiv j_1 - j_3 + m_2 = (j_1 - m_1) - (j_3 + m_3), \tag{11.4.23}$$

$$\alpha_2 \equiv j_2 - j_3 - m_1 = (j_2 + m_2) - (j_3 - m_3), \tag{11.4.24}$$

$$\beta_1 \equiv j_1 - m_1, \; \beta_2 \equiv j_2 + m_2, \; \beta_3 \equiv j_1 + j_2 - j_3, \tag{11.4.25}$$

$$f(x,y,z) \equiv \sqrt{\frac{(-x+y+z)!(x-y+z)!(x+y-z)!}{(x+y+z+1)!}}. \tag{11.4.26}$$

The function f vanishes unless

$$|j_1 - j_2| \leq j \leq j_1 + j_2.$$

Further details can be found in Biedenharn and Louck (1981).

Major progress in the understanding of the symmetries of the 3-j coefficients began when Regge (1958) arranged the nine non-negative integer parameters:

$$-j_1 + j_2 + j_3, j_1 - j_2 + j_3, j_1 + j_2 - j_3,$$

$$j_1 - m_1, j_2 - m_2, j_3 - m_3, j_1 + m_1, j_2 + m_2, j_3 + m_3,$$

into a 3×3 square symbol and represented the 3-j coefficient as

$$\begin{pmatrix} j_1 & j_2 & j_3 \\ m_1 & m_2 & m_3 \end{pmatrix} \rightarrow (a_{ik}) \quad i, k = 1, 2, 3, \tag{11.4.27a}$$

where

$$(a_{ik}) \equiv \begin{pmatrix} -j_1 + j_2 + j_3 & j_1 - j_2 + j_3 & j_1 + j_2 - j_3 \\ j_1 - m_1 & j_2 - m_2 & j_3 - m_3 \\ j_1 + m_1 & j_2 + m_2 & j_3 + m_3 \end{pmatrix}, \tag{11.4.27b}$$

and noted that

$$\sum_{k=1}^{3} a_{ik} = j_1 + j_2 + j_3 \quad \forall i = 1, 2, 3, \tag{11.4.28}$$

$$\sum_{i=1}^{3} a_{il} = j_1 + j_2 + j_3 \quad \forall l = 1, 2, 3. \tag{11.4.29}$$

Regge concluded that the 3-j coefficient has 72 symmetries, being invariant under 3! column permutations, 3! row permutations and a reflection about the diagonal of the 3×3 square symbol.

11.5 Postulates of quantum mechanics with spin

We can now give a thorough formulation of quantum mechanics when the effects of spin are included. To begin, note that the Pauli matrices (5.7.10)–(5.7.12) provide a particular example of a Clifford algebra. Their basic properties are as follows (hereafter we replace x, y, z by $1, 2, 3$ respectively).

(i) They form an algebra.

(ii) They obey the property

$$\sigma_j \, \sigma_k = \delta_{jk} \mathbb{I} + \mathrm{i} \varepsilon_{jkl} \, \sigma_l, \tag{11.5.1}$$

which implies

$$\sigma_j \, \sigma_k - \sigma_k \, \sigma_j = 2\mathrm{i} \varepsilon_{jkl} \, \sigma_l, \tag{11.5.2}$$

$$\sigma_j \, \sigma_k + \sigma_k \, \sigma_j = 2\delta_{jk} \mathbb{I}. \tag{11.5.3}$$

(iii) The triple $\left(\sigma_1, \sigma_2, \sigma_3\right) \equiv \vec{\sigma}$ is an operator with vector-valued expectation values.

(iv) One finds by direct calculation

$$\left(\vec{a} \cdot \vec{\sigma}\right)\left(\vec{b} \cdot \vec{\sigma}\right) = \left(\vec{a} \cdot \vec{b}\right)\mathbb{I} + \mathrm{i}\left(\vec{a} \wedge \vec{b}\right) \cdot \vec{\sigma}. \tag{11.5.4}$$

Thus, if one considers the rotation through an angle ω about a generic axis $\vec{n}$, which is image of

$$U_{\vec{n}}(\omega/2) \equiv \mathrm{e}^{-\mathrm{i}(\vec{\sigma}\cdot\vec{n})\omega/2}, \tag{11.5.5}$$

one finds

$$U_{\vec{n}}(\omega/2) = \cos\frac{\omega}{2}\ \mathbb{I} - \mathrm{i}\sin\frac{\omega}{2}\left(\vec{\sigma} \cdot \vec{n}\right). \tag{11.5.6}$$

Since double-valued representations of the rotation group exist, the next step is to consider the class of all *projective transformations* (i.e. up to a phase):

$$\mathcal{D}(R)\mathcal{D}(S) = \omega(R,S)\mathcal{D}(RS), \tag{11.5.7}$$

where

$$|\omega(R,S)| = 1. \tag{11.5.8}$$

Denoting by S_1, S_2 and S_3 the *generators* of $R \rightarrow \mathcal{D}(R)$, one can write

$$\mathcal{D}(n,\alpha) \equiv \mathcal{D}(R(n,\alpha)) = \mathrm{e}^{-\mathrm{i}\alpha\vec{n}\cdot\vec{S}}. \tag{11.5.9}$$

By virtue of (11.5.7), one recovers from (11.5.9) the commutation rules for angular momentum operators, i.e.

$$\left[S_j, S_k\right] = \mathrm{i}\hbar\varepsilon_{jkl}S_l. \tag{11.5.10}$$

We are thus led to say that the spin of a (quantum) particle behaves like an angular momentum. Note that spin *does not result from the translational motion of the particle*, and its magnitude can only take a fixed value. The basic postulates of a formalism which incorporates spin are thus as follows.

(a) Besides the orbital angular momentum $\vec{L}$, a spin angular momentum exists with components S_1, S_2, S_3. Such operators commute with position and momentum.

(b) The operator

$$\vec{J} \equiv \vec{L} + \vec{S} \tag{11.5.11}$$

plays the role of *total angular momentum*, and hence, by definition, commutes with all the operators invariant under rotations. Within this framework, wave functions are complex-valued maps

$$\psi : \mathbf{R}^3 \times \mathbf{C}^{2s+1} \rightarrow \mathbf{C}$$

which are linear in the second argument. Therefore we may consider maps

$$\psi : \mathbf{R}^3 \rightarrow \mathbf{C}^{2s+1},$$

where $\mathbf{C}^{2s+1}$ is identified with its dual vector space. Moreover, the operators become matrix-valued:

$$\hat{x}_k \, \psi = \begin{pmatrix} x_k & \ldots & 0 \\ \ldots & \ldots & \ldots \\ 0 & \ldots & x_k \end{pmatrix} \begin{pmatrix} \psi_1 \\ \ldots \\ \psi_{2s+1} \end{pmatrix}, \tag{11.5.12}$$

$$\hat{p}_k \, \psi = \begin{pmatrix} \frac{\hbar}{\mathrm{i}} \frac{\partial}{\partial x_k} & \ldots & 0 \\ \ldots & \ldots & \ldots \\ 0 & \ldots & \frac{\hbar}{\mathrm{i}} \frac{\partial}{\partial x_k} \end{pmatrix} \begin{pmatrix} \psi_1 \\ \ldots \\ \psi_{2s+1} \end{pmatrix}, \tag{11.5.13}$$

$$\hat{S}_k \, \psi = \begin{pmatrix} (S_k)_{1,1} & (S_k)_{1,2} & \ldots & (S_k)_{1,2s+1} \\ \ldots & \ldots & \ldots & \ldots \\ (S_k)_{2s+1,1} & (S_k)_{2s+1,2} & \ldots & (S_k)_{2s+1,2s+1} \end{pmatrix} \begin{pmatrix} \psi_1 \\ \ldots \\ \psi_{2s+1} \end{pmatrix}. \tag{11.5.14}$$

In other words, the wave function ψ is a column vector of functions belonging to $\mathcal{L}^2(\mathbf{R}^3)$, since the Hilbert space of quantum mechanics incorporating spin is isomorphic to the direct product of $\mathcal{L}^2(\mathbf{R}^3)$ with a finite-dimensional Hilbert space. If ψ undergoes a linear transformation under a rotation R, i.e.

$$\psi'(x') = D(R)\psi(x) \; x' = Rx, \tag{11.5.15}$$

then, on setting $\psi' = U(R)\psi$, one obtains the overall transformation law

$$(U(R)\psi)(x) = D(R)\psi(R^{-1}x), \tag{11.5.16}$$

and the composition of two rotations yields

$$(U(R_1)U(R_2)\psi)(x) = (D(R_1)D(R_2)\psi)((R_1R_2)^{-1}x). \tag{11.5.17}$$

Nothing prevents us from fulfilling Eq. (11.5.17) with the help of projective transformations defined by (11.5.7) and (11.5.8). For the group $SO(3)$, the analysis of section 5.7 shows that the phase factor $\sigma(R_1, R_2) = \pm 1$. The overall representation $R \rightarrow U(R)$ obeys the composition law

$$U(R_1R_2) = \sigma(R_1, R_2)U(R_1)U(R_2), \tag{11.5.18}$$

with exponentiated form

$$U(n,\alpha) = \mathrm{e}^{-\mathrm{i}\alpha\vec{n}\cdot\vec{J}}, \tag{11.5.19}$$

where $\vec{J}$ is given by (11.5.11). Once again, we stress that S_1, S_2, S_3 are the generators of the projective representation $R \to D(R)$, which maps the rotation R into the matrix

$$D(n,\vec{\alpha}) \equiv D(R(n,\alpha)) = \mathrm{e}^{-\mathrm{i}\alpha\vec{n}\cdot\vec{S}}. \tag{11.5.20}$$

11.6 Spin and Weyl systems

Recall from section 9.5 that a Weyl system is a continuous unitary map of a symplectic space into the set of unitary operators on a Hilbert space such that Eq. (9.5.11) holds, with W a strongly continuous function of z. It is clear that there we had a projective unitary representation of the Abelian group $S \equiv \mathbf{R}^{2n}$ with multiplier σ associated with the symplectic structure on S:

$$\sigma(x,y) = \mathrm{e}^{\frac{\mathrm{i}}{\hbar}\omega(x,y)}. \tag{11.6.1}$$

For the case of quantum mechanics with spin we consider the semi-direct group structure $SU(2) \otimes_\rho S$ where ρ is the action of $SU(2)$ on S preserving the symplectic structure. Moreover, the multiplier is provided by the one of S plus the $\sigma(R_1, R_2)$ of the previous section. The Hilbert space $\mathcal{H}$ carries a projective representation of the group $SU(2) \otimes_\rho S$. A specific realization is obtained by selecting any Lagrangian sub-space left invariant by the $SU(2)$-action and building on it square-integrable functions with values in $\mathbf{C}^{2s+1}$. A particular realization of $\mathcal{H}$ is therefore provided by

$$\mathcal{H} \equiv \mathcal{L}^2(\mathbf{R}^3) \otimes \mathbf{C}^{2s+1}. \tag{11.6.2}$$

For one-parameter groups one can always write

$$\sigma(R_1, R_2) = \mathrm{e}^{\mathrm{i}\Omega(R_1(s), R_2(t))}, \tag{11.6.3}$$

where $R_1(s)$ and $R_2(t)$ are such that

$$U(R_1(s)) = \mathrm{e}^{-\mathrm{i}sJ_1},\ U(R_2(t)) = \mathrm{e}^{-\mathrm{i}tJ_2}. \tag{11.6.4}$$

On writing the generators in the form

$$J_1 = \lambda_1^{\ k}\, j_k, \quad J_2 = \lambda_2^{\ l}\, j_l, \tag{11.6.5}$$

one obtains

$$U(R_1(s)) = \mathrm{e}^{-\mathrm{i}s\lambda_1^{\ k}\, j_k}, \tag{11.6.6}$$

$$U(R_2(t)) = \mathrm{e}^{-\mathrm{i}t\lambda_2^{\ l}\, j_l}. \tag{11.6.7}$$

For the rotation group, however, the phase factor reduces to ± 1 as we mentioned after Eq. (11.5.17).

11.7 Monopole harmonics

Recall from appendix 6.B that, in his attempt to understand the quantization of the electric charge, Dirac was led to assume that particles have both magnetic and electric charge, with corresponding magnetic and electric currents (Dirac 1931). The resulting macroscopic Maxwell equations would therefore read as

$$\text{curl}\, \vec{E} + \frac{1}{c}\frac{\partial \vec{B}}{\partial t} = -\frac{4\pi}{c}\vec{J}_m, \tag{11.7.1}$$

$$\text{div}\, \vec{B} = 4\pi \rho_m, \tag{11.7.2}$$

$$\text{curl}\, \vec{H} - \frac{1}{c}\frac{\partial \vec{D}}{\partial t} = \frac{4\pi}{c}\vec{J}_e, \tag{11.7.3}$$

$$\text{div}\, \vec{D} = 4\pi \rho_e. \tag{11.7.4}$$

Note now that under the *duality transformations*, defined by (with $\xi \in \mathbf{R}$)

$$\begin{pmatrix} \vec{E} \\ \vec{H} \end{pmatrix} = \begin{pmatrix} \cos\xi & \sin\xi \\ -\sin\xi & \cos\xi \end{pmatrix} \begin{pmatrix} \vec{E}' \\ \vec{H}' \end{pmatrix}, \tag{11.7.5}$$

$$\begin{pmatrix} \vec{D} \\ \vec{B} \end{pmatrix} = \begin{pmatrix} \cos\xi & \sin\xi \\ -\sin\xi & \cos\xi \end{pmatrix} \begin{pmatrix} \vec{D}' \\ \vec{B}' \end{pmatrix}, \tag{11.7.6}$$

$$\begin{pmatrix} \rho_e \\ \rho_m \end{pmatrix} = \begin{pmatrix} \cos\xi & \sin\xi \\ -\sin\xi & \cos\xi \end{pmatrix} \begin{pmatrix} \rho_e' \\ \rho_m' \end{pmatrix}, \tag{11.7.7}$$

$$\begin{pmatrix} \vec{J}_e \\ \vec{J}_m \end{pmatrix} = \begin{pmatrix} \cos\xi & \sin\xi \\ -\sin\xi & \cos\xi \end{pmatrix} \begin{pmatrix} \vec{J}_e' \\ \vec{J}_m' \end{pmatrix}, \tag{11.7.8}$$

the Maxwell equations for the transformed fields $\vec{E}', \vec{D}', \vec{B}', \vec{H}'$, with charge densities ρ_e', ρ_m' and current densities $\vec{J}_e', \vec{J}_m'$, remain formally analogous to Eqs. (11.7.1)–(11.7.4). Thus, what is really crucial is that for all particles the ratio of magnetic to electric charge is the same. If this is the case, one can perform a duality transformation to obtain vanishing values for the magnetic charge density ρ_m and magnetic current density $\vec{J}_m$ (Jackson 1975).

Note also that, by virtue of Eq. (11.7.2), one can no longer describe $\vec{B}$ as the curl of a vector potential $\vec{A}$ on the whole of $\mathbf{R}^3$. The correct mathematical description is instead obtained by considering a vector $\vec{B}$ such that

$$\operatorname{div} \vec{B} = 0 \quad \text{in } \mathbf{R}^3 - \{0\}, \tag{11.7.9}$$

i.e. in a space where not all closed surfaces can be shrunk continuously to a point without passing outside. The integer that classifies solutions of the Maxwell equations in $\mathbf{R}^3 - \{0\}$ is called the magnetic charge. An example of monopole solution with charge n is given below, with the notation of differential geometry:

$$F = \mathrm{d}A = \frac{n}{2r^3}\Big(x\,\mathrm{d}y \wedge \mathrm{d}z + y\,\mathrm{d}z \wedge \mathrm{d}x + z\,\mathrm{d}x \wedge \mathrm{d}y\Big), \tag{11.7.10}$$

$$\vec{B} = \frac{n}{2r^3}\vec{r}, \tag{11.7.11}$$

where $r \equiv \sqrt{x^2 + y^2 + z^2}$. Dirac pointed out that if magnetic monopoles exist, their charge is inversely proportional to the electric charge. It follows that, if monopoles exist, we can understand charge quantization!

The occurrence of monopoles, well defined mathematically, but for which the experimental evidence is still lacking, makes it possible to introduce the corresponding monopole harmonics. They are the next logical step after the spherical harmonics of section 5.6, and hence we focus on them hereafter. For this purpose, we study the wave function of a charged particle in the field of a magnetic monopole. Our presentation follows the work of Wu and Yang (1976), whereas, for further developments, the reader is referred to the work by Kazama *et al.* (1977), Dray (1985) and Weinberg (1994).

First it should be remarked that, since the space around a monopole is spherically symmetric and singularity-free, the wave function of a charged particle should also be singularity-free. On the other hand, cusps and discontinuities are found to occur because any choice of the vector potential $\vec{A}$ around the monopole must have singularities. One can overcome this difficulty after a careful consideration of a simpler problem, i.e. the choice of a coordinate system on the surface of a sphere. It is indeed well known that all possible choices have some singularities, whereas the geometry of the sphere is, clearly, singularity-free. To avoid introducing (fictitious) singularities in the coordinate system one can divide the sphere into more than one overlapping region, defining a singularity-free coordinate system in each region. Moreover, in the overlap one has a singularity-free coordinate transformation between the different coordinate
systems.

Similarly, one can divide the space outside a magnetic monopole into two regions R_1 and R_2, and define a vector potential $A_k^{(1)}$ in R_1 and a vector potential $A_k^{(2)}$ in R_2. On using spherical coordinates r, θ, φ *with the monopole at the origin*, one takes

$$\theta \in [0, \frac{\pi}{2} + \delta[, \ r > 0, \ \varphi \in [0, 2\pi[\text{ in } R_1, \tag{11.7.12}$$

$$\theta \in]\frac{\pi}{2} - \delta, \pi], \ r > 0, \ \varphi \in [0, 2\pi[\text{ in } R_2, \tag{11.7.13}$$

$$\theta \in]\frac{\pi}{2} - \delta, \frac{\pi}{2} + \delta[, \ r > 0, \ \varphi \in [0, 2\pi[\text{ in } R_1 \cap R_2, \tag{11.7.14}$$

where δ is chosen to lie in the interval $]0, \frac{\pi}{2}]$. The components of the vector potential are then assumed to be (g being the strength of the monopole)

$$A_r^{(1)} = A_\theta^{(1)} = 0, \tag{11.7.15}$$

$$A_\varphi^{(1)} = \frac{g}{r \sin \theta}(1 - \cos \theta), \tag{11.7.16}$$

$$A_r^{(2)} = A_\theta^{(2)} = 0, \tag{11.7.17}$$

$$A_\varphi^{(2)} = -\frac{g}{r \sin \theta}(1 + \cos \theta), \tag{11.7.18}$$

where, as usual, $A_r \equiv \vec{A} \cdot \vec{e}_r$, $A_\theta \equiv \vec{A} \cdot \vec{e}_\theta$, $A_\varphi \equiv \vec{A} \cdot \vec{e}_\varphi$ denote the projections of $\vec{A}$ in the three orthogonal directions. Such components in the two regions are related by the transformation formula

$$A_k^{(1)} = A_k^{(2)} + \frac{\mathrm{i}}{Ze} S \frac{\partial S^{-1}}{\partial x^k}, \tag{11.7.19}$$

where Ze is the charge of the particle and $S = \mathrm{e}^{2iq\varphi}$ is called the *transition function*, with $q = Zeg$. Remarkably, S is the gauge-transformation phase factor for changing from $A_k^{(2)}$ to $A_k^{(1)}$ in the overlap $R_1 \cap R_2$:

$$\psi^{(1)} = S \psi^{(2)}, \tag{11.7.20}$$

where $\psi^{(1)}$ and $\psi^{(2)}$ are the wave functions of a particle of charge Ze in the regions R_1 and R_2, respectively. In general, a function ξ taking the values ξ_1 in R_1 and ξ_2 in R_2, respectively, and satisfying the relation

$$\xi_1 = S \xi_2 = \mathrm{e}^{2iq\varphi} \xi_2 \quad \text{in } R_1 \cap R_2, \tag{11.7.21}$$

is called a *section*. Thus, at a deeper level, the wave function of a charged particle in the field of a magnetic monopole is a section (more precisely, a section of a complex line bundle).

The potential V in the Schrödinger equation for stationary states is taken to be spherically symmetric, and singularity-free for $r > 0$. Thus, on neglecting the effects of spin, one deals with the pair of equations

$$\frac{1}{2m}\left(\vec{p} - Ze\vec{A}^{(1)}\right)\psi^{(1)} + V\psi^{(1)} = E\psi^{(1)} \text{ in } R_1, \tag{11.7.22}$$

$$\frac{1}{2m}\left(\vec{p} - Ze\vec{A}^{(2)}\right)\psi^{(2)} + V\psi^{(2)} = E\psi^{(2)} \text{ in } R_2. \tag{11.7.23}$$

Equations (11.7.22) and (11.7.23) are compatible with the transformation rule (11.7.20), by virtue of (11.7.21).

If ξ is a section, then $x\xi$ and $\left(\vec{p} - Ze\vec{A}\right)\xi$ are also a section. One can thus regard $\vec{r}$ and $\vec{p} - Ze\vec{A}$ as operators on the Hilbert space of sections, with a scalar product of sections defined by

$$(\eta, \xi) \equiv \int \eta^* \xi \, \mathrm{d}^3 r. \tag{11.7.24}$$

This is consistent with the previous rules, because

$$\eta_2^* \xi_2 = \eta_1^* \xi_1 \text{ in } R_1 \cap R_2, \tag{11.7.25}$$

by virtue of (11.7.21). Of course, both $\vec{r}$ and $\vec{p} - Ze\vec{A}$ are Hermitian operators, but the non-trivial step is to build angular momentum operators. For this purpose, following Fierz (1944), one defines (cf. section 12.5)

$$\vec{G} \equiv \vec{r} \wedge \left(\vec{p} - Ze\vec{A}\right) - q\frac{\vec{r}}{r}, \tag{11.7.26}$$

the components of which are Hermitian operators on the Hilbert space of sections and obey the commutation relations

$$\left[G_j, G_k\right] = \mathrm{i}\hbar\varepsilon_{jkl}G_l. \tag{11.7.27}$$

Thus, G_x, G_y and G_z are, indeed, the angular momentum operators for a charged particle in the field of a magnetic monopole, with the monopole at the origin of the coordinate system. It should be stressed that both the Hilbert space of sections and the angular momentum operators are singularity-free.

Since r^2 commutes with G_x, G_y and G_z, one can diagonalize r^2 and study angular momentum operators for fixed r^2. Thus, one eventually studies sections of the form $\delta(r^2 - r_0^2)\xi$, where, hereafter, ξ is a section depending only on the angular coordinates θ and φ. Bearing in mind the general theory developed in section 11.1 we study, in our case, the functions $Y_{q,l,m}$ such that

$$G^2 Y_{q,l,m} = l(l+1)Y_{q,l,m}, \tag{11.7.28}$$

$$G_z Y_{q,l,m} = m Y_{q,l,m}, \qquad (11.7.29)$$

with the understanding that one deals with functions $Y^{(1)}_{q,l,m}$ in R_1 and $Y^{(2)}_{q,l,m}$ in R_2. These functions, *together*, make up what is called a *monopole harmonic* $Y_{q,l,m}$. They are, therefore, eigensections with

$$l = |q|, |q|+1, |q|+2, \ldots \quad m = -l, -l+1, \ldots, l,$$

and subject to the normalization condition

$$\int_0^{\pi} \sin\theta \, d\theta \int_0^{2\pi} |Y_{q,l,m}|^2 \, d\varphi = 1. \qquad (11.7.30)$$

In $R_1 \cap R_2$, one has $|Y^{(1)}_{q,l,m}|^2 = |Y^{(2)}_{q,l,m}|^2$. Moreover, for a fixed value of q, different monopole harmonics are orthogonal. The phases are chosen in such a way that Eq. (11.1.43) holds, i.e.

$$(G_x + iG_y) Y_{q,l,m} = \sqrt{(l-m)(l+m+1)} \, Y_{q,l,m+1}. \qquad (11.7.31)$$

The harmonics $Y^{(1)}_{q,l,m}$ are analytic in R_1, and the harmonics $Y^{(2)}_{q,l,m}$ are analytic in R_2.

To obtain the explicit form of the monopole harmonics, one can use the definition (11.7.26) to find

$$G^2 = \left[\vec{r} \wedge (\vec{p} - Ze\vec{A})\right]^2 + q^2, \qquad (11.7.32)$$

$$m Y^{(1)}_{q,l,m} = G_z Y^{(1)}_{q,l,m} = \left(-i\frac{\partial}{\partial\varphi} - q\right) Y^{(1)}_{q,l,m} \text{ in } R_1, \qquad (11.7.33)$$

$$m Y^{(2)}_{q,l,m} = G_z Y^{(2)}_{q,l,m} = \left(-i\frac{\partial}{\partial\varphi} + q\right) Y^{(2)}_{q,l,m} \text{ in } R_2. \qquad (11.7.34)$$

This implies that

$$Y^{(1)}_{q,l,m} = \Theta^{(1)}_{q,l,m}(\theta) e^{i(m+q)\varphi} \text{ in } R_1, \qquad (11.7.35)$$

$$Y^{(2)}_{q,l,m} = \Theta^{(2)}_{q,l,m}(\theta) e^{i(m-q)\varphi} \text{ in } R_2. \qquad (11.7.36)$$

By virtue of (11.7.21), one finds $\Theta^{(1)}_{q,l,m}(\theta) = \Theta^{(2)}_{q,l,m}(\theta)$ in $R_1 \cap R_2$. They are, in fact, the same function (Wu and Yang 1976). The explicit evaluation of the operator $\left[\vec{r} \wedge (\vec{p} - Ze\vec{A})\right]^2$ acting on $Y_{q,l,m}$ yields

$$[l(l+1) - q^2]\Theta_{q,l,m} = \left[-\frac{1}{\sin\theta}\frac{\partial}{\partial\theta}\sin\theta\frac{\partial}{\partial\theta} + \frac{1}{\sin^2\theta}(m + q\cos\theta)^2\right]\Theta_{q,l,m}. \qquad (11.7.37)$$

On defining $w \equiv \cos\theta$, this leads to (cf. Eq. (5.4.27))

$$[l(l+1)-q^2]\Theta = -(1-w^2)\Theta'' + 2w\Theta' + \frac{(m+qw)^2}{(1-w^2)}\Theta, \qquad (11.7.38)$$

where $w \in [-1,1]$, and the prime denotes differentiation with respect to w.

Note now that, since $Y_{q,l,m}$ is single-valued in each region, Eq. (11.7.36) implies that $m-q$ is an integer, and hence also $l-q$ is an integer. The function $\Theta_{q,l,-l}$ can be written as

$$\Theta_{q,l,-l} = N_{q,l}\sqrt{1-w}^{\,l-q}\,\sqrt{1+w}^{\,l+q}, \qquad (11.7.39)$$

with $l-|q|$ an integer ≥ 0, and

$$N_{q,l} \equiv \sqrt{\frac{(2l+1)!}{4\pi\, 2^{2l}(l-q)!(l+q)!}}. \qquad (11.7.40)$$

This property is easily checked by insertion of (11.7.40) into Eq. (11.7.38), whereas the factor (11.7.40) is obtained from the normalization condition (11.7.30) for $Y_{q,l,-l}$. As a next step, one has to apply repeatedly the spin-raising operator $(G_x + \mathrm{i}G_y)$ onto the monopole harmonics $Y_{q,l,-l}$. Hence one finds

$$Y^{(1)}_{q,l,m} = M_{q,l,m}(1-w)^{\alpha/2}(1+w)^{\beta/2}P_n^{\alpha,\beta}(w)\mathrm{e}^{\mathrm{i}(m+q)\varphi}, \qquad (11.7.41)$$

$$Y^{(2)}_{q,l,m} = Y^{(1)}_{q,l,m}\,\mathrm{e}^{-2\mathrm{i}q\varphi}, \qquad (11.7.42)$$

where

$$\alpha \equiv -q-m, \quad \beta \equiv q-m, \quad n \equiv l+m, \qquad (11.7.43)$$

$$M_{q,l,m} \equiv 2^m\sqrt{\frac{2l+1}{4\pi}\frac{(l-m)!(l+m)!}{(l-q)!(l+q)!}}, \qquad (11.7.44)$$

and $P_n^{\alpha,\beta}(w)$ are the Jacobi polynomials:

$$P_n^{\alpha,\beta}(w) \equiv \frac{(-1)^n}{2^n n!}(1-w)^{-\alpha}(1+w)^{-\beta}\frac{\mathrm{d}^n}{\mathrm{d}w^n}\Big[(1-w)^{\alpha+n}(1+w)^{\beta+n}\Big]. \qquad (11.7.45)$$

These polynomials are defined if $n, n+\alpha, n+\beta, n+\alpha+\beta$ are all integers ≥ 0. Further technical details can be found in the appendices A–C of Wu and Yang (1976).

Remark. From an abstract viewpoint we can say, to sum up some features of our chapter, that in terms of any realization of the Heisenberg algebra we construct realizations of any sub-algebra of the symplectic algebra. At the group level, the symplectic group is the automorphism group of the Heisenberg algebra. When the above sub-algebras correspond to compact groups, it is possible to realize the corresponding sub-algebras in terms of matrices. We deal then with finite-level quantum systems, i.e. quantum mechanics with a finite-dimensional Hilbert space. The Weyl map, considered as a projective unitary representation of the 'symplectic' vector group with the rotation group can be constructed, and a generalized version of the von Neumann theorem can be given. An instance of alternative realization of the angular momentum algebra is provided by the monopole harmonics.

11.8 Problems

11.P1. In spherical coordinates, the component of the orbital angular momentum along the z-axis takes the form $\frac{\hbar}{\mathrm{i}} \frac{\partial}{\partial \varphi}$. Study its eigenfunctions with

(i) Periodic boundary conditions:

$$u(0) = u(2\pi), \tag{11.8.1}$$

(ii) Anti-periodic boundary conditions:

$$u(0) = -u(2\pi). \tag{11.8.2}$$

Find the spectrum in these two cases, and interpret the result in the light of the general theory of angular momentum operators.

11.P2. Could we have values of the z-component of angular momentum which are not integers or half-odd? What would then be the expression for $\mathrm{e}^{2\pi \mathrm{i} J_3}$?

11.P3. Consider an isotropic harmonic oscillator in two dimensions. It is therefore possible to define two pairs of annihilation and creation operators, hereafter denoted by $(a_1, a_1^\dagger)$ and $(a_2, a_2^\dagger)$. These pairs can be used, in turn, to define the operators (11.2.3), (11.2.4) and (11.2.6).

(i) On using the operators J_+ and J_-, reconstruct the angular momentum operators J_1, J_2, J_3.

(ii) Derive the action of J_+, J_- and J_3 on the multiparticle states $|n_1, n_2\rangle$.

(iii) Evaluate the commutator $[J_k, J_l]$, for all $k, l = 1, 2, 3$, and re-express $J^2 \equiv \sum_{i=1}^{3} J_i^2$ by means of the number operator N.

(iv) On denoting by $|j, m\rangle$ the eigenstates of J^2 and J_3, re-express the quantum numbers j and m in terms of n_1 and n_2.

(v) Bearing in mind that the oscillator is taken to be isotropic, compute the eigenvalues of the Hamiltonian and their degeneracy.

11.P4. Consider a device with two input ports and two output ports, which interacts with electromagnetic radiation. One thus deals with 'input' pairs of annihilation and creation operators, $\left(a_1, a_1^\dagger\right)$, $\left(a_2, a_2^\dagger\right)$, jointly with corresponding 'output' pairs of annihilation and creation operators: $\left(b_1, b_1^\dagger\right)$ and $\left(b_2, b_2^\dagger\right)$.

(i) On defining the *scattering matrix* U by means of Eq. (11.3.9), derive the properties of the matrix U by imposing the commutation relations

$$\left[a_i, a_j\right] = \left[b_i, b_j\right] = 0, \tag{11.8.3}$$

$$\left[a_i^\dagger, a_j^\dagger\right] = \left[b_i^\dagger, b_j^\dagger\right] = 0, \tag{11.8.4}$$

$$\left[a_i, a_j^\dagger\right] = \left[b_i, b_j^\dagger\right] = \delta_{ij}. \tag{11.8.5}$$

(ii) In the particular case when U takes the form (11.3.11), derive the transformation properties of the angular momentum operators J_x, J_y and J_z, obtained from $J_+ \equiv a_1^\dagger a_2$ and $J_- \equiv a_2^\dagger a_1$.

(iii) Repeat the analysis (ii) when U is given by (11.3.15).

(iv) Use the above properties to describe in what sense such devices make it possible to measure rotations of angular momentum operators.

11.P5. Given a system with two input ports and two output ports, upon which an electromagnetic radiation is falling, one can associate to it annihilation operators a_1, a_2 in the input, with corresponding creation operators $a_1^\dagger$ and $a_2^\dagger$, and annihilation operators b_1, b_2 in the output, with corresponding creation operators $b_1^\dagger$ and $b_2^\dagger$. Consider now, in $\hbar = 1$ units, the raising and lowering operators defined by

$$K_+ \equiv a_1^\dagger a_2^\dagger \equiv K_1 + \mathrm{i}K_2, \tag{11.8.6}$$

$$K_- \equiv a_1 a_2 \equiv K_1 - \mathrm{i}K_2. \tag{11.8.7}$$

On requiring that

$$\left[K_-, K_+\right] = 2K_3, \tag{11.8.8}$$

find the form of the operators K_1, K_2, K_3, of the commutators

$$\left[K_1, K_2\right], \left[K_2, K_3\right], \left[K_3, K_1\right], \left[K_3, K_+\right], \left[K_3, K_-\right],$$

$$\left[J_3, K_1\right], \left[J_3, K_2\right], \left[J_3, K_3\right],$$

and of the Casimir invariant $K^2 \equiv K_3^2 - K_1^2 - K_2^2$.

As a next step, assuming that a device exists which transforms the modes according to the equation

$$\begin{pmatrix} b_1 \\ b_2^\dagger \end{pmatrix} = \begin{pmatrix} S_{11} & S_{12} \\ S_{21} & S_{22} \end{pmatrix} \begin{pmatrix} a_1 \\ a_2^\dagger \end{pmatrix}, \tag{11.8.9}$$

find all the equations obeyed by the elements of such matrix, by imposing the commutation relations

$$\left[a_i, a_j\right] = \left[a_i^\dagger, a_j^\dagger\right] = 0, \quad \left[a_i, a_j^\dagger\right] = \delta_{ij}, \tag{11.8.10}$$

$$\left[b_i, b_j\right] = \left[b_i^\dagger, b_j^\dagger\right] = 0, \quad \left[b_i, b_j^\dagger\right] = \delta_{ij}. \tag{11.8.11}$$

Is it possible to satisfy such conditions if S takes the form

$$S = \begin{pmatrix} \cosh(\beta/2) & -\mathrm{i}\sinh(\beta/2) \\ \mathrm{i}\sinh(\beta/2) & \cosh(\beta/2) \end{pmatrix}. \tag{11.8.12}$$

In the affirmative case, find the transformation law $K_{\text{out}} = A\, K_{\text{in}}$ for the triple

$$K \equiv \begin{pmatrix} K_1 \\ K_2 \\ K_3 \end{pmatrix}$$

of Hermitian operators. Is it possible to pick out, in the matrix A, a Lorentz boost?

11.P6. Prove that

$$\left[J_x^2, J_y^2\right] = \left[J_y^2, J_z^2\right] = \left[J_z^2, J_x^2\right], \tag{11.8.13}$$

and that these commutators vanish upon the states with $j = 0, \frac{1}{2}, 1$. When $j = 1$, find the basis which achieves simultaneous diagonalization of the operators J_x^2, J_y^2, J_z^2.

11.P7. On using the representation $\mathcal{D}^{(1)}$, compute the rotation operators

$$R(\vec{n}, \theta) = e^{-i(\vec{n} \cdot \vec{J})\theta/\hbar}. \tag{11.8.14}$$

11.P8. Let $|\psi\rangle$ be a spherically symmetric state for a particle. Prove that $(x_1 + ix_2)^l |\psi\rangle$ is an eigenstate of L^2 and L_z with eigenvalues $l(l+1)$ and l, respectively, in $\hbar = 1$ units.

11.P9. Let S_1, S_2, S_3 be the spin operators for $j = 1$. Show that they satisfy the Duffin–Kemmer commutation relations:

$$S_i S_j S_k + S_k S_j S_i = (\delta_{ij} S_k + \delta_{jk} S_i)\hbar^2. \tag{11.8.15}$$

Is this relation satisfied by the angular momentum operators

$$J_1, J_2, J_3$$

for $j = \frac{3}{2}$? From the characteristic equation

$$\left(J_j^2 - \frac{9}{4}\hbar^2\right)\left(J_j^2 - \frac{1}{4}\hbar^2\right) = 0 \tag{11.8.16}$$

try to deduce a fourth degree equation for the operators J_1, J_2, J_3.

11.P10. Use the identity

$$(2j_1 + 1)(2j_2 + 1) = \sum_{j=j_{\text{min}}}^{j_1+j_2} (2j + 1) \tag{11.8.17}$$

to prove that, in the addition of two triples of angular momentum operators, $j_{\text{min}} = |j_1 - j_2|$.

11.P11. On using lowering and raising operators, compute all Clebsch–Gordan coëfficients for the addition of two triples of angular momentum operators having quantum numbers $j_1 = 1, j_2 = \frac{1}{2}$. Try to repeat the calculation for larger values of j_1 and j_2.

12

Algebraic methods for eigenvalue problems

Quasi-exactly solvable differential operators are introduced. By definition, these operators can be written as a polynomial in the basis of a finite-dimensional Lie algebra. The concepts of equivalent operators and of conjugations which preserve the spectral problem are also introduced. The algebraic analysis of the Hamiltonians occurring in the hydrogen atom is then performed. This is obtained by means of a set of transformation operators. The general formalism of transformation operators, for which the action generates a family of isospectral Hamiltonians, following Darboux, is constructed in detail in one-dimensional problems. The Riccati equation emerges naturally within this framework. After this, the formalism of operators satisfying the $su(1,1)$ algebra is used to solve the Schrödinger equation in a central potential in the s-wave case. This yields yet another derivation of the Balmer formula, and points out a deep link with the formalism of $SU(1,1)$ interferometers. Last, the Runge–Lenz vector is introduced and used to derive the $su(2) \times su(2)/Z_2$ symmetry algebra of the Hamiltonian of the hydrogen atom.

12.1 Quasi-exactly solvable operators

In previous chapters we have considered finite-dimensional Lie algebras built out of creation and annihilation operators. In mathematical language they are finite-dimensional Lie algebras in the enveloping algebra (appendix 2.C) of the Heisenberg algebra (section 9.1). One may however consider situations where the starting algebra is not the Heisenberg algebra.

The von Neumann theorem (see chapter 9) makes it possible to consider quantum systems on the Hilbert space $\mathcal{H} = \mathcal{L}^2(\mathbf{R})$, with the operators realized as differential operators on $\mathbf{R}$. Indeed, the algebra of differential operators is generated by operators of the kind

$$T \equiv f(x)\frac{\partial}{\partial x} + \eta(x), \tag{12.1.1}$$

i.e. a vector field, jointly with an operator acting as multiplication by a function η. Our algebra is clearly associative, and hence the commutator defines a Lie algebra.

Once that quantization is carried out, with the differential operators being of any order, one deals with the eigenvalue problem

$$H|\psi\rangle = \lambda|\psi\rangle. \tag{12.1.2}$$

The problem of finding eigenvalues and eigenfunctions can be solved for a very limited number of systems (i.e. the harmonic oscillator, the hydrogen atom and a few more). As shown in chapter 7, one has to resort, in general, to perturbative methods. There exist, however, some operators which lie in between: they describe the so-called *quasi-exactly solvable systems* (Turbiner 1988, Ushveridze 1994). For these particular systems, a finite part of the spectrum can be determined by means of purely algebraic methods. In general, one studies a set of first-order differential operators:

$$J^a \equiv \sum_{i=1}^{d} \xi^{ai}(x)\frac{\partial}{\partial x^i} + \eta^a(x), \tag{12.1.3}$$

where $a \in \{1, \ldots, r\}$, and the coefficients ξ^{ai}, η^a are smooth functions of x. The operators J^a make it possible to build higher-order operators:

$$-T \equiv \sum_{a,b} C_{ab} J^a J^b + \sum_a C_a J^a + C_0, \tag{12.1.4}$$

for some constants C_{ab}, C_a, C_0. The crucial requirement is that

$$J^a J^b - J^b J^a = C^{ab}{}_m \, J^m, \tag{12.1.5}$$

where $C^{ab}{}_m$ are the structure constants of a finite-dimensional Lie algebra.

A Lie algebra of first-order differential operators is said to be quasi-exactly solvable if it is endowed with a finite-dimensional representation space (or *module*), here denoted by $W \subset C^\infty$. In other words, for all $\psi \in W$, one should find

$$J^a \psi \in W. \tag{12.1.6}$$

A differential operator is then said to be quasi-exactly solvable if it can be written as a polynomial in the base elements of the original Lie algebra.

By construction, $T(W) \subset W$, and hence the operator T reduces, on W, to a numerical matrix. Of course, it is not easy to determine whether a differential operator T can be expressed as a polynomial of generators of a Lie algebra. The naturally occurring problem is therefore that of reducing T to its 'normal form'. For this purpose, one has to give a precise definition of what is meant by equivalence of two differential operators.

Two differential operators $T(x)$ and $\overline{T}(\overline{x})$ are said to be equivalent if there exists a C^{∞} change of coordinates

$$\overline{x} = \varphi(x), \tag{12.1.7}$$

and a *conjugation*

$$\overline{T}(\overline{x}) \equiv \mathrm{e}^{f(x)}\, T(x)\, \mathrm{e}^{-f(x)}, \tag{12.1.8}$$

which transform T into $\overline{T}$. Such a transformation has two key properties.

(i) It is *canonical*, in that it preserves the Lie-algebra structure. Thus, if the original Lie algebra $\mathcal{A}_L$ is finite-dimensional, the transformed Lie algebra $\overline{\mathcal{A}}_L$ should be isomorphic to $\mathcal{A}_L$:

$$\overline{\mathcal{A}}_L \cong \mathcal{A}_L. \tag{12.1.9}$$

Moreover, if W is a finite-dimensional module, then the set

$$\overline{W} \equiv \left\{ \mathrm{e}^{f(x)} \psi(x)|_{x=\varphi^{-1}(\overline{x})} \ \ \psi \in W \right\} \tag{12.1.10}$$

should be a module for $\overline{\mathcal{A}}_L$. This means that we are considering transformations which preserve the property of the Lie algebra and of the operators of being quasi-exactly solvable.

(ii) The rule (12.1.8) preserves the spectral problem associated with the differential operator T, i.e. if

$$T\psi = \lambda\psi, \tag{12.1.11}$$

one then finds

$$\overline{T}\,\overline{\psi} = \lambda\,\overline{\psi}. \tag{12.1.12}$$

Indeed, one may check that (12.1.8) and (12.1.10) lead to

$$\overline{T}\,\overline{\psi} = \mathrm{e}^{f(x)} T \mathrm{e}^{-f(x)}\ \mathrm{e}^{f(x)}\psi = \mathrm{e}^{f(x)} T\psi = \lambda \mathrm{e}^{f(x)}\psi = \lambda\overline{\psi},$$

and hence the operators T and $\overline{T}$ have the same eigenvalues, provided that the eigenfunctions are transformed according to the rule $\overline{\psi} = \mathrm{e}^{f(x)}\psi$. Note, however, that canonical transformations do not preserve the norm of wave functions.

The problem of reduction to the normal form can thus be formulated by fixing some normal forms we are interested in. For example, one may fix the Hamiltonian to be of the form

$$H \equiv -\triangle + V(x), \tag{12.1.13}$$

and try to determine the class of operators equivalent to that defined in (12.1.13). For second-order differential operators on the real line:

$$D \equiv g(x)\frac{\mathrm{d}^2}{\mathrm{d}x^2} + h(x)\frac{\mathrm{d}}{\mathrm{d}x} + K(x), \tag{12.1.14}$$

it can be proved that any operator of this form is equivalent to that occurring in (12.1.13).

In one dimension, there is an essentially unique quasi-exactly solvable Lie algebra, in terms of first-order differential operators (n being an integer):

$$\widehat{g}_n \equiv \left\{\partial_x, x\partial_x, x^2\partial_x - nx, 1\right\} \cong gl(2). \tag{12.1.15}$$

For $\widehat{g}_n$, the corresponding module $W = \mathcal{P}_n$ consists of polynomials of degree $m \leq n$.

If one takes the basis

$$J^- = J_n^- \equiv \frac{\mathrm{d}}{\mathrm{d}x}, \tag{12.1.16}$$

$$J^0 = J_n^0 \equiv x\frac{\mathrm{d}}{\mathrm{d}x} - \frac{n}{2}, \tag{12.1.17}$$

$$J^+ = J_n^+ \equiv x^2\frac{\mathrm{d}}{\mathrm{d}x} - nx, \tag{12.1.18}$$

these operators can be shown to act as the generators of the $sl(2)$ algebra.

The most general form of second-order operator obeying (12.1.4) is

$$-T \equiv P(x)\frac{\mathrm{d}^2}{\mathrm{d}x^2} + Q(x)\frac{\mathrm{d}}{\mathrm{d}x} + R(x), \tag{12.1.19}$$

where $P(x), Q(x)$ and $R(x)$ are algebraic polynomials of degree 4, 3 and 2, respectively, the explicit expression of which depends on the constants C_{ab}, C_a and C_0 (see Eq. (12.1.4)).

12.2 Transformation operators for the hydrogen atom

In the course of studying bound states for the hydrogen atom in section 5.5, we arrived at a Hamiltonian operator which can be cast in the form

$$H_l \equiv p_\rho^2 + \frac{l(l+1)}{\rho^2} - \frac{2}{\rho}. \tag{12.2.1}$$

Here, we have rescaled the independent variable by considering the parameters

$$a_0 \equiv \frac{\hbar^2}{\mu e^2}, \quad \alpha \equiv \frac{Z}{a_0}, \tag{12.2.2}$$

and hence

$$\rho \equiv \alpha r, \tag{12.2.3}$$

$$p_\rho \equiv \frac{1}{\hbar\alpha} p_r = \frac{1}{\mathrm{i}} \left(\frac{\partial}{\partial \rho} + \frac{1}{\rho} \right). \tag{12.2.4}$$

The operators ρ and p_ρ obey the commutation rule

$$\left[\rho, p_\rho\right] \equiv \rho p_\rho - p_\rho \rho = \mathrm{i}. \tag{12.2.5}$$

We now consider the first-order operator

$$d_l \equiv \mathrm{i} p_\rho + \frac{l}{\rho} - \frac{1}{l}, \tag{12.2.6}$$

and its Hermitian conjugate

$$d_l^\dagger \equiv -\mathrm{i} p_\rho + \frac{l}{\rho} - \frac{1}{l}. \tag{12.2.7}$$

One thus finds

$$d_l \, d_l^\dagger = p_\rho^2 + \frac{l^2}{\rho^2} - \frac{2}{\rho} + \frac{1}{l^2} + \mathrm{i} l \left(p_\rho \frac{1}{\rho} - \frac{1}{\rho} p_\rho \right). \tag{12.2.8}$$

By virtue of (12.2.4) one finds, for any function f,

$$p_\rho \frac{1}{\rho} f = -\frac{\mathrm{i}}{\rho} \frac{\partial f}{\partial \rho}, \tag{12.2.9}$$

$$\mathrm{i} l \left[p_\rho \frac{1}{\rho} - \frac{1}{\rho} p_\rho \right] f = -\frac{l f}{\rho^2}. \tag{12.2.10}$$

These identities, combined with (12.2.1) and (12.2.8), lead to

$$d_l \, d_l^\dagger = H_{l-1} + \frac{1}{l^2}. \tag{12.2.11}$$

Moreover, the evaluation of $d_l^\dagger d_l$ leads to a second-order operator that differs from (12.2.8) for the sign multiplying the term in square brackets (see also Eq. (12.2.10)), i.e.

$$d_l^\dagger \, d_l = H_l + \frac{1}{l^2}. \tag{12.2.12}$$

The identities (12.2.11) and (12.2.12) can be used to derive some very useful formulae relating H_l to H_{l-1}. Indeed, one finds

$$H_l d_l^\dagger = \left(d_l^\dagger d_l - \frac{1}{l^2} \right) d_l^\dagger = d_l^\dagger \left(H_{l-1} + \frac{1}{l^2} \right) - \frac{1}{l^2} d_l^\dagger = d_l^\dagger H_{l-1}, \tag{12.2.13}$$

$$H_{l-1} d_l = \left(d_l d_l^\dagger - \frac{1}{l^2} \right) d_l = d_l H_l. \tag{12.2.14}$$

This means that the operators d_l and $d_l^\dagger$ act as transformation operators (see section 12.3) between the eigenfunctions of H_l and the eigenfunctions of H_{l-1}, and vice versa.

If we now denote by $|n, l\rangle$ the eigenfunctions of H_l belonging to the (negative) eigenvalue ε_n:

$$H_l|n, l\rangle = \varepsilon_n|n, l\rangle, \tag{12.2.15}$$

we find from (12.2.14) that

$$H_{l-1}d_l|n, l\rangle = d_l H_l|n, l\rangle = \varepsilon_n d_l|n, l\rangle. \tag{12.2.16}$$

In other words, $|n, l\rangle$ is an eigenstate of H_l belonging to the eigenvalue ε_n, and $d_l|n, l\rangle$ is an eigenstate of H_{l-1} belonging to the *same* eigenvalue. Furthermore, if we write, from (12.2.15), the eigenvalue equation

$$H_{l-1}|n, l-1\rangle = \varepsilon_n|n, l-1\rangle, \tag{12.2.17}$$

we find, by comparison with (12.2.16), that d_l acts as a lowering operator, in that one can write

$$d_l|n, l\rangle = C_{nl}|n, l-1\rangle. \tag{12.2.18}$$

The constant C_{nl} in (12.2.18) is evaluated by taking the norm of the vector $d_l|n, l\rangle$:

$$\begin{aligned} 0 \leq \langle n, l|d_l^\dagger d_l|n, l\rangle &= \langle n, l|H_l + \frac{1}{l^2}|n, l\rangle \\ &= \left(\varepsilon_n + \frac{1}{l^2}\right)\langle n, l|n, l\rangle = \left(\varepsilon_n + \frac{1}{l^2}\right) = |C_{nl}|^2, \end{aligned} \tag{12.2.19}$$

which implies

$$C_{nl} = \sqrt{\varepsilon_n + \frac{1}{l^2}}\, \mathrm{e}^{\mathrm{i}\gamma} = \sqrt{-\frac{1}{n^2} + \frac{1}{l^2}}\, \mathrm{e}^{\mathrm{i}\gamma}. \tag{12.2.20}$$

Hereafter we set $\gamma = 2n\pi$, with $n = 0, 1, 2, \ldots$.

The lowering action (12.2.18) does not go on forever. Indeed, if one takes the norm of the vector

$$|V_{n,l,k}\rangle \equiv d_{l+k}^\dagger\, d_{l+k-1}^\dagger \cdots d_{l+2}^\dagger\, d_{l+1}^\dagger|n, l\rangle, \tag{12.2.21}$$

which belongs to the eigenvalue ε_n of H_{l+k}, one finds the product

$$\left[\varepsilon_n + \frac{1}{(l+k)^2}\right]\left[\varepsilon_n + \frac{1}{(l+k-1)^2}\right]\cdots\left[\varepsilon_n + \frac{1}{(l+1)^2}\right] \geq 0. \tag{12.2.22}$$

It is now clear that an integer $k \geq 1$ exists such that the inequality (12.2.22) is saturated, since otherwise one would find at some stage

$$-\frac{1}{(l+k_0)^2} < \varepsilon_n < -\frac{1}{(l+k_0+1)^2}, \tag{12.2.23}$$

which implies that the product in (12.2.22) would contain at least one negative factor. This is impossible, since the norm of the vector (12.2.21) can never become negative.

Bearing in mind (12.2.18), jointly with the orthogonality properties of the eigenvectors $|n,l\rangle$, the only non-vanishing matrix elements of the operators d_l and $d_l^\dagger$ are found to be

$$\langle n, l-1|d_l|n,l\rangle = \langle n,l|d_l^\dagger|n,l-1\rangle = \frac{\sqrt{n^2-l^2}}{nl}. \tag{12.2.24}$$

When $l = n-1$, one finds

$$d_n^\dagger|n, n-1\rangle = 0. \tag{12.2.25}$$

This is the equation which *defines* the state belonging to $\varepsilon_n = -\frac{1}{n^2}$ and to $l = l_{\max} = n-1$. Starting from the vector $|n, n-1\rangle$, one can generate all states of the form $|n, n-k\rangle$, which correspond, for a given value of n, to the *positive* values of l which remain $\leq n-1$. By virtue of (12.2.18) and (12.2.20), one finds

$$|n, l-1\rangle = \frac{nl}{\sqrt{n^2-l^2}} d_l|n,l\rangle, \tag{12.2.26}$$

which leads to

$$|n, n-2\rangle = \frac{n(n-1)}{\sqrt{2n-1}} d_{n-1}|n, n-1\rangle, \tag{12.2.27}$$

$$|n, n-3\rangle = \frac{n(n-2)}{\sqrt{4(n-1)}} d_{n-2}|n, n-2\rangle, \tag{12.2.28}$$

and hence, by induction, to the general formula

$$|n, n-k-1\rangle = \frac{n^k(n-1)\cdots(n-k)}{\sqrt{k!(2n-1)\cdots(2n-k)}} d_{n-k}\cdots d_{n-1}|n, n-1\rangle. \tag{12.2.29}$$

12.3 Darboux maps: general framework

The calculations of the previous section are a particular case of a more general problem: how to generate families of isospectral Hamiltonians? For this purpose, one may apply the Darboux method (Darboux 1882), the foundations of which (Moutard 1875) are given by a theorem which, in modern language, can be stated as follows (Luban and Pursey 1986). Let ψ be the general solution of the Schrödinger equation

$$H\psi \equiv \left[-\frac{\mathrm{d}^2}{\mathrm{d}x^2} + V(x)\right]\psi(x) = E\psi(x). \tag{12.3.1}$$

If φ is a particular solution of Eq. (12.3.1) corresponding to an energy eigenvalue $\varepsilon \neq E$, then

$$\widetilde{\psi} \equiv \frac{1}{\varphi}\left(\psi \frac{\mathrm{d}\varphi}{\mathrm{d}x} - \frac{\mathrm{d}\psi}{\mathrm{d}x}\varphi\right) \tag{12.3.2}$$

is the general solution of the Schrödinger equation

$$\widetilde{H}\,\widetilde{\psi}(x) = E\,\widetilde{\psi}(x), \tag{12.3.3}$$

where

$$\widetilde{H} \equiv -\frac{\mathrm{d}^2}{\mathrm{d}x^2} + \widetilde{V}(x), \tag{12.3.4}$$

$$\widetilde{V}(x) \equiv V(x) - 2\frac{\mathrm{d}^2}{\mathrm{d}x^2}\log\varphi(x). \tag{12.3.5}$$

In other words, if two Hamiltonian operators H_A and H_B are given, one looks for a differential operator D, such that (Levitan 1987, Carinena *et al.* 1998)

$$H_B\,\mathrm{D} = \mathrm{D}\,H_A. \tag{12.3.6}$$

This makes it possible, in turn, to relate the eigenfunctions of H_A and H_B through the action of D. Here, we focus on one-dimensional problems, with

$$H_A = H_0 \equiv -\frac{\mathrm{d}^2}{\mathrm{d}x^2} + V_0, \tag{12.3.7}$$

$$H_B = H_1 \equiv -\frac{\mathrm{d}^2}{\mathrm{d}x^2} + V_1, \tag{12.3.8}$$

$$\mathrm{D} \equiv \frac{\mathrm{d}}{\mathrm{d}x} + G, \tag{12.3.9}$$

where V_0 and V_1 are the 'potential' functions which are assumed to be given, and G is a function the form of which will be determined by imposing the condition (12.3.6). In our case, this reads as

$$\left(-\frac{\mathrm{d}^2}{\mathrm{d}x^2} + V_1\right)\left(\frac{\mathrm{d}}{\mathrm{d}x} + G\right) f = \left(\frac{\mathrm{d}}{\mathrm{d}x} + G\right)\left(-\frac{\mathrm{d}^2}{\mathrm{d}x^2} + V_0\right) f, \tag{12.3.10}$$

for all functions f which are at least of class C^3. On imposing (12.3.10), one finds exact cancellation of the terms $-\frac{\mathrm{d}^3 f}{\mathrm{d}x^3}$ and $-G\frac{\mathrm{d}^2 f}{\mathrm{d}x^2}$, since they occur on both sides with the same sign. Hence one finds, for all f,

$$\left[\Big(-2G' + V_1 - V_0\Big)\frac{\mathrm{d}}{\mathrm{d}x} + \Big(-G'' - V_0' + (V_1 - V_0)G\Big)\right] f = 0, \tag{12.3.11}$$

which implies

$$2G' = V_1 - V_0, \tag{12.3.12}$$

$$-G'' - V_0' + (V_1 - V_0)G = 0, \tag{12.3.13}$$

by virtue of the arbitrariness of f. Note that, by virtue of Eq. (12.3.12), Eq. (12.3.13) takes the form

$$-G'' + 2G'G = V_0'. \tag{12.3.14}$$

Both sides of Eq. (12.3.14) are total derivatives, i.e.

$$\frac{\mathrm{d}}{\mathrm{d}x}\left(-G' + G^2\right) = \frac{\mathrm{d}}{\mathrm{d}x}V_0, \tag{12.3.15}$$

and this leads to

$$G^2 - G' = V_0 + C, \tag{12.3.16}$$

for some constant C. Thus, the desired function G is a solution of the system (12.3.12) and (12.3.16), where Eq. (12.3.16) is known as the Riccati equation (Riccati 1724, Carinena *et al.* 2000). To solve such an equation, it is convenient to consider a function φ with the property that

$$G = -\frac{\varphi'}{\varphi}. \tag{12.3.17}$$

What happens is that non-linear differential equations are, in general, very difficult to solve, and Eq. (12.3.17) provides an example of a transformation that reduces a non-linear problem to the solution of a linear equation. Indeed, from Eqs. (12.3.16) and (12.3.17) one finds

$$G^2 - G' = \frac{\varphi''}{\varphi} = V_0 + C, \tag{12.3.18}$$

which implies

$$H_0\, \varphi = -C\, \varphi. \tag{12.3.19}$$

This is a simple but deep result: one first has to find the eigenfunctions φ of H_0. After doing this, the desired function G is obtained from (12.3.17), and hence the *transformation operator* is

$$D = \frac{\mathrm{d}}{\mathrm{d}x} - \frac{\varphi'}{\varphi}. \tag{12.3.20}$$

From Eq. (12.3.12), the potential term for the Hamiltonian H_1 is then

$$V_1 = V_0 + 2\left(\frac{\varphi'^2}{\varphi^2} - \frac{\varphi''}{\varphi}\right). \tag{12.3.21}$$

The transformations (12.3.21) are known as the Darboux maps, and agree, of course, with Eq. (12.3.5). An equivalent form of the Riccati equation is then

$$G^2 + G' = V_1 + C. \tag{12.3.22}$$

For example, if

$$H_0 = -\frac{d^2}{dx^2}, \tag{12.3.23}$$

its eigenfunctions are $\varphi_0 = \cosh kx$, from which G is obtained as

$$G = -\frac{\varphi_0'}{\varphi_0} = -k \tanh kx. \tag{12.3.24}$$

In the Hamiltonian H_1, the potential is then

$$V_1 = 2G' = -\frac{2k}{(\cosh kx)^2}, \tag{12.3.25}$$

and the eigenfunctions of H_1 are obtained by applying the transformation operator,

$$D = \frac{d}{dx} - k \tanh kx, \tag{12.3.26}$$

to the eigenfunctions of H_0.

Note also that, in terms of the formal adjoint of D:

$$\mathrm{D}^\dagger \equiv -\frac{d}{dx} + G, \tag{12.3.27}$$

one finds, in general (cf. problem 6.P4),

$$\mathrm{D}\,\mathrm{D}^\dagger = H_1 + C, \tag{12.3.28}$$

$$\mathrm{D}^\dagger\,\mathrm{D} = H_0 + C. \tag{12.3.29}$$

It should be stressed that our analysis remains incomplete. In particular, a rigorous analysis of the self-adjointness problem for our operators, for a given form of the potentials, is completely lacking. For details and relations with old results by S. Lie, we refer the reader to the work in Carinena *et al.* (2000).

12.4 $SU(1,1)$ structures in a central potential

This section studies the deep link between the Schrödinger equation in a central potential, and the formalism of $su(1,1)$ algebras and $SU(1,1)$ interferometers. For simplicity, we study the s-wave case for a particle of unit mass in a Coulomb-like potential: $V(r) = -\frac{\alpha}{r}$. The corresponding Hamiltonian reads, in $\hbar = 1$ units:

$$H = \frac{1}{2}p^2 - \frac{\alpha}{r}. \tag{12.4.1}$$

We now introduce the following operators:

$$K_1 \equiv \frac{1}{2}\Big(rp^2 - r\Big), \tag{12.4.2}$$

$$K_2 \equiv rp, \tag{12.4.3}$$

$$K_3 \equiv \frac{1}{2}\Big(rp^2 + r\Big). \tag{12.4.4}$$

Bearing in mind that $[r,p] \equiv rp - pr = \mathrm{i}$, one finds that our operators obey the commutation relations

$$\Big[K_1, K_2\Big] = -\mathrm{i}K_3, \tag{12.4.5}$$

$$\Big[K_2, K_3\Big] = \mathrm{i}K_1, \tag{12.4.6}$$

$$\Big[K_3, K_1\Big] = \mathrm{i}K_2. \tag{12.4.7}$$

For example, the explicit calculation of $\Big[K_1, K_2\Big]$ yields

$$\begin{aligned}\Big[K_1, K_2\Big] &= \frac{1}{2}\Big[rp^2rp - rp(pr + \mathrm{i})p\Big] - \frac{\mathrm{i}}{2}\Big[r(pr + \mathrm{i}) - rpr\Big] \\ &= -\frac{\mathrm{i}}{2}\Big(rp^2 + r\Big) = -\mathrm{i}K_3.\end{aligned}$$

This is the first time that we meet the algebra defined by (12.4.5)–(12.4.7), which is called the $su(1,1)$ algebra. Our aim is to prove that it provides powerful tools for an elegant derivation of the energy spectrum. For this purpose, we point out that the stationary Schrödinger equation can be written as

$$r(H - E)\psi = 0, \tag{12.4.8}$$

where, from (12.4.2) and (12.4.4), one has

$$K_3 + K_1 = rp^2, \quad K_3 - K_1 = r.$$

Hence one finds the eigenvalue equation (we are interested in bound states only)

$$\left[\Big(\frac{1}{2} - E\Big)K_3 + \Big(\frac{1}{2} + E\Big)K_1 - \alpha\right]\psi = 0. \tag{12.4.9}$$

Remarkably, the operator $r(H-E)$ turns out to be a linear combination of operators obeying the commutation rules of the $su(1,1)$ algebra. We now try to transform Eq. (12.4.9) in such a way that it leads to a new eigenvalue equation involving the K_3 operator only. As a first step, we define a transformed 'wave function' (cf. section 12.1):

$$\overline{\psi} \equiv \mathrm{e}^{-\mathrm{i}\theta K_2}\,\psi, \tag{12.4.10}$$

where θ is a real-valued function of E, the form of which will be determined later. We now re-express ψ as $\mathrm{e}^{\mathrm{i}\theta K_2}\,\overline{\psi}$, and the resulting form of Eq. (12.4.9) is multiplied on the left by the operator $\mathrm{e}^{-\mathrm{i}\theta K_2}$. This leads to

$$\left[\left(\frac{1}{2}-E\right)\mathrm{e}^{-\mathrm{i}\theta K_2}K_3\mathrm{e}^{\mathrm{i}\theta K_2}+\left(\frac{1}{2}+E\right)\mathrm{e}^{-\mathrm{i}\theta K_2}K_1\mathrm{e}^{\mathrm{i}\theta K_2}-\alpha\right]\overline{\psi}=0. \tag{12.4.11}$$

At this stage, we consider the Taylor series defining the operators $\mathrm{e}^{-\mathrm{i}\theta K_2}$ and $\mathrm{e}^{\mathrm{i}\theta K_2}$, and take into account the commutation rules (12.4.5) and (12.4.6). One thus finds (cf. section 11.3)

$$\mathrm{e}^{-\mathrm{i}\theta K_2}\,K_3\,\mathrm{e}^{\mathrm{i}\theta K_2}=(\cosh\theta)K_3+(\sinh\theta)K_1, \tag{12.4.12}$$

$$\mathrm{e}^{-\mathrm{i}\theta K_2}\,K_1\,\mathrm{e}^{\mathrm{i}\theta K_2}=(\cosh\theta)K_1+(\sinh\theta)K_3. \tag{12.4.13}$$

Equation (12.4.11) now reads (since $\frac{1}{2}-E\neq 0$)

$$\left[\Big(\cosh\theta+\beta\sinh\theta\Big)K_3+\Big(\sinh\theta+\beta\cosh\theta\Big)K_1 -\frac{\alpha}{\left(\frac{1}{2}-E\right)}\right]\overline{\psi}=0, \tag{12.4.14}$$

where

$$\beta\equiv\frac{\left(\frac{1}{2}+E\right)}{\left(\frac{1}{2}-E\right)}. \tag{12.4.15}$$

On imposing that the function multiplying K_1 should vanish, one finds

$$\beta=-\tanh\theta, \tag{12.4.16}$$

which implies

$$\left[K_3-\frac{\alpha\cosh\theta}{\left(\frac{1}{2}-E\right)}\right]\overline{\psi}=0. \tag{12.4.17}$$

Bearing in mind the identity $(\cosh\theta)^2-(\sinh\theta)^2=1$, one can re-express $\cosh\theta$ as $\cosh\theta=\frac{1}{\sqrt{1-(\tanh\theta)^2}}$. This, jointly with the definition (12.4.15), makes it possible to express the coefficient in (12.4.17) as

$$\gamma\equiv\frac{\alpha\cosh\theta}{\left(\frac{1}{2}-E\right)}=\frac{\alpha}{\sqrt{-2E}}. \tag{12.4.18}$$

On the other hand, one can prove that, for any triple of Hermitian operators satisfying the $su(1,1)$ algebra, the K_3 operator obeys the eigenvalue equation

$$K_3\,\overline{\psi} = m\,\overline{\psi}, \tag{12.4.19}$$

where m is an integer. Combining (12.4.17)–(12.4.19) one thus arrives at the following formula for the spectrum of bound states in the s-wave sector:

$$E_m = -\frac{\alpha^2}{2m^2}. \tag{12.4.20}$$

At a deeper level, the formulae (12.4.12) and (12.4.13) correspond to the transformation formulae for K_1, K_2, K_3 in an $SU(1,1)$ interferometer (cf. the $SU(2)$ case in section 11.3), which is a device with two input ports and two output ports, relating ingoing and outgoing annihilation and creation operators according to the rule (cf. Eq. (11.3.9))

$$\begin{pmatrix} b_1 \\ b_2^\dagger \end{pmatrix} = \begin{pmatrix} S_{11} & S_{12} \\ S_{21} & S_{22} \end{pmatrix} \begin{pmatrix} a_1 \\ a_2^\dagger \end{pmatrix}. \tag{12.4.21}$$

The 2×2 matrix in (12.4.21), which is called, once again, the scattering matrix, can be chosen in such a way that ingoing and outgoing K_i operators get transformed indeed as

$$\begin{aligned} \begin{pmatrix} K_1 \\ K_2 \\ K_3 \end{pmatrix}_{\rm out} &= {\rm e}^{-i\theta K_2} \begin{pmatrix} K_1 \\ K_2 \\ K_3 \end{pmatrix} {\rm e}^{i\theta K_2} \\ &= \begin{pmatrix} \cosh\theta & 0 & \sinh\theta \\ 0 & 1 & 0 \\ \sinh\theta & 0 & \cosh\theta \end{pmatrix} \begin{pmatrix} K_1 \\ K_2 \\ K_3 \end{pmatrix}. \end{aligned} \tag{12.4.22}$$

The expert reader can recognize the occurrence of a Lorentz boost along the x-axis, if K_1 is viewed as the K_x operator.

12.5 The Runge–Lenz vector

The Hamiltonian of the hydrogen atom (hereafter, we use Cartesian coordinates):

$$H = \frac{1}{2m}\left(p_x^2 + p_y^2 + p_z^2\right) - \frac{e^2}{r}, \tag{12.5.1}$$

has a large symmetry algebra. In addition to the conserved angular momentum: $[H, \vec{L}] = 0$, there is also the Runge–Lenz vector $\vec{R}$, i.e. the Hermitian operator

$$\vec{R} \equiv \frac{1}{2m}\left(\vec{p} \times \vec{L} - \vec{L} \times \vec{p}\right) - \frac{e^2}{r}\vec{r}. \tag{12.5.2}$$

One thus finds (cf. Rabin 1995)

$$R_k = \frac{1}{m}\Big(r_k p_l p_l - r_l p_l p_k + i\hbar p_k\Big) - e^2 \frac{r_k}{r}, \tag{12.5.3}$$

where we have used the identities

$$\left(\vec{A} \times \vec{B}\right)_k = \varepsilon_{kij}\, A_i B_j, \tag{12.5.4}$$

$$\varepsilon_{ijk}\, \varepsilon_{kln} = \delta_{il}\, \delta_{jn} - \delta_{in}\, \delta_{jl}, \tag{12.5.5}$$

and the canonical commutation relations

$$r_k p_l - p_l r_k = \mathrm{i}\hbar\, \delta_{kl}, \tag{12.5.6}$$

with $r_1 = x, r_2 = y, r_3 = z$. The explicit calculation shows that $[H, \vec{R}] = 0$. Note that, in classical mechanics, the electron would follow an elliptic orbit with the origin at one focus, and the vector $\vec{R}$ points from the origin to the nearer vertex of the ellipse. Its length is proportional to the eccentricity of the orbit.

Other useful identities involving the Runge–Lenz vector are as follows (hereafter, we set $\hbar = 1$):

$$\vec{R} \cdot \vec{L} = \vec{L} \cdot \vec{R} = 0, \tag{12.5.7}$$

$$\Big[L_j, R_k\Big] = \mathrm{i}\varepsilon_{jkl} R_l, \tag{12.5.8}$$

$$|\vec{R}|^2 = e^4 + \frac{2}{m} H\Big(|\vec{L}|^2 + 1\Big), \tag{12.5.9}$$

$$\Big[R_j, R_k\Big] = -\frac{2\mathrm{i}}{m} H \varepsilon_{jkl} L_l. \tag{12.5.10}$$

Now, denoting by $D \subset \mathcal{L}^2(\mathbf{R}^3)$ the span of the eigenvectors of H having negative eigenvalues, one can define a Hermitian operator $\vec{K}$ with domain D by setting

$$\vec{K} \equiv \sqrt{-\frac{m}{2H}}\, \vec{R}. \tag{12.5.11}$$

At this stage, one can also define the operators

$$\vec{M} \equiv \frac{1}{2}(\vec{L} + \vec{K}), \quad \vec{N} \equiv \frac{1}{2}(\vec{L} - \vec{K}), \tag{12.5.12}$$

which obey the commutation relations

$$\Big[M_j, M_k\Big] = \mathrm{i}\varepsilon_{jkl} M_l, \tag{12.5.13}$$

$$\Big[N_j, N_k\Big] = \mathrm{i}\varepsilon_{jkl} N_l, \tag{12.5.14}$$

$$\left[M_j, N_k\right] = 0, \tag{12.5.15}$$

and the identities

$$|\vec{M}|^2 = |\vec{N}|^2 = \frac{1}{4}\left(|\vec{L}|^2 + |\vec{K}|^2\right), \tag{12.5.16}$$

$$H = -\frac{me^4}{2(4|\vec{M}|^2 + 1)}. \tag{12.5.17}$$

This means that the operators $\vec{M}$ and $\vec{N}$ acting in D generate a representation of the Lie algebra $su(2) \times su(2)$. Suppose now we decompose D as the direct sum of irreducible representations of this algebra. The Casimir operators $|\vec{M}|^2$ and $|\vec{N}|^2$ commute with the generators:

$$[M^2, M_k] = [M^2, N_k] = [N^2, N_k] = [N^2, M_k] = 0,$$

and are hence constant on each irreducible representation, by virtue of the Schur lemma. Thus, by virtue of Eq. (12.5.17), H is also constant on each irreducible representation, so these are the eigenspaces of H. Only those irreducible representations having $|\vec{M}|^2 = |\vec{N}|^2$ occur, and the value of each of these Casimir operators is $j(j+1)$ in the $(2j+1)$-dimensional irreducible representation of $su(2)$ having highest weight $j = 0, \frac{1}{2}, 1, \frac{3}{2}, \ldots$. The $su(2) \times su(2)$ symmetry of the hydrogen atom can be enlarged to $so(4,2)$ by means of suitable raising and lowering operators. For a detailed proof, we refer the reader to Wybourne (1974). The work in Sudarshan *et al.* (1965), Musto (1966) has studied the $O(4,1)$ symmetry for the hydrogen atom. In this as well as the treatment of $O(4,2)$ for hydrogen, one goes beyond the symmetries of the Hamiltonian.

12.6 Problems

12.P1. Consider a problem in a central potential in the s-wave case. Setting to 1, for simplicity, the reduced mass of the problem (think, for example, of the relative nucleus–electron motion in the hydrogen atom), the Hamiltonian takes the form

$$H = \frac{1}{2}p_r^2 - \frac{\alpha}{r}. \tag{12.6.1}$$

(i) Express H as a linear combination of the operators K_1, K_2, K_3, which obey the commutation rules (12.4.5)–(12.4.7).

(ii) Consider the transformation $\overline{\psi} \equiv e^{-i\theta K_2}\, \psi$ on the wave function, where θ is an unknown function (for the time being) of the eigenvalues E of H, and derive the Schrödinger equation for $\overline{\psi}$.

(iii) Find for which form of $\theta(E)$ the new Schrödinger equation reduces to

$$K_3\, \overline{\psi} = f_1(\alpha, E)\overline{\psi}, \tag{12.6.2}$$

and find the explicit form of $f_1(\alpha, E)$ and of the eigenvalues of the original problem.

(iv) Find for which form of $\theta(E)$ the new Schrödinger equation becomes

$$K_1 \overline{\psi} = f_2(\alpha, E) \overline{\psi}. \tag{12.6.3}$$

How does f_2 depend on α and E?

12.P2. Try to use the transformation operators of section 12.3 to study scattering from singular potentials in a central field. For this purpose, write D in the form

$$D = \frac{\mathrm{d}}{\mathrm{d}r} + G(r), \tag{12.6.4}$$

where, for $r \in]0, \infty[$, $G(r)$ admits a Laurent expansion

$$G(r) = \sum_{p=-\infty}^{\infty} b_p r^p, \tag{12.6.5}$$

and the potential V_1 is expressed by a Laurent series

$$V_1(r) = \sum_{n=-\infty}^{\infty} a_n r^n. \tag{12.6.6}$$

(i) By insertion of (12.6.5) and (12.6.6) into Eq. (12.3.22), prove that the coefficients b_p obey the non-linear algebraic system (Esposito 1998a)

$$(n+1)b_{n+1} + \sum_{p=-\infty}^{\infty} b_p b_{n-p} = a_n + C\delta_{n,0}. \tag{12.6.7}$$

(ii) If also $V_0(r)$ is represented by a Laurent expansion for $r \in]0, \infty[$:

$$V_0(r) = \sum_{n=-\infty}^{\infty} f_n r^n, \tag{12.6.8}$$

prove that

$$f_n = -(n+1)b_{n+1} + \sum_{p=-\infty}^{\infty} b_p b_{n-p} - C\delta_{n,0}. \tag{12.6.9}$$

(iii) In the particular case when $C = 0$ and, for some real k,

$$V_1(r) = \frac{k^2}{r^4} - \frac{2k}{r^3}, \tag{12.6.10}$$

$$V_0(r) = \frac{k^2}{r^4} + \frac{2k}{r^3}, \tag{12.6.11}$$

find $G(r)$.

13
From density matrix to geometrical phases

The density matrix ρ is defined and its basic properties are derived in detail: ρ is a Hermitian non-negative operator with unit trace, its eigenvalues are non-negative and lie in the closed interval $[0, 1]$; if ρ is a projector, it projects onto a one-dimensional sub-space, the trace of ρ^2 is ≤ 1, the equality holding only if ρ is a projector. A necessary and sufficient condition for ρ to be a projector is that all vectors in its definition are identical up to a phase. *Pure cases* are sets that can be described by a state vector, unlike *mixtures*. The density matrix of a mixture is not a projection operator, and this may be viewed as its distinguishing feature with respect to pure cases. The mean value of an observable can be re-expressed with the help of the density matrix, and a Schrödinger equation for density matrices holds. Last, but not least, *pure cases cannot unitarily evolve into mixtures.*

The applications of the formalism presented in this chapter deal with orientation of spin-$\frac{1}{2}$ particles, polarization of pairs of photons, thermal-equilibrium states. In the second part, quantum entanglement is presented, with emphasis on the existence of state vectors for entangled pairs of particles that belong to the tensor product of Hilbert spaces $\mathcal{H}_1 \otimes \mathcal{H}_2$, but are not of the simple product form $\psi_1 \otimes \psi_2$. The only correct interpretation of such states has a statistical nature, associated to series of repeated measurements performed by two observers on the constituents of the entangled pair. Hidden-variable theories, with the associated Bell inequalities, are introduced, and a way of using entangled quantum states is described which makes it possible to transfer the polarization state of a photon to another photon. This property provides the foundation of the modern investigations of quantum teleportation. The production of statistical mixtures is also studied in detail. The chapter ends with an analysis of geometrical phases and the Wigner theorem on symmetries in quantum mechanics.

13.1 The density matrix

Let $E_1, \ldots, E_\alpha, \ldots, E_\mu$ be μ sets of physical systems of the same type, and let us denote by N_α the number of elements of E_α and by $\hat{E}$ the set of all the

$$N = N_1 + \cdots + N_\alpha + \cdots + N_\mu$$

systems. Each E_α is described by a normalized ket vector $|\Phi_\alpha\rangle$, for which therefore $\langle \Phi_\alpha | \Phi_\alpha \rangle = 1$, where such vectors are not necessarily orthogonal to each other. One then defines

$$\rho \equiv \sum_{\alpha=1}^{\mu} |\Phi_\alpha\rangle \, \frac{N_\alpha}{N} \, \langle \Phi_\alpha| \tag{13.1.1}$$

as the *statistical operator* of the set $\hat{E}$, also called (hereafter) the *density matrix* associated to $\hat{E}$. Its basic properties are as follows (Gottfried 1966, D'Espagnat 1976).

(i) ρ is Hermitian.

(ii) ρ is non-negative:

$$\langle u|\rho|u\rangle \geq 0 \quad \forall \, |u\rangle. \tag{13.1.2}$$

(iii) ρ has unit trace. Indeed, by direct calculation one finds

$$\begin{aligned} \mathrm{Tr}(\rho) &= \sum_{\alpha,n} \frac{N_\alpha}{N} \langle \Phi_\alpha | n \rangle \langle n | \Phi_\alpha \rangle = \sum_{\alpha} \frac{N_\alpha}{N} \langle \Phi_\alpha | \sum_n |n\rangle \langle n | \Phi_\alpha \rangle \\ &= \sum_{\alpha} \frac{N_\alpha}{N} \langle \Phi_\alpha | \Phi_\alpha \rangle = 1. \end{aligned} \tag{13.1.3}$$

(iv) Each diagonal element of ρ, in every matrix representation, is non-negative. In particular, the eigenvalues of ρ are all non-negative.

(v) The eigenvalues p_n of ρ lie in the closed interval [0, 1]. This can be proved by pointing out that, if there were an m such that $p_m > 1$, the property of ρ of having unit trace would be violated, since all the eigenvalues of ρ are non-negative.

(vi) If ρ is a projection operator, it projects onto a one-dimensional subspace. Indeed, if ρ is a projector, so that $\rho^2 = \rho$, one has $p_n^2 = p_n$ with $p_n = 0, 1$. On the other hand, the property of unit trace implies that

$$\sum_{n=1}^{\infty} p_n = 1,$$

with $p_n = 0, 1 \ \forall n$. Thus, only one eigenvalue equals 1, and all the others vanish, which implies

$$\rho = \sum_{n=1}^{\infty} |n\rangle p_n \langle n| = |q\rangle \, p_q \, \langle q|. \tag{13.1.4}$$

(vii) The trace of ρ^2 satisfies the inequality $\text{Tr}(\rho^2) \leq 1$, where the equality holds only if ρ is a projection operator. To prove it, note first that no point of the hyperplane defined by the condition $\sum_{n=1}^{\infty} p_n = 1$, with non-negative coordinates p_n, lies outside the hypersphere $\text{Tr}(\rho^2) = \sum_{n=1}^{\infty} p_n^2 = 1$. One then finds, by subtracting the equations that define the hyperplane and the hypersphere,

$$\sum_{n=1}^{\infty} p_n(1 - p_n) = 0. \tag{13.1.5}$$

On the other hand, it is clear that

$$p_n \in [0, 1] \Longrightarrow p_n(1 - p_n) \geq 0, \tag{13.1.6}$$

and hence Eq. (13.1.5) is only satisfied if

$$p_n(1 - p_n) = 0 \quad \forall n. \tag{13.1.7}$$

On the other hand, the condition of unit trace for ρ, jointly with Eq. (13.1.7), implies that there exists only one m such that $p_m = 1$, whereas $p_n = 0, \ \forall n \neq m$. This completes the proof of our statement.

(viii) When ρ is written in the form (13.1.1), a necessary and sufficient condition for it to be a projection operator is that all ket vectors $|\Phi_\alpha\rangle$ are identical up to a phase factor, which implies that the summation defining ρ reduces to one term only. The sufficiency of the condition is obvious, and hence we focus on the proof of its necessity. For this purpose, we compute

$$\begin{aligned}
\text{Tr}(\rho^2) &= \text{Tr}\left[\sum_{\alpha=1}^{\mu} |\Phi_\alpha\rangle \frac{N_\alpha}{N} \langle \Phi_\alpha| \sum_{\beta=1}^{\mu} |\Phi_\beta\rangle \frac{N_\beta}{N} \langle \Phi_\beta|\right] \\
&= \text{Tr}\left[\sum_{\alpha,\beta} \frac{N_\alpha N_\beta}{N^2} \langle \Phi_\alpha|\Phi_\beta\rangle |\Phi_\alpha\rangle \langle \Phi_\beta|\right] \\
&= \sum_{\alpha,\beta,n} \frac{N_\alpha N_\beta}{N^2} \langle n|\Phi_\alpha\rangle \langle \Phi_\alpha|\Phi_\beta\rangle \langle \Phi_\beta|n\rangle \\
&= \sum_{\alpha,\beta,n} \frac{N_\alpha N_\beta}{N^2} \langle \Phi_\alpha|\Phi_\beta\rangle \langle \Phi_\beta|n\rangle \langle n|\Phi_\alpha\rangle
\end{aligned}$$

$$= \sum_{\alpha,\beta} \frac{N_\alpha N_\beta}{N^2} \langle \Phi_\alpha | \Phi_\beta \rangle \langle \Phi_\beta | \sum_n |n\rangle \langle n | \Phi_\alpha \rangle$$
$$= \sum_{\alpha,\beta} \frac{N_\alpha N_\beta}{N^2} \langle \Phi_\alpha | \Phi_\beta \rangle \langle \Phi_\beta | \Phi_\alpha \rangle$$
$$= \sum_{\alpha,\beta} \frac{N_\alpha N_\beta}{N^2} |\langle \Phi_\alpha | \Phi_\beta \rangle|^2. \tag{13.1.8}$$

But then, since we know that ρ^2 has a unit trace if it is a projector, we find from Eq. (13.1.8) the condition

$$\sum_{\alpha,\beta} N_\alpha N_\beta |\langle \Phi_\alpha | \Phi_\beta \rangle|^2 = N^2 = \sum_{\alpha,\beta} N_\alpha N_\beta,$$

and hence

$$\sum_{\alpha,\beta} N_\alpha N_\beta \left[1 - |\langle \Phi_\alpha | \Phi_\beta \rangle|^2\right] = 0. \tag{13.1.9}$$

This implies that one can write

$$|\Phi_\alpha\rangle = \lambda_{\alpha\beta} \, |\Phi_\beta\rangle \quad \text{with } |\lambda_{\alpha\beta}| = 1, \tag{13.1.10}$$

because then

$$|\langle \Phi_\alpha | \Phi_\beta \rangle|^2 = |\lambda_{\alpha\beta}|^2 |\langle \Phi_\beta | \Phi_\beta \rangle|^2 = |\lambda_{\alpha\beta}|^2 = 1. \tag{13.1.11}$$

By definition, a set E, which can be described by a single state vector $|\psi\rangle$ is called a *pure case*. The deep property is that, instead of describing E by means of the normalized ket $|\psi\rangle$, one can describe it equally well by means of the density matrix

$$\rho = |\psi\rangle \langle \psi|. \tag{13.1.12}$$

If A is an observable, one finds

$$\mathrm{Tr}(\rho A) = \sum_n \langle n|\psi\rangle \langle \psi|A|n\rangle = \sum_n \langle \psi|A|n\rangle \langle n|\psi\rangle$$
$$= \langle \psi|A \sum_n |n\rangle \langle n|\psi\rangle = \langle \psi|A|\psi\rangle, \tag{13.1.13a}$$

or, in other words, the mean value of the observable A is given by

$$\overline{A} = \mathrm{Tr}(\rho A). \tag{13.1.13b}$$

The projector onto the sub-space determined by the eigenvectors of A corresponding to the eigenvalue α_k reads as

$$P(\alpha_k) = \sum_r |\alpha_k, r\rangle \langle \alpha_k, r|. \tag{13.1.14}$$

Moreover, one finds

$$\begin{aligned}
\mathrm{Tr}\Big[\rho P(\alpha_k)\Big] &= \sum_{n,r} \langle n|\psi\rangle \, \langle\psi|\alpha_{k,r}\rangle \, \langle\alpha_{k,r}|n\rangle \\
&= \sum_r \langle\psi|\alpha_{k,r}\rangle \, \langle\alpha_{k,r}| \sum_n |n\rangle \, \langle n|\psi\rangle \\
&= \sum_r \langle\psi|\alpha_{k,r}\rangle \, \langle\alpha_{k,r}|\psi\rangle = \sum_r |\langle\alpha_{k,r}|\psi\rangle|^2, \quad (13.1.15)
\end{aligned}$$

which implies that $\mathrm{Tr}\Big[\rho P(\alpha_k)\Big]$ is the statistical frequency w_k with which one predicts that a measurement of A will give as a result the value α_k (see Eq. (9.11.5)).

On combining together all elements of a number of subsets E_α one can think of obtaining a *mixture*. However, this definition is a bit too vague. More precisely, if each E_α can be described by a state vector $|\Phi_\alpha\rangle$, and if the various $|\Phi_\alpha\rangle$ are not all identical up to a phase factor, the resulting mixture is not a pure case. Hereafter, we shall talk of mixtures only when they do not reduce to pure cases. Remarkably, one finds

$$\begin{aligned}
\overline{A} &= \frac{1}{N} \sum_\alpha N_\alpha \overline{A}_\alpha = \frac{1}{N} \sum_\alpha N_\alpha \mathrm{Tr}(\rho_\alpha A) \\
&= \frac{1}{N} \sum_\alpha N_\alpha \mathrm{Tr}\Big(|\Phi_\alpha\rangle \, \langle\Phi_\alpha|A\rangle\Big) = \mathrm{Tr}(\rho A). \qquad (13.1.16)
\end{aligned}$$

Moreover, it remains true that

$$w_k = \mathrm{Tr}\Big[\rho P(\alpha_k)\Big]. \qquad (13.1.17)$$

However, for mixtures,

$$\rho^2 \neq \rho. \qquad (13.1.18)$$

One has, in fact, to bear in mind that, by virtue of (viii), the density matrix is a projector if and only if all ket vectors $|\Phi_\alpha\rangle$ are identical up to a phase factor. But we have taken as a definition of mixtures just the case when the state vectors are not all identical up to a phase.

Last, but not least, a differential equation for density matrices can be derived. For this purpose, we rely on Eqs. (9.8.1) and (13.1.1), which imply that, by writing the time evolution of state vectors in the form (t_0 being the instant of time corresponding to the assignment of initial data)

$$|\Phi_\alpha(t)\rangle = U(t, t_0)|\Phi_\alpha(t_0)\rangle, \qquad (13.1.19)$$

one finds

$$\rho(t) = U(t, t_0)\rho(t_0)U^\dagger(t, t_0), \qquad (13.1.20)$$

and hence

$$
\begin{aligned}
i\hbar \frac{d\rho}{dt} &= i\hbar \left[\frac{\partial U(t,t_0)}{\partial t} \rho(t_0) U^\dagger(t,t_0) + U(t,t_0)\rho(t_0) \frac{\partial U^\dagger(t,t_0)}{\partial t} \right] \\
&= \Big[HU(t,t_0)\rho(t_0)U^\dagger(t,t_0) - U(t,t_0)\rho(t_0)U^\dagger(t,t_0)H \Big] \\
&= \Big[H\rho(t) - \rho(t)H \Big] = \Big[H, \rho(t) \Big].
\end{aligned} \tag{13.1.21}
$$

This is called the *quantum Liouville equation for the density matrix*, and it holds in the Schrödinger picture for state vectors. Note that if one deals with a pure case, so that the corresponding density matrix is initially a projector:

$$
\rho^2(t_0) = \rho(t_0),
$$

such a property is preserved by the time-evolution, because, by virtue of Eq. (13.1.20), one finds

$$
\begin{aligned}
\rho^2(t) &= U(t,t_0)\rho(t_0)U^\dagger(t,t_0)U(t,t_0)\rho(t_0)U^\dagger(t,t_0) \\
&= U(t,t_0)\rho^2(t_0)U^\dagger(t,t_0) = U(t,t_0)\rho(t_0)U^\dagger(t,t_0) \\
&= \rho(t).
\end{aligned} \tag{13.1.22}
$$

The derivation of this result is, as we have seen, quite simple, but its relevance is so great that we stress it again: in quantum mechanics of a closed system, *pure cases cannot unitarily evolve into mixtures.*

The advantage of formulating the rules of section 9.11 in terms of density matrices (rather than state vectors and their refinement, i.e. the space of rays in a Hilbert space), lies in the possibility of applying them directly not only to pure cases, but also to mixtures. Moreover, quantum statistical mechanics relies heavily on the density-matrix formalism.

13.2 Applications of the density matrix

This section outlines three of the many relevant applications of the density-matrix formalism. They are as follows (Fano 1957).

(i) **Orientation of spin-$\frac{1}{2}$ particles**. The orientation of spin is represented by a density matrix with two rows and two columns, corresponding to two pure states with opposite spin orientation (e.g. 'spin up' and 'spin down'). In a generic state, the degree and direction of spin orientation are indicated by the magnitude and direction of the vector

$$
\vec{P} \equiv \langle \vec{\sigma} \rangle = \mathrm{Tr}(\rho \vec{\sigma}), \tag{13.2.1}
$$

whose components are the mean values of the operators represented by the three Pauli matrices σ_x, σ_y and σ_z. A pure state with definite spin orientation has $P = 1$, a state with random orientation has $P = 0$. Particles with magnetic moment $\vec{\mu}$ in a magnetic field $\vec{H}$, have

$$\vec{P} = \frac{\mu \vec{H}}{\frac{1}{2}KT}, \tag{13.2.2}$$

under conditions of paramagnetic polarization. The knowledge of $\vec{P}$ is sufficient to determine the density matrix, because:

(1) As we know from section 5.7, every 2×2 Hermitian matrix can be expressed as a linear combination of $\sigma_x, \sigma_y, \sigma_z$ and of the unit matrix $\mathbb{1}$.

(2) The Pauli matrices have the properties

$$\mathrm{Tr}(\sigma_i) = 0, \tag{13.2.3}$$

$$\mathrm{Tr}(\sigma_i \sigma_k) = 2\delta_{ik}. \tag{13.2.4}$$

(3) The density matrix and the unit matrix have traces equal to 1 and 2, respectively, and hence

$$\rho = \frac{1}{2}\Big(\mathbb{1} + P_x\sigma_x + P_y\sigma_y + P_z\sigma_z\Big) = \frac{1}{2}\Big(\mathbb{1} + \vec{P}\cdot\vec{\sigma}\Big)$$
$$= \frac{1}{2}\begin{pmatrix} 1+P_z & P_x - \mathrm{i}P_y \\ P_x + \mathrm{i}P_y & 1-P_z \end{pmatrix}. \tag{13.2.5}$$

(ii) **Polarization of pairs of photons**. In the mutual annihilation of positrons and electrons by means of two-photon processes, the emitted γ-rays exhibit a polarization only when the two photons of a pair are annihilated simultaneously, but not when the photons are observed separately. The experimental results can be described in terms of the mean valuc $\langle D^{(A)} D^{(B)} \rangle$, where the operators D represent polarization analysers, which detect two photons A and B in coincidence, but not necessarily simultaneously. The matrix $D^{(A)}$ or $D^{(B)}$ is given by

$$D^{(A,B)} \equiv \frac{1}{2}\Big[(\varepsilon_{\mathrm{M}} + \varepsilon_{\mathrm{m}})\mathbb{1} + (\varepsilon_{\mathrm{M}} - \varepsilon_{\mathrm{m}})\vec{Q}\cdot\vec{\sigma}\Big]^{A,B}, \tag{13.2.6}$$

where $\vec{Q}$ is a unit vector, ε_{M} is the maximal efficiency and ε_{m} is the minimal efficiency. More precisely, $\vec{Q}$ is a vector of the Poincaré representation of polarizations.

The statement 'the photons of a pair have opposite polarizations' means that, for perfect analysers with $\varepsilon_{\mathrm{m}} = 0$, one would find

$$\langle D^{(A)} D^{(B)} \rangle = \frac{1}{4}\varepsilon_{\mathrm{M}}^{(A)}\ \varepsilon_{\mathrm{M}}^{(B)}\Big[1 - \vec{Q}^{(A)}\cdot\vec{Q}^{(B)}\Big], \tag{13.2.7}$$

which vanishes when $\vec{Q}^{(A)} = \vec{Q}^{(B)}$, and equals $\frac{1}{2}\varepsilon_{\rm M}^{(A)}\,\varepsilon_{\rm M}^{(B)}$ when $\vec{Q}^{(A)} = -\vec{Q}^{(B)}$, independent of the states of linear, circular or elliptic polarization corresponding to $\vec{Q}^{(A)}$. In a realistic application, with imperfect analysers, one finds instead

$$\langle D^{(A)}D^{(B)}\rangle = \frac{1}{4}\Big[(\varepsilon_{\rm M}+\varepsilon_{\rm m})^{(A)}(\varepsilon_{\rm M}+\varepsilon_{\rm m})^{(B)} - (\varepsilon_{\rm M}-\varepsilon_{\rm m})^{(A)}(\varepsilon_{\rm M}-\varepsilon_{\rm m})^{(B)}\vec{Q}^{(A)}\cdot\vec{Q}^{(B)}\Big]. \tag{13.2.8}$$

The polarization state of the pairs of photons is represented by a density matrix $\rho^{(AB)}$ with four rows and four columns, which can be always expressed as a linear combination of the 16 matrices

$$\mathbb{I}^{(A)}\mathbb{I}^{(B)},\ \mathbb{I}^{(A)}\sigma_x^{(B)}, \ldots \sigma_z^{(A)}\sigma_z^{(B)}.$$

The requirement that

$$\mathrm{Tr}\Big(\rho^{(AB)}D^{(A)}D^{(B)}\Big) = \langle D^{(A)}D^{(B)}\rangle \quad \forall\vec{\varepsilon}\ \text{and}\ \forall\vec{Q} \tag{13.2.9}$$

implies that

$$\rho^{(AB)} = \frac{1}{4}\Big[\mathbb{I}^{(A)}\mathbb{I}^{(B)} - \vec{\sigma}^{(A)}\cdot\vec{\sigma}^{(B)}\Big]. \tag{13.2.10}$$

This matrix represents the situation in a very condensed form, without referring to any specific type of polarization. The negative sign corresponds to the case when the photons have opposite polarizations. The density matrix for a single photon is obtained as a particular case, i.e.

$$\rho^{(A)} = \mathrm{Tr}_{\rm B}\rho^{(AB)} = \frac{1}{2}\mathbb{I}^{(A)}. \tag{13.2.11}$$

(iii) **Thermal-equilibrium states.** It is well known from statistical mechanics that the state of a system at temperature T is represented by an incoherent mixture of states with (unnormalized) statistical weights proportional to the Boltzmann distribution factor ${\rm e}^{-E_m/KT}$. To make sure that the sum of the weights for all the eigenstates equals 1, the weight of each state has to be equal to the factor ${\rm e}^{-E_m/KT}$, divided by the normalization factor (cf. Eq. (14.4.8))

$$Z(T) = \sum_l {\rm e}^{-E_l/KT}. \tag{13.2.12}$$

In other words, the density matrix is diagonal when the energy eigenstates are considered, and is given by

$$\rho_{mm'} = \frac{{\rm e}^{-E_m/KT}}{Z(T)}\delta_{mm'}, \tag{13.2.13}$$

in such a scheme. On using operator notation one can also write, in such a case,

$$\rho = \frac{e^{-H/KT}}{Z(T)} = \frac{e^{-H/KT}}{\mathrm{Tr}(e^{-H/KT})}. \tag{13.2.14}$$

13.3 Quantum entanglement

Let $\mathcal{U}$ and $\mathcal{V}$ be two quantum systems with Hilbert spaces $\mathcal{H}^{\mathcal{U}}$ and $\mathcal{H}^{\mathcal{V}}$, respectively, and let us assume that a system of type $\mathcal{U}$ and a system of type $\mathcal{V}$ collide for a time Δt. The preparation can be such that, before the collision, one has a state vector $|f_i\rangle \in \mathcal{H}^{\mathcal{U}}$ and a ket $|\varphi_0\rangle \in \mathcal{H}^{\mathcal{V}}$, and after the collision the ket $|f_i\rangle$ still lies in $\mathcal{H}^{\mathcal{U}}$ whereas one deals with a state vector $|\varphi_i\rangle$ in $\mathcal{H}^{\mathcal{V}}$. We are thus considering a process such that (D'Espagnat 1976)

$$|f_i\rangle\,|\varphi_0\rangle \rightarrow |f_i\rangle\,|\varphi_i\rangle, \tag{13.3.1}$$

or, more generally, (collision) processes for which

$$|f_i\rangle\,|\varphi_0\rangle \rightarrow |f_i'\rangle\,|\varphi_i\rangle, \tag{13.3.2}$$

where $|f_i'\rangle$ differs from $|f_i\rangle$, and the state vector of the composite system belongs to the tensor product $\mathcal{H}^{\mathcal{U}} \otimes \mathcal{H}^{\mathcal{V}}$.

However, among the phenomena of temporary interaction between quantum systems, those for which the property (13.3.2) holds constitute a very particular subset. For example, by virtue of the linearity of temporal evolution, one may consider processes such that

$$\Big(|f_i\rangle + |f_k\rangle\Big)|\varphi_0\rangle = |f_i\rangle\,|\varphi_0\rangle + |f_k\rangle\,|\varphi_0\rangle \rightarrow |f_i\rangle\,|\varphi_i\rangle + |f_k\rangle\,|\varphi_k\rangle. \tag{13.3.3}$$

In other words, quantum mechanics allows for state vectors of *entangled* systems *which are not tensor products of state vectors.* From a purely mathematical point of view, the concept is not so abstract, but has far reaching consequences for our understanding of the physical world. This is a first, clear indication that, when two quantum systems have interacted with each other in the past, *one cannot ascribe to each of them* any single state vector.

Indeed, several efforts have been produced, in the literature, to understand whether alternative descriptions exist of the final state of the two systems, which are compatible with the assignment of a state vector (and/or a well-defined set of physical properties) either of the systems $\mathcal{U}$ or $\mathcal{V}$ separately. According to the quantum-mechanical rules that we are teaching (cf. section 9.12), *it is impossible to consider each component, either $\mathcal{U}$ or $\mathcal{V}$, as a system with a complete set of well-defined properties.* If we fail to take this into account we may, following Einstein, Podolsky and Rosen (hereafter called EPR) argue that, if it is possible to predict with certainty, without affecting the system in any way, the value of a physical quantity, then *there exists an element of physical reality* which corresponds to this quantity (Einstein *et al.* 1935). An enlightening example of what this would imply is provided by the Bohm version of the EPR argument, which is as follows. A particle decays, producing two spin-$\frac{1}{2}$ particles, for which the total spin angular momentum $S_{\rm tot}$ vanishes. These particles then move away from each other, and two observers, A and B, measure the components of their spins along various directions. Since $S_{\rm tot} = 0$, if both observers measure the spin along a particular direction $\vec{u}$, and if the measurement performed by A yields $+\frac{\hbar}{2}$, then B must necessarily find $-\frac{\hbar}{2}$, and vice versa.

To recover the correct quantum-mechanical description of the phenomenon, let us assume, for simplicity, that the measurements are performed along the z-axes of the two observers, and let us recall that the spin part of the state vector of the two particles is the antisymmetric combination

$$\psi_{1,2} = \frac{1}{\sqrt{2}}\left(\psi_1^{\rm u} \otimes \psi_2^{\rm d} - \psi_1^{\rm d} \otimes \psi_2^{\rm u}\right), \tag{13.3.4}$$

where $\psi_1^{\rm u} \otimes \psi_2^{\rm d}$ is a state having spin $\frac{\hbar}{2}$ for particle 1 and spin $-\frac{\hbar}{2}$ for particle 2, and $\psi_1^{\rm d} \otimes \psi_2^{\rm u}$ is a state having spin $-\frac{\hbar}{2}$ for particle 1 and spin $\frac{\hbar}{2}$ for particle 2. The *up* and *down* states therefore obey the eigenvalue equations

$$S_z \psi^{\rm u} = \frac{\hbar}{2}\psi^{\rm u}, \tag{13.3.5}$$

$$S_z \psi^{\rm d} = -\frac{\hbar}{2}\psi^{\rm d}, \tag{13.3.6}$$

for particles 1 and 2. The state (13.3.4) is therefore an *entangled state*, the statistical interpretation of which is as follows (Isham 1995). If, in a series of repeated measurements performed by observer A, the pairs of particles are selected for which the measurement of his particle yields 'spin up', then, *with probability equal to 1, a series of measurements performed by B on his particle in this pair will yield 'spin down'.* Similarly, if A finds 'spin down', then, with probability equal to 1, B will find 'spin up'. One then says that the measurements performed by A, the effect of which is evaluated with the operator $S_z \otimes 1\!\!1$, lead to a reduction of the state vector (13.3.4) into either $\psi_1^{\rm u} \otimes \psi_2^{\rm d}$ or $\psi_1^{\rm d} \otimes \psi_2^{\rm u}$, depending on whether spin up or spin down is selected. This new state is an eigenstate of the operator $1\!\!1 \otimes S_z$ associated to the second particle, with an eigenvalue which is minus that found by A.

Even apart from arguing for definite values for the sub-systems, if we tried to assign probabilities we could do so. Both of the spins 1 and 2 have an equal probability of being up or down. But even such a probabilistic statement can be consistently made for any generic system for several successive times: in other words, generic quantum systems have no 'histories' for which we can assign probabilities. The kind of special histories for which we can assign probabilities consistently are called 'decoherent histories' (Gell-Mann and Hartle 1993) and may be used to have a classical statistical picture emerge for the quantum system. The generic class of consistent histories are very limited if we consider projection to well-defined states (one-dimensional projections).

13.4 Hidden variables and the Bell inequalities

After the paper of Einstein *et al.* (1935), several authors have considered the possibility that the standard quantum formalism is incomplete, and that the probabilistic nature of the results of measurements arises from the existence of some *hidden variables.* A common feature of the various hidden-variable theories is that, in any given quantum state ψ, any observable A possesses an objectively existing value determined by ψ and by the values of a set of hidden variables $\{\lambda_1, \ldots, \lambda_n\}$ belonging to some space Λ. One then writes, for this objective value,

$$A(\psi, \lambda_1, \ldots, \lambda_n).$$

Moreover, the existence of a probability density μ_ψ on Λ is assumed, such that the expected value of A in the state ψ reads

$$\langle A \rangle_\psi = \int_\Lambda \mu_\psi(\lambda_1, \ldots, \lambda_n) A(\psi, \lambda_1, \ldots, \lambda_n) \, \mathrm{d}\lambda_1 \cdots \mathrm{d}\lambda_n. \tag{13.4.1}$$

One then has to find the space Λ, the probability density μ_ψ and the value function A such that Eq. (13.4.1) may reproduce the predictions of standard quantum theory.

Within this framework, the Bell inequalities (Bell 1964, 1987) express some properties of a quantum theory where it would be meaningful to say that observables *have a definite value.* To obtain such a characterization, Bell studied what happens if two observers measure the spin of two particles along different axes. One observer has to consider unit vectors $\vec{a}$ and $\vec{a}'$, and the other observer deals with unit vectors $\vec{b}$ and $\vec{b}'$. A series of repeated measurements is made on a collection of systems for which the quantum state is described by the vector ψ in Eq. (13.3.4). Two key assumptions are made (Isham 1995).

(i) Each particle has a definite value of the projection of spin $\vec{S}$ along any direction at all times.

(ii) The value of any physical quantity is not affected by altering the position of remote measuring equipment.

For directions characterized by $\vec{a}$ and $\vec{b}$, the correlation between measurements performed by the two observers is defined by

$$C(\vec{a},\vec{b}) \equiv \lim_{N\to\infty} \frac{1}{N} \sum_{n=1}^{N} a_n b_n, \tag{13.4.2}$$

where a_n (respectively b_n) is $\frac{2}{\hbar}$ times the value of $\vec{a}\cdot\vec{S}$ (respectively $\vec{b}\cdot\vec{S}$) possessed by particle 1 (respectively 2) in the nth element of the collection. Now consider the quantity (Isham 1995)

$$g_n = a_n b_n + a_n b'_n + a'_n b_n - a'_n b'_n. \tag{13.4.3}$$

Since $a_n = \pm 1$ if $\vec{a}\cdot\vec{S} = \pm\frac{1}{2}\hbar$, and the same holds for b_n, each term in g_n takes the value $+1$ or -1. If $a_n = b_n = \pm 1 = a'_n = b'_n$, then $g_n = 2$. If $a_n = -b_n = \pm 1 = a'_n = -b'_n$, then $g_n = -2$. Thus, by construction, g_n can only take two values: ± 2, and one can write for its average value

$$\left| \frac{1}{N}\sum_{n=1}^{N} g_n \right| = \frac{1}{N}\left| \sum_{n=1}^{N} a_n b_n + \sum_{n=1}^{N} a_n b'_n + \sum_{n=1}^{N} a'_n b_n - \sum_{n=1}^{N} a'_n b'_n \right|. \tag{13.4.4}$$

Assumption (ii) plays a crucial role in Eq. (13.4.4), because it implies that a_n is the same independently of being multiplied by b_n or b'_n, i.e. independently of the direction along which the other observer is measuring the spin of particle 2. By virtue of (13.4.2) and (13.4.4) one obtains, in the limit as $N \to \infty$, the *Bell inequality*

$$\left| C(\vec{a},\vec{b}) + C(\vec{a},\vec{b}') + C(\vec{a}',\vec{b}) - C(\vec{a}',\vec{b}') \right| \leq 2. \tag{13.4.5}$$

On the other hand, according to 'orthodox' quantum mechanics, the correlation between spin measurements along axes with unit vectors $\vec{a}$ and $\vec{b}$ is

$$C(\vec{a},\vec{b}) \equiv \frac{\left(\psi, \vec{a}\cdot\vec{S}_{(1)} \otimes \vec{b}\cdot\vec{S}_{(2)}\psi\right)}{(\hbar/2)^2}, \tag{13.4.6}$$

where $\vec{S}_{(1)}$ and $\vec{S}_{(2)}$ are the spin operators for particles 1 and 2, respectively, and the tensor product of operators on Hilbert spaces is defined according to the rule

$$\left(A_1 \otimes A_2\right)\psi_1 \otimes \psi_2 \equiv (A_1\psi_1) \otimes (A_2\psi_2). \tag{13.4.7}$$

Since the correlation $C(\vec{a},\vec{b})$ can only depend on $\cos\theta_{ab} = \vec{a}\cdot\vec{b}$, one can assume that $\vec{a}$ points along the z-axis and that $\vec{b}$ lies in the x–z plane.

Then Eq. (13.4.6) leads to

$$C(\vec{a},\vec{b}) = \Big(\psi, \sigma_{1z} \otimes (\sigma_{2z} \cos\theta_{ab} + \sigma_{2x} \sin\theta_{ab})\psi\Big), \tag{13.4.8}$$

where

$$\sigma_{iz} = \begin{pmatrix} 1 & 0 \\ 0 & -1 \end{pmatrix} \quad i = 1, 2, \tag{13.4.9}$$

$$\sigma_{ix} = \begin{pmatrix} 0 & 1 \\ 1 & 0 \end{pmatrix} \quad i = 1, 2, \tag{13.4.10}$$

and ψ is given by Eq. (13.3.4), with

$$\psi_i^{\rm u} = \begin{pmatrix} 1 \\ 0 \end{pmatrix} \quad i = 1, 2, \tag{13.4.11}$$

$$\psi_i^{\rm d} = \begin{pmatrix} 0 \\ 1 \end{pmatrix} \quad i = 1, 2. \tag{13.4.12}$$

Therefore, the basic properties

$$\sigma_{iz}\ \psi_i^{\rm u} = \psi_i^{\rm u}, \tag{13.4.13}$$

$$\sigma_{iz}\ \psi_i^{\rm d} = -\psi_i^{\rm d}, \tag{13.4.14}$$

imply that

$$\begin{aligned} C(\vec{a},\vec{b}) &= \frac{1}{2}\cos\theta_{ab}\Big((\psi_1^{\rm u}\psi_2^{\rm d} - \psi_1^{\rm d}\psi_2^{\rm u}), (\sigma_{1z}\psi_1^{\rm u}\sigma_{2z}\psi_2^{\rm d} - \sigma_{1z}\psi_1^{\rm d}\sigma_{2z}\psi_2^{\rm u})\Big) \\ &= -\frac{1}{2}\cos\theta_{ab}\Big((\psi_1^{\rm u}\psi_2^{\rm d} - \psi_1^{\rm d}\psi_2^{\rm u}), (\psi_1^{\rm u}\psi_2^{\rm d} - \psi_1^{\rm d}\psi_2^{\rm u})\Big) \\ &= -\cos\theta_{ab}. \end{aligned} \tag{13.4.15}$$

In particular, if one further assumes that the unit vectors $\vec{a}, \vec{b}, \vec{a}', \vec{b}'$ all lie in the same plane, with $\vec{a}$ and $\vec{b}$ parallel to each other, and $\theta_{ab'} = \theta_{a'b} = \varphi$, the Bell inequality is satisfied provided that (Isham 1995)

$$|1 + 2\cos\varphi - \cos 2\varphi| \leq 2. \tag{13.4.16}$$

The left-hand side of such a condition, however, is larger than 2 when $\varphi \in]0, \frac{\pi}{2}[$. The resulting contradiction shows that the idea of systems possessing individual values for observables is incorrect, unless one is ready to accept the existence of a non-locality for which the formulation is indeed unclear and questionable (cf. Peres 1993, Isham 1995).

A new interpretation of the Bell inequalities has been proposed in Sudarshan and Rothman (1993). They stress that such inequalities are always derived assuming that local hidden-variable theories give a set of positive-definite probabilities for detecting a particle with a given spin orientation, while it is claimed that quantum mechanics cannot produce a set of such probabilities. However, they show that this is not the case if one allows for generalized, *non-positive-definite* 'master probability distributions'. Quantum mechanics and hidden-variable theories are therefore placed on a nearly equal footing, and the usual hidden-variable result might be viewed as merely the wrong answer to a quantum-mechanical problem; it is then not surprising that a set of non-quantum-mechanical rules gives a non-quantum-mechanical result.

Major progress in the experimental work on Bell inequalities began with the work of Alain Aspect and his collaborators, who studied the correlation between photons in a variety of configurations (Aspect *et al.* 1981, 1982a,b). Photons are much easier to use experimentally than electrons as they pass through air. The counterpart of electron spin along a particular axis are the polarization states of the photon. Details and implications of these experiments are discussed in Hughes (1989), Redhead (1989) and Peres (1993). Very recent experimental work on the Bell inequalities is described in Weihs *et al.* (1998), where the authors have observed a strong violation of a Bell inequality in an Einstein–Podolsky–Rosen-type experiment with independent observers.

It is usually accepted that probabilities are measures and therefore are normalized and non-negative. If we retain the normalization but relax the positivity of the probabilities it is indeed possible to give 'probabilities' to every history, be it decoherent or otherwise (Sudarshan and Rothman 1993). The use of even more general probabilities such as complex (or quaternionic) probabilities have been advanced in the literature. One can ask what to make of a negative probability or a complex probability. The traditional way of understanding probabilities is in terms of relative frequencies or of a quantification of prior knowledge. Neither of these allow negative, let alone complex probabilities. Neither sense of probability is appropriate for quantum mechanics. There is a third way of understanding probability as a distribution given its characteristic function. If $x_1, x_2, \ldots$ are random variables, the expectation value of any function of these variables is expressed in terms of the statistical state as a linear valuation on the function. Of these valuations the valuation of $e^{i(\lambda_1 x_1 + \lambda_2 x_2 + \cdots)}$ is special and is called the characteristic function

$$\chi(\lambda_1, \lambda_2, \ldots) \equiv \langle e^{i(\lambda_1 x_1 + \lambda_2 x_2 + \cdots)} \rangle. \tag{13.4.17}$$

The probability is *defined* as the inverse Fourier transform

$$P(x_1, x_2, \ldots) = (2\pi)^{-N} \chi(\lambda_1, \ldots, \lambda_N) \times e^{-i(\lambda_1 x_1 + \cdots + \lambda_N x_N)} \, d\lambda_1 \cdots d\lambda_N. \tag{13.4.18}$$

This concept was used by Wigner and more systematically by Moyal (see section 15.5).

13.5 Entangled pairs of photons

A highly non-trivial application of entangled quantum states has been studied recently in experiments involving three photons (see, in particular, Bouwmeester *et al.* 1997). The problem is so enlightening and so deep that we find it appropriate to describe its theoretical aspects here. In the experiment, a photon $F1$ carries the polarization that one wants to transfer, and the other two photons, $F2$ and $F3$, form a pair which is in an entangled state (cf. Eq. (13.3.4)). The photons $F1$ and $F2$ are then subject to a measurement in such a way that $F3$ acquires the polarization of the photon $F1$. This is obtained as follows.

Observer A has a particle in a certain quantum state ψ, and he wants to transfer to the particle of observer B the same quantum state. In general, their communication channel might not be good enough to preserve quantum coherence, or it might take too long a time for the particle to travel from A to B. Thus, one looks for a mechanism which *makes it possible to transfer the quantum state of particle A, but not the particle itself.* For example, a single photon might be polarized *horizontally*, with state vector $\psi^{\rm H}$, or *vertically*, with state vector $\psi^{\rm V}$, or might be, more generally, in a superposition of these two polarization states, i.e.

$$\psi = \alpha\, \psi^{\rm H} + \beta\, \psi^{\rm V}, \tag{13.5.1}$$

with $|\alpha|^2 + |\beta|^2 = 1$. Even more generally, for any two-state quantum system, one could have

$$\psi = \alpha\, \psi_0 + \beta\, \psi_1. \tag{13.5.2}$$

If we were to limit ourselves to let the photon pass through a beamsplitter that reflects horizontally or transmits vertically, the photon would be found in the reflected or transmitted beams with probabilities $|\alpha|^2$ and $|\beta|^2$, respectively. In such a case, the measurement process only has the effect of projecting the initial linear combination onto either $\psi^{\rm H}$ or $\psi^{\rm V}$. This is not quite what we want to obtain, but a more appropriate use of the projection postulate can be made to transfer the initial polarization state of a photon (Bennett *et al.* 1993). For this purpose, the key idea is to use the initial photon jointly with an entangled pair of photons, which are shared by the observers A and B. The photon $F1$, for which the polarization state is to be transferred, is in the initial state

$$\psi_1 = \alpha\, \psi_1^{\rm H} + \beta\, \psi_1^{\rm V}, \tag{13.5.3}$$

whereas the pair of photons $F2$ and $F3$, shared by A and B, are in the entangled state

$$\psi_{2,3}^{-} = \frac{1}{\sqrt{2}}\Big[\psi_2^{\rm H} \otimes \psi_3^{\rm V} - \psi_2^{\rm V} \otimes \psi_3^{\rm H}\Big]. \tag{13.5.4}$$

As should be clear from section 13.3, the entangled state $\psi_{2,3}^{-}$ does not contain any information on the individual particles which are its constituents, and it only shows that the photons $F2$ and $F3$ will be found in opposite polarization states. It has, however, the important property according to which, as soon as a measurement on $F2$ or $F3$ projects it onto the state $\psi^{\rm H}$, the state of $F3$ is forced to be $\psi^{\rm V}$, or vice versa. Thus, the quantum state of the other photon of the entangled pair is immediately determined, even though the two photons are separated by a very large distance.

At this stage, the crucial step consists in performing a measurement on the photons $F1$ and $F2$, in such a way that they are projected onto the entangled state

$$\psi_{1,2}^{-} = \frac{1}{\sqrt{2}} \Big[\psi_1^{\rm H} \otimes \psi_2^{\rm V} - \psi_1^{\rm V} \otimes \psi_2^{\rm H} \Big]. \tag{13.5.5}$$

Note that this is only one of the four possible entangled states into which any state of two photons can be decomposed. Its peculiar feature is the antisymmetry upon interchanging the photons $F1$ and $F2$ (hence the superscript $(-)$ is used to denote it). But then the desired phenomenon actually occurs. In other words, since the photons $F1$ and $F2$ are observed in the entangled state $\psi_{1,2}^{-}$ we know that, whatever the state of $F1$, the photon $F2$ must be in the state orthogonal to that of photon $F1$. On the other hand, since photons $F2$ and $F3$ had been initially prepared in the entangled state $\psi_{2,3}^{-}$ of Eq. (13.5.4), the state of photon $F2$ is also orthogonal to the state of photon $F3$. Thus, since the state of $F2$ is orthogonal to the states of both $F3$ and $F1$, the quantum state of $F3$ coincides, eventually, with the quantum state (13.5.3) of $F1$, in that

$$\psi_3 = \alpha\, \psi_3^{\rm H} + \beta\, \psi_3^{\rm V}. \tag{13.5.6}$$

It should be stressed that this result has been achieved at the price of destroying the quantum state (13.5.3), because photon $F1$ has become entangled with photon $F2$ by virtue of the measurement performed upon them. Some remarks are now in order.

(i) The transfer of *quantum information* from a particle $P1$ to a particle $P3$ (not necessarily photons) can take place over arbitrary distances. Hence it is called *quantum teleportation*. The work in Boschi *et al.* (1998) has realized quantum teleportation of unknown pure quantum states, obtaining results for the teleportation of a linearly polarized state and of an elliptically polarized state.

(ii) So far, on experimental grounds, *quantum entanglement* is known to survive over distances of the order of 10 kilometres.

(iii) It is not necessary for observer A to know where observer B is.

(iv) The initial quantum state of particle $P1$ can be completely unknown both to observer A and to anyone else. For example, particle $P1$ might be, itself, the member of an entangled pair of particles. In other words, it is now possible, at least in principle, *to transfer quantum entanglement* among particles.

(v) It is thus possible to test the Bell inequalities on particles that have not shared any common past.

(vi) It is also possible to test the *local realistic* nature of the physical world, by generating quantum entanglement between more than two particles which are spatially separated.

(vii) When the quantum state has been teleported from $P1$ to $P3$, the quantum state of $P1$ is no longer the same (this is the 'no cloning' restriction).

13.6 Production of statistical mixtures

After the analysis of entangled pairs of photons, it is now appropriate to study in more detail the quantum-mechanical description of composite systems. One then deals with a system S interacting with a system S', so that the Hilbert space of the composite system $S \cup S'$ is the tensor product

$$\mathcal{H}_{S \cup S'} = \mathcal{H}_S \otimes \mathcal{H}_{S'}. \tag{13.6.1}$$

One may assume that the initial state in the tensor product of $\mathcal{H}_S$ with $\mathcal{H}_{S'}$ is a pure state with density matrix $|\psi\rangle \langle\psi|$, although complete knowledge about it is not available. The situation is analogous to that encountered in the classical description: if enough information is not available, one has to use a statistical approach relying on a probability distribution in phase space. In a classical system in any pure state of the composite $S \cup S'$ the states in S and S' must be pure. The specific physical system under examination, however, occupies a definite point of phase space with definite values of coordinates and momenta.

On considering two orthonormal bases $\{|\phi_j\rangle\}$ and $\left\{|\phi'_j\rangle\right\}$ in $\mathcal{H}_S$ and $\mathcal{H}_{S'}$, respectively, one can always express a normalized state vector as the double sum

$$|\psi\rangle = \sum_{i,j} C_{ij} |\phi_i\rangle \, |\phi'_j\rangle. \tag{13.6.2}$$

On the other hand, for the mean value of any observable A of S one obtains the formula

$$\langle A \rangle_\psi \equiv \langle \psi | A | \psi \rangle = \sum_{i,j} \langle \phi_i | A | \phi_j \rangle \sum_k C_{ik}^* C_{jk}, \tag{13.6.3}$$

since A acts as the identity on $\mathcal{H}_{S'}$, and the basis $\left\{ |\phi_j'\rangle \right\}$ is orthonormal by hypothesis. Moreover, by virtue of Eq. (13.1.13b), one can express the statistical matrix on $\mathcal{H}_S$ in the form

$$\rho = \sum_{i,j} |\phi_i\rangle \; P_{ij} \; \langle \phi_j |, \tag{13.6.4}$$

where

$$P_{ij} = \sum_k C_{ik}^* C_{jk}. \tag{13.6.5}$$

The statistical matrix ρ is found to obey all properties defining a density matrix (see section 13.1).

Thus, insofar as only the observables of S are of interest, the pure state with density matrix $|\psi\rangle \; \langle \psi|$ behaves as a statistical mixture ρ for S'. This effect is due to our ignorance, complete and well acknowledged, concerning the part of S' upon the Hilbert space of which we therefore have to evaluate a trace. This is a partial trace $\mathrm{Tr}_{S'}$ as far as the whole Hilbert space $\mathcal{H}_{S \cup S'}$ is concerned:

$$\rho = \mathrm{Tr}_{S'} |\psi\rangle \; \langle \psi| \equiv \sum_j \langle \phi_j' | \psi \rangle \; \langle \psi | \phi_j' \rangle. \tag{13.6.6}$$

In general, such an operation does not completely get rid of the quantum coherence of the pure state in $\mathcal{H}_{S \cup S'}$, which becomes manifest through the occurrence of cross-terms (with $i \neq j$) in the expression (13.6.4). This happens because the basis $\{|\phi_j\rangle\}$ is not, in general, a basis of eigenvectors of ρ. Nor can one think of being able to choose it in such a form *a priori*, i.e. before taking the partial trace over $\mathcal{H}_{S'}$. At the very beginning there is not in fact any state, neither pure nor mixed, relative to the system S alone, but only a pure state $|\psi\rangle$ of the composite system $S \cup S'$.

Of course, once the partial trace over $\mathcal{H}_{S'}$ has been taken so that the density matrix ρ on $\mathcal{H}_S$ has been obtained, it is always possible to diagonalize ρ, which therefore reads as

$$\rho = \sum_j w_j p_j = \sum_j |\psi_j\rangle \; w_j \; \langle \psi_j |. \tag{13.6.7}$$

The P_{ij} coefficients of the diagonal terms (i.e. with $i = j$) in Eq. (13.6.4), which are, by construction, non-negative and properly normalized, i.e.

$$\sum_j P_{jj} = \sum_{i,j} |C_{ij}|^2 = 1, \tag{13.6.8}$$

are called the *populations* of pure states (with density matrix $|\phi_j\rangle\langle\phi_j|$), whereas the coefficients P_{ij} of mixed terms (i.e. with $i \neq j$) are called *residual coherences.* Note that the Schwarz inequality leads to

$$|P_{ij}|^2 \leq P_{ii}P_{jj}, \tag{13.6.9}$$

and hence $P_{ii} = 0$ implies that $P_{ij} = 0$ for all i and j, i.e. unoccupied states (with vanishing population) cannot give rise to interference phenomena with any other quantum state.

At the risk of repeating ourselves, let us look in greater detail at the previous properties. The general framework is a statistical ensemble (possibly inhomogeneous) of composite systems $S = S_1 \cup S_2$. Let $\rho^{(1,2)}$ be the statistical operator associated with the ensemble, and let us assume we are only interested in the physics of one of the two constituents, e.g. S_1. This means that only the observables of S_1 are taken into account. Let $A^{(1)}$ be an observable belonging to the algebra of observables of S_1, and let $\mathbb{I}^{(2)}$ be the identity operator for the algebra of observables of S_2. The mean value of $A^{(1)}$ is then obtained as

$$\langle A^{(1)}\rangle_\rho = \mathrm{Tr}^{(1,2)}\left[\left(A^{(1)} \otimes \mathbb{I}^{(2)}\right)\rho^{(1,2)}\right]. \tag{13.6.10}$$

Since the trace is independent of the basis chosen, we may choose a factorized basis. For a given basis $\{|\varphi_i\rangle\}$ in the Hilbert space $\mathcal{H}^{(1)}$ and $\{|\chi_j\rangle\}$ in the Hilbert space $\mathcal{H}^{(2)}$, the basis chosen in $\mathcal{H}^{(1)} \otimes \mathcal{H}^{(2)}$ therefore reads as

$$\{|\varphi_i\rangle \otimes |\chi_j\rangle\}.$$

This is why the formula (13.6.10) for the mean value of $A^{(1)}$ leads to

$$\begin{aligned}\langle A^{(1)}\rangle_\rho &= \sum_{i,j}\left(\langle\varphi_i^{(1)}| \otimes \langle\chi_j^{(2)}|\right)\left|\left(A^{(1)} \otimes \mathbb{I}^{(2)}\right)\rho^{(1,2)}\left(|\varphi_i^{(1)}\rangle \otimes |\chi_j^{(2)}\rangle\right)\right. \\ &= \sum_i \langle\varphi_i^{(1)}|A^{(1)}\sum_j\langle\chi_j^{(2)}|\rho^{(1,2)}|\chi_j^{(2)}\rangle_2|\varphi_i^{(1)}\rangle_1 \\ &= \sum_i \langle\varphi_i^{(1)}|A^{(1)}\rho^{(1)}|\varphi_i^{(1)}\rangle_1 = \mathrm{Tr}^{(1)}\left[A^{(1)}\rho^{(1)}\right],\end{aligned} \tag{13.6.11}$$

where we have used the definition

$$\rho^{(1)} \equiv \mathrm{Tr}^{(2)}\left[\rho^{(1,2)}\right] = \sum_r \langle\chi_r^{(2)}|\rho^{(1,2)}|\chi_r^{(2)}\rangle, \tag{13.6.12}$$

and $\rho^{(1)}$ turns out to be a statistical operator even when $\rho^{(1,2)}$ is a projection.

The analysis we have just performed becomes of particular interest when a homogeneous quantum ensemble of composite systems is considered. The homogeneity means that all members of the system are associated with the same state vector $\psi^{(1,2)}$. In such a case, $\rho^{(1,2)}$ coincides with the projector $P_{\psi^{(1,2)}}$ onto the one-dimensional sub-space associated with $\psi^{(1,2)}$, and hence turns out to be an idempotent operator. However, if we were interested in the physics of the sub-system S_1 only, we should use the reduced statistical operator, hence obtaining the partial trace of $\rho^{(1,2)}$ over $\mathcal{H}^{(2)}$. For example, let us assume that the state $\psi^{(1,2)}$ has the form (see Eq. (13.6.2))

$$\psi^{(1,2)} = \sum_{i,j} C_{ij} |\varphi_i^{(1)}\rangle \otimes |\chi_j^{(2)}\rangle, \tag{13.6.13}$$

where the coefficients C_{ij} obey the condition

$$\sum_{i,j} C_{ij} C_{ij}^* = 1. \tag{13.6.14}$$

On choosing the basis $\left\{\chi_r^{(2)}\right\}$ to evaluate the trace $P_{\psi^{(1,2)}}$ one then finds

$$\begin{aligned}
\rho^{(1)} &= \mathrm{Tr}^{(2)}\left[P_{\psi^{(1,2)}}\right] = \sum_k \langle \chi_k^{(2)} | P_{\psi^{(1,2)}} | \chi_k^{(2)} \rangle \\
&= \sum_k \langle \chi_k^{(2)} | \psi^{(1,2)} \rangle \, \langle \psi^{(1,2)} | \chi_k^{(2)} \rangle \\
&= \sum_k \sum_{i,j,l,m} C_{ij} C_{lm}^* \langle \chi_k^{(2)} | \chi_j^{(2)} \rangle \, \langle \chi_m^{(2)} | \chi_k^{(2)} \rangle \, |\varphi_i^{(1)}\rangle \, \langle \varphi_l^{(1)}| \\
&= \sum_k \sum_{i,j,l,m} C_{ij} C_{lm}^* \delta_{kj} \delta_{km} |\varphi_i^{(1)}\rangle \, \langle \varphi_l^{(1)}| \\
&= \sum_{i,l} \sum_k C_{ik} C_{lk}^* |\varphi_i^{(1)}\rangle \, \langle \varphi_l^{(1)}| = \sum_{i,l} |\varphi_i^{(1)}\rangle \, P_{il} \, \langle \varphi_l^{(1)}| \\
&= \sum_i \Big(\sum_j |C_{ij}|^2 \Big) |\varphi_i^{(1)}\rangle \, \langle \varphi_i^{(1)}| = \sum_i w_i P_i,
\end{aligned} \tag{13.6.15}$$

where $w_i \equiv \sum_j |C_{ij}|^2$ and $P_i \equiv |\varphi_i\rangle \, \langle \varphi_i|$, where $|\varphi_i\rangle$ are normalized. We have therefore obtained a statistical mixture for $S^{(1)}$ even though we started from a pure state of the composite system.

13.7 Pancharatnam and Berry phases

Given a wave function, by virtue of its physical interpretation as a probability amplitude, we may associate with it a ray in Hilbert space, and all bilinear quantities obtained from the wave function are independent of the absolute phase. If we consider a closed curve along which the wave function is evaluated, the phase of the wave function may or may not return to the original value. While the change of phase $\mathrm{d}\theta$ for an element of path is well defined, there is no unique θ of which $\mathrm{d}\theta$ is the differential. Whenever $\mathrm{d}\theta$ is a closed but not exact form, it may be referred to as a non-integrable phase. The most familiar case is the phase of a spinor wave function on rotation through a simple closed loop.

When there is a charged particle moving in a field-free region and we take it around a closed loop linked with a non-vanishing magnetic flux, the phase change is proportional to the flux. Pancharatnam discovered the contribution from such a phase in the passage of light along a closed triangle, and gave its geometrical interpretation.

Berry showed that non-integrable phases occur in composite systems in which one motion is 'fast' and the other one is 'slow' (Berry 1984). In such cases for the fast motion the slow variables may be considered in a quasistatic approximation to provide an effective potential. This potential can be a velocity-dependent vector potential. More generally if we consider a time-dependent Hamiltonian $H(t)$ and the corresponding Schrödinger equation

$$\mathrm{i}\hbar\frac{\mathrm{d}\psi}{\mathrm{d}t} = H(t)\psi(t) = E(t)\psi(t),$$

then $\psi(t)$ has the formal expression

$$\psi(t) = \Big[\exp -\frac{\mathrm{i}}{\hbar}\int_0^t H(t')\,\mathrm{d}t'\Big]\psi(0)$$

where the exponential is time ordered. We can write this in a more suggestive form using the following:

$$E(t) = (\psi(t), H(t)\psi(t)),$$

$$\psi^*\,\mathrm{d}\psi = \psi^*\dot{\psi}\,\mathrm{d}t = \frac{\mathrm{d}\psi}{\psi} = \mathrm{i}\,\mathrm{d}\varphi,$$

where $\mathrm{d}\varphi = (\mathrm{d}\varphi)^\dagger = \frac{1}{\mathrm{i}}\mathrm{d}(\log\psi)$ is the differential of a pure phase. The Schrödinger equation does not specify the phase of $\psi(t)$ apart from a dynamical factor $\mathrm{e}^{-\mathrm{i}E(t)}$. Using the construction of $\mathrm{d}\varphi$ we can obtain an additional contribution

$$\int \phi^*\,\mathrm{d}\phi = \int \phi^*(t)\dot{\phi}(t)\,\mathrm{d}t,$$

where (in $\hbar = 1$ units)

$$\psi(t) = \mathrm{e}^{-\mathrm{i}tE(t)}\phi(t).$$

Consequently, the total phase change between $\psi(0)$ and $\psi(t_0)$ is $\theta(0, t_0)$ given by

$$\mathrm{e}^{-\mathrm{i}\theta(0,t_0)} = \mathrm{e}^{-\int E(t)\,\mathrm{d}t}\,\mathrm{e}^{-\int \phi^*\,\mathrm{d}\phi}.$$

The first contribution

$$\theta_{\rm D} \equiv \int E(t)\,{\rm d}t$$

is dynamical in that it depends on the eigenvalues of the Hamiltonian; for a time-independent Hamiltonian it is simply Et. The other contribution is dependent on the manner in which the wave functions are chosen to have phases (note that the Schrödinger equation does not tell the absolute phase of the wave function). The second contribution

$$\theta_{\rm B} \equiv \int \phi^* \,{\rm d}\phi = \int \phi^* \dot{\phi}\,{\rm d}t$$

depends on the manner in which the phases are chosen and is called the Berry phase. The total phase is the sum of these two, i.e.

$$\theta = \theta_{\rm D} + \theta_{\rm B},$$

and may be called the Pancharatnam phase (Pancharatnam 1956).

Berry originally defined the phase for the cyclic situation $H(t_0) = H(0)$ and the phase was associated with the closed curve $\psi(t), 0 < t < t_0, \psi(t_0) = \psi(0)$, and he also assumed that the evolution is adiabatic so that a 'fast' sub-system may be treated using a time-dependent Hamiltonian. For example, for the electronic states in a molecule, the molecular motion is 'slow' and the fast electronic motion may be considered as being in a slowly varying 'external' potential.

However, as Aharonov and Anandan showed, we could dispense with the adiabaticity requirement but still require a cyclic evolution (Aharonov and Anandan 1987). The question arises: is cyclicity really necessary? In other words, can one define the total Pancharatnam phase for an open sequence of evolutions? This is easily done using the tools furnished above:

$$\frac{\psi(t)}{|\psi(t)|} = {\rm e}^{-{\rm i}\alpha(t)} = {\rm e}^{-{\rm i}\int_0^t E(t')\,{\rm d}t'}\,{\rm e}^{\int_0^t \psi^*(t')\dot{\psi}(t')\,{\rm d}t'},$$

which equals

$${\rm e}^{-{\rm i}\alpha_{\rm D}(t) - {\rm i}\alpha_{\rm G}},$$

where $\alpha_{\rm D}$ is the dynamical phase $\int E(t')\,{\rm d}t'$, while the 'geometrical phase' contribution $\alpha_{\rm G}$ is the generalization of Berry's phase for non-cyclic evolution. Note that $\alpha_{\rm G}$ (and hence $\varphi_{\rm G}$) is unaffected by a local phase change

$$\phi(t) \to {\rm e}^{{\rm i}p}\phi(t),$$

since this additional factor cancels in the computations; therefore it is called 'geometric'. Note also that $\alpha_{\rm G}$ or $\varphi_{\rm G}$ depend only on the path in the Hilbert space of vectors $\phi(t)$, and not on the Hamiltonian. From its construction it is clear that local phase transformations on $\phi(t)$ do not affect α_G. So they depend only on the rays of $\phi(t)$.

Here the geometry alluded to is that of 'curves' in Hilbert space, and the independence of α_G from local phase changes shows that it depends only on the sequence of rays in Hilbert space. If we take

$$\phi_1^*\phi_2 \cdot \phi_2^*\phi_3 \cdot \phi_3^*\phi_4 \cdots \phi_n^*\phi_1$$

this is an invariant under local phase transformations of $\phi_1, \phi_2, \ldots$ and gives a phase factor. When we take $\phi(t)$ continuously and multiply by dt, on integrating we obtain the 'geometrical phase'. This is a generalized Bargmann geometrical invariant (Bargmann 1964, Rabei *et al.* 1999).

13.7.1 More concerning non-integrable phases

We have come across a number of special instances where the manifold over which the Schrödinger equation is constructed may have non-trivial homotopy, i.e. when not all closed paths are contractible continuously to a point. If we consider two closed paths (loops) we could define the ordered product of these loops as the successive paths. This multiplication is associative, there is the trivial loop (which can be contracted to a point), and the inverse of a loop is the same loop traversed in the opposite sense. Thus, the loops fall into classes and these classes may be thought of as constituting a group. This is the fundamental group of the manifold. As an example, the homotopy group for spinors in $\mathbf{R}^3$ is Z_2. So is the homotopy of the complex-energy plane in which the momentum of a particle is considered. The first homotopy group need not be Abelian nor finite. On the other hand, if the orientation of an ellipsoid in $\mathbf{R}^3$ is considered as the configuration space, the homotopy group is the group of symmetries of the ellipsoid, which is finite but non-commutative. The homotopy group for the orientation of a generic rigid body is Z_2.

There are non-kinematic aspects inducing a non-trivial homotopy; the most familiar is the case of wave functions defined on $\mathbf{R}^3$ for a charged particle in a magnetic field. In this case any loop that encloses a magnetic flux acquires a phase depending on the number of times it loops around the magnetic flux and in which sense. The path does not have to be in any region where there is a magnetic flux! This startling conclusion follows from the gauge-invariant coupling of a charged particle to an external field:

$$H = \frac{1}{2m}(\vec{p} - e\vec{A})^2 - eA_0.$$

Then the wave function changes by the phase $\oint \vec{A} \cdot \mathrm{d}\vec{s}$. Although $\vec{A}$ is undefined up to a gauge, this loop integral is gauge-invariant. By virtue of the Stokes theorem it can be converted into a surface integral of the magnetic field, the flux. The phase would be unobservable if it were a multiple of 2π, and this leads to the notion of flux quantization. The non-trivial phase acquired by alternative paths for the wave function of a charged particle is called the Aharonov–Bohm effect (Aharonov and Bohm 1959).

In an apparently different context involving both nuclear and electron motion the electronic energy is so much more than the energy of the nuclear (ion) motion that the electronic energy levels can be computed considering the ion to be fixed. For the fast motions (of the electron) the energy depends on the position and momentum of the ion. Thus, when the electronic motions are integrated out, the ions find an additional potential energy, and the ion motion should be evaluated with this effective potential included. It turns out that the induced potentials involve a vector potential in addition to a scalar potential (Mead 1987). Whenever there is a vector potential there is the chance of an additional phase, and this in turn affects the energy of ion motion. Berry recognized that this is a situation in all complex-system periodic motions, and such a generic phase is the Berry phase previously introduced.

Independent of these developments, Pancharatnam investigated the phases of a light beam in a crystal that is transmitted from point A to point B directly or by an indirect route from A to C and from C to B. He found that there is a phase difference between the paths apart from the optical path length-caused phase difference. The Pancharatnam phase in a crystal can be made to modulate the frequency of light when the crystal undergoes rapid rotations around the beam axis.

13.8 The Wigner theorem and symmetries

In the course of investigating symmetries in quantum mechanics it becomes of crucial importance to understand the mathematical nature of the relationship between two descriptions used by two experimenters O and O' using differently oriented coordinate frames with a common origin. Let us assume that O uses the state vectors φ_α and φ_β to describe a pair of states with certain properties, while O' associates $\varphi_{\alpha'}$ and $\varphi_{\beta'}$ with states which, according to his description, have identical properties. Moreover, we assume that an invertible map θ exists between state vectors used by O and O', so that, if $c_1\varphi_1 + c_2\varphi_2$ is a state vector describing some state prepared by O, the corresponding state vector according to O' is $\theta(c_1\varphi_1 + c_2\varphi_2)$, and hence $\varphi_{\alpha'} = \theta\varphi_\alpha, \varphi_{\beta'} = \theta\varphi_\beta$. We assume that if O finds the value α for the observable $\hat{A}$ in the state φ_α, the probability of finding the value β immediately afterwards for the observable $\hat{B}$ in the state φ_β coincides with the corresponding probability evaluated by O', i.e.

$$|(\varphi_\beta,\varphi_\alpha)|^2 = |(\theta\varphi_\beta,\theta\varphi_\alpha)|^2, \tag{13.8.1a}$$

where unprimed observables $\hat{A}$ and $\hat{B}$ for O take the same values as primed observables $\hat{A}'$ and $\hat{B}'$ for O' in the transformed states $\theta\varphi_\alpha$ and $\theta\varphi_\beta$, respectively. Equation (13.8.1a) describes a one-to-one correspondence between rays for O and O', and hereafter we use it in the form

$$|(\varphi_\alpha,\varphi_\beta)| = |(\theta\varphi_\alpha,\theta\varphi_\beta)|. \tag{13.8.1b}$$

Let $\{u_n\}$ be an orthonormal basis prepared by O, while the equivalent orthonormal basis prepared by O' is denoted by $\{u_n'\}$. By hypothesis, the following completeness relations hold:

$$\sum_{n=1}^{\infty} u_n(u_n,\cdot) = \sum_{n=1}^{\infty} u_n'(u_n',\cdot) = \mathbb{1}. \tag{13.8.2}$$

The phases of these sets are completely arbitrary. If φ_α and its counterpart for O' are expanded according to Eq. (13.8.2), we are also free to choose the overall phases of φ_α and $\theta\varphi_\alpha$, since this does not affect the map between the ray containing φ_α and the ray containing $\theta\varphi_\alpha$. The Wigner theorem states that the phases can be chosen in such a way that the map θ is either unitary or anti-unitary. The former requirement means that θ preserves the scalar product and is linear (recall that our convention for the scalar product is anti-linear in the first argument, i.e. $(\varphi,\psi) \equiv \int_{\mathbf{R}^3} \varphi^*\psi \, d^3x$), i.e.

$$(\varphi_\alpha,\varphi_\beta) = (\theta\varphi_\alpha,\theta\varphi_\beta), \quad \theta\sum_{k=1}^{N} c_k\varphi_k = \sum_{k=1}^{N} c_k\theta\varphi_k, \tag{13.8.3}$$

while anti-unitarity changes the order of arguments in the scalar product and means anti-linearity, i.e.

$$(\varphi_\alpha,\varphi_\beta) = (\theta\varphi_\alpha,\theta\varphi_\beta)^*, \quad \theta\sum_{k=1}^{N} c_k\varphi_k = \sum_{k=1}^{N} c_k^*\theta\varphi_k. \tag{13.8.4}$$

Proof. We rely in part on the presentation in Gottfried (1966), and we first consider the state vector $\varphi_n \equiv u_1 + u_n$. Under the action of the map θ, it becomes

$$\begin{aligned}\theta\varphi_n &= \sum_{m=1}^{\infty} u_m'(u_m',\theta\varphi_n) \\ &= \sum_{m=1}^{\infty} u_m'\Big[(u_m',\theta u_1) + (u_m',\theta u_n)\Big],\end{aligned} \tag{13.8.5a}$$

where

$$(u'_m, \theta u_1) = e^{i\tau_m}(u_m, u_1) = e^{i\tau_m}\delta_{m1},$$

$$(u'_m, \theta u_n) = e^{i\tau_m}(u_m, u_n) = e^{i\tau_m}\delta_{mn},$$

and hence

$$\theta\varphi_n = e^{i\tau_1}u'_1 + e^{i\tau_n}u'_n. \tag{13.8.5b}$$

The phases of the basis $\{u'_n\}$ can be now redefined so as to absorb the phase factors $e^{i\tau_n}$, and eventually $\theta\varphi_n = u'_1 + u'_n$. Consider now the expansions of φ_α and $\theta\varphi_\alpha$, which imply

$$|c_n| = |c'_n| \tag{13.8.6}$$

by virtue of (13.8.1*b*). We can also obtain equations for the modulus of $c_1 + c_n$ and $c'_1 + c'_n$, because

$$(\varphi_n, \varphi_\alpha) = \sum_{m=1}^{\infty} c_m[(u_1, u_m) + (u_n, u_m)] = c_1 + c_n, \tag{13.8.7}$$

and similarly, $(\theta\varphi_n, \theta\varphi_\alpha) = c'_1 + c'_n$. We therefore have the chain of equalities

$$|(\varphi_n, \varphi_\alpha)| = |(\theta\varphi_n, \theta\varphi_\alpha)| = |c_1 + c_n| = |c'_1 + c'_n|$$

and squaring up the modulus and choosing $c_1 = c'_1$ we obtain the equation

$$(c_1 + c_n)(c_1^* + c_n^*) = (c_1 + c'_n)(c_1^* + {c'_n}^*), \tag{13.8.8}$$

where the squared modulus of c_1 and c_n occurs on both sides upon exploiting (13.8.6). In the resulting equation we multiply both sides by $x \equiv c'_n$ and obtain the algebraic equation

$$c_1^* x^2 - (c_1 c_n^* + c_1^* c_n)x + c_1|c_n|^2 = 0, \tag{13.8.9}$$

the roots of which are

$$x_+ = \frac{c_1 c_n^* + c_1^* c_n + \sqrt{(c_1 c_n^* - c_1^* c_n)^2}}{2c_1^*} = \frac{c_1}{c_1^*}c_n^*, \tag{13.8.10}$$

$$x_- = \frac{c_1 c_n^* + c_1^* c_n - \sqrt{(c_1 c_n^* - c_1^* c_n)^2}}{2c_1^*} = c_n. \tag{13.8.11}$$

The phase of the state vector φ_α can be now redefined so that c_1 becomes real-valued, and Eq. (13.8.10) eventually reads as $c'_n = c_n^*$, which corresponds to an anti-unitary map θ, for which

$$\theta\varphi_\alpha = \sum_{n=1}^{\infty} c_n^* u'_n, \tag{13.8.12}$$

while Eq. (13.8.11) corresponds to unitary transformations of φ_α.

Consider now another state vector

$$\varphi_\beta = \sum_{n=1}^{\infty} d_n u_n. \tag{13.8.13}$$

If both φ_α and φ_β are acted upon by a unitary map θ one finds

$$(\theta\varphi_\beta, \theta\varphi_\alpha) = \sum_{m,n=1}^{\infty} d_n^* c_m (u'_n, u'_m) = (\varphi_\beta, \varphi_\alpha), \tag{13.8.14}$$

in agreement with (13.8.1*b*), whereas if φ_α and φ_β were undergoing unitary and anti-unitary transformations, respectively, one would find

$$(\theta\varphi_\beta, \theta\varphi_\alpha) = \sum_{n=1}^{\infty} d_n c_n \neq (\varphi_\beta, \varphi_\alpha), \tag{13.8.15}$$

which violates the fundamental requirement (13.8.1b). Lastly, if both φ_α and φ_β undergo an anti-unitary transformation, one finds

$$(\theta\varphi_\beta, \theta\varphi_\alpha) = \sum_{m,n=1}^{\infty} d_n c_m^* (u'_n, u'_m) = (\varphi_\beta, \varphi_\alpha), \tag{13.8.16}$$

which again agrees with (13.8.1*b*). The desired proof is therefore completed (Wigner 1932a, 1959).

Note that an anti-unitary map θ can always be cast in the form

$$\theta = U\,K, \tag{13.8.17}$$

where U is a unitary operator, for which

$$UU^\dagger = U^\dagger U = \mathbb{I}, \tag{13.8.18}$$

and K denotes complex conjugation:

$$K\varphi_\alpha \equiv \varphi_\alpha^*. \tag{13.8.19}$$

Indeed, these formulae show that

$$\begin{aligned}(\theta\varphi_\beta, \theta\varphi_\alpha) &= \int_{\mathbf{R}^3} (UK\varphi_\beta)^* UK\varphi_\alpha \, \mathrm{d}^3x \\ &= \int_{\mathbf{R}^3} \varphi_\beta U^\dagger U \varphi_\alpha^* \, \mathrm{d}^3x = (\varphi_\alpha, \varphi_\beta).\end{aligned} \tag{13.8.20}$$

Anti-unitary symmetries are realized in nature by inversion of motion, more frequently called *time reversal* (Wigner 1932a). Let us assume that a physical system is described by the state vector φ_α at $t = 0$. After a short time interval δt it therefore evolves into

$$\varphi_\alpha(\delta t) = \left(\mathbb{I} - \frac{\mathrm{i}H}{\hbar}\delta t\right)\varphi_\alpha(0), \tag{13.8.21}$$

where H is the time-independent Hamiltonian operator. However, we can also apply the time-reversal operator T at $t = 0$ and then let the system evolve under the action of H. After the interval δt we then obtain the state vector $\left(\mathbb{I} - \frac{iH}{\hbar}\delta t\right) T\varphi_\alpha(0)$. If motion is symmetric under time reversal, such a state should coincide with $T\varphi_\alpha(-\delta t)$, i.e. we expect, for all $\varphi_\alpha(0)$, the condition

$$\left(\mathbb{I} - \frac{iH}{\hbar}\delta t\right) T\varphi_\alpha(0) = T\left(\mathbb{I} - \frac{iH}{\hbar}(-\delta t)\right)\varphi_\alpha(0), \tag{13.8.22}$$

which leads to the operator equation

$$-iHT = TiH. \tag{13.8.23}$$

If T were unitary, one would have $TiH = iHT$, and Eq. (13.8.23) would imply that T anti-commutes with the Hamiltonian. But this would allow for a negative energy spectrum for a free particle, whereas we know its energy spectrum is continuous and ranges over $\mathbf{R}_+ \cup \{0\}$. Under time reversal one finds that the momentum operator satisfies the condition

$$(\varphi_\alpha, \hat{p}\varphi_\alpha) = -(T\varphi_\alpha, \hat{p}T\varphi_\alpha), \tag{13.8.24a}$$

which can also be expressed in the form

$$T\hat{p}T^{-1} = -\hat{p}. \tag{13.8.24b}$$

The position operator is instead even under time reversal, in that

$$(\varphi_\alpha, \hat{x}\varphi_\alpha) = (T\varphi_\alpha, \hat{x}T\varphi_\alpha), \tag{13.8.25a}$$

which implies

$$T\hat{x}T^{-1} = \hat{x}. \tag{13.8.25b}$$

The anti-unitary nature of T makes it possible to preserve the canonical commutation relations, because position and momentum are even and odd as we just stated, while $Ti\hbar\mathbb{I} = -i\hbar T\mathbb{I}$.

Discrete symmetries described by unitary operators are instead parity, on the one hand, and translation on a lattice or finite rotations for a spherically symmetric problem, on the other hand.

13.9 A modern perspective on the Wigner theorem

We are now going to see that the geometrical phase introduced in section 13.7 makes it possible to gain a better understanding, and a more elegant proof, of the Wigner theorem. For this purpose, we begin by describing some aspects of ray space and its geometry.

Recall that rays of a Hilbert space $\mathcal{H}$ are equivalence classes of normalizable states of $\mathcal{H}$, differing only by multiplication by a non-vanishing complex number. The equivalence relation therefore states that two states $|\psi_1\rangle$ and $|\psi_2\rangle \in \mathcal{H} - \{0\}$ are equivalent:

$$|\psi_1\rangle \sim |\psi_2\rangle$$

if there exists $\alpha \in \mathbf{C} - \{0\}$ such that $|\psi_1\rangle = \alpha|\psi_2\rangle$. The ray space $\mathcal{R}$ is the quotient of $\mathcal{H} - \{0\}$ by this equivalence relation:

$$\mathcal{R} \equiv \mathcal{H} - \{0\} / \sim . \tag{13.9.1}$$

With the help of the natural projection $\Pi : \mathcal{H} - \{0\} \rightarrow \mathcal{R}$, one can map each normalizable state $|\psi\rangle$ to the ray ψ on which it lies. The concepts of *overlap* and *distance* of rays are quite important. The former is defined by

$$|\psi_1 \cdot \psi_2|^2 \equiv \frac{\langle\psi_1|\psi_2\rangle\langle\psi_2|\psi_1\rangle}{\langle\psi_1|\psi_1\rangle\langle\psi_2|\psi_2\rangle}. \tag{13.9.2}$$

By construction, the overlap takes values ≤ 1, and equals 1 if and only if $\psi_1 = \psi_2$. The *distance* is instead the map δ such that the square root of the overlap equals $\cos \frac{\delta(\psi_1,\psi_2)}{2}$, i.e.

$$|\psi_1 \cdot \psi_2| \equiv \cos \frac{\delta}{2}. \tag{13.9.3}$$

It is built in such a way that $\delta(\psi_1, \psi_2) = 0$ if and only if $\psi_1 = \psi_2$.

Given two non-orthogonal vectors $|A\rangle$ and $|B\rangle$, they are in phase if the inner product $\langle A|B\rangle$ is real and positive. Given a vector $|A\rangle$ and a ray $\mathbf{B}$, there is a unique vector $|B\rangle$ that is in phase with $|A\rangle$ and has the same size: $\langle B|B\rangle = \langle A|A\rangle$. The vector $|B\rangle$ is said to be the *Pancharatnam lift* of the ray $\mathbf{B}$, with the vector $|A\rangle$ as reference. Given three pairwise non-orthogonal rays $\mathbf{A}, \mathbf{B}, \mathbf{C}$ one can define the complex number

$$\delta_{ABC} \equiv \frac{\langle A|B\rangle\langle B|C\rangle\langle C|A\rangle}{\langle A|A\rangle\langle B|B\rangle\langle C|C\rangle} = \rho e^{i\beta}, \tag{13.9.4}$$

which depends only on the rays $\mathbf{A}, \mathbf{B}, \mathbf{C}$ and not on the representative vectors $|A\rangle, |B\rangle$ and $|C\rangle$. The phase β is the Pancharatnam excess phase, which is defined up to integer multiples of 2π.

An isometry of the ray space is a map $\mathbf{T} : \mathcal{R} \rightarrow \mathcal{R}$, which preserves distances. On writing $\psi' = \mathbf{T}\, \psi$ for the rays, the map $\mathbf{T}$ is an isometry if the overlaps are equal, i.e.

$$|\mathbf{A}' \cdot \mathbf{B}'| = |\mathbf{A} \cdot \mathbf{B}|. \tag{13.9.5}$$

The cosine of the Pancharatnam phase is an isometry invariant (Samuel 1997), and hence one can prove that

$$\delta'_{A'B'C'} = \chi(\delta_{ABC}), \tag{13.9.6}$$

where the function χ is defined by $\chi(\alpha) = \alpha$ or α^*. The following analysis depends crucially on the property (13.9.6).

Given a ray-space isometry $\mathbf{T}$, we now consider a map

$$T : \mathcal{H} - \{0\} \to \mathcal{H} - \{0\}$$

such that

$$\Pi(T|\psi\rangle) = \mathbf{T}(\Pi|\psi\rangle). \tag{13.9.7}$$

The map T is called the lift of the ray-space isometry $\mathbf{T}$. In general there are many such lifts, and one can make suitable requirements to pick out a relevant subset of the general set of lifts. For example, one may require continuity of T, but at that stage one is free to impose further restrictions on T. The Wigner theorem does just that. With the language of ray space, it can be stated as follows (Samuel 1997).

Wigner theorem. There exists a lift T of a given ray-space isometry $\mathbf{T}$ such that

(i) The inner product of vectors is preserved, i.e. $\langle \psi' | \psi' \rangle = \langle \psi | \psi \rangle$.

(ii) When extended to the Hilbert space $\mathcal{H}$ by $T|0\rangle = |0\rangle$, it preserves superpositions, i.e.

$$T\Big(|A\rangle + |B\rangle\Big) = T\,|A\rangle + T\,|B\rangle. \tag{13.9.8}$$

The lift T is unique up to an overall phase.

Thus, all ray-space isometries can be realized by maps on $\mathcal{H}$ satisfying the conditions (i) and (ii). To prove the Wigner theorem one has therefore to construct explicitly the map T, which we now do following the work in Samuel (1997). To begin, let $|u\rangle$ be any fixed vector in $\mathcal{H} - \{0\}$, $\mathbf{u}$ its ray with image

$$\mathbf{u}' = \mathbf{T}\,\mathbf{u}. \tag{13.9.9}$$

Let $|u'\rangle$ be an arbitrary vector corresponding to the ray $\mathbf{u}'$ and satisfying $\langle u' | u' \rangle = \langle u | u \rangle$, and define $T|u\rangle \equiv |u'\rangle$. Such a vector is arbitrary up to a phase. This is the only arbitrariness in the Samuel construction. Let

$$\mathcal{P} \equiv \{|\psi\rangle \in \mathcal{H} : \ \langle u | \psi \rangle = 0\} \tag{13.9.10}$$

be the set of elements in $\mathcal{H}$ orthogonal to $|u\rangle$. We also have to consider its complement $\mathcal{P}^{\rm c}$, i.e. the set of elements in $\mathcal{H}$ that are not orthogonal to $|u\rangle$:

$$\mathcal{P}^{\rm c} \equiv \{|\psi\rangle \in \mathcal{H} : \langle u | \psi \rangle \neq 0\}\,. \tag{13.9.11}$$

The idea is now to define the action of T on all elements of $\mathcal{P}^{\rm c}$ by using the Pancharatnam lift. Let $|\psi\rangle \in \mathcal{P}^{\rm c}$ be such an element. From

(13.9.5) it follows that $|(\psi'.\,\mathbf{u}')| \neq 0$. The vector $|\psi\rangle$ is mapped to the unique vector $|\psi'\rangle$, which satisfies the two conditions below. The first, i.e. $\langle\psi'|\psi'\rangle = \langle\psi|\psi\rangle$, determines the amplitude of $|\psi'\rangle$. The phase of $|\psi'\rangle$ can be chosen to satisfy the second condition, i.e.

$$\langle u'|\psi'\rangle = \chi(\langle u|\psi\rangle), \tag{13.9.12}$$

since $|\langle u'|\psi'\rangle| = |\langle u|\psi\rangle|$.

Equation (13.9.6), rewritten here in the form

$$\frac{\langle u'|A'\rangle\langle A'|B'\rangle\langle B'|u'\rangle}{\langle u'|u'\rangle\langle A'|A'\rangle\langle B'|B'\rangle} = \chi\left(\frac{\langle u|A\rangle\langle A|B\rangle\langle B|u\rangle}{\langle u|u\rangle\langle A|A\rangle\langle B|B\rangle}\right), \tag{13.9.13}$$

implies that if $|A\rangle$ and $|B\rangle$ are any two vectors in $\mathcal{P}^{\rm c}$, the vectors $|A'\rangle$ and $|B'\rangle$ satisfy

$$\langle A'|B'\rangle = \chi(\langle A|B\rangle). \tag{13.9.14}$$

The lift of the isometry $\mathbf{T}$ so realized preserves superpositions, because if $|\psi\rangle = |A\rangle + |B\rangle$ (with all vectors in $\mathcal{P}^{\rm c}$), the norm of the vector $|\phi'\rangle \equiv |\psi'\rangle - |A'\rangle - |B'\rangle$ vanishes. Indeed, one has

$$\begin{aligned}\langle\phi'|\phi'\rangle &= \langle\psi'|\psi'\rangle + \langle A'|A'\rangle + \langle B'|B'\rangle + \langle A'|B'\rangle + \langle B'|A'\rangle \\ &\quad - [\langle\psi'|A'\rangle + \langle A'|\psi'\rangle] - [\langle\psi'|B'\rangle + \langle B'|\psi'\rangle],\end{aligned} \tag{13.9.15}$$

where the three diagonal terms can be replaced by their unprimed versions, and the same can be done with the off-diagonal terms by exploiting (13.9.14). One then deals with the terms recombining into $\langle\phi|\phi\rangle$ which vanishes (we are indebted to J. Samuel for correspondence concerning his calculation). This leads to $|\psi'\rangle = |A'\rangle + |B'\rangle$, since the Hilbert space norm is positive-definite. Moreover, if $|A\rangle + |B\rangle = |C\rangle + |D\rangle$, again with all vectors in $\mathcal{P}^{\rm c}$, one can show that

$$|A'\rangle + |B'\rangle = |C'\rangle + |D'\rangle, \tag{13.9.16}$$

by computing the norm of the difference of both sides and again using Eq. (13.9.14).

It should be remarked that the sum $|A\rangle + |B\rangle$ need not be in $\mathcal{P}^{\rm c}$, and hence the action of T on elements of $\mathcal{P}$ can be defined by superposition. Any vector $|\Phi\rangle$ of $\mathcal{P}$ can be written as sum of elements in $\mathcal{P}^{\rm c}$. For instance,

$$|\Phi\rangle = \Big(|\Phi\rangle - |u\rangle\Big) + |u\rangle. \tag{13.9.17}$$

There are indeed many ways to express the vector $|\Phi\rangle$ as sum of elements of $\mathcal{P}^{\rm c}$, and it does not matter which of these ways one chooses. We have thus extended T on the whole Hilbert space $\mathcal{H}$ in such a way that conditions (i) and (ii) hold.

In the Samuel construction just outlined one first proves Eq. (13.9.6) as a geometric identity on the ray space. This result is then used as an input for constructing the lift T of the ray-space isometry $\mathbf{T}$ and showing that it does have the desired properties (i) and (ii). The proof of the Wigner theorem is then considerably simplified and acquires geometric nature. The Pancharatnam excess phase can either be preserved or reversed under isometries. The lift T is accordingly unitary or anti-unitary, respectively.

13.10 Problems

13.P1. Prove that the correlation defined in Eq. (13.4.6) can only depend on $\cos\theta_{ab}$.

13.P2. Prove that any state of two photons can be expressed as a linear combination of the four maximally entangled states

$$\psi_{12}^{+} \equiv \frac{1}{\sqrt{2}}\left(\psi_1^{\rm H} \otimes \psi_2^{\rm V} + \psi_1^{\rm V} \otimes \psi_2^{\rm H}\right), \tag{13.10.1}$$

$$\psi_{12}^{-} \equiv \frac{1}{\sqrt{2}}\left(\psi_1^{\rm H} \otimes \psi_2^{\rm V} - \psi_1^{\rm V} \otimes \psi_2^{\rm H}\right), \tag{13.10.2}$$

$$\phi_{12}^{+} \equiv \frac{1}{\sqrt{2}}\left(\psi_1^{\rm H} \otimes \psi_2^{\rm H} + \psi_1^{\rm V} \otimes \psi_2^{\rm V}\right), \tag{13.10.3}$$

$$\phi_{12}^{-} \equiv \frac{1}{\sqrt{2}}\left(\psi_1^{\rm H} \otimes \psi_2^{\rm H} - \psi_1^{\rm V} \otimes \psi_2^{\rm V}\right). \tag{13.10.4}$$

The projection of an arbitrary state of two photons onto such a basis is called a Bell-state measurement.

13.P3. In the experiment of section 13.5 on entangled pairs of photons, show how a three-photon state is related to the basis defined in (13.10.1)–(13.10.4).

13.P4. Try to apply the ideas of section 13.5 to give a theoretical description of an experiment where entangled states are transferred by means of quantum teleportation.

13.P5. Suppose that quantum information is passed from $P1$ to $P3$ to $P4$. Prove that such a 'teleportation' changes the original quantum states of $P1$ and $P3$.

13.P6. Prove that, for a spin-$\frac{1}{2}$ system, the time-reversal operator (Wigner 1932a) can be written in the form (η being an arbitrary phase factor)

$$T = \eta \, {\rm e}^{-{\rm i}\pi S_y/\hbar} K, \tag{13.10.5}$$

and hence its square is minus the identity. In the case of angular momentum quantum number j, prove that the eigenvalue of T^2 is $(-1)^{2j}$ (Sakurai 1985).

Part III
Selected topics

14

From classical to quantum statistical mechanics

Classical statistical mechanics tries to derive the macroscopic properties of matter starting from the mechanical laws that rule the behaviour of single particles. To describe equilibrium states, only observables accounting for correlations among states are considered, and only systems consisting of a large number of particles are taken into account. The observables are described by continuous functions on phase space, and the states are represented by linear assignments of a number to each observable. Within the framework of the canonical ensemble, one deals with mechanical systems in thermal equilibrium with a thermal reservoir, and the equilibrium state is reached as a result of the interaction with the external world. The external world may be really external, or equally well, the unrecognized internal degrees of freedom, like in the calculation of viscosity or thermal conductivity. If really only the external world is the cause we may expect some surface dependence while an internal unrecognized degree of freedom would have a volume effect, unless the interactions are long range. The microcanonical ensemble is instead introduced to study isolated mechanical systems, and the equilibrium is viewed as a temporal average, rather than as a limit. Attention is then focused on partition functions, the theorem of equipartition of energy and an elementary theory of specific heats.

In the second part, the Planck derivation of the law of black-body radiation is analysed, presenting in chronological order the Kirchhoff theorem, the Stefan law, the Wien displacement law, the Rayleigh–Jeans formula and the Planck hypothesis. Further topics discussed are the Einstein and Debye quantum models for specific heats of solids. These topics prepare the ground for the introduction of quantum statistical mechanics.

The third part is, in fact, devoted to the analysis of identical particles in quantum mechanics. Their distinguishing feature is the fact that they agree in all their intrinsic properties (i.e. all properties which are independent of the quantum state). Bose–Einstein and Fermi–Dirac statistics are studied in some detail. The former can be used to obtain a more fundamental derivation of the Planck radiation formula, and this task is accomplished in the last section.

14.1 Aims and main assumptions

The aim of statistical mechanics (e.g. Münster 1969, Morandi 1995) is to deduce the macroscopic properties of matter, e.g. those studied in thermodynamics, starting from the mechanical laws which rule the behaviour of single particles (material points possibly endowed with internal structure). Within this framework one has thus to pick out the dynamical quantities (described by functions on the phase space) which correspond to thermodynamical variables, and then check that the familiar laws of thermodynamics are indeed satisfied. One can also consider statistical states of 'small' systems realized by density matrices and then consider dynamical maps induced by outside systems. Stochastic evolution is natural for large systems but is by no means restricted to them. The basic and as yet not completely solved problem for non-equilibrium processes is to derive irreversible statistical processes from a reversible mechanics, which distinguish between the past and the future.

Some simple criteria can be easily stated. If the number N of elementary components is very large, the description of the system should not change in any essential way if the state of a relatively small number of components is altered (for example, when N is large, $\sqrt{N}$ is small compared with N). The resulting description of the state has a probabilistic nature, and a 'typical' sentence is the one according to which 'almost all configurations of elementary components studied so far have a given property'. For a system which is in a state of thermodynamical equilibrium, the same thermodynamical description should hold both for the whole system and for any 'sufficiently large' part of it. This leads to restrictions on the statistical distributions which can be used in statistical mechanics to describe states of thermodynamical equilibrium. Non-equilibrium thermodynamics can also be developed, and the basic problem is then to derive the irreversible statistical mechanics from a Hamiltonian description.

Starting from a Hamiltonian formalism, however, we will be interested in deriving a description of equilibrium states only. For this purpose, three fundamental assumptions are made.

(i) Only observables of a particular kind are taken into account, i.e. observables that elucidate only *correlations* among the states (position and momentum) of material points separated by a small distance. An example is provided by

$$F_N(q^1,\ldots,q^N;p_1,\ldots,p_N) \equiv \frac{1}{N}\sum_{i=1}^{N} f(q^i,p_i). \tag{14.1.1}$$

(ii) Only systems consisting of a large number N of particles are considered. Whenever appropriate, terms of order $O(1/N)$ will be neglected in the calculations. The thermodynamic limit is then taken, for which both the number of particles, N, and the volume, V, tend to ∞, while the ratio $\frac{V}{N}$ remains finite.

(iii) Only *equilibrium states* are studied. A state of the system is identified with a distribution on the phase space.

The *observables* are meant to be continuous functions on phase space, and *states* are linear assignments to each observable of a number, i.e. the result of the measurement of that particular observable in the given state (one can thus say that states are linear functionals). To the observables described by positive functions (called, hereafter, positive observables), the state ρ associates a non-negative number $\rho(f)$; in particular, to the observable represented by the function $f(q,p)=1$, one associates 1. The states are also required to be regular, in that

$$|f_n(z)| \leq 1, \quad \forall z, \quad \lim_{n\to\infty} \sup_z |f_n(z) - f(z)| = 0$$
$$\Longrightarrow \lim_{n\to\infty} \rho(f_n) = \rho(f). \tag{14.1.2}$$

States of a system can be thus identified with a regular (Borel) measure on phase space, normalized in such a way that the measure of the whole phase space is 1. Hence one deals with probability measures.

It should be stressed that the measures of statistical mechanics do not pick out a state characterized by the knowledge of positions and momenta of all its components, nor even a projection to a sub-space of the phase space. This is why the measures are actually called *statistical ensembles* (or, simply, ensembles). The measure of an open set $\mathcal{O}$ of phase space, describes the percentage of systems represented by points in $\mathcal{O}$ when one considers a collection of systems obtained by identical operations from the thermodynamical point of view (e.g. by heating up a gas under the same conditions of pressure and volume).

14.2 Canonical ensemble

Suppose one has a mechanical system, for which the Hamiltonian describes the interactions among the various parts of the system and with the external world. A measure is an *equilibrium measure* if it describes the state of the system at the end of an evolutionary process. Since one would like it to arrive at reasonable assumptions from which to derive an operational definition of an equilibrium state of a mechanical system, it is necessary to assume the existence of suitable mechanisms leading to 'equilibrium'. For this purpose, two schemes receive particular attention. The first goes under the name of *canonical ensemble*.

Within the framework of the canonical ensemble, the idea is to deal with a mechanical system in *thermal equilibrium with a thermal reservoir*. The equilibrium state is reached after a sufficiently long time, as a result of the interaction with the external world. However, although the external interaction is necessary for the system to reach equilibrium, one assumes that the measure representing such an equilibrium state depends on the interaction only through *parameters of a thermodynamical nature* and through the *Hamiltonian*. Moreover, the resulting measure μ should be such that, when restricted to thermodynamical functions depending only on M of the N points of the given system, it yields a state represented by a measure having the same form of μ, up to terms of order M^{-1}. This measure considered as a phase space density does not change with time for equilibrium states.

Under the above assumptions one can pick out, in simple cases, a measure on phase space which is absolutely continuous with respect to the Liouville measure, and having the form

$$\mu_T(z) = C\, e^{H(z)/T}, \tag{14.2.1}$$

where C is a normalization constant and T is the temperature of the thermal reservoir. Such a measure is called the Gibbs measure, it is nowhere vanishing in phase space and the corresponding ensemble is, by definition, the canonical ensemble. This concept is only of interest when N is very large.

14.3 Microcanonical ensemble

In the case of the *microcanonical ensemble* one studies isolated mechanical systems (hence the energy is constant), and the equilibrium is viewed as a temporal average (rather than as a limit). The underlying idea is that a thermodynamical process is quasi-static, and hence the measurements of quantities of physical interest take place over very long times (e.g. as compared with the mean time after which the position and momentum of a material point change by a considerable amount).

Thus, the values taken by the observables of the theory should be viewed as mean values (with respect to the time variable). If the measurement time T is very large, the mean value of the observable f

corresponding to the initial data

$$z^0 \equiv \left(q_0^1, \ldots, q_0^N, p_1^0, \ldots, p_N^0\right)$$

can be identified with

$$\overline{f}(z^0) \equiv \lim_{T\to\infty} \frac{1}{T} \int_0^T f(\phi(t, z^0))\, \mathrm{d}t, \tag{14.3.1}$$

where $\phi(t, \cdot)$ is the evolution described by the given Hamiltonian. If the surface Σ_E corresponding to the energy E is compact, and if the system is *ergodic* (a system which does not admit Lebesgue-measurable constants of motion besides the energy is called *ergodic*; an equivalent definition of ergodicity demands that the energy surface cannot be split into two positive-measure regions that are invariant under the temporal evolution), with equilibrium measure μ, for almost all the initial data one has

$$\overline{f}(z^0) = \int_{\Sigma_E} f \, \mathrm{d}\mu. \tag{14.3.2}$$

The equilibrium state is thus described by the measure μ. Nevertheless, it remains unknown whether the systems of material points of physical interest are ergodic. In other words, the choice of the measure μ as a measure describing thermodynamical equilibrium for a system of N material points with total energy E_N should be viewed as an *assumption.*

For mechanical systems which are ergodic, the equilibrium measure turns out to be the Lebesgue measure on Σ_E. The state described by such a measure is called the *microcanonical state* (or *microcanonical ensemble*). From the physical point of view, this represents what we can know, at a microscopic level, about the position and the momentum of a system of N particles of the same kind, when this is viewed as an isolated thermodynamical system, in equilibrium, with total energy E_N.

14.4 Partition function

Let us consider a system of N elements in the space Ω^N. Its energy is the extensive function

$$H(\widetilde{\omega}) \equiv \sum_{n=1}^{N} h(\omega_n), \widetilde{\omega} \equiv \{\omega_1, \ldots, \omega_N\}, \tag{14.4.1}$$

where h is a bounded and continuous function on Ω. One then has

$$\overline{h}(\mu) = \int_\Omega h(\omega)\, \mathrm{d}\mu(\omega), \tag{14.4.2}$$

and hence, for large values of N, the average energy $\frac{H_N}{N}$ is very likely to take the value $\overline{h}(\mu)$.

If one wants to study the behaviour of the system when the average energy has a fixed value h_1 different from $\overline{h}(\mu)$, one has to build, in Ω, a measure μ' that differs from μ, and such that

$$h_1 = \int_\Omega h(\omega)\, d\mu'(\omega), \tag{14.4.3}$$

and then build on Ω^N the corresponding measure $(\mu')^N$, and take the limit $N \to \infty$. The Gibbs formalism for canonical ensembles provides a systematic way to obtain this result.

Let us assume that all measures introduced on Ω are absolutely continuous with respect to the reference measure μ (in the physical applications, μ is the Liouville measure on phase space). Indeed, every probability measure ν which is absolutely continuous with respect to μ can be written in the form

$$\nu\, d\omega = e^{-\phi(\omega)}\, \mu(d\omega), \tag{14.4.4}$$

where $\phi(\omega)$ is a measurable function, and

$$\int_\Omega e^{-\phi(\omega)}\, \mu(d\omega) = 1. \tag{14.4.5}$$

In the mathematical literature, the function $e^{-\phi(\omega)}$ is called the Radon–Nykodim derivative of the measure ν with respect to the measure μ. The measure ν^N on Ω^N, obtained as a product of the measures ν, reads as

$$\nu^N = e^{-\sum_{n=1}^N \phi(\omega_n)}\, \mu^N. \tag{14.4.6}$$

The logarithm of the proportionality factor is hence an extensive variable. Since we want to vary the mean value of the energy by varying the function ϕ, we are led to take ϕ itself as being proportional to the energy, and we consider variations of the proportionality factor. Thus, for β a positive real number, one can write

$$\mu_\beta(d\omega) \equiv \frac{e^{-\beta h(\omega)}}{Z(\beta)} \mu(d\omega), \tag{14.4.7}$$

where

$$Z(\beta) \equiv \int_\Omega e^{-\beta h(\omega)}\, \mu(d\omega). \tag{14.4.8}$$

It is assumed that $h(\omega)$ is bounded from below. If $h(\omega)$ is also bounded from above β may also take negative values. The function Z is called the *partition function*, and turns out to provide a normalization factor which ensures that the measure μ_β is a probability measure. We note that β is the modulus of a distribution (Haas 1925, chapter 16) and is conjugate to the energy; it can be introduced as a Lagrange multiplier (negative values of β corresponding to negative temperatures) for the constant of energy conservation.

The probability measure μ_β^N in Ω^N, given by the product of N copies of μ_β, is called the *canonical measure* for the system of N elements, with the Hamiltonian

$$H_N \equiv \sum_{k=1}^{N} h(\omega_k)$$

and absolute temperature β^{-1} (the identification of β with $\frac{1}{KT}$ is actually a non-trivial step; only plausibility arguments can be given to justify this property).

14.5 Equipartition of energy

If a system of N material points is described by the canonical distribution, with Hamiltonian $H(q^1, \dots, q^N, p_1, \dots, p_N)$, and if *one assumes* that, for each value of the index n, one has

$$\lim_{|q_n| \to \infty} |q_n| \mathrm{e}^{-\beta H} = 0, \tag{14.5.1}$$

$$\lim_{|p_n| \to \infty} |p_n| \mathrm{e}^{-\beta H} = 0, \tag{14.5.2}$$

one finds, for all integer $n \in [1, N]$, and $\forall k = 1, 2, 3$,

$$\begin{aligned} &\int \mathrm{d}q^1 \cdots \mathrm{d}q^N \, \mathrm{d}p_1 \cdots \mathrm{d}p_N \; q_{n,k} \mathrm{e}^{-\beta H(q,p)} \frac{\partial H}{\partial q_{n,k}} \\ &\quad = \int \mathrm{d}q^1 \cdots \mathrm{d}q^N \, \mathrm{d}p_1 \cdots \mathrm{d}p_N \; p_{n,k} \mathrm{e}^{-\beta H(q,p)} \frac{\partial H}{\partial p_{n,k}} \\ &\quad = \frac{1}{\beta} = KT. \end{aligned} \tag{14.5.3}$$

To prove the equality (14.5.3) one has to bear in mind the identity

$$\mathrm{e}^{-\beta H(q,p)} \frac{\partial H}{\partial q_{n,k}} = -\frac{1}{\beta} \frac{\partial}{\partial q_{n,k}} \mathrm{e}^{-\beta H(q,p)}, \tag{14.5.4}$$

integrate by parts and then use (14.5.1) and (14.5.2). The result (14.5.3) expresses the generalized energy equipartition law.

In particular, if the Hamiltonian is a quadratic function of a set of (q, p) coordinates, with coefficients depending on the remaining coordinates,

$$H = \sum_{k=1}^{N} \Big(E_{q,k} + E_{p,k} \Big), \tag{14.5.5}$$

$$E_{q,k} \equiv \frac{1}{2} b_k q_k^2, \tag{14.5.6}$$

$$E_{p,k} \equiv \frac{1}{2} a_k p_k^2, \tag{14.5.7}$$

one finds

$$\overline{E}_{q,k} = \overline{E}_{p,k} = \frac{1}{2} KT. \tag{14.5.8}$$

Equation (14.5.8) expresses the *energy equipartition law*. Note that the scheme described by (14.5.5)–(14.5.7) is verified, for example, if q_k represents the displacement from an equilibrium position of a molecule acted upon by an elastic force, and if p_k is a component of the corresponding momentum. Note also that the assumption (14.5.1) may be replaced by the condition that the system should be confined in a cubic box with periodic boundary conditions. The explicit calculation shows indeed that such periodic boundary conditions are enough to ensure the vanishing of all terms resulting from the integration by parts.

14.6 Specific heats of gases and solids

The generalized equipartition law (14.5.3) can be used to evaluate the specific heats of perfect gases and solids. The resulting scheme is as follows.

(i) *Monatomic perfect gas.* Each molecule has an energy

$$h(q,p) = \frac{1}{2m} \left(p_x^2 + p_y^2 + p_z^2 \right), \tag{14.6.1}$$

and hence

$$\overline{h} = \frac{3}{2} KT, \tag{14.6.2}$$

and the average energy of a mole of the gas is

$$\overline{E} = \frac{3}{2} RT, \tag{14.6.3}$$

where $R = KN$ and N is the Avogadro number. If the gas is taken to be macroscopically at rest, the energy can be identified with the internal energy. The specific heat at constant volume V is then, per mole,

$$C_V \equiv \frac{d\overline{E}}{dT} = \frac{3}{2} R. \tag{14.6.4}$$

(ii) *Biatomic perfect gas.* Each molecule may be viewed as consisting of two material points of equal mass, linked by a rigid rod. Denoting by p

the momentum of the centre of gravity and by I the principal moment of the molecule with respect to the centre of gravity, one now has for the energy

$$h = \frac{1}{2m}\left(p_x^2 + p_y^2 + p_z^2\right) + \frac{1}{2I}\left(p_\theta^2 + \frac{1}{\sin^2\theta}p_\phi^2\right). \qquad (14.6.5)$$

By virtue of Eq. (14.5.3), this leads to

$$\overline{E} = \frac{5}{2}RT, \qquad (14.6.6)$$

$$C_V = \frac{5}{2}R. \qquad (14.6.7)$$

(iii) *Solids.* An useful model of a solid consists of N material points, with potential energy $U(x_1, \ldots, x_N)$. Let $\{x_n^0\}$ be an equilibrium configuration (e.g. the points of a crystal lattice). If the system performs small displacements from the equilibrium configuration, one can express the Hamiltonian as

$$H = \frac{1}{2}\sum_{n=1}^{N}\sum_{k=1}^{3} p_{n,k}^2 + \frac{1}{2}\sum_{n,m=1}^{N}\sum_{k,j=1}^{3}\left(\frac{\partial^2 U}{\partial x_{n,k}\partial x_{m,j}}\right)_{x=x^0} \times (x_{n,k} - x_{n,k}^0)(x_{m,j} - x_{m,j}^0). \qquad (14.6.8)$$

The resulting average energy is

$$\overline{E} = \frac{6}{2}(N-1)KT, \qquad (14.6.9)$$

where the degrees of freedom of the centre of gravity have been ruled out. Thus, for sufficiently large values of N, one finds the Dulong and Petit law

$$C_V = 3R. \qquad (14.6.10)$$

14.7 Black-body radiation

In this section, relying in part on Born (1969), we are aiming to derive the law of heat radiation, following Planck's method. We think of a box for which the walls are heated to a definite temperature T. The walls of the box send out energy to each other in the form of heat radiation, so that within the box there exists a radiation field. This electromagnetic field may be characterized by specifying the average energy density u, which in the case of equilibrium is the same for every internal point; if we split the radiation into its spectral components, we denote by $u_\nu \, d\nu$ the energy density of all radiation components for which the frequency falls in the interval between ν and $\nu + d\nu$. (The spectral density is not the only specification; we need to know the state of the entire radiation field including the photon multiplicity). Thus, the function u_ν extends over all frequencies from 0 to ∞, and represents a continuous spectrum. Note that, unlike individual atoms in rarefied gases, which emit line spectra, the molecules, which consist of a limited number of atoms, emit narrow 'bands', which are often resolvable. A solid represents an infinite number of vibrating systems of all frequencies, and hence emits an effectively continuous spectrum. But inside a black cavity all bodies emit a continuous spectrum characteristic of the temperature.

The first important property in our investigation is a theorem by Kirchhoff (1860), which states that the ratio of the emissive and absorptive powers of a body depends only on the *temperature* of the body, and not on its nature (recall that the *emissive power* is, by definition, the radiant energy emitted by the body per unit time, whereas the *absorptive power* is the fraction of the radiant energy falling upon it that the body absorbs). A *black body* is meant to be a body with absorptive power equal to unity, i.e. a body that absorbs all of the radiant energy that falls upon it. The radiation emitted by such a body, called *black-body radiation*, is therefore a function of the temperature alone, and it is important to know the spectral distribution of the intensity of this radiation. Any object inside the black cavity emits the same amount of radiant energy. We are now aiming to determine the law of this intensity, but before doing so it is instructive to describe in detail some arguments in the original paper by Kirchhoff (cf. Stewart 1858).

14.7.1 The Kirchhoff laws

The brightness $\mathcal{B}$ is the energy flux per unit frequency, per unit surface, for a given solid angle per unit time. Thus, if $\mathrm{d}E$ is the energy incident on a surface $\mathrm{d}S$ with solid angle $\mathrm{d}\Omega$ in a time $\mathrm{d}t$ with frequency $\mathrm{d}\nu$, one has (θ being the incidence angle)

$$\mathrm{d}E = \mathcal{B}\ \mathrm{d}\nu\ \mathrm{d}S\ \mathrm{d}\Omega\ \cos\theta\ \mathrm{d}t. \tag{14.7.1}$$

The brightness $\mathcal{B}$ is independent of position, direction and the nature of the material. This is proved as follows.

(i) $\mathcal{B}$ cannot depend on position, since otherwise two bodies absorbing energy at the same frequency and placed at different points P_1 and P_2 would absorb different amounts of energy, although they were initially at the same temperature T equal to the temperature of the cavity. One would then obtain the spontaneous creation of a difference of temperature, which would make it possible to realize a perpetual motion of the second kind. However, this is forbidden by thermodynamics.

(ii) $\mathcal{B}$ cannot depend on direction either. Let us insert into the cavity a mirror S of negligible thickness, and imagine we can move it along a direction parallel to its plane. In such a way no work is performed, and hence the equilibrium of radiation remains unaffected. Then let A and B be two bodies placed at different directions with respect to S and absorbing in the same frequency interval. If the amount of radiation incident upon B along the BS direction is smaller than that along the AS direction, bodies A and B attain spontaneously different temperatures, although they were initially in equilibrium at the same temperature! Thermodynamics forbids this phenomenon as well.

(iii) Once equilibrium is reached, $\mathcal{B}$ is also independent of the material the cavity is made of. Let the cavities C_1 and C_2 be made of different materials, and suppose they are at the same temperature and linked by a tube such that only radiation of frequency ν can pass through it. If $\mathcal{B}$ were different for C_1 and C_2 one would therefore obtain a non-vanishing energy flux through the tube. Thus, the two cavities would change their temperature spontaneously, against the second law of thermodynamics.

By virtue of (i)–(iii) equation (14.7.1) reads, more precisely, as

$$dE = \mathcal{B}(\nu, T)\, d\nu\, dS\, d\Omega\, \cos\theta\, dt. \tag{14.7.2}$$

Moreover, the energy absorbed by the surface element dS of the wall once equilibrium is reached is

$$dE_{\rm abs} = a_m(\nu, T, x)\, dE, \tag{14.7.3}$$

while the emitted energy is

$$dE_{\rm em} = e_m(\nu, T, x)\, d\nu\, dS\, d\Omega\, \cos\theta\, dt. \tag{14.7.4}$$

Under equilibrium conditions, the amounts of energy $dE_{\rm em}$ and $dE_{\rm abs}$ are equal, and hence

$$\frac{e_m(\nu, T, x)}{a_m(\nu, T, x)} = \mathcal{B} = \mathcal{B}(\nu, T). \tag{14.7.5}$$

Thus, the ratio of emissive and absorptive powers is equal to the brightness and hence can only depend on frequency and temperature, although both e_m and a_m can separately depend on the nature of materials.

As far as the production of black-body radiation is concerned, it has been proved by Kirchhoff that an enclosure (typically, an oven) at uniform temperature, in the wall of which there is a small opening, behaves as a black body. Indeed, all the radiation which falls on the opening from the outside passes through it into the enclosure, and is, after repeated reflection at the walls, completely absorbed by them. The radiation in the interior, and hence also the radiation which emerges again from the opening, should therefore possess exactly the spectral distribution of intensity, which is characteristic of the radiation of a black body.

14.7.2 Stefan and displacement laws

Remaining within the framework of thermodynamics and the electromagnetic theory of light, two laws can be deduced concerning the way in which black-body radiation depends on the temperature. First, the Stefan law (1879) states that the total emitted radiation is proportional to the fourth power of the temperature of the radiator. Thus, the hotter the body, the more it radiates. Secondly, Wien found the *displacement law* (1893), which states that the spectral distribution of the energy density is given by an equation of the form

$$u_\nu(\nu, T) = \nu^3 F(\nu/T), \tag{14.7.6}$$

where F is a function of the ratio of the frequency to the temperature, but cannot be determined more precisely with the help of thermodynamical methods. This formula can be proved by considering the radiation field in a volume V in the shape of a cube of edge length L with reflecting walls. The equilibrium radiation field will then consist of standing waves, and the condition that the electric field should vanish at the walls leads to the following relation for the frequency:

$$\left(\frac{\nu L}{c}\right)^2 = l^2 + m^2 + n^2, \tag{14.7.7}$$

where l, m, n are integers. If an adiabatic change of volume is performed, the quantities l, m and n being integers and hence unable to change infinitesimally will remain invariant. Under an adiabatic transformation the product νL is therefore invariant, or introducing the volume V instead of L:

$$\nu^3 V = \text{invariant} \tag{14.7.8}$$

under adiabatic transformation. The result can be proved to be independent of the shape of the volume.

However, it is more convenient to have a relation between ν and T, and for this purpose one has to consider the entropy of the radiation field. Classical electrodynamics tells us that the radiation pressure P is one-third of the total radiation energy density $u(T)$:

$$P = \frac{1}{3} u(T). \tag{14.7.9}$$

On combining Eq. (14.7.9) with the thermodynamic equation of state

$$\left(\frac{\partial U}{\partial V}\right)_T = T\left(\frac{\partial P}{\partial T}\right)_V - P, \tag{14.7.10}$$

and the relation $U = uV$, one obtains the differential equation

$$u = \frac{1}{3}T\frac{\mathrm{d}u}{\mathrm{d}T} - \frac{1}{3}u, \tag{14.7.11}$$

which is solved using the Stefan law:

$$u(T) = aT^4. \tag{14.7.12}$$

Equations (14.7.9) and (14.7.12), when combined with the thermodynamic Maxwell relation

$$\left(\frac{\partial S}{\partial V}\right)_T = \left(\frac{\partial P}{\partial T}\right)_V, \tag{14.7.13}$$

yield

$$S = \frac{4}{3}aT^3V. \tag{14.7.14}$$

By virtue of (14.7.8) and (14.7.14) one finds that, under an isentropic transformation, the ratio $\frac{\nu}{T}$ must be invariant. Moreover, since the resolution of a spectrum into its components is a reversible process, the entropy s per unit volume can be written as the sum of contributions $s_\nu(T)$ corresponding to different frequencies. Each of these terms, being a function of ν and with the entropy density corresponding to the specific frequency ν, can depend on ν and T only through the adiabatic invariant $\frac{\nu}{T}$, or (Ter Haar 1967)

$$s = \sum_\nu s(\nu/T). \tag{14.7.15}$$

Also the total energy density can be expressed by a sum:

$$u(T) = \sum_\nu \mathcal{U}_\nu(T), \tag{14.7.16}$$

and Eqs. (14.7.12) and (14.7.14) show that

$$s = \frac{4}{3}\frac{u}{T}, \tag{14.7.17}$$

and hence

$$\mathcal{U}_\nu(T) = Tf_1(\nu/T) = \nu f_2(\nu/T), \tag{14.7.18}$$

so that

$$u(T) = \sum_\nu \nu f_2(\nu/T) = \int_0^\infty \nu Z(\nu) f_2(\nu/T)\,\mathrm{d}\nu. \tag{14.7.19}$$

Such an equality is simple but non-trivial: summation over ν should be performed with the associated 'weight', and it should reduce to an integral over all values of ν from 0 to ∞ to recover agreement with the formula $u(T) = \int_0^\infty u_\nu(\nu, T)\, d\nu$, so that we write

$$\sum_\nu \cdot \rightarrow \int_0^\infty Z(\nu) \cdot d\nu.$$

This implies the following equation for the spectral distribution of energy density:

$$u_\nu(\nu, T) = \nu Z(\nu) f_2(\nu/T), \tag{14.7.20}$$

where $Z(\nu)\, d\nu$ is the number of frequencies in the radiation between ν and $\nu + d\nu$. By virtue of Eq. (14.7.7), this is proportional to the number of points with integral coordinates within the spherical shell between the spheres with radii $\nu L/c$ and $(\nu + d\nu)L/c$, from which it follows that

$$Z(\nu) = C\nu^2, \tag{14.7.21}$$

for some parameter C independent of ν. The laws expressed by (14.7.20) and (14.7.21) therefore lead to the Wien law (14.7.6) (Ter Haar 1967).

At this stage, however, it is still unclear why such a formula is called the 'displacement law'. The reason is as follows. It was found experimentally by Lummer and Pringsheim (see figure 14.1) that the intensity of the radiation from an incandescent body, maintained at a definite temperature, was represented, as a function of the wavelength, by a curve such that the product of the temperature T and the wavelength $\lambda_{\rm max}$ for which the intensity attains its maximum, is constant:

$$\lambda_{\rm max} T = \text{constant}. \tag{14.7.22}$$

The Wien law makes it possible to understand why Eq. (14.7.22) holds. Indeed, so far we have referred to the energy distribution as a function of the frequency ν, with u_ν representing the radiation energy in the frequency interval $d\nu$. The displacement law, however, refers to a graph showing the intensity distribution as a function of λ, so that we now deal with u_λ, representing the energy in the wavelength interval $d\lambda$. Of course, one has to require that

$$u_\nu\, d\nu = u_\lambda\, d\lambda. \tag{14.7.23}$$

Moreover, since $\lambda\nu = c$, one has, for the relation between $d\nu$ and λ,

$$\frac{|d\nu|}{\nu} = \frac{|d\lambda|}{\lambda}. \tag{14.7.24}$$

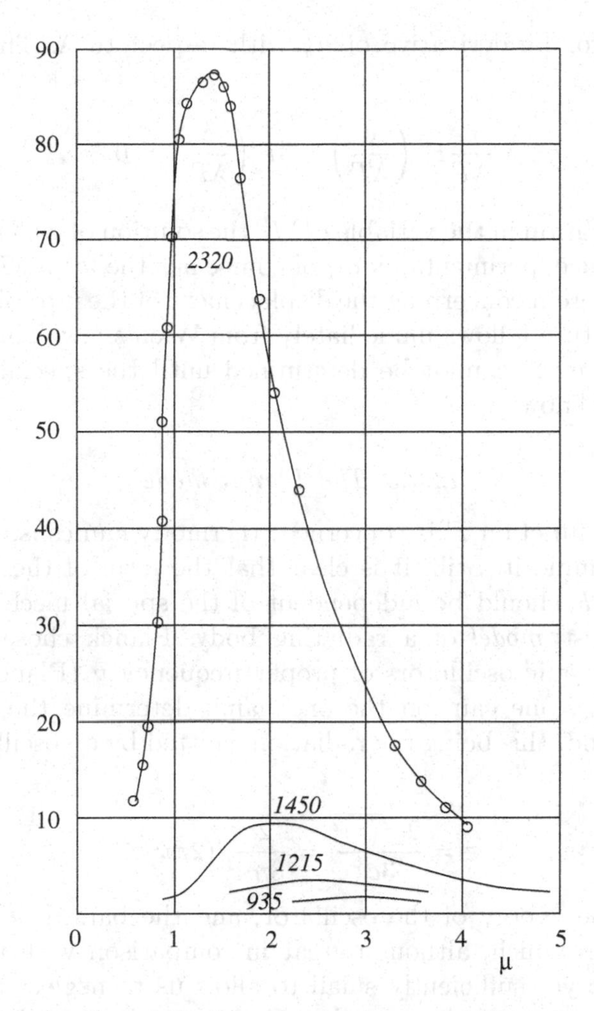

Fig. 14.1. Distribution of the intensity of thermal radiation as a function of wavelength according to the measurements of Lummer and Pringsheim. The y-axis corresponds to $u(\lambda, T) \times 10^{-11}$ in CGS units (Born 1969, by kind permission of professor Blin-Stoyle).

Thus, for the spectral distribution of energy expressed as a function of the wavelength, one finds

$$u_\lambda = \frac{c^4}{\lambda^5} F\left(\frac{c}{\lambda T}\right). \tag{14.7.25}$$

We are now in a position to prove the displacement law at once, by evaluating the wavelength for which u_λ is a maximum. For this purpose, we set to zero the derivative of u_λ with respect to λ. This eventually yields the equation

$$\frac{c}{\lambda T} F'\left(\frac{c}{\lambda T}\right) + 5F\left(\frac{c}{\lambda T}\right) = 0. \tag{14.7.26}$$

This is an equation in the variable $c/\lambda T$, the solution of which (we know it exists from the experimental data) of course has the form $\lambda T =$ constant. Thus, the theorem concerning the displacement of the intensity maximum with temperature follows immediately from Wien's law. The value of the constant, however, cannot be determined until the special form of the function F is known.

14.7.3 The Planck model

As far as the function F is concerned, thermodynamics is, by itself, unable to determine it. Still, it is clear that the form of the law given by the function F should be independent of the special mechanism. Thus, as the simplest *model* of a radiating body, Planck chose a collection of linear harmonic oscillators of proper frequency ν (Planck 1900). For each oscillator, one can, on the one hand, determine the energy radiated per second, this being the radiation emitted by an oscillating dipole, given by

$$\delta\varepsilon = \frac{2e^2\overline{(\ddot{r})^2}}{3c^3} = \frac{2e^2}{3mc^3}(2\pi\nu)^2\overline{\varepsilon}, \tag{14.7.27}$$

where ε is the energy of the oscillator, and the bars denote mean values over times which, although great in comparison with the period of vibration, are yet sufficiently small to allow us to neglect the radiation emitted. By virtue of the equation of motion one has $\ddot{r} = -(2\pi\nu)^2 r$, and

$$\overline{\varepsilon}_{\text{kin}} = \frac{1}{2}m\overline{\dot{r}^2} = \frac{1}{2}m\overline{(2\pi\nu r)^2} = \overline{\varepsilon}_{\text{pot}} = \frac{1}{2}\varepsilon. \tag{14.7.28}$$

On the other hand, it is well known from classical electrodynamics that the work done on the oscillator per second by a radiation field with the spectral energy density u_ν is (see appendix 14.A)

$$\delta W = \frac{\pi e^2}{3m}u_\nu. \tag{14.7.29}$$

When energy balance is achieved, these two amounts of energy should be equal. Hence one finds

$$u_\nu = \frac{8\pi\nu^2}{c^3}\overline{\varepsilon}. \tag{14.7.30}$$

It is thus clear that, if we know the mean energy of an oscillator, we also know the spectral intensity distribution of the cavity radiation.

The value of $\overline{\varepsilon}$, as determined by the theorem of equipartition of energy of classical statistical mechanics (section 14.5), would be $\overline{\varepsilon} = KT$. This happens because, according to a classical analysis, any term in the Hamiltonian that is proportional to the square of a coordinate or a momentum, contributes the amount $\frac{1}{2}KT$ to the mean energy. For the harmonic oscillator there are two such terms, and hence $\overline{\varepsilon}$ equals KT. Now if the classical mean value of the energy of the oscillator, as just determined, is inserted into the radiation formula (14.7.30), one obtains

$$u_\nu = \frac{8\pi\nu^2}{c^3} KT. \tag{14.7.31}$$

This is the Rayleigh–Jeans radiation formula proposed in Rayleigh (1900) and Jeans (1905) (actually Rayleigh forgot the polarization while Jeans corrected this). Some remarks are now in order.

(i) The Rayleigh–Jeans formula agrees with the Wien displacement law. This was expected to be the case, since the Wien law is deduced from thermodynamics, and hence should be of universal validity.

(ii) For long-wave components of the radiation, i.e. for small values of the frequency ν, the Rayleigh–Jeans equation reproduces the experimental intensity distribution very well.

(iii) For high frequencies, however, the formula (14.7.31) is completely wrong. It is indeed known from experiments that the intensity function reaches a maximum at a definite frequency and then decreases again. In contrast, Eq. (14.7.31) fails entirely to show this maximum, and instead describes a spectral intensity distribution that becomes infinite as the frequency ν tends to infinity. The same is true of the total energy of radiation, obtained by integrating u_ν over all values of ν from 0 to ∞: the integral diverges. We are facing, here, what is called in the literature the *ultraviolet catastrophe*.

To overcome this serious inconsistency, Planck *assumed* the existence of *discrete, finite quanta of energy*, here denoted by ε_0. According to this scheme, the energy of the oscillators can only take values that are integer multiples of ε_0, including 0. We are now going to see how this hypothesis leads to the so-called Planck radiation law. The essential point is, of course, the determination of the mean energy $\overline{\varepsilon}$. The derivation differs from that resulting from classical statistical mechanics only in replacing integrals by sums. The individual energy values occur again with the 'weight' given by the Boltzmann factor, but one should bear in mind that only the energy values $n\varepsilon_0$ are admissible, n being an integer greater than or equal to 0. In other words, the Planck hypothesis leads to the following expression for the mean energy (the parameter β being equal to $1/KT$):

$$\overline{\varepsilon} = \frac{\sum_{n=0}^{\infty} n\varepsilon_0 \mathrm{e}^{-\beta n\varepsilon_0}}{\sum_{n=0}^{\infty} \mathrm{e}^{-\beta n\varepsilon_0}} = -\frac{\mathrm{d}}{\mathrm{d}\beta} \log \sum_{n=0}^{\infty} \mathrm{e}^{-\beta n\varepsilon_0}. \tag{14.7.32}$$

At this stage, we recall an elementary property that follows from the analysis of the geometrical series:

$$\begin{aligned}\sum_{n=0}^{\infty} \mathrm{e}^{-\beta n\varepsilon_0} &= \lim_{N\to\infty} \sum_{n=0}^{N} (\mathrm{e}^{-\beta\varepsilon_0})^n \\ &= \lim_{N\to\infty} \frac{1-(\mathrm{e}^{-\beta\varepsilon_0})^{N+1}}{1-\mathrm{e}^{-\beta\varepsilon_0}} = \frac{1}{1-\mathrm{e}^{-\beta\varepsilon_0}}.\end{aligned} \tag{14.7.33}$$

The joint effect of Eqs. (14.7.32) and (14.7.33) is thus

$$\overline{\varepsilon} = \frac{\varepsilon_0 \mathrm{e}^{-\beta\varepsilon_0}}{1-\mathrm{e}^{-\beta\varepsilon_0}} = \frac{\varepsilon_0}{\mathrm{e}^{\beta\varepsilon_0}-1}. \tag{14.7.34}$$

Equations (14.7.30) and (14.7.34) therefore lead to the radiation formula

$$u_\nu = \frac{8\pi\nu^2}{c^3} \frac{\varepsilon_0}{\mathrm{e}^{\varepsilon_0/KT}-1}. \tag{14.7.35}$$

To avoid obtaining a formula that is inconsistent with the Wien displacement law, which, being derived from thermodynamics alone, is certainly valid, we have to assume that (the temperature being forced to appear only in the combination ν/T)

$$\varepsilon_0 = h\nu, \tag{14.7.36}$$

where h is the Planck constant that we know from chapter 1. Hence one eventually obtains the fundamental Planck radiation law:

$$u_\nu = \frac{8\pi h\nu^3}{c^3} \frac{1}{\mathrm{e}^{h\nu/KT}-1}. \tag{14.7.37}$$

Bose gave an independent derivation of the Planck formula by considering photons as strictly identical particles (Bose 1924a). This quantum derivation in 1924 before quantum theory was properly formulated (by Schrödinger, Heisenberg and Dirac) required a new method considering combinations of strictly interacting particles. This is the integer-spin analogue of the Pauli principle (section 14.9) and referred to as Bose statistics.

The radiation formula (14.7.37) is in very good agreement with all experimental results (see figure 14.2). In particular, for low values of the frequency, it reduces to the Rayleygh–Jeans formula (14.7.31), whereas,

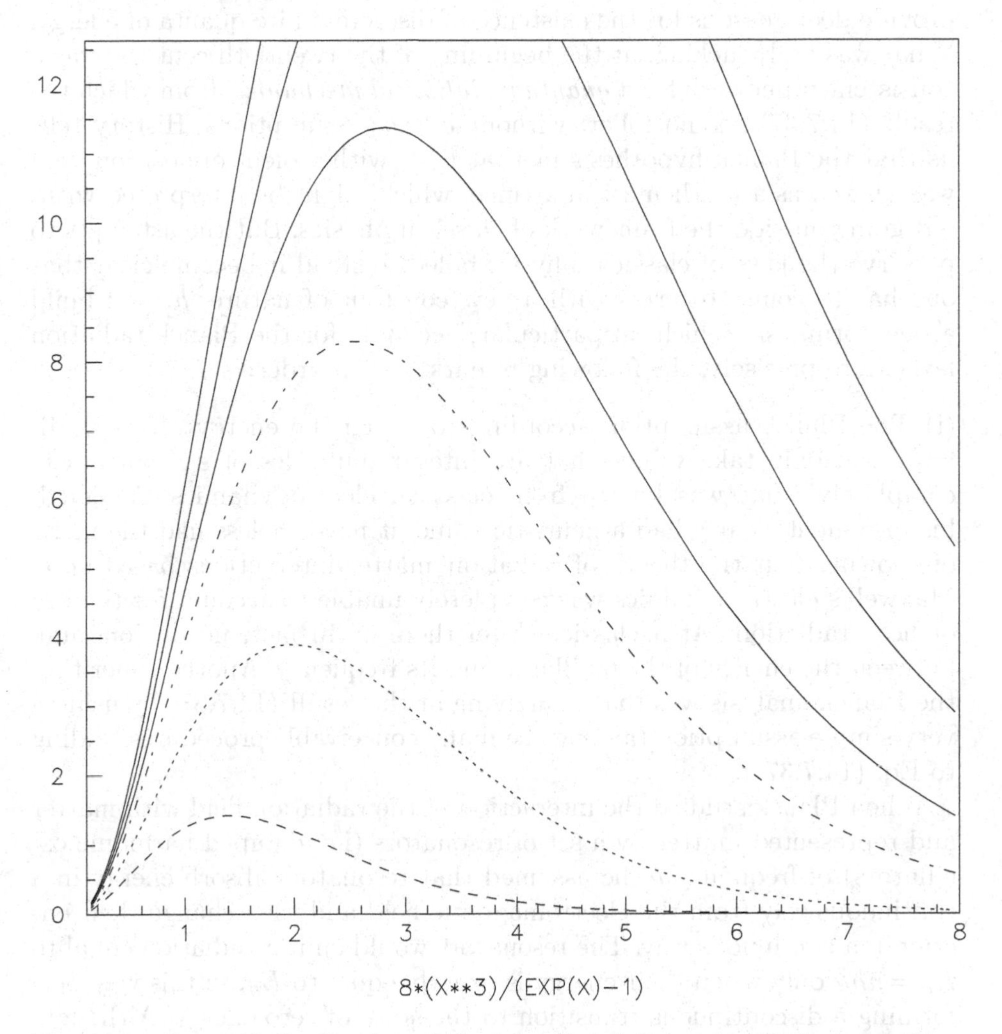

Fig. 14.2. Spectral distribution of the intensity of thermal radiation according to Planck, for temperatures between 2000 and 4500 K. The abscissa corresponds to wavelength. The lowest curve is obtained for $T = 2000$ K, and the following curves correspond to values of T equal to $2500, 3000, 3500, 4000$ and 4500 K, respectively. There is full agreement with the experimental results shown in figure 14.1.

as ν tends to ∞, it takes the approximate form

$$u_\nu \sim \frac{8\pi h\nu^3}{c^3} e^{-h\nu/KT}, \tag{14.7.38}$$

which agrees with a formula first derived, empirically, in Wien (1896), in the attempt to account for measurements in this region of the spectrum.

However, the derivation of the Planck radiation law given so far, following Planck's argument, is heuristic and unsatisfactory, since it does not provide deep reasons for the existence of discrete, finite quanta of energy. What was truly lacking, at the beginning of the twentieth century, was a consistent framework for a *quantum statistical mechanics*, from which the result (14.7.37) should follow without *ad hoc* assumptions. History tells us that the Planck hypothesis met, at first, with violent opposition, and was viewed as a mathematical artifice, which might be interpreted without going outside the framework of classical physics. But the attempts to preserve the laws of classical physics failed. Instead it became clear that one has to come to terms with a new constant of nature, h, and build a new formalism which, in particular, accounts for the Planck radiation law. More precisely, the following remarks are in order.

(i) The Planck assumption according to which the energy of the oscillator can only take values that are integer multiples of ε_0 contradicts completely what was known from classical electrodynamics. Although his argument clearly had a heuristic value, it nevertheless had the merit of showing that the theory of radiation–matter interactions based upon Maxwell's electrodynamics was completely unable to account for the law of heat radiation. At a classical level there is, in fact, no obvious link between the energy of the oscillator and its frequency. Another 'merit' of the Planck analysis was that of arriving at the result (14.7.37) by using a very simple assumption (among the many conceivable procedures leading to Eq. (14.7.37)).

When Planck studied the interaction of the radiation field with matter and represented matter by a set of resonators (i.e. damped harmonic oscillators) of frequency ν, he assumed that resonators absorb energy in a continuous way from the electromagnetic field and they change their energy in a continuous way. The resonators would emit a radiation equal to $E_n = nh\nu$ only when their energy is exactly equal to E_n, in this way performing a discontinuous transition to the state of zero energy. Although, with hindsight, we know that such a model is incorrect, we should acknowledge that it contains ideas which had a deep influence. (I) The different orbits (with $H = E_n$) divide phase space into regions of area h. (II) The average energy of a quantum state turns out to be $\left(n + \frac{1}{2}\right) h\nu$, hence leading to the concept of zero-point energy for the first time. (III)

The emission of radiation is viewed, for the first time, as a probabilistic process (Parisi 2001).

Planck did not realize that at equilibrium, in classical mechanics, the properties of the resonators do not affect the black-body radiation, but he had the merit of isolating the physically relevant points, where progress could be made, not paying attention to the possible contradictions (Parisi 2001).

(ii) Since the electromagnetic field inside the box interacts with a very large number of oscillators, it was suggested for some time that the particular 'collective' properties of matter, rather than energy exchanges with a single atom, can account for the Planck hypothesis without making it mandatory to give up the classical theory of electromagnetic phenomena. However the work in Einstein (1905, 1917), see below, proved that the energy of the electromagnetic field (and the associated linear momentum) is localized, and hence radiation–matter interactions are localized, and cannot be understood by appealing to collective properties of material media.

(iii) It should be stressed that no thermal equilibrium can ever be reached within a box with reflecting walls. The single monochromatic components of the electromagnetic field do not interact with each other, and hence no process can lead to thermal equilibrium in such a case. Fortunately, there are *no* perfect reflectors.

(iv) As far as the emission of radiation is concerned, all energy-balance arguments should take into account both induced emission and spontaneous emission. The former results from the interaction with an external field, whereas the latter may be due to energy acquired during previous collisions, or to previous interactions with an electromagnetic field.

14.7.4 The contributions of Einstein

In Einstein (1905) the author found that, in the region where the Wien law is valid, one can say that thermodynamically speaking monochromatic radiation consists of independent energy quanta of magnitude $h\nu$. To prove this, he applied thermodynamical concepts to the electromagnetic radiation, starting from the definition of temperature

$$\frac{1}{T} \equiv \left(\frac{\partial S}{\partial U}\right)_V = \frac{\partial \sigma}{\partial u_\nu}, \tag{14.7.39}$$

where the entropy density σ refers to a constant volume, and the same holds for u_ν. If the Wien law holds, i.e. for $h\nu >> KT$:

$$u_\nu = b\nu^3 e^{-h\nu/KT}, \tag{14.7.40}$$

Eq. (14.7.39) leads to

$$\frac{\partial \sigma}{\partial u_\nu} = -\frac{K}{h\nu} \log\left(\frac{u_\nu}{b\nu^3}\right), \tag{14.7.41}$$

which is solved by

$$\sigma(\nu, T) = -\frac{K u_\nu}{h\nu}\left[\log\left(\frac{u_\nu}{b\nu^3}\right) - 1\right]. \tag{14.7.42}$$

Thus, the entropy S in a volume V reads as

$$S = \sigma V = -\frac{K u_\nu V}{h\nu}\left[\log\left(\frac{u_\nu V}{bV\nu^3}\right) - 1\right] = -\frac{KE}{h\nu}\left[\log\left(\frac{E}{bV\nu^3}\right) - 1\right], \tag{14.7.43}$$

where $E = u_\nu V$ is the total energy of monochromatic radiation in a volume V. If the energy is kept fixed while the volume is expanded from V_0 to V, the resulting variation of entropy is

$$S - S_0 = \frac{KE}{h\nu} \log\left(\frac{V}{V_0}\right). \tag{14.7.44}$$

On the other hand, if the radiation is treated as an ideal gas undergoing an isothermal expansion, one can write for the variation of entropy another formula, i.e.

$$S - S_0 = \frac{1}{T}\int_{V_0}^{V} dQ = NK \int_{V_0}^{V} \frac{dy}{y} = NK \log \frac{V}{V_0}. \tag{14.7.45}$$

Equations (14.7.44) and (14.7.45) express the same variation of entropy at fixed energy, and tell us that monochromatic radiation of frequency $\nu >> \frac{KT}{h}$ behaves as a gas of N particles for which the total energy

$$E = Nh\nu. \tag{14.7.46}$$

Thus, each particle can be thought of as a photon of energy $h\nu$.

In Einstein (1917), the author obtained a profound and elegant derivation of the Planck radiation formula by considering the canonical distribution of statistical mechanics for molecules which can take only a discrete set of states $Z_1, Z_2, \ldots, Z_n, \ldots$ with energies $E_1, E_2, \ldots, E_n, \ldots$:

$$W_n = p_n e^{-E_n/kT}, \tag{14.7.47}$$

where W_n is the relative occurrence of the state Z_n, p_n is the statistical weight of Z_n and T is the temperature of the gas of molecules. On the one hand, a molecule might perform, without external stimulation, a transition from the state Z_m to the state Z_n while emitting the radiation energy $E_m - E_n$ of frequency ν. The probability $\mathrm{d}W$ for this process of *spontaneous emission* in the time interval $\mathrm{d}t$ is

$$\mathrm{d}W = A_{m \rightarrow n}\, \mathrm{d}t, \tag{14.7.48}$$

where $A_{m \rightarrow n}$ denotes a constant.

On the other hand, under the influence of a radiation density u_ν, a molecule can make a transition from the state Z_n to the state Z_m by absorbing the radiative energy $E_m - E_n$, and the probability law for this process is

$$\mathrm{d}W = B_{n \rightarrow m} u_\nu \, \mathrm{d}t. \tag{14.7.49}$$

Moreover, the radiation can also lead to a transition of the opposite kind, i.e. from state Z_m to state Z_n. The radiative energy $E_m - E_n$ is then freed according to the probability law

$$\mathrm{d}W = B_{m \rightarrow n} u_\nu \, \mathrm{d}t. \tag{14.7.50}$$

In these equations, $B_{n \rightarrow m}$ and $B_{m \rightarrow n}$ are yet other constants.

Einstein then looked for that particular radiation density u_ν which guarantees that the exchange of energy between radiation and molecules preserves the canonical distribution (14.7.47) of the molecules. This is achieved if and only if, on average, as many transitions from Z_m to Z_n take place as of the opposite type, i.e.

$$\Big(B_{m \rightarrow n} u_\nu + A_{m \rightarrow n} \Big) p_m \mathrm{e}^{-E_m/KT} = B_{n \rightarrow m} u_\nu p_n \mathrm{e}^{-E_n/KT}. \tag{14.7.51}$$

Einstein also assumed that the B constants are related by

$$p_m B_{m \rightarrow n} = p_n B_{n \rightarrow m}, \tag{14.7.52}$$

to ensure that as the temperature tends to infinity u_ν also tends to infinity, and hence found

$$u_\nu = \frac{A_{m \rightarrow n}/B_{m \rightarrow n}}{\mathrm{e}^{(E_m - E_n)/KT} - 1}, \tag{14.7.53}$$

which reduces to the Planck radiation formula (14.7.37), by virtue of the Wien displacement law (14.7.6), which implies

$$\frac{A_{m \rightarrow n}}{B_{m \rightarrow n}} = \alpha_1 \nu^3, \tag{14.7.54}$$

$$E_m - E_n = \alpha_2 \nu, \tag{14.7.55}$$

where α_1 and α_2 are constants, which cannot be fixed at this stage (Einstein 1917).

14.7.5 Dynamic equilibrium of the radiation field

While spontaneous emission was known for a long time in atomic physics, it was Einstein who emphasized its role and derived the Planck distribution of spectral energy on a *dynamic* basis as we have just seen, in contrast with the original Planck derivation. Einstein considered a two-level atom and monochromatic radiation of frequency $\nu = \frac{(E_2 - E_1)}{h}$. But in actual fact there are many frequencies, many species of atoms and many energy levels (and populations of these levels). This generic problem was posed and solved by Bose. In the briefest outlook his derivation observes that, like in Maxwell's derivation of the velocity distribution in kinetic theory, the various populations enter through appropriate Lagrange multipliers. Dynamic balance of the entire complex demands that for *every* frequency we have the law (14.7.37). In both Einstein's and Bose's derivations the atomic population in each level was proportional to the Boltzmann factor $e^{-E_n/KT}$. Other important work on black-body radiation can be found in Mandel (1963) and Mandel *et al.* (1964).

14.8 Quantum models of specific heats

As we know from (14.6.10) it is a well-established experimental property that, *at high temperatures*, a law holds, called the Dulong and Petit law, according to which the specific heat per mole is approximately 6 cal K^{-1} for all solids. The classical interpretation of such a property is as follows. In a solid, every atom may be viewed as a three-dimensional harmonic oscillator, since one thinks of the atom as quasi-elastically bound to a certain position of equilibrium. By virtue of the theorem of equipartition of energy of section 14.5, their mean total energy is hence $3KT$, so that a mole of the given substance possesses the energy $U = 3N_0KT = 3RT$. The specific heat at constant volume is therefore

$$C_V = \frac{\partial U}{\partial T} = 3R. \tag{14.8.1}$$

The experiments, however, show deviations from this classical formula, and the more firmly the atoms are bound to their equilibrium positions, the greater are the deviations.

The first attempt to overcome this inconsistency was due to Einstein. For this purpose, he proposed applying the expression obtained by Planck for the mean energy of a quantized oscillator. His idea was to consider a model of independent oscillators, *all of the same frequency*. This was unfortunate since it gave an exponentially decreasing specific heat. The resulting mean energy per mole should be

$$U = \frac{3N_0 h\nu}{\mathrm{e}^{h\nu/KT} - 1} = 3RT \frac{h\nu/KT}{\mathrm{e}^{h\nu/KT} - 1}. \tag{14.8.2}$$

At high temperatures, when the ratio $h\nu/KT$ is much smaller than 1, one finds

$$U \sim 3RT\Big[1 + \mathrm{O}(h\nu/KT)\Big], \tag{14.8.3}$$

and hence one recovers agreement with the property expressed by (14.8.1).

However, the assumption that the same frequency can be used for all oscillators is an extreme mathematical idealization, and the low-temperature behaviour of the resulting specific heat was incorrect. A first improvement is obtained by taking into account that the atoms in a lattice are very strongly coupled together. Thus, one cannot claim that the N_0 atoms in a crystal perform oscillations of the same frequency. One has instead to deal with a coupled system of $3N_0$ vibrations (corresponding to the $3N_0$ degrees of freedom of the N_0 atoms per mole). According to this improved scheme, one should express the energy in the form

$$U = \sum_{r=1}^{3N_0} \frac{h\nu_r}{\mathrm{e}^{h\nu_r/KT} - 1}, \tag{14.8.4}$$

where ν_r is the frequency corresponding to an individual vibration.

But such a description is still inappropriate. As pointed out by Debye, the frequencies ν_r are not the proper vibrations of the atom, but they correspond to the *frequencies of elastic waves* in the body (this implies that only waves for which the wavelength is much larger than the atomic separations are considered). The number $\mathrm{d}N$ of such vibrations in the frequency interval in between ν and $\nu + \mathrm{d}\nu$ in the volume V is

$$\mathrm{d}N = 4\pi V \left(\frac{2}{c_{\mathrm{t}}^3} + \frac{1}{c_{\mathrm{l}}^3} \right) \nu^2 \, \mathrm{d}\nu, \tag{14.8.5}$$

because both transverse waves (with speed c_{t}) and longitudinal waves (with speed c_{l}) can propagate in a crystal, the former having two degrees of freedom (two propagation directions exist, orthogonal to each other). Thus, on defining an average speed $\overline{c}$ by the equation

$$\frac{1}{\overline{c}^3} \equiv \frac{2}{c_{\mathrm{t}}^3} + \frac{1}{c_{\mathrm{l}}^3}, \tag{14.8.6}$$

one can write down for the mean energy the formula (Born 1969)

$$U = \int_0^{\nu_m} \frac{h\nu}{\mathrm{e}^{h\nu/KT} - 1} \frac{4\pi V}{\overline{c}^3} \nu^2 \, \mathrm{d}\nu. \tag{14.8.7}$$

At low temperature this gives a T^3 law for specific heat, which is the same law as for Planck's theory (which was formulated 16 years earlier!). What is non-trivial, here, is the existence of the maximum frequency $\nu_{\rm m}$, which is defined by the equation

$$3N_0 = \frac{4\pi V}{3\bar{c}^3}\nu_{\rm m}^3, \tag{14.8.8}$$

obtained upon integration of Eq. (14.8.5).

The mean energy, U, is eventually found to take the form

$$U = 3RT\frac{3}{x_{\rm m}^3}\int_0^{x_{\rm m}} \frac{x^3}{{\rm e}^x - 1}{\rm d}x, \tag{14.8.9}$$

where we have used Eqs. (14.8.7), (14.8.8), and the definition

$$x_{\rm m} \equiv \frac{h\nu_{\rm m}}{KT}. \tag{14.8.10}$$

Equation (14.8.9) is known as the Debye formula (Debye 1912), and gives a vastly better fit (see figure 14.3) than the Einstein model with a single frequency. We may also recognize that for $x_{\rm m} >> 1$ the Debye formula gives the same T^3 law as for the black body. It took two decades before this simple similarity was appreciated!

14.9 Identical particles in quantum mechanics

One knows from the work of Wigner, who studied the representations of the Poincaré group on a Hilbert space, that *elementary particles* can be identified with *unitary irreducible representations of the Poincaré group*, for which the Casimir operators (see appendix 2.B) represent the mass and spin of all known particles. The same result holds for projective irreducible representations of the Galilei group (Sudarshan and Mukunda 1974).

Now we study the problem of understanding under which conditions two elementary particles can be identified. This is part of a more general problem, aimed at understanding how to distinguish two physical systems. For this purpose, one might be tempted to measure the properties of the two systems. However, from what is known from measurement theory in quantum mechanics, it is clear that two systems cannot give the same results for all measurable properties if they are not in the same state. In other words, *only identical states* for the two systems can lead to identical results. This simple remark leads, therefore, to a deep distinction between *extrinsic properties*, which depend on the state, and *intrinsic properties*, which are independent of the state (Jauch 1973). Within this framework, it is therefore quite natural to say that two elementary

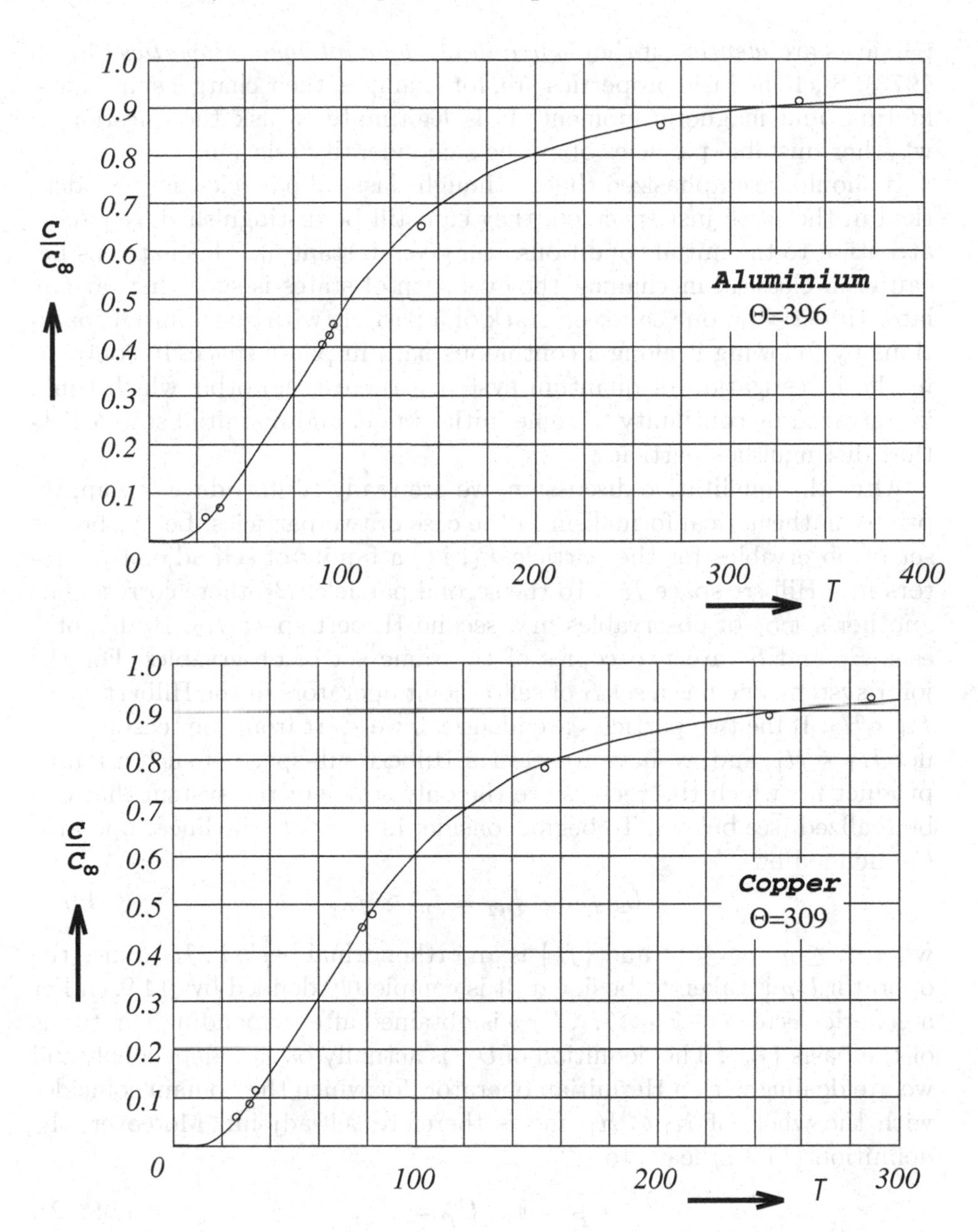

Fig. 14.3. Specific heat at low temperatures according to Debye. Small circles correspond to the observed points, whereas the continuous curve results from the Debye model (Born 1969, by kind permission of Professor Blin-Stoyle).

particles are *identical* if they *agree in all their intrinsic properties* (Jauch 1973). Such intrinsic properties are, for example, their charge, spin, mass, lifetime and magnetic moment. It is legitimate to ask the question of whether unstable particles obey the same identity relation.

It should be emphasized that, although classical particles may be identical in the sense just specified, they can still be distinguished by paying attention to the initial conditions at a given instant t_0. This happens because, in classical mechanics, the evolution of states is such that, at any later time $t > t_0$, one can keep track of a particle with given initial conditions by following it along a continuous path in phase space. In contrast, in the investigation of quantum systems, there is no orbit which could be retraced by continuity to some initial value, and no initial state exists that distinguishes particles.

After this qualitative discussion, we are ready to introduce the appropriate mathematical formalism in the case of two particles. Let S_1 be the set of observables for the particle P_1, i.e. a family of self-adjoint operators in a Hilbert space $\mathcal{H}_1$. To the second particle, P_2, there corresponds another set S_2 of observables in a second Hilbert space $\mathcal{H}_2$. By hypothesis, S_1 and S_2 are two copies of the same set of observables. For the joint system, one has a set S of self-adjoint operators in the Hilbert space $\mathcal{H}_1 \otimes \mathcal{H}_2$. If the two particles are identical, we start from the tensor product $\mathcal{H}_1 \otimes \mathcal{H}_1$, and we have to select a Hilbert sub-space of such a tensor product for which the vectors are the only states of the system that can be realized (see below). To begin, consider in $\mathcal{H}_1 \otimes \mathcal{H}_1$ the linear operator U_P defined by

$$U_P f_{n_1} \otimes f_{n_2} \equiv f_{n_2} \otimes f_{n_1}, \tag{14.9.1}$$

where $1 \leq n_1, n_2 \leq \infty$ and $\{f_n\}$ is an orthonormal basis in $\mathcal{H}_1$. Since the operator U_P is taken to be linear, it is completely defined by (14.9.1). For a generic vector $v \in \mathcal{H}_1 \otimes \mathcal{H}_1$, $U_P v$ is obtained after expanding v in terms of the basis $\{f_n\}$. The definition of U_P is actually basis-independent, and we are dealing with a Hermitian operator, for which the domain coincides with the whole of $\mathcal{H}_1 \otimes \mathcal{H}_1$ and is therefore self-adjoint. Moreover, the definition (14.9.1) leads to

$$U_P^2 = 1\!\!\mathrm{I}, \quad U_P^\dagger = U_P, \qquad (14.9.2),$$

which implies

$$U_P U_P^\dagger = U_P^\dagger U_P = 1\!\!\mathrm{I}. \tag{14.9.3}$$

Thus, U_P is unitary and has the spectrum ± 1. Since identical particles are indistinguishable, a vector $\xi \in \mathcal{H}_1 \otimes \mathcal{H}_1$ is a possible state of the composite system only if it is an eigenvector of the exchange operator U_P:

$$U_P \xi = c(\xi)\xi, \quad c(\xi) = \pm 1. \tag{14.9.4}$$

The state vectors of a system of two indistinguishable particles, with the corresponding Hilbert sub-space, are therefore restricted by the condition (14.9.4). Since the sum of eigenvectors belonging to different eigenvalues is no longer an eigenvector of a given linear operator, one finds that a sub-space $\mathcal{H}$ of $\mathcal{H}_1 \otimes \mathcal{H}_1$ can be the Hilbert space of state vectors for a system of two identical particles only if it coincides with the eigenspaces $\mathcal{H}_+$ and $\mathcal{H}_-$ of U_P belonging to the eigenvalues $+1$ and -1, respectively. It should be stressed that there are no *a priori* theoretical reasons for choosing $\mathcal{H}_+$ rather than $\mathcal{H}_-$.

The projectors onto the sub-spaces $\mathcal{H}_+$ and $\mathcal{H}_-$ are given by

$$\Pi_+ \equiv \frac{1}{2}\Big(\mathbb{I} + U_P\Big), \quad \Pi_- \equiv \frac{1}{2}\Big(\mathbb{I} - U_P\Big), \tag{14.9.5}$$

respectively. By virtue of (14.9.2) they are orthogonal:

$$\Pi_+\Pi_- = 0 = \Pi_-\Pi_+, \tag{14.9.6}$$

and hence $\mathcal{H}_+$ and $\mathcal{H}_-$ are orthogonal sub-spaces. Furthermore, they satisfy the completeness condition

$$\Pi_+ + \Pi_- = \mathbb{I}, \tag{14.9.7}$$

which implies that

$$\mathcal{H}_+ \oplus \mathcal{H}_- = \mathcal{H}_1 \otimes \mathcal{H}_1. \tag{14.9.8}$$

An orthonormal basis in $\mathcal{H}_+$ is given by the vectors

$$g^+_{n_1,n_2} \equiv \frac{1}{\sqrt{2}}\Big(f_{n_1} \otimes f_{n_2} + f_{n_2} \otimes f_{n_1}\Big) \quad \text{if } n_1 < n_2, \tag{14.9.9a}$$

$$g^+_{n_1,n_2} \equiv f_{n_1} \otimes f_{n_1} \quad \text{if } n_1 = n_2, \tag{14.9.9b}$$

whereas the orthonormal basis in $\mathcal{H}_-$ reads

$$g^-_{n_1,n_2} \equiv \frac{1}{\sqrt{2}}\Big(f_{n_1} \otimes f_{n_2} - f_{n_2} \otimes f_{n_1}\Big). \tag{14.9.10}$$

For the time being, we ask ourselves how self-adjoint operators in $\mathcal{H}_1 \otimes \mathcal{H}_1$ are related to observables of a system of two identical particles. Indeed, a self-adjoint operator in $\mathcal{H}_1 \otimes \mathcal{H}_1$ defines, by restriction to $\mathcal{H} = \mathcal{H}_+$ or $\mathcal{H}_-$, an observable of the system if and only if $\mathcal{H}$ remains unaffected. If Q denotes the projector onto $\mathcal{H}$, such a condition reads

$$AQv = QAQv, \quad \forall v \in \mathcal{H}_1 \otimes \mathcal{H}_1, \tag{14.9.11}$$

which implies the operator identity $AQ = QAQ$. Since both Q and A are Hermitian, the Hermitian conjugate of this identity leads to $QA = QAQ$, and hence

$$AQ = QA. \tag{14.9.12}$$

This is called the commutation property of A with Q. In other words, a self-adjoint operator in $\mathcal{H}_1 \otimes \mathcal{H}_1$ defines, by restriction, an observable of the system of two identical particles if and only if it commutes with the projector onto $\mathcal{H}$. Moreover, since we are considering a composite system of two particles only, Q reduces to Π_+ or Π_- (see Eq. (14.9.5)), and hence Eq. (14.9.12) leads to

$$AU_P = U_P A, \tag{14.9.13}$$

which implies (see Eq. (14.9.2))

$$U_P^2 A = A = U_P A U_P. \tag{14.9.14}$$

To interpret the condition (14.9.14), let us write A in the form

$$A = \sum_{A_1, A_2} A_1 \otimes A_2, \tag{14.9.15}$$

with A_1 and A_2 being self-adjoint operators in the Hilbert space $\mathcal{H}_1$ of single-particle state vectors. Hence one finds, from the definition (14.9.1),

$$\begin{aligned} U_P(A_1 \otimes A_2)U_P \xi_1 \otimes \xi_2 &= U_P(A_1 \otimes A_2)\xi_2 \otimes \xi_1 \\ &= U_P(A_1\xi_2) \otimes (A_2\xi_1) = (A_2\xi_1) \otimes (A_1\xi_2) \\ &= (A_2 \otimes A_1)\xi_1 \otimes \xi_2. \end{aligned} \tag{14.9.16}$$

Eventually this yields

$$U_P(A_1 \otimes A_2)U_P = A_2 \otimes A_1, \tag{14.9.17}$$

since the set of vectors of the kind $\xi_1 \otimes \xi_2$ is dense in $\mathcal{H}_1 \otimes \mathcal{H}_1$. Thus, the condition (14.9.14) expresses the symmetry of the self-adjoint operator A as a function of the single-particle observables.

On experimental grounds one knows that two identical particles are described by the Hilbert space $\mathcal{H}_+$ (Bose–Einstein statistics) if their spin is integer, and by the Hilbert space $\mathcal{H}_-$ (Fermi–Dirac statistics) if their spin is half-odd. This is the *symmetrization principle*, and can be extended to N identical particles. In particular, the state vector of N identical fermions is antisymmetric with respect to the exchange of *any* two arguments n_i and n_j, which implies that no non-trivial state vector exists if $n_i = n_j$. In other words, the *Pauli exclusion principle* is found to hold: in a quantum system, two identical fermions cannot share the same single-particle state (Pauli 1925).

In the case of several identical particles, it is convenient to denote by S_1 the observables for any one of the single particles. These are operators in the Hilbert space $\mathcal{H}_1$. For a system of n identical particles, the Hilbert space is then the n-fold tensor product

$$\mathcal{H}_n \equiv \mathcal{H}_1 \otimes \mathcal{H}_1 \otimes \cdots \otimes \mathcal{H}_1. \tag{14.9.18}$$

The set S of observables is a set of self-adjoint operators in the space $\mathcal{H}_n$. Since the particles are assumed to be identical, the elements of S are restricted to those particular linear operators that are symmetric under permutation of the n-particle variables. A general permutation changes 1 into i_1, 2 into $i_2, \ldots$, and hence can be represented by

$$P = \begin{pmatrix} 1 & 2 & \ldots\ldots & n \\ i_1 & i_2 & \ldots\ldots & i_n \end{pmatrix}.$$

To any permutation one associates a unitary operator U_P operating in $\mathcal{H}_n$. Its action on a vector f of the form

$$f = f_1 \otimes f_2 \otimes \cdots \otimes f_n \tag{14.9.19}$$

is given by (cf. Eq. (14.9.1))

$$U_P f \equiv f_{i_1} \otimes f_{i_2} \otimes \cdots \otimes f_{i_n}. \tag{14.9.20}$$

The operator U_P is then extended by linearity to the whole Hilbert space $\mathcal{H}_n$, hence leading to a *unitary representation of the permutation group.*

A vector f is *symmetric* if

$$U_P f = f \quad \forall P, \tag{14.9.21}$$

and it is *antisymmetric* if, instead,

$$U_P f = \delta_P f \quad \forall P, \tag{14.9.22}$$

where δ_P is the signature of the permutation P:

$$\delta_P = 1 \quad \text{if } P \text{ is even}, \tag{14.9.23a}$$

$$\delta_P = -1 \quad \text{if } P \text{ is odd}. \tag{14.9.23b}$$

It is then clear that the projection Π_+ on the symmetric sub-space is defined by (cf. first operator in (14.9.5))

$$\Pi_+ \equiv \frac{1}{n!} \sum_P U_P, \tag{14.9.24}$$

whereas the projection on the antisymmetric sub-space reads (cf. the second operator in (14.9.5))

$$\Pi_- \equiv \frac{1}{n!} \sum_P \delta_P U_P. \tag{14.9.25}$$

For identical particles, following what one knows in the two-particle case, one requires (see Eq. (14.9.13))

$$A U_P = U_P A, \quad \forall A \in S, \ \forall U_P. \tag{14.9.26}$$

Thus, the projections $\Pi_\pm$ defined in (14.9.24) and (14.9.25) commute with the observables, in that

$$\Pi_\pm A - A\Pi_\pm = 0, \quad \forall A \in S. \tag{14.9.27}$$

Every kind of particle falls either into the class characterized by

$$\Pi_+ = \mathbb{I}, \quad \Pi_- = 0,$$

or into the class having

$$\Pi_+ = 0, \quad \Pi_- = \mathbb{I}.$$

This means that the observables can only operate in one of the two spaces

$$\mathcal{H}_n^\pm \equiv \Pi_\pm \mathcal{H}_n. \tag{14.9.28}$$

The former corresponds to the Bose–Einstein statistics, and the latter to the Fermi–Dirac statistics. Without giving further details, we should however warn the general reader that the mathematical formalism allows for yet other statistics.

14.9.1 An application: helium atom

In a neutral helium atom, an electron is in the $1s$ ground state, while the other electron is in an excited state, with quantum numbers $n \geq 2$ and $l \geq 1$. Our aim is to evaluate the ionization energy for the (n, l) electron by making suitable approximations. The ideas and the various steps are as follows.

If the $1s$ electron is exposed to the full nuclear charge $2e$, but the (n, l) electron only interacts with the screened charge e, one can describe the two one-electron states by the solutions of the differential equations (Flügge 1971)

$$\left(-\frac{1}{2}\triangle_1 - \frac{2}{r_1}\right) u(r_1) = E_1 \, u(r_1), \tag{14.9.29}$$

$$\left(-\frac{1}{2}\triangle_2 - \frac{1}{r_2}\right) v_{n,l,m}(r_2) = E_n \, v_{n,l,m}(r_2), \tag{14.9.30}$$

where r_1 and r_2 are distances from the origin of the $1s$ and (n, l) electrons, respectively. In our problem, the bound states can be thus written as

$$u(r) = \sqrt{\frac{8}{\pi}}\, \mathrm{e}^{-2r}, \tag{14.9.31}$$

$$v_{n,l,m}(r) = R_{n,l}(r) Y_{l,m}(\theta, \varphi), \tag{14.9.32}$$

with energy eigenvalues

$$E_1 = -2, \tag{14.9.33}$$

$$E_n = -\frac{1}{2n^2}. \tag{14.9.34}$$

The states u and v are the building blocks in solving the Schrödinger equation for the two-electron problem:

$$\left(-\frac{1}{2}\triangle_1 - \frac{1}{2}\triangle_2 - \frac{2}{r_1} - \frac{2}{r_2} + \frac{1}{r_{12}}\right)\psi = E\psi, \tag{14.9.35}$$

where r_{12} is the distance among the two electrons. The main idea is to find approximate solutions of Eq. (14.9.35) by taking the symmetrized wave function

$$\psi = u(r_1)v_{n,l,m}(r_2) + \varepsilon v_{n,l,m}(r_1)u(r_2). \tag{14.9.36}$$

The choice $\varepsilon = 1$ corresponds to *para-helium*, for which the spins of the two electrons are antiparallel, so that the total spin s vanishes. The case $\varepsilon = -1$ corresponds instead to *ortho-helium*, for which the spins are parallel, so that $s = 1$. The states u and $v_{n,l,m}$ obey the normalization conditions $(u, u) = (v_{n,l,m}, v_{n,l,m}) = 1$, and the orthogonality condition $(u, v_{n,l,m}) = 0$. One thus finds (omitting, for simplicity of notation, the quantum numbers)

$$\left(u(r_1)v(r_2), \left(-\frac{1}{2}\triangle_1 - \frac{1}{2}\triangle_2 - \frac{2}{r_1} - \frac{2}{r_2} + \frac{1}{r_{12}}\right)\psi\right) = E(u(r_1)v(r_2), \psi), \tag{14.9.37}$$

where, bearing in mind (14.9.36), one has

$$\left(u(r_1)v(r_2), \left(-\frac{1}{2}\triangle_1 - \frac{2}{r_1}\right)u(r_1)v(r_2)\right) = E_1, \tag{14.9.38}$$

$$\left(u(r_1)v(r_2), \left(-\frac{1}{2}\triangle_1 - \frac{2}{r_1}\right)v(r_1)u(r_2)\right) = 0, \tag{14.9.39}$$

$$\left(u(r_1)v(r_2), \left(-\frac{1}{2}\triangle_2 - \frac{2}{r_2}\right)v(r_1)u(r_2)\right) = E_n - \left(v(r_2), \frac{1}{r_2}v(r_2)\right), \tag{14.9.40}$$

$$\left(u(r_1)v(r_2), \left(-\frac{1}{2}\triangle_2 - \frac{2}{r_2}\right)v(r_1)u(r_2)\right) = 0. \tag{14.9.41}$$

Now one can show that, for all one-electron states, a theorem fixes the value of the potential energy to be twice that of the total energy. More precisely, the *virial theorem* states that, if the potential of a system of particles is a homogeneous function of degree p, the temporal averages of the kinetic and position energies are related by the equation $2\overline{T} = p\overline{U}$. In particular, the position energy of a system of particles interacting through Coulomb forces is a homogeneous function of degree -1, and hence one finds $2\overline{T} = -\overline{U}$.

This makes it possible to evaluate the integral on the right-hand side of (14.9.40) as

$$\left(v(r_2), \frac{1}{r_2}v(r_2)\right) = -2E_n = \frac{1}{n^2}. \tag{14.9.42}$$

At this stage, to complete the evaluation of (14.9.37) we have to work out in detail the following integrals:

$$F \equiv \left(u(r_1)v(r_2), \frac{1}{r_{12}}u(r_1)v(r_2)\right), \tag{14.9.43}$$

$$G \equiv \left(u(r_1)v(r_2), \frac{1}{r_{12}}v(r_1)u(r_2)\right). \tag{14.9.44}$$

The integral G is usually referred to as the *exchange integral* of the electron–electron interaction. In terms of these integrals, the energy eigenvalue of the symmetrized wave function reads

$$E = E_1 + E_n - (-2E_n) + F + \varepsilon G = -2 - \frac{3}{2n^2} + F + \varepsilon G. \tag{14.9.45}$$

The exchange integral G contributes a negative term (see below) by virtue of the antisymmetry of the two-electron wave function. The next step is therefore the expansion in spherical harmonics of $\frac{1}{r_{12}}$, denoting by θ_{12} the angle between the position vectors $\vec{r}_1$ and $\vec{r}_2$ of the two electrons:

$$\frac{1}{r_{12}} = \frac{1}{r_2}\sum_{j=0}^{\infty}\left(\frac{r_1}{r_2}\right)^j P_j(\cos\theta_{12}) \quad \text{if } r_1 < r_2, \tag{14.9.46a}$$

$$\frac{1}{r_{12}} = \frac{1}{r_1}\sum_{j=0}^{\infty}\left(\frac{r_2}{r_1}\right)^j P_j(\cos\theta_{12}) \quad \text{if } r_1 > r_2. \tag{14.9.46b}$$

In the angular integration for F and G, one deals with

$$F_{\text{ang}} = \oint d\Omega_2 \, |Y_{l,m}(\theta_2,\varphi_2)|^2 \oint d\Omega_1 \, P_j(\cos\theta_{12}), \tag{14.9.47}$$

$$G_{\text{ang}} = \oint d\Omega_2 \, Y_{l,m}^*(\theta_2,\varphi_2) \oint d\Omega_1 \, Y_{l,m}(\theta_1,\varphi_1) P_j(\cos\theta_{12}), \tag{14.9.48}$$

because

$$(u(r_1)v(r_2))_{\text{ang}} = Y_{l,m}(\theta_2,\varphi_2), \tag{14.9.49}$$

$$(u^*(r_1)v^*(r_2))_{\text{ang}} = Y_{l,m}^*(\theta_2,\varphi_2), \tag{14.9.50}$$

$$(v(r_1)u(r_2))_{\text{ang}} = Y_{l,m}(\theta_1,\varphi_1). \tag{14.9.51}$$

One thus finds that $F_{\rm ang} = 4\pi\delta_{0,j}$, which implies

$$F = 4\pi \int_0^\infty {\rm d}r_2 \; r_2^2 |u(r_2)|^2 \left\{ \int_0^{r_2} {\rm d}r_1 \; \frac{r_1^2}{r_2} |R_{n,l}(r_1)|^2 \right.$$
$$\left. + \int_{r_2}^\infty {\rm d}r_1 \; r_1 |R_{n,l}(r_1)|^2 \right\}. \tag{14.9.52}$$

In the integration (14.9.48), it is appropriate to use the addition theorem for spherical harmonics:

$$P_j(\cos\theta_{12}) = \frac{4\pi}{(2j+1)} \sum_{\mu=-j}^{j} Y_{j,\mu}^*(\theta_1,\varphi_1) Y_{j,\mu}(\theta_2,\varphi_2). \tag{14.9.53}$$

This leads to

$$\oint {\rm d}\Omega_1 \, Y_{l,m}(\theta_1,\varphi_1) P_j(\cos\theta_{12}) = \frac{4\pi}{(2j+1)} \delta_{l,j} \sum_{\mu=-j}^{j} \delta_{m,\mu} Y_{j,\mu}(\theta_2,\varphi_2)$$
$$= \frac{4\pi}{(2l+1)} \delta_{l,j} Y_{l,m}(\theta_2,\varphi_2), \tag{14.9.54}$$

and hence

$$G_{\rm ang} = \frac{4\pi}{(2l+1)} \delta_{l,j}, \tag{14.9.55}$$

$$G = \frac{4\pi}{(2l+1)} \int_0^\infty {\rm d}r_2 \; r_2^2 u(r_2) R_{n,l}(r_2)$$
$$\times \left[\frac{1}{r_2} \int_0^{r_2} {\rm d}r_1 \; r_1^2 \left(\frac{r_1}{r_2}\right)^l u(r_1) R_{n,l}(r_1) \right.$$
$$\left. + \int_{r_2}^\infty {\rm d}r_1 \; r_1 \left(\frac{r_2}{r_1}\right)^l u(r_1) R_{n,l}(r_1) \right]. \tag{14.9.56}$$

Now we set $n = 2$ and $l = 1$, and we note that the term in curly brackets in (14.9.52), which is a function of r_2, can be written as

$$\frac{1}{24} \frac{1}{r_2} A + \frac{1}{24} B,$$

where

$$A \equiv \int_0^{r_2} y^4 {\rm e}^{-y} \, {\rm d}y, \quad B \equiv \int_{r_2}^\infty y^3 {\rm e}^{-y} \, {\rm d}y. \tag{14.9.57}$$

Repeated integration by parts yields

$$A = -r_2^4 {\rm e}^{-r_2} - 4r_2^3 {\rm e}^{-r_2} - 12 r_2^2 {\rm e}^{-r_2} - 24 r_2 {\rm e}^{-r_2} - 24 {\rm e}^{-r_2} + 24, \tag{14.9.58}$$

$$B = r_2^3 {\rm e}^{-r_2} + 3 r_2^2 {\rm e}^{-r_2} + 6 r_2 {\rm e}^{-r_2} + 6 {\rm e}^{-r_2}. \tag{14.9.59}$$

After the insertion of (14.9.58) and (14.9.59) into (14.9.52), some cancellations and simplifications occur, and one finds

$$F = \frac{4}{3} \sum_{i=1}^{5} F_i, \tag{14.9.60}$$

where

$$F_1 \equiv - \int_0^\infty y^4 e^{-5y} \, dy = -\frac{24}{3125}, \tag{14.9.61}$$

$$F_2 \equiv -6 \int_0^\infty y^3 e^{-5y} \, dy = -\frac{36}{625}, \tag{14.9.62}$$

$$F_3 \equiv -18 \int_0^\infty y^2 e^{-5y} \, dy = -\frac{36}{125}, \tag{14.9.63}$$

$$F_4 \equiv -24 \int_0^\infty y e^{-5y} \, dy = -\frac{24}{25}, \tag{14.9.64}$$

$$F_5 \equiv 24 \int_0^\infty y e^{-4y} \, dy = \frac{3}{2}. \tag{14.9.65}$$

Hence one obtains

$$F = \frac{778}{3125}. \tag{14.9.66}$$

Moreover, if $n = 2$ and $l = 1$, the similar term in (14.9.44) can be written as

$$\sqrt{\frac{8}{\pi}} \frac{1}{24} \left(\frac{1}{r_2^2} \widetilde{A} + r_2 \widetilde{B} \right),$$

where

$$\widetilde{A} \equiv \int_0^{r_2} y^4 e^{-\frac{5}{2}y} \, dy, \quad \widetilde{B} \equiv \int_{r_2}^\infty y e^{-\frac{5}{2}y} \, dy. \tag{14.9.67}$$

Repeated integration by parts yields

$$\begin{aligned} \widetilde{A} = & -\frac{2}{5} r_2^4 e^{-\frac{5}{2}r_2} - \frac{16}{25} r_2^3 e^{-\frac{5}{2}r_2} - \frac{96}{125} r_2^2 e^{-\frac{5}{2}r_2} \\ & - \frac{384}{625} r_2 e^{-\frac{5}{2}r_2} - \frac{768}{3125} e^{-\frac{5}{2}r_2} + \frac{768}{3125}, \end{aligned} \tag{14.9.68}$$

$$\widetilde{B} = \frac{2}{5} r_2 e^{-\frac{5}{2}r_2} + \frac{4}{25} e^{-\frac{5}{2}r_2}, \tag{14.9.69}$$

and hence G can be written in the form

$$G = \frac{4}{9} \sum_{i=1}^{5} G_i, \tag{14.9.70}$$

where

$$G_1 \equiv -\frac{12}{25}\int_0^\infty y^4 e^{-5y}\, dy = -\frac{288}{25}\frac{1}{3125}, \tag{14.9.71}$$

$$G_2 \equiv -\frac{96}{125}\int_0^\infty y^3 e^{-5y}\, dy = -\frac{576}{25}\frac{1}{3125}, \tag{14.9.72}$$

$$G_3 \equiv -\frac{384}{625}\int_0^\infty y^2 e^{-5y}\, dy = -\frac{768}{25}\frac{1}{3125}, \tag{14.9.73}$$

$$G_4 \equiv -\frac{768}{3125}\int_0^\infty y e^{-5y}\, dy = -\frac{768}{25}\frac{1}{3125}, \tag{14.9.74}$$

$$G_5 \equiv \frac{768}{3125}\int_0^\infty y e^{-\frac{5}{2}y}\, dy = \frac{3072}{25}\frac{1}{3125}. \tag{14.9.75}$$

One thus finds

$$G = \frac{896}{75}\frac{1}{3125}. \tag{14.9.76}$$

The results (14.9.45), (14.9.66) and (14.9.76) lead to

$$E = -2.126\,04 + \varepsilon \times 0.00382, \tag{14.9.77}$$

in atomic units. The ionization energy is the difference of E and the energy $E^+ = -2$ of He^+ in the ground state (in this case, one electron is still in the $1s$ state, and the other is removed):

$$I \equiv E^+ - E = 0.126\,04 - \varepsilon \times 0.00382. \tag{14.9.78}$$

This result may be re-expressed in electron volts:

$$I = \Bigl(3.429 - \varepsilon \times 0.104\Bigr)\ \mathrm{eV}. \tag{14.9.79}$$

The agreement between these theoretical values and the experiment is quite good. Interestingly, the ionization energy of para-helium turns out to be smaller than the ionization energy of ortho-helium. In physical terms, what happens is that the exchange integral, which results from the mutual repulsion of the two electrons, is positive, and is hence responsible for the state with space-symmetric wave function lying above the antisymmetric ortho-state:

$$I_{\varepsilon=1} = 3.325\ \mathrm{eV}, \tag{14.9.80}$$

$$I_{\varepsilon=-1} = 3.533\ \mathrm{eV}, \tag{14.9.81}$$

$$I_{\varepsilon=1} - I_{\varepsilon=-1} = -0.208\ \mathrm{eV}. \tag{14.9.82}$$

Thus, even though the potential has no explicit dependence on the spins, there is a different energy for the two spin orientations.

14.10 Bose–Einstein and Fermi–Dirac gases

A system consisting of n identical non-interacting particles satisfying the Bose–Einstein statistics is called, hereafter, a free Bose–Einstein gas (cf. Jauch 1973). We know from section 14.9 that its Hilbert space is $\mathcal{H}_n^+ \equiv \Pi_+ \mathcal{H}_n$. Let $\{\varphi_r\}$ be a complete orthonormal system of vectors in $\mathcal{H}_1$, with a corresponding complete system in $\mathcal{H}_n$ of the form

$$\varphi_{r_1} \otimes \varphi_{r_2} \otimes \cdots \otimes \varphi_{r_n}.$$

On acting upon it with Π_+, such a system is mapped into the sub-space $\mathcal{H}_n^+$ with vectors

$$\varphi[r_1 \cdots r_n] \equiv \Pi_+ \varphi_{r_1} \otimes \varphi_{r_2} \otimes \cdots \otimes \varphi_{r_n}. \tag{14.10.1}$$

Such vectors, however, do not have a unit norm. On denoting by n_1 the number of indices among the $r_1 \ldots r_n$ which are equal to 1, by n_2 the number of indices equal to $2, \ldots$, one finds that

$$\begin{aligned} \|\varphi[r_1 \cdots r_n]\|^2 &= \frac{1}{n!} \sum_P \delta_{r_1 r_{i_1}} \delta_{r_2 r_{i_2}} \cdots \delta_{r_n r_{i_n}} \\ &= \frac{n_1! n_2! \cdots n_r! \cdots}{n!}, \end{aligned} \tag{14.10.2}$$

with $n = n_1 + n_2 + \cdots + n_r + \cdots$. Thus, the complete orthonormal system of vectors in $\mathcal{H}_n$ consists of

$$\begin{aligned} \varphi(n_1 n_2 \cdots n_r \cdots) &\equiv \frac{\varphi[r_1 \cdots r_n]}{\|\varphi[r_1 \cdots r_n]\|} = \varphi_n \\ &= \sqrt{\frac{n!}{n_1! \cdots n_r! \cdots}}\, \Pi_+ \varphi_{r_1} \otimes \varphi_{r_2} \otimes \cdots \otimes \varphi_{r_n}. \end{aligned} \tag{14.10.3}$$

A representation of observables in the system described by (14.10.3) is called an *occupation-number representation*. The state $\varphi(n_1 n_2 \cdots n_r \cdots)$ has to satisfy the condition $\sum_i n_i < \infty$. Without this the space is not separable.

If the total number of particles is not fixed, the physical system can be described in a direct sum of Hilbert spaces called a *Fock space* (Fock 1932). By definition, a vector

$$F \in \mathcal{H}_0^+ \oplus \mathcal{H}_1^+ \oplus \cdots \oplus \mathcal{H}_n^+ \oplus \cdots$$

is a finite or infinite sequence of vectors $f_n \in \mathcal{H}_n^+$, subject to the condition of finite norm

$$\|F\|^2 \equiv \sum_{n=0}^{\infty} \|f_n\|^2 < \infty. \tag{14.10.4}$$

The scalar product of two vectors $F = \{f_n\}$ and $G = \{g_n\}$ is then defined by

$$(F, G) \equiv \sum_{n=0}^{\infty} (f_n, g_n). \tag{14.10.5}$$

Addition of vectors is defined by

$$F + G \equiv \{f_n + g_n\}, \tag{14.10.6}$$

and multiplication with scalars is obtained as

$$\lambda F \equiv \{\lambda f_n\}. \tag{14.10.7}$$

By virtue of (14.10.4)–(14.10.7), the set of all such sequences $F = \{f_n\}$ have the structure of a Hilbert space. In particular, the vector in the space $\mathcal{H}_0$ represents the system in the state with no particle, called the *vacuum state* and taken to be non-degenerate. The vectors defined in (14.10.3) may be considered as vectors in the Fock space $F = \{f_k\}$, with

$$f_k = 0 \quad \text{if } k \neq n, \quad f_k = \varphi_n \quad \text{if } k = n. \tag{14.10.8}$$

One can introduce a set of annihilation and creation operators in the Fock space, which make it possible to describe a Bose–Einstein gas in terms of harmonic oscillators:

$$\hat{a}_r \varphi(n_1 \cdots n_r \cdots) \equiv \sqrt{n_r}\, \varphi(n_1 \cdots n_r - 1 \cdots), \tag{14.10.9}$$

$$\hat{a}_r^\dagger \varphi(n_1 \cdots n_r \cdots) \equiv \sqrt{n_r + 1}\, \varphi(n_1 \cdots n_r + 1 \cdots). \tag{14.10.10}$$

One then finds that a free Bose–Einstein gas is kinematically indistinguishable from an independent collection of distinct harmonic oscillators. The Hamiltonian operator for a state in $\mathcal{H}_n$ is given by an operator of the form

$$H = H_1 + H_2 + \cdots + H_n + \cdots, \tag{14.10.11}$$

where

$$H_r \equiv I \otimes I \otimes \cdots \otimes H_0 \otimes \cdots . \tag{14.10.12}$$

The operator H_0 refers to a single boson, and is here taken to have a discrete spectrum only, with eigenvectors φ_r:

$$H_0 \varphi_r = \varepsilon_r \varphi_r. \tag{14.10.13}$$

One then finds

$$H\varphi(n_1 \cdots n_r \cdots) = \Big(n_1 \varepsilon_1 + \cdots + n_r \varepsilon_r + \cdots\Big) \varphi(n_1 \cdots n_r \cdots). \tag{14.10.14}$$

This means that the dynamical structure of a free Bose–Einstein gas is identical to that of a collection of harmonic oscillators. The result can be extended to a system of interacting identical bosons, which are put in correspondence with a system of interacting and distinct harmonic oscillators (cf. the Planck assumption in section 14.7).

A free Fermi–Dirac gas is, instead, a collection of n identical non-interacting particles satisfying the Fermi–Dirac statistics (Dirac 1926b, Fermi 1926). The Hilbert space for such systems is $\mathcal{H}_n^- \equiv \Pi_- \mathcal{H}_n$. For any complete orthonormal system $\{\varphi_n\} \in \mathcal{H}_1$, one defines another such system in $\mathcal{H}_n^-$ from the vectors (cf. Eq. (14.10.1))

$$\varphi[r_1 \cdots r_n] \equiv \Pi_- \varphi_{r_1} \otimes \varphi_{r_2} \otimes \cdots \otimes \varphi_{r_n}. \tag{14.10.15}$$

The natural basis in the occupation-number representation is now given by (cf. Eq. (14.10.3))

$$\begin{aligned}\varphi(n_1 n_2 \cdots n_r \cdots) &\equiv \frac{\varphi[r_1 \cdots r_n]}{\|\varphi[r_1 \cdots r_n]\|} \\ &= \sqrt{n!}\, \Pi_- \varphi_{r_1} \otimes \varphi_{r_2} \otimes \cdots \otimes \varphi_{r_n},\end{aligned} \tag{14.10.16}$$

because all the occupation numbers n_r are equal to 0 or 1 for fermionic particles. Unless the wave functions are antisymmetric this restriction is mode-dependent. The Fock space can be built in complete formal analogy to what has been done for bosonic particles, but bearing in mind that each f_n now belongs to $\mathcal{H}_n^-$. Annihilation and creation operators are instead defined by the rules (cf. Eqs. (14.10.9) and (14.10.10))

$$\hat{a}_r \varphi(n_1 \cdots n_r \cdots) \equiv (-1)^{S_r} n_r \varphi(n_1 \cdots n_r - 1 \cdots), \tag{14.10.17}$$

$$\hat{a}_r^\dagger \varphi(n_1 \cdots n_r \cdots) \equiv (-1)^{S_r} (1 - n_r) \varphi(n_1 \cdots n_r + 1 \cdots), \tag{14.10.18}$$

where

$$S_r \equiv \sum_{k=1}^{r-1} n_k. \tag{14.10.19}$$

Such operators obey *anticommutation relations*:

$$\{\hat{a}_r, \hat{a}_s\} = \left\{\hat{a}_r^\dagger, \hat{a}_s^\dagger\right\} = 0, \tag{14.10.20}$$

$$\left\{\hat{a}_r, \hat{a}_s^\dagger\right\} = \delta_{rs}, \tag{14.10.21}$$

where the anticommutator is defined by the relation (we here omit any discussion of domains and of unbounded operators)

$$\{A, B\} \equiv AB + BA. \tag{14.10.22}$$

The energy eigenvalues of a free Fermi–Dirac gas are formally analogous to what one finds in Eq. (14.10.14):

$$E = n_1 \varepsilon_1 + n_2 \varepsilon_2 + \cdots + n_r \varepsilon_r,$$

where $n_r = 0, 1 \; \forall r$, and $\sum_r n_r = n$. However, unlike a Bose–Einstein gas, the lowest eigenvalue is not simply $E_0 = n\varepsilon_1$, but is given by the finite sum

$$E_0 = \sum_{k=1}^{n} \varepsilon_k. \tag{14.10.23}$$

This form of the ground-state energy is responsible for the shell structure of the periodic system.

14.11 Statistical derivation of the Planck formula

We are now in a position to derive the Planck radiation formula (14.7.37) by assuming that light quanta are *completely indistinguishable*. At the mathematical level, the problem is therefore that of finding the number of *distinguishable arrangements* of Bose–Einstein particles in a sheet. The individual cells of the sheet may be denoted by $z_1, z_2, \ldots, z_{g_s}$, where g_s is the weight factor of the sheet. Moreover, we assume there are n_s particles in the sheet, denoted by $P_1, P_2, \ldots, P_{n_s}$. These particles should be distributed among the g_s cells of the sheet, and we may consider the elements z_i and P_j in an arbitrary order (Bose 1924a,b), e.g.

$$z_1 P_1 P_2 \; z_2 P_3 \; z_3 P_4 P_5 P_6 \; z_4 \; z_5 P_7 \ldots .$$

Following Born (1969), the particles standing in between two $z's$ are in each case *assumed* to be in the cell *to their left* in the sequence. Thus, in the above sequence, the particles P_1 and P_2 are in the cell z_1, the particle P_3 in the cell z_2, the particles P_4, P_5 and P_6 in the cell z_3, no particle is in z_4, whereas the particles $P_7 \ldots$ are in the cell z_5, and so on. This means that, to obtain all possible arrangements, we first set a z_k at the head of the sequence, which can be done in g_s different ways, and then write down the remaining $g_s - 1 + n_s$ letters in arbitrary order one after the other. The total number of such arrangements is

$$N_s = g_s(g_s + n_s - 1)!. \tag{14.11.1}$$

However, N_s is not yet the quantity we are interested in, because distributions derived from one another by merely permuting the *cells* among themselves, or the *particles* among themselves, do not represent different states, but one and the same state. The number of these permutations is

$$\mathcal{P}_s = g_s! n_s!. \tag{14.11.2}$$

Thus, if the Bose–Einstein statistics can be applied, the *number of distinguishable arrangements in the sheet s* is

$$A_s \equiv \frac{N_s}{\mathcal{P}_s} = \frac{g_s(g_s + n_s - 1)!}{g_s! n_s!} = \frac{(g_s + n_s - 1)!}{(g_s - 1)! n_s!}. \tag{14.11.3}$$

Lastly, one has to evaluate the number of distinguishable arrangements when there are n_k particles in the kth sheet with $k = 1, 2, \ldots$. This is equal to the product of the various A_s, i.e.

$$W = \prod_s A_s = \prod_s \frac{(g_s + n_s - 1)!}{(g_s - 1)! n_s!}, \tag{14.11.4}$$

which is the *probability of distribution* of the particles among the various sheets. The counting scheme is due to Bose (1924a,b).

We now determine the most probable distribution, subject to the supplementary condition which fixes the total energy of the set of light quanta:

$$\sum_s n_s \varepsilon_s = h \sum_s n_s \nu_s = E. \tag{14.11.5}$$

No restriction is instead put on the number of light quanta, since their number does not remain constant. We therefore require that

$$\frac{\partial}{\partial n_k}\Big[\log(W) - \beta E\Big] = 0, \tag{14.11.6}$$

where β is a multiplier (see below). If the numbers g_s and n_s are both large, one can use the Stirling formula

$$\log \Gamma(z) \sim z \log(z) - z + \frac{1}{2}\log(2\pi) + \frac{1}{12}\frac{1}{z} - \frac{1}{360}\frac{1}{z^3} + \mathrm{O}(z^{-5}), \tag{14.11.7}$$

which holds, in particular, if z is an integer, so that $\Gamma(z) = \Gamma(n) = (n-1)!$. On neglecting 1 with respect to g_s and n_s, as well as the constant and the negative powers and linear term in the asymptotic formula (14.11.7), one finds from (14.11.4) the expression

$$\log(W) \sim \sum_s \Big[(g_s + n_s)\log(g_s + n_s) - g_s \log(g_s) - n_s \log(n_s)\Big]. \tag{14.11.8}$$

By virtue of (14.11.5), (14.11.6) and (14.11.8), the problem of finding the maximum of $\log(W)$ is solved by the system of equations

$$\log(g_k + n_k) + 1 - \log(n_k) - 1 = \beta \varepsilon_k, \tag{14.11.9}$$

which implies

$$\log\left(\frac{g_k}{n_k} + 1\right) = \beta \varepsilon_k, \tag{14.11.10a}$$

and, eventually,

$$n_k = \frac{g_k}{\mathrm{e}^{\beta \varepsilon_k} - 1}. \tag{14.11.10b}$$

We can now omit, for simplicity of notation, the label k, and find the spectral distribution of energy density in the form

$$u_\nu \, d\nu = \frac{1}{c^3} \frac{8\pi h \nu^3 \, d\nu}{e^{\beta h\nu} - 1}. \tag{14.11.11}$$

So far, the parameter β remains undetermined. We can, however, use the fundamental equation for the entropy, S, in terms of the probability distribution, i.e.

$$S = K \log(W), \tag{14.11.12}$$

which, jointly with the equations

$$dQ = \sum_s \varepsilon_s \, dn_s, \tag{14.11.13}$$

$$d\log(W) = \beta \sum_s \varepsilon_s \, dn_s = \beta \, dQ, \tag{14.11.14}$$

$$dS = \frac{1}{T} dQ, \tag{14.11.15}$$

leads to

$$dS = K \, d\log(W) = K\beta \, dQ = \frac{1}{T} dQ, \tag{14.11.16}$$

and hence

$$\beta = \frac{1}{KT}. \tag{14.11.17}$$

In these formulae, dQ is the infinitesimal heat provided to the system by the variation of internal energy resulting from the new distribution of atoms in the various states, due to quantum transitions, and $\frac{1}{T}$, the inverse of the temperature, is the familiar integrating factor for dQ.

We have thus put on solid ground the derivation of the Planck radiation formula, which relies on the Bose–Einstein statistics for light quanta (the above computation is indeed due to Bose (1924a,b), while Einstein's work was merely a follow up in this respect), rather than a mixture of classical and quantum arguments, as in the original derivation of Planck (cf. section 14.7 and Planck 1991). It should be stressed, however, that the Planck contribution remains absolutely outstanding, even in hindsight: he (only) knew thermodynamics, classical electrodynamics, classical mechanics and the experimental properties of heat radiation. No quantum theory of harmonic oscillators was available, no quantum statistics, nor was the concept of photons well developed. The analysis of section 14.7 therefore shows that new physical ideas arise after quite a struggle, and rely on assumptions which are sometimes *ad hoc*, if not unjustifiable from the known physical principles. The present section is instead an example of how the moment frequently comes when the original results are derived from new general principles, which, in the meantime, have become well established, thanks to theoretical and experimental work.

14.12 Problems

14.P1. In the universe there exists a background electromagnetic radiation which is isotropic and the spectrum of which is similar to black-body radiation at a temperature $T \cong 3$ K. Find the density of photons and their mean energy.

14.P2. Write a short essay to describe how the specific heats of solids are measured in the laboratory.

Appendix 14.A
Towards the Planck formula

In section 14.7, we have made use of the formula (14.7.29) for the work δW done per second by a radiation field on an oscillator. Now we are going to prove it. For this purpose, we first perform a Fourier analysis of the electric field, from which we derive a relation for the spectral density of radiation. On the other hand, by studying the equation of motion of the oscillator subject to a time-varying electric field, one can derive a formula for the work δW. In comparison, one finds Eq. (14.7.29). The details are as follows. Usually Fourier transformations can be carried out for only absolutely integrable functions. In the case of a stationary situation (such as black-body radiation) the 'signal' $E_x(t)$ is not absolutely integrable since it is integrated over an infinite time. To deal with such cases Wiener (1930) introduced generalized harmonic analysis. The main idea is to evaluate the autocorrelation function

$$C(t-t') \equiv \langle E(t)E(t') \rangle$$

and take its Fourier transform. This spectrum is called the 'power spectrum' of the stationary process.

The radiation field is defined by specifying how the electric field depends on time. To ensure the convergence of the integrals that we are going to use, we assume that the radiation field is a function with compact support, and hence is non-vanishing only in the closed time interval $[0, T]$. The limit $T \to \infty$ may be taken eventually. The Fourier analysis of the x-component of the electric field is

$$E_x(t) = \int_{-\infty}^{\infty} f(\nu) e^{2\pi i \nu t} \, d\nu, \tag{14.A.1}$$

where the amplitudes $f(\nu)$ are determined by (the field having compact support)

$$f(\nu) \equiv \int_0^T E_x(t) e^{-2\pi i \nu t} \, dt. \tag{14.A.2}$$

By virtue of the reality of E_x, one has $f^*(\nu) = f(-\nu)$, and an analogous analysis may be performed for the y- and z-components. Thus, the total energy of the radiation field is given by

$$u = \frac{1}{8\pi} \overline{\left(E^2 + H^2 \right)} = \frac{1}{4\pi} \overline{E^2} = \frac{3}{4\pi} \overline{E_x^2}, \tag{14.A.3}$$

because the time averages (see the comments after (14.A.6b)) have the property

$$\overline{E_x^2} = \overline{E_y^2} = \overline{E_z^2}.$$

Now we can compute these time averages, starting from the identity

$$\overline{E_x^2} = \frac{1}{T}\int_0^T E_x^2\,\mathrm{d}t = \frac{1}{T}\int_0^T E_x\,\mathrm{d}t \int_{-\infty}^{\infty} f(\nu)\mathrm{e}^{2\pi\mathrm{i}\nu t}\,\mathrm{d}\nu \tag{14.A.4}$$

and changing the order of integration. This yields

$$\begin{aligned}\overline{E_x^2} &= \frac{1}{T}\int_{-\infty}^{\infty} f(\nu)\,\mathrm{d}\nu \int_0^T E_x \mathrm{e}^{2\pi\mathrm{i}\nu t}\,\mathrm{d}t = \frac{1}{T}\int_{-\infty}^{\infty} f(\nu)f^*(\nu)\,\mathrm{d}\nu \\ &= \frac{2}{T}\int_0^{\infty} |f(\nu)|^2\,\mathrm{d}\nu.\end{aligned} \tag{14.A.5}$$

Equations (14.A.3) and (14.A.5) imply that the total density of radiation reads as

$$u = \int_0^{\infty} u_\nu\,\mathrm{d}\nu = \frac{3}{2\pi T}\int_0^{\infty} |f(\nu)|^2\,\mathrm{d}\nu, \tag{14.A.6a}$$

so that the spectral density is

$$u_\nu = \frac{3}{2\pi T}|f(\nu)|^2. \tag{14.A.6b}$$

So far, we have presented the simplified argument given by Born in his book on atomic physics. However, the averaging process is a crucial point, and a number of comments are in order before we can continue our investigation. The calculation of the averages $\langle E^k(t)\rangle$, $\langle H^k(t)\rangle$, where the integer $k = 1, 2, \ldots$, makes it necessary to use the theory of electromagnetic signals. The physical problem consists of several atoms, which emit electromagnetic radiation over the whole range of frequencies $\nu \in]-\infty, \infty[$. The results that we need are as follows.

(i) Denoting by N the average number of events per unit time, and by $[0, T]$ the time interval during which the observations are performed, one finds

$$\langle E(t)\rangle = N\int_0^T E(\tau)\,\mathrm{d}\tau = NT\int_0^T \frac{E(\tau)}{T}\,\mathrm{d}\tau = 0. \tag{14.A.7}$$

One then says that the mean value of the electric field is equal to the product of the average number of events per unit time, N, with the time integral of E, or, equivalently, to the product of the average number of events, NT, with the temporal average of E. This mean value vanishes, because E is a rapidly varying function, represented, hereafter, by the infinite sum

$$\sum_{i=-\infty}^{\infty} \varepsilon(t - t_i),$$

where each ε is non-vanishing only in a finite interval.

(ii) In general, in the interval $[0, T]$ one detects a number M of distinct signals. Thus, for each component of the electric field, the mean value of $E^2(t)$ turns out to be (the index for the component is omitted, for simplicity of notation)

$$\langle E^2(t)\rangle = \sum_{i=1}^{M} N_i \int_{-\infty}^{\infty} \varepsilon^2(\tau)\,\mathrm{d}\tau = \sum_{i=1}^{M} N_i \int_{-\infty}^{\infty} |\tilde{\varepsilon}(\nu)|^2\,\mathrm{d}\nu, \tag{14.A.8}$$

where the Parseval formula has been used to obtain the second equality. The integrals in (14.A.8) do not exist for a steady beam, but following Norbert Wiener the 'power spectrum' can be obtained from the autocorrelation function which is square-integrable. Now although the atoms emit at different frequencies, one finds that, for all of them, $\varepsilon(\nu)$ may be approximated by a curve having the shape of a bell. More precisely, the behaviour of $\varepsilon(\nu)$ is well approximated by a curve which changes rapidly in the neighbourhood of some value ν_i, where it attains its maximum: $\varepsilon(\nu) = \varepsilon(|\nu - \nu_i|)$. Bearing in mind that the atoms may emit over the whole range of frequencies, so that, in (14.A.8), the sum

$$\sum_{i=1}^{M} N_i$$

should be replaced by the integral $\int_{-\infty}^{\infty} N(\nu')\, d\nu'$, one finds that

$$\langle E^2(t)\rangle = \int_{-\infty}^{\infty} \overline{N}(\nu) e_0(\nu)\, d\nu = \int_{-\infty}^{\infty} e(\nu)\, d\nu = 2\int_0^{\infty} e(\nu)\, d\nu, \tag{14.A.9}$$

where $\overline{N}(\nu)$ is the average number of events per unit time:

$$\overline{N}(\nu) \equiv \frac{\int_{-\infty}^{\infty} N(\nu')|\tilde{\varepsilon}(|\nu - \nu'|)|^2\, d\nu'}{\int_{-\infty}^{\infty} |\tilde{\varepsilon}(|\nu - \nu'|)|^2\, d\nu'}, \tag{14.A.10}$$

and $e_0(\nu)$ is the total energy of the signal corresponding to the frequency ν:

$$e_0(\nu) \equiv \int_{-\infty}^{\infty} |\tilde{\varepsilon}(|\nu - \nu'|)|^2 d\nu'. \tag{14.A.11}$$

It is thus crucial to appreciate, hereafter, that all mean values we refer to should be obtained, strictly, by combining the operation of a statistical average with the temporal average of the function under investigation. It is only upon considering the joint effect of these two operations that one obtains mean values that are independent of the instant of time when the measurements begin.

To complete our analysis, we have now to consider the vibrations of the linear harmonic oscillator under the action of the electric field. One can assume, for simplicity, that the oscillator vibrates only in the x-direction (Born 1969):

$$m\ddot{x} + ax = eE_x(t), \tag{14.A.12}$$

where $a \equiv 4\pi^2\nu_0^2 m$. The general solution of Eq. (14.A.12) consists of the general solution of the homogeneous equation plus a particular solution of the full equation. The former reads as

$$x_1(t) = x_0 \sin(2\pi\nu_0 t + \phi), \tag{14.A.13}$$

with x_0 and ϕ being arbitrary constants, while the latter is

$$x_2(t) = \frac{e}{2\pi\nu_0 m}\int_0^t E_x(\tau)\sin[2\pi\nu_0(t-\tau)]\, d\tau, \tag{14.A.14}$$

if the initial conditions are $x(0) = \dot{x}(0) = 0$. Of course, to derive Eq. (14.A.14) one has to apply the method of 'variation of parameters'. This requires that one should look for the solution of Eq. (14.A.12) in the form (here $\omega \equiv 2\pi\nu$)

$$x_2(t) = A_1(t)\cos(\omega t) + A_2(t)\sin(\omega t), \tag{14.A.15}$$

where A_1 and A_2 are two functions of the time variable subject to the conditions

$$\dot{A}_1\cos(\omega t) + \dot{A}_2\sin(\omega t) = 0, \tag{14.A.16}$$

$$-\omega\dot{A}_1\sin(\omega t) + \omega\dot{A}_2\cos(\omega t) = \frac{e}{m}E_x, \tag{14.A.17}$$

the solution of which is, by elementary integration,

$$A_1(t) = -\frac{e}{m\omega}\int_0^t E_x(\tau)\sin(\omega\tau)\, d\tau + A_1^0, \tag{14.A.18}$$

$$A_2(t) = \frac{e}{m\omega}\int_0^t E_x(\tau)\cos(\omega\tau)\, d\tau + A_2^0. \tag{14.A.19}$$

The initial conditions $x_2(0) = 0, \dot{x}_2(0) = 0$, imply that both A_1^0 and A_2^0 should vanish, and hence Eqs. (14.A.15), (14.A.18) and (14.A.19) lead to the result in the form (14.A.14).

We are now in a position to evaluate the work done by the field on the oscillator. More precisely, we are interested in the work per unit time (here $x(t) = x_1(t) + x_2(t)$)

$$\delta W = \frac{e}{T}\int_0^T \dot{x}(t)E_x(t)\, dt. \tag{14.A.20}$$

By construction, the contribution of x_1 to the integral (14.A.20) vanishes, and one finds

$$\delta W = \frac{e}{T}\frac{e}{m}\int_0^T E_x(t)\,\mathrm{d}t \int_0^t E_x(\tau)\cos[2\pi\nu_0(t-\tau)]\,\mathrm{d}\tau. \tag{14.A.21}$$

The integrand in (14.A.21) is a symmetric function of t and τ, and hence one finds

$$\delta W = \frac{e^2}{mT}\int_0^T E_x(\tau)\,\mathrm{d}\tau \int_\tau^T E_x(t)\cos[2\pi\nu_0(t-\tau)]\,\mathrm{d}t, \tag{14.A.22a}$$

which is re-expressed, after interchanging t and τ, as

$$\delta W = \frac{e^2}{mT}\int_0^T E_x(t)\,\mathrm{d}t \int_t^T E_x(\tau)\cos[2\pi\nu_0(t-\tau)]\,\mathrm{d}\tau. \tag{14.A.22b}$$

The comparison of Eqs. (14.A.21) and (14.A.22*b*) leads to the formula

$$\delta W = \frac{e^2}{2mT}\int_0^T E_x(t)\,\mathrm{d}t \left(\int_0^t + \int_t^T\right) E_x(\tau)\cos[2\pi\nu_0(t-\tau)]\,\mathrm{d}\tau. \tag{14.A.23}$$

Note that the cosine function is conveniently re-expressed in the form

$$\frac{1}{2}\left[\mathrm{e}^{2\pi\mathrm{i}\nu_0(t-\tau)} + \mathrm{e}^{-2\pi\mathrm{i}\nu_0(t-\tau)}\right],$$

and hence the work per unit time turns out to be

$$\begin{aligned}\delta W &= \frac{e^2}{4mT}\int_0^T E_x(t)\mathrm{e}^{2\pi\mathrm{i}\nu_0 t}\,\mathrm{d}t \int_0^T E_x(\tau)\mathrm{e}^{-2\pi\mathrm{i}\nu_0\tau}\,\mathrm{d}\tau \\ &+ \frac{e^2}{4mT}\int_0^T E_x(t)\mathrm{e}^{-2\pi\mathrm{i}\nu_0 t}\,\mathrm{d}t \int_0^T E_x(\tau)\mathrm{e}^{2\pi\mathrm{i}\nu_0\tau}\,\mathrm{d}\tau \\ &= \frac{e^2}{4mT}\left[f^*(\nu_0)f(\nu_0) + f(\nu_0)f^*(\nu_0)\right] \\ &= \frac{e^2}{2mT}|f(\nu_0)|^2. \end{aligned} \tag{14.A.24}$$

By virtue of Eqs. (14.A.6*b*) and (14.A.24) one finds the desired result,

$$\delta W = \frac{e^2}{2mT}\frac{2\pi T}{3}u_\nu = \frac{\pi e^2}{3m}u_\nu. \tag{14.A.25}$$

15
Lagrangian and phase-space formulations

We begin with a presentation of the Schwinger variational principle in quantum mechanics. In the second part, after a re-assessment of the problem of motion in quantum mechanics, it is shown that the Green kernel, which represents the probability amplitude of finding a particle in space at a given time, once its location at another time is known, can be represented by a sum over 'space–time paths'. According to this interpretation, one integrates the exponential of i times the classical action divided by $\hbar$, with a (formal) measure over all space–time paths matching the initial location x_i at time t_i and the final location x_f at time t_f. Such a way of evaluating the Green kernel is applied to any quadratic Lagrangian for a generic quantum system including, in particular, the harmonic oscillator and the free particle.

Lastly, we outline a formalism which involves ordinary functions of commuting variables, and exactly reproduces ordinary quantum mechanics. It works with a phase space endowed with commuting coordinates, so that one is dealing with quantum mechanics in phase space.

15.1 The Schwinger formulation of quantum dynamics

In both the Schrödinger formulation in terms of wave functions (with associated state vectors) and the Heisenberg formulation in terms of non-commuting matrices (with corresponding linear operators) the dynamical evolution is specified in terms of the Hamiltonian. This has motivated our presentation of classical dynamics, but is in marked contrast with the view according to which the Lagrangian and the action specify the dynamics. Hamilton's principle is formulated in terms of the action, the time integral of the Lagrangian. Why not in quantum mechanics?

This is what was formulated in Dirac (1933), where Dirac pointed out that in quantum mechanics the q, p variables at time t are still connected with the q, p variables at any other time T by a canonical transformation, and that a transformation function $\langle q_t|q_T\rangle$ exists connecting the two representations in which the q_t and q_T are diagonal respectively. He went on to show that

$$\langle q_t|q_T\rangle \text{ corresponds to } \exp\left(\mathrm{i}\int_T^t \mathcal{L}\,\mathrm{d}t/\hbar\right),$$

where $\mathcal{L}$ is the Lagrangian. In case T differs only infinitely little from t, he therefore obtained the result according to which

$$\langle q_{t+\mathrm{d}t}|q_t\rangle \text{ corresponds to } \exp(\mathrm{i}\mathcal{L}\,\mathrm{d}t/\hbar).$$

In other words, Dirac showed that the transformation function connecting the states at t and at $t+\Delta t$ may be viewed as

$$\langle \beta t|\alpha, t+\Delta t\rangle = \mathrm{i}\langle\beta|\,\triangle\,\mathcal{A}/\hbar|\alpha\rangle. \tag{15.1.1}$$

Moreover this transformation is transitive, i.e.

$$\sum_\beta \langle\alpha t|\beta t_1\rangle\langle\beta t_1|\alpha t_0\rangle = \langle\alpha t|\alpha t_0\rangle. \tag{15.1.2}$$

Starting with this Feynman developed a formalism leading to the transition amplitude

$$\langle t|t_0\rangle = \langle\beta t|\alpha t_0\rangle = \langle\beta|\exp\left(\mathrm{i}\int_{t_0}^t \mathcal{L}(t')\,\mathrm{d}t'/\hbar\right)|\alpha\rangle, \tag{15.1.3}$$

which is the path integral formulation. We describe this in detail in the following, but we first need to discuss the quantum action principle, which is central to the Schwinger approach to quantum mechanics (including quantum field theory).

If in classical dynamics we consider the Hamilton action principle, variations of dynamical variables and the time labels vanishing at the boundaries we obtain the Hamilton equations of motion. But what if extra variations were allowed? For this purpose, P. Weiss invented the extended action principle for classical dynamical systems (Weiss 1938). He considered general variations which do not vanish at the boundaries, while the time at the end point is also varied. Weiss demonstrated that the change in the action contains a surface term in terms of the changes of q and t, while the volume term gives the equations of motion. More precisely, the variation of the action is expressed in general by (the notation $\triangle$ is quite standard for total variations, but should not be confused with the Laplacian)

$$\begin{aligned}\triangle\mathcal{A} = &\int_{t_1}^{t_2}\sum_s\left(\frac{\partial\mathcal{L}}{\partial q^s} - \frac{\mathrm{d}}{\mathrm{d}t}\frac{\partial\mathcal{L}}{\partial v^s}\right)\delta q^s(t)\,\mathrm{d}t \\ &+\Big[\sum_s p_s\,\triangle q^s - H\,\triangle t\Big]_{t_1}^{t_2},\end{aligned} \tag{15.1.4}$$

where the total variation of position variables at the endpoints reads

$$\triangle q^s(t_i) \equiv \delta q^s(t_i) + v^s(t_i) \triangle t_i, \quad i = 1, 2. \tag{15.1.5}$$

If the path of integration is an actual dynamical path, the integral in (15.1.4) vanishes by virtue of the Euler–Lagrange equations and the variation of the action functional under general variations is not necessarily zero and reduces to

$$\triangle \mathcal{A} = \Big[-H \triangle t + \sum_s p_s \triangle q^s \Big]_{t_1}^{t_2}. \tag{15.1.6}$$

The variation (15.1.6) of the action depends only on the endpoints, and the coefficients of the variations are the canonical conjugates to t and q, respectively. In this Lagrangian formulation of the action principle, the dynamical path of a system is that path 'general variations' about which produce only endpoint contributions to $\triangle \mathcal{A}$ according to (15.1.6). Moreover, we may write

$$\delta F = \{F, \delta \mathcal{A}\}. \tag{15.1.7}$$

In this computation the variations of q and t are treated as numbers which have vanishing Poisson brackets with all dynamical variables.

In the hands of J. Schwinger the quantum action principle took the form of the variation of a matrix element in terms of the quantum variation $\delta \mathcal{A}$ (Schwinger 1951b):

$$\delta \langle \Phi_1 | \Phi_2 \rangle = \frac{\mathrm{i}}{\hbar} \langle \Phi_1 | \delta \mathcal{A} | \Phi_2 \rangle. \tag{15.1.8}$$

This is a succinct but powerful condensation of quantum dynamics. Schwinger then goes on to show that the Lagrangian and Hamiltonian equations are consistent, and the generator of changes in the field quantities is the change in the action. Since the dynamical variables in quantum theory do not commute with each other we must designate how the variations commute with the dynamical variables. For Bose field quantities Schwinger requires the variations $\delta\psi$ to commute with all variables. For Fermi field quantities the variations $\delta\psi$ anticommute with all fields.

Schwinger prefers to consider Lagrangians linear in the velocities, which give rise to first-order equations. For Bose or Fermi fields the kinematic terms in the Lagrangian are bilinear of the form

$$\mathcal{L}_{\text{kin}} = \psi^{\mathrm{T}} \Gamma_0 \frac{\partial}{\partial t} \psi; \quad \mathcal{A}_{\text{kin}} = \int \mathcal{L}_{\text{kin}} \, \mathrm{d}t. \tag{15.1.9}$$

Then the field variations are generated by $\delta \mathcal{A}$:

$$\delta \psi = \Big[\psi^{\mathrm{T}} \Gamma_0 \delta \psi, \psi \Big]. \tag{15.1.10}$$

For Bose fields $\delta\psi$ commutes and we obtain the commutator relations

$$\left[\psi^{\mathrm{T}}(x), \Gamma_0 \psi(y)\right] = \delta(x-y). \tag{15.1.11}$$

For Fermi fields $\delta\psi$ anticommutes to yield the anticommutation relations

$$\left\{\psi^{\mathrm{T}}(x), \Gamma_0 \psi(y)\right\}_+ = \delta(x-y). \tag{15.1.12}$$

Since the kinematic terms must be a scalar and since $\frac{\partial}{\partial t}$ is an antisymmetric operation, we find that the matrix Γ_0 must be antisymmetric for Bose fields and symmetric for Fermi fields. But since the scalar product of two tensor quantities is symmetric and of two spinor quantities is antisymmetric we immediately obtain the spin-statistics relation: tensor fields are Bose fields and spinor fields are Fermi fields. This fundamental connection is thus directly related to the geometry of spinors and tensors for rotations in three dimensions. Relativistic invariance is welcome but not necessary to deduce this relation (Duck and Sudarshan 1997).

When dynamical constraints obtain for singular Lagrangians the variations are not independent, and this lack of independence is to be taken into account to obtain the correct implementation of the Schwinger action principle.

15.2 Propagator and probability amplitude

We are familiar, from section 4.4, with the notion of the Green kernel of the Schrödinger equation. In general, this makes it possible to express the wave function in the form (see Eq. (4.4.25))

$$\psi(\vec{x}, t) = \int G(\vec{x} - \vec{x}', t - t') \psi(\vec{x}', t') \, \mathrm{d}^3 x'. \tag{15.2.1}$$

It is of course possible to formulate the problem of integrating the equations of motion by taking Eq. (15.2.1) as the starting point.

Indeed, let us consider a particle described by the wave function $\psi(x, t)$ (hereafter the arrows for vectors are omitted for simplicity of notation). The Schrödinger equation makes it possible to evaluate $\frac{\partial}{\partial t}\psi(x, t)$, i.e. the way in which the wave function varies in time. The equation therefore provides the temporal evolution of the wave function by using a differential point of view. It is, however, possible to adopt a more 'global' point of view that leads to the direct evaluation of $\psi(\widetilde{x}, t)$ starting from the knowledge of the wave function $\psi(x, t')$ at a previous time t', not necessarily close to t.

The consideration of this possibility can be suggested by electromagnetism. It is, in fact, possible to rely on the Maxwell equations, i.e. the

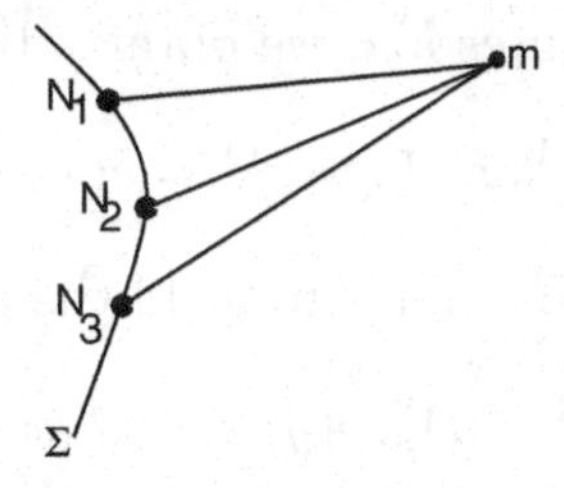

Fig. 15.1. According to a global viewpoint in electromagnetism, when a monochromatic field is known on a surface Σ, the field at any point $m \in M$ can be obtained by adding the fields propagated from all fictitious secondary sources located on Σ.

differential viewpoint, or the Huygens principle, and hence the global viewpoint, which makes it possible to evaluate directly, when a monochromatic field is known on a surface Σ, the field at any point of the manifold M (see figure 15.1). One then adds the fields propagated to the point $m \in M$ from the fictitious secondary sources $N_1, N_2, \ldots$ located on the surface Σ, for which the amplitudes and phases are determined from the values of the field at $N_1, N_2, \ldots$. It is therefore clear that an expression of the kind

$$\psi(\widetilde{x}, t_2) = \int G(\widetilde{x}, t_2; x, t_1)\psi(x, t_1)\, \mathrm{d}^3x \tag{15.2.2}$$

for $t_2 > t_1$ translates exactly the idea we have described. The possible physical interpretation is that the probability amplitude of finding the particle at m at time t_2 is obtained by adding all amplitudes 'propagated' from the secondary sources located on the surface Σ (having equation $t = t_1$) in space-time. The kernel G is the Green kernel (also called the Green function) of the Schrödinger equation, and it is possible to formulate the whole of quantum mechanics in terms of G.

Recall from section 9.8 that the propagator is the unitary operator $U(t, t_0)$ mapping the state vector at time t_0 into the state vector at time t. This operator obeys the first-order equation (9.8.1) with initial condition (9.8.2). This is equivalent to studying the integral equation

$$U(t, t_0) = \mathbb{1} - \frac{\mathrm{i}}{\hbar} \int_{t_0}^{t} H(t')U(t', t_0)\, \mathrm{d}t'. \tag{15.2.3}$$

On iterating the equation for $|\psi(t)\rangle$, i.e.

$$|\psi(t)\rangle = U(t, t')|\psi(t')\rangle = U(t, t')U(t', t'')|\psi(t'')\rangle, \tag{15.2.4}$$

we obtain

$$U(t_n, t_1) = U(t_n, t_{n-1})U(t_{n-1}, t_{n-2}) \cdots U(t_2, t_1). \tag{15.2.5}$$

By working at the 'infinitesimal level' we can write that

$$d|\psi(t)\rangle = |\psi(t+dt)\rangle - |\psi(t)\rangle = -\frac{i}{\hbar}H(t)|\psi(t)\rangle\, dt. \tag{15.2.6}$$

The state vector at time $t + dt$ is thus given by

$$|\psi(t+dt)\rangle = \left[\mathbb{1} - \frac{i}{\hbar}H(t)\, dt\right]|\psi(t)\rangle, \tag{15.2.7}$$

which implies, for the propagator,

$$U(t+dt, t) = \mathbb{1} - \frac{i}{\hbar}H(t)\, dt. \tag{15.2.8}$$

Starting from Eq. (15.2.4), here rewritten in the form

$$|\psi(t_2)\rangle = U(t_2, t_1)|\psi(t_1)\rangle, \tag{15.2.9}$$

we can determine the wave function $\psi(x_2, t_2)$ by evaluating (see Eq. (7.8.3))

$$\psi(x_2, t_2) = \langle x_2|\psi(t_2)\rangle, \tag{15.2.10}$$

which leads to

$$\psi(x_2, t_2) = \int \langle x_2|U(t_2, t_1)|x_1\rangle\langle x_1|\psi(t_1)\rangle\, d^3x_1. \tag{15.2.11}$$

By comparison of (15.2.2) and (15.2.11), the Green kernel is therefore found to be

$$G(x_2, t_2; x_1, t_1) = \langle x_2|U(t_2, t_1)|x_1\rangle. \tag{15.2.12}$$

This formula is of fundamental importance and specifies how to obtain the Green kernel from the propagator. Sometimes, in the terminology, one does not distinguish between the two, and it is precisely the previous set of equations which suggests calling G itself 'the propagator'. Following the ideas in Schwinger (1951a) such a formula has been extended to systems with an infinite number of degrees of freedom by obtaining various Green kernels as matrix elements of suitably defined operators between vectors of an abstract Hilbert space.

If we are only interested in the Green kernel for $t_2 > t_1$ we can write

$$G(x_2, t_2; x_1, t_1) \equiv \langle x_2|U(t_2, t_1)|x_1\rangle\theta(t_2 - t_1), \tag{15.2.13}$$

where θ is the Heaviside 'step function':

$$\theta(t_2 - t_1) = 1 \text{ if } t_2 > t_1, \tag{15.2.14a}$$

$$\theta(t_2 - t_1) = \frac{1}{2} \text{ if } t_2 = t_1, \tag{15.2.14b}$$

$$\theta(t_2 - t_1) = 0 \text{ if } t_2 < t_1. \tag{15.2.14c}$$

The introduction of the Heaviside function takes into account that the surface Σ propagates to the future only, and G is the 'retarded' Green kernel. The derivative of θ is a Dirac delta functional (see below).

15.2.1 Physical interpretation

The Green kernel $G(x_2, t_2; x_1, t_1)$ provides the probability amplitude that the particle starts at the point with coordinates (x_1, t_1) and arrives at (x_2, t_2). If one takes as an initial state at time t_1 a state localized at the point x_1:

$$|\psi(t_1)\rangle = |x_1\rangle, \tag{15.2.15}$$

the state vector at time t_2 becomes

$$|\psi(t_2)\rangle = U(t_2, t_1)|\psi(t_1)\rangle = U(t_2, t_1)|x_1\rangle. \tag{15.2.16}$$

The probability amplitude of finding the particle at point x_2 at time t_2 is therefore

$$\langle x_2|\psi(t_2)\rangle = \langle x_2|U(t_2, t_1)|x_1\rangle. \tag{15.2.17}$$

15.2.2 Green kernel in the energy representation

Let us assume that the Hamiltonian operator does not depend explicitly on time and has a purely discrete spectrum with eigenvectors φ_n. We are now going to consider again the analysis of section 4.4, but with the 'ket' and 'bra' notation and a more frequent use of Dirac delta functionals.

By virtue of the completeness relation

$$\sum_{n=0}^{\infty} |\varphi_n\rangle\langle\varphi_n| = \mathbb{I}, \tag{15.2.18}$$

the propagator can be written as

$$U(t_2, t_1) = \mathrm{e}^{-\mathrm{i}H(t_2-t_1)/\hbar} = \sum_{n=0}^{\infty} |\varphi_n\rangle \mathrm{e}^{-\mathrm{i}H(t_2-t_1)/\hbar} \langle\varphi_n|. \tag{15.2.19}$$

The right-hand side of Eq. (15.2.17) is hence equal to

$$\begin{aligned}\langle x_2|U(t_2, t_1)|x_1\rangle &= \sum_{n=0}^{\infty} \langle x_2|\varphi_n\rangle \mathrm{e}^{-\mathrm{i}H(t_2-t_1)/\hbar} \langle\varphi_n|x_1\rangle \\ &= \sum_{n=0}^{\infty} \varphi_n^*(x_1)\varphi_n(x_2)\mathrm{e}^{-\mathrm{i}H(t_2-t_1)/\hbar},\end{aligned} \tag{15.2.20}$$

and the Green kernel is obtained from (15.2.13) as

$$G(x_2, t_2; x_1, t_1) = \theta(t_2 - t_1) \sum_{n=0}^{\infty} \varphi_n^*(x_1)\varphi_n(x_2)\mathrm{e}^{-\mathrm{i}H(t_2-t_1)/\hbar}. \tag{15.2.21}$$

Since $\varphi_n(x_2)\mathrm{e}^{-\mathrm{i}E_n t_2/\hbar}$ is a solution of the Schrödinger equation

$$\left[\mathrm{i}\hbar\frac{\partial}{\partial t_2} - H\left(x_2, \frac{\hbar}{\mathrm{i}}\mathrm{grad}_2\right)\right] \varphi_n(x_2)\mathrm{e}^{-\mathrm{i}E_n t_2/\hbar} = 0, \quad (15.2.22)$$

we can use the identity

$$\frac{\partial}{\partial t_2}\theta(t_2 - t_1) = \delta(t_2 - t_1) \quad (15.2.23)$$

and hence write

$$\begin{aligned}\left[\mathrm{i}\hbar\frac{\partial}{\partial t_2} - H\left(x_2, \frac{\hbar}{\mathrm{i}}\mathrm{grad}_2\right)\right] &G(x_2, t_2; x_1, t_1)\\ &= \mathrm{i}\hbar\sum_{n=0}^{\infty}\varphi_n^*(x_1)\varphi_n(x_2)\mathrm{e}^{-\mathrm{i}E_n(t_2-t_1)/\hbar}\delta(t_2 - t_1)\\ &= \mathrm{i}\hbar\sum_{n=0}^{\infty}\varphi_n^*(x_1)\varphi_n(x_2)\delta(t_2 - t_1)\\ &= \mathrm{i}\hbar\sum_{n=0}^{\infty}\langle x_2|\varphi_n\rangle\langle\varphi_n|x_1\rangle\delta(t_2 - t_1)\\ &= \mathrm{i}\hbar\langle x_2|x_1\rangle\delta(t_2 - t_1)\\ &= \mathrm{i}\hbar\delta(x_2 - x_1)\delta(t_2 - t_1). \end{aligned} \quad (15.2.24)$$

Thus, the Green kernel obeys the equation (see Eq. (4.4.16a))

$$\left[\mathrm{i}\hbar\frac{\partial}{\partial t} - H\left(x, \frac{\hbar}{\mathrm{i}}\mathrm{grad}_x\right)\right] G(x, t; x_0, t_0) = \mathrm{i}\hbar\delta(x - x_0)\delta(t - t_0). \quad (15.2.25)$$

15.3 Lagrangian formulation of quantum mechanics

Let us consider a partition of the time interval $[t_1, t_2]$ into sub-intervals (see figure 15.2)

$$[t_1, t_{\alpha_1}],\ [t_{\alpha_1}, t_{\alpha_2}], \ldots, [t_{\alpha_N}, t_2],$$

in such a way that, by virtue of Eq. (15.2.5), the propagator $U(t_2, t_1)$ can be expressed in the form

$$\begin{aligned} U(t_2, t_1) &= \prod U(t_\alpha, t_\beta)\\ &= U(t_2, t_{\alpha_N})U(t_{\alpha_N}, t_{\alpha_{N-1}})\cdots U(t_{\alpha_1}, t_1). \end{aligned} \quad (15.3.1)$$

We now consider the right-hand side of Eq. (15.2.17), with the integration

$$\int \mathrm{d}^3x_{\alpha_N}\,|x_{\alpha_N}\rangle\langle x_{\alpha_N}|$$

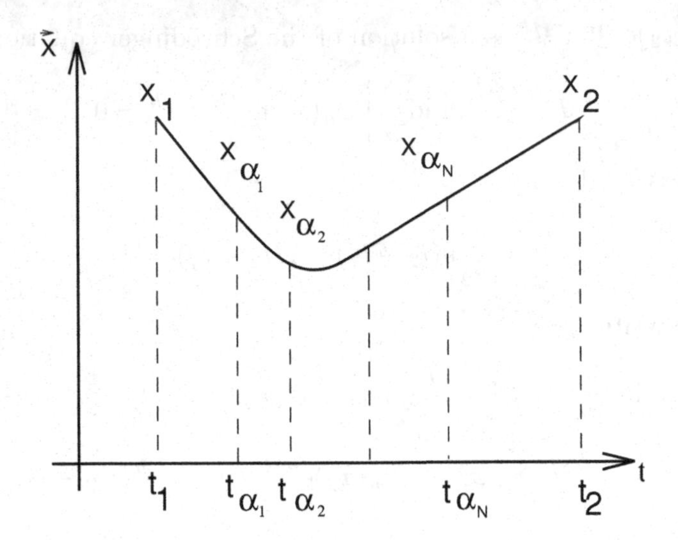

Fig. 15.2. In the analysis of a space-time path, the time interval $[t_1, t_2]$ is partitioned into sub-intervals.

to be inserted in between adjacent products.

Hereafter we shall need the formula

$$\langle q|p\rangle = \frac{1}{\sqrt{2\pi\hbar}} \mathrm{e}^{\mathrm{i}pq/\hbar}. \tag{15.3.2}$$

If we regard the Hamiltonian operator as depending on all position and momentum operators, with all momenta to the left and all positions to the right (this is called the standard ordering prescription), we can write

$$\langle p|H|q\rangle = h(p,q)\langle p|q\rangle = \frac{1}{\sqrt{2\pi\hbar}} \mathrm{e}^{-\mathrm{i}pq/\hbar} h(p,q). \tag{15.3.3}$$

We must use this mixed version so that H can be replaced by $h(p,q)$, assuming that the Hamiltonian is already standard ordered. If the Hamiltonian is independent of time, the propagator depends only on the difference $t_2 - t_1$, and hence

$$\langle q_2|U(t_2 - t_1)|q_1\rangle = \langle q_2|\mathrm{e}^{-\mathrm{i}H(t_2-t_1)/\hbar}|q_1\rangle = \langle q_2, t_2|q_1, t_1\rangle. \tag{15.3.4}$$

On the other hand, if $t_2 - t_1 = \Delta t$, we have

$$\langle p|U(t_1 + \Delta t, t_1)|q_1\rangle \cong \frac{1}{\sqrt{2\pi\hbar}} \mathrm{e}^{-\mathrm{i}pq_1/\hbar} \left[1 - \frac{\mathrm{i}}{\hbar} h(p, q_1)\Delta t\right]. \tag{15.3.5}$$

We can now insert the resolution of the identity

$$\int \mathrm{d}p\, |p\rangle\langle p| = \mathbb{I} \tag{15.3.6}$$

into the matrix element (15.3.4), so that

$$\begin{aligned}\langle q_2|U(t_2-t_1)|q_1\rangle &= \int \mathrm{d}p\, \langle q_2|p\rangle\langle p|U(t_2-t_1)|q_1\rangle \\ &\cong \frac{1}{2\pi\hbar}\int \mathrm{d}p\, \mathrm{e}^{\frac{\mathrm{i}}{\hbar}p(q_2-q_1)}\left[1-\frac{\mathrm{i}}{\hbar}h(p,q_1)\triangle t\right] \\ &\cong \frac{1}{2\pi\hbar}\int \mathrm{d}p\, \mathrm{e}^{\frac{\mathrm{i}}{\hbar}p(q_2-q_1)}\,\mathrm{e}^{-\frac{\mathrm{i}}{\hbar}h(p,q_1)\triangle t}. \qquad (15.3.7)\end{aligned}$$

Note that we must use the standard ordering (with p to the left and q to the right) so that $\langle p|H|q\rangle$ can be replaced by $h(p,q)$ which is essential in (15.3.3) and (15.3.5). On the other hand, if $\triangle t = \frac{t_2-t_1}{n}$, we can express the propagator as (see appendix 15.A)

$$U(t_2-t_1) = \left(\mathrm{e}^{-\frac{\mathrm{i}}{\hbar}H\triangle t}\right)^n, \qquad (15.3.8)$$

and inserting repeatedly the resolution of the identity with respect to position improper eigenfunctions:

$$\int \mathrm{d}q\, |q\rangle\langle q| = \mathbb{I}, \qquad (15.3.9)$$

we find

$$\begin{aligned}\langle q_2|U(t_2-t_1)|q_1\rangle &= \int \mathrm{d}\widetilde{q}_1 \cdots \int \mathrm{d}\widetilde{q}_{n-1}\langle \widetilde{q}_n|\mathrm{e}^{-\frac{\mathrm{i}}{\hbar}H\triangle t}|\widetilde{q}_{n-1}\rangle \\ \langle \widetilde{q}_{n-1}|\mathrm{e}^{-\frac{\mathrm{i}}{\hbar}H\triangle t}|\widetilde{q}_{n-2}\rangle &\cdots \langle \widetilde{q}_1|\mathrm{e}^{-\frac{\mathrm{i}}{\hbar}H\triangle t}|\widetilde{q}_0\rangle, \qquad (15.3.10)\end{aligned}$$

where $\widetilde{q}_0 \equiv q_1$ and $\widetilde{q}_N \equiv q_2$. Finally, repeated insertion of (15.3.6) into Eq. (15.3.10) yields

$$\begin{aligned}&\int \frac{\mathrm{d}p_n}{2\pi\hbar}\prod_{k=1}^{n-1}\frac{\mathrm{d}p_k\,\mathrm{d}q_k}{2\pi\hbar}\mathrm{e}^{\frac{\mathrm{i}}{\hbar}\sum_{k=1}^{n}[p_k(q_k-q_{k-1})-h(p_k,q_k)\triangle t]} \\ &= \langle q_2|U(t_2-t_1)|q_1\rangle. \qquad (15.3.11)\end{aligned}$$

On taking the limit as $n\to\infty$, the exponent in Eq. (15.3.11) becomes

$$\frac{\mathrm{i}}{\hbar}\int_{t_1}^{t_2}\mathrm{d}t\,[p(t)\dot{q}(t)-h(p,q)] = \frac{\mathrm{i}}{\hbar}\int_{t_1}^{t_2}L\,\mathrm{d}t.$$

Moreover, the corresponding measure can be (formally) written as

$$\frac{\mathrm{d}p_2}{2\pi\hbar}\prod_t \frac{\mathrm{d}p(t)\,\mathrm{d}q(t)}{2\pi\hbar},$$

with $q(t_1)=q_1$ and $q(t_2)=q_2$. The factor $\frac{\mathrm{d}p_2}{2\pi}$ occurs because there are $n-1$ integrations with respect to $\widetilde{q}_1,\ldots,\widetilde{q}_{n-1}$ and n integrations with respect to $p_1, \ldots, p_n$. The final result is normally written as

$$\langle q_2 | U(t_2 - t_1) | q_1 \rangle = \int \mathrm{D}q \, \mathrm{D}p \, \mathrm{e}^{\frac{\mathrm{i}}{\hbar} \int_{t_1}^{t_2} L \, \mathrm{d}t}, \tag{15.3.12}$$

where, in three dimensions,

$$\mathrm{D}q \, \mathrm{D}p = (2\pi\hbar)^{-3n} \prod_{k=1}^{n} \mathrm{d}q_k \prod_{k=0}^{n} \mathrm{d}p_k. \tag{15.3.13}$$

This is called the Feynman formula (Feynman 1948) for the probability amplitude. Note that

$$\mathrm{e}^{\frac{\mathrm{i}}{\hbar} \int \mathcal{L} \, \mathrm{d}t} = \mathrm{e}^{\frac{\mathrm{i}}{\hbar} S_{cl}},$$

and the stationarity points of the action clearly provide the dominant contribution to the integral.

One of the most interesting advantages of this Lagrangian formulation is the possibility of obtaining a relativistic generalization (since the Lagrangian is Lorentz invariant though the Hamiltonian is not), but we cannot present such a topic in a monograph mainly devoted to non-relativistic quantum theory. The Lagrangian approach is due to Feynman, who was inspired by early work in Dirac (1933), who pointed out the importance of the Lagrangian in quantum mechanics, and who also demonstrated the group property of the evolution operator. In Feynman (1948), the two basic postulates of a space-time approach to non-relativistic quantum mechanics were as follows.

(i) The probability that the path of a particle lies in a space-time region is the square of the modulus of a sum of contributions, each of which results from all possible paths in that region.

(ii) Every path has a *phase* proportional to the classical action for that path in $\hbar$ units: $\varphi(x(s)) \propto \mathrm{e}^{\frac{\mathrm{i}}{\hbar} S}$. The paths summed over in the expression of the Green kernel are all paths such that $q(t_1) = q_1$ and $q(t_2) = q_2$.

Interestingly, the path-integral approach also has a counterpart in classical mechanics, for which we refer the reader to the work in Gozzi and Regini (2000).

15.4 Green kernel for quadratic Lagrangians

It would now be instructive to see how the previous formulae lead to powerful tools for the evaluation of the Green kernel of the Schrödinger equation. For this purpose, we consider quadratic Lagrangians in one spatial dimension (in this case a term like $bx\dot{x}$ would be a total time derivative and therefore re-absorbed in the wave function). For any quadratic

Lagrangian

$$\mathcal{L} = \frac{1}{2}m\dot{x}^2 - \frac{1}{2}cx^2, \tag{15.4.1}$$

Eq. (15.3.11) makes it possible to express the corresponding Green kernel as the limit for $N \to \infty$ of

$$\left(m/2\pi i\hbar\varepsilon\right)^{N/2} \int_{y,t_0}^{z,t} dx_1 \cdots dx_{N-1}$$
$$\times \exp\left\{\frac{i\varepsilon}{\hbar}\sum_{j=0}^{N-1}\left[\frac{m}{2}\left(\frac{x_{j+1}-x_j}{\varepsilon}\right)^2 - \frac{1}{2}c(t_j)x_j^2\right]\right\}.$$

To evaluate such a Green kernel, we now set $x(s) = \overline{x}(s) + r(s)$, where $\overline{x}(s)$ represents the classical path and $r(s)$ is a fluctuation around $\overline{x}(s)$. Each point x_i can be thus expressed as

$$x_i = \overline{x}_i + r_i, \tag{15.4.2}$$

where $\overline{x}_i$ is the classical position at the time t_i. This leads to a Taylor expansion of the sum in the argument of the exponential in the form

$$\begin{aligned}\sum_{j=0}^{N-1} S(x_{j+1}, x_j) &= \sum_{j=0}^{N-1} S(\overline{x}_{j+1}, \overline{x}_j) \\ &\quad + \frac{1}{2}\sum_{k,l=0}^{N} \frac{\partial^2}{\partial x_k \partial x_l} \sum_{j=0}^{N-1} S(x_{j+1}, x_j)|_{\overline{x}_k, \overline{x}_l} r_k r_l \\ &= \sum_{j=0}^{N-1} S(\overline{x}_{j+1}, \overline{x}_j) \\ &\quad + \sum_{j=0}^{N-1}\left[\frac{m}{2\varepsilon}(r_{j+1}-r_j)^2 - \frac{1}{2}c(t_j)r_j^2\right].\end{aligned} \tag{15.4.3}$$

Note that terms involving the first derivatives of S at $\overline{x}_l$ vanish, since this is exactly the condition for these points to solve the classical equations of motion. Moreover, since S is a quadratic functional of the x_l variables, no higher-order derivatives contribute to the expansion (15.4.3). This implies that the Green kernel $G(z, t; y, t_0)$ reads

$$e^{\frac{i}{\hbar}S(\overline{x})} \lim_{N\to\infty} \left(m/2\pi i\hbar\varepsilon\right)^{N/2}$$
$$\times \int dr_1 \cdots dr_{N-1} \exp\left\{\frac{i\varepsilon}{\hbar}\sum_{j=0}^{N-1}\left[\frac{m}{2}\left(\frac{r_{j+1}-r_j}{\varepsilon}\right)^2 - \frac{1}{2}c(t_j)r_j^2\right]\right\}.$$

With our notation, $r_0 = r_N = 0$, since all paths have the endpoints $x(t_0) = y$ and $x(t) = z$. Hence one can write

$$G(z, t; y, t_0) = \mathrm{e}^{\frac{\mathrm{i}}{\hbar} S(\overline{x})} \widetilde{G}(0, t; 0, t_0). \tag{15.4.4}$$

Denoting by η the column vector $\eta = \begin{pmatrix} r_1 \\ \cdot \\ \cdot \\ r_{N-1} \end{pmatrix}$, the exponent in the Green kernel $\widetilde{G}$ can be written as $-\eta^{\mathrm{T}} \, \sigma \, \eta$, where σ is the matrix (hereafter $c(t_j) \equiv c_j$)

$$\sigma = \frac{m}{2\varepsilon\hbar \mathrm{i}} \begin{pmatrix} 2 & -1 & \cdot & \cdot & \cdot & \cdot & \cdot \\ -1 & 2 & -1 & \cdot & \cdot & \cdot & \cdot \\ \cdot & -1 & 2 & -1 & \cdot & \cdot & \cdot \\ \cdot & \cdot & \cdot & \cdot & \cdot & \cdot & \cdot \\ \cdot & \cdot & \cdot & \cdot & \cdot & \cdot & \cdot \\ \cdot & \cdot & \cdot & \cdot & \cdot & 2 & -1 \\ \cdot & \cdot & \cdot & \cdot & \cdot & -1 & 2 \end{pmatrix} + \frac{\mathrm{i}\varepsilon}{2\hbar} \mathrm{diag}\Big(c_1, \ldots, c_{N-1}\Big). \tag{15.4.5}$$

It is now possible to show that

$$\begin{aligned} \widetilde{G}(0, t; 0, t_0) &= \lim_{N \to \infty} \Big(m/2\pi \mathrm{i}\hbar\varepsilon\Big)^{N/2} \int \mathrm{d}\eta \, \mathrm{e}^{-\eta^{\mathrm{T}} \, \sigma \, \eta} \\ &= \lim_{N \to \infty} \left[\Big(m/2\pi \mathrm{i}\hbar\varepsilon\Big)^N \frac{\pi^{N-1}}{\det \sigma} \right]^{1/2}. \end{aligned} \tag{15.4.6}$$

What happens is that σ is the sum of a Hermitian matrix and a diagonal matrix. One can then put σ in diagonal form by means of a unitary matrix W:

$$\sigma = W^{\dagger} \, \sigma_{\mathrm{diag}} \, W.$$

By further setting $Z \equiv W\eta$, the integral with respect to η in (15.4.6) coincides with the integral with respect to Z, with σ replaced by σ_{diag}, i.e. $\eta^{\mathrm{T}} \sigma \eta = Z^{\mathrm{T}} W \sigma W^{\dagger} Z = Z^{\mathrm{T}} \sigma_{\mathrm{diag}} Z$.

It is now useful to define

$$f(t, t_0) \equiv \lim_{N \to \infty} \Big[\varepsilon (2\mathrm{i}\hbar\varepsilon/m)^{N-1} \det \sigma \Big] = \lim_{N \to \infty} \Big(\varepsilon \det \sigma'_{N-1} \Big). \tag{15.4.7}$$

This leads to (cf. Eq. (15.4.6))

$$\widetilde{G}(0, t; 0, t_0) = \left[\frac{m}{2\pi \mathrm{i}\hbar} \frac{1}{f(t, t_0)} \right]^{1/2}. \tag{15.4.8}$$

Let us now consider those sub-matrices in $\varepsilon\,\sigma'_{N-1}$ with only j rows and j columns, for which the determinant is denoted by $\varphi(t_j)$. They satisfy the recursive relation (hereafter $\varphi(t_0) = \varepsilon$)

$$\frac{\varphi(t_{j+1}) - 2\varphi(t_j) - \varphi(t_{j-1})}{\varepsilon^2} = -\frac{1}{m} c_{j+1} \varphi(t_j). \tag{15.4.9}$$

The limit as $\varepsilon \to 0$ of Eq. (15.4.9) implies

$$\frac{\mathrm{d}^2\varphi}{\mathrm{d}s^2} = -\frac{1}{m} c(s)\varphi, \tag{15.4.10}$$

and hence $f(t, t_0)$ obeys the equation

$$\left[m\frac{\partial^2}{\partial s^2} + c(s) \right] f(s, t_0) = 0, \tag{15.4.11}$$

subject to the initial conditions

$$f(t_0, t_0) = 0, \tag{15.4.12}$$

$$\frac{\partial f}{\partial s}(s = t_0) = 1. \tag{15.4.13}$$

15.4.1 Harmonic oscillator

Now we can solve in a few lines the problem for the harmonic oscillator. Indeed, the Lagrangian for a one-dimensional harmonic oscillator is a particular case of Eq. (15.4.1):

$$\mathcal{L} = \frac{m}{2}\left(\dot{x}^2 - \omega^2 x^2\right). \tag{15.4.14}$$

In the previous equations one should thus insert $c(s) = \omega^2 m$, and the solution of (15.4.11)–(15.4.13) reads

$$f(s, t_0) = \frac{1}{\omega} \sin[\omega(s - t_0)]. \tag{15.4.15}$$

Hereafter we set $s \equiv t$ and $T \equiv t - t_0$. By virtue of Eqs. (15.4.4), (15.4.8) and (15.4.15) one finds

$$G(z, t; y, t_0) = \left[\frac{m\omega}{2\pi \mathrm{i}\hbar(\sin \omega T)} \right]^{1/2} \mathrm{e}^{\frac{\mathrm{i}}{\hbar} S_{\mathrm{classical}}(y, t_0; z, t)}. \tag{15.4.16}$$

It is also instructive to evaluate the classical action by integration of (15.4.14) from t_0 to t. To simplify the calculation, we introduce

$$x \equiv A\,\mathrm{e}^{\mathrm{i}(\omega s + \varphi)}, \tag{15.4.17}$$

where A is real-valued. Hence one finds

$$\frac{S}{m} = \frac{\mathrm{i}}{2}\omega A^2\Big[\cos 2(\omega t+\varphi) + \mathrm{i}\sin 2(\omega t+\varphi) - \cos 2(\omega t_0+\varphi) - \mathrm{i}\sin 2(\omega t_0+\varphi)\Big]. \tag{15.4.18}$$

This implies

$$\mathrm{Re}\frac{S}{m} = \omega A\Big[y\sin(\omega t_0+\varphi) - z\sin(\omega t+\varphi)\Big], \tag{15.4.19}$$

where

$$y = A\cos(\omega t_0+\varphi), \tag{15.4.20}$$

$$z = A\cos(\omega t+\varphi). \tag{15.4.21}$$

We now use (15.4.20) and (15.4.21) to express

$$\tan\varphi = \frac{\left(\cos\omega t_0 - \frac{y}{z}\cos\omega t\right)}{\left(\sin\omega t_0 - \frac{y}{z}\sin\omega t\right)}. \tag{15.4.22}$$

Equations (15.4.22) and (15.4.19), jointly with the identity

$$\tan(\alpha+\beta) = \frac{(\tan\alpha + \tan\beta)}{[1-(\tan\alpha)(\tan\beta)]}, \tag{15.4.23}$$

lead to

$$S = \frac{m\omega}{\sin\omega T}\left[\left(y^2+z^2\right)(\cos\omega T) - 2yz\right]. \tag{15.4.24}$$

This formula for the classical action should be inserted in Eq. (15.4.16) for the Green kernel.

15.4.2 Free particle

For a free particle, c vanishes in Eq. (15.4.1), and hence one should take the limit as $\omega \rightarrow 0$ in Eq. (15.4.16). For this purpose, we multiply and divide by T within the square root and use the well-known limit

$$\lim_{\omega\rightarrow 0}\frac{\sin(\omega T)}{\omega T} = 1$$

to find

$$G(z,t;y,t_0) = \left(\frac{m}{2\pi\mathrm{i}\hbar T}\right)^{1/2} \mathrm{e}^{\frac{\mathrm{i}}{\hbar}S_{\mathrm{classical}}}, \tag{15.4.25}$$

for the Green kernel of a free particle in one dimension (see Eq. (4.7.26)).

15.5 Quantum mechanics in phase space

Quantum mechanics in the Schrödinger picture is formulated in terms of (suitable, square-integrable) functions on the configuration space. The operators of configuration space are represented by the same functions on the state space. On the other hand, the conjugate momentum variables are linear differential operators. Note that we could rewrite the Schrödinger picture in momentum space, and now the role of positions and momenta would be exchanged. In general, the Schrödinger picture requires a polarization of the dynamical variables into transversal Lagrangian sub-spaces. In contrast the Heisenberg picture formulation treats the whole set of dynamical variables but it makes use of a non-commutative algebra of operators. It would be desirable to have a formalism involving ordinary functions on phase space that faithfully reproduces quantum mechanics (not approximately but exactly). Such a formalism was initiated by Wigner and developed systematically by Moyal. It works with functions on phase space, hence we are dealing with quantum mechanics in phase space.

Wigner proposed the following question: is there a probability distribution in phase space which correctly deduces the expectation values for functions of coordinates and momenta? Since the operators of quantum mechanics do not commute one has to have a protocol in expressing such functions. Wigner chose the Weyl ordering in which $p^m q^n$ is made to correspond to the symmetrized product of m p-factors and n q-factors: that is as the coefficient of $\frac{\lambda^m \mu^n (m+n)!}{m!n!}$ in $(\lambda p + \mu q)^{m+n}$. (For example, $p^2 q^3$ when Weyl symmetrized denotes the terms which are the coefficients of $\lambda^2 \mu^3$ in the expansion of $(\lambda p + \mu q)^5$.) This could be more elegantly written using the power-series expansion for $\exp(\mathrm{i}\lambda p + \mathrm{i}\mu q)$, the Weyl operator that we have discussed extensively in chapter 9. We recall that unlike the monomials in p and q, which are unbounded operators, the Weyl operators are unitary and hence well defined. Wigner gave the formula (Wigner 1932b)

$$\langle p^m q^n{}_{\text{Weyl ordered}} \rangle = \int \int W(q,p) p^m q^n \, \mathrm{d}p \, \mathrm{d}q. \tag{15.5.1}$$

The distribution function W here has a characteristic function χ (see Eq. (13.4.17)) given by the expectation value of the Weyl operator $\mathrm{e}^{\mathrm{i}(\lambda \hat{p} + \mu \hat{q})}$ in the state described by ψ, and in $\hbar = 1$ units

$$W(q,p) \equiv \frac{1}{2\pi} \int_{-\infty}^{\infty} \psi^* \left(q + \frac{y}{2} \right) \mathrm{e}^{\mathrm{i}py} \psi \left(q - \frac{y}{2} \right) \mathrm{d}y. \tag{15.5.2}$$

At this stage, the distribution W is introduced as a generalization of the position probability distribution

$$P(x) = \psi^*(x)\psi(x)$$

and the momentum probability distribution

$$Q(p) = \varphi^*(p)\varphi(p).$$

In fact,

$$\int W(q,p)\,\mathrm{d}p = P(q), \tag{15.5.3}$$

$$\int W(q,p)\,\mathrm{d}q = Q(p). \tag{15.5.4}$$

Wigner introduced $W(q,p)$ for a wave function, but from its bilinearity in the wave function it is clear that instead of treating this as the distribution for a state, represented by the wave function $\psi(x)$, we could easily extend it to any density matrix $\rho = \sum_n \lambda_n \psi_n \psi_n^\dagger$ with the Wigner distribution

$$W = \sum_n \lambda_n W_n. \tag{15.5.5}$$

These distribution functions thus constitute a convex set and the pure state distributions form the extremal ones. The Wigner distribution function is integrable and square-integrable, and is analytic in p and q. But it is not non-negative for almost all wave functions. For example, only the ground state of the harmonic oscillator has a positive-definite Wigner distribution; all the excited states have an indefinite distribution.

The extension for the extremal distribution to convex linear combinations not only generalizes the distribution function to mixed states but it also enables us to look at it as a candidate to represent operators. This point was realized by Moyal in a fundamental contribution (Moyal 1949), who introduced a new calculus for these phase-space representatives for the operators, which relies upon the bracket

$$[f(q,p), g(q,p)] \equiv \mathrm{i}\hbar\,\{f,g\} + \sum_{k=2}^{\infty} \left(\frac{\mathrm{i}\hbar}{2}\right)^k \frac{[D_k(f,g) - D_k(g,f)]}{k!}, \tag{15.5.6}$$

where

$$D_k(f,g)(q,p) \equiv \frac{\partial^k f}{\partial q^k}\frac{\partial^k g}{\partial p^k} - \binom{k}{1}\frac{\partial^k f}{\partial q^{k-1}\partial p}\frac{\partial^k g}{\partial p^{k-1}\partial q} + \cdots + (-1)^k \frac{\partial^k f}{\partial p^k}\frac{\partial^k g}{\partial q^k}. \tag{15.5.7}$$

The bracket (15.5.6) has come to be known as the Moyal bracket (we have indicated its explicit dependence on $\hbar$). This bracket is bilinear, antisymmetric and satisfies the Jacobi identity. The usual derivation property

$$[f, g * h] = [f, g] * h + g * [f, h] \tag{15.5.8}$$

holds for the Moyal bracket provided $f * g$ is not the pointwise product of $f(q, p)$ with $g(q, p)$, but the Moyal product

$$(f * g)(q, p) \equiv fg + \frac{i\hbar}{2} \{f, g\} + \sum_{k=2}^{\infty} \left(\frac{i\hbar}{2}\right)^k \frac{1}{k!} D_k(f, g). \tag{15.5.9}$$

At a deeper level, Eq. (15.5.9) is found to be the asymptotic expansion of an integral formula for $f * g$, for which we refer the reader, for example, to the appendix of the work in Gracia-Bondia *et al.* (2002). This $*$-product is bilinear and associative but not commutative. In fact, the Moyal commutator bracket is the commutator of this associative product, i.e. (cf. Simoni et al. 1971)

$$[f, g]_{\text{Mb}} = (f * g - g * f). \tag{15.5.10}$$

The resulting equations of motion with Hamiltonian $H(q, p)$ can be cast in the form

$$\frac{\partial}{\partial t} f(q, p) = [f(q, p), H(q, p)]_{\text{Mb}}. \tag{15.5.11}$$

For any two functions f and g in which one function is quadratic or linear in (q, p), the Moyal bracket coincides with their Poisson bracket, but otherwise the two brackets are inequivalent.

With this formalism, states are described by functions on phase space that are projectors, i.e.

$$\rho * \rho = \rho, \tag{15.5.12}$$

the eigenvalue problem for the Hamiltonian H reads

$$H * \rho_E = E \rho_E, \tag{15.5.13}$$

and H admits the decomposition

$$H = \sum_E E \rho_E, \tag{15.5.14}$$

where

$$E = \frac{1}{2\pi\hbar} \int (H * \rho_E)(q, p) \, \mathrm{d}q \, \mathrm{d}p = \frac{1}{2\pi\hbar} \int H(q, p) \rho_E(q, p) \, \mathrm{d}q \, \mathrm{d}p. \tag{15.5.15}$$

The time evolution is ruled by the first-order differential equation

$$\mathrm{i}\hbar \frac{\mathrm{d}}{\mathrm{d}t} \mathrm{e}^{Ht} = H * \mathrm{e}^{Ht}, \tag{15.5.16}$$

where

$$e^{Ht} = \sum_{n=0}^{\infty} \frac{1}{n!} \left(-\frac{it}{\hbar}\right)^n (H*)^n, \tag{15.5.17}$$

having defined

$$(H*)^n \equiv H * H * \cdots * H \ (n \text{ factors}). \tag{15.5.18}$$

The operator e^{Ht} is then expressed through the Fourier–Dirichlet expansion

$$e^{Ht} = \sum_E \rho_E e^{-\frac{iEt}{\hbar}}. \tag{15.5.19}$$

The notion of the Wigner–Moyal bracket and the Moyal dynamical equations can be employed for more than one degree of freedom in a straightforward manner. It can also be applied to systems with a finite-dimensional state space, such as for example for spin systems. It should be emphasized that the Wigner–Moyal scheme is not a classical *approximation* but an *exact transcription* of quantum mechanics. The infinite series of derivatives in the Moyal associative product and in the Moyal bracket shows the sense in which quantum mechanics is *non-local*. In particular, a wave function at a point away from a potential is nevertheless affected by the potential, so that a potential acts not only within its support but elsewhere also.

15.5.1 *Operators vs. phase-space functions*

The Weyl expansion of a classical phase-space function $A(q,p)$ into a Fourier integral reads, in $\hbar = 1$ units (see Eq. (9.7.7)),

$$A(q,p) = \int d\lambda \int d\mu \, \chi(\lambda,\mu) e^{i(\lambda q + \mu p)}. \tag{15.5.20}$$

Weyl then defines the operator which corresponds to the exponential in the integrand of (15.5.20) as $e^{i(\lambda \hat{q} + \mu \hat{p})}$. The operator which corresponds to $A(q,p)$ is then given by (Hillery *et al.* 1984)

$$\mathcal{A}(\hat{q},\hat{p}) = \int d\lambda \int d\mu \, \chi(\lambda,\mu) e^{i(\lambda \hat{q} + \mu \hat{p})}. \tag{15.5.21}$$

If the classical phase-space function A goes over to the quantum operator $\mathcal{A}$, then the relation between them is that given by Wigner:

$$A(q,p) = \int dz \, e^{ipz} \langle q - \frac{z}{2} | \mathcal{A} | q + \frac{z}{2} \rangle. \tag{15.5.22}$$

Moreover, the expectation value of the result of the measurement of the operator $\mathcal{A}$, if carried out on a system in the state $|\psi\rangle$, is equal to the expectation value of the classical function $A(q,p)$, i.e.

$$\langle\psi|\mathcal{A}|\psi\rangle = \int \mathrm{d}q \int \mathrm{d}p \, W(q,p)A(q,p), \tag{15.5.23}$$

assuming that the system is described by the distribution function $W(q,p)$, which corresponds to the density matrix ρ.

The Moyal exponential bracket is the phase-space image of the usual ordered multiplication. It then follows that this Moyal multiplication is associative, and the phase-space functions now behave as the images of non-commuting quantum operators. A positive (non-negative) operator may be written as the positive linear combination of a number of Moyal sequences:

$$\mathcal{A}(q,p) = \sum_{\alpha} \mu_\alpha F_\alpha(q,p) * F_\alpha(q,p). \tag{15.5.24}$$

Alternatively, we may start by considering an even polynomial $\mathcal{P}$, which may be written in the form

$$\mathcal{P} = \sum_{\alpha} \mu_\alpha Q_\alpha * Q_\alpha, \tag{15.5.25}$$

with associated

$$P(q,p) = \sum_{\alpha} \mu_\alpha Q(q,p) * Q(q,p). \tag{15.5.26}$$

$\mathcal{A}$ is a positive operator if and only if

$$\mathrm{tr}(\mathcal{P}\mathcal{A}) \geq 0. \tag{15.5.27}$$

The pointwise positivity of $A(q,p)$ is not a guarantee of $\mathcal{A}$ being positive. If $A(q,p) \geq 0$ for all q,p but the support of $A(q,p)$ is smaller than $\frac{\hbar}{2}$ then $\mathcal{A}$ is not non-negative.

Given that $\mathcal{A}$ is an operator with image $A(q,p)$, we can look for its spectral decomposition. In particular, for the harmonic oscillator Hamiltonian in dimensionless form

$$\mathcal{H}(\hat{q},\hat{p}) = \frac{\hat{q}^2 + \hat{p}^2}{2}, \tag{15.5.28}$$

the corresponding phase-space function is given by

$$H(q,p) = \frac{q^2 + p^2}{2}. \tag{15.5.29}$$

This function has an integral representation of the kind

$$\frac{q^2 + p^2}{2} = \sum_{n=0}^{\infty} \left(n + \frac{1}{2}\right) \phi_n(q,p) * \phi_n(q,p). \tag{15.5.30}$$

Thus, the Hamiltonian operator must also satisfy ($\widetilde{\pi}_n$ being projection operators)

$$\mathcal{H}(\hat{q},\hat{p}) = \sum_{n=0}^{\infty} \left(n + \frac{1}{2}\right) \widetilde{\pi}_n, \tag{15.5.31}$$

which is the required spectral resolution.

We therefore see that the Moyal programme is a complete alternative treatment of quantum mechanics in terms of functions of commuting phase-space variables. The lack of commutation obtains from the associative non-commutative scalar product

$$(AB)(q,p) \equiv A(q,p) * B(q,p). \tag{15.5.32}$$

Similarly, positivity is defined by the projections $\pi_n(q,p)$ such that

$$\pi_n(q,p) * \pi_n(q,p) = \pi_n(q,p), \tag{15.5.33}$$

$$\int\!\!\int \pi_n(q,p)\, \mathrm{d}q\, \mathrm{d}p = 1, \tag{15.5.34}$$

$$\pi_n(q,p) * \pi_m(q,p) = \delta_{mn}\pi_n(q,p). \tag{15.5.35}$$

15.5.2 Quantum tomograms

The Wigner–Moyal distribution is not positive-definite, and several authors would like a way of expressing the content of the Wigner–Moyal distribution in terms of non-negative probability distributions. This is done in terms of quantum tomograms. For this purpose one makes use of the Radon transform of a planar distribution (cf. Radon 1917):

$$\tau_\rho(l,u,v) = \int \mathrm{d}q \wedge \mathrm{d}p\, \rho\,(q,p)\, \delta\,(l - uq - vp). \tag{15.5.36}$$

Given $\tau_\rho(l,u,v)$ we can recover the distribution $\rho(q,p)$ by means of the following formulae:

$$\tau_\rho(l,u,v) = \int \mathrm{d}q \wedge \mathrm{d}p \int \frac{\mathrm{d}\zeta}{2\pi} \rho\,(q,p)\, \mathrm{e}^{\mathrm{i}(uq+vp-l)\zeta}, \tag{15.5.37}$$

$$\rho(q,p) = \int \frac{\mathrm{d}u \wedge \mathrm{d}v}{(2\pi)^2} \int \mathrm{d}l\, \tau_\rho(l,u,v) \mathrm{e}^{\mathrm{i}(uq+vp-l)}. \tag{15.5.38}$$

The remarkable property of quantum tomograms is they are all non-negative, stemming from the positivity of the marginal distribution of Wigner–Moyal distribution with respect to any canonical variable. The transformation

$$p \to p' \equiv u'p - v'q, \quad q \to q' \equiv vp + uq \tag{15.5.39}$$

is canonical, hence the marginal distribution is positive. This is the *quantum tomogram.*

By the way this gives another way of stating the difference between classical and quantum mechanics: while the first is described by a single probability distribution $\rho(q,p)$ in phase space, quantum mechanics needs *all* tomograms $Q(l,u,v)$ to give a characterization in terms of only (positive) probability distributions.

There are yet other presentations of quantum theory. In particular, by virtue of the overcompleteness of coherent states studied in chapter 10, every density matrix

$$\rho = \sum_{n=0}^{\infty} \sum_{m=0}^{\infty} \rho(n,m)|n\rangle\langle m| \tag{15.5.40}$$

can be expressed in the form of the 'diagonal' representation (Sudarshan 1963)

$$\rho = \int \mathrm{d}^2 z \, \phi(z)|z\rangle\langle z|. \tag{15.5.41}$$

Equation (15.5.41) does not determine $\phi(z)$ uniquely, which is not necessarily an ordinary function (i.e. it may have a distributional nature), but it has a Fourier transform in the complex plane given by

$$N(\zeta) \equiv \int \phi(z) \mathrm{e}^{\zeta z^* - \zeta^* z} \, \mathrm{d}^2 z. \tag{15.5.42}$$

Interestingly, a distribution function K may be defined which, unlike the distribution function ϕ, is always an ordinary function. It can be defined as the expectation value of the density matrix in the state $|z\rangle$, i.e. (Husimi 1940, Kano 1965)

$$K(z) \equiv \frac{1}{\pi}\langle z|\rho|z\rangle = \frac{1}{\pi}\sum_{n,m} \rho(n,m) \mathrm{e}^{-|z|^2} \frac{(z^*)^n z^m}{\sqrt{n!m!}}. \tag{15.5.43}$$

The function K satisfies the conditions

$$K(z) \geq 0, \quad \int K(z) \, \mathrm{d}^2 z = 1, \tag{15.5.44}$$

and is particularly interesting since formally it plays the same role as a classical probability distribution function, when the expectation value of an operator given by an anti-normal ordered product of annihilation and creation operators is evaluated (see below). The Husimi–Kano distribution function $K(z)$ is the boundary value of the distribution function

$$R(z,z') \equiv \frac{1}{\pi}\langle z|\rho|z'\rangle, \tag{15.5.45}$$

in that $K(z) = R(z,z)$. The latter distribution function makes it possible to express the expectation value of a normal-ordered operator $\mathcal{A}$ according to

$$\mathrm{tr}(\rho\mathcal{A}) = \frac{1}{\pi}\int \mathrm{d}^2 z' \int \mathrm{d}^2 z \; R(z,z')\langle z'|\mathcal{A}|z\rangle. \tag{15.5.46}$$

If any operator is expressed in the 'normal-ordered' form $F_{\mathrm{N}}(a^\dagger, a)$, in which all annihilation operators come to the extreme right, one has

$$\langle F_{\mathrm{N}}\rangle_z = \int \phi(z) F_{\mathrm{N}}(z^*, z)\, \mathrm{d}^2 z, \tag{15.5.47}$$

in contrast to the expectation value of any anti-normal ordered form $F_{\mathrm{A}}(a, a^\dagger)$ with all the creation operators to the right:

$$\langle F_{\mathrm{A}}\rangle_z = \int K(z) F_{\mathrm{A}}(z, z^*)\, \mathrm{d}^2 z. \tag{15.5.48}$$

These representations are important in quantum-optics descriptions of coherence and interference. We also note that $\phi(z)$ is the anti-normal-ordered form for the density matrix ρ. (The discussion of the dynamics re-expressed in terms of coherent states is beyond the scope of this book.)

15.6 Problems

15.P1. Write an essay on the implementation of canonical transformations in the path-integral approach to non-relativistic quantum mechanics.

15.P2. Try to evaluate the path integral expressing the Green kernel of the Pauli equation.

15.P3. Find the relation between the Fourier transforms of the Husimi–Kano and $\phi(z)$ distribution functions (see Eqs. (15.5.42) and (15.5.43)).

Appendix 15.A
The Trotter product formula

The limit as $n \to \infty$ in Eq. (15.3.11) relies on a highly non-trivial result proved by Trotter according to which, if A and B are self-adjoint operators with domains $D(A)$ and $D(B)$, and if $A + B$ is essentially self-adjoint on the domain $D(A) \cap D(B)$, one has (Trotter 1959)

$$\mathrm{e}^{\mathrm{i}(A+B)t} = \lim_{n\to\infty} \left(\mathrm{e}^{\frac{\mathrm{i}t}{n}A}\,\mathrm{e}^{\frac{\mathrm{i}t}{n}B}\right)^n, \tag{15.A.1}$$

where the limit in (15.A.1) is a strong limit. Recall that, for maps $T : X \rightarrow Y$, with X and Y Banach spaces, the strong topology is that for which each map $E_x : T \rightarrow Tx \in Y$ is continuous for all $x \in X$. A sequence $\{T_\alpha\}$ then converges uniformly to the operator T if and only if

$$\|T_\alpha x - Tx\| \rightarrow 0 \quad \forall x \in X. \tag{15.A.2}$$

By virtue of (15.A.1), one can 'split' the kinetic (i.e. T) and potential (i.e. V) terms in the Hamiltonian operator, on taking the limit as $n \rightarrow \infty$ in Eq. (15.3.10), without relying on the approximate formula (15.3.7). In other words, one deals with infinitely many terms of the form

$$\langle q_{j+1}| \mathrm{e}^{-\frac{\mathrm{i}}{\hbar n} Tt} \, \mathrm{e}^{-\frac{\mathrm{i}}{\hbar n} Vt} |q_j\rangle = \int \mathrm{d}p \, \langle q_{j+1}| \mathrm{e}^{-\frac{\mathrm{i}}{\hbar n} Tt} |p\rangle \langle p| \mathrm{e}^{-\frac{\mathrm{i}}{\hbar n} Vt} |q_j\rangle$$

$$= \mathrm{e}^{-\frac{\mathrm{i}}{\hbar n} V(q_j)t} \int \mathrm{d}p \, \langle q_{j+1}| \mathrm{e}^{-\frac{\mathrm{i}}{\hbar n} \frac{p^2}{2m} t} |p\rangle \langle p|q_j\rangle.$$

16
Dirac equation and no-interaction theorem

The requirement of relativistic invariance is first used to derive the Dirac wave equation for the electron, with the associated γ-matrix formalism.

Relativistic invariance may indeed involve two different theoretical postulates: symmetry of the theory under the relativistic transformation group, and explicit transformation properties or *manifest invariance* of certain quantities. For a classical mechanical theory of a fixed number of particles, the Lorentz transformation formula can be assumed for the coordinates of the space-time events that comprise the world lines of the particles as defined by their positions as a function of time. The assumption of manifest invariance is expressed in terms of equations involving the Poisson brackets of the canonical position coordinates with the generators of the Lorentz group. For a theory of two particles, the only generators satisfying these latter equations jointly with the Poisson-bracket equations characteristic of Lorentz symmetry are those describing free-particle motion. In other words, the combined assumptions of Lorentz symmetry and Lorentz transformation of particle positions rule out any interaction. The need for a quantum theory of fields is therefore discussed, and important topics in the (relativistic) theory of spinor fields are eventually introduced.

16.1 The Dirac equation

In searching for relativistically invariant equations with the requirement that they should be of first order in time, it is crucial to realize that relativistic invariance requires that time and other coordinates should be treated on the same footing. Thus, to have a first-order differential operator in the time variable, the desired equation should also be of first order in the space coordinates. Moreover, our previous experience with the Pauli equation has shown that the expression $\vec{\sigma} \cdot \vec{p}$ is invariant under rotations if $\vec{p}$ transforms like a vector and $\vec{\sigma}$ as the infinitesimal generators of $SU(2)$. It is therefore natural to look for an expression like (Dirac 1928, 1958)

$$\gamma^0 p_0 + \gamma^1 p_1 + \gamma^2 p_2 + \gamma^3 p_3,$$

where $(p_0, \vec{p})$ is a 4-momentum and $(\gamma^0, \gamma^1, \gamma^2, \gamma^3)$ is a 4-vector, the components of which are matrices. γ transform like a vector under Poincaré transformations in such a way that the above symbol is Poincaré invariant, because (p_0, p_1, p_2, p_3) transforms like a covector.

Furthermore, invariance under translations on $\mathbf{R}^4$, which are the inhomogeneous part of the Poincaré group, requires that the components $\gamma^0, \gamma^1, \gamma^2$ and γ^3 should be independent of position. Lastly, the additional requirement we impose is that the square of the above expression should provide us with the symbol of the Klein–Gordon equation, motivated by analogy with the three-dimensional case, where $(\vec{\sigma} \cdot \vec{p})^2$ yields the Laplacian in three dimensions. All of this leads to the equation

$$\begin{aligned}
&(\gamma^0)^2 p_0^2 + (\gamma^1)^2 p_1^2 + (\gamma^2)^2 p_2^2 + (\gamma^3)^2 p_3^2 \\
&\quad + (\gamma^0\gamma^1 + \gamma^1\gamma^0) p_0 p_1 + (\gamma^0\gamma^2 + \gamma^2\gamma^0) p_0 p_2 \\
&\quad + (\gamma^0\gamma^3 + \gamma^3\gamma^0) p_0 p_3 + (\gamma^j\gamma^k + \gamma^k\gamma^j) p_j p_k \\
&= p_0^2 - p_1^2 - p_2^2 - p_3^2,
\end{aligned} \tag{16.1.1}$$

which implies the conditions

$$(\gamma^0)^2 = I, \quad (\gamma^j)^2 = -I \ (j = 1, 2, 3), \tag{16.1.2}$$

$$\gamma^0\gamma^j + \gamma^j\gamma^0 = 0, \tag{16.1.3}$$

$$\gamma^j\gamma^k + \gamma^k\gamma^j = 0, \quad j \neq k. \tag{16.1.4}$$

In looking for solutions of these equations in terms of matrices, one finds that they must have as order a multiple of 4, and that there exists a solution of order 4. In terms of Pauli matrices, one can write

$$\gamma^0 = \begin{pmatrix} \mathbb{I} & 0 \\ 0 & -\mathbb{I} \end{pmatrix}, \tag{16.1.5}$$

$$\gamma^j = \begin{pmatrix} 0 & \sigma_j \\ -\sigma_j & 0 \end{pmatrix}. \tag{16.1.6}$$

Therefore the 'square root' of the symbol of the Klein–Gordon equation becomes the symbol (here $p_\mu \equiv \mathrm{i}\frac{\partial}{\partial x^\mu}$)

$$\sigma(p_\mu) = \gamma_0 p_0 + \gamma_1 p_1 + \gamma_2 p_2 + \gamma_3 p_3 \tag{16.1.7}$$

of the Dirac equation

$$\left(\gamma_0 \frac{\partial}{\partial x^0} + \gamma_1 \frac{\partial}{\partial x^1} + \gamma_2 \frac{\partial}{\partial x^2} + \gamma_3 \frac{\partial}{\partial x^3}\right)\psi = \mathrm{i}m\psi, \tag{16.1.8}$$

where we have used the relations $\gamma^0 = -\gamma_0, \gamma^j = \gamma_j, p^j = p_j$, which result from the convention $\eta_{\mu\nu} = \text{diag}(-1,1,1,1)$ for the Minkowski metric. As a by-product, one finds that the wave functions acted upon by this operator should have four components, a further generalization with respect to the Pauli equation where the wave function had two components to accommodate the spin. Unlike the non-relativistic formalism with spin, one has two extra components which reflect the ability of the Dirac equation to describe negative-energy states (Dirac 1928, 1958).

Remark. The (γ^μ) are generators of an algebra called a Clifford algebra. Equations (16.1.2)–(16.1.4) define the structure constants and make it possible to express any product of a finite number of matrices in the form

$$a_0 \mathbb{I} + \sum_j a_j \gamma^j + \sum_{j<k} a_{jk} \gamma^j \gamma^k + \sum_{j<k<l} a_{jkl} \gamma^j \gamma^k \gamma^l + a_{0123} \gamma^0 \gamma^1 \gamma^2 \gamma^3.$$

16.1.1 Non-relativistic limit of the Dirac equation and antiparticles

The Dirac equation (16.1.8) for the spinor amplitude $\psi(\vec{x}, t)$ can be re-expressed in the more convenient form (note that $\hbar = c = 1$ units are used)

$$i\gamma^0 \frac{\partial \psi}{\partial t} = (\vec{\gamma} \cdot \vec{p} + m\mathbb{I})\psi. \tag{16.1.9}$$

This equation has four solutions for each momentum $\vec{p}$, two with positive energy $E = \sqrt{p^2 + m^2}$ and two with negative energy $-E = -\sqrt{p^2 + m^2}$. For either sign of the energy there are two orthogonal spin states, which are best described in terms of the helicities $\frac{\vec{\sigma} \cdot \vec{p}}{|\vec{p}|}$. The 'negative energy' solutions should be identified with the complex conjugates of positive energy solutions of the 'antiparticle', the positron for the electron.

Various representations of the γ-matrices are relevant for emphasizing different aspects. For very high energies or for theoretical considerations of higher dynamical structures the chiral representation (in which $\gamma^5 \equiv i\gamma^0\gamma^1\gamma^2\gamma^3$ is diagonal and all γ matrices are off-diagonal) is the most appropriate. In nuclear β-decay, for example, the positive chiral (negative helicity) parts of the various Fermi fields are coupled. Both electrodynamics and the kinematics of the field are most naturally described in this representation. On the other hand, the simplest representation for the discussion of particle–antiparticle conjugation is the Majorana representation in which all γ matrices are pure imaginary; the Dirac equation is thus real. If we need to have distinct antiparticles we may take a complex field obeying real equations (much as we would do for a spinless particle).

But for discussing the non-relativistic limit it is best to use the choice Dirac originally made in which γ^0 is real and diagonal while the γ^i

matrices (with $i = 1, 2, 3$) are off-diagonal with Pauli matrices in the off-diagonal blocks. The Dirac equation now reads as

$$\mathrm{i}\frac{\partial}{\partial t}\begin{pmatrix}\psi_{\mathrm{I}}\\ \psi_{\mathrm{II}}\end{pmatrix} = \begin{pmatrix}0 & \vec{\sigma}\cdot\vec{p}\\ \vec{\sigma}\cdot\vec{p} & 0\end{pmatrix}\begin{pmatrix}\psi_{\mathrm{I}}\\ \psi_{\mathrm{II}}\end{pmatrix} + \begin{pmatrix}\mathbb{I} & 0\\ 0 & -\mathbb{I}\end{pmatrix} m \begin{pmatrix}\psi_{\mathrm{I}}\\ \psi_{\mathrm{II}}\end{pmatrix}. \quad (16.1.10)$$

In the limit $\vec{p} \rightarrow \vec{0}$ we obtain simply

$$\mathrm{i}\frac{\partial \psi_{\mathrm{I}}}{\partial t} = m\psi_{\mathrm{I}}, \quad \mathrm{i}\frac{\partial \psi_{\mathrm{II}}}{\partial t} = -m\psi_{\mathrm{II}}. \quad (16.1.11)$$

One could go to the next approximation. Without approximation one has indeed

$$\mathrm{i}\frac{\partial \psi_{\mathrm{I}}}{\partial t} = \vec{\sigma}\cdot\vec{p}\psi_{\mathrm{II}} + m\psi_{\mathrm{I}}, \quad (16.1.12)$$

$$\mathrm{i}\frac{\partial \psi_{\mathrm{II}}}{\partial t} = \vec{\sigma}\cdot\vec{p}\psi_{\mathrm{I}} - m\psi_{\mathrm{II}}, \quad (16.1.13)$$

or

$$\psi_{\mathrm{II}} = \frac{\vec{\sigma}\cdot\vec{p}}{(E+m)}\psi_{\mathrm{I}}, \quad (16.1.14)$$

so that

$$\mathrm{i}\frac{\partial \psi_{\mathrm{I}}}{\partial t} = E\psi_{\mathrm{I}}. \quad (16.1.15)$$

This is called the Pauli form, where only the two-component spinor ψ_{I} and the Pauli matrices appear. We may then say that the Pauli equation has only positive energies and two spin (or helicity) states. In this sense the Pauli equation for spin $\frac{1}{2}$ does not differ from the Schrödinger–Klein–Gordon equation for spin 0:

$$\mathrm{i}\frac{\partial \psi}{\partial t} = (m^2 - \triangle)^{1/2}\psi, \quad (16.1.16)$$

$$\mathrm{i}\frac{\partial \psi^*}{\partial t} = -(m^2 - \triangle)^{1/2}\psi^*. \quad (16.1.17)$$

A more systematic method of arriving at a two-component positive-energy relativistic amplitude was discovered by Foldy and Wouthuysen. One makes the canonical transformation (Foldy and Wouthuysen 1950, Tani 1951)

$$\psi \rightarrow \left\{\exp\left[-\frac{\mathrm{i}}{2}\frac{\vec{\gamma}\cdot\vec{p}}{|\vec{p}|}\tan^{-1}\left(\frac{|\vec{p}|}{m}\right)\right]\right\}\psi = \widetilde{\psi} \quad (16.1.18)$$

to obtain

$$\mathrm{i}\frac{\partial}{\partial t}\widetilde{\psi} = \gamma^0\sqrt{p^2+m^2}\,\widetilde{\psi}. \quad (16.1.19)$$

In this case

$$\widetilde{\psi}^\dagger\widetilde{\psi} = \psi^\dagger\psi, \tag{16.1.20}$$

unlike for the Pauli equation, since $\widetilde{\psi}$ is related by a unitary transformation to ψ. The Pauli equation is an approximation working with only two components. The Foldy–Wouthuysen form still has four components with two helicities for either sign of the energy. If ψ were essentially complex we could have distinct antiparticles.

The kinematic term of the Lagrangian is now still $\widetilde{\psi}^\dagger\frac{\partial}{\partial t}\psi$ which is symmetric, and by the Schwinger action principle (see Eq. (15.1.10)) we obtain

$$\left\{\psi_r^\dagger, \psi_s\right\}_+ = \delta_{rs}. \tag{16.1.21}$$

So the spinstatistics relation (see subsection 16.7.2) still obtains. But since the Hamiltonian is only a function of p^2 there is no obvious need for $\widetilde{\psi}(p) \to \beta\eta\widetilde{\psi}(-p)$, but it is consistent ($\eta$ is a unimodular phase factor which is usually chosen to be ±1, but for the Majorana representation $\eta^2 = -1$ to preserve the reality of the spinor amplitude). (Unless $\psi(p) \to \beta\eta\psi(-p)$, omitting β would not satisfy a *local* transformation on ψ in the covariant form.) Therefore while the antiparticles with the *same* mass and opposite charge are obtained, the antiparticle–particle complex does not have intrinsic negative parity. Time reversal of the non-relativistic limit of the Foldy–Wouthuysen form is immediate.

When considering electromagnetic or other interactions, the Dirac equation allows simple forms like the gauge-invariant electromagnetic coupling. This same coupling appears extremely complicated with non-linear and derivative terms in the coupling. However, the Foldy–Wouthuysen form of the interacting neutron or proton gives an understanding of the anomalous moment, the Darwin term (proportional to $\text{div}\vec{E}$, which affects the s-states only) and the neutron–electron interaction which appears from the anomalous magnetic moment of the neutron when transcribed to the Foldy–Wouthuysen form, which has been clearly elucidated in Foldy (1950). In contrast, the Pauli reduction to two-component form gives confusing results such as an imaginary electric moment for the non-relativistic electron. We must therefore avoid considering the Pauli reduction of the non-relativistic Dirac electron wave equation as a valid framework.

16.2 Particles in mutual interaction

By and large we have described quantum particles which are free or interacting with an external potential (which may even be velocity dependent for charged particle in an electromagnetic field). We have not considered mutual interaction between particles. A mutual interaction potential between two particles in the non-relativistic case can be reduced to the motion of one particle (with reduced mass) moving in a potential, and the centre of mass moves with a constant total momentum (section 5.5). This is action-at-a-distance rather than action-by-contact. The only examples we have of action-by-contact is the coupling of particles with a field, say charged particles with an electromagnetic field or neutrons with lattice vibrations of a solid.

For classical non-relativistic (Galilean) mechanics, invariance of the potential under full Galilei transformation requires that the potential be a scalar made up of the relative coordinates and relative velocities. Similarly, for velocity-dependent vector potentials we expect it to transform as a vector coupled to the relative velocity, or as a suitable tensor in the relative variables for non-linear velocity-dependent interactions.

For quantum mechanics of interacting particles similar considerations obtain. We see that the Galilean group is realized projectively so that there is a superselection about the sum of the masses remaining invariant for any reaction, including those that change the particles like in nuclear β-decay. For the restrictions on interactions of spinning particles see Eisenbud and Wigner (1947) and Okubo and Marshak (1958).

16.3 Relativistic interacting particles. Manifest covariance

As soon as we consider relativistic systems, the problem is entirely different. Dirac defines a theory to be relativistically invariant if the ten generators furnish a representation of the Poincaré group (inhomogeneous Lorentz group). This is not difficult to do; but if we look at the commutation rules we see that, since the commutator of a boost and a displacement along the same axis gives the Hamiltonian, either the boost or the displacement should contain interaction terms.

For relativistic *classical* mechanics there is another property that is natural. This is the imposition of the world line conditions: according to this the trajectory should be a world line: $(q(t), t)$ transform as four coordinates under Lorentz transformations. If the boost generators are written $\vec{K}$, translations $\vec{P}$ then in a Hamiltonian theory the world-line condition demands that

$$\{K_j, q_k\} = q_j \{H, q_k\}.$$

This is automatically satisfied for free particles since

$$K_j = q_j H.$$

Currie *et al.* (1963) have analysed this world-line condition together with the Poincaré group commutation relations in a Hamiltonian theory, that is all the ten generators $\vec{K}, \vec{J}, \vec{P}, H$ are functions of p and q at the same time. They arrive at the startling conclusion that such theories contain no interaction. An outline of this 'no-interaction theorem' is given below for the case of two particles. It has been extended to an arbitrary number of particles by Leutwyler (1965). A Lagrangian proof of the no-interaction theorem has been furnished by Marmo *et al.* (1984).

16.4 The no-interaction theorem in classical mechanics

Classical dynamics contains as an idealization the concept of a spinless and structureless mass point. In the relativistic framework this may be viewed as an irreducible representation of the Poincaré group, the mass of the particle being the only non-trivial Casimir invariant of the representation. In the formulation of this point of view one works at the Hamiltonian level, exploiting the Poisson bracket structure to image the Lie algebra of the Poincaré group, and canonical transformations to image group elements. One can however start, equally well, with the notion of a one-parameter family of space-time points describing a straight-line world trajectory, with Newton's first law of motion obeyed.

On passing from a single free particle to a collection of free particles the Hamiltonian description remains valid, with the Poisson bracket realization of the Lie algebra of the Poincaré group using the sums of the individual particle contributions. The question we now consider is: can the particle world lines or space-time trajectories be made to transform properly under frame changes although influencing each other due to mutual interaction? If this were the case, we would say that the scheme can accommodate interactions.

The requirements of Poincaré invariance on the one hand, and the geometrical world-line transformation property on the other hand are distinct, and their joint effect leads to rather stringent limitations on the admissible dynamics. It may well happen that the individual response to a change of reference frame by each particle without reference to the overall system destroys any cohesion of the system.

The work in Currie *et al.* (1963) was the first to prove the no-interaction theorem of relativistic particle mechanics by relying on the phase-space formalism and Poisson brackets, whereas we are aiming to give a more intrinsic derivation, relying on the work in Marmo *et al.* (1984). One then finds that it is not the phase-space formulation which lies at the heart of the theorem, but the real cause resides in the geometric structures and the conflicting conditions formulated in terms of them.

In dealing with particle dynamics in a geometric framework, the carrier space is usually chosen to be either the tangent bundle TQ (as in the Newtonian and the Lagrangian formalism) or the cotangent bundle T^*Q

(as in the Hamiltonian formalism) on the configuration space Q. While the latter is suited for the study of canonical aspects of dynamics, the former, with the Lagrangian formalism, is more appropriate for the expression of relativistic invariance. The cotangent bundle is geometrically relevant by virtue of the existence of the natural 1-form $\theta \equiv p_r \, dq^r$ on it. On the tangent bundle TQ, however, there is no natural 1-form. One then tries to define a 1-form on TQ if a Lagrange function $\mathcal{L}$ is given on TQ. By exploiting the replacement (q, v being local coordinates on TQ) $p_r \rightarrow \frac{\partial \mathcal{L}}{\partial v^r}$, the natural 1-form is pulled back from the cotangent bundle to the tangent bundle of configuration space, in that one defines

$$\theta_{\mathcal{L}} \equiv \frac{\partial \mathcal{L}}{\partial v^r} \, dq^r. \tag{16.4.1}$$

By analogy with the passage from θ to ω on T^*Q, one obtains on TQ the 2-form

$$\omega_{\mathcal{L}} = d\theta_{\mathcal{L}} = \frac{\partial^2 \mathcal{L}}{\partial v^r \partial v^s} \, dv^s \wedge dq^r + \frac{1}{2} \left(\frac{\partial^2 \mathcal{L}}{\partial v^r \partial q^s} - \frac{\partial^2 \mathcal{L}}{\partial v^s \partial q^r} \right) dq^s \wedge dq^r. \tag{16.4.2}$$

Thus, the 2-form $\omega_{\mathcal{L}}$ is invertible if and only if

$$\det \left(\frac{\partial^2 \mathcal{L}}{\partial v^r \partial v^s} \right) \neq 0. \tag{16.4.3}$$

There are, however, physically relevant cases where the non-degeneracy condition (16.4.3) is violated. They will not be considered here both for simplicity and to avoid introducing the theory of constrained systems, which goes beyond the aims of the present monograph.

For our specific problem, the independent coordinates of configuration space are written as q_{aj}, where the indices $a, b, c, \ldots$ range from 1 to N and play the role of particle labels, while the indices $j, k, l, m, \ldots$ range from 1 to 3 and are Cartesian vector indices. Repeated indices are not summed over here, and every summation is explicitly indicated. Whenever we use the symbol v_{aj} for velocities, these are obtained in a second-order formalism by the derivative of position with respect to the physical time of an inertial observer. By virtue of (16.4.2), one has the relations

$$\omega_{\mathcal{L}}(\partial/\partial v_{aj}, \partial/\partial v_{bk}) = 0, \tag{16.4.4}$$

$$\omega_{\mathcal{L}}(\partial/\partial v_{aj}, \partial/\partial q_{bk}) = \omega_{\mathcal{L}}(\partial/\partial v_{bk}, \partial/\partial q_{aj}) = \partial^2 \mathcal{L}/\partial v_{aj} \partial v_{bk}, \tag{16.4.5}$$

$$\omega_{\mathcal{L}}(\partial/\partial q_{aj}, \partial/\partial q_{bk}) = \frac{1}{2}(\partial^2 \mathcal{L}/\partial q_{aj} \partial v_{bk} - \partial^2 \mathcal{L}/\partial q_{bk} \partial v_{aj}). \tag{16.4.6}$$

The dynamical vector field Γ, which obeys the Euler–Lagrange equations of motion, is a vector field on the tangent bundle of configuration space. It is given by the sum $\Gamma = \sum_a \Gamma^{(a)}$, where each term takes the form

$$\Gamma^{(a)} = \sum_j \left(v_{aj} \frac{\partial}{\partial q_{aj}} + A_{aj} \frac{\partial}{\partial v_{aj}} \right), \tag{16.4.7}$$

which is why Γ is said to take a second-order form. With our notation, A_{aj} are the accelerations. On defining the energy function on TQ:

$$E_{\mathcal{L}} \equiv i_\Gamma \theta_{\mathcal{L}} - \mathcal{L}, \tag{16.4.8}$$

the equations of motion become algebraic conditions on the dynamical vector field Γ, i.e.

$$i_\Gamma \omega_{\mathcal{L}} = -\mathrm{d}E_{\mathcal{L}}. \tag{16.4.9}$$

For simplicity, we assume that the Lagrangian is non-singular, which implies that the accelerations A_{aj} are uniquely determined functions on TQ, and the dynamics expressed by Γ exists unambiguously all over TQ. Moreover, every point of TQ is allowed to be chosen as a possible initial condition, which is necessary for a system of N particles. Equation (16.4.9) implies that $L_\Gamma \omega_{\mathcal{L}} = 0$. We similarly assume that the whole Lie algebra of the Poincaré group $\mathcal{P}$ is represented by vector fields on TQ obeying the equation $L_X \omega_{\mathcal{L}} = 0$. Thus, we assume the existence of vector fields $X_{P_j}, X_{J_j}, X_{K_j}$ generating spatial translations, spatial rotations and pure Lorentz transformations, respectively, and obeying

$$L_{X_{P_j}} \omega_{\mathcal{L}} = L_{X_{J_j}} \omega_{\mathcal{L}} = L_{X_{K_j}} \omega_{\mathcal{L}} = 0. \tag{16.4.10}$$

The 10 vector fields $\Gamma, X_{P_j}, X_{J_j}, X_{K_j}$ have to obey commutation relations corresponding to the Lie algebra of $\mathcal{P}$; of all these relations, the only ones needed explicitly in proving the no-interaction theorem are

$$\left[X_{K_j}, \Gamma \right] = X_{P_j}. \tag{16.4.11}$$

The forms of X_{P_j} and X_{J_j} are immediately obtained and read as

$$X_{P_j} = -\sum_a \frac{\partial}{\partial q_{aj}}, \tag{16.4.12}$$

$$X_{J_j} = -\sum_{akl} \varepsilon_{jkl} \left(q_{ak} \frac{\partial}{\partial q_{al}} + v_{ak} \frac{\partial}{\partial v_{al}} \right). \tag{16.4.13}$$

The structure of X_{K_j} will be derived shortly.

While translations and rotations have associated vector fields X_{P_j} and X_{J_j} which have the free-particle form, the vector field Γ for time translations must be different to account for interaction. The as yet undetermined boost vector fields should satisfy the commutation relations

$$\left[X_{K_j}, X_{P_k}\right] = \delta_{jk}\Gamma, \tag{16.4.14}$$

by virtue of the Poincaré algebra. Thus, if Γ includes an interaction contribution and X_{P_k} does not, then X_{K_j} must have an interaction contribution. This implies that the particle trajectories cannot transform by the familiar free-particle formulae, and the Lorentz transformation law is itself determined by the dynamics.

In the Hamiltonian formalism on T^*Q, a canonical transformation describing a symmetry of a system maps a state at a certain time into another state at the same time so as to preserve the equations of motion. For a relativistic system the pure Lorentz or boost generator gives rise to a canonical transformation mapping physical conditions at a certain time in one inertial frame onto physical conditions at the same value of time but in a Lorentz-transformed frame. It is on these grounds that, in the instant form of relativistic dynamics, the world-line condition was originally derived. In the language of the tangent bundle of configuration space, the world-line condition is the requirement that

$$L_{X_{K_j}} q_{ak} = q_{aj} v_{ak}. \tag{16.4.15}$$

This equation fixes the part of X_{K_j} involving $\frac{\partial}{\partial q}$. To fix the remaining $\frac{\partial}{\partial v}$ part, we apply L_Γ to Eq. (16.4.15) and use the commutation relation (16.4.11) and the forms of Γ and X_{P_j}:

$$\begin{aligned} L_{X_{K_j}} v_{ak} &= L_{X_{K_j}} L_\Gamma q_{ak} = L_{X_{P_j}} q_{ak} + L_\Gamma (q_{aj} v_{ak}) \\ &= -\delta_{jk} + v_{aj} v_{ak} + q_{aj} A_{ak}. \end{aligned} \tag{16.4.16}$$

By virtue of (16.4.15) and (16.4.16), X_{K_j} takes the form

$$X_{K_j} = \sum_{ak} \left[q_{aj} v_{ak} \frac{\partial}{\partial q_{ak}} + (v_{aj} v_{ak} - \delta_{jk} + q_{aj} A_{ak}) \frac{\partial}{\partial v_{ak}} \right]. \tag{16.4.17}$$

Bearing in mind the decomposition of the dynamical vector field, the previous formula becomes (see Eq. (16.4.7))

$$X_{K_j} = \sum_a q_{aj} \Gamma^{(a)} + \sum_{ak} (v_{aj} v_{ak} - \delta_{jk}) \frac{\partial}{\partial v_{ak}}. \tag{16.4.18}$$

Hence the boost generator is determined by the equations of motion to the same extent that the accelerations A_{aj} are. The three steps in the proof of the no-interaction theorem are now as follows (Marmo *et al.* 1984).

Step I. Application of the Lie derivative along X_{K_m} to the identity (16.4.4) and use of the properties (16.4.10) and (16.4.17) yields

$$(q_{am} - q_{bm}) \frac{\partial^2 \mathcal{L}}{\partial v_{aj} \partial v_{bk}} = 0. \tag{16.4.19}$$

This means that for distinct particles, for which $a \neq b$, the second derivative of the Lagrangian in Eq. (16.4.19) has to vanish, leading to the decomposition

$$\mathcal{L}(q; v) = \sum_a \mathcal{L}^{(a)}(q; v_a). \tag{16.4.20}$$

Step II. To the result of step I we apply first the operator L_Y, then $L_{X_{K_m}}$, compare the results, and derive some properties of the components of the 2-form $\omega_{\mathcal{L}}$ in Eq. (16.4.6). Since Eq. (16.4.19) is more conveniently expressed in the form

$$\omega_{\mathcal{L}} \left(\frac{\partial}{\partial q_{aj}}, \frac{\partial}{\partial v_{bk}} \right) = 0, \quad a \neq b, \tag{16.4.21}$$

we therefore find that

$$(q_{am} - q_{bm}) \omega_{\mathcal{L}} \left(\frac{\partial}{\partial q_{aj}}, \frac{\partial}{\partial q_{bk}} \right) = 0. \tag{16.4.22}$$

Hence the components of $\omega_{\mathcal{L}}$ in this equation should vanish whenever $a \neq b$. By exploiting Eq. (16.4.6) and the decomposition of the Lagrangian obtained in step I, we find the restriction

$$\frac{\partial^2 \mathcal{L}^{(a)}(q; v_a)}{\partial q_{bk} \partial v_{aj}} = \frac{\partial^2 \mathcal{L}^{(b)}(q; v_b)}{\partial q_{aj} \partial v_{bk}}, \quad a \neq b. \tag{16.4.23}$$

Thus, in any non-linear dependence of $\mathcal{L}^{(a)}$ on v_a, the position q_b for $b \neq a$ cannot occur, while in any linear dependence of $\mathcal{L}^{(a)}$ on v_a we have conditions imposed by Eq. (16.4.23). The only admissible linear terms in the Lagrangian would therefore read as

$$\sum_j \frac{\partial f(q)}{\partial q_{aj}} v_{aj},$$

and are here dropped because they amount to a total time derivative which does not affect the equations of motion. Step II then reduces the Lagrangian to the form

$$\mathcal{L}(q; v) = \sum_a \mathcal{L}_{\rm nl}^{(a)}(q_a; v_a) - V(q), \tag{16.4.24}$$

where the subscript 'nl'denotes non-linear functions of v_a and V is a sum of terms $V^{(a)}(q)$. The resulting decomposition of the 2-form $\omega_{\mathcal{L}}$ reads as $\omega_{\mathcal{L}} = \sum_a \omega_{\mathcal{L}}^{(a)}$, where

$$\omega_{\mathcal{L}}^{(a)} = \sum_j \mathrm{d}\left(\frac{\partial \mathcal{L}_{\mathrm{nl}}^{(a)}}{\partial v_{aj}}\right) \wedge \mathrm{d}q_{aj}. \tag{16.4.25}$$

Step III. To the result (16.4.22) of step II we apply L_Γ and $L_{X_{K_m}}$ and compare the results. This yields

$$(q_{am} - q_{bm}) \sum_l \omega_{\mathcal{L}}^{(a)}\left(\frac{\partial}{\partial q_{aj}}, \frac{\partial}{\partial v_{al}}\right) \frac{\partial A_{al}}{\partial q_{bk}} = 0, \tag{16.4.26}$$

which eventually leads to

$$\frac{\partial}{\partial q_{bk}} \sum_l \omega_{\mathcal{L}}^{(a)}\left(\frac{\partial}{\partial q_{aj}}, \frac{\partial}{\partial v_{al}}\right) A_{al} = 0, \quad a \neq b. \tag{16.4.27}$$

Such an equation makes it possible for the sum over l therein to depend on q_a, v_a, and possibly on q_b for $b \neq a$ through the accelerations A_{al}. It does not directly give us some conditions on the potential $V(q)$ in the Lagrangian, but this task is accomplished when we bear in mind that the dynamical equation (16.4.9) is taken to be solvable all over TQ. That equation now reads, in local coordinates,

$$-\frac{\partial \mathcal{L}_{\mathrm{nl}}^{(a)}}{\partial q_{aj}}(q_a; v_a) + \sum_k \left[\frac{\partial^2 \mathcal{L}_{\mathrm{nl}}^{(a)}(q_a; v_a)}{\partial v_{aj} \partial v_{ak}} A_{ak} + \frac{\partial^2 \mathcal{L}_{\mathrm{nl}}^{(a)}(q_a; v_a)}{\partial v_{aj} \partial q_{ak}} v_{ak}\right]$$
$$= -\frac{\partial V(q)}{\partial q_{aj}}. \tag{16.4.28}$$

Since the left-hand side of Eq. (16.4.28) is independent of q_b for $b \neq a$, while all positions and velocities can be treated as independent variables because solutions A_{aj} exist all over the tangent bundle of configuration space, the potential V is forced to be separable, i.e. $V(q) = \sum_a V^{(a)}(q_a)$. The Lagrangian now takes the completely separated form

$$\mathcal{L}(q; \dot{q}) = \sum_a \mathcal{L}_{\mathrm{nl}}^{(a)}(q_a; v_a) - V^{(a)}(q_a), \tag{16.4.29}$$

and the no-interaction theorem is proved, since we have assumed non-singularity of $\mathcal{L}$, and Eq. (16.4.28) shows that each particle moves independently of the others. By requiring Poincaré invariance, each $V^{(a)}$ is indeed found to vanish, while each $\mathcal{L}_{\rm nl}^{(a)}$ has to take the form (Balachandran *et al.* 1982b)

$$\mathcal{L}_{\rm nl}^{(a)}(q_a; v_a) = -m_a \sqrt{1 - \sum_j v_{aj} v_{aj}}. \tag{16.4.30}$$

For singular Lagrangians, some extra work is necessary, for which we refer the reader once more to the work in Marmo *et al.* (1984). In the course of arriving at the completely separated form of the Lagrangian, only the following features play an explicit role in the calculations: (i) the world-line condition (16.4.15); (ii) the form (16.4.12) for the translation generator X_{P_j}; (iii) the commutation relation (16.4.11); (iv) the annihilation of $\omega_{\mathcal{L}}$ by L_Γ and $L_{X_{K_m}}$. The original proof of the theorem for any finite number of particles, used as an essential assumption the existence of a Hamiltonian obtained from a non-singular Lagrangian. If this is the case, the proof presented here is much simpler than that relying on phase-space methods. On the other hand, we should stress that the separability of the Lagrangian can be proved without having to assume its non-singularity at all. It is then necessary to assume that the second-order dynamics described by the vector field Γ does exist everywhere on the tangent bundle of configuration space (Marmo *et al.* 1984), or, with the Bergmann language, that there are no secondary constraints.

Digression. In the proof of the no-interaction theorem the principle of manifest covariance and, hence, of a world line for each particle has been exploited. If this is relaxed it is very easy to construct relativistic interactions. In fact, Eddington attempted such constructions with particular reference to binary stars, but Pryce gave a more satisfactory formulation (Pryce 1948). It was revived in the context of quantum mechanics in Bakamjian and Thomas (1953) and by the work in Currie *et al.* (1963), Jordan *et al.* (1964). In this case the reduced mass is replaced by a scalar function of the relative coordinates. These satisfy the criterion set down by Dirac in his celebrated paper on forms of relativistic dynamics (Dirac 1949).

However, we are used to associating a world line with the trajectory of each particle which transforms in a geometric fashion when one changes observation from one frame to another. Though the manifest geometrical transformation has been demonstrated and applied to the free motion of a particle, we think it would be convenient if it held even under interaction. This geometric transformation law is given in its differential form in (16.4.16). What we find is that this requirement together with canonical dynamics constrain the dynamics to be trivial.

The way out of this difficulty is to weaken the constraints. Houtappel *et al.* (1965) formulated a scheme in which each particle interacts with every other particle lying inside its space-like region. In such a model we cannot set initial conditions and ask for the dynamical development nor can we assure energy–momentum conservation without including a 'field' energy and momentum which resides in space between particles. The same comments apply for the imaginative but largely incomplete formulation of action-at-a-distance theory in Wheeler and Feynman (1949). It can be shown that any such action-at-a-distance theory is best understood as a theory of particles interacting with fields locally (Sudarshan 1972).

16.5 Relativistic quantum particles

For a manifestly covariant treatment of quantum particles we must consider a relativistic wave equation. The Dirac equation (section 16.1) is the easiest and best known of these, though Schrödinger had considered the second-order differential equation, which is now called the Klein–Gordon equation. The Dirac equation describes particles with spin $\frac{1}{2}$; the Klein–Gordon equation describes spinless particles. For spin 1 we have the Duffin–Kemmer–Petiau first-order equation with 10 components (Petiau 1936, Duffin 1938, Kemmer 1939, Fischbach *et al.* 1973). For still higher spin we have complicated equations that contain second-class constraints and therefore create difficulties with describing interactions.

Even at the level of free particles, all of these relativistic wave equations contain the anomaly of having both positive- and negative-energy solutions. We can see this independently of the details of the equation by recognizing that any manifestly relativistic wave equation with finitely many components is automatically invariant under the complex Lorentz group. But if so the strong reflection $\vec{x} \to -\vec{x}, t \to -t$ is a real Lorentz transformation contained in the complex Lorentz group and it is connected with the identity in the complex group. But this purely geometric transformation takes positive energies to negative energies while the wave equation is invariant.

The existence of the negative-energy branch shows that such energies are unbounded from below. If these states are freely accessible the second law of thermodynamics would be violated. In the case of the Dirac equation, Dirac proposed the idea that the negative-energy states are all filled: by virtue of the Pauli exclusion principle no positive-energy electron can go to a negative-energy state. On the other hand, if there is an unoccupied level of negative energy ('holes in the physical vacuum') this would move like a bubble in water in a pond in the opposite direction to a pebble which goes down. In an electric field the hole would move like a particle of positive charge, now identified with the positron.

There are two points about this method of avoiding negative-energy particles: it takes us from the one-particle theory into a many particles (actually an infinite number) which can be created or annihilated. This would suggest that we should deal with the processes of creation and destruction of particles. Such a theory is the quantum theory of fields. We are thus driven to quantum fields rather than particles.

The other point is that a negative branch of the energy spectrum obtains for spin-0 and spin-1 particles, including the photon. No Pauli exclusion principle is available to invoke a negative-energy sea. The negative-energy sea is *not* the solution. Further, as noted by Friedrichs, the physical vacuum with all the negative-energy states filled is not Lorentz invariant (Friedrichs 1953).

16.6 From particles to fields

The correct interpretation of positive- and negative-energy states is that the positive-energy states have this wave function but the negative 'energy' states are the complex conjugate (or more generally the adjoint) of an anti-particle wave function. This does not discriminate between integer and half-odd spins. Rather, the information about the symmetry type of the many-particle wave functions is contained in the commutation properties of the operator coefficients, the creation and annihilation operators in the expansion of the relativistic amplitudes. Rather than being wave function expansion coefficients, they are operators which decrease (or increase) the number of particles. We have thus been tricked into a quantum theory of fields (Itzykson and Zuber 1985, Weinberg 1996, Wightman 1996).

What is inevitable, we must accept, and find merit in it. This is the situation in the quantum theory of fields, which becomes a quantum system with an infinite number of degrees of freedom. The investigation of 'radiative' corrections in quantum electrodynamics and the processes in particle physics have shown that correct and unambiguous predictions are obtained in agreement with experimental findings. This is all the more remarkable since in the process of calculating we encounter infinities which we have to sidestep skilfully. While the majority of physicists are resigned to this, Dirac makes the startling and unambiguous assertion that the theory must be fundamentally wrong (Dirac 1981).

Apart from such a judgement, the number of particle species has now grown so fast that perhaps we should attempt a new kind of theory which will give all the particles. But we are unable to even outline them in this basic text.

16.7 The Kirchhoff principle, antiparticles and QFT

In the study of the thermodynamics of radiation in equilibrium, the dynamic approach incorporates the continual emission and absorption of radiation by matter. Recall from section 14.7 that there is a connection between the emissivity and absorptivity of any material: when we see inside a cavity all materials glow equally when radiation is in equilibrium. This implies that the emissivity and absorptivity are proportional: this is the Kirchhoff principle, and it is an essential link in the demonstration that the radiation content in the interior of any cavity depends only on the temperature and is independent of the matter within the cavity. This in turn leads to the Wien displacement law which prescribes the scaling of the spectral distribution in black-body radiation.

When we study this problem with the quantum theory of light we have the emission and absorption within the eigenmodes of the electromagnetic field in the cavity. The standing modes of the electromagnetic field now become harmonic variables with well-defined frequencies. We have seen earlier that the coordinates and momenta of a quantum oscillator both contain the creation and annihilation operators in such a manner that we cannot separate the two parts in a 'local' manner. This, in turn, allows the probability of emission and absorption to be proportional to $n+1$ and n, respectively, where n is the number of photons of the corresponding eigenmode. In addition to the theory of the black-body spectrum, this principle is relevant for such topics as the Saha ionization law where atoms may partially ionize in the presence of radiation, such as in the mantle of a star.

The coupling of a harmonic oscillator to other degrees of freedom is in terms of the coordinate q; but the creation and annihilation operators are not separately and independently coupled. Such a coupling is also used in other contexts such as the coupling of cold neutrons to lattice vibrations.

When we come to relativistic quantum theory, the dynamical variables are relativistic fields: for the Dirac wave function we saw that there are positive- and negative-frequency solutions, and to consistently interpret these we must interpret the 'negative-energy' solutions as the complex conjugates of the wave functions with 'positive energy'. But this situation is not restricted to the Dirac equation. Both the Schrödinger–Klein–Gordon equation for spin-0 particles or the electromagnetic Maxwell equations also have both signs for the energy. (In fact, from the invariance of such wave equations under the *complex* Lorentz group we see that the operation of strong reflection which inverts both time and space is a symmetry of these equations. But the strong reflection takes a positive-energy solution to a negative-energy solution and vice versa: this is true of *any* relativistic wave equation with finite component wave functions.)

Thus, in complete analogy to the harmonic oscillator coordinate

$$q = \frac{a\mathrm{e}^{-\mathrm{i}\omega t} + a^\dagger \mathrm{e}^{\mathrm{i}\omega t}}{\sqrt{2\omega}}, \tag{16.7.1}$$

we expand a scalar field in the manner

$$\phi(\vec{x}, t) = \int \frac{\mathrm{d}^3 k}{\sqrt{2\omega}} \Big[a(k)\mathrm{e}^{\mathrm{i}\vec{k}\cdot\vec{x} - \mathrm{i}\omega t} + a^\dagger(k)\mathrm{e}^{-\mathrm{i}\vec{k}\cdot\vec{x} + \mathrm{i}\omega t}\Big], \tag{16.7.2}$$

with $\omega = +\sqrt{k^2 + m^2}$ the energy of the mode with momentum $\vec{k}$. The corresponding expansion of the conjugate field $\pi(\vec{x}, t)$, which is the time derivative of $\phi(\vec{x}, t)$, reads as

$$\pi(\vec{x}, t) = \int \mathrm{d}^3 k \sqrt{\frac{\omega}{2}} \mathrm{i}\Big[a^\dagger(k)\mathrm{e}^{-\mathrm{i}\vec{k}\cdot\vec{x} + \mathrm{i}\omega t} - a(k)\mathrm{e}^{\mathrm{i}\vec{k}\cdot\vec{x} - \mathrm{i}\omega t}\Big], \tag{16.7.3}$$

and the associated commutation law at equal times can be written in the form

$$[\phi(\vec{x}, t), \pi(\vec{y}, t)] = \mathrm{i}\delta(\vec{x} - \vec{y}). \tag{16.7.4}$$

The breakup of either the field or its time derivative into creation and annihilation operators is non-local: similar comments apply to the Maxwell field. For the Dirac field, with first-order equation in the spinor field $\psi_r(\vec{x}, t)$ we have to use the anticommutation rule:

$$\Big\{\psi_r(\vec{x}, t), \psi_s^\dagger(\vec{y}, t)\Big\}_+ = \delta(\vec{x} - \vec{y}). \tag{16.7.5}$$

Again, locality demands that we do not break up ψ into creation and annihilation parts and couple them independently. Thus, locality of the relativistic fields automatically incorporates the Kirchhoff principle, which asserts that emissivity and absorptivity should be proportional.

For the electromagnetic field there are both left and right circularly polarized photons, and the Hermitian fields $\vec{E}, \vec{B}, \vec{A}, A_0$ create and destroy the same particle. There are no distinct antiparticles. The photon is its own antiparticle. Similarly for the neutral scalar field (appropriate for the neutral pion) the particles and antiparticles are the same. If we want particle and antiparticles to be distinct we take two neutral scalar fields $\phi_1(\vec{x}, t)$ and $\phi_2(\vec{x}, t)$ and consider the complex field

$$\phi(\vec{x}, t) = \frac{1}{\sqrt{2}}[\phi_1(\vec{x}, t) + \mathrm{i}\phi_2(\vec{x}, t)]. \tag{16.7.6}$$

If the quanta of the complex field $\phi(\vec{x}, t)$ are identified as the particle, they will have a distinct antiparticle.

Similar considerations apply for the Dirac fields: we can have a spinor field for which the particles and antiparticles are the same. This is sometimes called the Majorana field. To get a familiar Dirac field with distinct particles and antiparticles we need a combination of two Majorana fields. This simple fact is often obscured by the freedom to choose different representations of the Dirac matrices. The spinor in Minkowski space-time is essentially real (that is the complex conjugate transforms in the same way as the original) in contrast to spinors in three dimensions. In the Majorana representation the γ matrices are all pure imaginary as we said before, and hence the Dirac equation becomes a purely real differential equation. Other representations of the γ matrices express the reality of the Majorana field in a more involved fashion.

The existence of distinct antiparticles implies that one is dealing with a symmetry group $SO(2)$ for which the covering group is the Abelian translation group. These transformations are called gauge transformations. For 'real' fields this degenerates into the two-element group Z_2, which has two representations. Some neutral fields transform by ± 1 under this; these are said to have particle conjugation parity ± 1. When the fields are complex, whether spin 0, spin $\frac{1}{2}$ or spin 1 the kinematic structure allows the symmetry of particle conjugation

$$\phi_1 \to +\phi_1, \quad \phi_2 \to -\phi_2, \quad \phi \to \phi^\dagger.$$

Any phase factor may be absorbed into the redefinition of the field.

There is an intimate connection between particle conjugation (C) and the selection of space (P) and time (T): the combined operation TCP is an invariance of any local relativistic theory, not only of its kinematics, since it is a real element of the complex Poincaré group. The breaking up of the field into creation and annihilation parts requires the use of projection operators that involve the energy $E = \sqrt{p^2 + m^2}$ and which is therefore highly non-local.

16.7.1 Time and space inversions in quantum mechanics

We know from chapter 9 that space inversion, or parity, is the coordinate transformation

$$\vec{x} \to -\vec{x}, \quad t \to t.$$

A *scalar* field under this transformation must be

$$\phi(\vec{x}, t) \to \phi'(\vec{x}, t) = \phi(-\vec{x}, t). \tag{16.7.7}$$

If the field is pseudo-scalar one has instead

$$\phi(\vec{x}, t) \to -\phi(-\vec{x}, t). \tag{16.7.8}$$

For the Schrödinger amplitude for a particle

$$\vec{p} \to -\vec{p}, \quad \vec{q} \to \vec{q}, \quad E \to E, \quad \vec{J} \to \vec{J},$$

so that

$$\psi(\vec{x},t) \to \eta\psi(-\vec{x},t), \tag{16.7.9}$$

where $\eta = \pm 1$ is the intrinsic parity. For a spinor field

$$\psi_r(\vec{x},t) \to \eta\beta_{rs}\psi_s(-\vec{x},t) = \psi'_r(\vec{x},t), \tag{16.7.10}$$

with β being introduced to guarantee that the transformed field also obeys the same equation. For a Majorana (real) particle we must choose $\eta = \pm\mathrm{i}$ since β_{rs} is pure imaginary. For a Dirac particle the choice of η is wider. It could be (Yang and Tiomno 1950)

$$\eta = \pm\mathrm{i},\ \pm 1,$$

where the ± 1 phase does not retain the reality properties of ψ. In the non-relativistic two-component form the wave function is automatically a complex spinor but the wave functions can transform as

$$\psi_a(\vec{x},t) \to \eta\psi_a(-\vec{x},t). \tag{16.7.11}$$

For the electromagnetic field one has the transformation laws

$$\vec{E}(\vec{x},t) \to -\vec{E}(-\vec{x},t), \quad \vec{B}(\vec{x},t) \to +\vec{B}(-\vec{x},t),$$

$$\vec{A}(\vec{x},t) \to -\vec{A}(-\vec{x},t), \quad A_0(\vec{x},t) \to +A_0(-\vec{x},t). \tag{16.7.12}$$

The Maxwell equations are invariant under (16.7.12).

The possibility of defining a space inversion transformation and the fact that the kinematics and equal-time anticommutation rules are invariant does not guarantee that parity is conserved in all interactions. In fact, for zero mass the Dirac equation splits into two parts, the right and left chiral components. The field must be complex since the chirality transformation involves an imaginary matrix (in the Majorana representation) and hence the chiral eigenstates are essentially complex. The same comment applies to the decomposition of the Maxwell field. It should also be recognized that the Poincaré group does *not* involve space inversion. Such a transformation is an extension of the Poincaré group.

The time inversion transformation is more subtle. If we merely change t to $-t$ the Schrödinger equation is not preserved. Moreover, we would like the following physical transformations:

$$\vec{p} \to -\vec{p}, \quad \vec{q} \to \vec{q}, \quad \vec{E} \to \vec{E}, \quad \vec{J} \to -\vec{J}.$$

These transformations are *not* canonical; consequently the corresponding time inversion cannot be a unitary transformation. The Wigner (weak) time inversion is the following *anti-unitary* transformation:

$$\psi(\vec{x},t) \to \psi^*(\vec{x},-t). \tag{16.7.13}$$

Under this transformation the Schrödinger equation is preserved. For the various relativistic equations one has

$$\phi(\vec{x},t) \to \phi^*(\vec{x},-t), \quad \psi(\vec{x},t) \to \psi^*(\vec{x},-t),$$

$$\vec{E}(\vec{x},t) \to \vec{E}(\vec{x},-t), \quad \vec{B}(\vec{x},t) \to -\vec{B}(\vec{x},-t).$$

Once more, any phase factor η_T can be absorbed by a redefinition of the fields. For the non-relativistic two-component spinor

$$\sigma_2 \sigma_k^* \sigma_2 = -\sigma_k, \tag{16.7.14}$$

so that the time reversal transformation is

$$\psi(\vec{x},t) \to \exp\left(\mathrm{i}\pi\frac{\sigma_2}{2}\right)\psi(\vec{x},-t). \tag{16.7.15}$$

Just as for parity, time inversion is a kinematic symmetry which holds but this does not imply that it will be a symmetry of the dynamics. The strong reflection which involves $\vec{x} \to -\vec{x}, t \to -t$ and particle conjugation is equivalent to the real element

$$\vec{x} \to -\vec{x}, \quad t \to -t$$

of the complex Poincaré group, provided that the equations are manifestly covariant. On one-particle wave functions T and C are anti-unitary, so the combination TCP is a unitary operation. But on quantized fields the particle conjugation is a unitary operation, so TCP is anti-unitary. TCP is automatically valid in any relativistic field theory.

16.7.2 The spin-statistics connection

In nature we observe that identical particles are really identical, and hence the interchange of any two of them should not change the density matrix. But on the wave functions this allows for the possibility of symmetry or antisymmetry:

$$\Psi(\vec{x}_1, ...; t) \to \mathcal{P}_{12}\Psi(\vec{x}_1, \vec{x}_2, ..., \vec{x}_N; t) = \pm\Psi(\vec{x}_2, \vec{x}_1, ..., \vec{x}_N; t). \tag{16.7.16}$$

The antisymmetric choice is appropriate for electrons and nucleons so as to satisfy the Pauli exclusion principle. For photons they obey Bose statistics. It is easy to see that the Michel identity (Michel 1964) holds, i.e.

$$\mathcal{P}_{12} = (-1)^{2s}. \tag{16.7.17}$$

Bose particles have integer spin and Fermi particles (obeying the exclusion principle) have half-odd spin. This connection is usually deduced from relativistic quantum field theory, but we shall discover that it is already true in non-relativistic quantum mechanics.

As a preliminary let us note that the configuration space of N identical particles is a multiply connected manifold. The point (x_j, x_k) and the point (x_k, x_j) are identified, so we find the space

$$\mathbf{R} \times \mathbf{R} \times \cdots \times \mathbf{R}/S_N,$$

where $\mathbf{R}$ is the manifold on which one-particle wave functions are defined. Unfortunately $\mathbf{R}^N/S_N$ is not a manifold, since two or more points x_1 to x_N may coincide. So the proper manifold is defined as

$$\left(\mathbf{R} \times \mathbf{R} \times \cdots \times \mathbf{R} - \mathcal{D}\right)/S_N,$$

where $\mathcal{D}$ is the set of points at which two or more of the points coincide. Indeed, $\mathcal{D}$ is precisely the subset of those points in the product that are fixed under the action of some element of S_N. This is a manifold. This manifold is multiply connected and the wave functions may acquire phase factors in going around a non-contractible loop. While these considerations tell us that wave functions may acquire phase factors, they do not tell us which species of particles prefer which phase (± 1).

To find a more restrictive result we consider the equations for the quantum field (not necessarily relativistic) expressed as first-order equations for several components. We therefore write the kinematic part of the Lagrangian in the form

$$\mathcal{L}_{\text{kin}} = \frac{1}{2}\left(\psi_r \Gamma_{rs} \frac{\partial \psi_s}{\partial t} - \frac{\partial \psi_r}{\partial t} \Gamma_{rs} \psi_s\right). \tag{16.7.18}$$

We want this term to be a scalar under rotations in three-dimensional space. The operation of time differentiation is anti-symmetric. But this term should be a scalar, bilinear in ψ. We know that the scalar product of two scalars (or two vectors or two tensors) is symmetric in the factors, but the scalar product of two spinors is anti-symmetric. So we deduce that Γ_{rs} is anti-symmetric for tensor fields and symmetric for spinor fields. This is already evident from the spin $0, \frac{1}{2}$ and 1 equations that we have considered.

If the components of ψ were to commute and Γ_{rs} was symmetric the Lagrangian density would reduce to a number; similarly if they were to anti-commute and Γ_{rs} was anti-symmetric. So we conclude that tensor fields alone can obey (equal-time) commutation relations, while spinor fields must obey (equal-time) anti-commutation relations. (If there is kinematic degeneracy we may be tempted to evade these restrictions by having Γ_{rs} anti-symmetric in the internal symmetry labels. But this is inconsistent since if we were to diagonalize an anti-symmetric matrix the trace would still be zero, and inconsistent with a Hilbert space realization.)

References

Abraham, R. and Marsden, J. E. (1985). *Foundations of Mechanics,* 2nd edn (Reading, Addison-Wesley).

Aharonov, Y. and Bohm, D. (1959). Significance of electromagnetic potentials in the quantum theory, *Phys. Rev.* **115,** 485–91.

Aharonov, Y. and Anandan, J. (1987). Phase change during a cyclic quantum evolution, *Phys. Rev. Lett.* **58,** 1593–6.

Airy, G. B. (1838). On the intensity of light in the neighbourhood of a caustic, *Trans. Camb. Phil. Soc.* **6,** 379–402.

Arnol'd, V. I. (1978). *Mathematical Methods of Classical Mechanics* (New York, Springer).

Aspect, A., Graingier, P. and Roger, G. (1981). Experimental tests of realistic local theories via Bell's theorem, *Phys. Rev. Lett.* **47,** 460–7.

Aspect, A., Dalibard, J. and Roger, G. (1982a). Experimental tests of Bell's inequalities using time-varying analysers, *Phys. Rev. Lett.* **49,** 1804–7.

Aspect, A., Graingier, P. and Roger, G. (1982b). Experimental realization of EPR Gedankenexperiment: a new violation of Bell's inequalities, *Phys. Rev. Lett.* **49,** 91–4.

Auletta, G. (2000). *Foundations and Interpretation of Quantum Mechanics* (Singapore, World Scientific).

Bakamjian, B. and Thomas, L. H. (1953). Relativistic particle dynamics. 2, *Phys. Rev.* **92,** 1300–10.

Baker Jr., G. A. (1958). Formulation of quantum mechanics based on the quasi-probability distribution induced on phase space, *Phys. Rev.* **109,** 2198–2206.

Balachandran, A. P., Marmo, G., Mukunda, N., Nilsson, J. S., Simoni, A., Sudarshan, E. C. G. and Zaccaria, F. (1982a). Relativistic-particle interactions. A third world view, *Nuovo Cimento* **A67,** 121–42.

Balachandran, A. P., Marmo, G. and Stern, A. (1982b). A Lagrangian approach to the no-interaction theorem, *Nuovo Cimento* **A69,** 175–186.

Balachandran, A. P., Marmo, G., Skagerstam, B. S. and Stern, A. (1983). *Gauge Symmetries and Fibre Bundles. Applications to Particle Dynamics* (Berlin, Springer).

Balachandran, A. P., Marmo, G., Mukunda, N., Nilsson, J. S., Simoni, A., Sudarshan, E. C. G. and Zaccaria, F. (1984). Unified geometrical approach to relativistic particle dynamics, *J. Math. Phys.* **25,** 167–76.

Balachandran, A. P., Marmo, G., Skagerstam, B. S. and Stern, A. (1991). *Classical Topology and Quantum States* (Singapore, World Scientific).

Balasubramanian, S. (1972). Sylvester's formula and the spin problem, *Am. J. Phys.* **40,** 918–9.

Balmer, J. J. (1885). Notiz über die spectrallinien des wasserstoffs, *Verh. d. Naturf. Ges. (Basel)* **7,** 548–60.

Bargmann, V. (1961). On a Hilbert space of analytic functions and an associated integral transform, 1, *Commun. Pure Appl. Math.* **14,** 187–214.

(1964). Note on Wigner's theorem on symmetry operations, *J. Math. Phys.* **5,** 862–8.

Barnes, J. F., Brascamp, H. J. and Lieb, E. H. (1976). Lower bounds for the ground-state energy of the Schrödinger equation using the sharp form of Young's inequality, in *Studies in Mathematical Physics,* ed. E.H. Lieb, B. Simon and A. S. Wightman (Princeton, Princeton University Press).

Bastai, A., Bertocchi, L., Fubini, S., Furlan G. and Tonin, M. (1963). On the treatment of singular Bethe–Salpeter equations, *Nuovo Cim.* **30,** 1512–31.

Bell, J. S. (1964). On the Einstein–Podolsky–Rosen paradox, *Physics* **1,** 195–200.

(1987). *Speakable and Unspeakable in Quantum Mechanics* (Cambridge, Cambridge University Press).

Beltrametti, E. G. and Cassinelli, G. (1976). Logical and mathematical structures of quantum mechanics, *Riv. Nuovo Cimento* **6,** 321–404.

(1981). *The Logic of Quantum Mechanics* (London, Addison-Wesley).

Bennett, C. H., Brassard, G., Crépeau, C., Jozsa, R., Peres, A. and Wootters, W. K. (1993). Teleporting an unknown quantum state via dual classic and Einstein–Podolsky–Rosen channels, *Phys. Rev. Lett.* **70,** 1895–9.

Berezin, F. A. and Shubin, M. A. (1991). *The Schrödinger Equation* (Dordrecht, Kluwer).

Berry, M. V. (1984). Quantal phase factors accompanying adiabatic changes, *Proc. Roy. Soc. Lond.* **A392,** 45–57.

Bethe, H. A. and Salpeter, E. E. (1957). *Quantum Mechanics of One and Two Electron Atoms* (Berlin, Springer).

Biedenharn, L. C. and Louck, J. D. (1981). *Angular Momentum in Quantum Physics, Encycl. of Maths. and Its Applications* (New York, Academic Press).

Bimonte, G. and Musto, R. (2003a). Comment on 'Quantitative wave-particle duality in multibeam interferometers', *Phys. Rev.* **A**, **67,** 066101.

(2003b). What is the concept of interferometric duality in multibeam interferometers?, QUANT-PH 0301017, preprint.

Bloch, F. (1928). Über die quantenmechanik der elektronen in kristallgittern, *Z. Phys.* **52,** 555–600.

Bohm, D. (1958). *Quantum Theory* (New York, Prentice-Hall).

Bohr, N. (1913). On the constitution of atoms and molecules, in *The Old Quantum Theory,* ed. D. Ter Haar (Glasgow, Pergamon Press).

Bolte, J. and Keppeler, S. (1999). A semiclassical approach to the Dirac equation, *Ann. Phys. (N.Y.)* **274,** 125–62.

Born, M. and Wolf, E. (1959). *Principles of Optics* (London, Pergamon Press).

Born, M. (1969). *Atomic Physics* (London, Blackie & Son).

Boschi, D., Branca, S., De Martini, F., Hardy, L. and Popescu, S. (1998). Experimental realization of teleporting an unknown pure quantum state via dual classical and Einstein–Podolsky–Rosen channels, *Phys. Rev. Lett.* **80,** 1121–1125.

Bose, S. N. (1924a). Plancks gesetz und lichtquantenhypothese, *Z. Phys.* **26,** 178–181.

(1924b). Wärmegleichgewicht und strahlungsfeld bei annesenheit von materie, *Z. Phys.* **27,** 384–93.

Bouwmeester, D., Pan Jan-Wei, Mattle, K., Eibl, M., Weinfurter, H. and Zeilinger, A. (1997). Experimental quantum teleportation, *Nature* **390,** 575–9.

Boya, L. J. and Sudarshan, E. C. G. (1996). Point interactions from flux conservation, *Int. J. Theor. Phys.* **35,** 1063–68.

Breit, G. and Wigner, E. P. (1936). Capture of slow neutrons, *Phys. Rev.* **49,** 519–31.

Caldirola, P., Cirelli, R. and Prosperi, G. M. (1982). *Introduzione alla Fisica Teorica* (Torino, UTET).

Calogero, F. (1967). *Variable Phase Approach to Potential Scattering* (New York, Academic Press).

Carinena, J. F., Marmo, G., Perelomov, A. M. and Ranada, M. F. (1998). Related operators and exact solutions of Schrödinger equations, *Int. J. Mod. Phys.* **A13,** 4913–29.

Carinena, J. F., Grabowski, J. and Marmo, G. (2000). *Lie–Scheffers Systems: A Geometric Approach* (Naples, Bibliopolis).

Cartan, E. (1938). *La Théorie des Spineurs I & II* (Paris, Hermann).

Casimir, H. B. G. (1948). On the attraction between two perfectly conducting plates, *Proc. Kon. Ned. Akad. Wetenschap.* **B51,** 793–5.

Cavallaro, S., Morchio, G. and Strocchi, F. (1999). A generalization of the Stone–von Neumann theorem to nonregular representations of the CCR-algebra, *Lett. Math. Phys.* **47,** 307–20.

Chiu, C. B., Sudarshan, E. C. G. and Bhamathi, G. (1992). Hamiltonian model for the Higgs resonance, *Phys. Rev.* **D45,** 884–91.

Clebsch, A. (1872). *Theorie der Binaren Algebraischen Formen* (Leipzig, Teubner).

Cohen-Tannoudji, C., Diu, B. and Laloe, F. (1977a). *Quantum Mechanics. I* (New York, Wiley).

(1977b). *Quantum Mechanics. II* (New York, Wiley).

Compton, A. H. (1923a). A quantum theory of the scattering of X-rays by light elements, *Phys. Rev.* **21,** 483–502.

(1923b). The spectrum of scattered X-rays, *Phys. Rev.* **22,** 409–13.

Currie, D. G., Jordan, T. F. and Sudarshan, E. C. G. (1963). Relativistic invariance and Hamiltonian theories of interacting particles, *Rev. Mod. Phys.* **35,** 350–75.

Daneri, A., Loinger, A. and Prosperi, G. M. (1962). Quantum theory of measurement and ergodicity conditions, *Nucl. Phys.* **33,** 297–319.

Darboux, G. (1882). Sur une proposition relative aux équations linéaires, *C. R. Acad. Sci. (Paris)* **94,** 1456–9.

Davisson, C. and Germer, L. H. (1927). Diffraction of electrons by a crystal of nickel, *Phys. Rev.* **30,** 705–40.

Davydov, A. S. (1981). *Quantum Mechanics* (Moscow, MIR).

De Alfaro, V. and Regge, T. (1965). *Potential Scattering* (Amsterdam, North-Holland).

de Broglie, L. (1923). Waves and quanta, *Nature* **112,** 540.

Debye, P. (1912). Les particularités des chaleurs specifiques à basse temperature, *Arch. Sci. Phys. Natur.* **33,** 256–58.

De Facio, B. and Hammer, C. L. (1974). Remarks on the Klauder phenomenon, *J. Math. Phys.* **15,** 1071–7.

De Filippo, S., Landi, G., Marmo, G. and Vilasi, G. (1989). Tensor fields defining a tangent bundle structure, *Ann. Poincaré Phys. Theor.* **50,** 205–18.

de la Madrid, R., Bohm, A. and Gadella, M. (2002). Rigged Hilbert space treatment of continuous spectrum, *Fortschr. Phys.* **50,** 185–216.

Dell'Antonio, G. (1996). *Elementi di Meccanica. I: Meccanica Classica* (Naples, Liguori).

D'Espagnat, B. (1976). *Conceptual Foundations of Quantum Mechanics* (New York, Benjamin).

Derezinski, J. (1993). Asymptotic completeness for *N*-particle long-range quantum systems, *Ann. Math.* **138,** 427–76.

DeWitt, B. S. (1952). Point transformations in quantum mechanics, *Phys. Rev.* **85,** 653–61.

(1965). *Dynamical Theory of Groups and Fields* (New York, Gordon and Breach).

DeWitt, B. S. and Graham, N. (1973). *The Many-Worlds Interpretation of Quantum Mechanics* (Princeton, Princeton University Press).

Dirac, P. A. M. (1926a). The fundamental equations of quantum mechanics, *Proc. R. Soc. Lond.* **A109,** 642–53.

(1926b). On the theory of quantum mechanics, *Proc. R. Soc. Lond.* **A112,** 661–77.

(1927). The quantum theory of emission and absorption of radiation, *Proc. R. Soc. Lond.* **A114,** 243–65.

(1928). The quantum theory of the electron, *Proc. R. Soc. Lond.* **A117,** 610–24.

(1931). Quantized singularities in the electromagnetic field, *Proc. R. Soc. Lond.* **A133,** 60–72.

(1933). The Lagrangian in quantum mechanics, *Physikalische Zeitschrift der Sowjetunion* **3,** 64–72.

(1949). Forms of relativistic dynamics, *Rev. Mod. Phys.* **21,** 392–9.

(1958). *The Principles of Quantum Mechanics* (Clarendon Press, Oxford).

(1981). Does renormalization make sense? in Conf. on Perturbative Quantum Chromodynamics, *AIP Conf. Proc.* **74,** 129–30.

Dolinszky, T. (1980). Scattering by singular potentials of the general type at high energies, *Nucl. Phys.* **A338,** 495–513.

Dray, T. (1985). The relationship between monopole harmonics and spin-weighted spherical harmonics, *J. Math. Phys.* **26,** 1030–3.

Dubin, D. A., Hennings, M. A. and Smith, T. B. (2000). *Mathematical Aspects of Weyl Quantization and Phase* (Singapore, World Scientific).

DuBridge, L. A. (1933). Theory of the energy distribution of photoelectrons, *Phys. Rev.* **43,** 727–41.

Duck, I. and Sudarshan, E. C. G. (1997). *Pauli and the Spin-Statistics Theorem* (Singapore, World Scientific).

Duck, I. and Sudarshan, E. C. G. (2000). *100 Years of Planck's Quantum* (Singapore, World Scientific).

Duffin, R. J. (1938). On the characteristic matrices of covariant systems, *Phys. Rev.* **54,** 1114.

Dyson, F. J. (1949). The radiation theories of Tomonaga, Schwinger and Feynman, *Phys. Rev.* **75,** 486–502.

Ehrenfest, P. (1927). Bemerkung über die angenäherte gültigkeit der klassischen mechanik innerhalb der quantenmechanik, *Z. Phys.* **45,** 455–7.

Ehrenpreis, L. (1954). Solution of some problems of division, *Amer. J. Math.* **76,** 883–903.

Einstein, A. (1905). On a heuristic point of view about the creation and conversion of light, in *The Old Quantum Theory,* ed. D. Ter Haar (Glasgow, Pergamon Press).

(1917). On the quantum theory of radiation, in *Sources of Quantum Mechanics,* ed. B. L. van der Waerden (New York, Dover).

Einstein, A., Podolsky, B. and Rosen, N. (1935). Can quantum-mechanical description of physical reality be considered complete?, *Phys. Rev.* **47,** 777–80.

Eisenbud, L. and Wigner, E. P. (1941). Invariant forms of interaction between nuclear particles, *Proc. Nat. Acad. Sci. U.S.* **27,** 281–9.

Elsasser, W. (1925). Bemerkungen zur quantenmechanik freier elektronen, *Naturwiss* **13,** 711.

Enss, V. (1978). Asymptotic completeness for quantum-mechanical potential scattering, *Commun. Math. Phys.* **61,** 285–91.

(1979). Asymptotic completeness for quantum-mechanical potential scattering. II: singular and long-range potentials, *Ann. Phys. (N.Y.)* **119,** 117–32.

Enss, V. (1983). Completeness of three-body quantum scattering, in *Dynamics and Processes,* ed. P. Blanchard and L. Streit (Berlin, Springer).

Ernst, D. J., Shakin, C. M. and Thaler, R. M. (1973). Separable representations of two-body interactions, *Phys. Rev.* **C8,** 46–52.

Esposito, G. (1998a). Scattering from singular potentials in quantum mechanics, *J. Phys.* **A31,** 9493–504.

(1998b). Complex parameters in quantum mechanics, *Found. Phys. Lett.* **11,** 535–47.

(2000). Non-Fuchsian singularities in quantum mechanics, *Found. Phys. Lett.* **13,** 29–40.

Fano, U. (1957). Description of states in quantum mechanics by density matrix and operator techniques, *Rev. Mod. Phys.* **29,** 74–93.

Fermi, E. (1926). Sulla quantizzazione del gas perfetto monoatomico, *Rend. Lincei* **3,** 145–9.

(1932). Quantum theory of radiation, *Rev. Mod. Phys.* **4,** 87–132.

(1950). *Nuclear Physics* (Chicago, The University of Chicago Press).

(1965). *Notes on Quantum Mechanics* (Chicago, The University of Chicago Press).

Feynman, R. P. (1948). Space–time approach to non-relativistic quantum mechanics, *Rev. Mod. Phys.* **20,** 367–87.

Fierz, M. (1944). Zur theorie magnetisch geladener teilchen, *Helv. Phys. Acta* **17,** 27–34.

Fischbach, E., Nieto, M. M. and Scott, C. K. (1973). Duffin–Kemmer–Petiau subalgebras: representations and applications, *J. Math. Phys.* **14,** 1760–74.

Flügge, S. (1971). *Practical Quantum Mechanics. I* (Berlin, Springer).

Fock, V. A. (1932). Konfigurationsraum und zweite quantelung, *Z. Phys.* **75,** 622–47.

(1978). *Fundamentals of Quantum Mechanics* (Moscow, MIR).

Foldy, L. L. (1950). A lower bound on the range of the triplet neutron–proton interaction, *Phys. Rev.* **78,** 636.

Foldy, L. L. and Wouthuysen, S. A. (1950). On the Dirac theory of spin $\frac{1}{2}$ particle and its non-relativistic limit, *Phys. Rev.* **78,** 29–36.

Fonda, L. and Newton, R. G. (1959). Threshold behavior of cross sections of charged particles., *Ann. Phys. (N.Y.)* **7,** 133–45.

(1960). Threshold effects in three-body channels, *Phys. Rev.* **119,** 1394–9.

Forsyth, A. R. (1959). *Theory of Differential Equations, Part III* (New York, Dover).

Franck, J. and Hertz, G. (1914). Über zusammenstösse zwischen elektronen und den molekülen des quecksilberdampfes und die ionisierungsspammung desselben, *Verh. d. Deutsch. Phys. Ges.* **16,** 457–67.

Frank, W. M., Land, D. J. and Spector, R. M. (1971). Singular potentials, *Rev. Mod. Phys.* **43,** 36–98.

Friedrichs, K. (1948). On the perturbation of continuous spectra, *Comm. Pure Appl. Math.* **1,** 361–406.

Friedrichs, K. O. (1953). *Mathematical Aspects of the Quantum Theory of Fields* (New York, Interscience).

Froman, N. and Froman, P. O. (1965). *JWKB Approximation* (Amsterdam, North-Holland).

Fubini, S. and Stroffolini, R. (1965). A new approach to scattering by singular potentials, *Nuovo Cim.* **37,** 1812–6.

Fuchs, L. (1866). Zur theorie der linearen differentialgleichungen mit veränderlichen koeffizienten, *J. Math.* **66,** 121–60.

Gamow, G. (1928). Zur quantentheorie des atomkernes, *Z. Phys.* **51,** 204–12.

Garabedian, P. R. (1964). *Partial Differential Equations* (New York, Chelsea).

Gell-Mann, M. and Hartle, J. B. (1993). Classical equations for quantum systems, *Phys. Rev.* **D47,** 3345–82.

Gel'fand, I. M. and Vilenkin, N. Y. (1964). *Generalized Functions. Vol. 4. Applications of Harmonic Analysis* (New York, Academic Press).

Gel'fand, I. M. and Shilov, G. E. (1967). *Generalized Functions. Vol. 3. Theory of Differential Equations* (New York, Academic Press).

Gerlach, W. and Stern, O. (1922). Der experimentelle nachweiss der richtungsquantelung im magnetfeld, *Z. Phys.* **9,** 349–52.

Gieres, F. (2000). Mathematical surprises and Dirac's formalism in quantum mechanics, *Rep. Prog. Phys.* **63,** 1893–931.

Glauber, R. J. (1963). Coherent and incoherent states of the radiation field, *Phys. Rev.* **131,** 2766–88.

Goldstein, H. (1980). *Classical Mechanics* (New York, Addison-Wesley).

Gordan, P. (1875). *Uber das Formensystem Binarer Formen* (Leipzig, Teubner).

Gordon, W. (1926). Comptoneffekt nach der Schrödingerschen theorie, *Z. Phys.* **40,** 117–33.

Gottfried, K. (1966). *Quantum Mechanics. I: Fundamentals* (New York, Benjamin).

Gozzi, E. and Regini, M. (2000). Addenda and corrections to work done on the path integral approach to classical mechanics, *Phys. Rev.* **D62,** 067702.

Gracia-Bondia, J. M., Lizzi, F., Marmo, G. and Vitale, P. (2002). Infinitely many star products to play with, *JHEP* **0204,** 026.

Gradshteyn, I. S. and Ryzhik, I. M. (1965). *Tables of Integrals, Series and Products* (New York, Academic Press).

Graeber, R. and Dürr, S. P. (1977). Singular potentials and the divergence problem in quantum field theory, *Nuovo Cim.* **A40,** 11–23.

Graffi, S. and Grecchi, V. (1978). Resonances in Stark effect and perturbation theory, *Commun. Math. Phys.* **62,** 83–96.

Grechko, L. G., Sugarov, V. I. and Tomasevich, O. F. (1977). *Problems in Theoretical Physics* (Moscow, MIR).

Grib, A. A., Mamaev, S. G. and Mostepanenko, V. M. (1994). *Vacuum Quantum Effects in Strong Fields* (St Petersburg, Friedmann Laboratory Publishing).

Guckenheimer, J. (1973). Catastrophes and partial differential equations, *Ann. Inst. Fourier (Grenoble)* **23,** 31–59.

Haas, A. E. (1910a). *Sitz. Ber. der Wiener Akad* **Abt IIa**, 119–44.

(1910b). Über eine neue theoretische methode zur bestimmung des elektrischen elementar quantums und des halbmessers des wasserstoffatoms, *Physikal. Zeitschr.* **11**, 537–8.

(1925). *Introduction to Theoretical Physics* (London, Constable & Company).

Hallwachs, W. (1888). Über den Einflu*β* des lichtes auf elektrostatisch geladene körper, *Ann. Phys.* **33**, 301–12.

Harrell, E. (1977). Singular perturbation potentials, *Ann. Phys. (N.Y.)* **105**, 379–406.

Hawking, S. W. and Ellis, G. F. R. (1973). *The Large-Scale Structure of Space-Time* (Cambridge, Cambridge University Press).

Hegerfeldt, G. C. (1994). Causality problems for Fermi's two-atom system, *Phys. Rev. Lett.* **72**, 596–9.

Heisenberg, W. (1925). Quantum-theoretical re-interpretation of kinematic and mechanical relations, *Z. Phys.* **33**, 879–93.

Hellinger, E. and Toeplitz, O. (1910). Grundlagen für eine theorie der unendlichen matrizen, *Math. Ann.* **69**, 289–330.

Hellwig, G. (1964). *Differential Operators of Mathematical Physics* (New York, Addison-Wesley).

Herzberg, G. (1944). *Atomic Spectra and Atomic Structure* (New York, Dover).

Hillery, M., O'Connell, R. F., Scully, M. O. and Wigner, E. P. (1984). Distribution functions in physics: fundamentals, *Phys. Rep.* **106**, 121–67.

Holland, P. R. (1993). *The Quantum Theory of Motion* (Cambridge, Cambridge University Press).

Holton, G. (2000). Millikan's struggle with theory, *Europhys. News* **31**, No 3, 12–14.

Hörmander, L. (1983). *The Analysis of Linear Partial Differential Operators. I. Distribution Theory and Fourier Analysis* (New York, Springer).

Houtappel, R. M. F., van Dam, H. and Wigner, E. P. (1965). The conceptual basis and use of the geometric invariance principles, *Rev. Mod. Phys.* **37**, 595–632.

Hughes, A. L. and DuBridge, L. A. (1932). *Photoelectric Phenomena* (New York, McGraw-Hill).

Hughes, R. (1989). *The Structure and Interpretation of Quantum Mechanics* (Harvard, Harvard University Press).

Hulthén, L. (1942). Über die eigenlosunger der Schrödinger-gleichung des deuterons, *Ark. Mat. Astron. Fys.* **A28**, Art. 5, 1–12.

(1943). On the virtual state of the deuteron, *Ark. Mat. Astron. Fys.* **B29**, Art. 1, 1–11.

Husimi, K. (1940). Some formal properties of the density matrix, *Proc. Phys. Math. Soc. Japan* **22**, 264–314.

Isham, C. J. (1989). *Lectures on Modern Differential Geometry* (Singapore, World Scientific).

(1995). *Lectures on Quantum Theory. Mathematical and Structural Foundations* (London, Imperial College Press).

Itzykson, C. and Zuber, J. B. (1985). *Quantum Field Theory* (New York, McGraw-Hill).

Jackson, J. D. (1975). *Classical Electrodynamics* (New York, Wiley).

Jauch, J. (1973). *Foundations of Quantum Mechanics* (London, Addison-Wesley).

Jeans, J. H. (1905). On the partition of energy between matter and aether, *Phil. Mag.* **10**, 91–98.

Jordan, T. F., MacFarlane, A. J. and Sudarshan, E. C. G. (1964). Hamiltonian model of Lorentz invariant particle interactions, *Phys. Rev.* **133**, B487–96.

José, J. V. and Saletan, E. J. (1998). *Classical Dynamics* (Cambridge, Cambridge University Press).

Jost, R. (1964). Poisson brackets (an unpedagogical lecture), *Rev. Mod. Phys.* **36**, 572–9.

Kano, Y. (1965). A new phase-space distribution function in the statistical theory of the electromagnetic field, *J. Math. Phys.* **6**, 1913–5.

Kato, T. (1995). *Perturbation Theory for Linear Operators* (Berlin, Springer).

Kazama, Y., Yang, C. N. and Goldhaber, A. S. (1977). Scattering of a Dirac particle with charge Ze by a fixed magnetic monopole, *Phys. Rev.* **D15**, 2287–99.

Kemmer, N. (1939). The particle aspect of meson theory, *Proc. R. Soc. Lond.* **A173**, 91–116.

Khalfin, L. A. (1958). Contribution to the decay theory of a quasi-stationary state, *JETP* **6**, 1053–63.

Khuri, N. N. and Pais, A. (1964). Singular potentials and peratization. I., *Rev. Mod. Phys.* **36**, 590–5.

Kirchhoff, G. (1860). On the relation between the radiating and absorbing powers of different bodies for light and heat, *Phil. Mag. (4)* **20**, 1–21.

Klauder, J. R. and Sudarshan, E. C. G. (1968). *Fundamentals of Quantum Optics* (New York, Benjamin).

Klauder, J. (1973). Field structure through models studies, *Acta Phys. Austriaca Supp.* **11**, 341–87.

Klauder, J. and Detwiler, L. (1975). Supersingular quantum perturbations, *Phys. Rev.* **D11**, 1436–41.

Klauder, J. (1997). Understanding quantization, *Found. Phys.* **27**, 1467–83.

(2000). *Beyond Conventional Quantization* (Cambridge, Cambridge University Press).

Klein, O. (1926). Quantentheorie und funfdimensionale relativitatstheorie, *Z. Phys.* **37**, 895–906.

Kummer, E. E. (1836). Über die hypergeometrische reihe $F(a, b; x)$, *J. reine angew. Math.* **15**, 39–83.

Kunihiro, T. (1998). Renormalization-group resummation of a divergent series of the perturbative wave functions of the quantum anharmonic oscillator, *Phys. Rev.* **57**, R2035–9.

Lagrange, J. L. (1788). *Mécanique Analytique* (Paris, Albert Blanchard).

Landau, L. D. (1930). Diamagnetismus der metalle, *Z. Phys.* **64**, 629–37.

Landau, L. D. and Lifshitz, E. M. (1958). *Course of Theoretical Physics. III: Quantum Mechanics, Non-Relativistic Theory* (Oxford, Pergamon Press).

(1960). *Course of Theoretical Physics. I: Mechanics* (Oxford, Pergamon Press).

Lee, T. D. (1954). Some special examples in renormalizable field theory, *Phys. Rev.* **95**, 1329–34.

Leutwyler, H. (1965). A no-interaction theorem in classical relativistic Hamiltonian particle mechanics, *Nuovo Cimento* **37**, 556–67.

Levinson, N. (1953). Certain explicit relationships between phase shift and scattering potential, *Phys. Rev.* **89**, 755–7.

Levitan, B. M. (1987). *Inverse Sturm–Liouville Problems* (Utrecht, VNU Science Press).

Lewis, G. N. (1926). The conservation of photons, *Nature* **118**, 874–5.

Lim, Y. K. (1998). *Problems and Solutions on Quantum Mechanics* (Singapore, World Scientific).

Lin, Q. G. (2000). On the partial wave amplitude of Coulomb scattering in three dimensions, *Am. J. Phys.* **68**, 1056–7.

Lippmann, B. A. and Schwinger, J. (1950). Variational principles for scattering processes, I, *Phys. Rev.* **79**, 469–80.

Lizzi, F., Miele, G. and Nicodemi, F. (1999). *Problemi di Istituzioni di Fisica Teorica* (University of Naples).

Luban, M. and Pursey, D. L. (1986). New Schrödinger equations for old: inequivalence of the Darboux and Abraham–Moses constructions, *Phys. Rev.* **D33**, 431–6.

Ludwig, G. (1968). *Wave Mechanics* (Oxford, Pergamon Press).

Ma, S. T. (1952). Bound states and the interaction representation, *Phys. Rev.* **87**, 652–5.

Ma, Z. Q. (1985). Proof of the Levinson theorem by the Sturm–Liouville theorem, *J. Math. Phys.* **26**, 1995–9.

Mackey, G. W. (1963). *Mathematical Foundations of Quantum Mechanics* (New York, Benjamin).

(1998). The relationship between classical mechanics and quantum mechanics, *Contemp. Math.* **214**, 91–109.

Majorana, E. (1938). *Lectures at the University of Naples* (Naples, Bibliopolis (1987)).

Malgrange, B. (1955). Existence et approximation des solutions des équations aux dérivées partielles et des equations de convolution, *Ann. Inst. Fourier (Grenoble)* **6**, 271–355.

Mandel, L. (1963). Intensity fluctuations of partially polarized light, *Proc. Phys. Soc. (London)* **81**, 1104–14.

Mandel, L., Sudarshan, E. C. G. and Wolf, E. (1964). Theory of photoelectric detection of light fluctuations, *Proc. Phys. Soc. (London)* **84**, 435–44.

Man'ko, V. I., Marmo, G., Sudarshan, E. C. G. and Zaccaria, F. (2000). Inner composition law of pure states as a purification of impure states, *Phys. Lett.* **A273**, 31–6.

Marmo, G., Mukunda, N. and Sudarshan, E. C. G. (1984). Relativistic particle dynamics. Lagrangian proof of the no-interaction theorem, *Phys. Rev.* **D30**, 2110–6.

Marmo, G., Saletan, E., Simoni, A. and Vitale, B. (1985). *Dynamical Systems, a Differential Geometric Approach to Symmetry and Reduction* (New York, Wiley).

Marmo, G. and Rubano, C. (1988). *Particle Dynamics on Fibre Bundles* (Naples, Bibliopolis).

Marshak, R. and Sudarshan, E. C. G. (1962). *Elementary Particles* (New York, Interscience).

Maslov, V. P. and Fedoriuk, M. V. (1981). *Semi-Classical Approximation in Quantum Mechanics* (Dordrecht, Reidel).

Mavromatis, H. A. (1987). *Exercises in Quantum Mechanics* (Dordrecht, Reidel).

Mead, C. A. (1987). Molecular Kramers degeneracy and nonabelian adiabatic phase factors, *Phys. Rev. Lett.* **59**, 161–4.

Mehra, J. and Rechenberg, H. (1982a). *The Historical Development of Quantum Theory, Vol. 1, Part 1* (New York, Springer).

(1982b). *The Historical Development of Quantum Theory, Vol. 1, Part 2* (New York, Springer).

(1982c). *The Historical Development of Quantum Theory, Vol. 2* (New York, Springer).

(1982d). *The Historical Development of Quantum Theory, Vol. 3* (New York, Springer).

(1982e). *The Historical Development of Quantum Theory, Vol. 4, Part 1* (New York, Springer).

(1982f). *The Historical Development of Quantum Theory, Vol. 4, Part 2* (New York, Springer).

(1987a). *The Historical Development of Quantum Theory, Vol. 5, Part 1* (New York, Springer).

(1987b). *The Historical Development of Quantum Theory, Vol. 5, Part 2* (New York, Springer).

Mehta, C. L. and Sudarshan, E. C. G. (1965). Relation between quantum and semiclassical description of optical coherence, *Phys. Rev.* **138**, B274–80.

Merzbacher, E. (1961). *Quantum Mechanics* (New York, Wiley).

Michel, L. (1964). Invariance in quantum mechanics and group extension, in *Group Theoretical Concepts and Methods in Elementary Particle Physics. Istanbul Summer School of Theoretical Physics,* ed. F. Gursey (New York, Gordon and Breach).

Millikan, R. A. (1916). A direct photoelectric determination of Planck's h, *Phys. Rev.* **7**, 355–88.

Misra, B. and Sudarshan, E. C. G. (1977). The Zeno's paradox in quantum theory, *J. Math. Phys.* **18**, 756–63.

Mitra, A. N. (1961). Analyticity of amplitudes and separable potentials, *Phys. Rev.* **123**, 1892–5.

Möller, C. (1945). General properties of the characteristic matrix in the theory of elementary particles, I, *Danske. Vid. Selsk. Mat.-Fys. Medd.* **23**, 1–48.

Morandi, G., Ferrario, C., Lo Vecchio, G., Marmo, G. and Rubano, C. (1990). The inverse problem in the calculus of variations and the geometry of the tangent bundle, *Phys. Rep.* **188**, 147–284.

Morandi, G. (1995). *Statistical Mechanics, an Intermediate Course* (Singapore, World Scientific).

Moutard, Th.F. (1875). Note sur les équations différentielles linéaires au second ordre, *C. R. Acad. Sci. (Paris)* **80**, 729–33.

Moyal, J. E. (1949). Quantum mechanics as a statistical theory, *Proc. Camb. Phil. Soc.* **45**, 99–124.

Mukunda, N. and Sudarshan, E. C. G. (1981). Form of relativistic dynamics with world lines, *Phys. Rev.* **D23**, 2210–17.

Münster, A. (1969). *Statistical Thermodynamics* (Berlin, Springer).

Musto, R. (1966). Generators of $O(4,1)$ for the quantum-mechanical hydrogen atom, *Phys. Rev.* **148**, 1274–5.

Newton, T. D. and Wigner, E. P. (1949). Localized states for elementary systems, *Rev. Mod. Phys.* **21**, 400–6.

Newton, R. G. and Fonda, L. (1960). Threshold discontinuities. Application to X-ray scattering, *Ann. Phys. (N.Y.)* **9**, 416–21.

Newton, I. (1999). *Philosophiae Naturalis Principia Mathematica* (Berkeley, University of California Press).

Okubo, S. and Marshak, R. E. (1958). Velocity dependence of the two-nucleon interaction, *Ann. Phys. (N.Y.)* **4**, 166–79.

Onofri, E. and Destri, C. (1996). *Istituzioni di Fisica Theorica* (Rome, La Nuova Italia Scientifica).

Pancharatnam, S. (1956). Generalized theory of interference, and its applications, *Proc. Ind. Acad. Sci.* **A44**, 247–62.

Parisi, G. (2001). Planck's legacy to statistical mechanics, COND-MAT 0101293, preprint.

Pars, L. A. (1965). *A Treatise on Analytical Dynamics* (London, Heinemann).

Paschen, F. and Back, E. (1912). Normale und anomale Zeemaneffekte, *Ann. d. Phys.* **39**, 897–932.

(1913). Normale und anomale Zeemaneffekte, *Nachtrag, Ann. Phys.* **40**, 960–70.

Pauli, W. (1925). Üben den zusammenhang des abschlusses der elektronengruppen im atom mit der komplexstruktur der spektren, *Z. Phys.* **31**, 765–83.

Perelomov, A. M. (1986). *Generalized Coherent States and Their Applications* (Berlin, Springer).

Peres, A. (1993). *Quantum Theory: Concepts and Methods* (Boston, Kluwer Academic).

Peter, F. and Weyl, H. (1927). Die vollständigkeit der primitiven dorstellungen einer geschlossenen kontinuierlichen gruppe, *Math. Ann.* **97**, 737–55.

Petiau, G. (1936). Contribution a la théorie des équations d'ondes corpusculaires, *Acad. R. Belg., Cl. Sci., Mem., Collect. N. 8* **16**, 1–116.

Phipps, T. E. and Taylor, J. B. (1927). The magnetic moment of the hydrogen atom, *Phys. Rev.* **29**, 309–20.

Picasso, L. E. (2000). *Lezioni di Meccanica Quantistica* (Pisa, ETS).

Planck, M. (1900). On the theory of the energy distribution law of the normal spectrum, *Verh. d. Deutsch. Phys. Ges.* **2**, 237–45.

(1991). *The Theory of Heat Radiation* (New York, Dover).

Poincaré, H. (1886). Sur les intégrales irregulières des équations linéaires, *Acta Mathematica* **8**, 295–344.

(1908). Sur les petits diviseurs dans la théorie de la Lune, *Bull. Astron.* **25**, 321–60.

Prugovecki, E. (1981). *Quantum Mechanics in Hilbert Space,* 2nd edn (New York, Academic Press).

Pryce, M. H. L. (1948). The mass centre in the restricted theory of relativity and its connection with the quantum theory of elementary particles, *Proc. R. Soc. Lond.* **A195**, 62–81.

Rabei, E. M., Arvind, Mukunda, N. and Simon, R. (1999). Bargmann invariants and geometric phases: a generalized connection, *Phys. Rev.* **A60**, 3397–409.

Rabin, J. M. (1995). Introduction to quantum field theory for mathematicians, in *Geometry and Quantum Field Theory,* ed. D. S. Freed and K. K. Uhlenbeck (American Mathematical Society, USA).

Radon, J. (1917). Über die bestimmung von funktionen durch ihre integralwerte längs gewisser mannigfaltigkeiten, *Sächs. Akad. Wiss. Leipzig, Math. Nat. Kl.* **69**, 262–77.

Rayleigh, J. W. S. (1900). Remarks upon the law of complete radiation, *Phil. Mag.* **49**, 539–40.

Redhead, M. (1989). *Incompleteness, Nonlocality, and Realism* (Oxford, Clarendon Press).

Reed, M. and Simon, B. (1975). *Methods of Modern Mathematical Physics. II: Fourier Analysis and Self-Adjointness* (New York, Academic Press).

(1978). *Methods of Modern Mathematical Physics. IV: Analysis of Operators* (New York, Academic Press).

(1979). *Methods of Modern Mathematical Physics. III: Scattering Theory* (New York, Academic Press).

(1980). *Methods of Modern Mathematical Physics. I: Functional Analysis* (revised and enlarged edition) (New York, Academic Press).

Regge, T. (1958). Symmetry properties of Clebsch–Gordon's coefficients, *Nuovo Cimento* **10**, 544–45.

Riccati, J. (1724). Animadversiones in aequationes differentiales secundi gradus, *Acta Erud. Suppl.* **8**, 66–73.

Robertson, H. P. (1930). A general formulation of the uncertainty principle and its classical interpretation, *Phys. Rev.* **35**, 667.

Rollnik, H. (1956). Streumaxima und gebundene Zustande, *Z. Phys.* **145**, 639–53.

Sakurai, J. J. (1985). *Modern Quantum Mechanics* (New York, Benjamin).

Samuel, J. (1997). The geometric phase and ray space isometries, *Pramana J. Phys.* **48**, 959–67.

Sartori, G. (1998). *Lezioni di Meccanica Quantistica* (Padova, Edizioni Libreria Cortina).

Schrödinger, E. (1926). Der stetige ubergang von der mikro-zur makromechanik, *Naturwiss.* **14**, 664–66.

(1930). Zum Heisenbergschen unschaerfeprinzip, *Ber. Kgl. Akad. Wiss.* **29**, 296–303.

(1977). *Collected Papers on Wave Mechanics* (New York, Chelsea).

Schwinger, J. (1951a). On gauge invariance and vacuum polarization, *Phys. Rev.* **82**, 664–79.

(1951b). The theory of quantized fields. I., *Phys. Rev.* **82**, 914–27.

Scully, M. O., Englert B. G. and Walther, H. (1991). Quantum optical tests of complementarity, *Nature* **351**, 111–6.

Scully, M. O. and Zubairy, M. S. (1997). *Quantum Optics* (Cambridge, Cambridge University Press).

Segal, I. E. (1959). Foundations of the theory of dynamical systems of infinitely many degrees of freedom. I., *Mat.-Fis. Medd. Dansk. Vid. Selsk.* **31 no 12**, 38 pp..

Sigal, I. M. and Soffer, A. (1987). The *N*-particle scattering problem: asymptotic completeness for short-range systems, *Ann. Math.* **126**, 35–108.

Simon, B. (1969). On positive eigenvalues of one-body Schrödinger operators, *Comm. Pure Appl. Math.* **22**, 531–38.

(1973a). Quadratic forms and Klauder's phenomenon: a remark on very singular perturbations, *J. Funct. Analysis* **14**, 295–98.

(1973b). Resonances in N-body quantum systems with dilatation analytic potentials and the foundations of time-dependent perturbation theory, *Ann. Math.* **97**, 247–74.

Simoni, A., Zaccaria, F. and Sudarshan, E. C. G. (1971). On kernels realizing associative multiplications between phase-space functions, *Nuovo Cim.* **B5**, 134–42.

Sommerfeld, A. (1964). *Partial Differential Equations in Physics* (New York, Academic Press).

Squires, G. L. (1995). *Problems in Quantum Mechanics* (Cambridge, Cambridge University Press).

Srinivasa Rao, K. and Rajeswari, V. (1993). *Quantum Theory of Angular Momentum. Selected Topics* (Delhi, Narosa Publishing House).

Stakgold, I. (1979). *Green's Functions and Boundary Value Problems* (New York, Wiley).

Stark, J. (1914). Beobachtungen über den effekt des elektrischen feldes auf spektrallinien, *Ann. Phys.* **43**, 965–82.

Stewart, B. (1858). An account of some experiments on radiant heat, involving an extension of Prévost's theory of exchanges, *Trans. R. Soc. Edin.* **22**, 1–20.

Stone, M. H. (1930). Linear transformations in Hilbert space III operational methods and group theory, *Proc. Nat. Acad. Sci. U.S.A.* **16**, 172–5.

(1932). On one-parameter unitary groups in Hilbert space, *Ann. Math.* **33**, 643–8.

Stroffolini, R. (1971). A constructive determination of the S-matrix for singular potentials, *Nuovo Cim.* **A2**, 793–828.

(2001). *Lezioni di Elettrodinamica* (Naples, Bibliopolis).

Sturm, C. (1836). Sur les équations différentielles linéaires du second ordre, *J. de Math.* **1**, 106–86.

Sudarshan, E. C. G. (1962). Relativistic particle interactions, in *1961 Brandeis Summer Institute Lectures in Theoretical Physics,* ed. (New York, Benjamin).

Sudarshan, E. C. G. (1963). Equivalence of semiclassical and quantum mechanical descriptions of statistical light beams, *Phys. Rev. Lett.* **10**, 277–9.

Sudarshan, E. C. G., Mukunda, N. and O'Raifeartaigh, L. (1965). Group theory of the Kepler problem, *Phys. Lett.* **19**, 322–26.

Sudarshan, E. C. G. (1972). Action-at-a-distance, *Fields Quanta* **2**, 175.

Sudarshan, E. C. G. and Mukunda, N. (1974). *Classical Dynamics: a Modern Perspective* (New York, Wiley).

Sudarshan, E. C. G., Chiu, C. B. and Gorini, V. (1978). Decaying states as complex energy eigenvectors in generalized quantum mechanics, *Phys. Rev.* **D18**, 2914–29.

Sudarshan, E. C. G. and Rothman, T. (1991). The two-slit interferometer re-examined, *Am. J. Phys.* **59**, 592–5.

(1993). A new interpretation of Bell's inequalities, *Int. J. Theor. Phys.* **32**, 1077–86.

Sudarshan, E. C. G. (1994). Quantum dynamics in dual spaces, *Phys. Rev.* **50**, 2006–26.

Sudarshan, E. C. G., Chiu, C. B. and Bhamathi, G. (1995). Generalized uncertainty relations and characteristic invariants for the multimode states, *Phys. Rev.* **A52**, 43–54.

Synge, J. L. (1954). *Geometrical Mechanics and de Broglie Waves* (Cambridge, Cambridge University Press).

Szebehely, V. (1967). *Theory of Orbits. The Restricted Problem of Three Bodies* (New York, Academic Press).

Tani, S. (1951). Connection between particle models and field theories, I. The case spin 1/2, *Prog. Theor. Phys.* **6**, 267–85.

Ter Haar, D. (1946). On the redundant zeros in the theory of the Heisenberg matrix, *Physica* **12**, 500–8.

(1967). *The Old Quantum Theory* (Glasgow, Pergamon Press).

(1975). *Problems in Quantum Mechanics,* 3rd edn (London, Pion).

Thirring, W. (1981). *A Course in Mathematical Physics. Vol. III: Quantum Mechanics of Atoms and Molecules* (Berlin, Springer).

Trotter, H. (1959). On the product of semigroups of operators, *Proc. Amer. Math. Soc.* **10**, 545–51.

Turbiner, A. V. (1988). Quasi-exactly-solvable problems and $sl(2)$ algebra, *Commun. Math. Phys.* **118**, 467–74.

Uhlenbeck, G. E. and Goudsmit, S. (1926). Spinning electrons and the structure of spectra, *Nature* **117**, 264–5.

Ushveridze, A. G. (1994). *Quasi-Exactly Solvable Models in Quantum Mechanics* (Bristol, IOP).

van der Waerden, B. L. (1968). *Sources of Quantum Mechanics* (New York, Dover).

Varadarajan, V. S. (1985). *Geometry of Quantum Theory* (Berlin, Springer-Verlag).

Varma, S. and Sudarshan, E. C. G. (1996). Quantum scattering theory in the light of an exactly solvable model with rearrangement collisions, *J. Math. Phys.* **37**, 1668–712.

von Neumann, J. (1931). Die eindeutigkeit der Schrödingerschen operatoren, *Math. Ann.* **104**, 570–8.

(1955). *Mathematical Foundations of Quantum Mechanics* (Princeton, Princeton University Press).

Watson, G. (1912). A theory of asymptotic series, *Philos. Trans. R. Soc. Lond.* **A211**, 279–313.

(1966). *A Treatise on the Theory of Bessel Functions* (Cambridge, Cambridge University Press).

Weihs, G., Jennewein, T., Simon, C., Weinfurter, H. and Zeilinger, A. (1998). Violation of Bell's inequality under strict Einstein locality conditions, *Phys. Rev. Lett.* **81**, 5039–43.

Weinberg, E. J. (1994). Monopole vector spherical harmonics, *Phys. Rev.* **D49**, 1086–92.

Weinberg, S. (1989). Testing quantum mechanics, *Ann. Phys. (N.Y.)* **194**, 336–86.

(1996). *The Quantum Theory of Fields. Vol. I* (Cambridge, Cambridge University Press).

Weiss, P. (1938). On the Hamilton–Jacobi theory and quantization of a dynamical continuum, *Proc. R. Soc. Lond.* **A169**, 102–19.

Weyl, H. (1927). Quantenmechanik und gruppentheorie, *Z. Phys.* **46**, 1–46.

(1931). *The Theory of Groups and Quantum Mechanics* (New York, Dover).

(1940). The method of orthogonal projection in potential theory, *Duke Math. J.* **7**, 411–44.

Wheeler, J. A. and Feynman, R. P. (1949). Classical electrodynamics in terms of direct interparticle action, *Rev. Mod. Phys.* **21**, 425–33.

Wheeler, J. A. and Zurek, W. H. (1983). *Quantum Theory and Measurement* (Princeton, Princeton University Press).

Whittaker, E. T. (1937). *A Treatise on the Analytical Dynamics of Particles and Rigid Bodies* (Cambridge, Cambridge University Press).

Wien, W. (1896). Über die energieverteilung im emissionsspektrum eines schwarzen körpers, *Ann. Phys.* **58**, 662–9.

Wiener, N. (1930). Generalized harmonic analysis, *Acta Math.* **55**, 117–258.

Wightman, A. S. and Schweber, S. S. (1955). Configuration space methods in relativistic quantum field theory. I, *Phys. Rev.* **98**, 812–37.

Wightman, A. S. (1996). How it was learned that quantized fields are operator-valued distributions, *Fortschr. Phys.* **44**, 143–78.

Wigner, E. P. (1932a). Über die operation der zeitumkehr in der quantenmechanik, *Gott. Nachr.* **31**, 546–59.

(1932b). On the quantum correction for thermodynamic equilibrium, *Phys. Rev.* **40,** 749–59.

(1959). *Group Theory and Its Application to the Quantum Mechanics of Atomic Spectra* (New York, Academic Press).

Wintner, A. (1947). The unboundedness of quantum-mechanical matrices, *Phys. Rev.* **71,** 738–9.

Witten, E. (1981). Dynamical breaking of supersymmetry, *Nucl. Phys.* **B188,** 513–54.

Wu, T. T. and Yang, C. N. (1976). Dirac monopole without strings: monopole harmonics, *Nucl. Phys.* **B107,** 365–80.

Wybourne, B. G. (1974). *Classical Groups for Physicists* (New York, Wiley).

Yamaguchi, Y. (1954). Two-nucleon problem when the potential is nonlocal but separable. I, *Phys. Rev.* **95,** 1628–34.

Yamaguchi, Y. and Yamaguchi, Y. (1954). Two-nucleon problem when the potential is nonlocal but separable. II, *Phys. Rev.* **95,** 1635–43.

Yang, C. N. and Tiomno, J. (1950). Reflection properties of spin-$\frac{1}{2}$ fields and a universal Fermi type interaction, *Phys. Rev.* **79,** 495–8.

Yosida, K. (1991). *Lectures on Differential and Integral Equations* (New York, Dover)

Yuen, H. P. (1976). Two-photon coherent states of the radiation field, *Phys. Rev.* **A13,** 2226–43.

Yurke, B., McCall, S. L. and Klauder, J. R. (1986). $SU(2)$ and $SU(1,1)$ interferometers, *Phys. Rev.* **A33,** 4033–54.

Zeeman, P. (1897a). On the influence of magnetism on the nature of the light emitted by a substance, *Phil. Mag.* **43,** 226–39.

(1897b). Doublets and triplets in the spectrum produced by external magnetic forces, *Phil. Mag.* **44,** 55–60, 255–9.

Index